Absolute-Value Inequalities

$|x - a| < b$ is equivalent to $-b < x - a < b$

$|x - a| > b$ is equivalent to $x - a > b$ and $x - a < -b$

The Number e

$e = 2.718281828459\ldots$

Definitions of the Number e

$$e = \lim_{x \to \infty} \left(1 + \frac{1}{x} \right)^x$$

$$e = \lim_{h \to 0} (1 + h)^{1/h}$$

Definition of a Logarithm

$y = \log_b x$ is equivalent to $x = b^y$

Natural Logarithm

$\log_e x = \ln x$

$y = \ln x$ is equivalent to $x = e^y$

Laws of Logarithms

$\log_b MN = \log_b M + \log_b N$

$$\log_b \frac{M}{N} = \log_b M - \log_b N$$

$\log_b M^c = c \log_b M$

Properties of Logarithms

$\log_b b = 1$

$\log_b 1 = 0$

$\log_b b^x = x$

$b^{\log_b x} = x$

Changing From Base b to Base e

$$\log_b x = \frac{\ln x}{\ln b}$$

Review of Graphs

To Find Intercepts

y-intercepts: Set $x = 0$ in an equation and solve for y

x-intercepts: Set $y = 0$ in an equation and solve for x

Vertex (h, k) of a Parabola

Complete the square in x for $f(x) = ax^2 + bx + c$ to obtain $f(x) = a(x - h)^2 + k$. Alternatively, compute the coordinates

$$\left(-\frac{b}{2a}, f\left(-\frac{b}{2a} \right) \right).$$

Even and Odd Functions

Even: $f(-x) = f(x)$;   Symmetry of graph: y-axis

Odd: $f(-x) = -f(x)$;   Symmetry of graph: origin

Rigid Transformations

Graph of $y = f(x)$ for $c > 0$:

$y = f(x) + c$, shifted up c units

$y = f(x) - c$, shifted down c units

$y = f(x + c)$, shifted left c units

$y = f(x - c)$, shifted right c units

$y = f(-x)$, reflection in y-axis

$y = -f(x)$, reflection in x-axis

Asymptotes

If the polynomial functions P and Q have no common factors, then the graph of a rational function

$$f(x) = \frac{P(x)}{Q(x)} = \frac{a_n x^n + \cdots + a_1 x + a_0}{b_m x^m + \cdots + b_1 x + b_0}$$

has a vertical asymptote where $Q(x) = 0$. The graph has the horizontal asymptote

$$y = a_n / b_m \text{ when } n = m,$$

and the horizontal asymptote

$$y = 0 \text{ when } n < m.$$

The graph has no horizontal asymptote when $n > m$. The graph has a slant asymptote when $n = m + 1$.

PRECALCULUS

with Calculus Previews

FIFTH EDITION

THE JONES & BARTLETT LEARNING SERIES IN MATHEMATICS

PRECALCULUS
with Calculus Previews
FIFTH EDITION

Dennis G. Zill

Jacqueline M. Dewar

Loyola Marymount University

JONES & BARTLETT
LEARNING

World Headquarters
Jones & Bartlett Learning
5 Wall Street
Burlington, MA 01803
978-443-5000
info@jblearning.com
www.jblearning.com

Jones & Bartlett Learning books and products are available through most bookstores and online booksellers. To contact Jones & Bartlett Learning directly, call 800-832-0034, fax 978-443-8000, or visit our website, www.jblearning.com.

Substantial discounts on bulk quantities of Jones & Bartlett Learning publications are available to corporations, professional associations, and other qualified organizations. For details and specific discount information, contact the special sales department at Jones & Bartlett Learning via the above contact information or send an email to specialsales@jblearning.com.

Production Credits
Chief Executive Officer: Ty Field
President: James Homer
SVP, Editor-in-Chief: Michael Johnson
SVP, Chief Technology Officer: Dean Fossella
SVP, Chief Marketing Officer: Alison M. Pendergast
Publisher, Higher Education: Cathleen Sether
Senior Acquisitions Editor: Timothy Anderson
Managing Editor: Amy Bloom
Director of Production: Amy Rose
Production Editor: Tiffany Sliter
Production Assistant: Alyssa Lawrence

Senior Marketing Manager: Andrea DeFronzo
V.P., Manufacturing and Inventory Control: Therese Connell
Rights & Photo Research Manager: Katherine Crighton
Rights & Photo Research Supervisor: Anna Genoese
Permissions & Photo Research Assistant: Lian Bruno
Composition: Aptara, Inc.
Cover and Title Page Design: Scott Moden
Cover and Title Page Image: Whirlpool galaxy: © Steve Nagy/Design Pics Inc./Alamy Images; Coma Cluster of galaxies: © Stocktrek Images, Inc./Alamy Images
Printing and Binding: Courier Corporation
Cover Printing: Courier Corporation

To order this product, use ISBN: 978-1-4496-4912-8

Library of Congress Cataloging-in-Publication Data
Zill, Dennis G.
    Precalculus with calculus previews / Dennis G. Zill and Jacqueline M. Dewar. — 5th ed.
        p. cm.
    Includes index.
    ISBN-13: 978-1-4496-4515-1 (casebound)
    ISBN-10: 1-4496-4515-1 (casebound)
  1. Precalculus—Textbooks.   I. Dewar, Jacqueline M.   II. Title.
    QA39.3.Z55 2012
    510—dc23
                                2011038351

6048

Printed in the United States of America
15 14 13 12 11   10 9 8 7 6 5 4 3 2 1

Contents

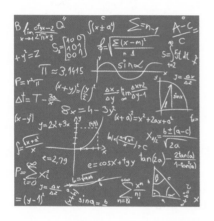

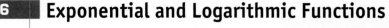

Preface

If you are familiar with our current companion volume, *Essentials of Precalculus with Calculus Previews*, *Fifth Edition*, you may know that for our 3-hour semester course in precalculus mathematics at Loyola Marymount University we have long favored a short text covering only what we consider to be basic material necessary for the successful completion of a course in calculus; a text that would allow time for instructors to work with their students to focus on strengthening their algebraic, logarithmic, and trigonometric skills.

This longer text, *Precalculus with Calculus Previews*, *Fifth Edition*, is a recognition of the needs of those instructors whose course syllabus contains topics not covered in *Essentials of Precalculus*, have more class time to cover extra material, or simply prefer to design their own course by being able to choose from a wider variety of topics.

☐ About This Text

Emphases As in previous editions, the focus of this revision is squarely on the function concept. We continue to provide applications and mathematical models culled from journals, newspapers, scientific texts, and even a calculus text. These applications in this revision span a wide variety of disciplines including astronomy, biology, business, chemistry, ecology, engineering, geology, medicine, meteorology, optics, and physics. Applied or "real-life" problems show students the power and usefulness of the mathematics they will learn in this course.

There is also a great emphasis on basics, especially algebra. Many times we have seen students in a calculus class perform an operation such as differentiation flawlessly, but fail to complete the problem because they had difficulty simplifying the resulting expression or solving a related equation. So, in this edition we continue to make an effort to strengthen algebraic skills. Marginal side notes and in-text annotations fill in the details of solutions of examples and convey additional information to the reader.

In this revision we have also increased the use of calculus terminology. We do not hesitate to use, where appropriate, words such as "continuous function," "discontinuity," "convergence of a sequence," "divergence," and so on. The idea is to give the student a good intuitive sense of what these words mean prior to their exposure to their formal definitions in calculus.

Building Functions from Words As teachers we know that the related rate and applied max-min, or optimization, problems can be a discouraging experience for some students of calculus. Typically, correctly interpreting the words of such a problem in order to set up an equation or a function presents the greatest challenge for many students. It follows then that it is appropriate to emphasize such material in a precalculus course. In Section 2.8 (Building a Function from Words), we begin by illustrating how to translate a verbal description into a symbolic representation of a function. We then present actual problems taken from *Calculus: Early Transcendentals, Fourth Edition* by Dennis G. Zill and Warren S. Wright (Jones & Bartlett Learning, 2011), and demonstrate how to decode the statement of the problem and transform those words into an objective function. We discuss the importance of drawing pictures, using variables to describe pertinent quantities, identifying a constraint between the variables, using a constraint to eliminate an extra variable, and observing that the domain of the objective function may not be the same as its implicit domain. To ensure that the focus is squarely on the process of fashioning a symbolic function from the words, we have chosen not to discuss how such optimization problems are actually solved.

Notes from the Classroom Selected sections of this text conclude with remarks called *Notes from the Classroom*. These remarks are aimed directly at the student and address a variety of student/textbook/classroom/calculus issues such as alternative terminology, reinforcement of important concepts, what material is or is not recommended for memorization, misinterpretations, common errors, solution procedures, calculators, and advice on the importance of neatness and organization.

Calculus Previews Every chapter in this text concludes with a section subtitled *Calculus Preview*. Each of these special sections is devoted to a single calculus concept (for example, *The Tangent Line Problem*) and the discussion is kept at a level easily within the reach of a precalculus student. The emphasis in these previews is *not* on the calculus; the calculus topic provides a framework and motivation for the precalculus mathematics we discuss. The focus in these sections is on the algebraic, logarithmic, and trigonometric manipulations that are necessary for the successful completion of typical calculus problems related to the *Calculus Preview* topic. Consequently the *Calculus Previews* are intended to be taught as part of a regular course in precalculus mathematics.

Final Examination Following the ten chapters of the text we present a list of 92 questions called the *Final Examination*. This "test" consists of mostly fill-in-the-blank and true/false questions. It was not our intention to emulate an actual final examination in a precalculus course, but rather our thought was to offer a vehicle for an informal wrap-up of the entire course. We suggest that, if time permits, a part of a class period be devoted to a discussion of these questions in order to help students prepare for their actual final examination and their subsequent transition to calculus. To facilitate the students' review, the answers of the *Final Examination* are given both in the *Student Resource Manual* as well as in the instructor's *Complete Solutions Manual*. Of course, the instructor is free to utilize this material in whatever manner he or she chooses (including ignoring it completely).

☐ **Updates to This New Edition** Those who have used the fourth edition of this text might notice that this new edition of *Precalculus* has put on a little weight. Here are some of the reasons why:

- Many new and interesting problems have been added throughout the text. As already mentioned, we have increased the number of applied problems as well as conceptual problems suitable for classroom discussion.

- At the request of some users of the previous edition, a discussion of relative extrema of a function has been added to Section 3.1 (Polynomial Functions).
- Section 3.5 (Approximating Real Zeros) is new to Chapter 3 (Polynomial and Rational Functions). In this section, the Intermediate Value Theorem and the root-finding method of interval bisection are introduced.
- Three new sections have been added to Chapter 4 (Trigonometric Functions). Section 4.5 (Verifying Trigonometric Identities) was inserted at the request of a user who felt his students needed more practice in manipulating and recognizing trigonometric functions and identities. Section 4.7 (Product-to-Sum and Sum-to-Product Formulas) was added because these formulas are arguably useful in calculus when the techniques of integration are studied. Also, the discussion of Simple Harmonic Motion has been expanded and now appears as its own section, Section 4.10.
- The discussion of two-dimensional vectors in Section 5.5 has been considerably lengthened. The discussion now includes the dot product of two vectors.
- We did not think that there were a sufficient number of problems devoted to solving equations involving exponential and logarithmic functions in the previous edition. Consequently we've added Section 6.3 (Exponential and Logarithmic Equations) to Chapter 6 (Exponential and Logarithmic Functions).
- Because the definition, graphs, and some of the properties of the hyperbolic functions are introduced, the *Calculus Preview* for Chapter 6 is now Section 6.5 (The Hyperbolic Functions).
- A new *Calculus Preview* has been added to the end of Chapter 7 (Conic Sections). In Section 7.5 (3-Space) the notions of distance, lines, planes, spheres, and vectors in 3-space are discussed. The cross product of two vectors is introduced in Exercises 7.5 as an application of a 3×3 determinant.
- We felt that the discussion of the polar coordinate system deserved its own chapter—it now appears as Chapter 8 (Polar Coordinates).
- Because of its similarity to shifted rectangular graphs, rotation of polar graphs is now discussed in Section 8.2 (Graphs of Polar Equations) and again in Section 8.3 (Conic Sections in Polar Coordinates).
- The *Calculus Preview* for Chapter 8 is Section 8.4 (Parametric Equations). This material appeared as the *Calculus Preview* for Chapter 7 in the previous edition.
- In Chapter 9 (Systems of Equations and Inequalities), Section 9.2 has been completely rewritten and renamed (Determinants and Cramer's Rule).
- Section 10.5 (Principles of Counting) and Section 10.6 (Introduction to Probability) are new to Chapter 10 (Sequences and Series).
- The *Final Examination* at the end of the text has been expanded and given a new format.
- A complete review of the complex number system can now be found in Appendix A (Complex Numbers).
- We feel that the brief new Appendix B (Descartes' Rule of Signs) could be used as a reading assignment to accompany Section 3.4 (Real Zeros of Polynomial Functions).
- Appendix C (Formulas from Geometry) includes a list of well-known (and not-so-well-known) formulas for area, circumference, volume, and surface area. The point of this addition is to supply some potentially useful information for Section 2.8 (Building a Function from Words).

☐ **Supplements for the Instructor** The following materials are available online, at go.jblearning.com/precalc5e.

Complete Solutions Manual (CSM) provides worked-out solutions for every problem in the text as well as answers to all questions in the Final Examination.

Computerized Testing System for both Mac OS® and Windows® computer operating systems. This testing system allows instructors to create customized quizzes and tests. The questions and answers are sorted by chapter and can be easily installed on a computer. Publisher-supplied .rtf files can also be uploaded to the instructor's learning management system.

PowerPoint Image Bank feature all labeled figures as they appear in the text. This useful tool allows instructors to easily display and discuss figures and problems that appear in the textbook.

WebAssign® developed by instructors for instructors, is a premier independent online teaching and learning environment, guiding several million students through their academic careers since 1997. With WebAssign, instructors can create and distribute algorithmic assignments using selected questions specific to this textbook. Instructors can also grade, record, and analyze student responses and performance instantly; offer more practice exercises, quizzes, and homework; and upload additional resources to share and communicate with their students seamlessly, such as the PowerPoint image bank and the test items supplied by Jones & Bartlett Learning's Computerized Testing System.

eBook Format offers an added convenience. This complete textbook is now available in eBook format for instructors and students through WebAssign.

CourseSmart® is a new way for instructors and students to access this textbook in digital format, anytime from anywhere. Jones & Bartlett Learning has partnered with CourseSmart to make many of our leading mathematics textbooks available in the CourseSmart eTextbook store. For more information on CourseSmart Editions please visit

www.jblearning.com/elearning/econtent/coursesmart/

Please contact your Jones & Bartlett Learning Account Specialist for information on, and access to, online demonstrations of the supplements and services described above.

☐ **Supplements for the Student**

Student Resource Manual (SRM) prepared by Warren S. Wright and Carol D. Wright. This manual continues to be popular with students using any one of the Zill series of mathematics textbooks. Unlike the traditional student solutions manual, where a selected subset of the problems are worked out, the *SRM* is divided into five parts: *Algebra Topics, Use of a Calculator, Basic Skills, Selected Solutions*, and *Answers to the Final Examination*. In *Algebra Topics*, selected topics from algebra (such as multiplication of an inequality by an unknown, implicit conditions in a word problem, Pascal's triangle, factoring techniques, binomial expansions, rationalizations of numerators and denominators, adding symbolic fractions, long division of polynomials, synthetic division of polynomials, factorial notation, and so on) are reviewed because of their relevance to calculus. Because we do not discuss how to use technology within the text proper, we have devoted the section *Use of a Calculator* to the review of graphing calculator essentials. In *Selected Solutions*, a detailed solution of every third problem in the exercise sets is given.

Answers to the Final Examination is a list of answers for all questions in the *Final Examination*.

Available in both print and online formats, this student manual can be purchased separately or ordered bundled with the textbook at a substantial savings.

Student Companion Website is available at go.jblearning.com/precalc5e. This online tutorial learning center can be accessed at any time during the term. The resources are tied directly to the text and include: Practice Quizzes, an Online Glossary of Key Terms, and Animated Flashcards.

Exploring Mathematics: Solving Problems with the TI-84 Plus Graphing Calculator by Jeffery M. Gervasi, EdD, Porterville College, Porterville, CA is a helpful graphing calculator manual and can be ordered through the bookstore or online directly from Jones & Bartlett Learning.

WebAssign Access Card can be bundled with this text or purchased separately by the student online at www.webassign.net.

eBook with Course Access Card can be bundled with this text or purchased separately by the student online at www.webassign.net.

CourseSmart is a new way for students to access college textbooks in digital format, anytime from anywhere. Jones & Bartlett Learning has partnered with CourseSmart to make this textbook available in the CourseSmart eTextbook store.

For students, this CourseSmart Edition has many features designed to make studying more efficient such as highlighting, online search, note taking, and print capabilities. For more information on purchasing this CourseSmart Edition please visit
www.jblearning.com/elearning/econtent/coursesmart/

▮ To the Student

After teaching collegiate mathematics for many years, we have seen almost every type of student, from a budding genius who invented his own calculus, to students who struggled to master the most rudimentary mechanics of the subject. Frequently the source of difficulty in calculus can be traced to weak algebra skills and an inadequate background in trigonometry. Calculus builds immediately on your prior knowledge and skills and there is much new ground to be covered. Consequently there is very little time to review precalculus mathematics in the calculus classroom. So those who teach calculus must assume that you can factor, simplify and solve equations, solve inequalities, handle absolute values, use a calculator, correctly apply the laws of exponents, find equations of lines, plot points, sketch basic graphs, and apply important trigonometric and logarithmic identities. The ability to do algebra and trigonometry, work with exponentials and logarithms, and sketch by hand basic graphs quickly and accurately are keys to success in a calculus course.

In this text, we have tried to give you as much help as possible within the confines of the printed page using such features as marginal annotations, arrow annotations within examples, notes of caution, *Notes from the Classroom*, and the *Final Examination*. The many marginal and in-text annotations provide additional information or further explanation of the steps in the solution of an example. The *Student Resource Manual* (described earlier) was written just for you. It contains review material not found in the text, extra examples, information on calculators, solutions of problems, and answers to the *Final Examination.*

Those of us who teach and write mathematics texts strive to communicate clearly *how* to do mathematics. This text reflects our philosophy that a mathematics text for the beginning college/university-level student should be readable, straightforward, and loaded with motivation. The principal reason for studying precalculus is to become well prepared for calculus. To show you how the material covered in this text is essential for success in calculus, we end each chapter with a section called *Calculus Preview*. In each of these previews, a calculus topic provides a framework and motivation for precalculus mathematics and shows you how this mathematics is a vital part of the calculus problem.

Finally, we caution you that *learning* mathematics is not like learning how to ride a bicycle, that once learned, the ability sticks for a lifetime. Mathematics is more like learning another language or learning to play a musical instrument; it requires time and effort to memorize basic formulas and to understand when and how to apply them, and most importantly, it requires a lot of practice to develop and maintain proficiency. Even experienced musicians practice the fundamental scales before playing their instrument. So, ultimately, you the student can learn mathematics (that is, make it stick) only through the hard work of doing mathematics.

We wish you the best of luck in this preparatory course and in your subsequent study of calculus.

Acknowledgments

We are deeply indebted to our editor Tim Anderson and to the staffs of the editorial, production, and marketing departments at Jones & Bartlett Learning. Another job well done; kudos to everyone. Last, but certainly not least, we thank our production editor, Tiffany Sliter. It was a pleasure working with someone who has not only an expertise in all things related to book production, but as an unexpected bonus, an impressive knowledge of mathematics. We greatly appreciated her hard work, cooperation, never-ending patience with our multitude of tweaks to the final manuscript, and especially her good sense of humor.

We would also like to thank the students and instructors for their feedback on the previous editions as well as Melanie Fulton for the creation of the Test Bank.

To conclude, any mathematical or typographical errors that you may find in the text belong to the authors. We apologize for them in advance. In order to correct the errors expeditiously, please send them directly to our editor at: tanderson@jblearning.com.

Dennis G. Zill   Jacqueline M. Dewar

About the Cover

Imagine that, after reading and successfully completing all of the problems and review exercises in this text, you decide to reward yourself with a vacation to visit the galaxies shown on this book's cover. We suggest that you set aside a lot of time for your trip! Even if you could travel at the speed of light (approximately 671,000,000 miles per hour), it would take you about 31 million years to reach the Whirlpool Galaxy (M51A or NGC 5194) and its companion galaxy, M51B or NGC 5195.

The main spiral structure you see was discovered in 1773 by noted French astronomer Charles Messier and later named the Whirlpool Galaxy. The spherical, yellowish structure that seems to be attached to one of the arms of the spiral is actually a distinctly separate galaxy discovered in 1781 by Messier's assistant, Pierre Méchain (M51B/NGC 5195). Although these two galaxies appear to be intertwined, the image you see here taken by NASA's Hubble Space Telescope has helped modern astronomers to determine that M51B/NGC 5195 has been passing through and behind the Whirlpool Galaxy for hundreds of millions of years. In fact, the Whirlpool Galaxy's spiral structure is believed to have been formed by its gravitational interaction with this smaller galaxy.

After your 31 million year trip, you might find viewing the galaxies firsthand quite a different experience than looking at the image on this book's cover. The Hubble Space Telescope is capable of gathering a broad range of wavelengths from visible light through near-infrared light. The image you see has been enhanced and created in color to highlight nonvisible aspects of the galaxy such as gases and dust.

Perhaps instead of going to the Whirlpool Galaxy, you can take a short trip to Earth's nearest star. Traveling at the speed of light, you'll get there in about 8 minutes. Just don't forget your sunscreen.

1 Inequalities, Equations, and Graphs

Chapter Outline

The Real Line

≡ **Introduction** In calculus you will study quantities described by real numbers. Therefore, we begin with a review of the set of real numbers using the terminology and notation you will encounter in calculus.

☐ **Real Number System** Recall that the set R of **real numbers** consists of numbers that are either **rational** or **irrational**. Rational numbers are numbers of the form a/b, where a and $b \neq 0$ are integers. For example, -3, $-\frac{1}{2}$, $\frac{2}{3}$, 5, and $\frac{127}{4}$ are rational numbers. Irrational numbers are numbers that are not rational, that is, they are numbers that cannot be expressed as a quotient of integers. For example, $\sqrt{2}$ and π are irrational numbers. Every real number can also be written as a decimal. A rational number can be expressed either as a *terminating decimal*, such as $\frac{1}{8} = 0.125$, or a *nonterminating and repeating decimal*, such as $\frac{1}{3} = 0.333\ldots$. Repeating decimals, such as $0.666\ldots$ and $8.545454\ldots$, are often written as $0.\overline{6}$ and $8.\overline{54}$, respectively, where the bar indicates the digit or block of digits that repeat. An irrational number is always a *nonterminating and nonrepeating decimal* such as $\sqrt{2} = 1.41421\ldots$ or $\pi = 3.14159\ldots$. The following chart summarizes the relationship between the principal sets of real numbers.

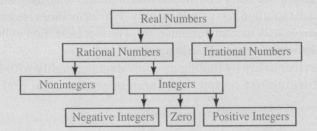

☐ **The Number Line** The set R of real numbers can be put into a one-to-one correspondence with the set of points on a line. As a consequence, we can visualize or represent real numbers as points on a *horizontal line* called the **number line** or **coordinate line**. The point chosen to represent the number 0 is called the **origin**. The direction to the right of 0 is said to be the **positive direction** on the number line; the direction to the left of 0 is the **negative direction**. Real numbers corresponding to points to the right of 0 are called **positive numbers** and numbers corresponding to points to the left of 0 are **negative numbers**. As indicated in **FIGURE 1.1.1**, the number 0 is considered to be neither positive nor negative. From here on, we will not distinguish between a point on the number line and the number that corresponds to this point.

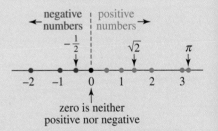

FIGURE 1.1.1 The real number line

☐ **Inequalities** The number line is useful in demonstrating order relations between two real numbers a and b. As shown in **FIGURE 1.1.2**, we say that the number a is **less than** the number b, and write $a < b$, whenever the number a lies to the left of the number b on the number line. Equivalently, because the number b lies to the right of a on the number line we say that b is **greater than** a and write $b > a$. For example, $4 < 9$ is the same as $9 > 4$. We also use the notation $a \leq b$ if the number a is either **less than**

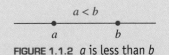

FIGURE 1.1.2 a is less than b

or **equal** to the number b. Similarly, $b \geq a$ means b is **greater than or equal** to a. For example, $2 \leq 5$ since $2 < 5$. Also, $4 \geq 4$ because $4 = 4$. For any two real numbers a and b, exactly *one* of the following is true:

$$a < b, \qquad a = b, \qquad \text{or} \qquad a > b.$$

The symbols $<$, $>$, $\geq$, and $\leq$ are called **inequality symbols** and expressions such as $a < b$ or $b \geq a$ are called **inequalities**. The inequality $a > 0$ means the number a lies to the right of the number 0 on the number line, and so a is **positive**. We signify that a number a is **negative** by the inequality $a < 0$. Because the inequality $a \geq 0$ means a is either greater than 0 (positive) or equal to 0 (which is neither positive nor negative), we say that a is **nonnegative**.

☐ **Solving Inequalities** We are interested in solving various kinds of inequalities containing a variable. If a real number a is substituted for the variable x in an inequality such as

$$8x + 4 < 16 + 5x, \tag{1}$$

and if the result is a true statement, then a is said to be a **solution** of the inequality. For example, -2 is a solution of (1) because if x is replaced by -2, then the resulting inequality $8(-2) + 4 < 16 + 5(-2)$ simplifies to the true statement $-12 < 6$. The word *solve* means that we are to find the set of *all* solutions of an inequality such as (1). This set is called the **solution set** of the inequality. Two inequalities are said to be **equivalent** if they have exactly the same solution set. The representation of the solution set on the number line is the **graph** of the inequality.

We solve an inequality by finding an equivalent inequality with obvious solutions. The following list summarizes three operations that yield equivalent inequalities.

THEOREM 1.1.1 Properties of Inequalities

Suppose a and b are real numbers and c is a nonzero real number. Then the inequality $a < b$ is equivalent to:

   (*i*) $a + c < b + c$,
   (*ii*) $ac < bc$,   for $c > 0$,
   (*iii*) $ac > bc$,   for $c < 0$.

Property (*iii*) of Theorem 1.1.1 is frequently forgotten. In words, (*iii*) states that:

> *If an inequality is multiplied by a negative number, then the direction of the resulting inequality is reversed.*

For example, if we multiply the inequality $-2 < 5$ by -3 then the *less than* symbol is changed to a *greater than* symbol:

$$-2(-3) > 5(-3) \qquad \text{or} \qquad 6 > -15.$$

EXAMPLE 1	Solving the Inequality (1)

Solve $8x + 4 < 16 + 5x$.

Solution We solve the inequality by using the properties of inequalities to obtain a sequence of equivalent inequalities:

$$8x + 4 < 16 + 5x$$
$$8x + 4 - 4 < 16 + 5x - 4 \quad \leftarrow \text{by } (i) \text{ of Theorem 1.1.1}$$
$$8x < 12 + 5x$$
$$8x - 5x < 12 + 5x - 5x \quad \leftarrow \text{by } (i) \text{ of Theorem 1.1.1}$$
$$3x < 12$$
$$\left(\tfrac{1}{3}\right)3x < \left(\tfrac{1}{3}\right)12 \quad \leftarrow \text{by } (ii) \text{ of Theorem 1.1.1}$$
$$x < 4.$$

Using set-builder notation, the solution set is $\{x | x \text{ real and } x < 4\}$.   ≡

☐ **Interval Notation** The solution set in Example 1 is graphed on the number line in **FIGURE 1.1.3** as a colored arrow over the line pointing to the left. In the figure, the right parenthesis at 4 indicates that the number 4 is *not* included in the solution set. Because the solution set extends indefinitely to the left—the negative direction—the inequality $x < 4$ can also be written as $-\infty < x < 4$, where ∞ is the infinity symbol. In other words, the solution set of the inequality $x < 4$ is

$$\{x | x \text{ real and } x < 4\} = \{x | -\infty < x < 4\}.$$

Using **interval notation** this set of real numbers is written $(-\infty, 4)$ and is an example of an **unbounded interval**. Table 1.1.1 summarizes various inequalities and their solution sets, as well as interval notations, names, and graphs. In each of the first four entries

FIGURE 1.1.3 Solution set in Example 1 in interval notation is $(-\infty, 4)$

TABLE 1.1.1	Inequalities and Intervals

Inequality	Solution Set	Interval Notation	Name	Graph
$a < x < b$	$\{x | a < x < b\}$	(a, b)	Open interval	
$a \leq x \leq b$	$\{x | a \leq x \leq b\}$	$[a, b]$	Closed interval	
$a < x \leq b$	$\{x | a < x \leq b\}$	$(a, b]$	Half-open interval	
$a \leq x < b$	$\{x | a \leq x < b\}$	$[a, b)$	Half-open interval	
$a < x$	$\{x | a < x < \infty\}$	(a, ∞)		
$a \leq x$	$\{x | a \leq x < \infty\}$	$[a, \infty)$	Unbounded intervals	
$x < b$	$\{x | -\infty < x < b\}$	$(-\infty, b)$		
$x \leq b$	$\{x | -\infty < x \leq b\}$	$(-\infty, b]$		
$-\infty < x < \infty$	$\{x | -\infty < x < \infty\}$	$(-\infty, \infty)$		

of the table, the numbers a and b are called the **endpoints** of the interval. As a set, the **open interval**

$$(a, b) = \{x \mid a < x < b\}$$

does not include either endpoint, whereas the **closed interval**

$$[a, b] = \{x \mid a \le x \le b\}$$

includes both endpoints. Note, too, that the graph of the last interval in Table 1.1.1, which extends indefinitely both to the left and to the right, is the entire real number line. In calculus the interval notation $(-\infty, \infty)$ is generally used to represent the set R of real numbers.

A word of caution is in order as you peruse Table 1.1.1. The **infinity symbols** $-\infty$ ("minus infinity") and ∞ ("infinity") do not represent real numbers and should *never* be manipulated arithmetically like a number. The infinity symbols are merely notational devices: $-\infty$ and ∞ are used to indicate unboundedness in the negative direction and in the positive direction, respectively. Thus when using interval notation, the symbols $-\infty$ and ∞ can never appear next to a square bracket. For example, the expression $(2, \infty]$ is meaningless.

An inequality of the form $a < x < b$ is sometimes referred to as a **simultaneous inequality** because the number x is *between* the numbers a and b. In other words, $x > a$ *and* simultaneously $x < b$. For example, the real numbers that satisfy $2 < x < 5$ is the intersection of the intervals defined by the inequalities $2 < x$ and $x < 5$. Recall that the **intersection** of two sets A and B, written $A \cap B$, is the set of elements that are in A *and* in B—in other words, the elements that are common to both sets. As illustrated in **FIGURE 1.1.4** by the overlapping arrows extending indefinitely to the right and to the left, the solution set of the inequality $2 < x < 5$ can be written as the intersection $(2, \infty) \cap (-\infty, 5) = (2, 5)$.

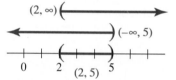

FIGURE 1.1.4 The numbers in $(2, 5)$ are the numbers common to both $(2, \infty)$ and $(-\infty, 5)$

▊ EXAMPLE 2   Solving a Simultaneous Inequality

Solve $-2 \le 1 - 2x < 3$.

Solution As previously discussed, one way of proceeding is to solve two inequalities:

$$-2 \le 1 - 2x \qquad \text{and} \qquad 1 - 2x < 3$$

and then take the intersection of the two solution sets. A faster method is to solve both of the inequalities simultaneously in the following manner:

$$-2 \le 1 - 2x < 3$$
$$-1 - 2 \le -1 + 1 - 2x < -1 + 3 \qquad \leftarrow \text{by } (i) \text{ of Theorem 1.1.1}$$
$$-3 \le -2x < 2.$$

We isolate the variable x in the middle of the last simultaneous inequality by multiplying by $-\frac{1}{2}$:

$$\left(-\tfrac{1}{2}\right)(-3) \ge \left(-\tfrac{1}{2}\right)(-2x) > \left(-\tfrac{1}{2}\right)2 \qquad \leftarrow \text{by } (iii) \text{ of Theorem 1.1.1}$$
$$\tfrac{3}{2} \ge x > -1,$$

where we note that multiplication by the negative number has reversed the direction of the inequalities. To express this inequality in interval notation, we first rewrite it with the leftmost number on the number line on the left side of the inequality: $-1 < x \le \frac{3}{2}$. The solution set of the last inequality is the half-open interval $\left(-1, \frac{3}{2}\right]$; the square bracket on the right signifies that $\frac{3}{2}$ is included in the solution set. The graph of this interval is given in **FIGURE 1.1.5**. ≡

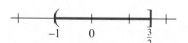

FIGURE 1.1.5 Solution set in Example 2

CHAPTER 1 INEQUALITIES, EQUATIONS, AND GRAPHS

☐ **Sign-Chart Method** In Examples 1 and 2 we solved **linear inequalities** in one variable x, that is, inequalities that can be put into one of the forms

$$ax + b < 0, \quad ax + b \leq 0, \quad ax + b > 0, \quad ax + b \geq 0,$$

where $a \neq 0$. In the next several examples we illustrate the **sign-chart method** used in calculus for solving **nonlinear inequalities**. A nonlinear inequality in one variable x is simply an inequality that is not linear. For example, $x^2 \geq -2x + 15$ is a nonlinear inequality because of the presence of the x^2 term. The two properties of real numbers given next are fundamental to constructing a sign chart of an inequality.

THEOREM 1.1.2 Sign Properties of Products

(*i*) The product of two real numbers is **positive** if and only if the numbers have the same signs, that is, either $(+)(+)$ or $(-)(-)$.

(*ii*) The product of two real numbers is **negative** if and only if the numbers have opposite signs, that is, $(+)(-)$ or $(-)(+)$.

Here are some of the basic steps of the sign-chart method illustrated in the next example.

Guidelines for the Sign Chart Method

- Use the properties of inequalities to recast the given inequality into a form where all variables and nonzero constants are on the same side of the inequality symbol and the number 0 on the other side.
- Then, if possible, factor the expression involving the variables and constants into linear factors $ax + b$.
- Mark the number line at the points where the factors are zero. These points divide the number line into intervals.
- In each of these intervals, determine the sign of each factor and the corresponding sign of the product using (*i*) and (*ii*) of Theorem 1.1.2.

EXAMPLE 3 Solving a Nonlinear Inequality

Solve $x^2 \geq -2x + 15$.

Solution We begin by rewriting the inequality with all terms to the left of the inequality symbol and 0 to the right. By (*i*) of Theorem 1.1.1,

$$x^2 \geq -2x + 15 \quad \text{is equivalent to} \quad x^2 + 2x - 15 \geq 0.$$

Factoring, the last expression is the same as $(x + 5)(x - 3) \geq 0$.

Then we indicate on the number line where each factor is 0—in this case, $x = -5$ and $x = 3$. As shown in FIGURE 1.1.6, this divides the number line into three disjoint, or nonintersecting, intervals: $(-\infty, -5)$, $(-5, 3)$, and $(3, \infty)$. Note, too, that since the given inequality requires the product to be nonnegative, that is, "greater than or *equal to* 0," the numbers -5 and 3 are two solutions. Next, we must determine the signs of the factors $x + 5$ and $x - 3$ on each of the three intervals. We are looking for those intervals on which the two factors are either both positive or both negative, for then their product will be positive. Since the linear factors $x + 5$ and $x - 3$ cannot change signs within these intervals, it suffices to obtain the sign of each factor at just *one* test value

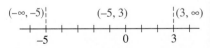

FIGURE 1.1.6 Three disjoint intervals

◀ See (*i*) of Theorem 1.1.2

chosen from inside each interval. For example, on the interval $(-\infty, -5)$, if we use $x = -10$, then

Interval	$(-\infty, -5)$	
Sign of $x + 5$	$-$	$\leftarrow$ at $x = -10, x + 5 = -10 + 5 < 0$
Sign of $x - 3$	$-$	$\leftarrow$ at $x = -10, x - 3 = -10 - 3 < 0$
Sign of $(x + 5)(x - 3)$	$+$	$\leftarrow (-)(-)$ is $(+)$

Continuing in this manner for the remaining two intervals we get the sign chart in FIGURE 1.1.7. As can be seen from the third line of this figure, the product $(x + 5)(x - 3)$ is nonnegative on either of the unbounded intervals $(-\infty, -5]$ or $[3, \infty)$.

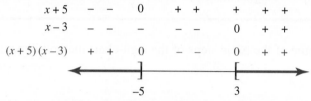

FIGURE 1.1.7 Sign chart for Example 3

Because the solution set in Example 3 consists of two nonintersecting, or **disjoint**, intervals it cannot be expressed as a single interval. The best we can do is to write the solution set as the union of the two intervals. Recall that the **union** of two sets A and B, written $A \cup B$, is the set of elements that are in either A or in B, or in both. Thus the solution set in Example 3 can be written $(-\infty, -5] \cup [3, \infty)$.

■ EXAMPLE 4   Solving a Nonlinear Inequality

Solve $(x - 4)^2(x + 8)^3 > 0$.

Solution Since the given inequality already has the form appropriate for the sign-chart method (a factored expression to the left of the inequality symbol and 0 to the right), we begin by finding the numbers where each factor is 0, in this case, $x = 4$ and $x = -8$. We place these numbers on the number line and determine three intervals. Then in each interval we consider the signs of the powers of each linear factor. Because of the even power, we see that $(x - 4)^2$ is never negative. However, because of the odd power, $(x + 8)^3$ has the same sign as the factor $x + 8$. Observe that the numbers $x = 4$ and $x = -8$ are not solutions of the inequality because of the "greater than" symbol. Therefore, as we see in FIGURE 1.1.8, the solution set is $(-8, 4) \cup (4, \infty)$.

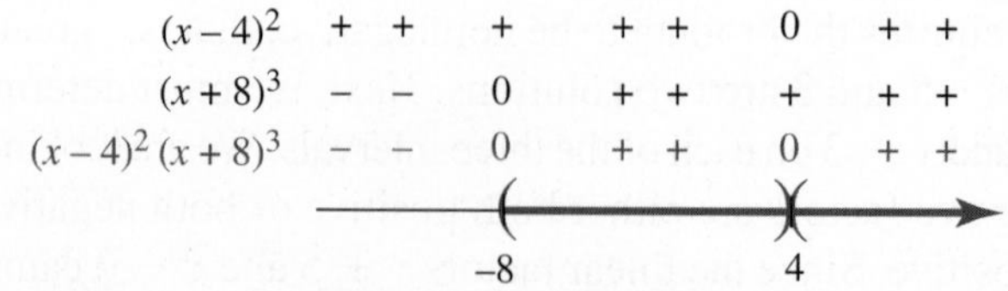

$$
\begin{array}{c}
(x-4)^2 \quad + \quad + \quad + \quad\quad + \quad + \quad\quad 0 \quad\quad + \quad + \\
(x+8)^3 \quad - \quad - \quad\quad 0 \quad\quad + \quad + \quad\quad + \quad\quad + \quad + \\
(x-4)^2\,(x+8)^3 \quad - \quad - \quad\quad 0 \quad\quad + \quad + \quad\quad 0 \quad\quad + \quad +
\end{array}
$$

FIGURE 1.1.8 Sign chart for Example 4

CHAPTER 1 INEQUALITIES, EQUATIONS, AND GRAPHS

EXAMPLE 5
Solving a Nonlinear Inequality

Solve $x \le 3 - \dfrac{6}{x + 2}$.

Solution We begin by rewriting the inequality with all variables and nonzero constants to the left and 0 to the right of the inequality sign,

$$x - 3 + \frac{6}{x + 2} \le 0.$$

Next we put the terms over a common denominator,

$$\frac{(x - 3)(x + 2) + 6}{x + 2} \le 0 \quad \text{and simplify to} \quad \frac{x(x - 1)}{x + 2} \le 0. \qquad (2)$$

One thing we *don't do* is clear the denominator by multiplying the inequality by $x + 2$. See Problem 70 in Exercises 1.1.

Now the numbers that make the three linear factors in the last expression equal to 0 are $-2, 0,$ and 1. On the number line these three numbers determine four intervals. As a result of the "less than or *equal to* 0," we see that 0 and 1 are members of the solution set. However, -2 is excluded from the solution set since substituting this value into the fractional expression results in a zero denominator (making the fraction undefined). As we can see from the sign chart in FIGURE 1.1.9, the solution set is $(-\infty, -2) \cup [0, 1]$.

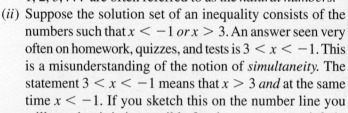

FIGURE 1.1.9 Sign chart for Example 5

NOTES FROM THE CLASSROOM

(*i*) Terminology used in mathematics often varies from teacher to teacher and from textbook to textbook. For example, inequalities using the symbols $<$ or $>$ are sometimes called *strict* inequalities, whereas inequalities using $\le$ or $\ge$ are called *nonstrict*. As another example, the *positive integers* $1, 2, 3, \ldots$ are often referred to as the *natural numbers*.

(*ii*) Suppose the solution set of an inequality consists of the numbers such that $x < -1$ *or* $x > 3$. An answer seen very often on homework, quizzes, and tests is $3 < x < -1$. This is a misunderstanding of the notion of *simultaneity*. The statement $3 < x < -1$ means that $x > 3$ *and* at the same time $x < -1$. If you sketch this on the number line you will see that it is impossible for the same x to satisfy both inequalities. The best we can do in rewriting "$x < -1$ or $x > 3$" is to use the union of intervals $(-\infty, -1) \cup (3, \infty)$.

(*iii*) Here is another frequent error: The notation $a < x > b$ is meaningless. If, say, we have $x > -2$ *and* $x > 6$, then only the numbers $x > 6$ satisfy *both* conditions.

(*iv*) In the classroom we frequently hear the response "positive" when in reality the student means "nonnegative." Question: x under the square root sign $\sqrt{x}$ must be positive, right? Raise your hand if you agree. Invariably, lots of hands go up. Correct answer: x must be nonnegative, that is, $x \ge 0$. Don't forget that $\sqrt{0} = 0$.

In Problems 1–6, write the given statement as an inequality.

1. $a + 2$ is positive
2. $4y$ is negative
3. $a + b$ is nonnegative
4. a is less than -3
5. $2b + 4$ is greater than or equal to 100
6. $c - 1$ is less than or equal to 5

In Problems 7–14, write the given inequality using interval notation and then graph the interval.

7. $x < 0$
8. $0 < x < 5$
9. $x \geq 5$
10. $-1 \leq x$
11. $8 < x \leq 10$
12. $-5 < x \leq -3$
13. $-2 \leq x \leq 4$
14. $x > -7$

In Problems 15–18, write the given interval as an inequality.

15. $[-7, 9]$
16. $[1, 15)$
17. $(-\infty, 2)$
18. $[-5, \infty)$

In Problems 19–34, solve the given linear inequality. Write the solution set using interval notation. Graph the solution set.

19. $x + 3 > -2$
20. $3x - 9 < 6$
21. $\frac{3}{2}x + 4 \leq 10$
22. $5 - \frac{5}{4}x \geq -4$
23. $\frac{3}{2} - x > x$
24. $-(1 - x) \geq 2x - 1$
25. $2 + x \geq 3(x - 1)$
26. $-7x + 3 \leq 4 - x$
27. $-\frac{20}{3} < \frac{2}{3}x < 4$
28. $-3 \leq -x < 2$
29. $-7 < x - 2 < 1$
30. $3 < x + 4 \leq 10$
31. $7 < 3 - \frac{1}{2}x \leq 8$
32. $100 + x \leq 41 - 6x \leq 121 + x$
33. $-1 \leq \dfrac{x - 4}{4} < \frac{1}{2}$
34. $2 \leq \dfrac{4x + 2}{-3} \leq 10$

In Problems 35–58, solve the given nonlinear inequality. Write the solution set using interval notation. Graph the solution set.

35. $x^2 - 9 < 0$
36. $x^2 \geq 16$
37. $x(x - 5) \geq 0$
38. $4x^2 + 7x < 0$
39. $x^2 - 8x + 12 < 0$
40. $(3x + 2)(x - 1) \leq 0$
41. $9x \geq 2x^2 - 18$
42. $4x^2 > 9x + 9$
43. $(x + 1)(x - 2)(x - 4) < 0$
44. $(1 - x)\left(x + \frac{1}{2}\right)(x - 3) \leq 0$
45. $(x^2 - 1)(x^2 - 4) \leq 0$
46. $(x - 1)^2(x + 3)(x - 5) \geq 0$
47. $\dfrac{5}{x + 8} < 0$
48. $\dfrac{10}{x^2 + 2} > 0$
49. $\dfrac{5}{x} \geq -1$
50. $\dfrac{x - 3}{x + 2} < 0$
51. $\dfrac{x + 1}{x - 1} + 2 > 0$
52. $\dfrac{x - 2}{x + 3} \leq 1$
53. $\dfrac{x(x - 1)}{x + 5} \geq 0$
54. $\dfrac{(1 + x)(1 - x)}{x} \leq 0$

55. $\dfrac{x^2 - 2x + 3}{x + 1} \leq 1$

56. $\dfrac{x}{x^2 - 16} > 0$

57. $\dfrac{2}{x + 3} - \dfrac{1}{x + 1} < 0$

58. $\dfrac{4x + 5}{x^2} \geq \dfrac{4}{x + 5}$

59. If 7 times a number is decreased by 6, the result is less than 50. What can be determined about the number?

60. The sides of a square are extended to form a rectangle. As shown in **FIGURE 1.1.10**, one side is extended 2 inches and the other side is extended 5 inches. If the area of the resulting rectangle is less than 130 in.², what are the possible lengths of a side of the original square?

61. A polygon is a closed figure made by joining line segments. For example, a *triangle* is a three-sided polygon. Shown in **FIGURE 1.1.11** is an eight-sided polygon called an *octagon*. A *diagonal* of a polygon is defined to be a line segment that joins any two nonadjacent vertices. The number of diagonals d in a polygon with n sides is given by $d = \frac{1}{2}(n - 1)n - n$. For what polygons will the number of diagonals exceed 35?

62. The total number N of dots in a triangular array with n rows is given by the formula $N = \frac{1}{2}n(n + 1)$. See **FIGURE 1.1.12**. How many rows can the array have if the total number of dots is to be less than 5050?

FIGURE 1.1.10 Rectangle in Problem 60

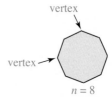

FIGURE 1.1.11 Octagon in Problem 61

$n = 1$ • $n = 2$ • •

$n = 3$ • • • $n = 4$ • • • •

FIGURE 1.1.12 Triangular arrays of dots in Problem 62

Miscellaneous Applications

63. Flower Garden A rectangular flower bed is to be twice as long as it is wide. If the area enclosed must be greater than 98 m², what can you conclude about the width of the flower bed?

64. Fever The relationship between degrees Celsius T_C and degrees Fahrenheit T_F is given by $T_F = \frac{9}{5}T_C + 32$. A person is considered to have a fever if he or she has an oral temperature greater than 98.6°F. What temperatures on the Celsius scale indicate a fever?

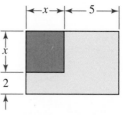

Oral thermometer

65. Parallel Resistors A 5-ohm resistor and a variable resistor are placed in parallel. The resulting resistance is given by $R_T = \dfrac{5R}{5 + R}$. Determine the values of the variable resistor R for which the resulting resistance R_T will be greater than 2 ohms.

66. What Goes Up ... With the aid of calculus it is easy to show that the height s of a projectile launched straight upward from an initial height s_0 with an initial velocity v_0 is given by $s = -\frac{1}{2}gt^2 + v_0 t + s_0$, where t is in seconds and $g = 32$ ft/s². If a toy rocket is shot straight upward from ground level, then $s_0 = 0$. If its initial velocity is 72 ft/s, during what time interval will the rocket be more than 80 ft above the ground?

67. Linear Depreciation The value V of a new car, which cost $50,000 initially, when depreciated linearly over 20 years is given by $V = 50,000(1 - x/20)$, where x represents years. Determine the values of x such that $0 < V < 20,000$.

68. Pulse Rate The pulse rate of a healthy person while engaged in aerobic exercises can vary widely. To obtain the maximum beneficial effect from the exercises, the pulse rate P_R should be maintained in a certain interval $[a, b]$. For jogging, one mathematical model determines the endpoints a and b of that interval subtracting the jogger's age from 220 and multiplying the result by 0.70 and 0.85, respectively. Write the desired interval for the pulse rate of a 40-year old jogger as a simultaneous inequality.

For Discussion

69. Discuss how you might determine the set of numbers for which the given expression is a real number.

    (a) $\sqrt{2x - 3}$ (b) $\sqrt{4 - 10x}$ (c) $\sqrt{x(x - 5)}$ (d) $\dfrac{1}{\sqrt{x + 2}}$

    Carry out your ideas.

70. In Example 5, explain why one should not multiply the last expression in (2) by $x + 2$.

71. (a) If $0 < a < b$, then use the properties of inequalities to show that $a^2 < b^2$.
    (b) If $a < b$, then explain why, in general, $a^2 < b^2$ is *not* true.

72. If $0 < a < b$, then use the properties of inequalities to show that $0 < \sqrt{a} < \sqrt{b}$. [*Hint*: One way of proceeding is to use the factorization $b - a = (\sqrt{b} + \sqrt{a})(\sqrt{b} - \sqrt{a}) > 0$.]

73. If a and b are real numbers, then the number $(a + b)/2$ is called the **arithmetic mean**, or **average**, of a and b. Use the properties of inequalities to show that if $a < b$, then $a < \dfrac{a + b}{2} < b$.

74. If a and b are positive real numbers, then the number $\sqrt{ab}$ is called the **geometric mean** of a and b. Use the properties of inequalities to show that if $0 < a < b$, then $a < \sqrt{ab} < b$.

75. If $0 < a < b$, then show that the geometric mean of a and b is less than the arithmetic mean of a and b, that is, $\sqrt{ab} < \dfrac{a + b}{2}$. See Problems 73 and 74.

76. Using the definition in Problem 74 as a model, how would you define the geometric mean of three positive numbers a, b, and c? Of n positive numbers?

77. Do a little bit of research and find an application where the geometric mean of positive real numbers is used rather than the arithmetic mean of the numbers. You might have to use Problem 76.

1.2 Absolute Value

≡ **Introduction** We can use the number line to picture distance. As shown in FIGURE 1.2.1, the distance between the number 0 and the number 3 is 3, and the distance between -3 and 0 is also 3. In general, for any *positive* real number x, the distance between x and 0 is x. If x represents a *negative* number, then the distance between x and 0 is $-x$. The concept of distance from a number on the number line to the number 0 is described by the **absolute value** of that number.

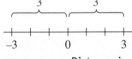

FIGURE 1.2.1 Distance is 3 units

DEFINITION 1.2.1 Absolute Value

For any real number x, the **absolute value** of x, denoted by $|x|$, is

$$|x| = \begin{cases} x, & \text{if } x \geq 0 \\ -x, & \text{if } x < 0. \end{cases} \tag{1}$$

Be careful. It is a common mistake to think that the symbol $-x$ represents a negative quantity simply because of the presence of the minus sign. If a symbol x represents a negative number (that is, $x < 0$), then $-x$ is a positive number. For example, if $x = -10 < 0$, then $|x| = -x = -(-10) = 10$.

As our first example shows, the symbol x in (1) is a placeholder. Other quantities can be placed inside the absolute value symbols $|\,|$.

EXAMPLE 1   Absolute Value

Write $|x - 5|$ without absolute value symbols.

Solution Wherever the symbol x appears in (1) we replace it by $x - 5$:

$$|x - 5| = \begin{cases} x - 5, & \text{if } x - 5 \geq 0 \\ -(x - 5), & \text{if } x - 5 < 0. \end{cases}$$

Let's consider each part of the foregoing definition separately. First, the inequality $x - 5 \geq 0$ means that $x \geq 5$. Therefore,

$$|x - 5| = x - 5 \quad \text{if} \quad x \geq 5.$$

Check this result (that is, $x - 5$ is nonnegative) by substituting numbers such as 5, 8, and 10. Next, $x - 5 < 0$ means that $x < 5$. In this case,

distributive law
$$\downarrow \qquad \qquad \downarrow$$
$$|x - 5| = -(x - 5) = -x + 5 \quad \text{if} \quad x < 5.$$

Again, you should convince yourself that this is correct (that is, $-x + 5$ is positive) by substituting a few numbers, such as 2 and -3. $\equiv$

As illustrated in Figure 1.2.1, for any real number x and its negative $-x$, the distance to 0 is the same. That is, $|x| = |-x|$. This is one property in a list of properties of the absolute value that is given next.

THEOREM 1.2.1   Properties of Absolute Values

(i) $|a| = |-a|$
(ii) $|a| = 0$ if and only if $a = 0$
(iii) $|ab| = |a||b|$
(iv) $\left|\dfrac{a}{b}\right| = \dfrac{|a|}{|b|}, \quad b \neq 0$
(v) $|a + b| \leq |a| + |b|$ (**Triangle inequality**)

For example, by virtue of property (iii) of Theorem 1.2.1 we can rewrite the expression $|-2x|$ as $|-2||x| = 2|x|$.

☐ **Distance Again** If we wish to find the distance between any two numbers on the real number line, then all we have to do is subtract the leftmost number from the rightmost number. For example, the distance between 10 and -2 is

rightmost number   leftmost number
$$\downarrow \qquad\qquad\qquad \downarrow$$
$$10 - (-2) = 12.$$

As we saw in the introduction, the distance between -3 and 0 is $0 - (-3) = 3$. If an absolute value is used to define the distance, then we do not have to worry about the order of subtraction.

DEFINITION 1.2.2   Distance Between Two Numbers

If a and b are any two numbers on the number line, the **distance** between a and b is

$$d(a, b) = |b - a|. \qquad (2)$$

Using the properties of absolute values,

by property *(iii)* of Theorem 1.2.1
$$\downarrow \qquad \qquad \downarrow \quad \downarrow$$
$$|b - a| = |(-1)(a - b)| = |-1|\,|a - b| = |a - b|,$$

and so we have $d(a, b) = d(b, a)$. For example, the distance between $\sqrt{2}$ and 3 is

$$d(\sqrt{2}, 3) = |3 - \sqrt{2}| = 3 - \sqrt{2}$$

because $3 > \sqrt{2}$ or $3 - \sqrt{2} > 0$, or

$$d(3, \sqrt{2}) = |\sqrt{2} - 3| = -(\sqrt{2} - 3) = 3 - \sqrt{2}$$

because $\sqrt{2} < 3$ or $\sqrt{2} - 3 < 0$.

☐ **Midpoint** Suppose a and b represent two distinct numbers on the number line such that $a < b$. The **midpoint** m of the line segment between the numbers a and b is given by the average of the two endpoints of the interval $[a, b]$, that is

$$m = \frac{a + b}{2}. \qquad (3)$$

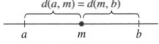

$$d(a, m) = d(m, b)$$

FIGURE 1.2.2 Midpoint m between a and b

As shown in **FIGURE 1.2.2**, (3) is easy to verify by using (2) to show that $d(a, m) = d(m, b)$.

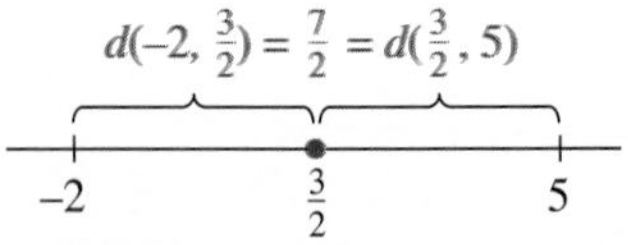

$$d\left(-2, \tfrac{3}{2}\right) = \tfrac{7}{2} = d\left(\tfrac{3}{2}, 5\right)$$

FIGURE 1.2.3 Midpoint in Example 2

EXAMPLE 2   **Midpoint**

From (3), the midpoint of the line segment joining the numbers -2 and 5 is

$$\frac{(-2) + 5}{2} = \frac{3}{2}.$$

See **FIGURE 1.2.3**.   ≡

☐ **Equations** Since *(i)* of Theorem 1.2.1 implies that $|-6| = |6| = 6$, we can conclude that the simple equation $|x| = 6$ has two solutions, either $x = -6$ or $x = 6$. In general, if a is a positive real number, then

$$|x| = a \qquad \text{if and only if} \qquad x = a \qquad \text{or} \qquad x = -a. \qquad (4)$$

EXAMPLE 3   **An Absolute-Value Equation**

Solve **(a)** $|5x - 3| = 8$ **(b)** $|x - 4| = -3$.

Solution **(a)** In (4) the symbol x is a placeholder for any quantity. By replacing x by $5x - 3$, the given equation is equivalent to two equations

$$5x - 3 = 8 \qquad \text{or} \qquad 5x - 3 = -8.$$

We solve each of these. From $5x - 3 = 8$, we obtain

$$5x = 11 \quad \text{which implies} \quad x = \frac{11}{5}.$$

From $5x - 3 = -8$, we have

$$5x = -5 \quad \text{which implies} \quad x = -1.$$

Therefore, the solutions are $\frac{11}{5}$ and -1.

(b) Since the absolute value of a real number is always nonnegative, there is no solution to an equation such as $|x - 4| = -3$. ≡

□ **Absolute-Value Inequalities** Many important applications of inequalities involve absolute values. We have just seen that $|x|$ represents the distance along the number line between the number x and the number 0. Thus the inequality $|x| < a$, where $a > 0$, means that the distance between x and 0 is less than a. We can see in FIGURE 1.2.4(a) that this is the set of real numbers x such that $-a < x < a$. On the other hand, $|x| > a$ means that the distance between x and 0 is greater than a. In Figure 1.2.4(b), we see that these are the numbers that satisfy either $x > a$ or $x < -a$. These graphical observations suggest two additional properties of absolute value.

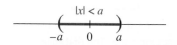

(a) The distance between x and 0 is *less* than a

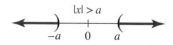

(b) The distance between x and 0 is *greater* than a

FIGURE 1.2.4 Graphical interpretation of (*i*) and (*ii*) of Theorem 1.2.2

THEOREM 1.2.2 Absolute-Value Inequalities

Let a be a positive real number.

   (*i*) $\ |x| < a \ $ if and only if $\ -a < x < a$.
   (*ii*) $|x| > a \ $ if and only if $\ x > a$ or $x < -a$.

    Properties (*i*) and (*ii*) of Theorem 1.2.2 also hold with the inequality symbols $<$ and $>$ are replaced by $\leq$ and $\geq$, respectively.

EXAMPLE 4　　**Two Absolute-Value Inequalities**

(a) From (*i*) of Theorem 1.2.2, the absolute-value inequality $|x| < 1$ is equivalent to the simultaneous inequality $-1 < x < 1$.

(b) From (*ii*) of Theorem 1.2.2, the absolute-value inequality $|x| \geq 5$ is equivalent to two inequalities: $x \geq 5$ or $x \leq -5$. ≡

EXAMPLE 5　　**Two Absolute-Value Inequalities**

Solve **(a)** $|3x - 7| < 1$　　**(b)** $|2x - 5| \leq 0$.

Solution (a) As in Example 3, the symbol x in the inequality $|x| < a$ is simply a placeholder for other quantities. If we replace x by $3x - 7$ and a by the number 1, then (*i*) of Theorem 1.2.2 yields the simultaneous inequality

$$-1 < 3x - 7 < 1$$

which we solve in the usual manner (see Example 2 in Section 1.1):

$$-1 + 7 < 3x - 7 + 7 < 1 + 7$$
$$6 < 3x < 8$$
$$\left(\tfrac{1}{3}\right)6 < \left(\tfrac{1}{3}\right)3x < \left(\tfrac{1}{3}\right)8$$
$$2 < x < \tfrac{8}{3}.$$

FIGURE 1.2.5 Solution set in Example 5

The solution set is the open interval $\left(2, \tfrac{8}{3}\right)$ shown in **FIGURE 1.2.5**.

(b) Since the absolute value of any expression is never negative, the values of x that satisfy the inequality $\leq$ are those for which $|2x - 5| = 0$. By *(ii)* of Theorem 1.2.1 we conclude that $2x - 5 = 0$. Hence the only solution is $\tfrac{5}{2}$. ≡

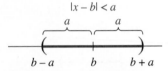

FIGURE 1.2.6 The distance between x and b is *less* than a

An absolute-value inequality such as $|x - b| < a$ can also be interpreted in terms of distance along the number line. Since $|x - b|$ is distance between x and b, the inequality $|x - b| < a$ is satisfied by all real numbers x whose distance between x and b is less than a. This interval is shown in **FIGURE 1.2.6**. Note that when $b = 0$ we get *(i)* of Theorem 1.2.2. Similarly, the set of numbers satisfying $|x - b| > a$ are the numbers x whose distance between x and b is greater than a.

EXAMPLE 6 An Absolute-Value Inequality

Solve $\left|4 - \tfrac{1}{2}x\right| \geq 7$.

Solution If we replace x and a in $|x| \geq a$ by $4 - \tfrac{1}{2}x$ and 7, respectively, then we see from *(ii)* of Theorem 1.2.2 that $\left|4 - \tfrac{1}{2}x\right| \geq 7$ is equivalent to the two different inequalities

$$4 - \tfrac{1}{2}x \geq 7 \qquad \text{or} \qquad 4 - \tfrac{1}{2}x \leq -7.$$

We solve each of these inequalities separately. First, we solve

$$4 - \tfrac{1}{2}x \geq 7$$
$$-\tfrac{1}{2}x \geq 3$$
$$x \leq -6. \quad \leftarrow \begin{cases} \text{multiplication by } -2 \text{ reverses} \\ \text{the direction of the inequality} \end{cases}$$

In interval notation the solution set of this inequality is $(-\infty, -6]$. Next, we solve

$$4 - \tfrac{1}{2}x \leq -7$$
$$-\tfrac{1}{2}x \leq -11$$
$$(-2)\left(-\tfrac{1}{2}\right)x \geq (-2)(-11) \quad \leftarrow \begin{cases} \text{multiplication by } -2 \text{ reverses} \\ \text{the direction of the inequality} \end{cases}$$
$$x \geq 22.$$

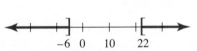

FIGURE 1.2.7 Solution set in Example 6

In interval notation the solution set is $[22, \infty)$.

Since the two intervals are disjoint, the solution set is the union of intervals: $(-\infty, -6] \cup [22, \infty)$. The graph of this solution set is shown in **FIGURE 1.2.7**. ≡

Note in Figure 1.2.4(a) that the number 0 is the midpoint of the solution interval for $|x| < a$ and in Figure 1.2.6 that the number b is the midpoint of the solution interval for the inequality $|x - b| < a$. With this in mind, work through the next example.

CHAPTER 1 INEQUALITIES, EQUATIONS, AND GRAPHS

| EXAMPLE 7 | Constructing an Inequality |

Find an inequality of the form $|x - b| < a$ for which the open interval $(4, 8)$ is its solution set.

Solution The midpoint of the interval $(4, 8)$ is $m = \dfrac{4 + 8}{2} = 6$. The distance between the midpoint m and one of the endpoints of the interval is $d(m, 8) = |8 - 6| = 2$. Therefore the required inequality is $|x - 6| < 2$. ≡

1.2 Exercises Answers to selected odd-numbered problems begin on page ANS-1.

In Problems 1–6, write the given quantity without the absolute value symbols.

1. $|\pi - 4|$ **2.** $|\sqrt{5} - 3|$
3. $|8 - \sqrt{63}|$ **4.** $|\sqrt{5} - 2.3|$
5. $|-6| - |-2|$ **6.** $||-3| - |10||$

In Problems 7–12, write the given expression without the absolute value symbols.

7. $|h|$, if h is negative   **8.** $|-h|$, if h is negative
9. $|x - 6|$, if $x < 6$ **10.** $|2x - 1|$, if $x \geq \frac{1}{2}$
11. $|x - y| - |y - x|$ **12.** $\dfrac{|x - y|}{|y - x|}, x \neq y$

In Problems 13–16, write the expression $|x - 2| + |x - 5|$ without the absolute value symbols if x is in the given interval.

13. $(-\infty, 1)$ **14.** $(7, \infty)$ **15.** $(3, 4]$ **16.** $[2, 5]$

In Problems 17–20, write the expression $|x + 1| - |x - 3|$ without the absolute value symbols if x is in the given interval.

17. $[-1, 3)$ **18.** $(0, 1)$ **19.** (π, ∞) **20.** $(-\infty, -5)$

In Problems 21–24, find the distance between the given numbers and find the midpoint of the line segment between them.

21. $3, 7$ **22.** $-100, 255$ **23.** $-\frac{3}{2}, \frac{3}{2}$ **24.** $-\frac{1}{4}, \frac{7}{4}$

In Problems 25–28, m is the midpoint of the line segment joining a (the left endpoint) and b (the right endpoint). Use the given conditions to find the indicated quantities.

25. $m = 5, d(a, m) = 3$; a and b **26.** $m = -1, d(m, b) = 2$; a and b
27. $a = 4, d(a, m) = \pi$; m and b **28.** $a = 10, d(m, b) = 5$; m and b

In Problems 29–34, solve the given equation.

29. $|4x - 1| = 2$ **30.** $|5v - 4| = 7$
31. $|\frac{1}{4} - \frac{3}{2}y| = 1$ **32.** $|2 - 16t| = 0$
33. $\left|\dfrac{x}{x - 1}\right| = 2$ **34.** $\left|\dfrac{x + 1}{x - 2}\right| = 4$

In Problems 35–46, solve the given inequality. Write the solution set using interval notation. Graph the solution set.

35. $|-5x| < 4$

36. $|3x| > 18$

37. $|3 + x| > 7$

38. $|x - 4| \leq 9$

39. $|2x - 7| \leq 1$

40. $|5 - \frac{1}{3}x| < \frac{1}{2}$

41. $|x + \sqrt{2}| \geq 1$

42. $|6x + 4| > 4$

43. $\left|\dfrac{3x - 1}{-4}\right| < 2$

44. $\left|\dfrac{2 - 5x}{3}\right| \geq 5$

45. $|x - 5| < 0.01$

46. $|x - (-2)| < 0.001$

In Problems 47–50, proceed as in Example 7 and find an inequality $|x - b| < a$ or $|x - b| > a$ for which the given interval is its solution set.

47. $(-3, 11)$

48. $(1, 2)$

49. $(-\infty, 1) \cup (9, \infty)$

50. $(-\infty, -3) \cup (13, \infty)$

In Problems 51 and 52, find an inequality whose solution is the set of real numbers x satisfying the given condition. Express each set using interval notation.

51. Greater than or equal to 2 units from -3

52. Less than $\frac{1}{2}$ unit from 3.5

Miscellaneous Applications

53. Comparing Ages Bill and Mary's ages, A_B and A_M, differ by at most 3 years. Write this fact as an inequality using absolute value symbols.

54. Survival Your score on the first exam is 72%. The midterm grade is the average of the first exam score with the midterm exam score. If the B range is from 80% to 89%, what score should you obtain on the midterm exam so that your mid-semester grade is B?

55. Weight of Coffee The weight w of the coffee in cans filled by a food processing company satisfies

$$\left|\frac{w - 12}{0.05}\right| \leq 1,$$

where w is measured in ounces. Determine the interval in which w lies.

56. Weight of Cans A grocery scale is designed to be accurate to within 0.25 oz. If two identical cans of soup placed on the scale have a combined weight of 33.15 oz, what are the largest and smallest possible weights of one of the cans?

For Discussion

57. Discuss how you might solve the given inequality and equation.

(a) $\left|\dfrac{x + 5}{x - 2}\right| \leq 3$

(b) $|5 - x| = |1 - 3x|$

Carry out your ideas.

58. The distance between the number x and 5 is $|x - 5|$.

(a) In words, describe the graphical interpretation of the inequalities
$0 < |x - 5|$ and $0 < |x - 5| < 3$.

(b) Solve each inequality in part (a) and write each solution set using interval notation.

59. (a) Interpret $|x - 3|$ as distance between the numbers x and 3. Sketch on the number line the set of real numbers that satisfy $2 < |x - 3| < 5$.

(b) Now solve the simultaneous inequality $2 < |x - 3| < 5$ by first solving $|x - 3| < 5$ and then $2 < |x - 3|$. Take the intersection of the two solution sets and compare with your sketch in part (a).

60. Here is a statement you may encounter in the beginning of a course in calculus. Express the following statement as best you can in words:

> *For every $\epsilon > 0$ there exists a $\delta > 0$ such that $|y - L| < \epsilon$ whenever $0 < |x - a| < \delta$.*

Do not use the symbols $>$, $<$, or $|\ |$. The symbols ϵ and δ are the Greek letters epsilon and delta and represent real numbers.

1.3 The Rectangular Coordinate System

≡ **Introduction** In Section 1.1 we saw that each real number can be associated with exactly one point on the number, or coordinate, line. We now examine a correspondence between points in a plane and ordered pairs of real numbers.

☐ **Coordinate Plane** A **rectangular coordinate system** is formed by two perpendicular number lines that intersect at the point corresponding to the number 0 on each line. This point of intersection is called the **origin** and is denoted by the symbol O. The horizontal and vertical number lines are called the ***x*-axis** and the ***y*-axis**, respectively. These axes divide the plane into four regions, called **quadrants**, which are numbered as shown in FIGURE 1.3.1(a). As we can see in FIGURE 1.3.1(b), the scales on the *x*- and *y*-axes need not be the same. Throughout this text, if tick marks are *not* labeled on the coordinates axes, as in Figure 1.3.1(a), then you may assume that one tick corresponds to one unit. A plane containing a rectangular coordinate system is called an ***xy*-plane**, a **coordinate plane**, or simply **2-space**.

René Descartes

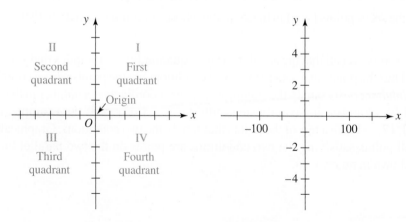

(a) Four quadrants   (b) Different scales on *x*- and *y*-axes

FIGURE 1.3.1 Coordinate plane

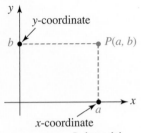

FIGURE 1.3.2 Point with coordinates (a, b)

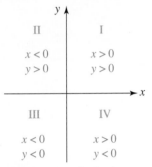

FIGURE 1.3.3 Algebraic signs of coordinates in the four quadrants

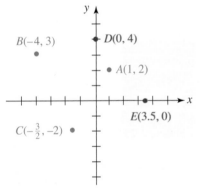

FIGURE 1.3.4 Plots of five points in Example 1

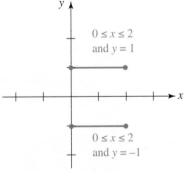

FIGURE 1.3.5 Set of points in Example 2

The rectangular coordinate system and the xy-plane are also called the **Cartesian coordinate system** and the **Cartesian plane** after the famous French mathematician and philosopher **René Descartes** (1596–1650).

☐ **Coordinates of a Point** Let P represent a point in the coordinate plane. We associate an ordered pair of real numbers with P by drawing a vertical line from P to the x-axis and a horizontal line from P to the y-axis. If the vertical line intersects the x-axis at the number a and the horizontal line intersects the y-axis at the number b, we associate the **ordered pair** of real numbers (a, b) with the point. Conversely, to each ordered pair (a, b) of real numbers there corresponds a point P in the plane. This point lies at the intersection of the vertical line through a on the x-axis and the horizontal line passing through b on the y-axis. Hereafter we will refer to an ordered pair as a **point** and denote it by either $P(a, b)$ or (a, b).* The number a is the **x-coordinate** of the point and the number b is the **y-coordinate** of the point and we say that P has **coordinates** (a, b). For example, the coordinates of the origin are $(0, 0)$. See **FIGURE 1.3.2**.

The algebraic signs of the x-coordinate and the y-coordinate of any point (x, y) in each of the four quadrants are indicated in **FIGURE 1.3.3**. Points on either of the two axes are not considered to be in any quadrant. Because a point on the x-axis has the form $(x, 0)$, an equation that describes the x-axis is $y = 0$. Similarly, a point on the y-axis has the form $(0, y)$ and so an equation of the y-axis is $x = 0$. When we locate a point in the coordinate plane corresponding to an ordered pair of numbers and represent it using a solid dot, we say that we **plot** or **graph** the point.

EXAMPLE 1   **Plotting Points**

Plot the points $A(1, 2)$, $B(-4, 3)$, $C\left(-\frac{3}{2}, -2\right)$, $D(0, 4)$, and $E(3.5, 0)$. Specify the quadrant in which each point lies.

Solution The five points are plotted in the coordinate plane in **FIGURE 1.3.4**. Point A lies in the first quadrant (quadrant I), B in the second quadrant (quadrant II), and C is in the third quadrant (quadrant III). Points D and E, which lie on the y- and x-axes, respectively, are not in any quadrant. ≡

EXAMPLE 2   **Plotting Points**

Sketch the set of points (x, y) in the xy-plane whose coordinates satisfy both $0 \leq x \leq 2$ and $|y| = 1$.

Solution First, recall that the absolute-value equation $|y| = 1$ implies that $y = -1$ or $y = 1$. Thus the points that satisfy the given conditions are the points whose coordinates (x, y) *simultaneously* satisfy the conditions: each x-coordinate is a number in the closed interval $[0, 2]$ and each y-coordinate is either $y = -1$ or $y = 1$. For example, $(1, 1)$, $\left(\frac{1}{2}, -1\right)$, $(2, -1)$ are a few of the points that satisfy the two conditions. Graphically, the set of all points satisfying the two conditions are points on the two parallel line segments shown in **FIGURE 1.3.5**. ≡

_____

*This is the same notation used to denote an open interval. It should be clear from the context of the discussion whether we are considering a point (a, b) or an open interval (a, b).

CHAPTER 1 INEQUALITIES, EQUATIONS, AND GRAPHS

EXAMPLE 3   **Regions Defined by Inequalities**

Sketch the set of points (x, y) in the xy-plane whose coordinates satisfy each of the following conditions. **(a)** $xy < 0$ **(b)** $|y| \geq 2$

Solution **(a)** From (*ii*) of the sign properties of products in Section 1.1, we know that a product of two real numbers x and y is negative when one of the numbers is positive and the other is negative. Thus, $xy < 0$ when $x > 0$ and $y < 0$ *or* when $x < 0$ and $y > 0$. We see from Figure 1.3.3 that $xy < 0$ for all points (x, y) in the second and fourth quadrants. Hence we can represent the set of points for which $xy < 0$ by the shaded regions in **FIGURE 1.3.6**. The coordinate axes are shown as dashed lines to indicate that the points on these axes are not included in the solution set.

(b) In Section 1.2 we saw that $|y| \geq 2$ means that either $y \geq 2$ or $y \leq -2$. Since x is not restricted in any way it can be any real number, and so the points (x, y) for which

$$y \geq 2 \quad \text{and} \quad -\infty < x < \infty \qquad \text{or} \qquad y \leq -2 \quad \text{and} \quad -\infty < x < \infty$$

can be represented by the two shaded regions in **FIGURE 1.3.7**. We use solid lines to represent the boundaries $y = -2$ and $y = 2$ of the region to indicate that the points on these boundaries are included in the solution set.

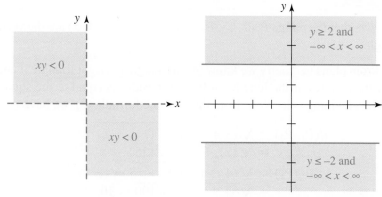

FIGURE 1.3.6 Region in the xy-plane satisfying the condition in (a) of Example 3

FIGURE 1.3.7 Region in the xy-plane satisfying the condition in (b) of Example 3   ≡

☐ **Distance Formula** Suppose $P_1(x_1, y_1)$ and $P_2(x_2, y_2)$ are two distinct points in the xy-plane that are not on a vertical line or on a horizontal line. As a consequence, P_1, P_2, and $P_3(x_1, y_2)$ are vertices of a right triangle, as shown in **FIGURE 1.3.8**. The length of the side P_3P_2 is $|x_2 - x_1|$ and the length of the side P_1P_3 is $|y_2 - y_1|$. If we denote the length of P_1P_2 by d, then

$$d^2 = |x_2 - x_1|^2 + |y_2 - y_1|^2 \tag{1}$$

by the Pythagorean theorem. Since the square of any real number is equal to the square of its absolute value, we can replace the absolute value signs in (1) with parentheses. The distance formula given next follows immediately from (1).

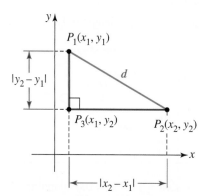

FIGURE 1.3.8 Distance between points P_1 and P_2

THEOREM 1.3.1 Distance Formula

The **distance** between any two points $P_1(x_1, y_1)$ and $P_2(x_2, y_2)$ in the xy-plane is given by

$$d(P_1, P_2) = \sqrt{(x_2 - x_1)^2 + (y_2 - y_1)^2}. \qquad (2)$$

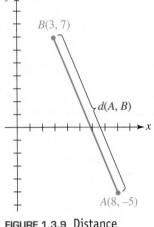

FIGURE 1.3.9 Distance between two points in Example 4

Although we derived this equation for two points not on a vertical or horizontal line, (2) holds in these cases as well. Also, because $(x_2 - x_1)^2 = (x_1 - x_2)^2$, it makes no difference which point is used first in the distance formula, that is, $d(P_1, P_2) = d(P_2, P_1)$.

EXAMPLE 4　　　Distance Between Two Points

Find the distance between the points $A(8, -5)$ and $B(3, 7)$.

Solution From (2), with A and B playing the parts of P_1 and P_2:

$$\begin{aligned} d(A, B) &= \sqrt{(3 - 8)^2 + (7 - (-5))^2} \\ &= \sqrt{(-5)^2 + (12)^2} = \sqrt{169} = 13. \end{aligned}$$

The distance d is illustrated in FIGURE 1.3.9.　　≡

EXAMPLE 5　　　Three Points Form a Triangle

Determine whether the points $P_1(7, 1)$, $P_2(-4, -1)$, and $P_3(4, 5)$ are the vertices of a right triangle.

Solution From plane geometry we know that a triangle is a right triangle if and only if the sum of the squares of the lengths of two of its sides is equal to the square of the length of the remaining side. Now, from the distance formula (2), we have

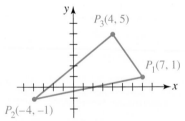

FIGURE 1.3.10 Triangle in Example 5

$$\begin{aligned} d(P_1, P_2) &= \sqrt{(-4 - 7)^2 + (-1 - 1)^2} \\ &= \sqrt{121 + 4} = \sqrt{125}, \\ d(P_2, P_3) &= \sqrt{(4 - (-4))^2 + (5 - (-1))^2} \\ &= \sqrt{64 + 36} = \sqrt{100} = 10, \\ d(P_3, P_1) &= \sqrt{(7 - 4)^2 + (1 - 5)^2} \\ &= \sqrt{9 + 16} = \sqrt{25} = 5. \end{aligned}$$

Since

$$[d(P_3, P_1)]^2 + [d(P_2, P_3)]^2 = 25 + 100 = 125 = [d(P_1, P_2)]^2,$$

we conclude that P_1, P_2, and P_3 are the vertices of a right triangle with the right angle at P_3. See FIGURE 1.3.10.　　≡

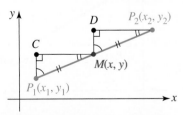

FIGURE 1.3.11 M is the midpoint of the line segment joining P_1 and P_2

□ **Midpoint Formula** In Section 1.2 we saw that the midpoint of a line segment between two numbers a and b on the number line is the average, $(a + b)/2$. In the xy-plane, each coordinate of the midpoint M of a line segment joining two points $P_1(x_1, y_1)$ and $P_2(x_2, y_2)$ shown in FIGURE 1.3.11 is the average of the corresponding coordinates of the endpoints of the intervals $[x_1, x_2]$ and $[y_1, y_2]$.

To prove this, we note in Figure 1.3.11 that triangles P_1CM and MDP_2 are congruent because corresponding angles are equal and $d(P_1, M) = d(M, P_2)$. Hence, $d(P_1, C) = d(M, D)$, or $y - y_1 = y_2 - y$. Solving the last equation for y gives

$y = \dfrac{y_1 + y_2}{2}$. Similarly, $d(C, M) = d(D, P_2)$, so that $x - x_1 = x_2 - x$, and therefore $x = \dfrac{x_1 + x_2}{2}$. We have proved the following result.

THEOREM 1.3.2 Midpoint Formula

The coordinates of the **midpoint** M of the line segment joining the points $P_1(x_1, y_1)$ and $P_2(x_2, y_2)$ are given by

$$M = \left(\frac{x_1 + x_2}{2}, \frac{y_1 + y_2}{2}\right). \tag{3}$$

EXAMPLE 6   **Midpoint of a Line Segment**

Find the coordinates of the midpoint of the line segment joining $A(-2, 5)$ and $B(4, 1)$.

Solution From the midpoint formula (3), the coordinates of the midpoint M are given by

$$\left(\frac{-2 + 4}{2}, \frac{5 + 1}{2}\right) \quad \text{or} \quad (1, 3).$$

This point is indicated in color in **FIGURE 1.3.12**. $\equiv$

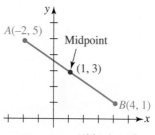

FIGURE 1.3.12 Midpoint of line segment in Example 6

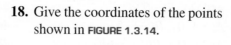

1.3 Exercises   Answers to selected odd-numbered problems begin on page ANS-2.

In Problems 1–4, plot the given points.

1. $(2, 3)$, $(4, 5)$, $(0, 2)$, $(-1, -3)$ **2.** $(1, 4)$, $(-3, 0)$, $(-4, 2)$, $(-1, -1)$
3. $(-\frac{1}{2}, -2)$, $(0, 0)$, $(-1, \frac{4}{3})$, $(3, 3)$ **4.** $(0, 0.8)$, $(-2, 0)$, $(1.2, -1.2)$, $(-2, 2)$

In Problems 5–16, determine the quadrant in which the given point lies if (a, b) is in quadrant I.

5. $(-a, b)$ **6.** $(a, -b)$ **7.** $(-a, -b)$ **8.** (b, a)
9. $(-b, a)$ **10.** $(-b, -a)$ **11.** (a, a) **12.** $(b, -b)$
13. $(-a, -a)$ **14.** $(-a, a)$ **15.** $(b, -a)$ **16.** $(-b, b)$

17. Plot the points given in Problems 5–16 if (a, b) is the point shown in **FIGURE 1.3.13**.

18. Give the coordinates of the points shown in **FIGURE 1.3.14**.

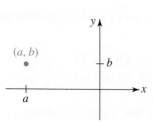

FIGURE 1.3.13 Point (a, b) in Problem 17

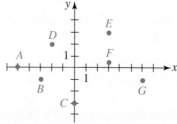

FIGURE 1.3.14 Points A–G in Problem 18

19. The points $(-2, 0)$, $(-2, 6)$, and $(3, 0)$ are vertices of a rectangle. Find the fourth vertex.

20. Describe the set of all points (x, x) in the coordinate plane. The set of all points $(x, -x)$.

In Problems 21–26, sketch the set of points (x, y) in the xy-plane whose coordinates satisfy the given conditions.

21. $xy = 0$ **22.** $xy > 0$

23. $|x| \leq 1$ and $|y| \leq 2$ **24.** $x \leq 2$ and $y \geq -1$

25. $|x| > 4$ **26.** $|y| \leq 1$

In Problems 27–32, find the distance between the given points.

27. $A(1, 2)$, $B(-3, 4)$ **28.** $A(-1, 3)$, $B(5, 0)$

29. $A(2, 4)$, $B(-4, -4)$ **30.** $A(-12, -3)$, $B(-5, -7)$

31. $A\left(-\frac{3}{2}, 1\right)$, $B\left(\frac{5}{2}, -2\right)$ **32.** $A\left(-\frac{5}{3}, 4\right)$, $B\left(-\frac{2}{3}, -1\right)$

In Problems 33–36, determine whether the points A, B, and C are vertices of a right triangle.

33. $A(8, 1)$, $B(-3, -1)$, $C(10, 5)$ **34.** $A(-2, -1)$, $B(8, 2)$, $C(1, -11)$

35. $A(2, 8)$, $B(0, -3)$, $C(6, 5)$ **36.** $A(4, 0)$, $B(1, 1)$, $C(2, 3)$

37. Determine whether the points $A(0, 0)$, $B(3, 4)$, and $C(7, 7)$ are vertices of an isosceles triangle.

38. Find all points on the y-axis that are 5 units from the point $(4, 4)$.

39. Consider the line segment joining $A(-1, 2)$ and $B(3, 4)$.
    (a) Find an equation that expresses the fact that a point $P(x, y)$ is equidistant from A and from B.
    (b) Describe geometrically the set of points described by the equation in part (a).

40. Use the distance formula to determine whether the points $A(-1, -5)$, $B(2, 4)$, and $C(4, 10)$ lie on a straight line.

41. Find all points each with x-coordinate 6 such that the distance from each point to $(-1, 2)$ is $\sqrt{85}$.

42. Which point, $\left(1/\sqrt{2}, 1/\sqrt{2}\right)$ or $(0.25, 0.97)$, is closer to the origin?

In Problems 43–48, find the midpoint M of the line segment joining the points A and B.

43. $A(4, 1)$, $B(-2, 4)$ **44.** $A\left(\frac{2}{3}, 1\right)$, $B\left(\frac{7}{3}, -3\right)$

45. $A(-1, 0)$, $B(-8, 5)$ **46.** $A\left(\frac{1}{2}, -\frac{3}{2}\right)$, $B\left(-\frac{5}{2}, 1\right)$

47. $A(2a, 3b)$, $B(4a, -6b)$ **48.** $A(x, x)$, $B(-x, x + 2)$

In Problems 49–52, find the point B if M is the midpoint of the line segment joining points A and B.

49. $A(-2, 1)$, $M\left(\frac{3}{2}, 0\right)$ **50.** $A\left(4, \frac{1}{2}\right)$, $M\left(7, -\frac{5}{2}\right)$

51. $A(5, 8)$, $M(-1, -1)$ **52.** $A(-10, 2)$, $M(5, 1)$

53. Find the distance from the midpoint M_1 of the line segment joining $A(-1, 3)$ and $B(3, 5)$ to the midpoint M_2 of the line segment joining $C(4, 6)$ and $D(-2, -10)$.

54. Find all points on the *x*-axis that are 3 units from the midpoint of the line segment joining $(5, 2)$ and $(-5, -6)$.

55. The *x*-axis is the perpendicular bisector of the line segment through $A(2, 5)$ and $B(x, y)$. Find *x* and *y*.

56. Consider the line segment joining the points $A(0, 0)$ and $B(6, 0)$. Find a point $C(x, y)$ in the first quadrant such that *A, B,* and *C* are vertices of an equilateral triangle.

57. Find the points $P_1(x_1, y_1)$, $P_2(x_2, y_2)$, and $P_3(x_3, y_3)$ on the line segment joining $A(3, 6)$ and $B(5, 8)$ that divide the line segment into four equal parts.

Miscellaneous Applications

58. Going to Chicago Kansas City and Chicago are not directly connected by an interstate highway, but each city is connected to St. Louis and Des Moines. See FIGURE 1.3.15. Des Moines is approximately 40 mi east and 180 mi north of Kansas City, St. Louis is approximately 230 mi east and 40 mi south of Kansas City, and Chicago is approximately 360 mi east and 200 mi north of Kansas City. Assume that this part of the Midwest is a flat plane and that the connecting highways are straight lines. Which route from Kansas City to Chicago, through St. Louis or through Des Moines, is shorter?

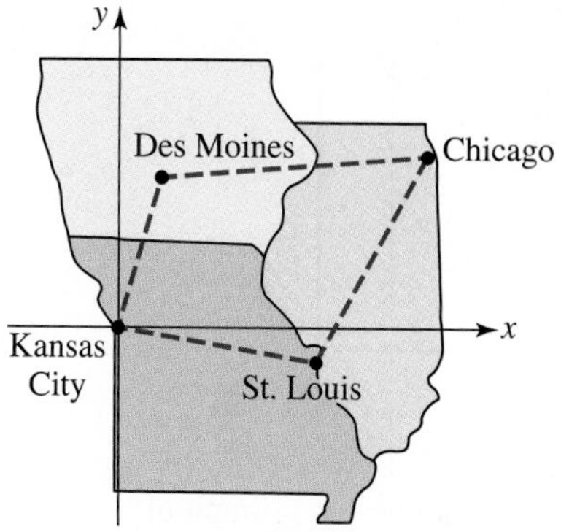

FIGURE 1.3.15 Map for Problem 58

For Discussion

59. The points $A(1, 0)$, $B(5, 0)$, $C(4, 6)$, and $D(8, 6)$ are vertices of a parallelogram. Discuss: How can it be shown that the diagonals of the parallelogram bisect each other? Carry out your ideas.

60. The points $A(0, 0)$, $B(a, 0)$, and $C(a, b)$ are vertices of a right triangle. Discuss: How can it be shown that the midpoint of the hypotenuse is equidistant from the vertices? Carry out your ideas.

1.4 Circles and Graphs

≡ **Introduction** An **equation in two variables**, say *x* and *y,* is simply a mathematical statement that asserts two quantities involving these variables are equal. In the fields of the physical sciences, engineering, and business, equations are a means of communication. For example, if a physicist wants to tell someone how far a rock dropped from a great height travels in a certain time *t,* he/she will write $s = 16t^2$. A mathematician will look at $s = 16t^2$ and immediately classify it as a certain *type* of equation. The classification of an equation carries with it information about properties shared by all equations of that kind. The remainder of this text is devoted to examining different kinds of equations involving two variables and studying their properties. Here is a sample of some of the equations you will see:

$$x = 1, \quad x^2 + y^2 = 1, \quad y = x^2, \quad y = \sqrt{x},$$
$$y = 5x - 1, \quad y = x^3 - 3x, \quad y = 2^x, \quad y = \ln x, \tag{1}$$
$$y = \sin x, \quad y^2 = x - 1, \quad \frac{x^2}{4} + \frac{y^2}{9} = 1, \quad \tfrac{1}{2}x^2 - y^2 = 1.$$

A **solution** of an equation in two variables x and y is an ordered pair of numbers (a, b) that yields a true statement when $x = a$ and $y = b$ are substituted into the equation. For example, $(-2, 4)$ is a solution of the equation $y = x^2$ because

$$\overset{\underset{\downarrow}{y = 4}}{4} = \overset{\underset{\downarrow}{x = -2}}{(-2)^2}$$

is a true statement. We also say that the coordinates $(-2, 4)$ **satisfy** the equation. The set of all solutions of an equation is called its **solution set**. Two equations are said to be **equivalent** if they have the same solution set. For example, we will see in Example 4 of this section that the equation $x^2 + y^2 + 10x - 2y + 17 = 0$ is equivalent to $(x + 5)^2 + (y - 1)^2 = 3^2$.

In the list given in (1), you might object that the first equation $x = 1$ does not involve two variables. It is a matter of interpretation! Because there is no explicit y dependence in the equation, $x = 1$ can be interpreted to mean the set

$$\{(x, y) \mid x = 1, \text{ where } y \text{ is any real number}\}.$$

The solutions of $x = 1$ are then ordered pairs $(1, y)$, where you are free to choose y arbitrarily so long as it is a real number. For example, $(1, 0)$ and $(1, 3)$ are solutions of the equation $x = 1$. The **graph** of an equation is the visual representation in the coordinate plane of the set of points whose coordinates (a, b) satisfy the equation. The graph of $x = 1$ is the vertical line shown in FIGURE 1.4.1.

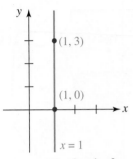

FIGURE 1.4.1 Graph of equation $x = 1$

☐ **Circles** The distance formula discussed in Section 1.3 can be used to define a set of points in the coordinate plane. One such important set is defined as follows.

DEFINITION 1.4.1   Circle

A **circle** is the set of all points $P(x, y)$ in the coordinate plane that are a given fixed distance r, called the **radius**, from a given fixed point C, called the **center**.

If the center has coordinates $C(h, k)$, then from the preceding definition a point $P(x, y)$ lies on a circle of radius r if and only if

$$d(P, C) = r \qquad \text{or} \qquad \sqrt{(x - h)^2 + (y - k)^2} = r.$$

Since $(x - h)^2 + (y - k)^2$ is always nonnegative, we obtain an equivalent equation when both sides are squared. We conclude that a circle of radius r and center $C(h, k)$ has the equation

$$(x - h)^2 + (y - k)^2 = r^2. \tag{2}$$

In FIGURE 1.4.2 we have sketched a typical graph of an equation of the form given in (2). Equation (2) is called the **standard form** of the equation of a circle. We note that the symbols h and k in (2) represent real numbers and as such can be positive, zero, or negative. When $h = 0$ and $k = 0$, we see that the standard form of the equation of a circle with center at the origin is

$$x^2 + y^2 = r^2. \tag{3}$$

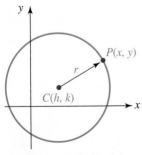

FIGURE 1.4.2 Circle with radius r and center (h, k)

CHAPTER 1 INEQUALITIES, EQUATIONS, AND GRAPHS

See **FIGURE 1.4.3**. When $r = 1$ we say that (2) is an equation of a **unit circle**. For example, $x^2 + y^2 = 1$ is an equation of a unit circle centered at the origin.

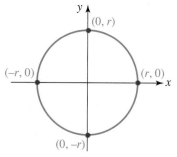

FIGURE 1.4.3 Circle with radius r and center (0, 0)

EXAMPLE 1   Center and Radius

Find the center and radius of the circle whose equation is

$$(x - 8)^2 + (y + 2)^2 = 49. \tag{4}$$

Solution To obtain the standard form of the equation, we rewrite (4) as

$$(x - 8)^2 + (y - (-2))^2 = 7^2.$$

From this last form we identify $h = 8$, $k = -2$, and $r = 7$. Thus the circle is centered at $(8, -2)$ and has radius 7. ≡

EXAMPLE 2   Equation of a Circle

Find an equation of the circle with center $C(-5, 4)$ with radius $\sqrt{2}$.

Solution Substituting $h = -5$, $k = 4$, and $r = \sqrt{2}$ in (2), we obtain

$$(x - (-5))^2 + (y - 4)^2 = (\sqrt{2})^2 \quad \text{or} \quad (x + 5)^2 + (y - 4)^2 = 2. \ ≡$$

EXAMPLE 3   Equation of a Circle

Find an equation of the circle with center $C(4, 3)$ and passing through $P(1, 4)$.

Solution With $h = 4$ and $k = 3$, we have from (2)

$$(x - 4)^2 + (y - 3)^2 = r^2. \tag{5}$$

Since the point $P(1, 4)$ lies on the circle as shown in **FIGURE 1.4.4**, its coordinates must satisfy equation (5). That is,

$$(1 - 4)^2 + (4 - 3)^2 = r^2 \quad \text{or} \quad 10 = r^2.$$

Thus the required equation in standard form is $(x - 4)^2 + (y - 3)^2 = 10$. ≡

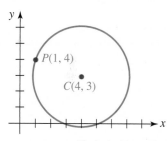

FIGURE 1.4.4 Circle in Example 3

☐ **Completing the Square** If the terms $(x - h)^2$ and $(y - k)^2$ are expanded and the like terms grouped together, an equation of a circle in standard form can be written as

$$x^2 + y^2 + ax + by + c = 0. \tag{6}$$

Of course in this last form the center and radius are not apparent. To reverse the process—in other words, to go from (6) to the standard form (2)—we must **complete the square** in both x and y. Recall from algebra that adding $(a/2)^2$ to an expression such as $x^2 + ax$ yields $x^2 + ax + (a/2)^2$, which is the perfect square $(x + a/2)^2$. By rearranging the terms in (6),

$$(x^2 + ax \quad) + (y^2 + by \quad) = -c,$$

and then adding $(a/2)^2$ and $(b/2)^2$ to *both* sides of the last equation,

$$\left(x^2 + ax + \left(\frac{a}{2}\right)^2\right) + \left(y^2 + by + \left(\frac{b}{2}\right)^2\right) = \left(\frac{a}{2}\right)^2 + \left(\frac{b}{2}\right)^2 - c,$$

◀ The terms in color added inside the parentheses on the left-hand side are also added to the right-hand side of the equality. This new equation is equivalent to (6).

we obtain the standard form of the equation of a circle:

$$\left(x + \frac{a}{2}\right)^2 + \left(y + \frac{b}{2}\right)^2 = \frac{1}{4}(a^2 + b^2 - 4c).$$

You should *not* memorize the last equation; we strongly recommend that you work through the process of completing the square each time.

▮ EXAMPLE 4　　　　Completing the Square

Find the center and radius of the circle whose equation is

$$x^2 + y^2 + 10x - 2y + 17 = 0. \tag{7}$$

Solution To find the center and radius we rewrite equation (7) in the standard form (2). First, we rearrange the terms,

$$(x^2 + 10x\quad) + (y^2 - 2y\quad) = -17.$$

Then, we complete the square in *x* and *y* by adding, in turn, $(10/2)^2$ in the first set of parentheses and $(-2/2)^2$ in the second set of parentheses. Proceed carefully here because we must add these numbers to both sides of the equation:

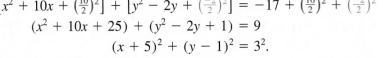

$$(x^2 + 10x + 25) + (y^2 - 2y + 1) = 9$$
$$(x + 5)^2 + (y - 1)^2 = 3^2.$$

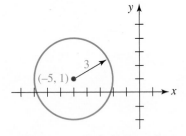

FIGURE 1.4.5 Circle in Example 4

From the last equation we see that the circle is centered at $(-5, 1)$ and has radius 3. See **FIGURE 1.4.5**.　　　　　　　　　　　　　　　　　　≡

　　It is possible that an expression for which we must complete the square has a leading coefficient other than 1. For example,

Note:

$$3x^2 + 3y^2 - 18x + 6y + 2 = 0$$

is an equation of circle. As in Example 4, we start by rearranging the equation:

$$(3x^2 - 18x\quad) + (3y^2 + 6y\quad) = -2.$$

Now, however, we must do one extra step before attempting completion of the square, that is, we must divide both sides of the equation by 3 so that the coefficients of x^2 and y^2 are each 1:

$$(x^2 - 6x\quad) + (y^2 + 2y\quad) = -\tfrac{2}{3}.$$

At this point we can now add the appropriate numbers within each set of parentheses *and* to the right-hand side of the equality. You should verify that the resulting standard form is $(x - 3)^2 + (y + 1)^2 = \frac{28}{3}$.

□ **Semicircles**　If we solve (3) for *y* we get $y^2 = r^2 - x^2$ or $y = \pm\sqrt{r^2 - x^2}$. This last expression is equivalent to two equations, $y = \sqrt{r^2 - x^2}$ and $y = -\sqrt{r^2 - x^2}$. In like manner if we solve (3) for *x* we obtain $x = \sqrt{r^2 - y^2}$ and $x = -\sqrt{r^2 - y^2}$.

　　By convention, the symbol $\sqrt{}$ denotes a nonnegative quantity, thus the *y*-values defined by an equation such as $y = \sqrt{r^2 - x^2}$ are nonnegative. The graphs of the four equations highlighted in color are, in turn, the upper half, lower half, right half, and left half of the circle shown in Figure 1.4.3. Each graph in **FIGURE 1.4.6** is called a **semicircle**.

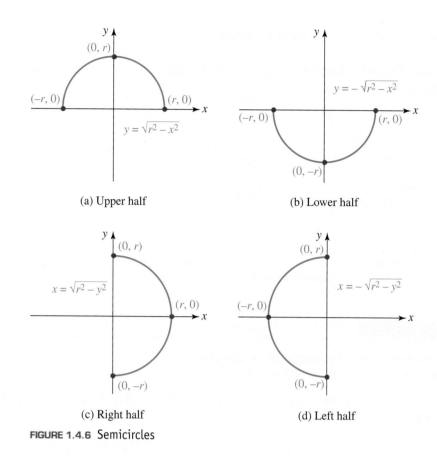

(a) Upper half   (b) Lower half

(c) Right half   (d) Left half

FIGURE 1.4.6 Semicircles

☐ **Inequalities** One last point about circles: On occasion we encounter problems where we must sketch the set of points in the xy-plane whose coordinates satisfy inequalities such as $x^2 + y^2 < r^2$ or $x^2 + y^2 \geq r^2$. The equation $x^2 + y^2 = r^2$ describes the set of points (x, y) whose distance to the origin $(0, 0)$ is exactly r. Therefore the inequality $x^2 + y^2 < r^2$ describes the set of points (x, y) whose distance to the origin is less than r. In other words, the points (x, y) whose coordinates satisfy the inequality $x^2 + y^2 < r^2$ are in the *interior* of the circle. Similarly, the points (x, y) whose coordinates satisfy $x^2 + y^2 \geq r^2$ lie either *on* the circle or are *exterior* to it. Inequalities such as these will be considered in greater detail in Section 9.4.

☐ **Graphs** It is difficult to read a newspaper, read a science or business text, surf the Internet, or even watch the news on TV without seeing graphical representations of data. It may even be impossible to get past the first page in a mathematics text without seeing some kind of graph. So many diverse quantities are connected by means of equations, and so many questions about the behavior of the quantities linked by the equation can be answered by means of a graph, that the ability to graph equations quickly and accurately—like the ability to do algebra quickly and accurately— is high on the list of skills essential to your success in a course in calculus. For the rest of this section we are going to talk about graphs in general, and more specifically about two important aspects of graphs of equations.

☐ **Intercepts** Locating the points at which the graph of an equation crosses the coordinate axes can be helpful when sketching a graph by hand. The **x-intercepts** of a graph of an equation are the points at which the graph crosses the x-axis. Since every point on the x-axis has y-coordinate 0, the x-coordinates of these points (if there are any) can be found from the given equation by setting $y = 0$ and solving for x. In turn, the **y-intercepts** of the graph of an equation are the points at which its graph crosses the y-axis. The y-coordinates of these points can found by setting $x = 0$ in the equation and solving for y. See **FIGURE 1.4.7**.

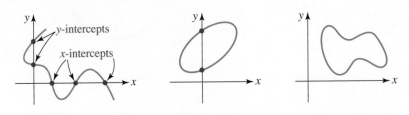

(a) Five intercepts   (b) Two y-intercepts   (c) Graph has no intercepts

FIGURE 1.4.7 Intercepts of a graph

EXAMPLE 5   Intercepts

Find the intercepts of the graphs of the equations
(a) $x^2 - y^2 = 9$ **(b)** $y = 2x^2 + 5x - 12$.

Solution (a) To find the x-intercepts we set $y = 0$ and solve the resulting equation $x^2 = 9$ for x:

$$x^2 - 9 = 0 \quad \text{or} \quad (x + 3)(x - 3) = 0$$

gives $x = -3$ and $x = 3$. The x-intercepts of the graph are the points $(-3, 0)$ and $(3, 0)$. To find the y-intercepts we set $x = 0$ and solve $-y^2 = 9$ or $y^2 = -9$ for y. Because there are no real numbers whose square is negative we conclude the graph of the equation does not cross the y-axis.

(b) Setting $y = 0$ yields $2x^2 + 5x - 12 = 0$. This is a quadratic equation and can be solved either by factoring or by the quadratic formula. Factoring gives

$$(x + 4)(2x - 3) = 0,$$

and so $x = -4$ and $x = \frac{3}{2}$. The x-intercepts of the graph are the points $(-4, 0)$ and $\left(\frac{3}{2}, 0\right)$. Now, setting $x = 0$ in the equation $y = 2x^2 + 5x - 12$ immediately gives $y = -12$. The y-intercept of the graph is the point $(0, -12)$. $\equiv$

EXAMPLE 6   Example 4 Revisited

Let's return to the circle in Example 4 and determine its intercepts from the equation in (7). Setting $y = 0$ in $x^2 + y^2 + 10x - 2y + 17 = 0$ and using the quadratic formula to solve $x^2 + 10x + 17 = 0$ shows the x-intercepts of this circle are $\left(-5 - 2\sqrt{2}, 0\right)$ and $\left(-5 + 2\sqrt{2}, 0\right)$. If we let $x = 0$, then the quadratic formula shows that the roots of the equation $y^2 - 2y + 17 = 0$ are complex numbers. As seen in Figure 1.4.5, the circle does not cross the y-axis. $\equiv$

☐ **Symmetry** A graph can also possess symmetry. You may already know that the graph of the equation $y = x^2$ is called a *parabola*. **FIGURE 1.4.8** shows that the graph of $y = x^2$ is symmetric with respect to the y-axis since the portion of the graph that lies in the second quadrant is the *mirror image* or *reflection* of that portion of the graph in the first

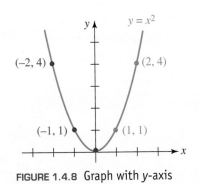

FIGURE 1.4.8 Graph with y-axis symmetry

CHAPTER 1 INEQUALITIES, EQUATIONS, AND GRAPHS

quadrant. In general, a graph is **symmetric with respect to the y-axis** if whenever (x, y) is a point on the graph, $(-x, y)$ is also a point on the graph. Note in Figure 1.4.8 that the points $(1, 1)$ and $(2, 4)$ are on the graph. Because the graph possesses y-axis symmetry, the points $(-1, 1)$ and $(-2, 4)$ must also be on the graph. A graph is said to be **symmetric with respect to the x-axis** if whenever (x, y) is a point on the graph, $(x, -y)$ is also a point on the graph. Finally, a graph is **symmetric with respect to the origin** if whenever (x, y) is on the graph, $(-x, -y)$ is also a point on the graph. FIGURE 1.4.9 illustrates these three types of symmetries.

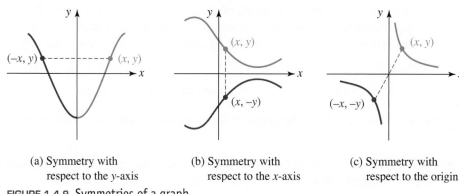

(a) Symmetry with
respect to the y-axis

(b) Symmetry with
respect to the x-axis

(c) Symmetry with
respect to the origin

FIGURE 1.4.9 Symmetries of a graph

Observe that the graph of the circle given in Figure 1.4.3 possesses all three of these symmetries.

As a practical matter we would like to know whether a graph possesses any symmetry in advance of plotting it. This can be done by applying the following tests to the equation that defines the graph.

THEOREM 1.4.1 Tests for Symmetry

The graph of an equation is symmetric with respect to:

    (*i*)   the **y-axis** if replacing x by $-x$ results in an equivalent equation;
    (*ii*) the **x-axis** if replacing y by $-y$ results in an equivalent equation;
    (*iii*) the **origin** if replacing x and y by $-x$ and $-y$ results in an equivalent equation.

The advantage of using symmetry in graphing should be apparent: If, say, the graph of an equation is symmetric with respect to the x-axis, then we need only produce the graph for $y \geq 0$ since points on the graph for $y < 0$ are obtained by taking the mirror images, through the x-axis, of the points in the first and second quadrants.

EXAMPLE 7   **Test for Symmetry**

By replacing x by $-x$ in the equation $y = x^2$ and using $(-x)^2 = x^2$, we see that

$$y = (-x)^2 \quad \text{is equivalent to} \quad y = x^2.$$

By (*i*) of Theorem 1.4.1 this proves what is apparent in Figure 1.4.8; the graph of $y = x^2$ is symmetric with respect to the y-axis. ≡

EXAMPLE 8   **Intercepts and Symmetry**

Determine the intercepts and any symmetry for the graph of

$$x + y^2 = 10. \tag{8}$$

Solution *Intercepts*: Setting $y = 0$ in equation (8) immediately gives $x = 10$. The graph of the equation has a single x-intercept, $(10, 0)$. When $x = 0$, we get $y^2 = 10$, which implies that $y = -\sqrt{10}$ or $y = \sqrt{10}$. Thus there are two y-intercepts, $\left(0, -\sqrt{10}\right)$ and $\left(0, \sqrt{10}\right)$.

Symmetry: If we replace x by $-x$ in the equation $x + y^2 = 10$ we get $-x + y^2 = 10$. This is not equivalent to equation (8). You should also verify that replacing x and y by $-x$ and $-y$ in (8) does not yield an equivalent equation. However, if we replace y by $-y$, we find that

$$x + (-y)^2 = 10 \quad \text{is equivalent to} \quad x + y^2 = 10.$$

Thus, by (*ii*) of Theorem 1.4.1 the graph of the equation is symmetric with respect to the x-axis.

Graph: In the graph of the equation given in FIGURE 1.4.10, the intercepts are indicated and the x-axis symmetry should be apparent. ≡

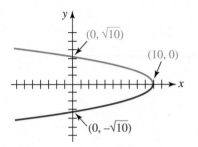

FIGURE 1.4.10 Graph of equation in Example 8

1.4   **Exercises**   Answers to selected odd-numbered problems begin on page ANS–2.

In Problems 1–6, find the center and the radius of the given circle. Sketch its graph.

1. $x^2 + y^2 = 5$ **2.** $x^2 + y^2 = 9$
3. $x^2 + (y - 3)^2 = 49$ **4.** $(x + 2)^2 + y^2 = 36$
5. $\left(x - \frac{1}{2}\right)^2 + \left(y - \frac{3}{2}\right)^2 = 1$ **6.** $(x + 3)^2 + (y - 5)^2 = 25$

In Problems 7–14, complete the square in x and y to find the center and the radius of the given circle.

7. $x^2 + y^2 + 8y = 0$ **8.** $x^2 + y^2 - 6x = 0$
9. $x^2 + y^2 + 2x - 4y - 4 = 0$ **10.** $x^2 + y^2 - 18x - 6y - 10 = 0$
11. $x^2 + y^2 - 20x + 16y + 128 = 0$ **12.** $x^2 + y^2 + 3x - 16y + 63 = 0$
13. $2x^2 + 2y^2 + 4x + 16y + 1 = 0$ **14.** $\frac{1}{2}x^2 + \frac{1}{2}y^2 + \frac{5}{2}x + 10y + 5 = 0$

In Problems 15–24, find an equation of the circle that satisfies the given conditions.

15. center $(0, 0)$, radius 1
16. center $(1, -3)$, radius 5
17. center $(0, 2)$, radius $\sqrt{2}$
18. center $(-9, -4)$, radius $\frac{3}{2}$
19. endpoints of a diameter at $(-1, 4)$ and $(3, 8)$
20. endpoints of a diameter at $(4, 2)$ and $(-3, 5)$
21. center $(0, 0)$, graph passes through $(-1, -2)$
22. center $(4, -5)$, graph passes through $(7, -3)$
23. center $(5, 6)$, graph tangent to the x-axis
24. center $(-4, 3)$, graph tangent to the y-axis

In Problems 25–28, sketch the semicircle defined by the given equation.

25. $y = \sqrt{4 - x^2}$

26. $x = 1 - \sqrt{1 - y^2}$

27. $x = \sqrt{1 - (y - 1)^2}$

28. $y = -\sqrt{9 - (x - 3)^2}$

29. Find an equation for the upper half of the circle $x^2 + (y - 3)^2 = 4$. Repeat for the right half of the circle.

30. Find an equation for the lower half of the circle $(x - 5)^2 + (y - 1)^2 = 9$. Repeat for the left half of the circle.

In Problems 31–34, sketch the set of points in the xy-plane whose coordinates satisfy the given inequality.

31. $x^2 + y^2 \geq 9$

32. $(x - 1)^2 + (y + 5)^2 \leq 25$

33. $1 \leq x^2 + y^2 \leq 4$

34. $x^2 + y^2 > 2y$

In Problems 35 and 36, find the x- and y-intercepts of the given circle.

35. the circle with center $(3, -6)$ and radius 7

36. the circle $x^2 + y^2 + 5x - 6y = 0$

In Problems 37–62, find any intercepts of the graph of the given equation. Determine whether the graph of the equation possesses symmetry with respect to the x-axis, y-axis, or origin. Do not graph.

37. $y = -3x$

38. $y - 2x = 0$

39. $-x + 2y = 1$

40. $2x + 3y = 6$

41. $x = y^2$

42. $y = x^3$

43. $y = x^2 - 4$

44. $x = 2y^2 - 4$

45. $y = x^2 - 2x - 2$

46. $y^2 = 16(x + 4)$

47. $y = x(x^2 - 3)$

48. $y = (x - 2)^2(x + 2)^2$

49. $x = -\sqrt{y^2 - 16}$

50. $y^3 - 4x^2 + 8 = 0$

51. $4y^2 - x^2 = 36$

52. $\dfrac{x^2}{25} + \dfrac{y^2}{9} = 1$

53. $y = \dfrac{x^2 - 7}{x^3}$

54. $y = \dfrac{x^2 - 10}{x^2 + 10}$

55. $y = \dfrac{x^2 - x - 20}{x + 6}$

56. $y = \dfrac{(x + 2)(x - 8)}{x + 1}$

57. $y = \sqrt{x} - 3$

58. $y = 2 - \sqrt{x + 5}$

59. $y = |x - 9|$

60. $x = |y| - 4$

61. $|x| + |y| = 4$

62. $x + 3 = |y - 5|$

In Problems 63–66, state all the symmetries of the given graph.

63.

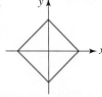

FIGURE 1.4.11 Graph for Problem 63

64.

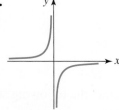

FIGURE 1.4.12 Graph for Problem 64

65.

FIGURE 1.4.13 Graph for
Problem 65

66.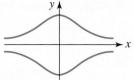

FIGURE 1.4.14 Graph for
Problem 66

In Problems 67–72, use symmetry to complete the given graph.

67. The graph is symmetric
with respect to the *y*-axis.

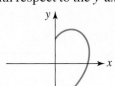

FIGURE 1.4.15 Graph for
Problem 67

68. The graph is symmetric
with respect to the *x*-axis.

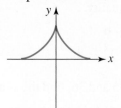

FIGURE 1.4.16 Graph for
Problem 68

69. The graph is symmetric
with respect to the origin.

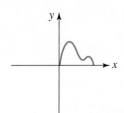

FIGURE 1.4.17 Graph for
Problem 69

70. The graph is symmetric
with respect to the *y*-axis.

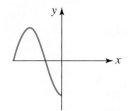

FIGURE 1.4.18 Graph for
Problem 70

71. The graph is symmetric with
respect to the *x*- and *y*-axes.

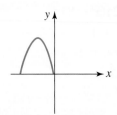

FIGURE 1.4.19 Graph for
Problem 71

72. The graph is symmetric
with respect to the origin.

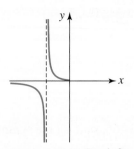

FIGURE 1.4.20 Graph for
Problem 72

For Discussion

73. Determine whether the following statement is true or false. Defend your answer.

If a graph has two of the three symmetries defined on page 29, *then the graph
must necessarily possess the third symmetry.*

74. (a) The radius of the circle in FIGURE 1.4.21(a) is r. What is its equation in standard form?

(b) The center of the circle in FIGURE 1.4.21(b) is (h, k). What is its equation in standard form?

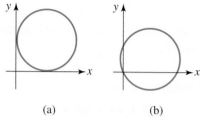

(a)   (b)

FIGURE 1.4.21 Graph for Problem 74

75. Discuss whether the following statement is true or false:

Every equation of the form $x^2 + y^2 + ax + by + c = 0$ is a circle.

76. Find the areas of the shaded regions in FIGURE 1.4.22.

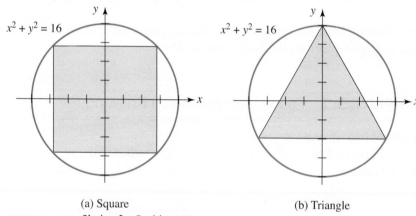

(a) Square   (b) Triangle

FIGURE 1.4.22 Circles for Problem 76

77. Show that the triangle in part (b) of Problem 76 is an equilateral triangle.

1.5 Algebra and Limits

$\int$**Calculus** ≡ **Introduction** A calculus problem often consists of a sequence
PREVIEW of steps, where most of the steps are algebra and only the last few—
sometimes just the last step—involve calculus. The discussion that
follows focuses on one kind of calculus problem: the computation of a certain type of *limit*.
Although we give a brief and intuitive introduction to the notion of a limit, the thrust of the
discussion is an overview of the type of algebra frequently encountered in such problems.

☐ **Algebraic Expressions** In this section we are concerned only with **fractional
expressions**. It suffices to think of a fractional expression as a quotient of two
algebraic expressions.* Roughly, an **algebraic expression** is one that is the result

---

*At this point we are excluding trigonometric functions, logarithms, and exponentials. See
Chapters 4 and 6.

of performing a finite number of additions, subtractions, multiplications, divisions, or roots on a collection of variables and real numbers. For example, some algebraic expressions in a single variable x are

$$5x^3 - 3x + 1, \qquad \frac{2x^2 - 18}{x + 3}, \qquad \text{and} \qquad x + \sqrt{x - 5}.$$

An area of algebra that causes difficulties in working calculus problems is the manipulation of fractional expressions.

☐ **Factoring** When the distributive law

$$a(b + c) = ab + ac$$

is read right to left,

$$ab + ac = a(b + c),$$

we say that the expression $ab + ac$ has been **factored**. We will see in Chapter 3 that factoring plays an important role in solving equations, as well as in graphing. But in the present context we are concerned only with using factoring to simplify fractional expressions.

The following three factorization formulas are important and are used as a matter of course throughout various fields of mathematics.

THEOREM 1.5.1   Factorizations Worth Knowing

Difference of two squares: $a^2 - b^2 = (a - b)(a + b)$	(1)
Difference of two cubes: $a^3 - b^3 = (a - b)(a^2 + ab + b^2)$	(2)
Sum of two cubes: $a^3 + b^3 = (a + b)(a^2 - ab + b^2)$	(3)

The symbols a and b in (1)–(3) are placeholders. For example, the expression $x^4 - 16$ is of the form given in (1). With the identifications $a = x^2$ and $b = 4$, we have

$$x^4 - 16 = (x^2)^2 - 4^2 = (x^2 - 4)(x^2 + 4). \qquad (4)$$

Since the factor $x^2 - 4$ is also the difference of two squares, (4) continues as

$$x^4 - 16 = (x^2 - 4)(x^2 + 4) = (x - 2)(x + 2)(x^2 + 4). \qquad (5)$$

The factorization in (5) is as far as we can go using real numbers and integer exponents; the sum of two squares $x^2 + 4$ does not factor using real numbers. As another example, consider the expression $2x^2 - 3$. Since any positive real number can be written as the square of its square root we have $2 = (\sqrt{2})^2$ and $3 = (\sqrt{3})^2$, and so from (1) the expression $2x^2 - 3$ factors in the following manner:

$$2x^2 - 3 = \overset{\underset{\displaystyle a}{\downarrow}}{(\sqrt{2}x)^2} - \overset{\underset{\displaystyle b}{\downarrow}}{(\sqrt{3})^2} = (\sqrt{2}x - \sqrt{3})(\sqrt{2}x + \sqrt{3}).$$

We use factoring and the cancellation property to simplify a fractional expression.

Cancellation Property: If a, b, and c are ▶ real numbers, then
$$\frac{ac}{bc} = \frac{a}{b}, \quad c \neq 0.$$

EXAMPLE 1   **Factoring and Canceling**

Simplify **(a)** $\dfrac{x^2 - 1}{x - 1}$ **(b)** $\dfrac{x + 3}{x^2 - 4x - 21}$.

Solution (a) From (1) we see

$$\frac{x^2 - 1}{x - 1} = \frac{(x - 1)(x + 1)}{x - 1} = x + 1.$$

The cancellation of $x - 1$ in the foregoing expression is only valid for $x \neq 1$. For $x = 1$ we would be dividing by 0.

(b) We look for factors $x - a$ and $x - b$ such that

$$x^2 - 4x - 21 = (x - a)(x - b).$$

This implies $ab = -21$, so a and b must be factors of -21 whose sum is $-(a + b) = -4$. The usual trial and error procedure leads to $a = 7$ and $b = -3$. Therefore,

$$\frac{x + 3}{x^2 - 4x - 21} = \frac{x + 3}{(x + 3)(x - 7)} = \frac{1}{x - 7}, \qquad x \neq -3. \qquad \equiv$$

☐ **Binomial Expansion** A two-term algebraic expression $a + b$ is called a **binomial**. You undoubtedly have worked problems where you had to expand powers of binomials such as $(a + b)^2$ and $(a + b)^3$. This occurs so often in mathematics courses that we recommend that you memorize the expansions given in (6) and (7) below.

THEOREM 1.5.2   Binomial Expansions Worth Knowing

Expansions of $(a + b)^n$ for $n = 2$ and $n = 3$ are, respectively,

$$(a + b)^2 = a^2 + 2ab + b^2 \qquad\qquad (6)$$
$$(a + b)^3 = a^3 + 3a^2b + 3ab^2 + b^3 \qquad\qquad (7)$$

Of course, formulas (6) and (7) work just as well for a binomial in the form of a difference $a - b$. Simply treat $a - b$ as the sum $a + (-b)$ and replace the symbol b in (6) and (7) with $-b$:

$$(a - b)^2 = (a + (-b))^2 = a^2 + 2a(-b) + (-b)^2 = a^2 - 2ab + b^2,$$

and

$$(a - b)^3 = (a + (-b))^3 = a^3 + 3a^2(-b) + 3a(-b)^2 + (-b)^3$$
$$= a^3 - 3a^2b + 3ab^2 - b^3.$$

There are ways of remembering how to obtain the coefficients in the expansion of higher powers such as $(a + b)^4$. **Pascal's triangle** is one such way and is reviewed in Section 10.4.

EXAMPLE 2   **Binomial Expansion**

Simplify $\dfrac{(7 + h)^2 - 49}{h}$.

Solution We use the expansion of $(a + b)^2$ given in (6) with $a = 7$ and $b = h$:

$$\frac{(7 + h)^2 - 49}{h} = \frac{(7^2 + 2(7)h + h^2) - 49}{h}$$

$$= \frac{49 + 14h + h^2 - 49}{h} \qquad \leftarrow 49 - 49 = 0$$

$$= \frac{h(14 + h)}{h} \quad \leftarrow \text{cancel the } h\text{'s}$$

$$= 14 + h, \quad h \neq 0. \qquad\qquad\qquad \equiv$$

☐ **Addition of Fractional Expressions** Combining two or more fractional expressions, or simplification of a complex fraction where the numerator or denominator is itself a fraction, can be particularly troublesome for some students.

> ### EXAMPLE 3 — Addition of Fractions

Write as one fraction $\dfrac{10x}{2x^2 + 3x - 2} - \dfrac{4}{x + 2} + \dfrac{8}{2x - 1}$.

Solution Because $2x^2 + 3x - 2 = (2x - 1)(x + 2)$, the least common denominator of the three terms is $(2x - 1)(x + 2)$. Therefore, we multiply the second term by $(2x - 1)/(2x - 1)$ and the third term by $(x - 2)/(x - 2)$:

$$\frac{10x}{(2x - 1)(x + 2)} - \frac{4}{x + 2}\frac{2x - 1}{2x - 1} + \frac{8}{2x - 1}\frac{x + 2}{x + 2}.$$

Adding numerators and simplifying gives

$$\frac{10x - 4(2x - 1) + 8(x + 2)}{(2x - 1)(x + 2)} = \frac{10x - 8x + 4 + 8x + 16}{(2x - 1)(x + 2)}$$

$$= \frac{10x + 20}{(2x - 1)(x + 2)}$$

$$= \frac{10(x + 2)}{(2x - 1)(x + 2)}$$

$$= \frac{10}{2x - 1}.$$

The cancellation of $x + 2$ is permissible provided $x \neq -2$. $\qquad\qquad \equiv$

We will illustrate the simplification of a complex fraction in Example 9.

☐ **Rationalization** You may have learned **rationalization of a denominator** in a previous mathematics course. Recall that rationalization of a denominator consists of multiplying an expression by a factor equal to 1 with the intent of clearing a radical from a denominator. For example, to rationalize the denominator in $1/\sqrt{2}$ we multiply the fraction by $\sqrt{2}/\sqrt{2}$:

$$\overset{\text{fraction is equal to 1}}{\underset{\downarrow}{}}$$

$$\frac{1}{\sqrt{2}} = \frac{1}{\sqrt{2}}\frac{\sqrt{2}}{\sqrt{2}} = \frac{\sqrt{2}}{(\sqrt{2})^2} = \frac{\sqrt{2}}{2}.$$

There is no rule in mathematics that says only denominators must be rationalized. There are times in calculus when we are interested in rationalization not only of denominators but numerators as well. The next example uses the factorization of the difference of two squares in a slightly different manner. For $a > 0$ and $b > 0$, we can write $(\sqrt{a})^2 = a$, $(\sqrt{b})^2 = b$, and so we can write $a - b = (\sqrt{a})^2 - (\sqrt{b})^2$. It then follows from (1) that

$$a - b = (\sqrt{a} - \sqrt{b})(\sqrt{a} + \sqrt{b}). \qquad\qquad (8)$$

A variation of (8) is $a^2 - b = (a - \sqrt{b})(a + \sqrt{b})$. Thus if a numerator or denominator of a fractional expression contains a binomial term that includes at least one radical, such as

$$a - \sqrt{b}, a + \sqrt{b}, \sqrt{a} - b, \sqrt{a} + b, \sqrt{a} - \sqrt{b}, \text{ or } \sqrt{a} + \sqrt{b},$$

we multiply the numerator and the denominator of the fraction by the corresponding **conjugate factor**

$$a + \sqrt{b}, a - \sqrt{b}, \sqrt{a} + b, \sqrt{a} - b, \sqrt{a} + \sqrt{b}, \text{ or } \sqrt{a} - \sqrt{b}.$$

For example, to rationalize the denominator of $3/(\sqrt{2} - \sqrt{5})$ we use (8) to write

$$\frac{3}{\sqrt{2} - \sqrt{5}} = \frac{3}{\sqrt{2} - \sqrt{5}} \overset{\substack{\text{fraction is equal to 1} \\ \downarrow}}{\underset{\underset{\text{conjugate factor of denominator}}{\uparrow}}{\frac{\sqrt{2} + \sqrt{5}}{\sqrt{2} + \sqrt{5}}}} = \frac{3(\sqrt{2} + \sqrt{5})}{(\sqrt{2})^2 - (\sqrt{5})^2}$$

$$= \frac{3(\sqrt{2} + \sqrt{5})}{-3} = -(\sqrt{2} + \sqrt{5}).$$

■ EXAMPLE 4   Rationalization of a Numerator

Rationalize the numerator in $\dfrac{\sqrt{4 + x} - 2}{x}$.

Solution Think of the numerator as $\sqrt{a} - b$ where $a = 4 + x$ and $b = 2$. Because the conjugate factor of $\sqrt{a} - b$ is $\sqrt{a} + b$, we are able to clear the radical in the numerator by multiplying the numerator and denominator of the given fractional expression by $\sqrt{4 + x} + 2$:

$$\frac{\sqrt{4 + x} - 2}{x} = \frac{\sqrt{4 + x} - 2}{x} \frac{\sqrt{4 + x} + 2}{\sqrt{4 + x} + 2} = \frac{(\sqrt{4 + x})^2 - 2^2}{x(\sqrt{4 + x} + 2)}$$

$$= \frac{4 + x - 4}{x(\sqrt{4 + x} + 2)} = \frac{x}{x(\sqrt{4 + x} + 2)}.$$

After canceling the x's in the numerator and the denominator in the last term the rationalization is complete:

$$\frac{\sqrt{4 + x} - 2}{x} = \frac{1}{\sqrt{4 + x} + 2}, \qquad x \neq 0. \qquad\qquad \equiv$$

☐ **Limits—The Calculus Connection** Consider the fractional algebraic expression $\dfrac{x^2 - 1}{x - 1}$. Observe that this fraction cannot be evaluated at $x = 1$ because substituting 1 into the expression results in the undefined quantity 0/0. However, the fractional expression can be evaluated at any other real number; in particular, it can be evaluated at numbers that are very *close* to 1. The numerical values of the fractional expression given in the following two tables are easily obtained using the simplification in part (a) of Example 1:

$$\frac{x^2 - 1}{x - 1} = x + 1, \qquad \text{for } x \neq 1.$$

x	0.9	0.99	0.999
$\dfrac{x^2 - 1}{x - 1}$	1.9	1.99	1.999

x	1.1	1.01	1.001
$\dfrac{x^2 - 1}{x - 1}$	2.1	2.01	2.001

(9)

☐ **Arrow Notation** The discussion of the limit concept is facilitated by using a special notation. If we let the **arrow symbol** → represent the word *approach*, then the symbolism

$x \to a^-$ *indicates that x approaches a number a from the left,*

that is, through numbers that are less than a, and

$x \to a^+$ *indicates that x approaches a number a from the right,*

that is, through numbers that are greater than a. Finally, the notation

$x \to a$ *signifies that x approaches a number a from both sides,*

in other words, from the left and the right sides of a on the number line. In the left-hand table in (9) we are letting $x \to 1^-$, and in the right-hand table $x \to 1^+$. Each table in (9) shows that the fractional expression $\dfrac{x^2 - 1}{x - 1}$ is close to the number 2 when x is close to 1, that is,

$$\frac{x^2 - 1}{x - 1} \to 2 \text{ as } x \to 1^- \qquad \text{and} \qquad \frac{x^2 - 1}{x - 1} \to 2 \text{ as } x \to 1^+. \qquad (10)$$

We say that 2 is the **limit** of $\dfrac{x^2 - 1}{x - 1}$ as x approaches 1 and write

$$\lim_{x \to 1} \frac{x^2 - 1}{x - 1} = 2. \qquad (11)$$

Before proceeding any further, we should make it clear that *a limit of an expression need not exist*. In the next two tables, consider $1/x$ as x approaches zero:

$x \to 0^-$	-0.1	-0.01	-0.001
$1/x$	-10	-100	-1000

$x \to 0^+$	0.1	0.01	0.001
$1/x$	10	100	1000

As can be seen in the tables, as x gets closer and closer to 0, the values of $1/x$ are becoming larger and larger in absolute value. In other words, $1/x$ is becoming unbounded. In this case we write

$$\frac{1}{x} \to -\infty \text{ as } x \to 0^- \qquad \text{and} \qquad \frac{1}{x} \to \infty \text{ as } x \to 0^+,$$

where ∞ is the infinity symbol. We say that $\lim\limits_{x \to 0} 1/x$ does not exist.

Suppose the symbol $f(x)$ denotes an expression involving a single variable x and that the symbols a and L represent real numbers. If, as illustrated in (10),

$$f(x) \to L \text{ as } x \to a^- \qquad \text{and} \qquad f(x) \to L \text{ as } x \to a^+,$$

then we say that $\lim\limits_{x \to a} f(x)$ **exists** and write

$$\lim_{x \to a} f(x) = L.$$

In calculus, you will not be asked to find a limit by constructing tables of numerical values, although you surely will be asked to construct such tables because they are useful in convincing yourself of either the existence or the nonexistence of a limit.

(See Problems 47 and 48 in Exercises 1.5.) Limits are either found or are proved to exist using analytical methods, in many cases using proven laws or properties of limits. Because it is not our goal to delve into theoretical or geometrical interpretations of a limit, and because we want to make the point that the calculus part of *some* problems is often the least significant part of the solution, we will accept three results from calculus without proof: If a and c are real numbers, then

$$\lim_{x \to a} c = c, \qquad \lim_{x \to a} x = a, \qquad \text{and} \qquad \lim_{x \to a} x^n = a^n, \tag{12}$$

where n is a positive integer. For example, (12) allows us to write*

$$\lim_{x \to 3} (5x + 4) = 5(3) + 4 = 19$$

and

$$\lim_{x \to 3} (2x^2 + x + 1) = 2(3)^2 + 3 + 1 = 22.$$

In the preceding line we used $\lim_{x \to 3} x = 3$, $\lim_{x \to 3} x^2 = 9$, $\lim_{x \to 3} 4 = 4$, and $\lim_{x \to 3} 1 = 1$.

The limit concept is the foundation of calculus, and one kind of limit is of particular significance in calculus: the limit of a fractional expression where *both* the numerator and the denominator are approaching 0. Such a limit is said to have the **indeterminate form 0/0**. For example, in view of the results in (12), $\lim_{x \to 1} (x - 1) = 0$ and $\lim_{x \to 1} (x^2 - 1) = 0$. Therefore $\lim_{x \to 1} \dfrac{x^2 - 1}{x - 1}$ has the indeterminate form 0/0. Of course, not all limit problems have this indeterminate form, but because of their importance (see Section 2.9), the limits in the remaining five examples, as well as *all* the limits in Exercises 1.5, have the form 0/0. Moreover, for simplicity we will only consider limits that actually exist.

We now show you how to find $\lim_{x \to 1} \dfrac{x^2 - 1}{x - 1}$ without the help of numerical tables:

algebra from Example 1(a)
↓

$$\lim_{x \to 1} \frac{x^2 - 1}{x - 1} = \lim_{x \to 1} (x + 1) = 1 + 1 = 2.$$

Done! The intermediate steps were all algebra performed to rewrite the expression in a more tractable form, a form where the actual limit can be computed with minimal effort.

EXAMPLE 5   Example 1 Revisited

Find $\lim_{x \to -3} \dfrac{x + 3}{x^2 - 4x - 21}$.

Solution This is the fractional expression in part (b) of Example 1. Observe that as $x \to -3$, the given limit has the indeterminate form 0/0. Now, using the algebraic simplification of this expression done in Example 1, we find that

algebra from Example 1(b)
↓

$$\lim_{x \to -3} \frac{x + 3}{x^2 - 4x - 21} = \lim_{x \to -3} \frac{1}{x - 7} = \frac{1}{-10} = -\frac{1}{10}. \qquad \equiv$$

EXAMPLE 6   Example 2 Revisited

Find $\lim_{h \to 0} \dfrac{(7 + h)^2 - 49}{h}$.

---

*We are actually using several other properties of limits here. However, we do not feel this is the place to discuss all the properties of the limit concept.

Solution Using the algebra from Example 2,

$$\lim_{h \to 0} \frac{(7 + h)^2 - 49}{h} = \lim_{h \to 0} (14 + h) = 14.$$ ≡

| EXAMPLE 7 | Example 3 Revisited |

Find $\displaystyle \lim_{x \to -2} \left[\frac{10x}{2x^2 + 3x - 2} - \frac{4}{x + 2} + \frac{8}{2x - 1} \right]$.

Solution Were this a calculus course, you should observe that the first and second terms are of the form $1/0$ as $x \to -2$. You may think that this is the situation $\infty - \infty$, and so gives 0. No. Remember we never treat ∞ as we would a number. The observation that the given algebraic expression contains these undefined quantities should trigger the idea that combining the fractions into *one* fractional expression would be a way to proceed. After carrying out the algebra, as done in Example 3, you would then finish the problem as follows:

$$\lim_{x \to -2} \left[\frac{10x}{2x^2 + 3x - 2} - \frac{4}{x + 2} + \frac{8}{2x - 1} \right] \overset{\text{algebra}}{=} \lim_{x \to -2} \frac{10}{2x - 1} = \frac{10}{-5} = -2.$$ ≡

| EXAMPLE 8 | Example 4 Revisited |

Find $\displaystyle \lim_{x \to 0} \frac{\sqrt{4 + x} - 2}{x}$.

Solution Using the algebra from Example 4, we find

$$\lim_{x \to 0} \frac{\sqrt{4 + x} - 2}{x} \overset{\text{algebra}}{=} \lim_{x \to 0} \frac{1}{\sqrt{4 + x} + 2} = \frac{1}{\sqrt{4} + 2} = \frac{1}{2 + 2} = \frac{1}{4}.$$ ≡

When finding the value of a limit, the algebra can be done as a side problem (as we have done in Examples 1–4), and then making use of your work, completing the problem as we have illustrated in Examples 5–8. In our last example, we combine the algebra with computing the limit. We recommend that you work through this example rather than just read it.

| EXAMPLE 9 | Limit of a Complex Fraction |

Find $\displaystyle \lim_{x \to 0} \frac{\dfrac{1}{(2 + x)^3} - \dfrac{1}{8}}{x}$.

Solution The given expression is an example of a complex fraction, that is, a quotient where either the numerator or denominator is a fractional expression. We begin by finding a common denominator in the numerator:

$$\lim_{x \to 0} \frac{\dfrac{1}{(2 + x)^3} - \dfrac{1}{8}}{x} = \lim_{x \to 0} \frac{\dfrac{1}{(2 + x)^3} \dfrac{8}{8} - \dfrac{1}{8} \dfrac{(2 + x)^3}{(2 + x)^3}}{x}$$

$$= \lim_{x \to 0} \frac{\dfrac{8 - (2 + x)^3}{8(2 + x)^3}}{x}$$

To continue we use the expansion of $(a + b)^3$ given in (7) with $a = 2$ and $b = x$:

$$\lim_{x\to 0} \frac{\dfrac{1}{(2+x)^3} - \dfrac{1}{8}}{x} = \lim_{x\to 0} \frac{\dfrac{8 - (2^3 + 3x(2)^2 + 3x^2(2) + x^3)}{8(2+x)^3}}{x}$$

$$= \lim_{x\to 0} \frac{\dfrac{8 - 8 - 12x - 6x^2 - x^3}{8(2+x)^3}}{x} \quad \leftarrow 8 - 8 = 0$$

$$= \lim_{x\to 0} \frac{\dfrac{-12x - 6x^2 - x^3}{8(2+x)^3}}{x}.$$

Because the x in the denominator of the last complex fraction is equivalent to the fraction $x/1$ we invert and multiply:

$$\lim_{x\to 0} \frac{\dfrac{1}{(2+x)^3} - \dfrac{1}{8}}{x} = \lim_{x\to 0} \frac{\dfrac{-12x - 6x^2 - x^3}{8(2+x)^3}}{\dfrac{x}{1}}$$

> ◀ Recall that division of fractions is converted into multiplication of fractions:
> $$\frac{\dfrac{a}{b}}{\dfrac{c}{d}} = \frac{a}{b} \times \frac{d}{c} = \frac{ad}{bc}, \quad bc \neq 0$$

$$= \lim_{x\to 0} \frac{-12x - 6x^2 - x^3}{8(2+x)^3} \frac{1}{x}$$

$$= \lim_{x\to 0} \frac{x(-12 - 6x - x^2)}{8(2+x)^3} \frac{1}{x} \quad \leftarrow \left\{ \begin{array}{l} \text{factor } x \text{ from numerator} \\ \text{and cancel } x\text{'s} \end{array} \right.$$

$$= \lim_{x\to 0} \frac{-12 - 6x - x^2}{8(2+x)^3}.$$

Finally, we have

$$\lim_{x\to 0} \frac{\dfrac{1}{(2+x)^3} - \dfrac{1}{8}}{x} = \lim_{x\to 0} \frac{-12 - 6x - x^2}{8(2+x)^3} = \frac{-12}{8 \cdot 2^3} = -\frac{3}{16},$$

since $\lim\limits_{x\to 0} x = 0$ and $\lim\limits_{x\to 0} x^2 = 0$ by (12). ≡

NOTES FROM THE CLASSROOM

(i) On tests we see students carrying out the expansion of $(a + b)^3$ by brute force, multiplying out $(a + b)(a + b)(a + b)$. This procedure is not recommended; it is slow and you are prone to errors. Instead, you should memorize (6) and (7).

(ii) In *any* mathematics course—not just calculus—do not erase or leave out important steps of your work. Most mathematics instructors want to see all work. Presenting that work in a neat and orderly fashion is also to your advantage. Finally, in the case of a limit problem such as Example 9, be sure to write down the symbol $\lim\limits_{x\to a}$ at each step. For example, we frequently see *incorrect* statements like this:

$$\lim_{x\to 1} \frac{x - 1}{x^2 - 1} = \frac{1}{x + 1} = \frac{1}{2}$$

on students' papers. The *correct* version of the preceding line is

$$\lim_{x \to 1} \frac{x-1}{x^2-1} = \lim_{x \to 1} \frac{1}{x+1} = \frac{1}{2}.$$

1.5 Exercises
Answers to selected odd-numbered problems begin on page ANS–3.

In Problems 1–12, use factorization to simplify the given expression in part (a). Then, if instructed, find the indicated limit in part (b).

1. (a) $\dfrac{x^2 - 25}{x - 5}$ (b) $\displaystyle\lim_{x \to 5} \dfrac{x^2 - 25}{x - 5}$

2. (a) $\dfrac{y - 3}{y^2 - 9}$ (b) $\displaystyle\lim_{y \to 3} \dfrac{y - 3}{y^2 - 9}$

3. (a) $\dfrac{x^2 - 7x + 6}{x - 1}$ (b) $\displaystyle\lim_{x \to 1} \dfrac{x^2 - 7x + 6}{x - 1}$

4. (a) $\dfrac{2x + 10}{x^2 + 7x + 10}$ (b) $\displaystyle\lim_{x \to -5} \dfrac{2x + 10}{x^2 + 7x + 10}$

5. (a) $\dfrac{x^2 + x - 6}{x^2 - 5x + 6}$ (b) $\displaystyle\lim_{x \to 2} \dfrac{x^2 + x - 6}{x^2 - 5x + 6}$

6. (a) $\dfrac{x^2 - 8x}{x^2 - 6x - 16}$ (b) $\displaystyle\lim_{x \to 8} \dfrac{x^2 - 8x}{x^2 - 6x - 16}$

7. (a) $\dfrac{x^3 - 1}{x - 1}$ (b) $\displaystyle\lim_{x \to 1} \dfrac{x^3 - 1}{x - 1}$

8. (a) $\dfrac{x^2 - 4}{x^3 + 8}$ (b) $\displaystyle\lim_{x \to -2} \dfrac{x^2 - 4}{x^3 + 8}$

9. (a) $\dfrac{x^3 - 1}{x^2 + 3x - 4}$ (b) $\displaystyle\lim_{x \to 1} \dfrac{x^3 - 1}{x^2 + 3x - 4}$

10. (a) $\dfrac{x^5 + 2x^4 + x^3}{x^4 - 2x^2 + 1}$ (b) $\displaystyle\lim_{x \to -1} \dfrac{x^5 + 2x^4 + x^3}{x^4 - 2x^2 + 1}$

11. (a) $\dfrac{x^3 + 3x^2 + 3x + 1}{x^4 + x^3 + x + 1}$ (b) $\displaystyle\lim_{x \to -1} \dfrac{x^3 + 3x^2 + 3x + 1}{x^4 + x^3 + x + 1}$

12. (a) $\dfrac{x^4 - 5x^3 + 4x - 20}{x^4 - 5x^3 + x - 5}$ (b) $\displaystyle\lim_{x \to 5} \dfrac{x^4 - 5x^3 + 4x - 20}{x^4 - 5x^3 + x - 5}$

In Problems 13–20, use binomial expansion to simplify the given expression in part (a). Then, if instructed, find the indicated limit in part (b).

13. (a) $\dfrac{(2 + h)^2 - 4}{h}$ (b) $\displaystyle\lim_{h \to 0} \dfrac{(2 + h)^2 - 4}{h}$

14. (a) $\dfrac{5 - 5(h + 1)^2}{h}$ (b) $\displaystyle\lim_{h \to 0} \dfrac{5 - 5(h + 1)^2}{h}$

15. (a) $\dfrac{(2x + 1)^2 - 9}{x - 1}$ (b) $\displaystyle\lim_{x \to 1} \dfrac{(2x + 1)^2 - 9}{x - 1}$

16. (a) $\dfrac{2(x-1)^2 - 4(x-1) - 6}{x}$ **(b)** $\displaystyle\lim_{x\to 0} \dfrac{2(x-1)^2 - 4(x-1) - 6}{x}$

17. (a) $\dfrac{(1+x)^3 - 1}{x}$ **(b)** $\displaystyle\lim_{x\to 0} \dfrac{(1+x)^3 - 1}{x}$

18. (a) $\dfrac{(x+1)^3 + (x-1)^3}{x}$ **(b)** $\displaystyle\lim_{x\to 0} \dfrac{(x+1)^3 + (x-1)^3}{x}$

19. (a) $\dfrac{2(h+1)^3 - 5(h+1)^2 + 3}{h}$ **(b)** $\displaystyle\lim_{h\to 0} \dfrac{2(h+1)^3 - 5(h+1)^2 + 3}{h}$

20. (a) $\dfrac{(x+2)^4 - 16}{x}$ **(b)** $\displaystyle\lim_{x\to 0} \dfrac{(x+2)^4 - 16}{x}$

In Problems 21–26, use addition of algebraic fractions to simplify the given expression in part (a). Then, if instructed, find the indicated limit in part (b).

21. (a) $\dfrac{1}{x-2} - \dfrac{6}{x^2 + 2x - 8}$ **(b)** $\displaystyle\lim_{x\to 2} \left[\dfrac{1}{x-2} - \dfrac{6}{x^2 + 2x - 8} \right]$

22. (a) $\dfrac{x^2 + 3x - 1}{x} + \dfrac{1}{x}$ **(b)** $\displaystyle\lim_{x\to 0} \left[\dfrac{x^2 + 3x - 1}{x} + \dfrac{1}{x} \right]$

23. (a) $\dfrac{1}{x-10} - \dfrac{20}{x^2 - 100}$ **(b)** $\displaystyle\lim_{x\to 10} \left[\dfrac{1}{x-10} - \dfrac{20}{x^2 - 100} \right]$

24. (a) $\dfrac{1}{x}\left[\dfrac{1}{9} - \dfrac{1}{x+9} \right]$ **(b)** $\displaystyle\lim_{x\to 0} \dfrac{1}{x}\left[\dfrac{1}{9} - \dfrac{1}{x+9} \right]$

25. (a) $\dfrac{\dfrac{1}{(2+h)^2} - \dfrac{1}{4}}{h}$ **(b)** $\displaystyle\lim_{h\to 0} \dfrac{\dfrac{1}{(2+h)^2} - \dfrac{1}{4}}{h}$

26. (a) $\dfrac{1}{t-1}\left[\dfrac{1}{(t+3)^2} - \dfrac{1}{16} \right]$ **(b)** $\displaystyle\lim_{t\to 1} \dfrac{1}{t-1}\left[\dfrac{1}{(t+3)^2} - \dfrac{1}{16} \right]$

In Problems 27–34, use rationalization to simplify the given expression in part (a). Then, if instructed, find the indicated limit in part (b).

27. (a) $\dfrac{\sqrt{x} - 3}{x - 9}$ **(b)** $\displaystyle\lim_{x\to 9} \dfrac{\sqrt{x} - 3}{x - 9}$

28. (a) $\dfrac{\dfrac{1}{\sqrt{x}} - \dfrac{1}{\sqrt{2}}}{x - 2}$ **(b)** $\displaystyle\lim_{x\to 2} \dfrac{\dfrac{1}{\sqrt{x}} - \dfrac{1}{\sqrt{2}}}{x - 2}$

29. (a) $\dfrac{x}{\sqrt{7 + x} - \sqrt{7}}$ **(b)** $\displaystyle\lim_{x\to 0} \dfrac{x}{\sqrt{7 + x} - \sqrt{7}}$

30. (a) $\dfrac{\sqrt{u + 4} - 3}{u - 5}$ **(b)** $\displaystyle\lim_{u\to 5} \dfrac{\sqrt{u + 4} - 3}{u - 5}$

31. (a) $\dfrac{25 - t}{5 - \sqrt{t}}$ **(b)** $\displaystyle\lim_{t\to 25} \dfrac{25 - t}{5 - \sqrt{t}}$

32. (a) $\dfrac{1}{h}\left[1 - \dfrac{1}{\sqrt{1 + h}} \right]$ **(b)** $\displaystyle\lim_{h\to 0} \dfrac{1}{h}\left[1 - \dfrac{1}{\sqrt{1 + h}} \right]$

33. (a) $\dfrac{4y^2}{\sqrt{y^2 + y + 1} - \sqrt{y + 1}}$ **(b)** $\displaystyle\lim_{y\to 0} \dfrac{4y^2}{\sqrt{y^2 + y + 1} - \sqrt{y + 1}}$

34. (a) $\dfrac{9t^2}{t + 2 - 2\sqrt{t + 1}}$ **(b)** $\displaystyle\lim_{t\to 0} \dfrac{9t^2}{t + 2 - 2\sqrt{t + 1}}$

Miscellaneous Calculus-Related Problems

In Problems 35–40, the given algebraic expression is an unsimplified answer to a calculus problem. Simplify the expression.

35. $\dfrac{x + \dfrac{1}{x} - a - \dfrac{1}{a}}{x - a}$

36. $\dfrac{\dfrac{3}{(x + 1)^2} - \dfrac{3}{(a + 1)^2}}{x - a}$

37. $(3x^2 + 4x - 1)(4)(2x - 3)^3(2) + (2x - 3)^4(6x + 4)$

38. $(12x - 1)^{1/3}(2)(x^2 - 1)(2x) + (x^2 - 1)^2(\tfrac{1}{3})(12x - 1)^{-2/3}(12)$

39. $\dfrac{2x(-4x + 6)^{1/2} - x^2(\tfrac{1}{2})(-4x + 6)^{-1/2}(-4)}{[(-4x + 6)^{1/2}]^2}$

40. $\dfrac{1}{2}\left(\dfrac{2x - 1}{4x + 1}\right)^{-\frac{1}{2}} \cdot \dfrac{(4x + 1)2 - (2x - 1)4}{(4x + 1)^2}$

In Problems 41–46, the given equation is a partial answer to a calculus problem. Solve the equation for the symbol y'.

41. $3y^2y' - y - xy' = x$

42. $y' = 2(x - y)(1 - y')$

43. $2yy' + 2x = y'$

44. $2xy^2 + x^2(2y)y' - 2 = -3y'$

45. $\dfrac{(x - y)(1 + y') - (x + y)(1 - y')}{(x - y)^2} = 1$

46. $\dfrac{1}{1 + x^2y^2}(xy' + y) = 2xyy' + y^2$

Calculator/Computer Problems

In Problems 47 and 48, use a calculator or computer to estimate the given limit by completing each table. Round the entries in each table to eight decimal places.

47. $\displaystyle\lim_{x \to 1} \dfrac{x^3 - 1}{\sqrt[3]{x} - 1}$;

$x \to 0^+$	1.1	1.01	1.001	1.0001	1.00001
$\dfrac{x^3 - 1}{\sqrt[3]{x} - 1}$					

$x \to 0^-$	0.9	0.99	0.999	0.9999	0.99999
$\dfrac{x^3 - 1}{\sqrt[3]{x} - 1}$					

48. $\displaystyle\lim_{x \to 0} (1 + x)^{1/x}$;

$x \to 0^+$	0.1	0.01	0.001	0.0001	0.00001
$(1 + x)^{1/x}$					

$x \to 0^-$	-0.1	-0.01	-0.001	-0.0001	-0.00001
$(1 + x)^{1/x}$					

For Discussion

In Problems 49 and 50, discuss what algebra is necessary to evaluate the given limit. Carry out your ideas.

49. $\displaystyle\lim_{x \to 1} \dfrac{x - 1}{x^8 - 1}$

50. $\displaystyle\lim_{x \to 0} \dfrac{\sqrt[3]{x + 27} - 3}{x}$

Review Exercises Answers to selected odd-numbered problems begin on page ANS-3.

A. Fill in the Blanks

In Problems 1–20, fill in the blanks.

1. An inequality with $(-\infty, 9]$ as its solution set is _____.
2. The solution set of the inequality $-3 < x \leq 8$ as an interval is _____.
3. If the point (a, b) lies in quadrant IV, then (b, a) lies in quadrant _____.
4. The point $(x, -3x)$ in the second quadrant that is 5 units from $(2, -1)$ is _____.
5. If the graph of an equation contains the point $(2, 3)$ and is symmetric with respect to the x-axis, then the graph also contains the point _____.
6. If the graph of an equation contains the point $(-1, 6)$ and is symmetric with respect to the origin, then the graph also contains the point _____.
7. An equation of a circle with center $(-2, -5)$ and radius 6 is _____.
8. If $|2 - x| = 15$, then $x =$ _____.
9. The distance from the midpoint of the line segment joining $(4, -6)$ and $(-2, 0)$ to the origin is _____.
10. The graph of $y = 2|x| - 5$ is symmetric with respect to _____.
11. The intercepts of the graph of $y = 2|x| - 5$ are _____.
12. The circle $x^2 - 16x + y^2 = 0$ is symmetric with respect to _____.
13. The center and radius of the circle $x^2 - 16x + y^2 = 0$ are _____.
14. The intercepts of the circle $(x - 1)^2 + (y - 2)^2 = 10$ are _____.
15. Two points on the circle $x^2 + y^2 = 25$ with the same x-coordinate -3 are _____.
16. The graph of $y = -\sqrt{100 - x^2}$ is a _____.
17. The inequality _____ describes the set of points in the xy-plane outside the circle $x^2 + y^2 = 36$.
18. If $\left(a, a + \sqrt{3}\right)$ lies on the graph of $y = 2x$, then $a =$ _____.
19. The set of real numbers x whose distance between x and $\sqrt{2}$ is greater than 3 is defined by the absolute-value inequality _____.
20. A point (x, y) in the xy-plane whose coordinates satisfy $xy < 0$ lies in quadrant(s) _____ or _____.

B. True/False

In Problems 1–20, answer true or false.

1. The word *nonnegative* means the same as the word *positive*. _____
2. The number 0 is neither positive nor negative. _____
3. -3 is not greater than -1. _____
4. If $a < b$, then $b - a$ is a positive number. _____
5. If $a < b$, then $a^2 < b^2$. _____
6. For any real number a, $-a \leq a$. _____
7. If $a < 0$, then $\dfrac{a}{-a} < 0$. _____
8. If $a^2 < a$, then $a < 1$. _____
9. If x is a negative number, then $-x$ is a positive number. _____
10. The solution set of $|4x - 6| \geq -1$ is $(-\infty, \infty)$. _____
11. $|-3t + 6| = 3|t - 2|$ _____
12. The point $(5, 0)$ is in quadrant I. _____
13. The point $(-3, 7)$ is in quadrant III. _____

14. The distance between the points $(0, 0)$ and $(3, 6)$ is 9. _____
15. To find y-intercepts of the graph of an equation we let $x = 0$ and solve for y. _____
16. There is no point on the circle $x^2 + y^2 - 10x + 22 = 0$ with x-coordinate 2. _____
17. A circle whose equation can be put into the form $x^2 + y^2 + ax + by = 0$ must pass through the origin. _____
18. The points $(0, 0)$, $(a, 0)$, $a > 0$, and $(0, b)$, $b < 0$, are vertices of a right triangle. _____
19. The graph of the equation $x^2 y + 4y = x$ is symmetric with respect to the origin. _____
20. The inequality $\dfrac{100}{x^2 + 64} \leq 0$ has no solution. _____

C. Review Exercises

In Problems 1–4, assume that $0 < a < b$. Compare the given expressions using inequality symbols.

1. a^2 and ab **2.** $-a$ and $-b$ **3.** a and $a + b$ **4.** $\dfrac{1}{a}$ and $\dfrac{1}{a + b}$

In Problems 5–10, fill in the blank with either an appropriate inequality symbol or a number.

5. If $x - 10 > 5$, then $x +$ _____ > 25.
6. If $x - 2 \leq 7$, then x _____ 9.
7. If $-\frac{1}{3}x \geq 4$, then x _____ -12.
8. If $3x - 6 \leq 4x - 4$, then x _____ 2.
9. If $-2 \leq 1 - x \leq 5$, then _____ $\leq x \leq$ _____.
10. If $-3 < x < 9$, then _____ $< -2x <$ _____.
11. On the number line, $m = 5$ is the midpoint of the line segment joining the number a (left endpoint) and the number b (right endpoint). Use the fact that $d(a, b) = 2$ to find a and b.
12. In the xy-plane, find an equation that describes the set of points (x, y) that are equidistant from $(0, 5)$ and $(x, -5)$.

In Problems 13–16, describe the given interval on the real number line using **(a)** an inequality, **(b)** interval notation.

13.

FIGURE 1.R.1 Graph for Problem 13

14.

FIGURE 1.R.2 Graph for Problem 14

15.
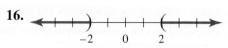
FIGURE 1.R.3 Graph for Problem 15

16.

FIGURE 1.R.4 Graph for Problem 16

In Problems 17–30, solve the given inequality. Write the solution set using interval notation.

17. $2x - 5 \geq 6x + 7$ **18.** $\frac{1}{4}x - 3 < \frac{1}{2}x + 1$
19. $-4 < x - 8 < 4$ **20.** $7 \leq 3 - 2x < 11$
21. $|x| > 10$ **22.** $|-6x| \leq 42$
23. $|3x - 4| < 5$ **24.** $|5 - 2x| \geq 7$
25. $3x \geq 2x^2 - 5$ **26.** $x^2 > 6x - 9$

27. $x^3 > x$

28. $(x^2 - x)(x^2 + x) \leq 0$

29. $\dfrac{1}{x} + x > 2$

30. $\dfrac{2x - 6}{x - 1} \geq 1$

In Problems 31–34, find equations of two different circles so that each circle satisfies the given conditions.

31. center in the first quadrant and graph is tangent to both the x- and y-axes

32. x-intercepts of the graph are $(-6, 0)$ and $(-2, 0)$

33. passes through the origin and center is on the negative y-axis

34. center on the x-axis and graph is tangent to the horizontal line through the point $(0, 3)$

35. Lens Equation The lens equation

$$\frac{1}{f} = \frac{1}{d_o} + \frac{1}{d_i},$$

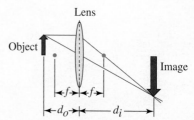

FIGURE 1.R.5 Convex lens in Problem 35

discovered by Carl Friedrich Gauss in 1841, relates the distance d_o from an object to a thin convex lens (in meters) to the distance d_i from the lens to its image (in meters), where f is the focal length of the lens and $d_o > f$. See FIGURE 1.R.5. If $f = 0.30$ m, then what distances d_o correspond to $d_i > 0.5$ m? Write the solution as a simultaneous inequality.

36. Solve the inequality $|x - 3| + |x + 1| < 10$. Write the solution set using interval notation.

In Problems 37–40, simplify the given expression in part (a). Then, if instructed, find the indicated limit in part (b).

37. (a) $\dfrac{2x - 1}{4x^2 - 1}$

(b) $\displaystyle\lim_{x \to \frac{1}{2}} \dfrac{2x - 1}{4x^2 - 1}$

38. (a) $\dfrac{x^2 - 6x + 5}{x - 5}$

(b) $\displaystyle\lim_{x \to 5} \dfrac{x^2 - 6x + 5}{x - 5}$

39. (a) $\dfrac{x^2 - 16}{\sqrt{x} - 2}$

(b) $\displaystyle\lim_{x \to 4} \dfrac{x^2 - 16}{\sqrt{x} - 2}$

40. (a) $\dfrac{1}{h}\left(\dfrac{1}{3 + h} - \dfrac{1}{3}\right)$

(b) $\displaystyle\lim_{h \to 0} \dfrac{1}{h}\left(\dfrac{1}{3 + h} - \dfrac{1}{3}\right)$

2 Functions

Chapter Outline

2.1 Functions and Graphs

≡ **Introduction** Using the objects and the persons around us, it is easy to make up a rule of correspondence that associates, or pairs, the members, or elements, of one set with the members of another set. For example, to each social security number there is a person, to each car registered in the state of California there is a license plate number, to each book there corresponds at least one author, to each state there is a governor, and so on. A natural correspondence occurs between a set of 20 students and a set of, say, 25 desks in a classroom when each student selects and sits in a different desk. In mathematics we are interested in a special type of correspondence, a *single-valued correspondence*, called a function.

Student/desk correspondence

DEFINITION 2.1.1 Function

A **function** from a set X to a set Y is a rule of correspondence that assigns to each element x in X exactly one element y in Y.

In the student/desk correspondence above suppose the set of 20 students is the set X and the set of 25 desks is the set Y. This correspondence is a function from the set X to the set Y provided no student sits in two desks at the same time.

☐ **Terminology** A function is usually denoted by a letter such as f, g, or h. We can then represent a function f from a set X to a set Y by the notation $f: X \rightarrow Y$. The set X is called the **domain** of f. The set of corresponding elements y in the set Y is called the **range** of the function. For our student/desk function, the set of students is the domain and the set of 20 desks actually occupied by the students constitutes the range. Notice that the range of f need not be the entire set Y. The unique element y in the range that corresponds to a selected element x in the domain X is called the **value** of the function at x, or the **image** of x, and is written $f(x)$. The latter symbol is read "f of x" or "f at x," and we write $y = f(x)$.* See FIGURE 2.1.1. Since the value of y depends on the choice of x, y is called the **dependent variable**; x is called the **independent variable**. Unless otherwise stated, we will assume hereafter that the sets X and Y consist of real numbers.

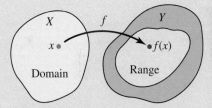

FIGURE 2.1.1 Domain and range of a function f

EXAMPLE 1   The Squaring Function

The rule for squaring a real number is given by the equation $y = x^2$ or $f(x) = x^2$. The values of f at $x = -5$ and $x = \sqrt{7}$ are obtained by replacing x, in turn, by the numbers -5 and $\sqrt{7}$:

$$f(-5) = (-5)^2 = 25 \quad \text{and} \quad f(\sqrt{7}) = (\sqrt{7})^2 = 7. \qquad \equiv$$

Occasionally for emphasis we will write a function using parentheses in place of the symbol x. For example, we can write the squaring function $f(x) = x^2$ as

$$f(\ \ ) = (\ \ )^2. \qquad (1)$$

---

*Many instructors like to call x the *input* of the function and $f(x)$ the *output*.

This illustrates the fact that x is a *placeholder* for any number in the domain of the function $y = f(x)$. Thus, if we wish to evaluate (1) at, say, $3 + h$, where h represents a real number, we put $3 + h$ into the parentheses and carry out the appropriate algebra:

See (6) of Section 1.5. ▶

$$f(3 + h) = (3 + h)^2 = 9 + 6h + h^2.$$

If a function f is defined by means of a formula or an equation, then typically the domain of $y = f(x)$ is not expressly stated. We will see that we can usually deduce the domain of $y = f(x)$ either from the structure of the equation or from the context of the problem.

■ EXAMPLE 2   Domain and Range

In Example 1, since any real number x can be squared and the result x^2 is another real number, $f(x) = x^2$ is a function from R to R, that is, $f: R \rightarrow R$. In other words, the domain of f is the set R of real numbers. Using interval notation, we also write the domain as $(-\infty, \infty)$. The range of f is the set of nonnegative real numbers or $[0, \infty)$; this follows from the fact that $x^2 \geq 0$ for every real number x.   ≡

□ **Domain of a Function** As mentioned earlier, the domain of a function $y = f(x)$ that is defined by a formula is usually not specified. Unless stated or implied to the contrary, it is understood that:

> *The domain of a function f is the largest subset of the set of real numbers for which $f(x)$ is a real number.*

This set is sometimes referred to as the **implicit domain** of the function. For example, we cannot compute $f(0)$ for the reciprocal function $f(x) = 1/x$ since $1/0$ is not a real number. In this case we say that f is **undefined** at $x = 0$. Since every nonzero real number has a reciprocal, the domain of $f(x) = 1/x$ is the set of real numbers except 0. By the same reasoning, the function $g(x) = 1/(x^2 - 4)$ is not defined at either $x = -2$ or $x = 2$, and so its domain is the set of real numbers with -2 and 2 excluded. The square root function $h(x) = \sqrt{x}$ is not defined at $x = -1$ because $\sqrt{-1}$ is not a real number. In order for $h(x) = \sqrt{x}$ to be defined in the real number system we must require the **radicand**, in this case simply x, to be nonnegative. From the inequality $x \geq 0$ we see that the domain of the function h is the interval $[0, \infty)$.

■ EXAMPLE 3   Domain and Range

Determine the domain and range of $f(x) = 4 + \sqrt{x - 3}$.

Solution The radicand $x - 3$ must be nonnegative. By solving the inequality $x - 3 \geq 0$ we get $x \geq 3$, and so the domain of f is $[3, \infty)$. Now, since the symbol $\sqrt{}$ denotes the nonnegative square root of a number, $\sqrt{x - 3} \geq 0$ for $x \geq 3$ and consequently $4 + \sqrt{x - 3} \geq 4$. The smallest value of $f(x)$ occurs at $x = 3$ and is $f(3) = 4 + \sqrt{0} = 4$. Moreover, because $x - 3$ and $\sqrt{x - 3}$ increase as x takes on increasing larger values, we conclude that $y \geq 4$. Consequently the range of f is $[4, \infty)$.   ≡

■ EXAMPLE 4   Domain of f

Determine the domain of $f(x) = \sqrt{x^2 + 2x - 15}$.

Solution As in Example 3, the expression under the radical symbol—the radicand—must be nonnegative, that is, the domain of f is the set of real numbers x for which $x^2 + 2x - 15 \geq 0$ or $(x - 3)(x + 5) \geq 0$. We have already solved the last inequality by means of a sign chart in Example 3 of Section 1.1. The solution set of the inequality $(-\infty, -5] \cup [3, \infty)$ is also the domain of f.   ≡

EXAMPLE 5 Domains of Two Functions

Determine the domain of the given function.

(a) $g(x) = \dfrac{1}{\sqrt{x^2 + 2x - 15}}$ **(b)** $h(x) = \dfrac{5x}{x^2 - 3x - 4}$

Solution A function that is given by a fractional expression is not defined at the x-values for which its denominator is equal to 0.

(a) The expression under the radical is the same as in Example 4. Since $x^2 + 2x - 15$ is in the denominator we must have $x^2 + 2x - 15 \neq 0$. This excludes $x = -5$ and $x = 3$. In addition, since $x^2 + 2x - 15$ appears under a radical, we must have $x^2 + 2x - 15 > 0$ for all other values of x. Thus the domain of the function g is the union of two open intervals $(-\infty, -5) \cup (3, \infty)$.

(b) Since the denominator of $h(x)$ factors,

$$x^2 - 3x - 4 = (x + 1)(x - 4)$$

we see that $(x + 1)(x - 4) = 0$ for $x = -1$ and $x = 4$. In contrast to the function in part (a), these are the *only* numbers for which h is not defined. Hence, the domain of the function h is the set of real numbers with $x = -1$ and $x = 4$ excluded. ≡

Using interval notation, the domain of the function h in part (b) of Example 5 can be written as

$$(-\infty, -1) \cup (-1, 4) \cup (4, \infty).$$

As an alternative to this ungainly union of disjoint intervals, this domain can also be written using set-builder notation as $\{x \mid x \neq -1 \text{ and } x \neq 4\}$.

☐ **Graphs** A function is often used to describe phenomena in fields such as science, engineering, and business. In order to interpret and utilize data, it is useful to display this data in the form of a graph. The graph of a function f is the graph of the set of ordered pairs $(x, f(x))$, where x is in the domain of f. In the xy-plane an ordered pair $(x, f(x))$ is a point, so that the graph of a function is a set of points. If a function is defined by an equation $y = f(x)$, then the graph of f is the graph of the equation. To obtain points on the graph of an equation $y = f(x)$, we judiciously choose numbers $x_1, x_2, x_3, \ldots$ in its domain, compute $f(x_1), f(x_2), f(x_3), \ldots$, plot the corresponding points $(x_1, f(x_1))$, $(x_2, f(x_2)), (x_3, f(x_3)), \ldots$, and then connect these points with a curve. See **FIGURE 2.1.2**. Keep in mind that:

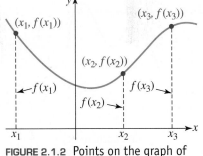

FIGURE 2.1.2 Points on the graph of an equation $y = f(x)$

- a value of x is a directed distance from the y-axis, and
- a function value $f(x)$ is a directed distance from the x-axis.

☐ **End Behavior** A word about the figures in this text is in order. With a few exceptions, it is usually impossible to display the complete graph of a function, and so we often display only the more important features of the graph. In **FIGURE 2.1.3(a)**, notice that the graph goes down on its left and right sides. Unless indicated to the contrary, we may assume that there are no major surprises beyond what we have shown and the graph simply continues in the manner indicated. The graph in Figure 2.1.3(a) indicates the so-called **end behavior** or **global behavior** of the function: For a point (x, y) on the graph, the values of the y-coordinate become unbounded in magnitude in the downward or negative direction as the x-coordinate becomes unbounded in magnitude in both the negative and positive directions on the number line. It is convenient to describe this end behavior using the arrow symbols

$$y \to -\infty \text{ as } x \to -\infty \quad \text{and} \quad y \to -\infty \text{ as } x \to \infty.$$

Recall from Section 1.5 that the symbol $\to$ is read "approaches." Thus, for example, $y \to -\infty$ as $x \to \infty$ is read "y approaches negative infinity as x approaches infinity." More will be said about the concept of global behavior of a function in Chapter 3. If a graph terminates at either its right or left end, we will indicate this by a dot when clarity demands it. See FIGURE 2.1.4. We will use a solid dot to represent the fact that the endpoint is included on the graph and an open dot to signify that the endpoint is not included on the graph.

☐ **Vertical Line Test** From the definition of a function we know that for each x in the domain of f there corresponds only one value $f(x)$ in the range. This means a vertical line that intersects the graph of a function $y = f(x)$ (this is equivalent to choosing an x) can do so in at most one point. Conversely, if *every* vertical line that intersects a graph of an equation does so in at most one point, then the graph is the graph of a function. The last statement is called the **vertical line test** for a function. See Figure 2.1.3(a). On the other hand, if *some* vertical line intersects a graph of an equation more than once, then the graph is not that of a function. See Figures 2.1.3(b) and 2.1.3(c). When a vertical line intersects a graph in several points, the same number x corresponds to different values of y in contradiction to the definition of a function.

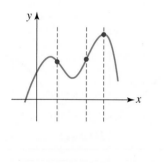

(a) Function

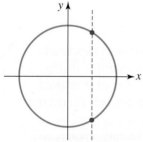

(b) Not a function

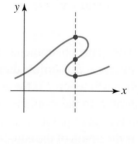

(c) Not a function

FIGURE 2.1.3 Vertical line test

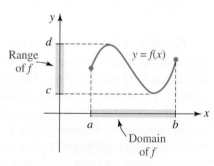

FIGURE 2.1.4 Domain and range interpreted graphically

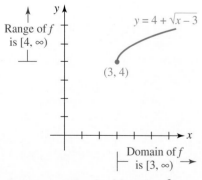

FIGURE 2.1.5 Graph of function f in Example 6

If you have an accurate graph of a function $y = f(x)$ it is often possible to *see* the domain and range of f. In Figure 2.1.4 assume that the colored curve is the entire, or complete, graph of some function f. The domain of f then is the interval $[a, b]$ on the x-axis and the range is the interval $[c, d]$ on the y-axis.

EXAMPLE 6   Example 3 Revisited

From the graph of $f(x) = 4 + \sqrt{x - 3}$ given in FIGURE 2.1.5, we can see that the domain and range of f are, respectively, the interval $[3, \infty)$ on the x-axis and the interval $[4, \infty)$ on the y-axis. This agrees with the results in Example 3.   ≡

As shown in Figure 2.1.3(b), a circle is not the graph of a function. Actually, an equation such as $x^2 + y^2 = 9$ defines (at least) two functions of x. If we solve this equation for y in terms of x we get $y = \pm\sqrt{9 - x^2}$. Because of the single-valued convention of the $\sqrt{}$ sign, both equations $y = \sqrt{9 - x^2}$ and $y = -\sqrt{9 - x^2}$ define functions. As we saw in Section 1.4, the first equation defines an *upper semicircle* and the second defines a *lower semicircle*. From the graphs shown in FIGURE 2.1.6, the domain of $y = \sqrt{9 - x^2}$ is the interval $[-3, 3]$ on the x-axis and the range is the interval $[0, 3]$ on the y-axis; the domain and range of $y = -\sqrt{9 - x^2}$ are $[-3, 3]$ and $[-3, 0]$, respectively.

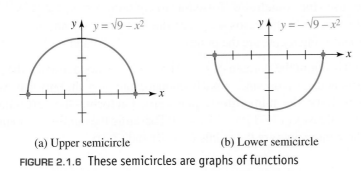

(a) Upper semicircle   (b) Lower semicircle

FIGURE 2.1.6 These semicircles are graphs of functions

☐ **Intercepts** To graph a function defined by an equation $y = f(x)$, it is usually a good idea to first determine whether the graph of f has any intercepts. Recall that all points on the y-axis are of the form $(0, y)$. Thus, if 0 is the domain of a function f, the **y-intercept** is the point on the y-axis whose y-coordinate is $f(0)$, in other words, $(0, f(0))$. See FIGURE 2.1.7(a). Similarly, all points on the x-axis have the form $(x, 0)$. This means that to find the **x-intercepts** of the graph of $y = f(x)$, we determine the values of x that make $y = 0$. That is, we must solve the equation $f(x) = 0$ for x. A number c for which

$$f(c) = 0$$

is referred to as either a **zero** of the function f or a **root** (or **solution**) of the equation $f(x) = 0$. The *real* zeros of a function f are the x-coordinates of the x-intercepts of the graph of f. In Figure 2.1.7(b), we have illustrated a function that has three zeros $x_1, x_2,$ and x_3 because $f(x_1) = 0$, $f(x_2) = 0$, and $f(x_3) = 0$. The corresponding three x-intercepts are the points $(x_1, 0)$, $(x_2, 0)$, and $(x_3, 0)$. Of course, the graph of the function may have no intercepts. This case is illustrated in Figure 2.1.5.

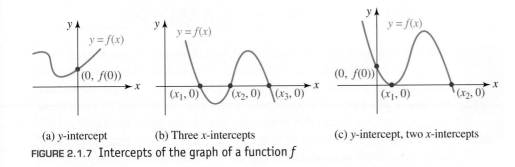

(a) y-intercept   (b) Three x-intercepts   (c) y-intercept, two x-intercepts

FIGURE 2.1.7 Intercepts of the graph of a function f

A graph does not necessarily have to *cross* a coordinate axis at an intercept; a graph could simply be tangent to, or *touch*, an axis. In Figure 2.1.7(c) the graph of $y = f(x)$ is tangent to the x-axis at $(x_1, 0)$. Also, the graph of a function f can have at most one y-intercept since, if 0 is the domain of f, there can correspond only one y-value, namely, $y = f(0)$.

◀ More will be said about this in Chapter 3.

EXAMPLE 7   Intercepts

Find, if possible, the x- and y-intercepts of the given function.

(a) $f(x) = x^2 + 2x - 2$ **(b)** $f(x) = \dfrac{x^2 - 2x - 3}{x}$

Solution (a) Since 0 is in the domain of f, $f(0) = -2$ is the y-coordinate of the y-intercept of the graph of f. The y-intercept is the point $(0, -2)$. To obtain the x-intercepts we must determine whether f has any real zeros, that is, real solutions of the equation $f(x) = 0$. Since the left-hand side of the equation $x^2 + 2x - 2 = 0$ has no obvious factors, we use the quadratic formula to obtain $x = \frac{1}{2}(-2 \pm \sqrt{12})$. Since $\sqrt{12} = \sqrt{4 \cdot 3} = 2\sqrt{3}$ the zeros of f are the irrational numbers $-1 - \sqrt{3}$ and $-1 + \sqrt{3}$. The x-intercepts are the points $(-1 - \sqrt{3}, 0)$ and $(-1 + \sqrt{3}, 0)$.

(b) Because 0 is not in the domain of f ($f(0) = -3/0$ is not defined), the graph of f possesses no y-intercept. Now since f is a fractional expression, the only way we can have $f(x) = 0$ is to have the numerator equal zero. Factoring the left-hand side of $x^2 - 2x - 3 = 0$ gives $(x + 1)(x - 3) = 0$. Therefore the numbers -1 and 3 are the zeros of f. The x-intercepts are the points $(-1, 0)$ and $(3, 0)$.   ≡

☐ **Approximating Zeros** Even when it is obvious that the graph of a function $y = f(x)$ possesses x-intercepts it is not always a straightforward matter to solve the equation $f(x) = 0$. In fact, it is *impossible* to solve some equations exactly; some times the best we can do is to **approximate** the zeros of the function. One way of doing this is to obtain a very accurate graph of f.

EXAMPLE 8   Approximate Intercepts

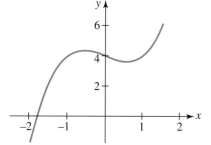

FIGURE 2.1.8 Approximate x-intercept in Example 8

With the aid of a graphing utility the graph of the function $f(x) = x^3 - x + 4$ is given in FIGURE 2.1.8. From $f(0) = 4$ we see that the y-intercept is $(0, 4)$. As we see in the figure, there appears to be only one x-intercept with x-coordinate close to -1.7 or -1.8. But there is no convenient way of finding the roots of the equation $x^3 - x + 4 = 0$. We can however approximate the real root of this equation with the aid of the *find root* feature of either a graphing calculator or computer algebra system. We find that $x \approx -1.796$ and so the approximate x-intercept is $(-1.796, 0)$. As a check, note that the function value

$$f(-1.796) = (-1.796)^3 - (-1.796) + 4 \approx 0.0028$$

is nearly 0.   ≡

NOTES FROM THE CLASSROOM

When sketching the graph of a function, you should never resort to plotting a lot of points by hand. That is something a graphing calculator or a computer algebra system (CAS) does so well. On the other hand, you should not become dependent on a calculator to obtain a graph. Believe it or not, there are precalculus and calculus instructors who do not allow the use of graphing calculators on quizzes or tests. Usually there is no objection to your using calculators or computers as an aid in checking homework problems, but in the classroom instructors want to see the product of your own mind, namely, the ability to analyze. So you are strongly encouraged to develop your graphing skills to the point where you are able to quickly sketch by hand the graph of a function from a basic familiarity of types of functions and by plotting a minimum of well-chosen points such as intercepts.

Exercises Answers to selected odd-numbered problems begin on page ANS–3.

In Problems 1–6, find the indicated function values.

1. If $f(x) = x^2 - 1$; $f(-5), f(-\sqrt{3}), f(3)$, and $f(6)$
2. If $f(x) = -2x^2 + x$; $f(-5), f(-\frac{1}{2}), f(2)$, and $f(7)$
3. If $f(x) = \sqrt{x+1}$; $f(-1), f(0), f(3)$, and $f(5)$
4. If $f(x) = \sqrt{2x+4}$; $f(-\frac{1}{2}), f(\frac{1}{2}), f(\frac{5}{2})$, and $f(4)$
5. If $f(x) = \dfrac{3x}{x^2+1}$; $f(-1), f(0), f(1)$, and $f(\sqrt{2})$
6. If $f(x) = \dfrac{x^2}{x^3-2}$; $f(-\sqrt{2}), f(-1), f(0)$, and $f(\frac{1}{2})$

In Problems 7 and 8, find

$$f(x), f(2a), f(a^2), f(-5x), f(2a+1), f(x+h)$$

for the given function f and simplify as much as possible.

7. $f(\ ) = -2(\ )^2 + 3(\ )$ **8.** $f(\ ) = (\ )^3 - 2(\ )^2 + 20$

9. For what values of x is $f(x) = 6x^2 - 1$ equal to 23?
10. For what values of x is $f(x) = \sqrt{x} - 4$ equal to 4?

In Problems 11–20, find the domain of the given function f.

11. $f(x) = \sqrt{4x-2}$ **12.** $f(x) = \sqrt{15-5x}$
13. $f(x) = \dfrac{10}{\sqrt{1-x}}$ **14.** $f(x) = \dfrac{2x}{\sqrt{3x-1}}$
15. $f(x) = \dfrac{2x-5}{x(x-3)}$ **16.** $f(x) = \dfrac{x}{x^2-1}$
17. $f(x) = \dfrac{1}{x^2-10x+25}$ **18.** $f(x) = \dfrac{x+1}{x^2-4x-12}$
19. $f(x) = \dfrac{x}{x^2-x+1}$ **20.** $f(x) = \dfrac{x^2-9}{x^2-2x-1}$

In Problems 21–26, use the sign-chart method to find the domain of the given function f.

21. $f(x) = \sqrt{25-x^2}$ **22.** $f(x) = \sqrt{x(4-x)}$
23. $f(x) = \sqrt{x^2-5x}$ **24.** $f(x) = \sqrt{x^2-3x-10}$
25. $f(x) = \sqrt{\dfrac{3-x}{x+2}}$ **26.** $f(x) = \sqrt{\dfrac{5-x}{x}}$

In Problems 27–30, determine whether the graph in the figure is the graph of a function.

27.

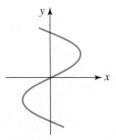

FIGURE 2.1.9 Graph for Problem 27

28.

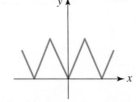

FIGURE 2.1.10 Graph for Problem 28

2.1 Functions and Graphs   55

29.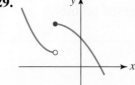

FIGURE 2.1.11 Graph for
Problem 29

30.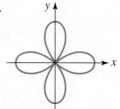

FIGURE 2.1.12 Graph for
Problem 30

In Problems 31–34, use the graph of the function f given in the figure to find its
domain and range.

31.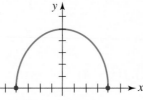

FIGURE 2.1.13 Graph for
Problem 31

32.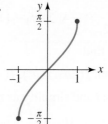

FIGURE 2.1.14 Graph for
Problem 32

33.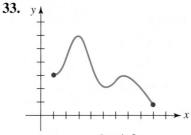

FIGURE 2.1.15 Graph for
Problem 33

34.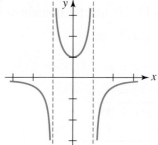

FIGURE 2.1.16 Graph for
Problem 34

In Problems 35–42, find the real zeros of the given function f.

35. $f(x) = 5x + 6$ **36.** $f(x) = -2x + 9$
37. $f(x) = x^2 - 5x + 6$ **38.** $f(x) = x^2 - 2x - 1$
39. $f(x) = x(3x - 1)(x + 9)$ **40.** $f(x) = x^3 - x^2 - 2x$
41. $f(x) = x^4 - 1$ **42.** $f(x) = 2 - \sqrt{4 - x^2}$

In Problems 43–50, find the x- and y-intercepts, if any, of the graph of the given
function f. Do not graph.

43. $f(x) = \frac{1}{2}x - 4$ **44.** $f(x) = x^2 - 6x + 5$
45. $f(x) = 4(x - 2)^2 - 1$ **46.** $f(x) = (2x - 3)(x^2 + 8x + 16)$
47. $f(x) = \dfrac{x^2 + 4}{x^2 - 16}$ **48.** $f(x) = \dfrac{x(x + 1)(x - 6)}{x + 8}$
49. $f(x) = \frac{3}{2}\sqrt{4 - x^2}$ **50.** $f(x) = \frac{1}{2}\sqrt{x^2 - 2x - 3}$

In Problems 51 and 52, find two functions $y = f_1(x)$ and $y = f_2(x)$ defined by the
given equation. Find the domain of the functions f_1 and f_2.

51. $x = y^2 - 5$ **52.** $x^2 - 4y^2 = 16$

In Problems 53–58, find f if $f(2) = -3$ in each case.

53. $f(x) = 4x + k$

54. $f(x) = -2x^2 + kx$

55. $f(x) = kx^3 - x + 1$

56. $f(x) = x^4 + x^3 + kx^2 - x - 1$

57. $f(x) = \dfrac{2x - k}{x}$

58. $f(x) = \dfrac{x + k}{x - k}$

In Problems 59 and 60, use the graph of the function f given in the figure to estimate the values of $f(-3), f(-2), f(-1), f(1), f(2),$ and $f(3)$. Estimate the y-intercept.

59.

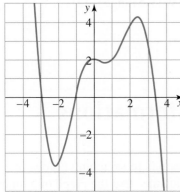

FIGURE 2.1.17 Graph for Problem 59

60.

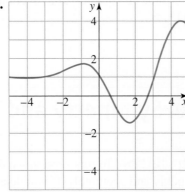

FIGURE 2.1.18 Graph for Problem 60

In Problems 61 and 62, use the graph of the function f given in the figure to estimate the values of $f(-2), f(-1.5), f(0.5), f(1), f(2),$ and $f(3.2)$. Estimate the x-intercepts.

61.

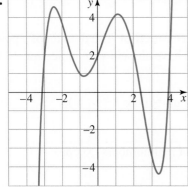

FIGURE 2.1.19 Graph for Problem 61

62.

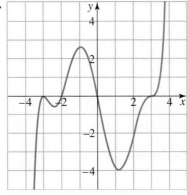

FIGURE 2.1.20 Graph for Problem 62

Miscellaneous Calculus-Related Problems

63. In calculus some of the functions that you will encounter have as their domain the set of positive integers n. The **factorial function** $f(n) = n!$ is defined as the product of the first n positive integers, that is,

$$f(n) = n! = 1 \cdot 2 \cdot 3 \cdots (n - 1) \cdot n.$$

(a) Evaluate $f(2), f(3), f(5),$ and $f(7)$.

(b) Show that $f(n + 1) = f(n) \cdot (n + 1)$.

(c) Simplify $f(n + 2)/f(n)$.

64. Another function of a positive integer n gives the sum of the first n squared positive integers:

$$S(n) = \tfrac{1}{6}n(n + 1)(2n + 1) = 1^2 + 2^2 + \cdots + n^2.$$

(a) Find the value of the sum $1^2 + 2^2 + \cdots + 99^2 + 100^2$.

(b) Find n such that $300 < S(n) < 400$. [*Hint*: Use a calculator.]

For Discussion

65. Determine an equation of a function $y = f(x)$ whose domain is (a) $[3, \infty)$, (b) $(3, \infty)$.

66. Determine an equation of a function $y = f(x)$ whose range is (a) $[3, \infty)$, (b) $(3, \infty)$.

2.2 Symmetry and Transformations

≡ **Introduction** In this section we discuss two aids in sketching graphs of functions quickly and accurately. If you determine in advance that the graph of a function possesses *symmetry*, then you can cut your work in half. In addition, sketching a graph of a complicated-looking function is expedited if you recognize that the required graph is actually a *transformation* of the graph of a simpler function. This latter graphing aid is based on your prior knowledge of the graphs of some basic functions.

☐ **Power Functions** A function of the form

$$f(x) = x^n,$$

where n represents a real number, is called a **power function**. The domain of a power function depends on the power n. For example, we have already seen in Section 2.1 for $n = 2, n = \tfrac{1}{2}$, and $n = -1$, respectively, that:

- the domain of $f(x) = x^2$ is the set R of real numbers or $(-\infty, \infty)$,
- the domain of $f(x) = x^{1/2} = \sqrt{x}$ is $[0, \infty)$,
- the domain of $f(x) = x^{-1} = \dfrac{1}{x}$ is the set R of real numbers except $x = 0$.

Simple power functions, or modified versions of these functions, occur so often in problems in calculus that you do not want to spend valuable time plotting their graphs. We suggest that you know (memorize) the short catalogue of graphs of power functions given in FIGURE 2.2.1 on the next page. You might already know that the graph in part (a) of that figure is a **line** and the graph in part (b) is called a **parabola**.

☐ **Symmetry** In Section 1.4 we discussed symmetry of a graph with respect to the y-axis, the x-axis, and the origin. Of those three types of symmetries, the graph of a function f can be symmetric with respect to the y-axis or with respect to the origin, but the graph of a nonzero function f *cannot* be symmetric with respect to the x-axis. Before proceeding with the discussion of symmetry of graphs of functions we need the following definition.

Can you explain why the graph of a function ▶ cannot have symmetry with respect to the x-axis? See Problem 47 in Exercises 2.2.

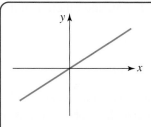

(a) $n = 1$, $f(x) = x$

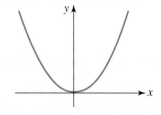

(b) $n = 2$, $f(x) = x^2$

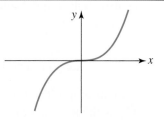

(c) $n = 3$, $f(x) = x^3$

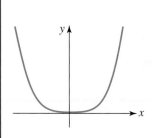

(d) $n = 4$, $f(x) = x^4$

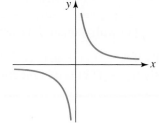

(e) $n = -1$, $f(x) = x^{-1} = \dfrac{1}{x}$

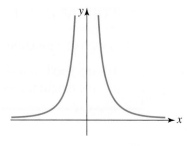

(f) $n = -2$, $f(x) = x^{-2} = \dfrac{1}{x^2}$

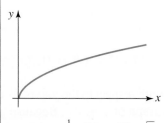

(g) $n = \frac{1}{2}$, $f(x) = x^{1/2} = \sqrt{x}$

(h) $n = \frac{1}{3}$, $f(x) = x^{1/3} = \sqrt[3]{x}$

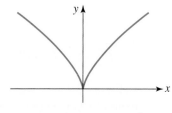

(i) $n = \frac{2}{3}$, $f(x) = x^{2/3} = \sqrt[3]{x^2}$

FIGURE 2.2.1 Brief catalogue of power functions $f(x) = x^n$ for various n

DEFINITION 2.2.1   Even and Odd Functions

(*i*) A function f with domain X is said to be an **even function** if
$f(-x) = f(x)$ for every x in X.

(*ii*) A function f with domain X is said to be an **odd function** if
$f(-x) = -f(x)$ for every x in X.

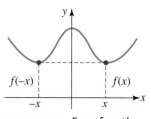

FIGURE 2.2.2 Even function

The graphical interpretation of Definition 2.2.1 is illustrated in FIGURES 2.2.2 and 2.2.3. In Figure 2.2.2, observe that if f is an even function and

$$f(x) \qquad\qquad f(-x) = f(x)$$
$$\downarrow \qquad\qquad \downarrow$$

(x, y) is a point on its graph, then necessarily $(-x, y)$

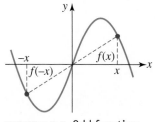

FIGURE 2.2.3 Odd function

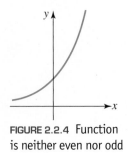

FIGURE 2.2.4 Function is neither even nor odd

is also on its graph. Similarly we see in Figure 2.2.3, that if f is an odd function and

$$\overset{f(x)}{\downarrow} \qquad\qquad\qquad\qquad \overset{f(-x)=-f(x)}{\downarrow}$$

(x, y) is a point on its graph, then necessarily $(-x, -y)$

is on its graph. The function whose graph is given in FIGURE 2.2.4 is neither even or odd. Using the information on page 29 of Section 1.4, we summarize these observations in terms of symmetry of the graph in the next theorem.

THEOREM 2.2.1   Tests for Symmetry

The graph of a function $y = f(x)$ with domain X is symmetric with respect to:

(*i*) the **y-axis** if and only if $y = f(x)$ is an even function, or
(*ii*) the **origin** if and only if $y = f(x)$ is odd function.

EXAMPLE 1   **Even and Odd Functions**

(a) $f(x) = x^{2/3}$ is an even function since by (*i*) of Definition 2.2.1 and the laws of exponents

$$\overset{\text{cube root of } -1 \text{ is } -1}{\downarrow}$$

$$f(-x) = (-x)^{2/3} = (-1)^{2/3}x^{2/3} = \left(\sqrt[3]{-1}\right)^2 x^{2/3} = (-1)^2 x^{2/3} = x^{2/3} = f(x).$$

In Figure 2.2.1(i), we see that the graph of f is symmetric with respect to the y-axis. For example, since $f(8) = 8^{2/3} = 4$, $(8, 4)$ is a point on the graph of $y = x^{2/3}$. Because f is an even function, $f(-8) = f(8)$ implies $(-8, 4)$ is on the same graph.

(b) $f(x) = x^3$ is an odd function since by (*ii*) of Definition 2.2.1,

$$f(-x) = (-x)^3 = (-1)^3 x^3 = -x^3 = -f(x).$$

Inspection of Figure 2.2.1(c) shows that the graph of f is symmetric with respect to the origin. For example, since $f(1) = 1$, $(1, 1)$ is a point on the graph of $y = x^3$. Because f is an odd function, $f(-1) = -f(1)$ implies $(-1, -1)$ is on the same graph.

(c) $f(x) = x^3 + 1$ is neither even nor odd. From

$$f(-x) = (-x)^3 + 1 = -x^3 + 1$$

we see that $f(-x) \neq f(x)$, and $f(-x) \neq -f(x)$. The graph of f has neither y-axis nor origin symmetry.   ≡

The graphs in Figure 2.2.1, with part (g) the only exception, possess either y-axis or origin symmetry. The functions in Figures 2.2.1(b), (d), (f), and (i) are even, whereas the functions in Figures 2.2.1(a), (c), (e), and (h) are odd.

Often we can sketch the graph of a function by applying a certain transformation to the graph of a simpler function (such as those given in Figure 2.2.1). We are going to consider two kinds of graphical transformations, rigid and nonrigid.

☐ **Rigid Transformations** A **rigid transformation** of a graph is one that changes only the *position* of the graph in the *xy*-plane but not its shape. We have already examined this concept briefly in the discussion of the circle in Section 1.4. For example, the circle $(x - 2)^2 + (y - 3)^2 = 1$ with center $(2, 3)$ and radius $r = 1$ has *exactly* the same shape as the circle $x^2 + y^2 = 1$ with center at the origin. We can think of the graph of $(x - 2)^2 + (y - 3)^2 = 1$ as the graph of $x^2 + y^2 = 1$ shifted horizontally two units to the right followed by an upward vertical shift of three units. For the graph of a function $y = f(x)$ we examine four kinds of shifts or translations.

THEOREM 2.2.2   Vertical and Horizontal Shifts

Suppose $y = f(x)$ is a function and c is a positive constant. Then the graph of

(i) $y = f(x) + c$ is the graph of f shifted vertically **up** c units,
(ii) $y = f(x) - c$ is the graph of f shifted vertically **down** c units,
(iii) $y = f(x + c)$ is the graph of f shifted horizontally to the **left** c units,
(iv) $y = f(x - c)$ is the graph of f shifted horizontally to the **right** c units.

Consider the graph of a function $y = f(x)$ given in FIGURE 2.2.5. The shifts of this graph described in (*i*)–(*iv*) of Theorem 2.2.2 are the graphs in red in parts (a)–(d) of Figure 2.2.6. If (x, y) is a point on the graph of $y = f(x)$ and the graph of *f* is shifted, say, upward by $c > 0$ units, then $(x, y + c)$ is a point on the new graph. In general, the *x*-coordinates do not change as a result of a vertical shift. See FIGURES 2.2.6(a) and 2.2.6(b). Similarly, in a horizontal shift the *y*-coordinates of points on the shifted graph are the same as on the original graph. See Figures 2.2.6(c) and 2.2.6(d).

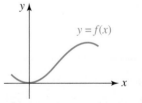

FIGURE 2.2.5 Graph of $y = f(x)$

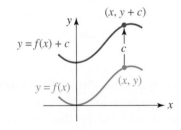

(a) Vertical shift up

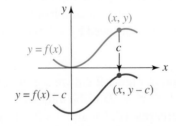

(b) Vertical shift down

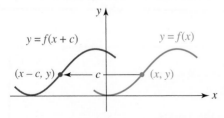

(c) Horizontal shift left

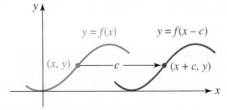

(d) Horizontal shift right

FIGURE 2.2.6 Vertical and horizontal shifts of the graph of $y = f(x)$ by an amount $c > 0$

EXAMPLE 2   Vertical and Horizontal Shifts

The graphs of $y = x^2 + 1$, $y = x^2 - 1$, $y = (x + 1)^2$, and $y = (x - 1)^2$ are obtained from the graph of $f(x) = x^2$ in FIGURE 2.2.7(a) by shifting this graph, in turn, 1 unit up (Figure 2.2.7(b)), 1 unit down (Figure 2.2.7(c)), 1 unit to the left (Figure 2.2.7(d)), and 1 unit to the right (Figure 2.2.7(e)).

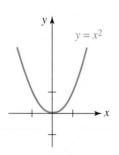 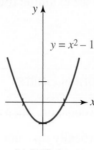 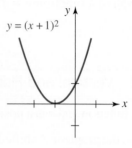 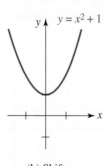 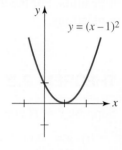

| (a) Starting point | (b) Shift up | (c) Shift down | (d) Shift left | (e) Shift right |

FIGURE 2.2.7 Shifted graphs in Example 2

☐ **Combining Shifts** In general, the graph of a function

$$y = f(x \pm c_1) \pm c_2, \tag{3}$$

where c_1 and c_2 are positive constants, combines a horizontal shift (left or right) with a vertical shift (up or down). For example, the graph of $y = f(x - c_1) + c_2$ is the graph of $y = f(x)$ shifted c_1 units to the right and then c_2 units up.

▶ The order in which the shifts are done is irrelevant. We could do the upward shift first followed by the shift to the right.

EXAMPLE 3   Graph Shifted Vertically and Horizontally

Graph $y = (x + 1)^2 - 1$.

Solution From the preceding paragraph we identify in (3) the form $y = f(x + c_1) - c_2$ with $c_1 = 1$ and $c_2 = 1$. Thus, the graph of $y = (x + 1)^2 - 1$ is the graph of $f(x) = x^2$ shifted 1 unit to the left followed by a downward shift of 1 unit. The graph is given in FIGURE 2.2.8.

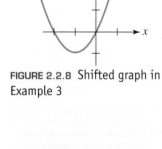

FIGURE 2.2.8 Shifted graph in Example 3

From the graph in Figure 2.2.8 we see immediately that the range of the function $y = (x + 1)^2 - 1 = x^2 + 2x$ is the interval $[-1, \infty)$ on the y-axis. Note also that the graph has x-intercepts $(0, 0)$ and $(-2, 0)$; you should verify this by solving $x^2 + 2x = 0$. Also, if you reexamine Figure 2.1.5 in Section 2.1 you will see that the graph of $y = 4 + \sqrt{x - 3}$ is the graph of the square root function $f(x) = \sqrt{x}$ (Figure 2.2.1(g)) shifted 3 units to the right and then 4 units up.

Another way of rigidly transforming a graph of a function is by a **reflection** in a coordinate axis.

Reflection or mirror image in a horizontal axis

THEOREM 2.2.3   Reflections

Suppose $y = f(x)$ is a function. Then the graph of

   (*i*) $y = -f(x)$ is the graph of f reflected in the **x-axis**,
   (*ii*) $y = f(-x)$ is the graph of f reflected in the **y-axis**.

In part (a) of FIGURE 2.2.9 we have reproduced the graph of a function $y = f(x)$ given in Figure 2.2.5. The reflections of this graph described in (i)–(ii) of Theorem 2.2.3 are illustrated in Figures 2.2.9(b) and 2.2.9(c). If (x, y) denotes a point on the graph of $y = f(x)$, then the point $(x, -y)$ is on the graph of $y = -f(x)$, and $(-x, y)$ is on the graph of $y = f(-x)$. Each of these reflections is a mirror image of the graph of $y = f(x)$ in the respective coordinate axis.

Reflection or mirror image in a vertical axis

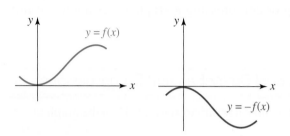

(a) Starting point   (b) Reflection in x-axis   (c) Reflection in y-axis

FIGURE 2.2.9 Reflections in the coordinate axes

EXAMPLE 4   Reflections

Graph **(a)** $y = -\sqrt{x}$ **(b)** $y = \sqrt{-x}$

Solution The starting point is the graph of $f(x) = \sqrt{x}$ given in FIGURE 2.2.10(a).
(a) The graph of $y = -\sqrt{x}$ is the reflection of the graph of $f(x) = \sqrt{x}$ in the x-axis. Observe in Figure 2.2.10(b) that since $(1, 1)$ is on the graph of f, the point $(1, -1)$ is on the graph of $y = -\sqrt{x}$.

(b) The graph of $y = \sqrt{-x}$ is the reflection of the graph of $f(x) = \sqrt{x}$ in the y-axis. Observe in Figure 2.2.10(c) that since $(1, 1)$ is on the graph of f, the point $(-1, 1)$ is on the graph of $y = \sqrt{-x}$. The function $y = \sqrt{-x}$ looks a little strange, but bear in mind that its domain is determined by the requirement that $-x \geq 0$, or equivalently $x \leq 0$, and so the reflected graph is defined on the interval $(-\infty, 0]$.

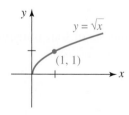 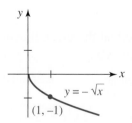 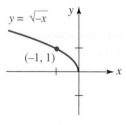

(a) Starting point   (b) Reflection in x-axis   (c) Reflection in y-axis

FIGURE 2.2.10 Graphs in Example 4   ≡

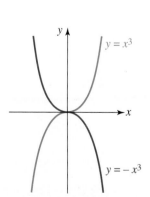

FIGURE 2.2.11 Reflection (red) of an odd function (blue) in y-axis

If a function f is even, then $f(-x) = f(x)$ shows that a reflection in the y-axis would give precisely the same graph. If a function is odd, then from $f(-x) = -f(x)$ we see that a reflection of the graph of f in the y-axis is identical to the graph of f reflected in the x-axis. In FIGURE 2.2.11 the blue curve is the graph of the odd function $f(x) = x^3$; the red curve is the graph of $y = f(-x) = (-x)^3 = -x^3$. Notice that if the blue curve is reflected in either the y-axis or the x-axis, we get the red curve.

☐ **Nonrigid Transformations** If a function f is multiplied by a constant $c > 0$ the shape of the graph is changed but retains, *roughly*, its original shape. The graph of $y = cf(x)$ is the graph of $y = f(x)$ distorted vertically; the graph of f is either stretched (or elongated) vertically or is compressed (or flattened) vertically depending on the value of c. Stretching or compressing a graph are examples of **nonrigid transformations**.

THEOREM 2.2.4   Vertical Stretches and Compressions

Suppose $y = f(x)$ is a function and c a positive constant. Then the graph of $y = cf(x)$ is the graph of f

  (i)   vertically stretched by a factor of c units if $c > 1$,
  (ii) vertically compressed by a factor of c units if $0 < c < 1$.

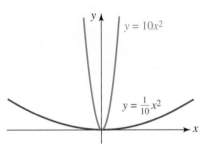

FIGURE 2.2.12 Vertical stretch (red) of the graph of $f(x) = x$ (blue)

If (x, y) represents a point on the graph of f, then the point (x, cy) is on the graph of cf. The graphs of $y = x$ and $y = 3x$ are compared in FIGURE 2.2.12; the y-coordinate of a point on the graph of $y = 3x$ is 3 times as large as the y-coordinate of the point with the same x-coordinate on the graph of $y = x$. The comparison of the graphs of $y = 10x^2$ (blue graph) and $y = \frac{1}{10}x^2$ (red graph) in FIGURE 2.1.13 is a little more dramatic; the graph of $y = \frac{1}{10}x^2$ exhibits considerable vertical flattening, especially in a neighborhood of the origin. Note that c is positive in this discussion. To sketch the graph of $y = -10x^2$ we think of it as $y = -(10x^2)$, which means we first stretch the graph of $y = x^2$ vertically by a factor of 10 units and then reflect that graph in the x-axis.

The next example illustrates shifting, reflecting, and stretching of a graph.

FIGURE 2.2.13 Vertical stretch (blue) and vertical compression (red) of the graph of $f(x) = x^2$

▮ **EXAMPLE 5**   **Combining Transformations**

Graph $y = 2 - 2\sqrt{x - 3}$.

Solution You should recognize that the given function consists of four transformations of the basic function $f(x) = \sqrt{x}$:

We start with the graph of $f(x) = \sqrt{x}$ in FIGURE 2.2.14(a). Then stretch this graph vertically by a factor of 2 to obtain $y = 2\sqrt{x}$ in Figure 2.2.14(b). Reflect this second graph in the x-axis to obtain $y = -2\sqrt{x}$ in Figure 2.2.14(c). Shift this third graph 3 units to the right to obtain $y = -2\sqrt{x - 3}$ in Figure 2.2.14(d). Finally, shift the fourth graph upward 2 units to obtain $y = 2 - 2\sqrt{x - 3}$ in Figure 2.2.14(e). Note that the point $(0, 0)$ on the graph of $f(x) = \sqrt{x}$ remains fixed in the vertical stretch and the reflection in the x-axis, but under the first (horizontal) shift $(0, 0)$ moves to $(3, 0)$ and under the second (vertical) shift $(3, 0)$ moves to $(3, 2)$.

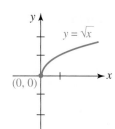

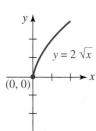

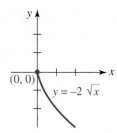

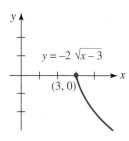

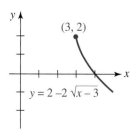

(a) Starting point   (b) Vertical stretch   (c) Reflection in x-axis   (d) Shift right   (e) Shift up

FIGURE 2.2.14 Graph of function in Example 5

≡

| **2.2** | Exercises | Answers to selected odd-numbered problems begin on page ANS-3. |

In Problems 1–10, use (1) and (2) to determine whether the given function $y = f(x)$ is even, odd, or neither even nor odd. Do not graph.

1. $f(x) = 4 - x^2$

2. $f(x) = x^2 + 2x$

3. $f(x) = x^3 - x + 4$

4. $f(x) = x^5 + x^3 + x$

5. $f(x) = 3x - \dfrac{1}{x}$

6. $f(x) = \dfrac{x}{x^2 + 1}$

7. $f(x) = 1 - \sqrt{1 - x^2}$

8. $f(x) = \sqrt[3]{x^3 + x}$

9. $f(x) = |x^3|$

10. $f(x) = x|x|$

In Problems 11–14, classify the function $y = f(x)$ whose graph is given as even, odd, or neither even nor odd.

11.

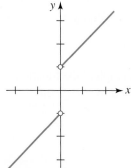

FIGURE 2.2.15 Graph for Problem 11

12.

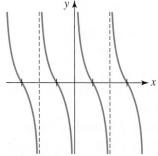

FIGURE 2.2.16 Graph for Problem 12

13.

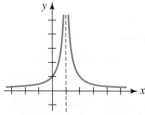

FIGURE 2.2.17 Graph for Problem 13

14.

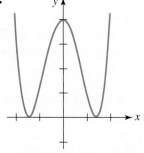

FIGURE 2.2.18 Graph for Problem 14

2.2 Symmetry and Transformations

In Problems 15–18, complete the graph of the given function $y = f(x)$ if **(a)** f is an even function and **(b)** f is an odd function.

15.

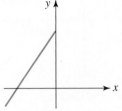

FIGURE 2.2.19 Graph for Problem 15

16.

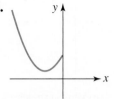

FIGURE 2.2.20 Graph for Problem 16

17.

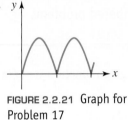

FIGURE 2.2.21 Graph for Problem 17

18.

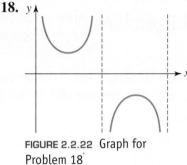

FIGURE 2.2.22 Graph for Problem 18

In Problems 19 and 20, suppose that $f(-2) = 4$ and $f(3) = 7$. Determine $f(2)$ and $f(-3)$.

19. If f is an even function.
20. If f is an odd function.

In Problems 21 and 22, suppose that $g(-1) = -5$ and $g(4) = 8$. Determine $g(1)$ and $g(-4)$.

21. If g is an odd function.
22. If g is an even function.

In Problems 23–32, the points $(-2, 1)$ and $(3, -4)$ are on the graph of the function $y = f(x)$. Find the corresponding points on the graph obtained by the given transformations.

23. the graph of f shifted up 2 units
24. the graph of f shifted down 5 units
25. the graph of f shifted to the left 6 units
26. the graph of f shifted to the right 1 unit
27. the graph of f shifted up 1 unit and to the left 4 units
28. the graph of f shifted down 3 units and to the right 5 units
29. the graph of f reflected in the y-axis
30. the graph of f reflected in the x-axis
31. the graph of f stretched vertically by a factor of 15 units
32. the graph of f compressed vertically by a factor of $\frac{1}{4}$ unit, then reflected in the x-axis

In Problems 33–36, use the graph of the function $y = f(x)$ given in the figure to graph the following functions.

(a) $y = f(x) + 2$　　　　　　**(b)** $y = f(x) - 2$
(c) $y = f(x + 2)$　　　　　　**(d)** $y = f(x - 5)$
(e) $y = -f(x)$　　　　　　　**(f)** $y = f(-x)$

33.

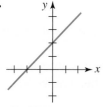

FIGURE 2.2.23 Graph for Problem 33

34.

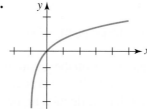

FIGURE 2.2.24 Graph for Problem 34

35.

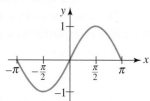

FIGURE 2.2.25 Graph for Problem 35

36.

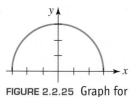

FIGURE 2.2.26 Graph for Problem 36

In Problems 37 and 38, use the graph of the function $y = f(x)$ given in the figure to graph the following functions.

(a) $y = f(x) + 1$ **(b)** $y = f(x) - 1$
(c) $y = f(x + \pi)$ **(d)** $y = f(x - \pi/2)$
(e) $y = -f(x)$ **(f)** $y = f(-x)$
(g) $y = 3f(x)$ **(h)** $y = -\frac{1}{2}f(x)$

37.

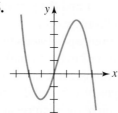

FIGURE 2.2.27 Graph for Problem 37

38.

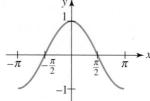

FIGURE 2.2.28 Graph for Problem 38

In Problems 39–42, find the equation of the final graph after the given transformations are applied to the graph of $y = f(x)$.

39. the graph of $f(x) = x^3$ shifted up 5 units and right 1 unit
40. the graph of $f(x) = x^{2/3}$ stretched vertically by a factor of 3 units, then shifted right 2 units
41. the graph of $f(x) = x^4$ reflected in the x-axis, then shifted left 7 units
42. the graph of $f(x) = 1/x$ reflected in the y-axis, then shifted left 5 units and down 10 units

In Problems 43–46, describe in words how the graph of the first function is obtained from the graph of the second function using rigid and nonrigid transformations. Carefully graph the first function.

43. $y = -1 + 2\sqrt{-x + 2};\ y = \sqrt{x}$ **44.** $y = 2 + \frac{1}{2}(-x)^3;\ y = x^3$

45. $y = 2 - \dfrac{2}{x - 1};\ y = \dfrac{1}{x}$ **46.** $y = -1 - \dfrac{1}{(x - 2)^2};\ y = \dfrac{1}{x^2}$

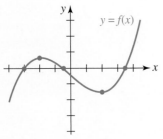

FIGURE 2.2.29 Graph for Problem 49

For Discussion

47. Explain why the graph of a nonzero function cannot be symmetric with respect to the *x*-axis.

48. What points, if any, on the graph of $y = f(x)$ remain fixed, that is, the same on the resulting graph after a vertical stretch or compression? After a reflection in the *x*-axis? After a reflection in the *y*-axis?

49. Copy the graph of $y = f(x)$ in FIGURE 2.2.29 on a piece of paper. By paying close attention to the five blue dots on the graph of $y = f(x)$ draw a representative graph of a vertical stretch and a vertical compression defined by **(a)** $y = cf(x), c > 1$ and **(b)** $y = cf(x), 0 < c < 1$. [*Hint*: See the first question in Problem 48.]

50. Discuss the relationship between the graphs of $y = f(x)$ and $y = f(|x|)$.

51. Discuss the relationship between the graphs of $y = f(x)$ and $y = f(cx)$, where $c > 0$ is a constant. Consider two cases: $0 < c < 1$ and $c > 1$.

52. Review the graphs of $y = x$ and $y = 1/x$ in Figure 2.2.1. Then discuss how to obtain the graph of the reciprocal $y = 1/f(x)$ from the graph of $y = f(x)$. Sketch the graph of $y = 1/f(x)$ for the function f whose graph is given in Figure 2.2.26.

2.3 Linear Functions

≡ **Introduction** The notion of a line plays an important role in the study of differential calculus. There are three types of lines in the *xy*- or Cartesian plane: horizontal lines, vertical lines, and slant or oblique lines. We will see in this section that an equation of each of these lines stems from a **linear equation in two variables**

$$Ax + By + C = 0, \tag{1}$$

where A, B, and C are real constants. The characteristic that gives (1) its name *linear* is that the variables x and y appear only to the first power. We will refer back to (1) when we review lines and their equations, but let's note the cases of special interest:

$$A = 0, B \neq 0, \text{ gives } y = -\frac{C}{B}, \tag{2}$$

$$A \neq 0, B = 0, \text{ gives } x = -\frac{C}{A}, \tag{3}$$

$$A \neq 0, B \neq 0, \text{ gives } y = -\frac{A}{B}x - \frac{C}{B}. \tag{4}$$

The first and the third of these three equations define functions. By relabeling $-C/B$ in (2) as b we get a constant function.

DEFINITION 2.3.1 Constant Function

A **constant function** $y = f(x)$ is a function of the form

$$y = b, \tag{5}$$

where b is a constant.

The **domain** of a constant function is the set of real numbers $(-\infty, \infty)$. In the definition of a function we are pairing each real number x with the same value of y, that is, (x, b). In our student/desk example of a function in Section 2.2 this is equivalent to having all the students in a classroom sit in one desk. On the other hand, the equation in (3) does not define a function. We cannot have one student (the fixed value of x) sit in all the desks in a classroom.

By relabeling $-A/B$ and $-C/B$ in (4) as a and b, respectively, we get the form of a linear function.

DEFINITION 2.3.2 Linear Function

A **linear function** $y = f(x)$ is a function of the form
$$f(x) = ax + b, \qquad (6)$$
where $a \neq 0$ and b are constants.

The **domain** of a linear function is the set of real numbers $(-\infty, \infty)$.

☐ **Graphs** Since the graphs of constant and linear functions are straight lines, it is appropriate that we will review equations of all lines. We begin with the recollection from plane geometry that through any two distinct points (x_1, y_1) and (x_2, y_2) in the plane there passes only one line L. If $x_1 \neq x_2$, then the number

$$m = \frac{y_2 - y_1}{x_2 - x_1} \qquad (7)$$

is called the **slope** of the line determined by these two points. It is customary to call $y_2 - y_1$ the **change in y** or the **rise** of the line; $x_2 - x_1$ is the **change in x** or the **run** of the line. Therefore (7) is

$$m = \frac{\text{rise}}{\text{run}}.$$

See FIGURE 2.3.1(a). Any pair of distinct points on a line will determine the same slope. To see why this is so, consider the two similar right triangles in Figure 2.3.1(b). Since we know that the ratios of corresponding sides in similar triangles are equal we have

$$\frac{y_2 - y_1}{x_2 - x_1} = \frac{y_4 - y_3}{x_4 - x_3}.$$

Hence the slope of a line is independent of the choice of points on the line.

In FIGURE 2.3.2 we compare the graphs of lines with positive, negative, zero, and undefined slopes. In Figure 2.3.2(a) we see, reading the graph from left to right, that a line with positive slope ($m > 0$) rises as x increases. Figure 2.3.2(b) shows that a line with negative slope ($m < 0$) falls as x increases. If (x_1, y_1) and (x_2, y_2) are points on a horizontal line, then $y_1 = y_2$ and so its rise is $y_2 - y_1 = 0$. Hence from (7) the slope is zero ($m = 0$). See Figure 2.3.2(c). If (x_1, y_1) and (x_2, y_2) are points on a vertical line, then $x_1 = x_2$ and so its run is $x_2 - x_1 = 0$. In this case we say that the slope of the line is **undefined** or that the line has no slope. See Figure 2.3.2(d).

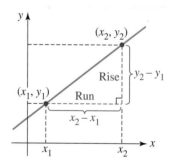

(a) Rise and run

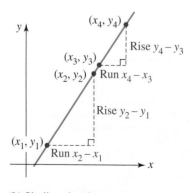

(b) Similar triangles

FIGURE 2.3.1 Slope of a line

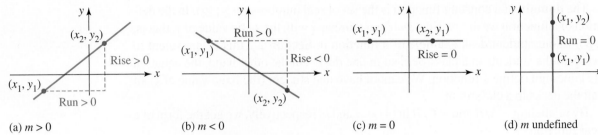

(a) $m > 0$ (b) $m < 0$ (c) $m = 0$ (d) m undefined

FIGURE 2.3.2 Lines with slope (a)–(c); line with no slope (d)

So when you see the phrase *line with slope* in a discussion you know that vertical lines are excluded.

☐ **Point-Slope Equation** We are now in a position to find an equation of a line L. To begin, suppose L has slope m and that (x_1, y_1) is on the line. If (x, y) represents any other point on L, then (7) gives

$$m = \frac{y - y_1}{x - x_1}.$$

Multiplying both sides of the last equality by $x - x_1$ gives an important equation.

THEOREM 2.3.1   Point-Slope Equation of a Line

The **point-slope equation** of the line through (x_1, y_1) with slope m is

$$y - y_1 = m(x - x_1). \tag{8}$$

EXAMPLE 1   Point-Slope Equation

Find an equation of the line with slope 6 and passing through $\left(-\frac{1}{2}, 2\right)$.

Solution Letting $m = 6, x_1 = -\frac{1}{2}$, and $y_1 = 2$ we obtain from (8)

$$y - 2 = 6\left[x - \left(-\frac{1}{2}\right)\right].$$

Simplifying gives $y - 2 = 6\left(x + \frac{1}{2}\right)$ or $y = 6x + 5$.   ≡

EXAMPLE 2   Point-Slope Equation

Find an equation of the line passing through the points $(4, 3)$ and $(-2, 5)$.

Solution First we compute the slope of the line through the points. From (7),

$$m = \frac{5 - 3}{-2 - 4} = \frac{2}{-6} = -\frac{1}{3}.$$

The distributive law $a(b + c) = ab + ac$ is the source of many errors on students' papers. A common error goes something like this:

$$-(2x - 3) = -2x - 3.$$

The correct result is:

$$\begin{aligned} -(2x - 3) &= (-1)(2x - 3) \\ &= (-1)2x - (-1)3 \\ &= -2x + 3. \end{aligned}$$

► The point-slope equation (8) then gives

the distributive law
↓   ↓
$$y - 3 = -\frac{1}{3}(x - 4) \qquad \text{or} \qquad y = -\frac{1}{3}x + \frac{13}{3}.$$ ≡

☐ **Slope-Intercept Equation** Any line with slope (that is, any line that is not vertical) must cross the y-axis. If this y-intercept is $(0, b)$, then with $x_1 = 0, y_1 = b$, the point-slope form (8) gives $y - b = m(x - 0)$. The last equation simplifies to the next result.

THEOREM 2.3.2 Slope-Intercept Equation of a Line

The **slope-intercept equation** of the line with slope m and y-intercept $(0, b)$ is

$$y = mx + b. \qquad (9)$$

☐ **Family of Lines** For $m \neq 0$, (8) and (9) give us the form of the linear function in (6). The coefficient a in (6) is, of course, the slope m of the line. When $b = 0$ in (9), the equation $y = mx$ represents a **family of lines** that pass through the origin $(0, 0)$. In FIGURE 2.3.3 we have drawn a few of the members of that family.

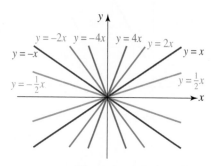

FIGURE 2.3.3 Lines through the origin are $y = mx$

EXAMPLE 3 Example 2 Revisited

We can also use the slope-intercept from (9) to obtain the equation of the line through two points in Example 2. As in that example, we start by finding the slope $m = -\frac{1}{3}$. The equation of the line is then $y = -\frac{1}{3}x + b$. Substituting the coordinates of either point $(4, 3)$ or $(-2, 5)$ into the last equation enables us to determine b. If we use $x = 4$ and $y = 3$, then $3 = -\frac{1}{3} \cdot 4 + b$ and so $b = 3 + \frac{4}{3} = \frac{13}{3}$. The equation of the line is $y = -\frac{1}{3}x + \frac{13}{3}$. ≡

☐ **Horizontal and Vertical Lines** We saw in Figure 2.3.2(c) that a horizontal line has slope $m = 0$. An equation of a horizontal line passing through a point (a, b) can be obtained from (8), that is, $y - b = 0(x - a)$. The **equation of a horizontal line** is then

$$y = b. \qquad (10)$$

We have already seen this in (5) and in (2) where $-C/B$ played the part of the symbol b. A vertical line through (a, b) has undefined slope and all points on the line have the same x-coordinate. The **equation of a vertical line** is then

$$x = a. \qquad (11)$$

Equation (11) is (3) with $-C/A$ replaced by the symbol a.

EXAMPLE 4 Vertical and Horizontal Lines

Find equations for the vertical and horizontal lines through $(3, -1)$. Graph these lines.

Solution Any point on the vertical line through $(3, -1)$ has x-coordinate 3. The equation of this line is then $x = 3$. Similarly, any point on the horizontal line through $(3, -1)$ has y-coordinate -1. The equation of this line is $y = -1$. Both lines are graphed in FIGURE 2.3.4. Don't forget, only $y = -1$ is a function. ≡

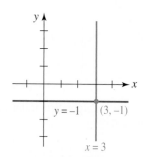

FIGURE 2.3.4 Horizontal and vertical lines in Example 4

☐ **Parallel and Perpendicular Lines** Suppose L_1 and L_2 are two distinct lines with slope. This assumption means that both L_1 and L_2 are nonvertical lines. Then necessarily L_1 and L_2 are either parallel or they intersect. If the lines intersect at a right angle they are said to be perpendicular. We can determine whether two lines are parallel or are perpendicular by examining their slopes.

Parallel lines

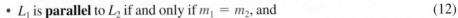

THEOREM 2.3.3   Slopes of Parallel and Perpendicular Lines

If L_1 and L_2 are lines with slopes m_1 and m_2, respectively, then

- L_1 is **parallel** to L_2 if and only if $m_1 = m_2$, and (12)
- L_1 is **perpendicular** to L_2 if and only if $m_1 m_2 = -1$. (13)

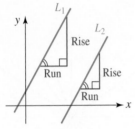

FIGURE 2.3.5 Parallel lines

There are several ways of proving the two parts of Theorem 2.3.3. The proof of (12) can be obtained using similar right triangles, as in FIGURE 2.3.5, and the fact that the ratios of corresponding sides in such triangles are equal. We leave the justification of (13) as an exercise. See Problem 64 in Exercises 2.3. Note that the condition $m_1 m_2 = -1$ implies that $m_2 = -1/m_1$, that is, the slopes are negative reciprocals of each other. A horizontal line $y = b$ and a vertical line $x = a$ are perpendicular, but the latter is a line with no slope.

EXAMPLE 5   Parallel Lines

The linear equations $3x + y = 2$ and $6x + 2y = 15$ can be rewritten in the slope-intercept forms

$$y = -3x + 2 \qquad \text{and} \qquad y = -3x + \tfrac{15}{2},$$

respectively. As noted in color in the preceding line the slope of each line is -3. Therefore the lines are parallel. The graphs of these equations are shown in FIGURE 2.3.6. ≡

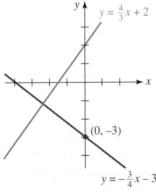

FIGURE 2.3.6 Parallel lines in Example 5

$y = -3x + 2$ $y = -3x + \tfrac{15}{2}$

EXAMPLE 6   Perpendicular Lines

Find an equation of the line through $(0, -3)$ that is perpendicular to the graph of $4x - 3y + 6 = 0$.

Solution We express the given linear equation in slope-intercept form:

$$4x - 3y + 6 = 0 \qquad \text{implies} \qquad 3y = 4x + 6.$$

Dividing by 3 gives $y = \tfrac{4}{3}x + 2$. This line, whose graph is given in blue in FIGURE 2.3.7, has slope $\tfrac{4}{3}$. The slope of any line perpendicular to it is the negative reciprocal of $\tfrac{4}{3}$, namely, $-\tfrac{3}{4}$. Since $(0, -3)$ is the y-intercept of the required line, it follows from (9) that its equation is $y = -\tfrac{3}{4}x - 3$. The graph of the last equation is the red line in FIGURE 2.3.7. ≡

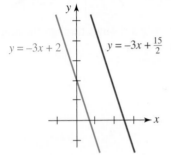

$y = \tfrac{4}{3}x + 2$

$(0, -3)$

$y = -\tfrac{3}{4}x - 3$

FIGURE 2.3.7 Perpendicular lines in Example 6

☐ **Graphs** As mentioned in the earlier sections of this chapter, when graphing an equation it is always a good habit to try to find x- and y-intercepts of its graph. Except in the cases of horizontal and vertical lines, and lines through the origin, a line will have distinct x- and y-intercepts. Of course, that is all we need to draw a line: two points.

EXAMPLE 7   Graph of a Linear Equation

Graph the linear equation $3x - 2y + 8 = 0$.

Solution There is no need to rewrite the linear equation in the form $y = mx + b$. We simply find the intercepts.

y-intercept: Setting $x = 0$ gives $-2y + 8 = 0$ or $y = 4$. The y-intercept is $(0, 4)$.
x-intercept: Setting $y = 0$ gives $3x + 8 = 0$ or $x = -\tfrac{8}{3}$. The x-intercept is $\left(-\tfrac{8}{3}, 0\right)$.

As shown in FIGURE 2.3.8, the line is drawn through the two intercepts $(0, 4)$ and $\left(-\tfrac{8}{3}, 0\right)$. ≡

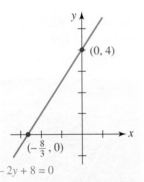

$(0, 4)$

$\left(-\tfrac{8}{3}, 0\right)$

$3x - 2y + 8 = 0$

FIGURE 2.3.8 Graph of equation in Example 7

CHAPTER 2 FUNCTIONS

□ **Increasing-Decreasing Functions** We have just seen in Figures 2.3.2(a) and 2.3.2(b) that if $a > 0$ (which, as we have just seen plays the part of m) the values of a linear function $f(x) = ax + b$ increase as x increases, whereas for $a < 0$, the values $f(x)$ decrease as x increases. The notions of increasing and decreasing can be extended to *any* function. The ability to determine intervals over which a function f is either increasing or decreasing plays an important role in applications of calculus.

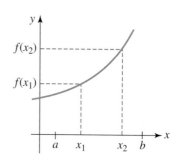

(a) $f(x_1) < f(x_2)$

DEFINITION 2.3.3 Increasing/Decreasing

Suppose $y = f(x)$ is a function defined on an interval $[a, b]$, and x_1 and x_2 are any two numbers in the interval such that $x_1 < x_2$. Then the function f is

- **increasing** on the interval if $f(x_1) < f(x_2)$,   (14)
- **decreasing** on the interval if $f(x_1) > f(x_2)$.   (15)

In FIGURE 2.3.9(a) the function f is increasing on the interval $[a, b]$, whereas f is decreasing on $[a, b]$ in Figure 2.3.9(b). A linear function $f(x) = ax + b$ increases on the interval $(-\infty, \infty)$ for $a > 0$ and decreases on the interval $(-\infty, \infty)$ for $a < 0$.

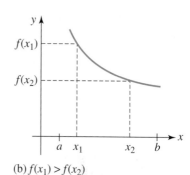

(b) $f(x_1) > f(x_2)$

FIGURE 2.3.9 Increasing function in (a); decreasing function in (b)

□ **Points of Intersection** We are often interested in finding the points where the graphs of two functions intersect. The x-intercepts of the graph of a function f can be interpreted as the points where the graph of f intersects the graph of the constant function $y = 0$. In general, at a point P of intersection of the graphs of two functions f and g, the coordinates (x, y) of P must satisfy both equations $y = f(x)$ and $y = g(x)$, and so $f(x) = g(x)$.

EXAMPLE 8   **Intersecting Lines**

Find the point where the two lines in Figure 2.3.7 intersect.

Solution We equate $y = \frac{4}{3}x + 2$ and $y = -\frac{3}{4}x - 3$ and solve for x:

$$\frac{4}{3}x + 2 = -\frac{3}{4}x - 3$$
$$\left(\frac{4}{3} + \frac{3}{4}\right)x = -5$$
$$\frac{25}{12}x = -5$$
$$x = -\frac{12}{5}.$$

By substituting $x = -\frac{12}{5}$ into either equation we find that $y = -\frac{6}{5}$. The point of intersection of the lines is then $\left(-\frac{12}{5}, -\frac{6}{5}\right)$.   ≡

2.3   Exercises   Answers to selected odd-numbered problems begin on page ANS-4.

In Problems 1–6, find the slope of the line through the given points. Graph the line through the points.

1. $(3, -7), (1, 0)$ **2.** $(-4, -1), (1, -1)$
3. $(5, 2), (4, -3)$ **4.** $(1, 4), (6, -2)$
5. $(-1, 2), (3, -2)$ **6.** $\left(8, -\frac{1}{2}\right), \left(2, \frac{5}{2}\right)$

In Problems 7 and 8, use the graph of the given line to estimate its slope.

7.

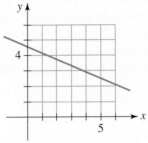

FIGURE 2.3.10 Graph for Problem 7

8.

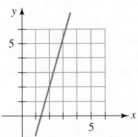

FIGURE 2.3.11 Graph for Problem 8

In Problems 9–16, find the slope and the x- and y-intercepts of the given line. Graph the line.

9. $3x - 4y + 12 = 0$

10. $\frac{1}{2}x - 3y = 3$

11. $2x - 3y = 9$

12. $-4x - 2y + 6 = 0$

13. $2x + 5y - 8 = 0$

14. $\frac{y}{2} - \frac{x}{10} - 1 = 0$

15. $y + \frac{2}{3}x = 1$

16. $y = 2x + 6$

In Problems 17–22, find an equation of the line through $(1, 2)$ with the indicated slope.

17. $\frac{2}{3}$

18. $\frac{1}{10}$

19. 0

20. -2

21. -1

22. undefined

In Problems 23–36, find an equation of the line that satisfies the given conditions.

23. through $(2, 3)$ and $(6, -5)$
24. through $(5, -6)$ and $(4, 0)$
25. through $(8, 1)$ and $(-3, 1)$
26. through $(2, 2)$ and $(-2, -2)$
27. through $(-2, 0)$ and $(-2, 6)$
28. through $(0, 0)$ and (a, b)
29. through $(-2, 4)$ parallel to $3x + y - 5 = 0$
30. through $(1, -3)$ parallel to $2x - 5y + 4 = 0$
31. through $(5, -7)$ parallel to the y-axis
32. through the origin parallel to the line through $(1, 0)$ and $(-2, 6)$
33. through $(2, 3)$ perpendicular to $x - 4y + 1 = 0$
34. through $(0, -2)$ perpendicular to $3x + 4y + 5 = 0$
35. through $(-5, -4)$ perpendicular to the line through $(1, 1)$ and $(3, 11)$
36. through the origin perpendicular to every line with slope 2

37. Find the coordinates of the point P shown in FIGURE 2.3.12.
38. A line through $(2, 4)$ has slope 8. Without finding an equation of the line, determine whether the point $(1, -5)$ is on the line.

FIGURE 2.3.12 Lines in Problem 37

In Problems 39–42, determine which of the given lines are parallel to each other and which are perpendicular to each other.

39. (a) $3x - 5y + 9 = 0$

(b) $5x = -3y$

(c) $-3x + 5y = 2$

(d) $3x + 5y + 4 = 0$

(e) $-5x - 3y + 8 = 0$

(f) $5x - 3y - 2 = 0$

CHAPTER 2 FUNCTIONS

40. (a) $2x + 4y + 3 = 0$ (b) $2x - y = 2$
(c) $x + 9 = 0$ (d) $x = 4$
(e) $y - 6 = 0$ (f) $-x - 2y + 6 = 0$

41. (a) $3x - y - 1 = 0$ (b) $x - 3y + 9 = 0$
(c) $3x + y = 0$ (d) $x + 3y = 1$
(e) $6x - 3y + 10 = 0$ (f) $x + 2y = -8$

42. (a) $y + 5 = 0$ (b) $x = 7$
(c) $4x + 6y = 3$ (d) $12x - 9y + 7 = 0$
(e) $2x - 3y - 2 = 0$ (f) $3x + 4y - 11 = 0$

In Problems 43 and 44, find a linear function (6) that satisfies both of the given conditions.

43. $f(-1) = 5, f(1) = 6$ **44.** $f(-1) = 1 + f(2), f(3) = 4f(1)$

In Problems 45–48, find the point of intersection of the graphs of the given linear functions. Sketch both lines.

45. $f(x) = -2x + 1, g(x) = 4x + 6$ **46.** $f(x) = 2x + 5, g(x) = \frac{3}{2}x + 5$
47. $f(x) = 4x + 7, g(x) = \frac{1}{3}x + \frac{10}{3}$ **48.** $f(x) = 2x - 10, g(x) = -3x$

In Problems 49 and 50, for the given linear function compute the quotient

$$\frac{f(x + h) - f(x)}{h},$$

where h is a constant.

49. $f(x) = -9x + 12$ **50.** $f(x) = \frac{4}{3}x - 5$

51. Find an equation of the red line L shown in FIGURE 2.3.13 if an equation of the blue curve is $y = x^2 + 1$.

52. A tangent line L to a circle at a point P on the circle is perpendicular to the line through P and the center of the circle. Find an equation of the red line L shown in FIGURE 2.3.14.

Miscellaneous Applications

53. Thermometers The functional relationship between degrees Celsius T_C and degrees Fahrenheit T_F is linear.
(a) Express T_F as a function of T_C if $(0°C, 32°F)$ and $(60°C, 140°F)$ are on the graph of T_F.
(b) Show that $100°C$ is equivalent to the Fahrenheit boiling point $212°F$. See FIGURE 2.3.15.

54. Thermometers—Continued The functional relationship between degrees Celsius T_C and temperature measured in Kelvin units T_K is linear.
(a) Express T_K as a function of T_C if $(0°C, 273\,K)$ and $(27°C, 300\,K)$ are on the graph of T_K.
(b) Express the boiling point of water $100°C$ in Kelvin units. See Figure 2.3.15.
(c) Absolute zero is defined as $0\,K$. What is $0\,K$ in degrees Celsius?
(d) Express T_K as a linear function of T_F.
(e) What is $0\,K$ in degrees Fahrenheit?

55. Simple Interest In simple interest, the amount A accrued over time is the linear function $A = P + Prt$, where P is the principal, t is measured in years, and r is the annual interest rate (expressed as a decimal). Compute A after 20 years if the principal is $P = 1000$ and the annual interest rate is 3.4%. At what time is $A = 2200$?

56. Linear Depreciation Straight line, or linear, depreciation consists of an item losing all its initial worth of A dollars over a period of n years by an amount A/n

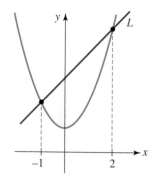

FIGURE 2.3.13 Graphs for Problem 51

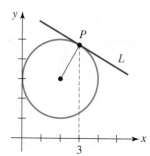

FIGURE 2.3.14 Circle and tangent line in Problem 52

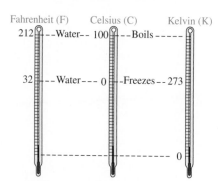

FIGURE 2.3.15 Thermometers in Problems 53 and 54

each year. If an item costing $20,000 when new is depreciated linearly over 25 years, determine a linear function giving its value V after x years, where $0 \le x \le 25$. What is the value of the item after 10 years?

For Discussion

57. Consider the linear function $f(x) = \frac{5}{2}x - 4$. If x is changed by 1 unit, how many units will y change? If x is changed by 2 units? If x is changed by n (n a positive integer) units?
58. Consider the interval $[x_1, x_2]$ and the linear function $f(x) = ax + b, a \ne 0$. Show that

$$f\left(\frac{x_1 + x_2}{2}\right) = \frac{f(x_1) + f(x_2)}{2}$$

and interpret this result geometrically for $a > 0$.
59. How would you find an equation of the line that is the perpendicular bisector of the line segment through $\left(\frac{1}{2}, 10\right)$ and $\left(\frac{3}{2}, 4\right)$?
60. Using only the concepts of this section, how would you prove or disprove that the triangle with vertices $(2, 3)$, $(-1, -3)$, and $(4, 2)$ is a right triangle?
61. Using only the concepts of this section, how would you prove or disprove that the quadrilateral with vertices $(0, 4)$, $(-1, 3)$, $(-2, 8)$, and $(-3, 7)$ is a parallelogram?
62. If C is an arbitrary real constant, the equation $2x - 3y = C$ defines a family of lines. Choose four different values of C and plot the corresponding lines on the same coordinate axes. What is true about the lines that are members of this family?
63. Find the equations of the lines through $(0, 4)$ that are tangent to the circle $x^2 + y^2 = 4$.
64. To prove (13) you have to prove two things, the "only if" and the "if" parts of the theorem.
    (a) In FIGURE 2.3.16, without loss of generality, we have assumed that two perpendicular lines, $y = m_1 x$, $m_1 > 0$ and $y = m_2 x$, $m_2 < 0$, intersect at the origin. Use the information in the figure to prove the "only if" part:

    If L_1 and L_2 are perpendicular lines with slopes m_1 and m_2, then $m_1 m_2 = -1$.

    (b) Reverse your argument in part (a) to prove the "if" part:

    If L_1 and L_2 are lines with slopes m_1 and m_2 such that $m_1 m_2 = -1$, then L_1 and L_2 are perpendicular.

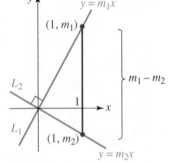

FIGURE 2.3.16 Lines through origin in Problem 64

2.4 Quadratic Functions

≡ **Introduction** The squaring function $y = x^2$ that played an important role in Section 2.2 is a member of a family of functions called **quadratic functions**.

DEFINITION 2.4.1 Quadratic Function

A **quadratic function** $y = f(x)$ is a function of the form

$$f(x) = ax^2 + bx + c, \tag{1}$$

where $a \ne 0, b$, and c are constants.

The **domain** of a quadratic function f is the set of real numbers $(-\infty, \infty)$.

□ **Graphs** The graph of any quadratic function is called a **parabola**. The graph of a quadratic function has the same basic shape of the squaring function $y = x^2$ shown in FIGURE 2.4.1. In the examples that follow we will see that the graphs of quadratic functions (1) are simply transformations of the graph of $y = x^2$:

- The graph of $f(x) = ax^2$, $a > 0$, is the graph of $y = x^2$ **stretched** vertically when $a > 1$, and **compressed** vertically when $0 < a < 1$.
- The graph of $f(x) = ax^2$, $a < 0$, is the graph of $y = ax^2$, $a > 0$, **reflected** in the x-axis.
- The graph of $f(x) = ax^2 + bx + c$, $b \neq 0$, is the graph of $y = ax^2$ **shifted** horizontally or vertically.

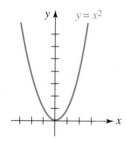

FIGURE 2.4.1 Graph of simplest parabola

From the first two items in the bulleted list, we conclude that the graph of a quadratic function opens upward (as in Figure 2.4.1) if $a > 0$ and opens downward if $a < 0$.

EXAMPLE 1　　　Stretch, Compression, and Reflection

(a) The graphs of $y = 4x^2$ and $y = \frac{1}{10}x^2$ are, respectively, a vertical stretch and a vertical compression of the graph of $y = x^2$. The graphs of these functions are shown in FIGURE 2.4.2(a); the graph of $y = 4x^2$ is shown in red, the graph of $y = \frac{1}{10}x^2$ is green, and the graph of $y = x^2$ is blue.

(b) The graphs of $y = -4x^2$, $y = -\frac{1}{10}x^2$, $y = -x^2$ are obtained from the graphs of the functions in part (a) by reflecting their graphs in the x-axis. See Figure 2.4.2(b).

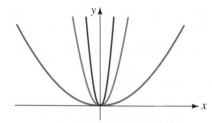

(a) Red graph is a vertical stretch of blue graph; green graph is a vertical compression of blue graph

(b) Reflections in x-axis

FIGURE 2.4.2 Graphs of quadratic functions in Example 1　　≡

□ **Vertex and Axis** If the graph of a quadratic function opens upward $a > 0$ (or downward $a < 0$), the lowest (highest) point (h, k) on the parabola is called its **vertex**. All parabolas are symmetric with respect to a vertical line through the vertex (h, k). The line $x = h$ is called the **axis of symmetry** or simply the **axis** of the parabola. See FIGURE 2.4.3.

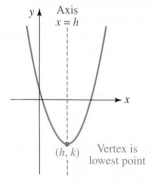
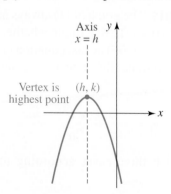

(a) $y = ax^2 + bx + c$, $a > 0$　　　(b) $y = ax^2 + bx + c$, $a < 0$

FIGURE 2.4.3 Vertex and axis of a parabola

□ **Standard Form** The vertex (h, k) of a parabola can be determined by recasting the equation $f(x) = ax^2 + bx + c$ into the **standard form**

$$f(x) = a(x - h)^2 + k. \tag{2}$$

See Section 1.4. ▶

The form (2) is obtained from the equation (1) by completing the square in x. Completing the square in (1) starts with factoring the number a from all terms involving the variable x:

$$f(x) = ax^2 + bx + c$$
$$= a\left(x^2 + \frac{b}{a}x\right) + c.$$

Within the parentheses we add and subtract the square of one-half the coefficient of x:

$$\text{square of } \tfrac{b}{2a}$$
$$\downarrow$$

$$f(x) = a\left(x^2 + \frac{b}{a}x + \frac{b^2}{4a^2} - \frac{b^2}{4a^2}\right) + c \quad \leftarrow \text{terms in color add to 0}$$
$$= a\left(x^2 + \frac{b}{a}x + \frac{b^2}{4a^2}\right) - \frac{b^2}{4a} + c \quad \leftarrow \text{note that } a \cdot \left(-\frac{b^2}{4a^2}\right) = -\frac{b^2}{4a} \tag{3}$$
$$= a\left(x + \frac{b}{2a}\right)^2 + \frac{4ac - b^2}{4a}$$

The last expression is equation (2) with the identifications $h = -b/2a$ and $k = (4ac - b^2)/4a$. If $a > 0$, then necessarily $a(x - h)^2 \geq 0$. Hence $f(x)$ in (2) is a minimum when $(x - h)^2 = 0$, that is, for $x = h$. A similar argument shows that if $a < 0$ in (2), $f(x)$ is a maximum value for $x = h$. Thus (h, k) is the vertex of the parabola. The equation of the axis of the parabola is $x = h$ or $x = -b/2a$.

If $a > 0$, then the function f in (2) is decreasing on the interval $(-\infty, h]$ and increasing on the interval $[h, \infty)$. If $a < 0$, we have just the opposite, that is, f is increasing on $(-\infty, h]$ followed by decreasing on $[h, \infty)$.

We strongly suggest that you *do not memorize* the result in the last line of (3), but practice completing the square each time. However, if memorization is permitted by your instructor to save time, then the vertex can be found by computing the coordinates of the point

$$\left(-\frac{b}{2a}, f\left(-\frac{b}{2a}\right)\right). \tag{4}$$

□ **Intercepts** The graph of (1) always has a **y-intercept** since $f(0) = c$, and so the y-intercept is $(0, c)$. To determine whether the graph has **x-intercepts** we must solve the equation $f(x) = 0$. The last equation can be solved either by factoring or by using the quadratic formula. Recall that a quadratic equation $ax^2 + bx + c = 0$, $a \neq 0$, has the solutions

$$x_1 = \frac{-b - \sqrt{b^2 - 4ac}}{2a}, \qquad x_2 = \frac{-b + \sqrt{b^2 - 4ac}}{2a}.$$

We distinguish three cases according to the algebraic sign of the discriminant $b^2 - 4ac$.

- If $b^2 - 4ac > 0$, then there are two distinct real solutions x_1 and x_2. The parabola crosses the x-axis at the points $(x_1, 0)$ and $(x_2, 0)$.

- If $b^2 - 4ac = 0$, then there is a single real solution x_1. The vertex of the parabola is located on the x-axis at $(x_1, 0)$. The parabola is tangent to, or touches, the x-axis at this point.
- If $b^2 - 4ac < 0$, then there are no real solutions. The parabola does not cross the x-axis.

As the next example shows, a reasonable sketch of a parabola can be obtained by plotting the intercepts and the vertex.

EXAMPLE 2   Graph Using Intercepts and Vertex

Graph $f(x) = x^2 - 2x - 3$.

Solution Since $a = 1 > 0$ we know that the parabola will open upward. From $f(0) = -3$ we get the y-intercept $(0, -3)$. To see whether there are any x-intercepts we solve $x^2 - 2x - 3 = 0$. By factoring

$$(x + 1)(x - 3) = 0,$$

we find the solutions $x = -1$ and $x = 3$. The x-intercepts are $(-1, 0)$ and $(3, 0)$. To locate the vertex we complete the square:

$$f(x) = (x^2 - 2x + 1) - 1 - 3 = (x^2 - 2x + 1) - 4.$$

Thus the standard form is $f(x) = (x - 1)^2 - 4$. With the identifications $h = 1$ and $k = -4$, we conclude that the vertex is $(1, -4)$. Using this information we draw a parabola through these four points as shown in FIGURE 2.4.4.

One last observation. By finding the vertex we automatically determine the range of a quadratic function. In our current example, $y = -4$ is the smallest number in the range of f and so the range of f is the interval $[-4, \infty)$ on the y-axis. $\equiv$

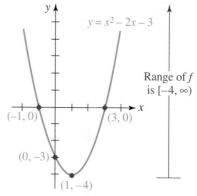

FIGURE 2.4.4 Parabola in Example 2

EXAMPLE 3   Vertex Is the x-intercept

Graph $f(x) = -4x^2 + 12x - 9$.

Solution The graph of this quadratic function is a parabola that opens downward because $a = -4 < 0$. To complete the square we start by factoring -4 from the two x-terms:

$$\begin{aligned}
f(x) &= -4x^2 + 12x - 9 \\
&= -4(x^2 - 3x) - 9 \\
&= -4\left(x^2 - 3x + \frac{9}{4} - \frac{9}{4}\right) - 9 \\
&= -4\left(x^2 - 3x + \frac{9}{4}\right) - 9 + 9 \\
&= -4\left(x^2 - 3x + \frac{9}{4}\right).
\end{aligned}$$

Thus the standard form is $f(x) = -4\left(x - \frac{3}{2}\right)^2$. With $h = \frac{3}{2}$ and $k = 0$ we see that the vertex is $\left(\frac{3}{2}, 0\right)$. The y-intercept is $(0, f(0)) = (0, -9)$. Solving $-4x^2 + 12x - 9 = 0$, we find that there is only one x-intercept, namely, $\left(\frac{3}{2}, 0\right)$. Of course, this was to be expected because the vertex $\left(\frac{3}{2}, 0\right)$ is on the x-axis. As shown in FIGURE 2.4.5 a rough sketch can be obtained from these two points alone. The parabola is tangent to the x-axis at $\left(\frac{3}{2}, 0\right)$. $\equiv$

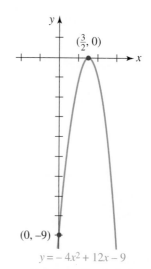

FIGURE 2.4.5 Parabola in Example 3

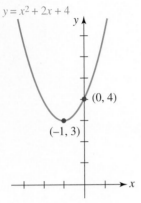

$y = x^2 + 2x + 4$

(0, 4)

(−1, 3)

FIGURE 2.4.6 Parabola in
Example 4

| EXAMPLE 4 | Using (4) to Find the Vertex |

Graph $f(x) = x^2 + 2x + 4$.

Solution The graph is a parabola that opens upward because $a = 1 > 0$. For the sake of illustration we will use (4) this time to find the vertex. With $b = 2$, $-b/2a = -2/2 = -1$ and

$$f(-1) = (-1)^2 + 2(-1) + 4 = 3,$$

the vertex is $(-1, f(-1)) = (-1, 3)$. Now the y-intercept is $(0, f(0)) = (0, 4)$ but the quadratic formula shows that the equation $f(x) = 0$ or $x^2 + 2x + 4 = 0$ has no real solutions. Therefore the graph has no x-intercepts. Since the vertex is above the x-axis and the parabola opens upward, the graph must lie entirely above the x-axis. See FIGURE 2.4.6. ≡

☐ **Graphs by Transformations** The standard form (2) clearly describes how the graph of any quadratic function is constructed from the graph of $y = x^2$ starting with a non-rigid transformation followed by two rigid transformations:

- $y = ax^2$ is the graph of $y = x^2$ stretched or compressed vertically.
- $y = a(x - h)^2$ is the graph of $y = ax^2$ shifted $|h|$ units horizontally.
- $y = a(x - h)^2 + k$ is the graph of $y = a(x - h)^2$ shifted $|k|$ units vertically.

FIGURE 2.4.7 illustrates the horizontal and vertical shifting in the case where $a > 0, h > 0$, and $k > 0$.

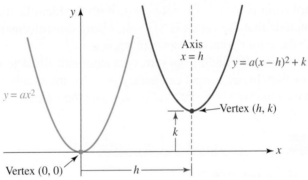

Axis
$x = h$

$y = a(x - h)^2 + k$

$y = ax^2$

Vertex (h, k)

k

Vertex $(0, 0)$

h

FIGURE 2.4.7 The red graph is obtained by shifting the blue graph h units to the right and k units upward.

| EXAMPLE 5 | Horizontally Shifted Graphs |

Compare the graphs of **(a)** $y = (x - 2)^2$ and **(b)** $y = (x + 3)^2$.

Solution The blue dashed graph in FIGURE 2.4.8 is the graph of $y = x^2$. Matching the given functions with (2) shows in each case that $a = 1$ and $k = 0$. This means that neither graph undergoes a vertical stretch or compression, and neither graph is shifted vertically.

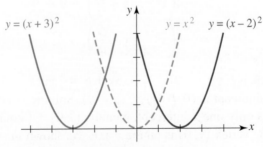

$y = (x + 3)^2$ $y = x^2$ $y = (x - 2)^2$

FIGURE 2.4.8 Shifted graphs (red and green) in
Example 5

80   CHAPTER 2 FUNCTIONS

(a) With the identification $h = 2$, the graph of $y = (x - 2)^2$ is the graph of $y = x^2$ shifted horizontally 2 units to the right. The vertex $(0, 0)$ for $y = x^2$ becomes the vertex $(2, 0)$ for $y = (x - 2)^2$. See the red graph in Figure 2.4.8.

(b) With the identification $h = -3$, the graph of $y = (x + 3)^2$ is the graph of $y = x^2$ shifted horizontally $|-3| = 3$ units to the left. The vertex $(0, 0)$ for $y = x^2$ becomes the vertex $(-3, 0)$ for $y = (x + 3)^2$. See the green graph in Figure 2.4.8. ≡

EXAMPLE 6   Shifted Graph

Graph $y = 2(x - 1)^2 - 6$.

Solution The graph is the graph of $y = x^2$ stretched vertically upward, followed by a horizontal shift to the right of 1 unit, followed by a vertical shift downward of 6 units. In FIGURE 2.4.9, you should note how the vertex $(0, 0)$ on the graph of $y = x^2$ is moved to $(1, -6)$ on the graph of $y = 2(x - 1)^2 - 6$ as a result of these transformations. You should also follow how the point $(1, 1)$ shown in Figure 2.4.9(a) ends up as $(2, -4)$ in Figure 2.4.9(d).

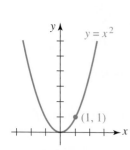 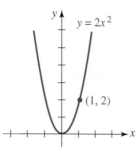 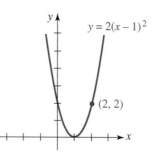 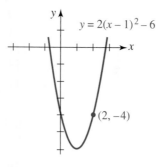

(a) Basic parabola   (b) Vertical stretch   (c) Horizontal shift   (d) Vertical shift

FIGURE 2.4.9 Graphs in Example 6   ≡

☐ **Quadratic Inequalities** Graphs can be of help in solving certain inequalities when a sign chart is not useful because the quadratic does not factor conveniently. For example, the quadratic function in Example 6 is equivalent to $y = 2x^2 - 4x - 4$. Were we required to solve the inequality $2x^2 - 4x - 4 \geq 0$, we see in Figure 2.4.9(d) that $y \geq 0$ to the left of the x-intercept on the negative x-axis and to the right of the x-intercept on the positive x-axis. The x-coordinates of these intercepts, obtained by solving $2x^2 - 4x - 4 = 0$ by the quadratic formula, are $1 - \sqrt{3}$ and $1 + \sqrt{3}$. Thus the solution of $2x^2 - 4x - 4 \geq 0$ is $(-\infty, 1 - \sqrt{3}] \cup [1 + \sqrt{3}, \infty)$.

☐ **Freely Falling Object** Suppose an object, such as a ball, is either thrown straight upward (downward) or simply dropped from an initial height s_0. Then if the positive direction is taken to be upward, the height $s(t)$ of the object above ground is given by the quadratic function

$$s(t) = -\tfrac{1}{2}gt^2 + v_0t + s_0, \tag{5}$$

where g is the acceleration due to gravity (32 ft/s^2 or 9.8 m/s^2), v_0 is the initial velocity imparted to the object, and t is time measured in seconds. See FIGURE 2.4.10. If the object is dropped, then $v_0 = 0$. An assumption in the derivation of (5), a straightforward exercise in integral calculus, is that the motion takes place close to the surface of Earth and so the retarding effects of air resistance is ignored. Also, the velocity of the object while it is in the air is given by the linear function

$$v(t) = -gt + v_0. \tag{6}$$

See Problems 49–52 in Exercises 2.4.

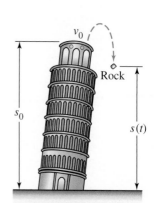

FIGURE 2.4.10 Rock thrown upward from an initial height s_0

In Problems 1–6, sketch the graph of the given function f.

1. $f(x) = 2x^2$

2. $f(x) = -2x^2$

3. $f(x) = 2x^2 - 2$

4. $f(x) = 2x^2 + 5$

5. $f(x) = -2x^2 + 1$

6. $f(x) = -2x^2 - 3$

In Problems 7–18, consider the quadratic function f.
(a) Find all intercepts of the graph of f.
(b) Express the function f in standard form.
(c) Find the vertex and axis of symmetry.
(d) Sketch the graph of f.

7. $f(x) = x(x + 5)$

8. $f(x) = -x^2 + 4x$

9. $f(x) = (3 - x)(x + 1)$

10. $f(x) = (x - 2)(x - 6)$

11. $f(x) = x^2 - 3x + 2$

12. $f(x) = -x^2 + 6x - 5$

13. $f(x) = 4x^2 - 4x + 3$

14. $f(x) = -x^2 + 6x - 10$

15. $f(x) = -\frac{1}{2}x^2 + x + 1$

16. $f(x) = x^2 - 2x - 7$

17. $f(x) = x^2 - 10x + 25$

18. $f(x) = -x^2 + 6x - 9$

In Problems 19 and 20, find the maximum or the minimum value of the function f.
Give the range of the function f.

19. $f(x) = 3x^2 - 8x + 1$

20. $f(x) = -2x^2 - 6x + 3$

In Problems 21–24, find the largest interval on which the function f is increasing and the largest interval on which f is decreasing.

21. $f(x) = \frac{1}{3}x^2 - 25$

22. $f(x) = -(x + 10)^2$

23. $f(x) = -2x^2 - 12x$

24. $f(x) = x^2 + 8x - 1$

In Problems 25–30, describe in words how the graph of the given function can be obtained from the graph of $y = x^2$ by rigid or nonrigid transformations.

25. $f(x) = (x - 10)^2$

26. $f(x) = (x + 6)^2$

27. $f(x) = -\frac{1}{3}(x + 4)^2 + 9$

28. $f(x) = 10(x - 2)^2 - 1$

29. $f(x) = (-x - 6)^2 - 4$

30. $f(x) = -(1 - x)^2 + 1$

In Problems 31–36, the given graph is the graph of $y = x^2$ shifted/reflected in the xy-plane. Write an equation of the graph.

31.

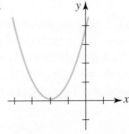

FIGURE 2.4.11 Graph for Problem 31

32.

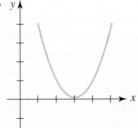

FIGURE 2.4.12 Graph for Problem 32

33.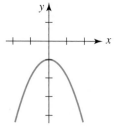

FIGURE 2.4.13 Graph for Problem 33

34.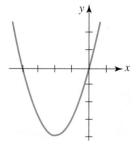

FIGURE 2.4.14 Graph for Problem 34

35.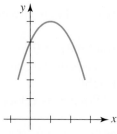

FIGURE 2.4.15 Graph for Problem 35

36.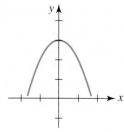

FIGURE 2.4.16 Graph for Problem 36

In Problems 37 and 38, find a quadratic function $f(x) = ax^2 + bx + c$ that satisfies the given conditions.

37. f has the values $f(0) = 5, f(1) = 10, f(-1) = 4$
38. graph passes through $(2, -1)$, zeros of f are 1 and 3

In Problems 39 and 40, find a quadratic function in standard form $f(x) = a(x - h)^2 + k$ that satisfies the given conditions.

39. the vertex of the graph of f is $(1, 2)$, graph passes through $(2, 6)$
40. the maximum value of f is 10, axis of symmetry is $x = -1$, and y-intercept is $(0, 8)$

In Problems 41–44, sketch the region in the xy-plane that is bounded between the graphs of the given functions. Find the points of intersection of the graphs.

41. $y = -x + 4, \quad y = x^2 + 2x$
42. $y = 2x - 2, \quad y = 1 - x^2$
43. $y = x^2 + 2x + 2, \quad y = -x^2 - 2x + 2$
44. $y = x^2 - 6x + 1, \quad y = -x^2 + 2x + 1$

45. Find the maximum value of $f(x) = -x + 6\sqrt{x} + 10$. [*Hint*: See Problems 19 and 20. Let $t = \sqrt{x}$.]
46. Consider the graphs shown in FIGURE 2.4.17. Find the points on both graphs for $1 \le x \le 6$ such that the vertical distance d between the graphs, indicated by the black dashed line, is a maximum. What is the maximum vertical distance?
47. (a) Express the square of the distance d from the point (x, y) on the graph of $y = 2x$ to the point $(5, 0)$ shown in FIGURE 2.4.18 as a function of x.
(b) Use the function in part (a) to find the point (x, y) that is closest to $(5, 0)$.
48. As shown in FIGURE 2.4.19 on page 84, an arrow that is shot at a $45°$ angle with the horizontal travels along a parabolic arc defined by the equation $y = ax^2 + x + c$. Use the fact that the arrow is launched at a vertical height of 6 ft and travels a horizontal distance of 200 ft to find the coefficients a and c. What is the maximum height attained by the arrow?

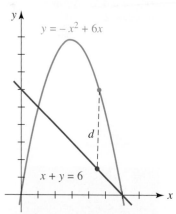

FIGURE 2.4.17 Graphs for Problem 46

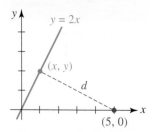

FIGURE 2.4.18 Distance in Problem 47

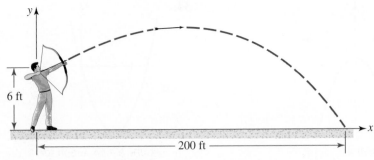

FIGURE 2.4.19 Arrow in Problem 48

49. An arrow is shot vertically upward with an initial velocity of 64 ft/s from a point 6 ft above the ground. See FIGURE 2.4.20.
   (a) Find the height $s(t)$ and the velocity $v(t)$ of the arrow at time $t \geq 0$.
   (b) What is the maximum height attained by the arrow? What is the velocity of the arrow at the time the arrow attains its maximum height?
   (c) At what time does the arrow fall back to the 6-ft level? What is its velocity at this time?

50. The height above ground of a toy rocket launched upward from the top of a building is given by $s(t) = -16t^2 + 96t + 256$.
   (a) What is the height of the building?
   (b) What is the maximum height attained by the rocket?
   (c) Find the time when the rocket strikes the ground.

51. A ball is dropped from the roof of a building that is 122.5 meters above ground level.
   (a) What is the height and velocity of the ball at $t = 1$ s?
   (b) At what time does the ball hit the ground?
   (c) What is the impact velocity of the ball when it hits the ground?

52. A few years ago a newspaper in the Midwest reported that an escape artist was planning to jump off a bridge into the Mississippi River wearing 70 lb of chains and manacles. The newspaper article stated that the height of the bridge was 48 ft and predicted that the escape artist's impact velocity on hitting the water would be 85 mi/h. Assuming that he simply dropped from the bridge, then his height (in feet) and velocity (in feet/second) t seconds after jumping off the bridge are given by the functions $s(t) = -16t^2 + 48$ and $v(t) = -32t$, respectively. Determine whether the newspaper's estimate of his impact velocity was accurate.

Miscellaneous Applications

53. Spread of a Disease One model for the spread of a flu virus assumes that within a population of P persons the rate at which a disease spreads is jointly proportional to the number D of persons already carrying the disease and the number $P - D$ of persons not yet infected. Mathematically, the model is given by the quadratic function

$$R(D) = kD(P - D),$$

where $R(D)$ is the rate of spread of the flu virus (in cases per day) and $k > 0$ is a constant of proportionality.
   (a) Show that if the population P is a constant, then the disease spreads most rapidly when exactly one-half the population is carrying the flu.
   (b) Suppose that in a town of 10,000 persons, 125 are sick on Sunday, and 37 new cases occur on Monday. Estimate the constant k.
   (c) Use the result of part (b) to estimate the number of new cases on Tuesday. [*Hint*: The number of persons carrying the flu on Monday is $162 = 125 + 37$.]
   (d) Estimate the number of new cases on Wednesday, Thursday, Friday, and Saturday.

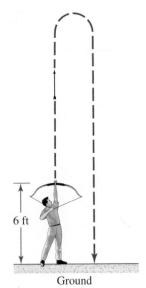

6 ft

Ground

FIGURE 2.4.20 Arrow in Problem 49

Spreading a virus

CHAPTER 2 FUNCTIONS

For Discussion

54. In Problems 50 and 52, what is the domain of the function $s(t)$? [*Hint*: It is *not* $(-\infty, \infty)$.]

55. On the Moon the acceleration due to gravity is one-sixth the acceleration due to gravity on Earth. If a ball is tossed vertically upward from the surface of the Moon, would it attain a maximum height six times that on Earth when the same initial velocity is used? Defend your answer.

56. Suppose the quadratic function $f(x) = ax^2 + bx + c$ has two distinct real zeros. How would you prove that the x-coordinate of the vertex is the midpoint of the line segment between the x-coordinates of the intercepts? Carry out your ideas.

57. Carefully graph the quadratic function $f(x) = x^2 - 4x + 3$. Use a calculator or CAS if necessary.

(a) Simplify $m(x) = \dfrac{f(x) - f(1)}{x - 1}$.

(b) Use $m(x)$ in part (a) to find $m(-5)$ and $m(4)$.

(c) The numbers $m(-5)$ and $m(4)$ in part (b) represent the slopes of two lines. Find the equations of these lines and then graph the lines superimposed on the graph of $f(x)$. What are these lines called?

2.5 Piecewise-Defined Functions

≡ **Introduction** A function f may involve two or more expressions or formulas, with each formula defined on different parts of the domain of f. A function defined in this manner is called a **piecewise-defined function**. For example,

$$f(x) = \begin{cases} x^2, & x < 0 \\ x + 1, & x \geq 0 \end{cases}$$

is not two functions, but a single function in which the rule of correspondence is given in two pieces. In this case, one piece is used for the negative real numbers ($x < 0$) and the other part on the nonnegative numbers ($x \geq 0$); the domain of f is the union of the intervals $(-\infty, 0) \cup [0, \infty) = (-\infty, \infty)$. For example, since $-4 < 0$, the rule indicates that we square the number:

$$f(-4) = (-4)^2 = 16;$$

on the other hand, since $6 \geq 0$ we add 1 to the number:

$$f(6) = 6 + 1 = 7.$$

☐ **Postage Stamp Function** The USPS first-class mailing rates for a letter, a card, or a package provide a real-world illustration of a piecewise-defined function. As of this writing, the postage for sending a letter in a standard-size envelope by first-class mail depends on its weight in ounces:

$$\text{Postage} = \begin{cases} \$0.44, & 0 < \text{weight} \leq 1 \text{ ounce} \\ \$0.61, & 1 < \text{weight} \leq 2 \text{ ounces} \\ \$0.78, & 2 < \text{weight} \leq 3 \text{ ounces,} \\ \vdots \\ \$2.92, & 12 < \text{weight} \leq 13 \text{ ounces.} \end{cases} \qquad (1)$$

The rule in (1) is called the **postage stamp function** P and consists of 13 pieces (letters over 13 ounces are sent priority mail). A value $P(w)$ is one of thirteen constants; the constant changes depending on the weight w (in ounces) of the letter. For example,

$$P(0.5) = \$0.44, P(1.7) = \$0.61, P(2.2) = \$0.78, P(2.9) = \$0.78,$$
$$\text{and } P(12.1) = \$2.92.$$

The domain of the function P is the union of the intervals:

$$(0, 1] \cup (1, 2] \cup (2, 3] \cup \cdots \cup (12, 13] = (0, 13].$$

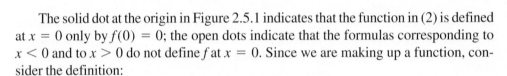

EXAMPLE 1   **Graph of a Piecewise-Defined Function**

Graph the piecewise-defined function

$$f(x) = \begin{cases} -1, & x < 0, \\ 0, & x = 0, \\ x + 1, & x > 0. \end{cases} \qquad (2)$$

Solution Although the domain of f consists of all real numbers $(-\infty, \infty)$, each piece of the function is defined on a different part of this domain. We draw

- the horizontal line $y = -1$ for $x < 0$,
- the point $(0, 0)$ for $x = 0$, and
- the line $y = x + 1$ for $x > 0$.

The graph is given in FIGURE 2.5.1. $\equiv$

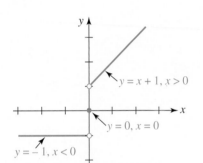

FIGURE 2.5.1 Graph of piecewise-defined function in Example 1

The solid dot at the origin in Figure 2.5.1 indicates that the function in (2) is defined at $x = 0$ only by $f(0) = 0$; the open dots indicate that the formulas corresponding to $x < 0$ and to $x > 0$ do not define f at $x = 0$. Since we are making up a function, consider the definition:

$$g(x) = \begin{cases} -1, & x \le 0, \\ x + 1, & x > 0. \end{cases} \qquad (3)$$

The graph of g shown in FIGURE 2.5.2 is very similar to the graph of (2), but (2) and (3) are not the same function since $f(0) = 0$ but $g(0) = -1$.

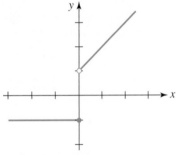

FIGURE 2.5.2 Graph of function g defined in (3)

☐ **Greatest Integer Function** We consider next a piecewise-defined function that is similar to the "postage stamp" function (1) in that both are examples of *step functions*; each function is constant on an interval and then jumps to another constant value on the next abutting interval. This new function, which has many notations, will be denoted here by $f(x) = [\![x]\!]$, and is defined by the rule

$$[\![x]\!] = n, \quad \text{where } n \text{ is an integer satisfying } n \le x < n + 1. \qquad (4)$$

The function f is called the **greatest integer function** because (4), translated into words, means that:

$f(x)$ is the greatest integer n that is less than or equal to x.

For example,

$$f(6) = 6 \text{ since } 6 \le x = 6, \qquad\qquad f(-1.5) = -2 \text{ since } -2 \le x = -1.5,$$
$$f(0.4) = 0 \text{ since } 0 \le x = 0.4, \qquad\quad f(7.6) = 7 \text{ since } 7 \le x = 7.6,$$
$$f(\pi) = 3 \text{ since } 3 \le x = \pi, \qquad\qquad f(-\sqrt{2}) = -2 \text{ since } -2 \le x = -\sqrt{2},$$

and so on. The domain of f is the set of real numbers and consists of the union of an infinite number of disjoint intervals; in other words, $f(x) = [\![x]\!]$ is a piecewise-defined function given by

$$f(x) = [\![x]\!] = \begin{cases} \vdots & \\ -2, & -2 \le x < -1 \\ -1, & -1 \le x < 0 \\ 0, & 0 \le x < 1 \\ 1, & 1 \le x < 2 \\ 2, & 2 \le x < 3 \\ \vdots & \end{cases} \tag{5}$$

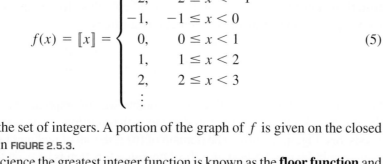

FIGURE 2.5.3 Greatest integer function

The range of f is the set of integers. A portion of the graph of f is given on the closed interval $[-2, 5]$ in FIGURE 2.5.3.

In computer science the greatest integer function is known as the **floor function** and is denoted by $f(x) = \lfloor x \rfloor$. See Problems 47, 48, and 53 in Exercises 2.5.

 **EXAMPLE 2**　　　　**Shifted Graph**

Graph $y = [\![x - 2]\!]$.

Solution The function is $y = f(x - 2)$, where $f(x) = [\![x]\!]$. Thus the graph in Figure 2.5.3 is shifted horizontally 2 units to the right. Note in Figure 2.5.3 that if n is an integer, then $f(n) = [\![n]\!] = n$. But in FIGURE 2.5.4, for $x = n$, $y = n - 2$. ≡

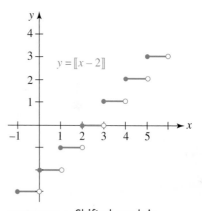

FIGURE 2.5.4 Shifted graph in Example 2

☐ **Continuous Functions** The graph of a **continuous function** has no holes, finite gaps, or infinite breaks. While the formal definition of continuity of a function is an important topic of discussion in calculus, in this course it suffices to think in informal terms. A continuous function is often characterized by saying that its graph can be drawn "without lifting pencil from paper." Parts (a)–(c) of FIGURE 2.5.5 illustrate functions that are *not* continuous, or **discontinuous**, at $x = 2$. The function

$$f(x) = \frac{x^2 - 4}{x - 2} = x + 2, \quad x \ne 2,$$

in FIGURE 2.5.5(a) has a hole in its graph (there is no point $(2, f(2))$); the function $f(x) = \dfrac{|x - 2|}{x - 2}$ in Figure 2.5.5(b) has a finite gap or jump in its graph at $x = 2$; the function $f(x) = \dfrac{1}{x - 2}$ in Figure 2.5.5(c) has an infinite break in its graph at $x = 2$. The function $f(x) = x^3 - 3x + 2$ is continuous; its graph given in Figure 2.5.5(d) has no holes, gaps, or infinite breaks.

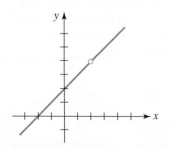
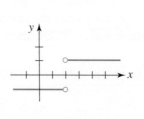
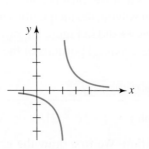
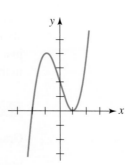

(a) Hole in graph　　　(b) Finite gap in graph　　　(c) Infinite break in graph　　　(d) No holes, gaps, or breaks
FIGURE 2.5.5 Discontinuous functions (a)–(c); continuous function (d)

You should be aware that constant functions, linear functions, and quadratic functions are continuous. Piecewise-defined functions can be continuous or discontinuous. The functions given in (2), (3), and (4) are discontinuous.

☐ **Absolute-Value Function** The function $y = |x|$, called the **absolute-value function**, appears frequently in the study of calculus. To obtain the graph, we graph its two pieces consisting of perpendicular half lines:

$$y = |x| = \begin{cases} -x, & \text{if } x < 0 \\ x, & \text{if } x \geq 0. \end{cases} \tag{6}$$

See FIGURE 2.5.6(a). Since $y \geq 0$ for all x, another way of graphing (6) is simply to sketch the line $y = x$ and then reflect in the x-axis that portion of the line that is below the x-axis. See Figure 2.5.6(b). The domain of (6) is the set of real numbers $(-\infty, \infty)$, and as is seen in Figure 2.5.6(a), the absolute-value function is an even function, decreasing on the interval $(-\infty, 0)$, increasing on the interval $(0, \infty)$, and is continuous.

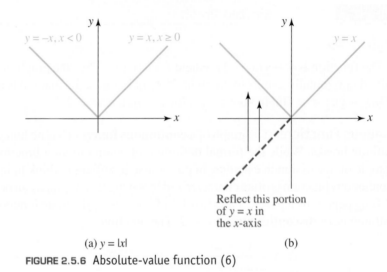

(a) $y = |x|$ (b)

Reflect this portion
of $y = x$ in
the x-axis

FIGURE 2.5.6 Absolute-value function (6)

In some applications we are interested in the graph of the absolute value of an arbitrary function $y = f(x)$; in other words, $y = |f(x)|$. Since $|f(x)|$ is nonnegative for all numbers x in the domain of f, the graph of $y = |f(x)|$ does not extend below the x-axis. Moreover, the definition of the absolute value of $f(x)$,

$$|f(x)| = \begin{cases} -f(x), & \text{if } f(x) < 0 \\ f(x), & \text{if } f(x) \geq 0, \end{cases} \tag{7}$$

shows that we must negate $f(x)$ whenever $f(x)$ is negative. There is no need to worry about solving the inequalities in (7); to obtain the graph of $y = |f(x)|$, we can proceed just as we did in Figure 2.5.6(b): Carefully draw the graph of $y = f(x)$ and then reflect in the x-axis all portions of the graph that are below the x-axis.

EXAMPLE 3   Absolute Value of a Function

Graph $y = |-3x + 2|$.

Solution We first draw the graph of the linear function $f(x) = -3x + 2$. Note that since the slope is negative, f is decreasing and its graph crosses the x-axis at $\left(\frac{2}{3}, 0\right)$. We dash the graph for $x > \frac{2}{3}$ since that portion is below the x-axis. Finally, we reflect that portion upward in the x-axis to obtain the solid blue v-shaped graph in FIGURE 2.5.7. Since $f(x) = x$ is a simple linear function, it is not surprising that the graph of the absolute value of any linear function $f(x) = ax + b, a \neq 0$, will result in a graph similar to that of the absolute-value function shown in Figure 2.5.6(a). ≡

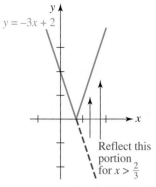

FIGURE 2.5.7 Graph of function in Example 3

EXAMPLE 4   Absolute Value of a Function

Graph $y = |-x^2 + 2x + 3|$.

Solution As in Example 3, we begin by drawing the graph of the function $f(x) = -x^2 + 2x + 3$ by finding its intercepts $(-1, 0), (3, 0), (0, 3)$ and, since f is a quadratic function, its vertex $(1, 4)$. Observe in FIGURE 2.5.8(a) that $y < 0$ for $x < -1$ and for $x > 3$. These portions of the graph of f are reflected in the x-axis to obtain the graph of $y = |-x^2 + 2x + 3|$ given in Figure 2.5.8(b).

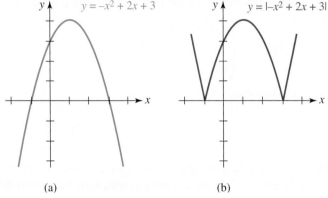

(a)   (b)

FIGURE 2.5.8 Graphs of functions in Example 4   ≡

2.5   Exercises   Answers to selected odd-numbered problems begin on page ANS–6.

In Problems 1–4, find the indicated values of the given piecewise-defined function f.

1. $f(x) = \begin{cases} \dfrac{x^2 - 4}{x - 2}, & x \neq 2 \\ 4, & x = 2 \end{cases}$; $f(0), f(2), f(-7)$

2. $f(x) = \begin{cases} \dfrac{x^4 - 1}{x^2 - 1}, & x \neq \pm 1 \\ 3, & x = -1 \\ 5, & x = 1 \end{cases}$; $f(-1), f(1), f(3)$

3. $f(x) = \begin{cases} x^2 + 2x, & x \geq 1 \\ -x^3, & x < 1 \end{cases}$; $f(1), f(0), f(-2), f(\sqrt{2})$

4. $f(x) = \begin{cases} 0, & x < 0 \\ x, & 0 < x < 1 \\ x + 1, & x \geq 1 \end{cases}$; $f(-\frac{1}{2}), f(\frac{1}{3}), f(4), f(6.2)$

5. If the piecewise-defined function f is defined by

$$f(x) = \begin{cases} 1, & x \text{ a rational number} \\ 0, & x \text{ an irrational number,} \end{cases}$$

find each of the following function values.

(a) $f\left(\frac{1}{3}\right)$ (b) $f(-1)$ (c) $f(\sqrt{2})$

(d) $f(1.\overline{12})$ (e) $f(5.72)$ (f) $f(\pi)$

6. What is the y-intercept of the graph of the function f in Problem 5?

7. Determine the values of x for which the piecewise-defined function

$$f(x) = \begin{cases} x^3 + 1, & x < 0 \\ x^2 - 2, & x \geq 0, \end{cases}$$

is equal to the given number.

(a) 7   (b) 0   (c) -1

(d) -2 (e) 1   (f) -7

8. Determine the values of x for which the piecewise-defined function

$$f(x) = \begin{cases} x + 1, & x < 0 \\ 2, & x = 0 \\ x^2, & x > 0, \end{cases}$$

is equal to the given number.

(a) 1   (b) 0   (c) 4

(d) $\frac{1}{2}$ (e) 2   (f) -4

In Problems 9–34, sketch the graph of the given piecewise-defined function. Find any x- and y-intercepts of the graph. Give any numbers at which the function is discontinuous.

9. $y = \begin{cases} -x, & x \leq 1 \\ -1, & x > 1 \end{cases}$ **10.** $y = \begin{cases} x - 1, & x < 0 \\ x + 1, & x \geq 0 \end{cases}$

11. $y = \begin{cases} -3, & x < -3 \\ x, & -3 \leq x \leq 3 \\ 3, & x > 3 \end{cases}$ **12.** $y = \begin{cases} -x^2 - 1, & x < 0 \\ 0, & x = 0 \\ x^2 + 1, & x > 0 \end{cases}$

13. $y = [\![x + 2]\!]$ **14.** $y = 2 + [\![x]\!]$

15. $y = -[\![x]\!]$ **16.** $y = [\![-x]\!]$

17. $y = |x + 3|$ **18.** $y = -|x - 4|$

19. $y = 2 - |x|$ **20.** $y = -1 - |x|$

21. $y = -2 + |x + 1|$ **22.** $y = 1 - \frac{1}{2}|x - 2|$

23. $y = -|5 - 3x|$ **24.** $y = |2x - 5|$

25. $y = |x^2 - 1|$ **26.** $y = |4 - x^2|$

27. $y = |x^2 - 2x|$ **28.** $y = |-x^2 - 4x + 5|$

29. $y = |\,|x| - 2|$ **30.** $y = |\sqrt{x} - 2|$

31. $y = |x^3 - 1|$ **32.** $y = |[\![x]\!]|$

33. $y = \begin{cases} 1, & x < 0 \\ |x - 1|, & 0 \leq x \leq 2 \\ 1, & x > 2 \end{cases}$ **34.** $y = \begin{cases} -x, & x < 0 \\ 1 - |x - 1|, & 0 \leq x \leq 2 \\ x - 2, & x > 2 \end{cases}$

35. Without graphing, give the range of the function $f(x) = (-1)^{[\![x]\!]}$.

36. Compare the graphs of $y = 2[\![x]\!]$ and $y = [\![2x]\!]$.

CHAPTER 2 FUNCTIONS

In Problems 37–40, find a piecewise-defined formula for the function f whose graph is given. Assume that the domain of f is $(-\infty, \infty)$.

37.

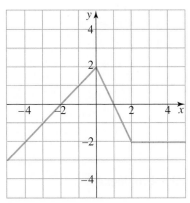

FIGURE 2.5.9 Graph for Problem 37

38.

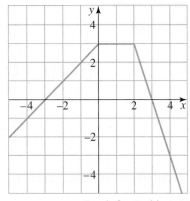

FIGURE 2.5.10 Graph for Problem 38

39.

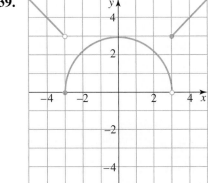

FIGURE 2.5.11 Graph for Problem 39

40.

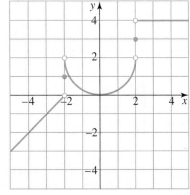

FIGURE 2.5.12 Graph for Problem 40

In Problems 41 and 42, sketch the graph of $y = |f(x)|$.

41. f is the function whose graph is given in Figure 2.5.9.
42. f is the function whose graph is given in Figure 2.5.10.

In Problems 43 and 44, use the definition of absolute value and express the given function f as a piecewise-defined function.

43. $f(x) = \dfrac{|x|}{x}$

44. $f(x) = \dfrac{x - 3}{|x - 3|}$

In Problems 45 and 46, find the value of the constant k such that the given piecewise-defined function f is continuous at $x = 2$. That is, the graph of f has no holes, gaps, or breaks in its graph at $x = 2$.

45. $f(x) = \begin{cases} \frac{1}{2}x + 1, & x \le 2 \\ kx, & x > 2 \end{cases}$

46. $f(x) = \begin{cases} kx + 2, & x < 2 \\ x^2 + 1, & x \ge 2 \end{cases}$

47. The **ceiling function** $g(x) = \lceil x \rceil$ is defined to be the least integer n that is greater than or equal to x. Fill in the blanks.

$$g(x) = \lceil x \rceil = \begin{cases} \vdots & \\ \underline{\hspace{2cm}}, & -3 < x \le -2 \\ \underline{\hspace{2cm}}, & -2 < x \le -1 \\ \underline{\hspace{2cm}}, & -1 < x \le 0 \\ \underline{\hspace{2cm}}, & 0 < x \le 1 \\ \underline{\hspace{2cm}}, & 1 < x \le 2 \\ \underline{\hspace{2cm}}, & 2 < x \le 3 \\ \vdots & \end{cases}$$

48. Graph the ceiling function $g(x) = \lceil x \rceil$ defined in Problem 47.

For Discussion

In Problems 49–52, describe in words how the graphs of the given functions differ. [*Hint*: Factor and cancel.]

49. $f(x) = \dfrac{x^2 - 9}{x - 3}$, $\qquad g(x) = \begin{cases} \dfrac{x^2 - 9}{x - 3}, & x \ne 3 \\ 4, & x = 3 \end{cases}$,

$h(x) = \begin{cases} \dfrac{x^2 - 9}{x - 3}, & x \ne 3 \\ 6, & x = 3 \end{cases}$

50. $f(x) = -\dfrac{x^2 - 7x + 6}{x - 1}$, $\qquad g(x) = \begin{cases} -\dfrac{x^2 - 7x + 6}{x - 1}, & x \ne 1 \\ 8, & x = 1 \end{cases}$,

$h(x) = \begin{cases} -\dfrac{x^2 - 7x + 6}{x - 1}, & x \ne 1 \\ 5, & x = 1 \end{cases}$

51. $f(x) = \dfrac{x^4 - 1}{x^2 - 1}$, $\qquad g(x) = \begin{cases} \dfrac{x^4 - 1}{x^2 - 1}, & x \ne 1 \\ 0, & x = 1 \end{cases}$,

$h(x) = \begin{cases} \dfrac{x^4 - 1}{x^2 - 1}, & x \ne 1 \\ 2, & x = 1 \end{cases}$

52. $f(x) = \dfrac{x^3 - 8}{x - 2}$, $\qquad g(x) = \begin{cases} \dfrac{x^3 - 8}{x - 2}, & x \ne 2 \\ 5, & x = 2 \end{cases}$,

$h(x) = \begin{cases} \dfrac{x^3 - 8}{x - 2}, & x \ne 2 \\ 12, & x = 2 \end{cases}$

53. Using the notion of a reflection of a graph in an axis, express the ceiling function $g(x) = \lceil x \rceil$ in terms of the floor function $f(x) = \lfloor x \rfloor$ (see page 87).

54. Discuss how to graph the function $y = |x| + |x - 3|$. Carry out your ideas.

2.6 Combining Functions

≡ **Introduction** Two functions f and g can be combined in several ways to create new functions. In this section we will examine two such ways in which functions can be combined: through arithmetic operations, and through the operation of function composition.

☐ **Arithmetic Combinations** Two functions can be combined through the familiar four arithmetic operations of addition, subtraction, multiplication, and division.

DEFINITION 2.6.1 Arithmetic Combinations

If f and g are two functions, then the **sum** $f + g$, the **difference** $f - g$, the **product** fg, and the **quotient** f/g are defined as follows:

$$(f + g)(x) = f(x) + g(x) \qquad (1)$$
$$(f - g)(x) = f(x) - g(x) \qquad (2)$$
$$(fg)(x) = f(x)g(x) \qquad (3)$$
$$\left(\frac{f}{g}\right)(x) = \frac{f(x)}{g(x)}, \quad \text{provided } g(x) \neq 0 \qquad (4)$$

■ EXAMPLE 1　　　**Sum, Difference, Product, and Quotient**

Consider the functions $f(x) = x^2 + 4x$ and $g(x) = x^2 - 9$. From (1)–(4) we can produce four new functions:

$$(f + g)(x) = f(x) + g(x) = (x^2 + 4x) + (x^2 - 9) = 2x^2 + 4x - 9,$$
$$(f - g)(x) = f(x) - g(x) = (x^2 + 4x) - (x^2 - 9) = 4x + 9,$$
$$(fg)(x) = f(x)g(x) = (x^2 + 4x)(x^2 - 9) = x^4 + 4x^3 - 9x^2 - 36x,$$

and
$$\left(\frac{f}{g}\right)(x) = \frac{f(x)}{g(x)} = \frac{x^2 + 4x}{x^2 - 9}. \qquad ≡$$

☐ **Domain of an Arithmetic Combination** When combining two functions arithmetically it is necessary that both f and g be defined at a same number x. Hence the **domain** of the functions $f + g, f - g$, and fg is the set of real numbers that are *common* to both domains, that is, the domain is the *intersection* of the domain of f with the domain of g. In the case of the quotient f/g, the domain is also the intersection of the two domains, *but* we must also exclude any values of x for which the denominator $g(x)$ is zero. In Example 1 the domain of f and the domain of g is the set of real numbers $(-\infty, \infty)$, and so the domain of $f + g, f - g$, and fg is also $(-\infty, \infty)$. However, since $g(-3) = 0$ and $g(3) = 0$, the domain of the quotient $(f/g)(x)$ is $(-\infty, \infty)$ with $x = 3$ and $x = -3$ excluded, in other words, $(-\infty, -3) \cup (-3, 3) \cup (3, \infty)$. In summary, if the domain of f is the set X_1 and the domain of g is the set X_2, then:

- the domain of $f + g, f - g$, and fg is $X_1 \cap X_2$, and
- the domain of f/g is the set $\{x | x \in X_1 \cap X_2, g(x) \neq 0\}$.

■ EXAMPLE 2　　　**Domain of** $f + g$

By solving the inequality $1 - x \geq 0$, it is seen that the domain of $f(x) = \sqrt{1 - x}$ is the interval $(-\infty, 1]$. Similarly, the domain of the function $g(x) = \sqrt{x + 2}$ is the interval $[-2, \infty)$. Hence, the domain of the sum

$$(f + g)(x) = f(x) + g(x) = \sqrt{1 - x} + \sqrt{x + 2}$$

is the intersection $(-\infty, 1] \cap [-2, \infty)$. You should verify this result by sketching these intervals on the number line that this intersection, or the set of numbers common to both domains, is the closed interval $[-2, 1]$.

≡

☐ **Composition of Functions** Another method of combining functions f and g is called **function composition**. To illustrate the idea, let's suppose that for a given x in the domain of g the function value $g(x)$ is a number in the domain of the function f. This means we are able to evaluate f at $g(x)$, in other words, $f(g(x))$. For example, suppose $f(x) = x^2$ and $g(x) = x + 2$. Then for $x = 1$, $g(1) = 3$, and since 3 is the domain of f, we can write $f(g(1)) = f(3) = 3^2 = 9$. Indeed, for these two particular functions it turns out that we can evaluate f at any function value $g(x)$, that is,

$$f(g(x)) = f(x + 2) = (x + 2)^2.$$

The resulting function, called the composition of f and g, is defined next.

DEFINITION 2.6.2   Function Composition

If f and g are two functions, then the **composition** of f and g, denoted by $f \circ g$, is the function defined by

$$(f \circ g)(x) = f(g(x)). \qquad (5)$$

The **composition** of g and f, denoted by $g \circ f$, is the function defined by

$$(g \circ f)(x) = g(f(x)). \qquad (6)$$

When computing a composition such as $(f \circ g)(x) = f(g(x))$, be sure to substitute $g(x)$ for every x that appears in $f(x)$. See part (a) of the next example.

EXAMPLE 3   **Two Compositions**

If $f(x) = x^2 + 3x - 1$ and $g(x) = 2x^2 + 1$, find **(a)** $(f \circ g)(x)$ and **(b)** $(g \circ f)(x)$.

Solution (a) For emphasis we replace x by the set of parentheses ( ) and write f in the form

$$f(\ ) = (\ )^2 + 3(\ ) - 1.$$

Thus to evaluate $(f \circ g)(x)$ we fill each set of parentheses with $g(x)$. We find

$$\begin{aligned}
(f \circ g)(x) = f(g(x)) &= f(2x^2 + 1) \\
&= (2x^2 + 1)^2 + 3(2x^2 + 1) - 1 \qquad \leftarrow \begin{cases} \text{use } (a + b)^2 = a^2 + 2ab + b^2 \\ \text{and the distributive law} \end{cases} \\
&= 4x^4 + 4x^2 + 1 + 3 \cdot 2x^2 + 3 \cdot 1 - 1 \\
&= 4x^4 + 10x^2 + 3.
\end{aligned}$$

(b) In this case write g in the form

$$g(\ ) = 2(\ )^2 + 1.$$

Then

$$\begin{aligned}
(g \circ f)(x) = g(f(x)) &= g(x^2 + 3x - 1) \\
&= 2(x^2 + 3x - 1)^2 + 1 \qquad \leftarrow \text{use } (a + b + c)^2 = ((a + b) + c)^2 \\
&= 2(x^4 + 6x^3 + 7x^2 - 6x + 1) + 1 \qquad = (a + b)^2 + 2(a + b)c + c^2 \text{ etc.} \\
&= 2 \cdot x^4 + 2 \cdot 6x^3 + 2 \cdot 7x^2 - 2 \cdot 6x + 2 \cdot 1 + 1 \\
&= 2x^4 + 12x^3 + 14x^2 - 12x + 3.
\end{aligned}$$

≡

Parts (a) and (b) of Example 3 illustrate that function composition is not commutative. That is, in general

$$f \circ g \neq g \circ f.$$

The next example shows that a function can be composed with itself.

EXAMPLE 4   *f* Composed with *f*

If $f(x) = 5x - 1$, then the composition $f \circ f$ is given by

$$(f \circ f)(x) = f(f(x)) = f(5x - 1) = 5(5x - 1) - 1 = 25x - 6.$$ ≡

EXAMPLE 5   Writing a Function as a Composition

Express $F(x) = \sqrt{6x^3 + 8}$ as the composition of two functions f and g.

Solution If we define f and g as $f(x) = \sqrt{x}$ and $g(x) = 6x^3 + 8$, then

$$F(x) = (f \circ g)(x) = f(g(x)) = f(6x^3 + 8) = \sqrt{6x^3 + 8}.$$ ≡

There are other solutions to Example 5. For instance, if the functions f and g are defined by $f(x) = \sqrt{6x + 8}$ and $g(x) = x^3$, then observe $(f \circ g)(x) = f(x^3) = \sqrt{6x^3 + 8}$.

☐ **Domain of a Composition** As stated in the introductory example to this discussion, to evaluate the composition $(f \circ g)(x) = f(g(x))$ the number $g(x)$ must be in the domain of f. For example, the domain $f(x) = \sqrt{x}$ is $x \geq 0$, and the domain of $g(x) = x - 2$ is the set of real numbers $(-\infty, \infty)$. Observe that we cannot evaluate $f(g(1))$ because $g(1) = -1$ and -1 is not in the domain of f. In order to substitute $g(x)$ into $f(x)$, $g(x)$ must satisfy the inequality that defines the domain of f, namely, $g(x) \geq 0$. This last inequality is the same as $x - 2 \geq 0$ or $x \geq 2$. The domain of the composition $f(g(x)) = \sqrt{g(x)} = \sqrt{x - 2}$ is $[2, \infty)$, which is only a portion of the original domain $(-\infty, \infty)$ of g. In general:

- The domain of the composition $f \circ g$ consists of the numbers x in the domain of g such that $g(x)$ is in the domain of f.   ◀ Read this sentence several times.

EXAMPLE 6   Domain of a Composition

Consider the function $f(x) = \sqrt{x - 3}$. From the requirement that $x - 3 \geq 0$ we see that whatever number x is substituted into f must satisfy $x \geq 3$. Now suppose $g(x) = x^2 + 2$ and we want to evaluate $f(g(x))$. Although the domain of g is the set of all real numbers, in order to substitute $g(x)$ into $f(x)$ we require that x be a number in that domain so that $g(x) \geq 3$. From FIGURE 2.6.1 we see that the last inequality is satisfied whenever $x \leq -1$ or $x \geq 1$. In other words, the domain of the composition

$$f(g(x)) = f(x^2 + 2) = \sqrt{(x^2 + 2) - 3} = \sqrt{x^2 - 1}$$

is $(-\infty, -1] \cup [1, \infty)$.   ≡

In certain applications a quantity y is given as a function of a variable x which in turn is a function of another variable t. By means of function composition we can express y as a function of t. The next example illustrates the idea; the symbol V plays the part of y and r plays the part of x.

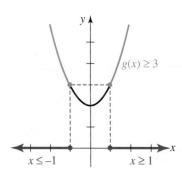

FIGURE 2.6.1 Domain of $(f \circ g)(x)$ in Example 6

Weather balloon

EXAMPLE 7 Inflating a Balloon

A weather balloon is being inflated with a gas. If the radius of the balloon is increasing at a rate of 5 cm/s, express the volume of the balloon as a function of time t in seconds.

Solution Let's assume that as the balloon is inflated, its shape is that of a sphere. If r denotes the radius of the balloon, then $r(t) = 5t$. Since the volume of a sphere is $V = \frac{4}{3}\pi r^3$, the composition is $(V \circ r)(t) = V(r(t)) = V(5t)$ or

$$V = \frac{4}{3}\pi(5t)^3 = \frac{500}{3}\pi t^3.$$

≡

☐ **Postscript** The rigid and nonrigid transformations that were studied in Section 2.2 are examples of the operations on functions just discussed. For $c > 0$, a constant, the rigid transformations defined by $y = f(x) + c$ and $y = f(x) - c$ are the *sum* and *difference* of the function $f(x)$ and the constant function $g(x) = c$. The nonrigid transformation $y = cf(x)$ is the *product* of $f(x)$ and the constant function $g(x) = c$. The rigid transformations defined by $y = f(x + c)$ and $y = f(x - c)$ are *compositions* of $f(x)$ with the linear functions $g(x) = x + c$ and $g(x) = x - c$, respectively.

| 2.6 | Exercises | Answers to selected odd-numbered problems begin on page ANS–7. |

In Problems 1–8, find the functions $f + g$, $f - g$, fg, and f/g, and give their domains.

1. $f(x) = x^2 + 1$, $g(x) = 2x^2 - x$

2. $f(x) = x^2 - 4$, $g(x) = x + 3$

3. $f(x) = x$, $g(x) = \sqrt{x - 1}$

4. $f(x) = x - 2$, $g(x) = \dfrac{1}{x + 8}$

5. $f(x) = 3x^3 - 4x^2 + 5x$, $g(x) = (1 - x)^2$

6. $f(x) = \dfrac{4}{x - 6}$, $g(x) = \dfrac{x}{x - 3}$

7. $f(x) = \sqrt{x + 2}$, $g(x) = \sqrt{5 - 5x}$

8. $f(x) = \dfrac{1}{x^2 - 9}$, $g(x) = \dfrac{\sqrt{x + 4}}{x}$

9. Fill in the table.

x	0	1	2	3	4
$f(x)$	−1	2	10	8	0
$g(x)$	2	3	0	1	4
$(f \circ g)(x)$					

10. Fill in the table where g is an odd function.

x	0	1	2	3	4
$f(x)$	−2	−3	0	−1	−4
$g(x)$	9	7	−6	−5	13
$(g \circ f)(x)$					

In Problems 11–14, find the functions $f \circ g$ and $g \circ f$ and give their domains.

11. $f(x) = x^2 + 1, \quad g(x) = \sqrt{x - 1}$

12. $f(x) = x^2 - x + 5, \quad g(x) = -x + 4$

13. $f(x) = \dfrac{1}{2x - 1}, \quad g(x) = x^2 + 1$

14. $f(x) = \dfrac{x + 1}{x}, \quad g(x) = \dfrac{1}{x}$

In Problems 15–20, find the functions $f \circ g$ and $g \circ f$.

15. $f(x) = 2x - 3, \quad g(x) = \frac{1}{2}(x + 3)$ **16.** $f(x) = x - 1, \quad g(x) = x^3$

17. $f(x) = x + \dfrac{1}{x^2}, \quad g(x) = \dfrac{1}{x}$ **18.** $f(x) = \sqrt{x - 4}, \quad g(x) = x^2$

19. $f(x) = x + 1, \quad g(x) = x + \sqrt{x - 1}$ **20.** $f(x) = x^3 - 4, \quad g(x) = \sqrt[3]{x + 3}$

In Problems 21–24, find $f \circ f$ and $f \circ (1/f)$.

21. $f(x) = 2x + 6$ **22.** $f(x) = x^2 + 1$

23. $f(x) = \dfrac{1}{x^2}$ **24.** $f(x) = \dfrac{x + 4}{x}$

In Problems 25 and 26, find $(f \circ g \circ h)(x) = f(g(h(x)))$.

25. $f(x) = \sqrt{x}, \quad g(x) = x^2, \quad h(x) = x - 1$

26. $f(x) = x^2, \quad g(x) = x^2 + 3x, \quad h(x) = 2x$

27. For the functions $f(x) = 2x + 7, g(x) = 3x^2$, find $(f \circ g \circ g)(x)$.

28. For the functions $f(x) = -x + 5, g(x) = -4x^2 + x$, find $(f \circ g \circ f)(x)$.

In Problems 29 and 30, find $(f \circ f \circ f)(x) = f(f(f(x)))$.

29. $f(x) = 2x - 5$ **30.** $f(x) = x^2 - 1$

In Problems 31–34, find functions f and g such that $F(x) = f \circ g$.

31. $F(x) = (x^2 - 4x)^5$ **32.** $F(x) = \sqrt{9x^2 + 16}$

33. $F(x) = (x - 3)^2 + 4\sqrt{x - 3}$ **34.** $F(x) = 1 + |2x + 9|$

In Problems 35 and 36, sketch the graphs of the compositions $f \circ g$ and $g \circ f$.

35. $f(x) = |x| - 2, \quad g(x) = |x - 2|$ **36.** $f(x) = [\![x - 1]\!], \quad g(x) = |x|$

37. Consider the function $y = f(x) + g(x)$, where $f(x) = x$ and $g(x) = -[\![x]\!]$. Fill in the blanks and then sketch the graph of the sum $f + g$ on the indicated intervals.

$$ y = \begin{cases} \vdots \\ \underline{\hspace{2cm}}, & -3 \leq x < -2 \\ \underline{\hspace{2cm}}, & -2 \leq x < -1 \\ \underline{\hspace{2cm}}, & -1 \leq x < 0 \\ \underline{\hspace{2cm}}, & 0 \leq x < 1 \\ \underline{\hspace{2cm}}, & 1 \leq x < 2 \\ \underline{\hspace{2cm}}, & 2 \leq x < 3 \\ \vdots \end{cases} $$

38. Consider the function $y = f(x) + g(x)$, where $f(x) = |x|$ and $g(x) = [\![x]\!]$. Proceed as in Problem 37 and then sketch the graph of the sum $f + g$.

In Problems 39 and 40, sketch the graph of the sum $f + g$.

39. $f(x) = |x - 1|$, $g(x) = |x|$ **40.** $f(x) = x$, $g(x) = |x|$

In Problems 41 and 42, sketch the graph of the product fg.

41. $f(x) = x$, $g(x) = |x|$ **42.** $f(x) = x$, $g(x) = [\![x]\!]$

In Problems 43 and 44, sketch the graph of the reciprocal $1/f$.

43. $f(x) = |x|$ **44.** $f(x) = x - 3$

Miscellaneous Calculus-Related Problems

In Problems 45 and 46,

(a) find the points of intersection of the graphs of the given functions,
(b) find the vertical distance d between the graphs on the interval I determined by the x-coordinates of their points of intersection,
(c) use the concept of a vertex of a parabola to find the maximum value of d on the interval I.

45.

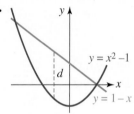

FIGURE 2.6.2 Graph for Problem 45

46.

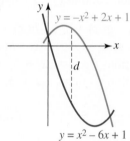

FIGURE 2.6.3 Graph for Problem 46

Miscellaneous Applications

47. For the Birds A birdwatcher sights a bird 100 ft due east of her position. If the bird is flying due south at a rate of 500 ft/min, express the distance d from the birdwatcher to the bird as a function of time t. Find the distance 5 minutes after the sighting. See FIGURE 2.6.4.

48. Bacteria A certain bacteria when cultured grows in a circular shape. The radius of the circle, measured in centimeters, is given by

$$r(t) = 4 - \frac{4}{t^2 + 1},$$

where time t is measured in hours.

(a) Express the area covered by the bacteria as a function of time t.
(b) Express the circumference of the area covered as a function of time t.

For Discussion

49. Suppose $f(x) = x^2 + 1$ and $g(x) = \sqrt{x}$. Discuss: Why is the domain of

$$(f \circ g)(x) = f(g(x)) = (\sqrt{x})^2 + 1 = x + 1$$

not $(-\infty, \infty)$?

FIGURE 2.6.4 Birdwatcher in Problem 47

Bird-watcher 100 ft Bird's path d

     CHAPTER 2 FUNCTIONS

50. Suppose $f(x) = \dfrac{2}{x - 1}$ and $g(x) = \dfrac{5}{x + 3}$. Discuss: Why is the domain of

$$(f \circ g)(x) = f(g(x)) = \frac{2}{g(x) - 1} = \frac{2}{\dfrac{5}{x + 3} - 1} = \frac{2x + 6}{2 - x}$$

not $\{x \mid x \neq 2\}$?

51. Find the error in the following reasoning: If $f(x) = 1/(x - 2)$ and $g(x) = 1/\sqrt{x + 1}$, then

$$\left(\frac{f}{g}\right)(x) = \frac{1/(x - 2)}{1/\sqrt{x + 1}} = \frac{\sqrt{x + 1}}{x - 2} \quad \text{and so} \quad \left(\frac{f}{g}\right)(-1) = \frac{\sqrt{0}}{-3} = 0.$$

52. Suppose $f_1(x) = \sqrt{x + 2}, f_2(x) = \dfrac{x}{\sqrt{x(x - 10)}}$, and $f_3(x) = \dfrac{x + 1}{x}$. What is the domain of the function $y = f_1(x) + f_2(x) + f_3(x)$?

53. Suppose $f(x) = x^3 + 4x, g(x) = x - 2$, and $h(x) = -x$. Discuss: Without actually graphing, how are the graphs of $f \circ g, g \circ f, f \circ h$, and $h \circ f$ related to the graph of f?

54. The domain of each piecewise-defined function,

$$f(x) = \begin{cases} x, & x < 0 \\ x + 1, & x \geq 0, \end{cases}$$

$$g(x) = \begin{cases} x^2, & x \leq -1 \\ x - 2, & x > -1, \end{cases}$$

is $(-\infty, \infty)$. Discuss how to find $f + g, f - g$, and fg. Carry out your ideas.

55. Discuss how the graph of $y = \frac{1}{2}\{f(x) + |f(x)|\}$ is related to the graph of $y = f(x)$. Illustrate your ideas using $f(x) = x^2 - 6x + 5$.

56. Discuss:
   (a) Is the sum of two even functions f and g even?
   (b) Is the sum of two odd functions f and g odd?
   (c) Is the product of an even function f with an odd function g even, odd, or neither?
   (d) Is the product of an odd function f with an odd function g even, odd, or neither?

57. The product fg of two linear functions with real coefficients, $f(x) = ax + b$ and $g(x) = cx + d$, is a quadratic function. Discuss: Why must the graph of this quadratic function have at least one x-intercept?

58. Make up two different functions f and g so that the domain of $F(x) = f \circ g$ is $[-2, 0) \cup (0, 2]$.

59. Suppose $y = f(x)$ is a function with domain the set X. Discuss: Is the domain of the composition $f \circ f$ also X?

60. The **Heaviside function**

$$U(x - a) = \begin{cases} 0, & x < a \\ 1, & x \geq a, \end{cases}$$

is frequently combined with other functions by either addition of multiplication. Given that $f(x) = x^2$, compare the graphs of $y = f(x - 3)$ and $y = f(x - 3) U(x - 3)$.

61. If U is the Heaviside function defined in Problem 60, then sketch the following functions.
   (a) $y = 2U(x - 1) + U(x - 2)$ **(b)** $y = U\left(x + \frac{1}{2}\right) - U\left(x - \frac{1}{2}\right)$

62. Find an equation for the function f illustrated in FIGURE 2.6.5 in terms of the Heaviside function $U(x - a)$. [*Hint*: Think addition.]

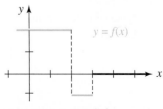

FIGURE 2.6.5 Graph for Problem 62

2.7 ■ Inverse Functions

≡ Introduction Recall that a function f is a rule of correspondence that assigns to each value x in its domain X, a single or unique value y in its range. This rule does not preclude having the same number y associated with several *different* values of x. For example, for $f(x) = x^2 + 1$, the value $y = 5$ occurs at either $x = -2$ or $x = 2$. On the other hand, for the function $g(x) = x^3$, the value $y = 64$ occurs only at $x = 4$. Indeed, for every value y in the range of $g(x) = x^3$, there corresponds only one value of x in the domain. Functions of this last kind are given a special name.

DEFINITION 2.7.1 One-to-One Function

A function f is said to be **one-to-one** if each number in the range of f is associated with exactly one number in its domain X.

☐ Horizontal Line Test Interpreted geometrically, this means that a horizontal line ($y = $ constant) can intersect the graph of a one-to-one function in at most one point. Furthermore, if *every* horizontal line that intersects the graph of a function does so in at most one point, then the function is necessarily one-to-one. A function is *not* one-to-one if *some* horizontal line intersects its graph more than once.

EXAMPLE 1 Horizontal Line Test

The graphs of the functions $f(x) = x^2 + 1$ and $g(x) = x^3$, and a horizontal line $y = c$ intersecting the graphs of f and g, are shown in **FIGURE 2.7.1**. Figure 2.7.1(a) indicates that there are two numbers x_1 and x_2 in the domain of f for which $f(x_1) = f(x_2) = c$. Inspection of Figure 2.7.1(b) shows that for every horizontal line $y = c$ intersecting the graph, there is only one number x_1 in the domain of g such that $g(x_1) = c$. Hence the function f is not one-to-one, whereas the function g is one-to-one. ≡

A one-to-one function can be defined in several different ways. Based on the preceding discussion, the following statement should make sense.

> *A function f is **one-to-one** if and only if $f(x_1) = f(x_2)$ implies $x_1 = x_2$ for all x_1 and x_2 in the domain of f.*   (1)

Stated in a negative way, (1) indicates that a function f is *not* one-to-one if different numbers x_1 and x_2 (that is, $x_1 \neq x_2$) can be found in the domain of f such that $f(x_1) = f(x_2)$. You will see this formulation of the one-to-one concept when we solve certain kinds of equations in Chapter 5.

Consider (1) as a way of determining whether a function f is one-to-one without the benefit of a graph.

EXAMPLE 2 Checking for One-to-One

(a) Consider the function $f(x) = x^4 - 8x + 6$. Now $0 \neq 2$ but observe that $f(0) = f(2) = 6$. Therefore f is not one-to-one.

(b) Consider the function $f(x) = \dfrac{1}{2x - 3}$, and let x_1 and x_2 be numbers in the domain of f. If we assume $f(x_1) = f(x_2)$, that is, $\dfrac{1}{2x_1 - 3} = \dfrac{1}{2x_2 - 3}$, then by taking the

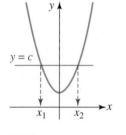

$y = x^2 + 1$

(a) Not one-to-one

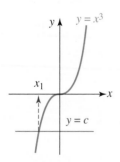

$y = x^3$

$y = c$

(b) One-to-one

FIGURE 2.7.1 Two types of functions in Example 1

100   CHAPTER 2 FUNCTIONS

reciprocal of both sides we see

$$2x_1 - 3 = 2x_2 - 3 \qquad \text{implies} \qquad 2x_1 = 2x_2 \qquad \text{or} \qquad x_1 = x_2.$$

We conclude from (1) that f is one-to-one. ≡

☐ **Inverse of a One-to-One Function** Suppose f is a one-to-one function with domain X and range Y. Since every number y in Y corresponds to precisely one number x in X, the function f must actually determine a "reverse" function f^{-1} whose domain is Y and range is X. As shown in FIGURE 2.7.2, f and f^{-1} must satisfy

$$f(x) = y \qquad \text{and} \qquad f^{-1}(y) = x. \tag{2}$$

The equations in (2) are actually the compositions of the functions f and f^{-1}:

$$f(f^{-1}(y)) = y \qquad \text{and} \qquad f^{-1}(f(x)) = x. \tag{3}$$

The function f^{-1} is called the **inverse** of f or the **inverse function** for f. Following the convention that each domain element be denoted by the symbol x, the first equation in (3) is rewritten as $f(f^{-1}(x)) = x$. We summarize the results in (3).

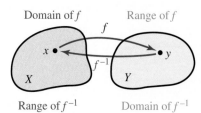

FIGURE 2.7.2 Functions f and f^{-1}

DEFINITION 2.7.2 Inverse Function

Let f be a one-to-one function with domain X and range Y. The **inverse** of f is the function f^{-1} with domain Y and range X for which

$$f(f^{-1}(x)) = x \text{ for every } x \text{ in } Y, \tag{4}$$

and
$$f^{-1}(f(x)) = x \text{ for every } x \text{ in } X. \tag{5}$$

Of course, if a function f is not one-to-one, then it has no inverse function.

EXAMPLE 3 Verifying an Inverse

Verify that the inverse of the one-to-one function $f(x) = \frac{1}{2}x + 7$ is $g(x) = 2x - 14$.

Solution First observe that the domain and range of both functions is the entire set of real numbers $(-\infty, \infty)$. Now we can use (4) and (5).

First from (4) we see that

$$f(g(x)) = f(2x - 14) = \frac{1}{2}(2x - 14) + 7 = x - 7 + 7 = x$$

for every real number x. Similarly, from (5)

$$g(f(x)) = g(\tfrac{1}{2}x + 7) = 2(\tfrac{1}{2}x + 7) - 14 = x + 14 - 14 = x$$

for every real number x. This shows that $g = f^{-1}$ ≡

☐ **Properties** Before we actually examine methods for finding the inverse of a one-to-one function f, let's list some important properties about f and its inverse f^{-1}.

THEOREM 2.7.1 Properties of Inverse Functions

(*i*) The domain of $f^{-1} = $ range of f.
(*ii*) The range of $f^{-1} = $ domain of f.
(*iii*) $y = f(x)$ is equivalent to $x = f^{-1}(y)$.
(*iv*) An inverse function f^{-1} is one-to-one.
(*v*) The inverse of f^{-1} is f; that is, $(f^{-1})^{-1} = f$.
(*vi*) The inverse of f is unique.

☐ **First Method for Finding f^{-1}** We will consider two ways of finding the inverse of a one-to-one function f. Both methods require that you solve an equation; the first method begins with (4).

EXAMPLE 4   **Inverse of a Function**

(a) Find the inverse of $f(x) = \dfrac{1}{2x - 3}$.

(b) Find the domain and range of f^{-1}. Find the range of f.

Solution (a) We proved in part (b) of Example 2 that f is one-to-one. To find the inverse of f using (4), we must substitute $f^{-1}(x)$ wherever x appears in f and then set the expression $f(f^{-1}(x))$ equal to x:

solve this equation for $f^{-1}(x)$
↓

$$f(f^{-1}(x)) = \boxed{\frac{1}{2f^{-1}(x) - 3} = x}$$

By taking the reciprocal of both sides of the equation in the outline box we get,

$$2f^{-1}(x) - 3 = \frac{1}{x}$$

$$2f^{-1}(x) = 3 + \frac{1}{x} = \frac{3x + 1}{x}. \quad \leftarrow \text{ common denominator}$$

Dividing both sides of the last equation by 2 yields the inverse of f:

$$f^{-1}(x) = \frac{3x + 1}{2x}.$$

(b) Inspection of f reveals that its domain is the set of real numbers except $\frac{3}{2}$, that is, $\{x \mid x \neq \frac{3}{2}\}$. Moreover, from the inverse just found we see that the domain of f^{-1} is $\{x \mid x \neq 0\}$. Because range of f^{-1} = domain of f we then know that the range of f^{-1} is $\{y \mid y \neq \frac{3}{2}\}$. From domain of f^{-1} = range of f we have also discovered that the range of f is $\{y \mid y \neq 0\}$.

☐ **Second Method for Finding f^{-1}** The inverse of a function f can be found in a different manner. If f^{-1} is the inverse of f, then $x = f^{-1}(y)$. Thus we need only do the following two things:

- Solve $y = f(x)$ for the symbol x in terms of y (if possible). This gives $x = f^{-1}(y)$.
- Relabel the variable x as y and the variable y as x. This gives $y = f^{-1}(x)$.

EXAMPLE 5   **Inverse of a Function**

Find the inverse of $f(x) = x^3$.

Solution In Example 1 we saw that this function was one-to-one. To begin, we rewrite the function as $y = x^3$. Solving for x then gives $x = y^{1/3}$. Next we relabel variables to obtain $y = x^{1/3}$. Thus $f^{-1}(x) = x^{1/3}$ or equivalently $f^{-1}(x) = \sqrt[3]{x}$.   ≡

Finding the inverse of a one-to-one function $y = f(x)$ is sometimes difficult and at times impossible. For example, it can be shown that the function $f(x) = x^3 + x + 3$ is one-to-one and so has an inverse f^{-1}, but solving the equation $y = x^3 + x + 3$ for x is difficult for everyone (including your instructor). Nevertheless since f only involves positive integer powers of x, its domain is $(-\infty, \infty)$. If you investigate f

graphically you are led to the fact that the range of f is also $(-\infty, \infty)$. Consequently, the domain and range of f^{-1} are $(-\infty, \infty)$. Even though we don't know f^{-1} explicitly, it makes complete sense to talk about values such as $f^{-1}(3)$ and $f^{-1}(5)$. In the case of $f^{-1}(3)$ note that $f(0) = 3$. This means that $f^{-1}(3) = 0$. Can you figure out the value of $f^{-1}(5)$?

□ **Graphs of f and f^{-1}** Suppose that (a, b) represents any point on the graph of a one-to-one function f. Then $f(a) = b$ and

$$f^{-1}(b) = f^{-1}(f(a)) = a$$

implies that (b, a) is a point on the graph of f^{-1}. As shown in FIGURE 2.7.3(a), the points (a, b) and (b, a) are reflections of each other in the line $y = x$. This means that the line $y = x$ is the perpendicular bisector of the line segment from (a, b) to (b, a). Because each point on one graph is the reflection of a corresponding point on the other graph, we see in Figure 2.7.3(b) that the graphs of f^{-1} and f are **reflections** of each other in the line $y = x$. We also say that the graphs of f^{-1} and f are symmetric with respect to the line $y = x$.

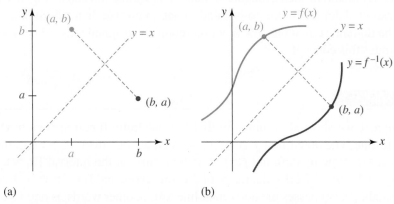

(a)   (b)

FIGURE 2.7.3 Graphs of f and f^{-1} are reflections in the line $y = x$.

EXAMPLE 6   Graphs of f and f^{-1}

In Example 5 we saw that the inverse of $y = x^3$ is $y = x^{1/3}$. In FIGURES 2.7.4(a) and 2.7.4(b) we show the graphs of these functions; in Figure 2.7.4(c) the graphs are superimposed on the same coordinate system to illustrate that the graphs are reflections of each other in the line $y = x$.

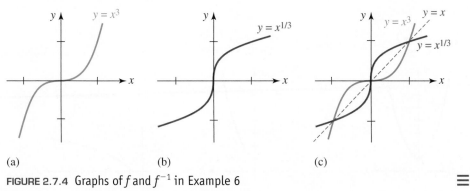

(a)   (b)   (c)

FIGURE 2.7.4 Graphs of f and f^{-1} in Example 6

Every linear function $f(x) = ax + b$, $a \neq 0$, is one-to-one.

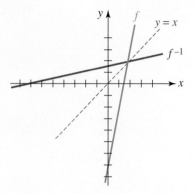

FIGURE 2.7.5 Graphs of f and f^{-1} in Example 7

EXAMPLE 7   Inverse of a Function

Find the inverse of the linear function $f(x) = 5x - 7$.

Solution Since the graph of $y = 5x - 7$ is a nonhorizontal line, it follows from the horizontal line test that f is a one-to-one function. To find f^{-1}, solve $y = 5x - 7$ for x:

$$5x = y + 7 \quad \text{implies} \quad x = \frac{1}{5}y + \frac{7}{5}.$$

Relabeling variables in the last equation gives $y = \frac{1}{5}x + \frac{7}{5}$. Therefore $f^{-1}(x) = \frac{1}{5}x + \frac{7}{5}$. The graphs of f and f^{-1} are compared in **FIGURE 2.7.5**.   ≡

Every quadratic function $f(x) = ax^2 + bx + c, a \neq 0$, is not one-to-one.

☐ **Restricted Domains** For a function f that is not one-to-one, it may be possible to restrict its domain in such a manner so that the new function consisting of f defined on this restricted domain is one-to-one and so has an inverse. In most cases we want to restrict the domain so that the new function retains its original range. The next example illustrates this concept.

EXAMPLE 8   Restricted Domain

In Example 1 we showed graphically that the quadratic function $f(x) = x^2 + 1$ is not one-to-one. The domain of f is $(-\infty, \infty)$, and as seen in **FIGURE 2.7.6(a)**, the range of f is $[1, \infty)$. Now by defining $f(x) = x^2 + 1$ only on the interval $[0, \infty)$, we see two things in Figure 2.7.6(b): the range of f is preserved and $f(x) = x^2 + 1$ confined to the domain $[0, \infty)$ passes the horizontal line test, in other words, is one-to-one. The inverse of this new one-to-one function is obtained in the usual manner. Solving $y = x^2 + 1$ implies

$$x^2 = y - 1 \quad \text{and} \quad x = \pm\sqrt{y - 1} \quad \text{and so} \quad y = \pm\sqrt{x - 1}.$$

The appropriate algebraic sign in the last equation is determined from the fact that the domain and range of f^{-1} are $[1, \infty)$ and $[0, \infty)$, respectively. This forces us to choose $f^{-1}(x) = \sqrt{x - 1}$ as the inverse of f. See Figure 2.7.6(c).

$y = x^2 + 1$
on $(-\infty, \infty)$

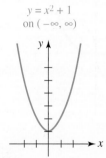

$y = x^2 + 1$
on $[0, \infty)$

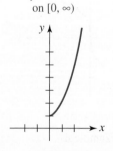

$y = \sqrt{x - 1}$
on $[1, \infty)$

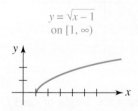

(a) Not a one-to-one function   (b) One-to-one function   (c) Inverse of function in part (b)

FIGURE 2.7.6 Inverse function in Example 8   ≡

In Problems 1–6, the graph of a function f is given. Use the horizontal line test to determine whether f is one-to-one.

1.

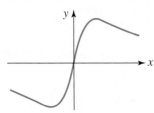

FIGURE 2.7.7 Graph for Problem 1

2.

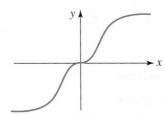

FIGURE 2.7.8 Graph for Problem 2

3.

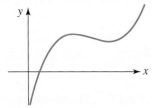

FIGURE 2.7.9 Graph for Problem 3

4.

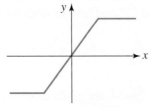

FIGURE 2.7.10 Graph for Problem 4

5.

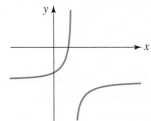

FIGURE 2.7.11 Graph for Problem 5

6.

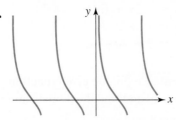

FIGURE 2.7.12 Graph for Problem 6

In Problems 7–10, sketch the graph of the given piecewise-defined function f to determine whether it is one-to-one.

7. $f(x) = \begin{cases} x - 2, & x < 0 \\ \sqrt{x}, & x \geq 0 \end{cases}$

8. $f(x) = \begin{cases} -\sqrt{-x}, & x < 0 \\ \sqrt{x}, & x \geq 0 \end{cases}$

9. $f(x) = \begin{cases} -x - 1, & x < 0 \\ x^2, & x \geq 0 \end{cases}$

10. $f(x) = \begin{cases} x^2 + x, & x < 0 \\ x^2 - x, & x \geq 0 \end{cases}$

In Problems 11–14, proceed as in Example 2(a) to show that the given function f is *not* one-to-one.

11. $f(x) = x^2 - 6x$

12. $f(x) = (x - 2)(x + 1)$

13. $f(x) = \dfrac{x^2}{4x^2 + 1}$

14. $f(x) = |x + 10|$

In Problems 15–18, proceed as in Example 2(b) to show that the given function f is one-to-one.

15. $f(x) = \dfrac{2}{5x + 8}$

16. $f(x) = \dfrac{2x - 5}{x - 1}$

17. $f(x) = \sqrt{4 - x}$

18. $f(x) = \dfrac{1}{x^3 + 1}$

In Problems 19–24, proceed as in Example 3 and verify that the inverse of the one-to-one function f is the function g by showing $f(g(x)) = x$ and $g(f(x)) = x$.

19. $f(x) = x + 5$; $g(x) = x - 5$ **20.** $f(x) = 5x - 10$; $g(x) = \frac{1}{5}x + 2$

21. $f(x) = \dfrac{1}{x^3}$; $g(x) = \dfrac{1}{\sqrt[3]{x}}$ **22.** $f(x) = \sqrt[3]{\frac{1}{3}x + 9}$; $g(x) = 3x^3 - 27$

23. $f(x) = \dfrac{1}{x - 4}$; $g(x) = \dfrac{1}{x} + 4$ **24.** $f(x) = \dfrac{x - 3}{x + 1}$; $g(x) = \dfrac{x + 3}{1 - x}$

In Problems 25 and 26, the given function f is one-to-one. Without finding f^{-1} find its domain and range.

25. $f(x) = 4 + \sqrt{x}$ **26.** $f(x) = 5 - \sqrt{x + 8}$

In Problems 27 and 28, the given function f is one-to-one. The domain and range of f is given. Find f^{-1} and give its domain and range.

27. $f(x) = \dfrac{2}{\sqrt{x}}$, $x > 0, y > 0$ **28.** $f(x) = 2 + \dfrac{3}{\sqrt{x}}$, $x > 0, y > 2$

In Problems 29–34, the given function f is one-to-one. Find f^{-1}. Sketch the graph of f and f^{-1} on the same coordinate axes.

29. $f(x) = -2x + 6$ **30.** $f(x) = -2x + 1$
31. $f(x) = x^3 + 2$ **32.** $f(x) = 1 - x^3$
33. $f(x) = 2 - \sqrt{x}$ **34.** $f(x) = \sqrt{x - 7}$

In Problems 35–38, the given function f is one-to-one. Find f^{-1}. Proceed as in Example 4(b) and find the domain and range of f^{-1}. Then find the range of f.

35. $f(x) = \dfrac{1}{2x - 1}$ **36.** $f(x) = \dfrac{2}{5x + 8}$

37. $f(x) = \dfrac{7x}{2x - 3}$ **38.** $f(x) = \dfrac{1 - x}{x - 2}$

In Problems 39–42, the given function f is one-to-one. Without finding f^{-1}, find the point on the graph of f^{-1} corresponding to the indicated value of x in the domain of f.

39. $f(x) = 2x^3 + 2x$; $x = 2$ **40.** $f(x) = 8x - 3$; $x = 5$

41. $f(x) = x + \sqrt{x}$; $x = 9$ **42.** $f(x) = \dfrac{4x}{x + 1}$; $x = \frac{1}{2}$

In Problems 43 and 44, sketch the graph of f^{-1} from the graph of f.

43. 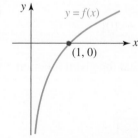 **44.**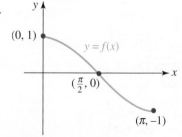

FIGURE 2.7.13 Graph for Problem 43   FIGURE 2.7.14 Graph for Problem 44

CHAPTER 2 FUNCTIONS

In Problems 45 and 46, sketch the graph of f from the graph of f^{-1}.

45.

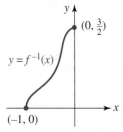

FIGURE 2.7.15 Graph for Problem 45

46.

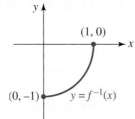

FIGURE 2.7.16 Graph for Problem 46

In Problems 47–50, the function f is not one-to-one on the given domain but is one-to-one on the restricted domain (the second interval). Find the inverse of the one-to-one function and give its domain. Sketch the graph of f on the restricted domain and the graph of f^{-1} on the same coordinate axes.

47. $f(x) = 4x^2 + 2$, $(-\infty, \infty)$; $[0, \infty)$ **48.** $f(x) = (3 - 2x)^2$, $(-\infty, \infty)$; $[\frac{3}{2}, \infty)$
49. $f(x) = \frac{1}{2}\sqrt{4 - x^2}$, $[-2, 2]$; $[0, 2]$ **50.** $f(x) = \sqrt{1 - x^2}$, $[-1, 1]$; $[0, 1]$

51. If the functions f and g have inverses, then it can be proved that

$$(f \circ g)^{-1} = g^{-1} \circ f^{-1}.$$

Verify this property for the one-to-one functions $f(x) = x^3$ and $g(x) = 4x + 5$.
52. It can be shown that the equation $y = \sqrt[3]{x} - \sqrt[3]{y}$ defines a one-to-one function $y = f(x)$. Without finding f, find f^{-1}.

Discussion Problems

53. Suppose f is a continuous function that is increasing (or decreasing) for all x in its domain. Explain why f is necessarily one-to-one.
54. Explain why the graph of a one-to-one function f can have at most one x-intercept.
55. The function $f(x) = |2x - 4|$ is not one-to-one. How should the domain of f be restricted so that the new function has an inverse? Find f^{-1} and give its domain and range. Sketch the graph of f on the restricted domain and the graph of f^{-1} on the same coordinate axes.
56. What property do the one-to-one functions $y = f(x)$ shown in FIGURES 2.7.17(a) and 2.7.17(b) have in common? Find two more explicit functions with this same property. Be very explicit about what this property has to do with f^{-1}.
57. The piecewise-defined function

$$f(x) = \begin{cases} x + 1, & -4 \le x < 0 \\ \frac{1}{4}(x + 4), & 0 \le x \le 4, \end{cases}$$

shown in FIGURE 2.7.18 is one-to-one. Sketch the graph of f^{-1}.

58. Find f^{-1} for the piecewise-defined function f in Problem 57. Give the domain and range of f^{-1}.
59. Suppose a function f is one-to-one. For c a constant, we know that the graph of the function $g(x) = f(x) + c$ is the graph of f shifted vertically and so g is also a one-to-one function. What is the inverse of the function g? Describe the graph of g^{-1} in terms of the graph of f^{-1}.
60. Suppose that f is one-to-one and that the domain of f and f^{-1} is $(-\infty, \infty)$. If $f^{-1}(5) = 10$, find x such that $3 + f(2x - 4) = 8$.

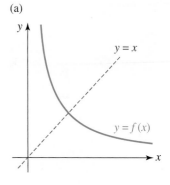

(a)

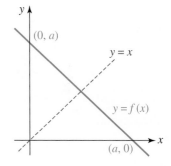

(b)

FIGURE 2.7.17 Graphs for Problem 56

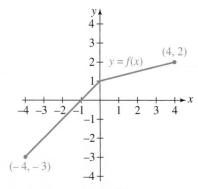

FIGURE 2.7.18 Graphs for Problem 57

2.8 Building a Function from Words

≡ **Introduction** In calculus there will be several instances when you will be expected to translate the words that describe a problem into mathematical symbols and then set up or construct either an *equation* or a *function*.

In this section we focus on problems that involve functions. We begin with a verbal description about the product of two numbers.

EXAMPLE 1   Product of Two Numbers

The sum of two nonnegative numbers is 15. Express the product of one and the square of the other as a function of one of the numbers.

Solution We first represent the two numbers by the symbols x and y and recall that "nonnegative" means that $x \geq 0$ and $y \geq 0$. The first sentence then says that $x + y = 15$; this is *not* the function we are seeking. The second sentence describes the function we want; it is called "the product." Let's denote "the product" by the symbol P. Now P is the product of one of the numbers, say, x and the square of the other, that is, y^2:

$$P = xy^2. \tag{1}$$

No, we are not finished because P is supposed to be a "function of *one* of the numbers." We now use the fact that the numbers x and y are related by $x + y = 15$. From this last equation we substitute $y = 15 - x$ into (1) to obtain the desired result:

$$P(x) = x(15 - x)^2. \tag{2} \quad \equiv$$

Here is a symbolic summary of the analysis of the problem given in Example 1:

$$x + y = 15$$

let the numbers be $x \geq 0$ and $y \geq 0$ P

The sum of two nonnegative numbers is 15. Express the product of $\qquad$ (3)

x y^2 use x

one and the square of the other as a function of one of the numbers.

Notice that the second sentence is vague about which number is squared. This means that it really doesn't matter; (1) could also be written as $P = yx^2$. Also, we could have used $x = 15 - y$ in (1) to arrive at $P(y) = (15 - y)y^2$. In a calculus setting it would not have mattered whether we worked with $P(x)$ or with $P(y)$ because by finding *one* of the numbers we automatically find the other from the equation $x + y = 15$. This last equation is commonly called a **constraint**. A constraint not only defines the relationship between the variables x and y but often puts a limitation on how x and y can vary. As we see in the next example, the constraint helps in determining the domain of the function that you have just constructed.

EXAMPLE 2   Example 1 Continued

What is the domain of the function $P(x)$ in (2)?

Solution Taken out of the context of the statement of the problem in Example 1, one would have to conclude from the discussion on page 50 of Section 2.1 that the domain of the function

$$P(x) = x(15 - x)^2 = 225x - 30x^2 + x^3$$

is the set of real numbers $(-\infty, \infty)$. *But* in the context of the original problem, the numbers were to be nonnegative. From the requirement that $x \geq 0$ *and* $y = 15 - x \geq 0$ we get $x \geq 0$ and $x \leq 15$, which means that x must satisfy the simultaneous inequality $0 \leq x \leq 15$. Using interval notation, the domain of the product function P in (2) is the closed interval $[0, 15]$. ≡

Another way of looking at the conclusion of Example 2 is this: The constraint $x + y = 15$ dictates that $y = 15 - x$. Thus *if* x were allowed to be larger than 15 (say, $x = 17.5$), then $y = 15 - x$ would be a negative number, which contradicts the initial assumption that $y \geq 0$.

☐ **Optimization Problems** For the remainder of this section we are going to examine "word problems" taken directly from a calculus text. These problems, variously called "optimization problems" or "applied maximum and minimum problems," consist of two parts, the "precalculus part" where you set up the function to be optimized and the "calculus part" where you perform calculus-specific operations on the function that you have just found to find its maximum or minimum value. The calculus part is usually identifiable by words such as "maximum (or minimum)," "least," "greatest," "large as possible," "find the dimensions," and so on. For example, the actual statement of Example 1 as it appears in a calculus text is:

> *Find two nonnegative numbers whose sum is* 15 *such that the product of one and the square of the other is a maximum.*

The big hurdle for many students is separating out the words that define the function to be optimized from all the words contained in the statement of the problem.

Before proceeding with the examples, you are encouraged to read the *Notes from the Classroom* at the end of this section.

The next example describes a geometric problem that asks for a "largest rectangle." Remember, you are not expected to work the entire problem by trying to actually find the "largest rectangle," which you would do in a calculus course. Right now your only job is to pick out the words, as we illustrated in (3), that tell you what the function is and then construct it using the variables introduced. In calculus, the function to be optimized is called the **objective function**.

◀ Please note.

EXAMPLE 3   Largest Rectangle

Find the objective function in the following calculus problem:

> *A rectangle has two vertices <u>on the</u> x-axis and two vertices on the semicircle whose equation is* $y = \sqrt{25 - x^2}$. *See* FIGURE 2.8.1(a). *Find the dimensions of the largest rectangle.*

Solution In calculus the words "largest rectangle" mean that we are seeking *the* rectangle, of the many that can be drawn in the semicircle, that has the greatest or maximum *area*. Hence, the function we must construct is the area A of the rectangle. If (x, y), $x > 0$, $y > 0$, denotes the vertex of the rectangle on the circle in the first quadrant, then as shown in Figure 2.8.1(b) the area A is length × width, or

$$A = (2x) \times y = 2xy. \qquad (4)$$

The constraint in this problem is the equation $y = \sqrt{25 - x^2}$ of the semicircle. We use the constraint equation to eliminate y in (4) and obtain the area of the rectangle or the objective function,

$$A(x) = 2x\sqrt{25 - x^2}. \qquad (5)$$

This ends the "precalculus part" of the problem.

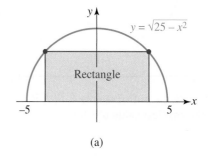

(a)

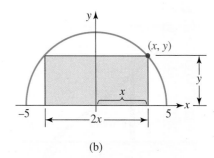

(b)

FIGURE 2.8.1 Rectangle in Example 3

The next step would be calculus procedures to determine the value of x for which the objective function $A(x)$ takes on its largest value. ☰

Were we again to consider the function $A(x)$ out of the context of the problem in Example 3, its domain would be $[-5, 5]$. Because we assumed that $x > 0$ the domain of $A(x)$ in (4) is actually the open interval $(0, 5)$. But in calculus we would use the closed interval $[0, 5]$ even though when $x = 0$ and $x = 5$ the area would be $A(0) = 0$ and $A(5) = 0$, respectively. Do not worry about this last technicality.

■ EXAMPLE 4   Least Amount of Fencing

Find the objective function and its domain in the following calculus problem:

A rancher intends to mark off a rectangular plot of land that will have an area of 1000 m^2. The plot will be fenced and divided into two equal portions by an additional fence parallel to two sides. Find the dimensions of the land that require the least amount of fence.

Solution Your drawing should be a rectangle with a line drawn down its middle, similar to that given in FIGURE 2.8.2. As shown in the figure, let $x > 0$ be the length of the rectangular plot of land and let $y > 0$ denote its width. The function we seek is the "amount of fence." If the symbol F represents this amount, then the sum of the lengths of the *five* portions—two horizontal and three vertical—of the fence is

$$F = 2x + 3y. \tag{6}$$

But the fenced-in land is to have an area of 1000 m^2, and so x and y must be related by the constraint $xy = 1000$. From the last equation we get $y = 1000/x$ which can be used to eliminate y in (6). Thus, the amount of fence F as a function of x is $F(x) = 2x + 3(1000/x)$ or

$$F(x) = 2x + \frac{3000}{x}. \tag{7}$$

Since x represents a physical dimension that satisfies $xy = 1000$, we conclude that it is positive. But other than that, there is no restriction on x. Thus, unlike the previous example, the objective function (7) is not defined on a closed interval. The domain of $F(x)$ is $(0, \infty)$. ☰

As can be seen from the graph of (7) given in FIGURE 2.8.3, F has a minimum at some value of x, say $x = c$. With a graphing calculator or computer we can approximate c and $F(c)$, but with calculus we can find their exact values.

If a problem involves triangles, you should study the problem carefully and determine whether the Pythagorean theorem, similar triangles, or trigonometry is applicable.

■ EXAMPLE 5   Shortest Ladder

Find the objective function and its domain in the following calculus problem:

A 10-ft wall stands 5 ft from a building. Find the length of the shortest ladder, supported by the wall, that reaches from the ground to the building.

Solution The words "shortest ladder" indicate that we want a function that describes the length of the ladder. Let L denote this length. With x and y defined in FIGURE 2.8.4, we see that there are two right triangles, the larger triangle has three sides with

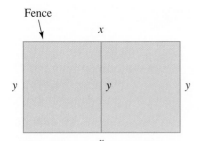

Fence

FIGURE 2.8.2 Rectangular plot of land in Example 4

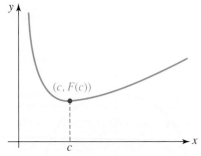

FIGURE 2.8.3 In Example 4, $F(c)$ is the smallest value of F for $x > 0$

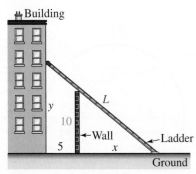

FIGURE 2.8.4 Ladder in Example 5

CHAPTER 2 FUNCTIONS

lengths L, y, and $x + 5$, and the smaller triangle has two sides of lengths x and 10. Now the ladder is the hypotenuse of the larger right triangle, so by the Pythagorean theorem,

$$L^2 = (x + 5)^2 + y^2. \tag{8}$$

The right triangles in Figure 2.8.4 are similar because they both contain a right angle and share the common acute angle the ladder makes with the ground. We then use the fact that the ratios of corresponding sides of similar triangles are equal. This enables us to write

$$\frac{y}{x + 5} = \frac{10}{x} \quad \text{so that} \quad y = \frac{10(x + 5)}{x}.$$

Using the last result, (8) becomes

$$L^2 = (x + 5)^2 + \left(\frac{10(x + 5)}{x}\right)^2$$

$$= (x + 5)^2\left(1 + \frac{100}{x^2}\right) \quad \leftarrow \text{factoring } (x + 5)^2$$

$$= (x + 5)^2\left(\frac{x^2 + 100}{x^2}\right) \quad \leftarrow \text{common denominator}$$

Taking the square root gives us L as a function of x,

$$L(x) = \frac{x + 5}{x}\sqrt{x^2 + 100}. \quad \leftarrow \begin{cases} \text{square root of a product} \\ \text{is the product of the square roots} \end{cases}$$

The domain of the objective function $L(x)$ is $(0, \infty)$. ≡

▮ EXAMPLE 6 ▮   Closest Point

Find the objective function and its domain in the following calculus problem:

> *Find the point in the first quadrant on the circle $x^2 + y^2 = 1$ that is closest to the point $(2, 4)$.*

Solution Let (x, y) denote the point in the first quadrant on the circle closest to $(2, 4)$ and let d represent the distance from (x, y) to $(2, 4)$. See **FIGURE 2.8.5**. Then from the distance formula, (2) of Section 1.3,

$$d = \sqrt{(x - 2)^2 + (y - 4)^2} = \sqrt{x^2 + y^2 - 4x - 8y + 20}. \tag{9}$$

The constraint in this problem is the equation of the circle $x^2 + y^2 = 1$. From this we can immediately replace $x^2 + y^2$ in (9) by the number 1. Moreover, using the constraint to write $y = \sqrt{1 - x^2}$ allows us to eliminate y in (9). Thus the distance d as a function of x is:

$$d(x) = \sqrt{21 - 4x - 8\sqrt{1 - x^2}}. \tag{10}$$

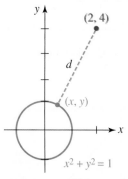

FIGURE 2.8.5 Distance d in Example 6

Since (x, y) is a point on the circle in the first quadrant it suffices to let the variable x range from 0 to 1, that is, the domain of the objective function in (10) is the closed interval $[0, 1]$. ≡

When we get to the sections in a calculus text devoted to word problems students often react with groans, ambivalence, and dismay. While not guaranteeing anything, the following suggestions might help you to get through the problems in Exercises 2.8.

- At least try to develop a positive attitude. Try to be neat and organized.
- Read the problem slowly. Then read the problem several more times.
- Pay attention to words such as "maximum," "least," "greatest," and "closest" because they may provide a clue about the nature of the function you are seeking. For example, if a problem asks for "closest," then the function you are seeking most probably involves *distance*; if a problem asks for "least material," then the function you want may be *surface area*. See Problems 35 and 42 in Exercises 2.8.
- Whenever possible, sketch a curve or a picture and identify given quantities in your sketch. Keep your sketch simple.
- Introduce variables and note any constraint or relationship between the variables (such as $x + y = 15$ in Example 1).
- Identify the domain of the function just constructed. Keep in mind that if the problem mentions "dimensions" then the variables representing those quantities must be nonnegative.

| 2.8 | Exercises | Answers to selected odd-numbered problems begin on page ANS-8. |

In Problems 1–26, proceed as in Example 1 and translate the words into an appropriate function. Give the domain of the function.

1. The product of two positive numbers is 50. Express their sum as a function of one of the numbers.
2. Express the sum of a nonzero number and its reciprocal as a function of the number.
3. The sum of two nonnegative numbers is 1. Express the sum of the square of one and twice the square of the other as a function of one of the numbers.
4. Let m and n be positive integers. The sum of two nonnegative numbers is S. Express the product of the mth power of one and the nth power of the other as a function of one of the numbers.
5. A rectangle has a perimeter of 200 in. Express the area of the rectangle as a function of the length of one of its sides.
6. A rectangle has an area of 400 in^2. Express the perimeter of the rectangle as a function of the length of one of its sides.
7. Express the area of the rectangle shaded in FIGURE 2.8.6 as a function of x.
8. Express the length of the line segment containing the point $(2, 4)$ shown in FIGURE 2.8.7 as a function of x.
9. Express the distance from a point (x, y) on the graph of $x + y = 1$ to the point $(2, 3)$ as a function of x.

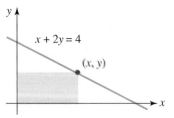

FIGURE 2.8.6 Rectangle in Problem 7

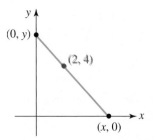

FIGURE 2.8.7 Line segment in Problem 8

10. Express the distance from a point (x, y) on the graph of $y = 4 - x^2$ to the point $(0, 1)$ as a function of x.

11. Express the perimeter of a square as a function of its area A.

12. Express the area of a circle as a function of its diameter d.

13. Express the diameter of a circle as a function of its circumference C.

14. Express the volume of a cube as a function of the area A of its base.

15. Express the area of an equilateral triangle as a function of its height h.

16. Express the area of an equilateral triangle as a function of the length s of one of its sides.

17. A wire of length x is bent into the shape of a circle. Express the area of the circle as a function of x.

18. A wire of length L is cut x units from one end. One piece of the wire is bent into a square and the other piece is bent into a circle. Express the sum of the areas as a function of x.

19. A tree is planted 30 ft from the base of a street lamp that is 25 ft tall. Express the length of the tree's shadow as a function of its height.

20. The frame of a kite consists of six pieces of lightweight plastic. The outer frame of the kite consists of four precut pieces, two pieces of length 2 ft, and two pieces of length 3 ft. Express the area of the kite as a function of x, where $2x$ is the length of the horizontal cross bar piece shown in FIGURE 2.8.8.

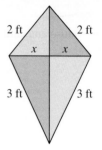

FIGURE 2.8.8 Kite in Problem 20

21. A company wants to construct an open rectangular box with a volume of 450 in³ so that the length of its base is 3 times its width. Express the surface area of the box as a function of the width.

22. A conical tank, with vertex down, has a radius of 5 ft and a height of 15 ft. Water is pumped into the tank. Express the volume of the water as a function of its depth. [*Hint*: The volume of a cone is $V = \frac{1}{3}\pi r^2 h$. Although the tank is a three-dimensional object, examine it in cross section as a two-dimensional triangle.]

23. Car A passes point O heading east at a constant rate of 40 mi/h; car B passes the same point 1 hour later heading north at a constant rate of 60 mi/h. Express the distance between the cars as a function of time t, where t is measured starting when car B passes point O.

24. At time $t = 0$ (measured in hours), two airliners with a vertical separation of 1 mile pass each other going in opposite directions. If the planes are flying horizontally at rates of 500 mi/h and 550 mi/h:
(a) Express the horizontal distance between them as a function of t.
    [*Hint*: distance = rate × time.]
(b) Express the diagonal distance between them as a function of t.

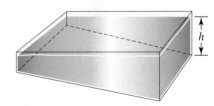

FIGURE 2.8.9 Swimming pool in Problem 25

25. The swimming pool shown in FIGURE 2.8.9 is 3 ft deep at the shallow end, 8 ft deep at the deepest end, 40 ft long, 30 ft wide, and the bottom is an inclined plane. Water is pumped into the pool. Express the volume of the water in the pool as a function of height h of the water above the deep end. [*Hint*: The volume will be a piecewise-defined function with domain defined by $0 \le h \le 8$.]

26. USPS regulations for parcel post stipulate that the length plus girth (the perimeter of one end) of a package must not exceed 108 inches. Express the volume of the package as a function of the width x shown in FIGURE 2.8.10.

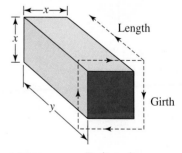

FIGURE 2.8.10 Package in Problem 26

In Problems 27–48, proceed as in Examples 3–5 and find the objective function for the given calculus problem. Give the domain of the objective function, but *do not* attempt to solve the problem.

27. Find a number that exceeds its square by the greatest amount.

28. Of all rectangles with perimeter 20 inches, find the one with the shortest diagonal.

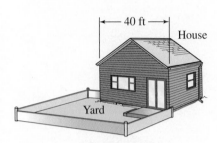

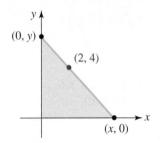

FIGURE 2.8.11 House and yard in Problem 32

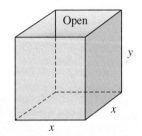

FIGURE 2.8.12 Line segment in Problem 34

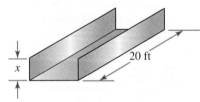

FIGURE 2.8.13 Box in Problem 35

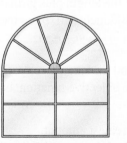

FIGURE 2.8.14 Rain gutter in Problem 37

29. A rectangular plot of land will be fenced into three equal portions by two dividing fences parallel to two sides. If the area to be enclosed is 4000 m², find the dimensions of the land that require the least amount of fence.

30. A rectangular plot of land will be fenced into three equal portions by two dividing fences parallel to two sides. If the total fence to be used is 8000 m, find the dimensions of the land that has the greatest area.

31. A rancher wishes to build a rectangular corral with an area of 128,000 ft² with one side along a straight river. The fencing along the river costs $1.50 per foot, whereas along the other three sides the fencing costs $2.50 per foot. Find the dimensions of the corral so that the cost of construction is a minimum. [*Hint*: Along the river the cost of x ft of fence is $1.50x$.]

32. A rectangular yard is to be enclosed with a fence by attaching it to a house whose length is 40 feet. See FIGURE 2.8.11. The amount of fencing to be used is 160 feet. Find the dimensions of the yard so that the greatest area is enclosed.

33. Consider all rectangles that have the same perimeter p. (Here p represents a constant.) Of these rectangles, show that the one with the largest area is a square.

34. Find the vertices $(x, 0)$ and $(0, y)$ of the shaded triangular region in FIGURE 2.8.12 so that its area is a minimum.

35. (a) An open rectangular box is to be constructed with a square base and a volume of 32,000 cm³. Find the dimensions of the box that require the least amount of material. See FIGURE 2.8.13.

   (b) If the rectangular box in part (a) is closed, find the dimensions that require the least amount of material.

36. A closed rectangular box is to be constructed with a square base. The material for the top costs $2 per square foot whereas the material for the remaining sides costs $1 per square foot. If the total cost to construct each box is $36, find the dimensions of the box of greatest volume that can be made.

37. A rain gutter with a rectangular cross section is made from a 1 ft × 20 ft piece of metal by bending up equal amounts from the 1-ft side. See FIGURE 2.8.14. How should the metal be bent up on each side in order to make the capacity of the gutter a maximum? [*Hint*: Capacity = volume.]

38. A Norman window consists of a rectangle surmounted by a semicircle as shown in FIGURE 2.8.15. If the total perimeter of the window is 10 m, find the dimensions of the window with the largest area.

39. A printed page will have 2-in. margins of white space on the sides and 1-in. margins of white space on the top and bottom. The area of the printed portion is 32 in². Determine the dimensions of the page so that the least amount of paper is used.

40. Find the dimensions of the right circular cylinder with greatest volume that can be inscribed in a right circular cone of radius 8 in. and height 12 in. See FIGURE 2.8.16.

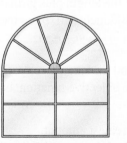

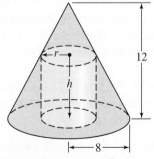

FIGURE 2.8.15 Norman window in Problem 38

FIGURE 2.8.16 Inscribed cylinder in Problem 40

41. Find the maximum length L of a thin board that can be carried horizontally around the right-angle corner shown in FIGURE 2.8.17. [*Hint*: Use similar triangles.]

42. A juice can is to be made in the form of a right circular cylinder and have a volume of 32 in.^{3}. See FIGURE 2.8.18. Find the dimensions of the can so that the least amount of material is used in its construction. [*Hint*: Material = total surface area of can = area of top + area of bottom + area of lateral side. If the circular top and bottom covers are removed and the cylinder is cut straight up its side and flattened out, the result is the rectangle shown in Figure 2.8.18(c).]

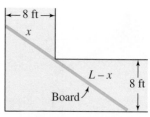

FIGURE 2.8.17 Board in Problem 41

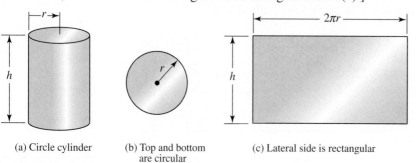

(a) Circle cylinder

(b) Top and bottom are circular

(c) Lateral side is rectangular

FIGURE 2.8.18 Juice can in Problem 42

43. The lateral side of a cylinder is to be made from a rectangle of flimsy sheet plastic. Because the plastic material cannot support itself, a thin, stiff wire is embedded in the material as shown in FIGURE 2.8.19(a). Find the dimensions of the cylinder of largest volume that can be constructed if the wire has a fixed length L. [*Hint*: There are two constraints in this problem. In Figure 2.8.19(b), the circumference of a circular end of the cylinder is y.]

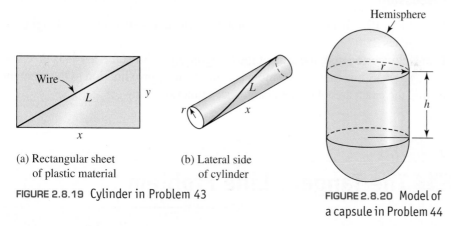

(a) Rectangular sheet of plastic material

(b) Lateral side of cylinder

FIGURE 2.8.19 Cylinder in Problem 43

Hemisphere

FIGURE 2.8.20 Model of a capsule in Problem 44

44. Many medications are packaged in capsules as shown in the accompanying photo. Assume that a capsule is formed by adjoining two hemispheres to the ends of a right circular cylinder as shown in FIGURE 2.8.20. If the total volume of the capsule is to be 0.007 in.^{3}, find the dimensions of the capsule so that the least amount of material is used in its construction. [*Hint*: The volume of a sphere is $\frac{4}{3}\pi r^3$ and its surface area is $4\pi r^2$.]

Capsule

45. A 20-ft long water trough has ends in the form of isosceles triangles with sides that are 4 ft long. See FIGURE 2.8.21. Determine the dimension across the top of the triangular end so that the volume of the trough is a maximum. [*Hint*: A *right cylinder* is not necessarily a *circular cylinder* where the top and bottom are circles. The top and bottom of a right cylinder are the same but could be a triangle, a pentagon, a trapezoid, and so on. The volume of a right cylinder is the area of the base $\times$ the height.]

46. Some birds fly more slowly over water than over land. A bird flies at constant rates 6 km/h over water and 10 km/h over land. Use the information in

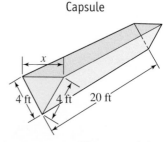

FIGURE 2.8.21 Water trough in Problem 45

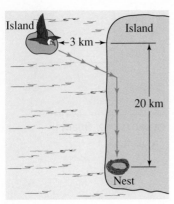

FIGURE 2.8.22 The bird in Problem 46

FIGURE 2.8.22 to find the path the bird should take to minimize the total flying time between the shore of one island and its nest on the shore of another island. [*Hint*: distance = rate × time.]

47. In a race a woman is required to swim from a floating dock *A* to the beach and, without stopping, swim from the beach out to another floating dock *C*. The distances are shown in FIGURE 2.8.23. She estimates that she can swim from dock *A* to the beach at a constant rate of 3 mi/h and out from the beach to dock *C* at a rate of 2 mi/h. Where should she touch the beach in order to minimize the total swimming time from *A* to *C*?

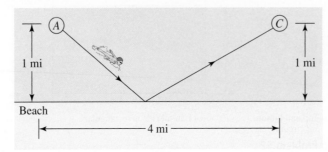

FIGURE 2.8.23 Swimmer in Problem 47

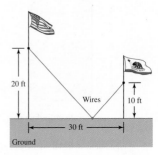

FIGURE 2.8.24 Flag poles in Problem 48

48. Two flag poles are secured by wires that are attached at a single point between the poles. See FIGURE 2.8.24. Where should the point be located to minimize the total length of wire used?

For Discussion

49. In Problem 19, what happens to the length of the tree's shadow as its height approaches 25 ft?

50. In an engineering text, the area of the octagon shown in FIGURE 2.8.25 is given as $A = 3.31r^2$. Show that this formula is actually an approximation to the area; that is, find the exact area *A* of the octagon as a function of *r*.

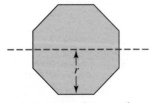

FIGURE 2.8.25 Octagon in Problem 50

2.9 The Tangent Line Problem

∫ **Calculus PREVIEW** ≡ **Introduction** In a calculus course you will study many different things, but roughly, the subject "calculus" is divided into two broad but related areas known as **differential calculus** and **integral calculus**. The discussion of each of these topics invariably begins with a motivating problem involving the graph of a function. Differential calculus is motivated by the problem

Find a tangent line to the graph of a function f,

whereas integral calculus is motivated by the problem

Find the area under the graph of a function f.

The first problem will be addressed in this section; the second problem will be discussed in Section 3.7.

☐ **Tangent Line to a Graph** The word *tangent* stems from the Latin verb *tangere*, meaning "to touch." You might remember from the study of plane geometry that a tangent to a circle is a line L that intersects, or touches, the circle in exactly one point P. See FIGURE 2.9.1. It is not quite as easy to define a tangent line to the graph of a function f. The idea of *touching* carries over to the notion of a tangent line to the graph of a function, but the idea of *intersecting the graph in one point* does not carry over.*

☐ **Using Secant Lines** Suppose $y = f(x)$ is a continuous function. If, as shown in FIGURE 2.9.2, f possesses a line L tangent to its graph at a point P, then what is the equation of this line? To answer this question, we need the coordinates of P and the slope $m_{\tan}$ of L. The coordinates of P pose no difficulty, since a point on the graph of a function f is obtained by specifying a value of x in the domain of f. The coordinates of the point of tangency at $x = a$ are then $(a, f(a))$. As a means of approximating the slope $m_{\tan}$, we can readily find the slopes $m_{\sec}$ of *secant lines* that pass through the point P and any other point Q on the graph. See FIGURE 2.9.3.

☐ **Definition of a Tangent Line** If P has coordinates $(a, f(a))$ and if Q has coordinates $(a + h, f(a + h))$, then as shown in FIGURE 2.9.4, the slope of the secant line through P and Q is

$$m_{\sec} = \frac{\text{rise}}{\text{run}} = \frac{f(a + h) - f(a)}{(a + h) - a}$$

or

$$m_{\sec} = \frac{f(a + h) - f(a)}{h}. \tag{1}$$

The expression on the right-hand side of the equality in (1) is called a **difference quotient**. When we let h take on values that are closer and closer to zero, that is, as $h \to 0$, the sequence of points $Q(a + h, f(a + h))$ move along the curve closer and closer to the point $P(a, f(a))$. Intuitively, we expect the secant lines to approach the tangent line L, and that $m_{\sec} \to m_{\tan}$ as $h \to 0$. Using the idea of a limit introduced in Section 1.5 we write $m_{\tan} = \lim_{h \to 0} m_{\sec}$. We summarize this discussion using the difference quotient (1).

DEFINITION 2.9.1 Tangent Line with Slope

Let $y = f(x)$ be continuous at the number a. If the limit,

$$m_{\tan} = \lim_{h \to 0} \frac{f(a + h) - f(a)}{h}, \tag{2}$$

exists, then the **tangent line** to the graph of f at $(a, f(a))$ is that line passing through the point $(a, f(a))$ with slope $m_{\tan}$.

Just like the problems discussed in Section 1.5, observe that the limit in (2) has the indeterminate form $0/0$ as $h \to 0$.

We are not going to delve into any theoretical details about when the limit (2) exists or does not exist—that discussion properly belongs in a calculus course. So to simplify the discussion, we will drop the phase "provided the limit exists." For this course it suffices simply to be aware of the fact that the limit (2) may not exist for certain values of a See Problem 43 in Exercises 2.9.

It is very likely that early on in your calculus course you will be asked to compute the limit of a difference quotient such as (2). The computation of (2) is essentially a *four-step*

---

* We leave the discussion of the many subtleties and questions surrounding the tangent line problem to a course in calculus.

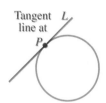

FIGURE 2.9.1 Tangent line L touches a circle at point P

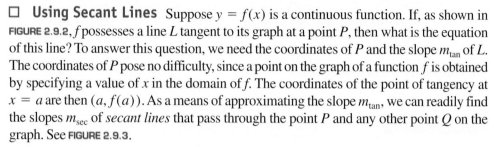

FIGURE 2.9.2 Tangent line L to a graph at point P

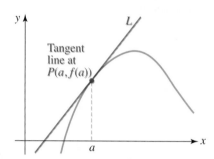

FIGURE 2.9.3 Slopes of secant lines approximate the slope $m_{\tan}$ of L

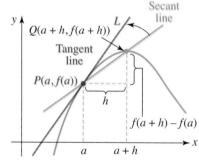

FIGURE 2.9.4 Secant lines swing into the tangent line L as $h \to 0$

process, and three of these steps involve only precalculus mathematics: algebra and trigonometry. Getting over the hurdles of algebraic or trigonometric manipulations in these first three steps is your primary goal. If done accurately, the fourth step, or the calculus step, may be the easiest part of the problem. In preparation for calculus we recommend that you be able to carry out the calculation of (2) for functions involving

review $(a + b)^n$ for $n = 2$ and 3 ▶

review adding symbolic fractions ▶

review rationalization of numerators and ▶ denominators

- positive integer powers of x such as x^n for $n = 1, 2$, and 3,
- division of functions such as $\dfrac{1}{x}$ and $\dfrac{x}{x + 1}$, and
- radicals such as $\sqrt{x}$.

See Problems 1–10 in Exercises 2.9.

▮ EXAMPLE 1   The Four-Step Process

Find the slope of the tangent line to the graph of $y = x^2 + 2$ at $x = 1$.

Solution We first compute the difference quotient in (2) with the identification that $a = 1$.

(*i*) The initial step is the computation of $f(a + h)$. Because functions can be complicated, it might help in this step to think of x wherever it appears in the function $f(x)$ as a set of parentheses ( ). For the given function we write $f(\ ) = (\ )^2 + 2$. The idea is to substitute $1 + h$ into those parentheses and carry out the required algebra:

$$\begin{aligned} f(1 + h) &= (1 + h)^2 + 2 \\ &= (1 + 2h + h^2) + 2 \\ &= 3 + 2h + h^2. \end{aligned}$$

(*ii*) The computation of the difference $f(a + h) - f(a)$ is the most important step. It is imperative that you simplify this step as much as possible. Here is a tip: In many of the problems that you will be required to do in calculus you will be able to factor h from the difference $f(a + h) - f(a)$. To begin, compute $f(a)$, which in this case is $f(1) = 1^2 + 2 = 3$. Next, you can use the result from the preceding step:

$$\begin{aligned} f(1 + h) - f(1) &= 3 + 2h + h^2 - 3 \\ &= 2h + h^2 \\ &= h(2 + h). \qquad \text{← notice the factor of } h \end{aligned}$$

(*iii*) The computation of the difference quotient $\dfrac{f(a + h) - f(a)}{h}$ is now straightforward. Again, we use the results from the preceding step:

$$\frac{f(1 + h) - f(1)}{h} = \frac{h(2 + h)}{h} = 2 + h. \qquad \text{← cancel the } h\text{'s}$$

(*iv*) The calculus step is now easy. From (2) we have

$$m_{\tan} = \lim_{h \to 0} \overset{\displaystyle \text{from the result in (\textit{iii})}}{\frac{f(1 + h) - f(1)}{h}} = \lim_{h \to 0} (2 + h) = 2.$$

The slope of the tangent line to the graph of $y = x^2 + 2$ at $(1, 3)$ is 2.   ≡

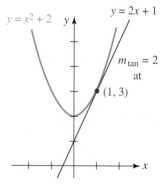

FIGURE 2.9.5 Tangent line in Example 2

| EXAMPLE 2 | **Equation of Tangent Line** |

Find an equation of the tangent line whose slope was found in Example 1.

Solution We know a point $(1, 3)$ and a slope $m_{tan} = 2$, and so from the point-slope equation of a line we find

$$y - 3 = 2(x - 1) \quad \text{or} \quad y = 2x + 1.$$

Observe that the last equation is consistent with the x- and y-intercepts of the red line in FIGURE 2.9.5. ☰

☐ **The Derivative** As you inspect Figure 2.9.5, imagine tangent lines at various points on the graph of $f(x) = x^2 + 2$. This particular function is known to have a tangent line at every point on its graph. The tangent lines to the left of the origin have negative slope, the tangent line at $(0, 2)$ has zero slope, and the tangent lines to the right of the origin have positive slope (as seen in Example 1). In other words, for a function f the value of m_{tan} at a point $(a, f(a))$ depends on the choice of the number a. Roughly speaking, there is at most *one* value of m_{tan} for each number a in the domain of a function f. More specifically, m_{tan} is itself a *function* with a domain that is a subset of the domain of the function f. Furthermore, it is usually possible to obtain a formula for this *slope function*. This is accomplished by computing the limit of the difference quotient $\dfrac{f(x + h) - f(x)}{h}$ as $h \to 0$. We then substitute a value of x *after* the limit has been found. The slope function derived in this manner from f is said to be the **derivative of f** and (instead of m_{tan}) is denoted by the symbol f'.

DEFINITION 2.9.2 The Derivative

The **derivative** of a function $y = f(x)$ is the function f' defined by

$$f'(x) = \lim_{h \to 0} \frac{f(x + h) - f(x)}{h}. \tag{3}$$

| EXAMPLE 3 | **Example 1 Revisited** |

Compute the derivative of $f(x) = x^2 + 2$.

Solution We proceed exactly as in Example 1 except that we find $f(x + h)$ instead of $f(1 + h)$. In the first three steps we calculate the difference quotient; in steps (*ii*) and (*iii*) we use the results in the preceding step. In step (*iv*) we compute the limit of the difference quotient.

(*i*) $f(x + h) = (x + h)^2 + 2 = x^2 + 2xh + h^2 + 2$

(*ii*) $f(x + h) - f(x) = x^2 + 2xh + h^2 + 2 - (x^2 + 2)$
$$= x^2 + 2xh + h^2 + 2 - x^2 - 2$$
$$= 2xh + h^2$$
$$= h(2x + h)$$

(*iii*) $\dfrac{f(x + h) - f(x)}{h} = \dfrac{h(2x + h)}{h} = 2x + h \quad \leftarrow \text{cancel } h\text{'s}$

(*iv*) From (3) the derivative of f is the limit as $h \to 0$ of the result in (*iii*). During the process of shrinking h smaller and smaller, x is held fixed. Hence

$$f'(x) = \lim_{h \to 0}(2x + h) = 2x.$$

So now we have two functions; from $f(x) = x^2 + 2$ we have obtained the derivative $f'(x) = 2x$. When evaluated at a number x, the function f gives the y-coordinate of a point on the graph and the derived function f' gives the slope of the tangent line at that point. We have already seen in Example 1 that $f(1) = 3$ and $f'(1) = 2$.   ☰

With the aid of the derivative $f'(x) = 2x$ we can find slopes at other points on the graph of $f(x) = x^2 + 2$. For example,

at $x = 0$, $\begin{cases} f(0) = 2 \\ f'(0) = 0 \end{cases}$ ← point of tangency is $(0, 2)$
← slope of tangent line at $(0, 2)$ is $m = 0$

at $x = -3$, $\begin{cases} f(-3) = 11 \\ f'(-3) = -6 \end{cases}$ ← point of tangency is $(-3, 11)$
← slope of tangent line at $(-3, 11)$ is $m = -6$

The fact that $f'(0) = 0$ means that the tangent line to the graph of f is horizontal at the point $(0, 2)$.

▐ EXAMPLE 4   **Derivative of a Function**

Compute the derivative of $f(x) = 2x^3 - 4x + 5$.

Solution
(*i*) The function is $f(\ \ ) = 2(\ \ )^3 - 4(\ \ ) + 5$ and so

$$f(x + h) = 2(x + h)^3 - 4(x + h) + 5.$$

The algebra here is a bit more complicated than in the previous example. We will use the binomial expansion for $(a + b)^3$ and the distributive law. Continuing,

See (7) on page 35. ▶

$$f(x + h) = 2(x^3 + 3x^2h + 3xh^2 + h^3) - 4(x + h) + 5$$
$$= 2x^3 + 6x^2h + 6xh^2 + 2h^3 - 4x - 4h + 5 \quad \leftarrow \begin{cases} \text{two applications} \\ \text{of the distributive law} \end{cases}$$

(*ii*) As mentioned previously, in this step we are looking for a factor of h:

$$f(x + h) - f(x) = 2x^3 + 6x^2h + 6xh^2 + 2h^3 - 4x$$
$$- 4h + 5 - (2x^3 - 4x + 5)$$
$$= 2x^3 + 6x^2h + 6xh^2 + 2h^3 - 4x$$
$$- 4h + 5 - 2x^3 + 4x - 5 \quad \leftarrow \text{terms in color add to 0}$$
$$= 6x^2h + 6xh^2 + 2h^3 - 4h$$
$$= h(6x^2 + 6xh + 2h^2 - 4) \quad \leftarrow \text{factor out } h$$

(*iii*) We use the last result:

$$\frac{f(x + h) - f(x)}{h} = \frac{h(6x^2 + 6xh + 2h^2 - 4)}{h} \quad \leftarrow \text{cancel } h\text{'s}$$
$$= 6x^2 + 6xh + 2h^2 - 4$$

(*iv*) From (3) and the preceding step the derivative of f is

$$f'(x) = \lim_{h \to 0}(6x^2 + 6xh + 2h^2 - 4) = 6x^2 - 4.$$ ☰

EXAMPLE 5   **Equation of Tangent Line**

Find an equation of the tangent line to the graph of $f(x) = 2/x$ at $x = 2$.

Solution We start by finding the derivative of f. In the second of the four steps we will have to combine two symbolic fractions by means of a common denominator.

(*i*) $f(x + h) = \dfrac{2}{x + h}$

(*ii*) $f(x + h) - f(x) = \dfrac{2}{x + h} - \dfrac{2}{x}$

$\qquad = \dfrac{2}{x + h}\dfrac{x}{x} - \dfrac{2}{x}\dfrac{x + h}{x + h}$ ← a common denominator is $x(x + h)$

$\qquad = \dfrac{2x - 2x - 2h}{x(x + h)}$ ← $2x - 2x = 0$

$\qquad = \dfrac{-2h}{x(x + h)}$ ← there is the factor of h

(*iii*) The last result is to be divided by h, or more precisely $\dfrac{h}{1}$. We invert and multiply by $\dfrac{1}{h}$:

$$\dfrac{f(x + h) - f(x)}{h} = \dfrac{\dfrac{-2h}{x(x + h)}}{\dfrac{h}{1}} = \dfrac{-2h}{x(x + h)}\dfrac{1}{h} = \dfrac{-2}{x(x + h)}$$ ← cancel h's

(*iv*) From (3) the derivative of f is

$$f'(x) = \lim_{h \to 0} \dfrac{-2}{x(x + h)} = \dfrac{-2}{x^2}.$$

We are now in a position to find an equation of the tangent line at the point corresponding to $x = 2$. From $f(2) = 2/2 = 1$, we get the point of tangency $(2, 1)$. Then from the derivative $f'(x) = -2/x^2$ we see that $f'(2) = -2/4$, and so the slope of the tangent line at $(2, 1)$ is $-\frac{1}{2}$. From the point-slope equation of a line, the tangent line is

$$y - 1 = -\tfrac{1}{2}(x - 2) \qquad \text{or} \qquad y = -\tfrac{1}{2}x + 2.$$

The graph of $y = 2/x$ is the graph of $y = 1/x$ stretched vertically. (See Figure 2.2.1(e).) The tangent line at $(2, 1)$ is shown in red in FIGURE 2.9.6. ≡

□ **Alternative Definition** There is an alternative definition of the derivative. If we let $x = a + h$ in (2), then $h = x - a$. Consequently the slope of the secant line through $P(a, f(a))$ and $Q(x, f(x))$, as shown in FIGURE 2.9.7, is $\dfrac{f(x) - f(a)}{x - a}$. As $h \to 0$ we must have $x \to a$, and so the derivative (3) takes on the form

$$f'(a) = \lim_{x \to a} \dfrac{f(x) - f(a)}{x - a}. \qquad (4)$$

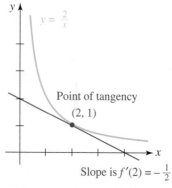

FIGURE 2.9.6 Tangent line in Example 5

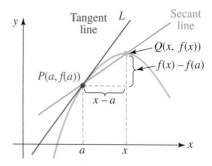

FIGURE 2.9.7 Secant line and tangent line at $(a, f(a))$

EXAMPLE 6   **Using (4)**

Use (4) to compute the derivative of $f(x) = 4x^2 - 5x + 9$.

Review (1) and (2) of Section 1.5. ▶

Solution We use the four-step process exactly as in Examples 3 and 4. The algebra is a slightly different; the analogue of the tip in (*ii*) of Example 1 is that we look for the factor $x - a$ in the difference $f(x) - f(a)$. Thus step (*ii*) will often require factoring the difference of two squares, the difference of two cubes, and so on.

(*i*) $f(a) = 4a^2 - 5a + 9$

(*ii*) $f(x) - f(a) = 4x^2 - 5x + 9 - (4a^2 - 5a + 9)$
$$= 4x^2 - 5x + 9 - 4a^2 + 5a - 9$$
$$= 4x^2 - 5x - 4a^2 + 5a \qquad \leftarrow \begin{cases} \text{regroup terms in} \\ \text{preparation for factoring} \end{cases}$$
$$= 4x^2 - 4a^2 - 5x + 5a$$
$$= 4(x^2 - a^2) - 5(x - a) \qquad \leftarrow \begin{cases} \text{first term is the} \\ \text{difference of two squares} \end{cases}$$
$$= 4(x - a)(x + a) - 5(x - a) \qquad \leftarrow \text{notice the factor of } x - a$$
$$= (x - a)[4(x + a) - 5]$$
$$= (x - a)(4x + 4a - 5)$$

(*iii*) $\dfrac{f(x) - f(a)}{x - a} = \dfrac{(x - a)(4x + 4a - 5)}{x - a} \qquad \leftarrow \text{cancel } x - a$
$$= 4x + 4a - 5$$

(*iv*) In the limit process indicated in (4), a is held fixed. Hence

$$f'(a) = \lim_{x \to a}(4x + 4a - 5) = 8a - 5. \qquad \leftarrow \text{the limit of } 4x \text{ as } x \to a \text{ is } 4a \qquad \equiv$$

As you can see in (4) and the final line in Example 6, the derivative comes out a function of the symbol a rather than x, that is, $f'(a) = 8a - 5$. As a consequence, (4) is not used as often as (3) to compute a derivative. See Problems 33–40 in Exercises 2.9. Nevertheless, (4) is important because it is convenient to use in some theoretical aspects of differential calculus.

2.9 **Exercises** Answers to selected odd-numbered problems begin on page ANS-8.

In Problems 1–10, proceed as in Example 1.

(a) Compute the difference quotient $\dfrac{f(a + h) - f(a)}{h}$ at the given value of a.

(b) Then, if instructed, compute $m_{\tan} = \lim_{h \to 0} \dfrac{f(a + h) - f(a)}{h}$.

(c) Use the result of part (b) to find an equation of the tangent line at the point of tangency.

1. $f(x) = x^2 - 6, a = 3$ 2. $f(x) = -3x^2 + 10, a = -1$
3. $f(x) = x^2 - 3x, a = 1$ 4. $f(x) = -x^2 + 5x - 3, a = -2$
5. $f(x) = -2x^3 + x, a = 2$ 6. $f(x) = 8x^3 - 4, a = \frac{1}{2}$
7. $f(x) = \dfrac{1}{2x}, a = -1$ 8. $f(x) = \dfrac{4}{x - 1}, a = 2$
9. $f(x) = \sqrt{x}, a = 4$ 10. $f(x) = \dfrac{1}{\sqrt{x}}, a = 1$

CHAPTER 2 FUNCTIONS

In Problems 11–26, proceed as in Examples 3 and 4.

(a) Compute the difference quotient $\dfrac{f(x + h) - f(x)}{h}$ for the given function.

(b) Then, if instructed, compute the derivative $f'(x) = \lim\limits_{h \to 0} \dfrac{f(x + h) - f(x)}{h}$.

11. $f(x) = 10$

12. $f(x) = -3x + 8$

13. $f(x) = -4x^2$

14. $f(x) = x^2 - x$

15. $f(x) = 3x^2 - x + 7$

16. $f(x) = 2x^2 + x - 1$

17. $f(x) = x^3 + 5x - 4$

18. $f(x) = 2x^3 + x^2$

19. $f(x) = \dfrac{1}{4 - x}$

20. $f(x) = \dfrac{3}{2x - 4}$

21. $f(x) = \dfrac{x}{x - 1}$

22. $f(x) = \dfrac{2x + 3}{x + 5}$

23. $f(x) = x + \dfrac{1}{x}$

24. $f(x) = \dfrac{1}{x^2}$

25. $f(x) = 2\sqrt{x}$

26. $f(x) = \sqrt{2x + 1}$

In Problems 27–32, use the appropriate derivatives obtained in Problems 11–26. For the given function, find the point of tangency and slope of the tangent line at the indicated value of x. Find an equation of the tangent line at that point.

27. $f(x) = 3x^2 - x + 7, \quad x = 2$

28. $f(x) = x^2 - x, \quad x = 3$

29. $f(x) = x^3 + 5x - 4, \quad x = 1$

30. $f(x) = 2x^3 + x^2, \quad x = -\frac{1}{2}$

31. $f(x) = x + \dfrac{1}{x}, \quad x = \frac{1}{2}$

32. $f(x) = \dfrac{3}{2x - 4}; \quad x = -1$

In Problems 33–40, proceed as in Example 6.

(a) Compute the difference quotient $\dfrac{f(x) - f(a)}{x - a}$ for the given function.

(b) Then, if instructed, compute the derivative $f'(a) = \lim\limits_{x \to a} \dfrac{f(x) - f(a)}{x - a}$.

33. $f(x) = 3x^2 + 1$

34. $f(x) = x^2 - 8x - 3$

35. $f(x) = 10x^3$

36. $f(x) = x^4$

37. $f(x) = \dfrac{1}{x}$

38. $f(x) = \dfrac{3x - 1}{x}$

39. $f(x) = \sqrt{7x}$

40. $f(x) = -\sqrt{x + 9}$

For Discussion

In Problems 41 and 42, use either (3) or (4) to compute the derivative of the given function. Find the points on the graph of f at which $f'(x) = 0$. Interpret your answers geometrically.

41. $f(x) = x^3 - 3x^2 - 9x$

42. $f(x) = x^4 - \frac{4}{3}x^3 + 2$

43. Use (2) to show that the graph of $f(x) = |x|$ possesses no tangent line at the point $(0, 0)$.

44. Use either (3) or (4) to compute the derivative of $f(x) = x^{1/3}$. [*Hint*: Recall from Section 1.5, $a^3 - b^3 = (a - b)(a^2 + ab + b^2)$.]

45. What is the tangent line to the graph of a linear function $f(x) = ax + b$?

46. If $f'(x) > 0$ for every x in an interval, then what can be said about f on the interval? If $f'(x) < 0$ for every x in an interval, then what can be said about f on the interval? [*Hint*: Draw a graph.]

47. If f is an even function and if (x, y) is on the graph of f, then $(-x, y)$ is also on the graph of f. How are the slopes of the tangent lines at (x, y) and $(-x, y)$ related?

48. If f is an odd function and if (x, y) is on the graph of f, then $(-x, -y)$ is also on the graph of f. How are the slopes of the tangent lines at (x, y) and $(-x, -y)$ related?

49. Consider the semicircle whose equation is $f(x) = \sqrt{1 - x^2}$. Discuss: How can the derivative $f'(x)$ be found using only the geometric fact that the radius of a circle is perpendicular to the tangent line at a point (x, y) on the circle?

50. Consider the semicircle whose equation is $f(x) = \sqrt{1 - x^2}$. Use (3) to find the derivative $f'(x)$ and compare your result with that in Problem 49.

51. Find an equation of the tangent line, shown in red in FIGURE 2.9.8, to the graph of $y = f(x)$ at point P. What are $f(-3)$ and $f'(-3)$?

52. Find an equation of the tangent line, shown in red in FIGURE 2.9.9, to the graph of $y = f(x)$ at point P. What is $f'(3)$? What is the y-intercept of the tangent line?

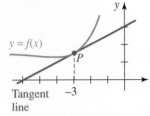

FIGURE 2.9.8 Graph for Problem 51

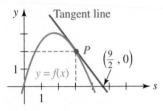

FIGURE 2.9.9 Graph for Problem 52

CHAPTER 2 | **Review Exercises** Answers to selected odd-numbered problems begin on page ANS-8.

A. Fill in the Blanks_____

In Problems 1–32, fill in the blanks.

1. If $f(x) = \dfrac{2x^3 - 1}{x^2 + 2}$, then $\left(\frac{1}{2}, \underline{\hspace{1cm}}\right)$ is a point on the graph of f.

2. If $f(x) = \dfrac{Ax}{10x - 2}$ and $f(2) = 3$, then $A = \underline{\hspace{1cm}}$.

3. The domain of the function $f(x) = \dfrac{1}{\sqrt{5 - x}}$ is $\underline{\hspace{1cm}}$.

4. The range of the function $f(x) = |x| - 10$ is $\underline{\hspace{1cm}}$.

5. The zeros of the function $f(x) = \sqrt{x^2 - 2x}$ are $\underline{\hspace{1cm}}$.

6. If the graph of f is symmetric with respect to the y-axis, $f(-x) = \underline{\hspace{1cm}}$.

7. The lines $2x - 5y = 1$ and $kx + 3y + 3 = 0$ are parallel if $k = \underline{\hspace{1cm}}$.

8. The x- and y-intercepts of the line $-4x + 3y - 48 = 0$ are $\underline{\hspace{1cm}}$.

9. The graph of a linear function for which $f(-2) = 0$ and $f(0) = -3$ has slope $m = \underline{\hspace{1cm}}$.

10. An equation of a line through $(1, 2)$ that is perpendicular to $y = 3x - 5$ is $\underline{\hspace{1cm}}$.

11. The x- and y-intercepts of the parabola $f(x) = x^2 - 2x - 1$ are $\underline{\hspace{1cm}}$.

12. The range of the function $f(x) = -x^2 + 6x - 21$ is $\underline{\hspace{1cm}}$.

13. The quadratic function $f(x) = ax^2 + bx + c$ for which $f(0) = 7$ and whose only x-intercept is $(-2, 0)$ is $f(x) = \underline{\hspace{1cm}}$.

14. If $f(x) = x + 2$ and $g(x) = x^2 - 2x$, then $(f \circ g)(-1) = \underline{\hspace{1cm}}$.

15. The vertex of the graph of $f(x) = x^2$ is $(0, 0)$. Therefore, the vertex of the graph of $y = -5(x - 10)^2 + 2$ is $\underline{\hspace{1cm}}$.

16. Given that $f^{-1}(x) = \sqrt{x - 4}$ is the inverse of a one-to-one function f, and without finding f, the domain of f is _____ and range of f is _____.

17. The x-intercept of a one-to-one function f is $(5, 0)$, and so the y-intercept of f^{-1} is _____.

18. The inverse of $f(x) = \dfrac{x - 5}{2x + 1}$ is $f^{-1} = $ _____.

19. The point $(a, 16a)$ lies on the graph of

$$f(x) = \begin{cases} 4x - 3, & x < 0 \\ x^3, & 0 \le x \le 1 \\ x^2 + 64, & x > 1 \end{cases}$$

for $a = $ _____.

20. For $f(x) = [\![x + 2]\!] - 4, f(-5.3) = $ _____.

21. If the entire graph of a one-to-one function f lies in the fourth quadrant, then graph of f^{-1} lies in the _____ quadrant.

22. The point $(3, 1)$ lies on the graph of a one-to-one function f. If $f^{-1}(2x) = 3$, then $x = $ _____.

In Problems 23–32, refer to FIGURE 2.R.1. Use approximation if necessary.

23. The domain of f is _____.
24. The range of f is _____.
25. x-intercepts of the graph of f are _____.
26. f is decreasing on the intervals _____.
27. f is increasing on the intervals _____.
28. $f(x) > 0$ on the intervals _____.
29. $f(x) < 0$ on the intervals _____.
30. $f(1) = $ _____.
31. The greatest function value on the interval $[2, 6]$ is _____.
32. If $f(x) = 0.5$, then $x = $ _____.

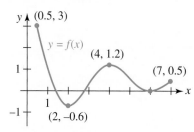

FIGURE 2.R.1 Graph for Problems 23–32

B. True/False

In Problems 1–22, answer true or false.

1. The points $(0, 3)$, $(2, 2)$, and $(6, 0)$ are collinear. _____
2. The graph of a function can have only one y-intercept. _____
3. If f is a function such that $f(a) = f(b)$, then $a = b$. _____
4. No nonzero function f can be symmetric with respect to the x-axis. _____
5. The domain of $f(x) = (x - 1)^{1/3}$ is $(-\infty, \infty)$. _____
6. If $f(x) = x$ and $g(x) = \sqrt{x + 2}$, then the domain of g/f is $[-2, \infty)$. _____
7. A function f is one-to-one if it never takes on the same value twice. _____
8. Two lines with positive slopes cannot be perpendicular. _____
9. The equation of a vertical line through $(2, -5)$ is $x = 2$. _____
10. A point of intersection of the graphs of f and f^{-1} must lie on the line $y = x$. _____

11. The one-to-one function $f(x) = 1/x$ has the property that $f = f^{-1}$. _____

12. The function $f(x) = 2x^2 + 16x - 2$ decreases on the interval $[-7, -2]$. _____

13. No even function defined on the interval $(-a, a)$, $a > 0$, can be one-to-one. _____

14. All odd functions are one-to-one. _____

15. If a function f is one-to-one, then $f^{-1}(x) = \dfrac{1}{f(x)}$. _____

16. If f is an increasing function on an interval containing $x_1 < x_2$, then $f(x_1) < f(x_2)$. _____

17. The function $f(x) = |x| - 1$ is decreasing on the interval $[0, \infty)$. _____

18. For function composition, $f \circ (g + h) = f \circ g + f \circ h$. _____

19. If the y-intercept for the graph of a function f is $(0, 1)$, then the y-intercept for the graph of $y = 4 - 3f(x)$ is $(0, 1)$. _____

20. For any function f, $f(x_1 + x_2) = f(x_1) + f(x_2)$. _____

21. The graph of $y = x^2 + 4x + 4$ is the graph of $f(x) = x^2$ shifted horizontally to the right. _____

22. The graph of $y = \sqrt{3 + x}$ is the graph of $f(x) = \sqrt{3 - x}$ reflected in the y-axis. _____

C. Review Exercises

In Problems 1 and 2 , identify two functions f and g so that $h = f \circ g$.

1. $h(x) = \dfrac{(3x - 5)^2}{x^2}$

2. $h(x) = 4(x + 1) - \sqrt{x + 1}$

3. Write the equation of each new function if the graph of $f(x) = x^3 - 2$ is
   (a) shifted to the left 3 units.
   (b) shifted down 5 units.
   (c) shifted to the right 1 unit and up 2 units.
   (d) reflected in the x-axis.
   (e) reflected in the y-axis.
   (f) vertically stretched by a factor of 3.

4. **FIGURE 2.R.2** shows the graph of a function f whose domain is $(-\infty, \infty)$. Sketch the graph of the following functions.

   (a) $y = f(x) - \pi$ **(b)** $y = f(x - 2)$

   (c) $y = f(x + 3) + \dfrac{\pi}{2}$ **(d)** $y = -f(x)$

   (e) $y = f(-x)$ **(f)** $y = 2f(x)$

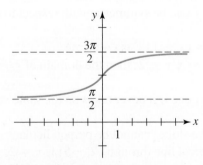

FIGURE 2.R.2 Graph for Problem 4

In Problems 5 and 6, use the graph of the one-to-one function f in Figure 2.R.2.

5. Give the domain and range of f^{-1}.
6. Sketch the graph of f^{-1}.

7. Express $y = x - |x| + |x - 1|$ as a piecewise-defined function. Sketch the graph of the function.
8. Sketch the graph of the function $y = [\![x]\!] + [\![-x]\!]$. Give the numbers at which the function is discontinuous.

In Problems 9 and 10, by examining the graph of the function f give the domain of the function g.

9. $f(x) = x^2 - 6x + 10$, $\quad g(x) = \sqrt{x^2 - 6x + 10}$

10. $f(x) = -x^2 + 7x - 6$, $\quad g(x) = \dfrac{1}{\sqrt{-x^2 + 7x - 6}}$

In Problems 11 and 12, $f(x) = \dfrac{1}{x + 1}$, $g(x) = \dfrac{5}{x - 2}$. Give the domain of the indicated composition.

11. $f \circ g$ **12.** $g \circ f$

In Problems 13 and 14, the given function f is one-to-one. Find f^{-1}.

13. $f(x) = (x + 1)^3$ **14.** $f(x) = x + \sqrt{x}$

15. Express the area of the shaded region in FIGURE 2.R.3 as a function of h.
16. Determine a quadratic function that describes the parabolic arch shown in FIGURE 2.R.4.

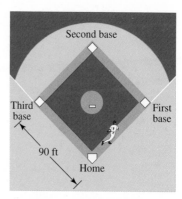

FIGURE 2.R.5 Cube in Problem 17

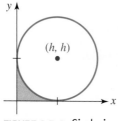

FIGURE 2.R.3 Circle in Problem 15

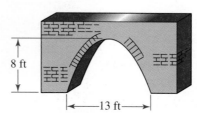

FIGURE 2.R.4 Arch in Problem 16

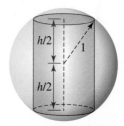

FIGURE 2.R.6 Inscribed cylinder in Problem 18

17. The diameter d of a cube is the distance between opposite vertices as shown in FIGURE 2.R.5. Express the diameter d as a function of the length s of a side of the cube by first expressing the length y of the diagonal in Figure 2.R.5 as a function of s.

18. A circular cylinder of height h is inscribed in a sphere of radius 1 as shown in FIGURE 2.R.6. Express the volume of the cylinder as a function of h.

19. A baseball diamond is a square that is 90 ft on a side. See FIGURE 2.R.7. After a player hits a home run, he jogs around the bases at a rate of 6 ft/s.
   (a) As the player jogs between home base and first base, express his distance from home base as a function of time t, where $t = 0$ corresponds to the time he left home base—that is, $0 \le t \le 15$.
   (b) As the player jogs between home base and first base, express his distance from second base as a function of time t, where $0 \le t \le 15$.

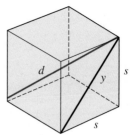

FIGURE 2.R.7 Baseball player in Problem 19

20. Consider the four circles shown in FIGURE 2.R.8. Express the area of the shaded region between them as a function of h.

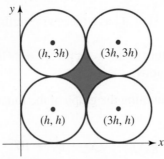

FIGURE 2.R.8 Circles in Problem 20

In Problems 21–24, find the objective function for the given calculus problem. Do **not** actually attempt to solve the problem.

21. Find the minimum value of the sum of 20 times a positive number and 5 times the reciprocal of that number.

22. A rancher wants to use 100 m of fence to construct a diagonal fence connecting two existing walls that meet at a right angle. How should this be done so that the area enclosed by the walls and fence is a maximum?

23. The running track shown as the black outline curve in FIGURE 2.R.9 is to consist of two parallel straight parts and two semicircular parts. The length of the track is to be 2 km. Find the design of the track so that the rectangular plot of land enclosed by the track is a maximum.

FIGURE 2.R.9 Running track in Problem 23

24. A pipeline is to be constructed from a refinery across a swamp to storage tanks. See FIGURE 2.R.10. The cost of construction is $25,000 per mile over the swamp and $20,000 per mile over land. How should the pipeline be made so that the cost of construction is a minimum?

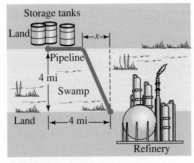

FIGURE 2.R.10 Pipeline in Problem 24

In Problems 25–28, compute $f'(x) = \lim\limits_{h \to 0} \dfrac{f(x+h) - f(x)}{h}$ for the given function.

Find an equation of the tangent line to the graph of f at the indicated value of x.

25. $f(x) = -3x^2 + 16x + 12, \quad x = 2$ **26.** $f(x) = x^3 - x^2, \quad x = -1$

27. $f(x) = \dfrac{-1}{2x^2}, \quad x = \frac{1}{2}$ **28.** $f(x) = x + 4\sqrt{x}, \quad x = 4$

In Problems 29 and 30, use of the derivative to find the points on the graph of the function f where the tangent line is horizontal.

29. f in Problem 26 **30.** f in Problem 28

3 Polynomial and Rational Functions

Chapter Outline

☰ **Introduction** In Chapter 2 we graphed functions such as $y = 3$, $y = 2x - 1$, $y = 5x^2 - 2x + 4$, and $y = x^3$. These functions, in which the variable x is raised to a *nonnegative integer power*, are examples of a more general type of function called a **polynomial function**. Our goal in this section is to examine some of the properties of polynomial functions and to present some general guidelines for graphing such functions. First we state the formal definition of a polynomial function.

DEFINITION 3.1.1 Polynomial Function

A **polynomial function** $y = f(x)$ is a function of the form

$$f(x) = a_n x^n + a_{n-1} x^{n-1} + \cdots + a_2 x^2 + a_1 x + a_0, \qquad (1)$$

where the coefficients $a_n, a_{n-1}, \ldots, a_2, a_1, a_0$ are real numbers and n is a nonnegative integer.

The **domain** of any polynomial function f is the set of all real numbers $(-\infty, \infty)$.

The following functions are *not* polynomial functions:

$$\overset{\text{not a nonnegative integer}}{\underset{\downarrow}{}} \qquad \qquad \overset{\text{not a nonnegative integer}}{\underset{\downarrow}{}}$$
$$y = 5x^2 - 3x^{-1} \qquad \text{and} \qquad y = 2x^{1/2} - 4.$$

The function

$$\overset{\text{nonnegative integer powers}}{\underset{\downarrow \quad \downarrow \quad \downarrow \quad \downarrow \quad \downarrow}{}}$$
$$y = 8x^5 - \tfrac{1}{2}x^4 - 10x^3 + 7x^2 + 6x + 4$$

is a polynomial, where we interpret the number 4 as the coefficient of x^0. Since 0 is a nonnegative integer, a constant function such as $y = 3$ is a polynomial function because it is the same as $y = 3x^0$.

☐ **Degree** Polynomial functions are classified by their degree. The highest power of x in a polynomial is said to be its **degree**. So if $a_n \neq 0$, then we say that $f(x)$ in (1) has **degree** n. The number a_n in (1) is called the **leading coefficient** and a_0 is called the **constant term** of the polynomial. For example,

$$\overset{\text{degree}}{\underset{\downarrow}{}}$$
$$f(x) = 3x^5 - 4x^3 - 3x + 8,$$
$$\underset{\text{leading coefficient}}{\uparrow} \qquad \qquad \underset{\text{constant term}}{\uparrow}$$

is a polynomial function of degree 5. We have already studied special polynomial functions in Sections 2.3 and 2.4. Polynomial functions of degrees $n = 0$, $n = 1$, and $n = 2$ are, respectively,

$$f(x) = a_0, \qquad\qquad \textbf{constant function} \Big\}$$
$$f(x) = a_1 x + a_0, \qquad\quad \textbf{linear function} \quad\Big\} \quad \text{Section 2.3}$$
$$f(x) = a_2 x^2 + a_1 x + a_0, \quad \textbf{quadratic function} \big\} \quad \text{Section 2.4}$$

Polynomials of degrees $n = 3$, $n = 4$, and $n = 5$ are, in turn, commonly referred to as **cubic**, **quartic**, and **quintic functions**. The constant function $f(x) = 0$ is called the **zero polynomial**.

☐ **Graphs** Recall that the graph of a constant function $f(x) = a_0$ is a **horizontal line**, the graph of a linear function $f(x) = a_1x + a_0$ is a **line with slope** $m = a_1$, and the graph of a quadratic function $f(x) = a_2x^2 + a_1x + a_0$ is a **parabola**. See Sections 2.3 and 2.4. Such descriptive statements cannot be made about the graph of a higher-degree polynomial function. What is the shape of the graph of a fifth-degree polynomial function? It turns out that the graph of a polynomial function of degree $n \geq 3$ can have several possible shapes. In general, graphing a polynomial function f of degree $n \geq 3$ often demands the use of either calculus or a graphing utility. However, we will see in the discussion that follows that by determining

- shifting,
- end behavior,
- symmetry,
- intercepts, and
- local behavior

of the function we can, in some instances, quickly sketch a reasonable graph of a higher-degree polynomial function while keeping point-plotting to a minimum. Before elaborating on each of these concepts we return to the notion of a power function first introduced in Section 2.2.

☐ **Power Function** A special case of the power function (see Section 2.2) is the **single-term polynomial function** or **monomial**,

$$f(x) = x^n, \quad n \text{ a positive integer.} \tag{2}$$

The graphs of f for degrees $n = 1, 2, 3, 4, 5$, and 6 are given in FIGURE 3.1.1. The interesting fact about (2) is that all the graphs for n odd are basically the same. The notable characteristics are that the graphs are symmetric about the origin and become increasingly flatter near the origin as the degree n increases. See Figures 3.1.1(a)–3.1.1(c). A similar observation is true for the graphs of (2) for n even, except, of course, the graphs are symmetric with respect to the y-axis. See Figures 3.1.1(d)–3.1.1(f).

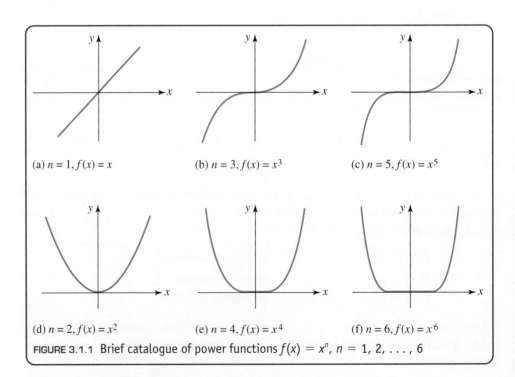

(a) $n = 1, f(x) = x$ (b) $n = 3, f(x) = x^3$ (c) $n = 5, f(x) = x^5$

(d) $n = 2, f(x) = x^2$ (e) $n = 4, f(x) = x^4$ (f) $n = 6, f(x) = x^6$

FIGURE 3.1.1 Brief catalogue of power functions $f(x) = x^n$, $n = 1, 2, \ldots, 6$

□ **Shifted Graphs** Recall from Section 2.2 that for $c > 0$, the graphs of polynomial functions of the form

$$y = ax^n + c, \qquad y = ax^n - c$$
and
$$y = a(x + c)^n, \qquad y = a(x - c)^n$$

can be obtained by vertical and horizontal shifts of the graph of $y = ax^n$. Also, if the leading coefficient a is positive, the graph of $y = ax^n$ is either a vertical stretch or a vertical compression of the graph of the basic single-term polynomial function $f(x) = x^n$. When a is negative we also carry out a reflection in the x-axis.

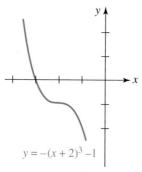

| EXAMPLE 1 | **Graphing a Shifted Polynomial Function** |

The graph of $y = -(x + 2)^3 - 1$ is the graph of $f(x) = x^3$ reflected in the x-axis, shifted 2 units to the left, and then shifted vertically downward 1 unit. First review Figure 3.1.1(b) and then see FIGURE 3.1.2. ≡

FIGURE 3.1.2 Reflected and shifted graph in Example 1

□ **End Behavior** The knowledge of the shape of a single-term polynomial function $f(x) = x^n$ is important for another reason. First, examine the computer-generated graphs given in FIGURES 3.1.3 and 3.1.4. Although the graph in Figure 3.1.3 certainly resembles the graphs in Figures 3.1.1(b) and 3.1.1(c), and the graph in FIGURE 3.1.4 resembles the graphs in Figure 3.1.1(d)–(f), the functions graphed in these two figures are *not* power functions $f(x) = x^n$, n odd, or $f(x) = x^n$, n even. We will not tell you at this point what the specific functions are except to say that they were both graphed on the interval $[-1000, 1000]$. The point is this: the function whose graph is given in Figure 3.1.3 could be almost *any* polynomial function

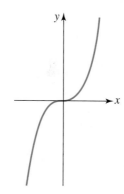

FIGURE 3.1.3 Mystery graph #1

$$f(x) = a_n x^n \boxed{+ \, a_{n-1}x^{n-1} + \cdots + a_1 x + a_0} \tag{3}$$

$a_n > 0$, of *odd* degree n, $n = 3, 5, \ldots$ when graphed on $[-1000, 1000]$. Similarly, the graph in Figure 3.1.4 could be that of any polynomial function given in (1), with $a_n > 0$, of *even* degree n, $n = 2, 4, \ldots$ when graphed on a large interval around the origin. As the next theorem indicates, the terms enclosed in the colored rectangle in (3) are irrelevant when we look at a graph of a polynomial globally—that is, for $|x|$ large. How a polynomial function f behaves when $|x|$ is very large is said to be its **end behavior**.

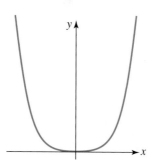

FIGURE 3.1.4 Mystery graph #2

THEOREM 3.1.1 End Behavior

For $|x|$ very large, that is, for $x \to -\infty$ and $x \to \infty$, the graph of the polynomial function $f(x) = a_n x^n + a_{n-1}x^{n-1} + \cdots + a_2 x^2 + a_1 x + a_0$, resembles the graph of $y = a_n x^n$.

To see why the graph of a polynomial function such as $f(x) = -2x^3 + 4x^2 + 5$ resembles the graph of the single-term polynomial $y = -2x^3$ when $|x|$ is large, let's factor out the highest power of x, that is, x^3:

both these terms become
negligible when $|x|$ is large
↓ ↓

$$f(x) = x^3\left(-2 + \frac{4}{x} + \frac{5}{x^3}\right). \tag{4}$$

By letting $|x|$ increase without bound, both $4/x$ and $5/x^3$ can be made as close to 0 as we want. Thus when $|x|$ is large, the values of the function f in (4) are closely approximated by the values of $y = -2x^3$. For example, for $x = 1000$ we see that

$$f(1000) = -2(1000)^3 + 4(1000)^2 + 5 = -1{,}995{,}999{,}995$$

whereas $\qquad y = -2(1000)^3 = -2{,}000{,}000{,}000.$

There can be only four types of end behavior for a polynomial function f. Although two of the end behaviors are already illustrated in Figures 3.1.3 and 3.1.4, we include them again in the pictorial summary given in Figure 3.1.5. To interpret the arrows in FIGURE 3.1.5 let's examine Figure 3.1.5(a). The position and direction of the left arrow (left arrow points down) indicates that as $x \to -\infty$, the values $f(x)$ are negative and large in magnitude. Stated another way, the graph is heading downward as $x \to -\infty$. Similarly, the position and direction of the right arrow (right arrow points up) indicates that the graph is heading upward as $x \to \infty$.

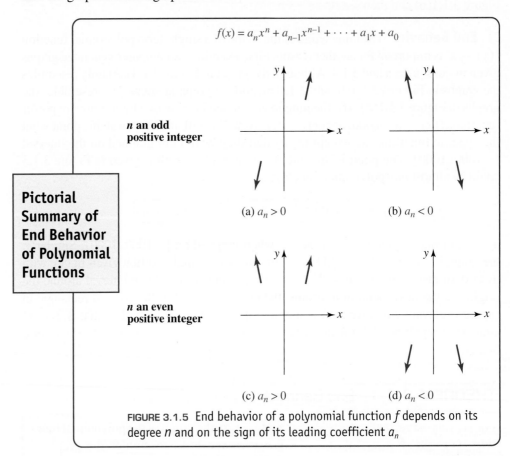

Pictorial Summary of End Behavior of Polynomial Functions

FIGURE 3.1.5 End behavior of a polynomial function f depends on its degree n and on the sign of its leading coefficient a_n

☐ **Relative Extrema** The gaps between the arrows in Figure 3.1.5 correspond to some interval around the origin. In these gaps the graph of f exhibits **local behavior**, in other words, the graph of f shows the characteristics of a polynomial function of a particular degree. This local behavior includes the x- and y-intercepts of the graph, the behavior of the graph at an x-intercept, the turning points of the graph, and observable symmetry of the graph (if any). A **turning point** is a point $(c, f(c))$ at which the graph of a polynomial function f changes direction, that is, the function f changes from increasing to decreasing or vice versa. The graph of a polynomial function of degree n can have up to $n - 1$ turning points. In calculus a turning point corresponds to a **relative**, or **local**, **extremum** of a function f. A relative extremum of f is classified as either a **maximum** or a **minimum**. This leads to the following definition.

DEFINITION 3.1.2 Relative Extremum

(*i*) A number $f(c)$ is a **relative maximum** of a function f if $f(x) \leq f(c)$ for every x in some open interval (a, b) that contains c.

(*ii*) A number $f(c)$ is a **relative minimum** of a function f if $f(x) \geq f(c)$ for every x in some open interval (a, b) that contains c.

If $(c, f(c))$ is a turning point of a polynomial function, then in some interval (a, b) containing c the function value $f(c)$ is either the *largest* (relative maximum) or the *smallest* (relative minimum) function value in the interval. If $f(c)$ is a relative maximum, then the graph of a polynomial function f must change from increasing immediately to the left of c to decreasing immediately to the right of c, whereas if $f(c)$ is a relative minimum the function f changes from decreasing to increasing at c. The graph in FIGURE 3.1.6 shows a graph of a function f with two relative extrema; $f(c_1)$ is a relative maximum in the interval (a_1, b_1) and $f(c_2)$ is a relative minimum in the interval (a_2, b_2).

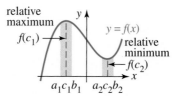

FIGURE 3.1.6 Two relative extrema of f

☐ **Symmetry** It is easy to tell by inspection those polynomial functions whose graphs possess symmetry with respect to either the y-axis or the origin. The words "even" and "odd" functions have special meaning for polynomial functions. Recall that an even function is one for which $f(-x) = f(x)$ and an odd function is one for which $f(-x) = -f(x)$. These two conditions hold for polynomial functions in which all the powers of x are even integers and odd integers, respectively. For example,

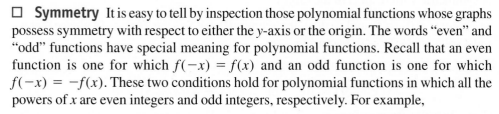

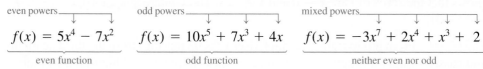

A function such as $f(x) = 3x^6 - x^4 + 6$ is an even function because the obvious powers are even integers; the constant term 6 is actually $6x^0$, and 0 is an even nonnegative integer.

☐ **Intercepts** The graph of every polynomial function f passes through the y-axis since $x = 0$ is the domain of the function. The y-intercept is the point $(0, f(0))$. Recall that a number c is a **zero** of a function f if $f(c) = 0$. In this discussion we assume c is a real zero. If $x - c$ is a factor of a polynomial function f, that is, $f(x) = (x - c)q(x)$ where $q(x)$ is another polynomial, then clearly $f(c) = 0$ and the corresponding point on the graph is $(c, 0)$. Thus the real zeros of a polynomial function are the x-coordinates of the x-intercepts of its graph. If $(x - c)^m$ is a factor of f, where $m > 1$ is a positive integer, and $(x - c)^{m+1}$ is *not* a factor of f, then c is said to be a **repeated zero**, or more precisely, a **zero of multiplicity** m. For example, $f(x) = x^2 - 10x + 25$ is equivalent to $f(x) = (x - 5)^2$. Hence 5 is a repeated zero or a zero of multiplicity 2. When $m = 1$, c is called a **simple zero**. For example, $-\frac{1}{3}$ and $\frac{1}{2}$ are simple zeros of $f(x) = 6x^2 - x - 1$ since f can be written as $f(x) = 6\left(x + \frac{1}{3}\right)\left(x - \frac{1}{2}\right)$. The behavior of the graph of f at an x-intercept $(c, 0)$ depends on whether c is a simple zero or a zero of multiplicity $m > 1$, where m is either an even or an odd integer.

(a) Simple zero

(b) Zero of odd multiplicity $m = 3, 5, \ldots$

(c) Zero of even multiplicity $m = 2, 4, \ldots$

FIGURE 3.1.7 x-intercepts of a polynomial function f

- If c is a simple zero, then the graph of f passes directly through the x-axis at $(c, 0)$. See FIGURE 3.1.7(a).
- If c is a zero of odd multiplicity $m = 3, 5, \ldots$, then the graph of f passes through the x-axis but is flattened at $(c, 0)$. See Figure 3.1.7(b).
- If c is a zero of even multiplicity $m = 2, 4, \ldots$, then the graph of f is tangent to, or touches, the x-axis at $(c, 0)$. See Figure 3.1.7(c).

In the case when c is either a simple zero or a zero of odd multiplicity $m = 3, 5, \ldots, f(x)$ changes sign *at* $(c, 0)$, whereas if c is a zero of even multiplicity $m = 2, 4, \ldots, f(x)$ does not change sign at $(c, 0)$. We note that depending on the sign of the leading coefficient of the polynomial, the graphs in Figure 3.1.7 could be reflected in the x-axis. For example, at a zero of even multiplicity the graph of f could be tangent to the x-axis from below that axis.

■ EXAMPLE 2 Graphing a Polynomial Function

Graph $f(x) = x^3 - 9x$.

Solution Here are some of the things we look at to sketch the graph of f:

End Behavior: By ignoring all terms but the first, we see that the graph of f resembles the graph of $y = x^3$ for large $|x|$. That is, the graph goes down to the left as $x \to -\infty$ and up to the right as $x \to \infty$, as illustrated in Figure 3.1.5(a).

Symmetry: Since all the powers are odd integers, f is an odd function. The graph of f is symmetric with respect to the origin.

Intercepts: $f(0) = 0$, and so the y-intercept is $(0, 0)$. Setting $f(x) = 0$, we see that we must solve $x^3 - 9x = 0$. Factoring

$$\overset{\text{difference of two squares}}{\overset{\downarrow}{x(x^2 - 9) = 0}} \quad \text{or} \quad x(x - 3)(x + 3) = 0$$

shows that the zeros of f are $x = 0$ and $x = \pm 3$. The x-intercepts are $(0, 0)$, $(-3, 0)$, and $(3, 0)$.

The Graph: From left to right, the graph rises (f is increasing) from the third quadrant and passes straight through $(-3, 0)$ since -3 is a simple zero. Although the graph is rising as it passes through this intercept it must turn back downward (f decreasing) at some point in the second quadrant to get through the intercept $(0, 0)$. Since the graph is symmetric with respect to the origin, its behavior is just the opposite in the first and fourth quadrants. See FIGURE 3.1.8. ≡

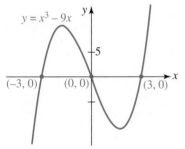

FIGURE 3.1.8 Graph of function in Example 2

In Example 2, the graph of f has two turning points. On the interval $[-3, 0]$ there is a relative maximum of f and on the interval $[0, 3]$ there is a relative minimum of f. We made no attempt to locate the corresponding turning points precisely; this is something that would, in general, require techniques from calculus. The best we can do using precalculus mathematics to refine the graph is to resort to plotting additional points on the intervals of interest. By the way, $f(x) = x^3 - 9x$ is the function whose graph on the interval $[-1000, 1000]$ is given in Figure 3.1.3.

■ EXAMPLE 3 Graphing a Polynomial Function

Graph $f(x) = (1 - x)(x + 1)^2$.

Solution Multiplying out, f is the same as $f(x) = -x^3 - x^2 + x + 1$.

End Behavior: From the preceding line we see that the graph of f resembles the graph of $y = -x^3$ for large $|x|$, just the opposite of the end behavior of the function in Example 2. See Figure 3.1.5(b).

Symmetry: As we see from $f(x) = -x^3 - x^2 + x + 1$, there are both even and odd powers of x present. Hence f is neither even nor odd; its graph possesses no y-axis or origin symmetry.

Intercepts: $f(0) = 1$ so the y-intercept is $(0, 1)$. From the given factored form of $f(x)$, we see that $(-1, 0)$ and $(1, 0)$ are the x-intercepts.

The Graph: From left to right, the graph falls (f decreasing) from the second quadrant and then, because -1 is a zero of multiplicity 2, the graph is tangent to the x-axis at $(-1, 0)$. The graph then rises (f increasing) as it passes through the y-intercept $(0, 1)$. At some point within the interval $[-1, 1]$ the graph turns downward (f decreasing) and, since 1 is a simple zero, passes through the x-axis at $(1, 0)$, heading downward into the fourth quadrant. See **FIGURE 3.1.9**. ≡

In Example 3, there are again two turning points. It should be clear that the point $(-1, 0)$ is a turning point (f changes from decreasing to increasing at -1) and $f(-1) = 0$ is a relative minimum of f. There is a turning point (f changes from increasing to decreasing at this point) somewhere within the interval $[-1, 1]$ and the function value at this point is a relative maximum of f.

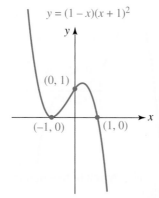

FIGURE 3.1.9 Graph of function in Example 3

■ EXAMPLE 4   **Zeros of Multiplicity Two**

Graph $f(x) = x^4 - 4x^2 + 4$.

Solution Before proceeding, note that the right-hand side of f is a perfect square. That is, $f(x) = (x^2 - 2)^2$. Since $x^2 - 2 = (x - \sqrt{2})(x + \sqrt{2})$, by the laws of exponents we can write

$$f(x) = (x - \sqrt{2})^2(x + \sqrt{2})^2. \tag{5}$$

End Behavior: Inspection of $f(x)$ shows that its graph resembles the graph of $y = x^4$ for large $|x|$. That is, the graph goes up to the left as $x \to -\infty$ and up to the right as $x \to \infty$, as shown in Figure 3.1.5(c).

Symmetry: Because $f(x)$ contains only even powers of x, it is an even function and so its graph is symmetric with respect to the y-axis.

Intercepts: $f(0) = 4$, so the y-intercept is $(0, 4)$. Inspection of (5) shows the x-intercepts are $(-\sqrt{2}, 0)$ and $(\sqrt{2}, 0)$.

The Graph: From left to right, the graph falls from the second quadrant and then, because $-\sqrt{2}$ is a zero of multiplicity 2, the graph touches the x-axis at $(-\sqrt{2}, 0)$. The graph then rises from this point to the y-intercept $(0, 4)$. We then use the y-axis symmetry to finish the graph in the first quadrant. See **FIGURE 3.1.10**. ≡

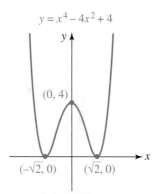

FIGURE 3.1.10 Graph of function in Example 4

In Example 4, the graph of f has three turning points. From the even multiplicity of the zeros, along with the y-axis symmetry, it can be deduced that the x-intercepts $(-\sqrt{2}, 0)$ and $(\sqrt{2}, 0)$ are turning points and $f(-\sqrt{2}) = 0$ and $f(\sqrt{2}) = 0$ are relative minima, and that the y-intercept $(0, 4)$ is a turning point and $f(0) = 4$ is a relative maximum.

■ EXAMPLE 5   **Zero of Multiplicity Three**

Graph $f(x) = -(x + 4)(x - 2)^3$.

Solution *End Behavior*: Inspection of f shows that its graph resembles the graph of $y = -x^4$ for large $|x|$. This end behavior of f is shown in Figure 3.1.5(d).

Symmetry: The function f is neither even nor odd. It is straightforward to show that $f(-x) \neq f(x)$ and $f(-x) \neq -f(x)$.

Intercepts: $f(0) = (-4)(-2)^3 = 32$, so the y-intercept is $(0, 32)$. From the factored form of $f(x)$, we see that $(-4, 0)$ and $(2, 0)$ are the x-intercepts.

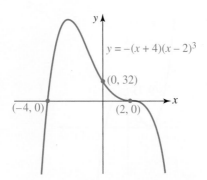

FIGURE 3.1.11 Graph of function in Example 5

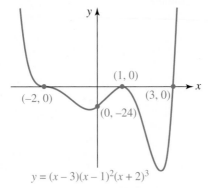

$y = (x-3)(x-1)^2(x+2)^3$

FIGURE 3.1.12 Graph of function in Example 6

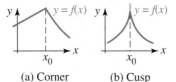

(a) Corner   (b) Cusp

FIGURE 3.1.13 The graph of a polynomial function cannot have corners or cusps

The Graph: From left to right, the graph rises from the third quadrant and then, because -4 is a simple zero, the graph of f passes directly through the x-axis at $(-4, 0)$. Somewhere within the interval $[-4, 0]$ the function f must change from increasing to decreasing to enable its graph to pass through the y-intercept $(0, 32)$. After its graph passes through the y-intercept, the function f continues to decrease but, since 2 is a zero of multiplicity 3, its graph flattens as it passes through $(2, 0)$, heading downward into the fourth quadrant. See **FIGURE 3.1.11**. $\equiv$

Note in Example 5 that since f is of degree 4, its graph could have up to three turning points. But as can be seen from Figure 3.1.11, the graph of f possesses only one turning point and at this point the function value is a relative maximum.

EXAMPLE 6   Zeros of Multiplicity Two and Three

Graph $f(x) = (x-3)(x-1)^2(x+2)^3$.

Solution The function f is of degree 6 and so its end behavior resembles the graph of $y = x^6$ for large $|x|$. See Figure 3.1.5(c). Also, the function f is neither even nor odd; its graph possesses no y-axis or origin symmetry. The y-intercept is $(0, f(0)) = (0, -24)$. From the factors of f we see that x-intercepts of the graph are $(-2, 0)$, $(1, 0)$, and $(3, 0)$. Since -2 is a zero of multiplicity 3, the graph of f is flattened as it passes through $(-2, 0)$. Since 1 is a zero of multiplicity 2, the graph of f is tangent to the x-axis at $(1, 0)$. Since 3 is a simple zero, the graph of f passes directly through the x-axis at $(3, 0)$. Putting all these facts together we obtain the graph in **FIGURE 3.1.12**. $\equiv$

In Example 6, since the function f is of degree 6 its graph could have up to five turning points. But as the graph in Figure 3.1.12 shows, there are only three turning points. At two of these points the unknown function values are relative minima; at the remaining point $(1, 0)$, the function value $f(1) = 0$ is a relative maximum.

☐ **Continuous Function** As is apparent from the graphs presented in this section a polynomial function is **continuous everywhere**, that is, continuous on the interval $(-\infty, \infty)$. Recall from the discussion of continuity on page 87 of Section 2.5 that this means the graph of a polynomial function f can have no holes, finite gaps, or infinite breaks in it. Moreover, a polynomial function f is a **smooth function** which means that its graph does not contain any sharp corners or cusps. In **FIGURE 3.1.13(a)** the point $(x_0, f(x_0))$ is a corner of the graph of f whereas $(x_0, f(x_0))$ is a cusp in Figure 3.1.13(b). For example, the functions $f(x) = |x|$ and $f(x) = x^{2/3}$ are continuous everywhere but are not smooth functions; the graph of $f(x) = |x|$ has a corner at the origin whereas the graph of $f(x) = x^{2/3}$ has a cusp at the origin. See Figures 2.5.6(a) and 2.2.1(i).

3.1   Exercises   Answers to selected odd-numbered problems begin on page ANS-9.

In Problems 1–8, proceed as in Example 1 and use transformations to sketch the graph of the given polynomial function.

1. $y = x^3 - 3$
                    2. $y = -(x+2)^3$
3. $y = (x-2)^3 + 2$
         4. $y = 3 - (x+2)^3$
5. $y = (x-5)^4$
               6. $y = x^4 - 1$
7. $y = 1 - (x-1)^4$
       8. $y = 4 + (x+1)^4$

     CHAPTER 3 POLYNOMIAL AND RATIONAL FUNCTIONS

In Problems 9–12, determine whether the given polynomial function f is even, odd, or neither even nor odd. Do not graph.

9. $f(x) = -2x^3 + 4x$

10. $f(x) = x^6 - 5x^2 + 7$

11. $f(x) = x^5 + 4x^3 + 9x + 1$

12. $f(x) = x^3(x + 2)(x - 2)$

In Problems 13–18, match the given graph with one of the polynomial functions in (a)–(f).

(a) $f(x) = x^2(x - 1)^2$

(b) $f(x) = -x^3(x - 1)$

(c) $f(x) = x^3(x - 1)^3$

(d) $f(x) = -x(x - 1)^3$

(e) $f(x) = -x^2(x - 1)$

(f) $f(x) = x^3(x - 1)^2$

13.

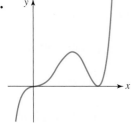

FIGURE 3.1.14 Graph for Problem 13

14.

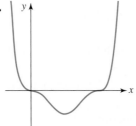

FIGURE 3.1.15 Graph for Problem 14

15.

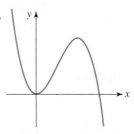

FIGURE 3.1.16 Graph for Problem 15

16.

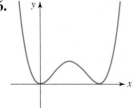

FIGURE 3.1.17 Graph for Problem 16

17.

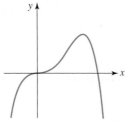

FIGURE 3.1.18 Graph for Problem 17

18.

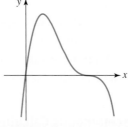

FIGURE 3.1.19 Graph for Problem 18

In Problems 19–22, construct a polynomial function f that has the given properties. There is no unique answer.

19. f is of degree 4, its graph is symmetric with respect to the y-axis, y-intercept is $(0, -6)$

20. f is of degree 5, 0 is a zero of multiplicity 3, its graph is symmetric with respect to the origin

21. f has four real zeros, 1 is a simple zero, -3 is zero of multiplicity 2, behaves like $y = -7x^4$ for large values of $|x|$

22. f is of degree 6, has four real zeros, 2 is a zero of multiplicity 3, behaves like $y = 2x^6$ for large values of $|x|$, $f(0) = 8$

In Problems 23–44, proceed as in Example 2 and sketch the graph of the given polynomial function f.

23. $f(x) = x^3 - 4x$

24. $f(x) = 9x - x^3$

25. $f(x) = -x^3 + x^2 + 6x$

26. $f(x) = x^3 + 7x^2 + 12x$

27. $f(x) = (x + 1)(x - 2)(x - 4)$

28. $f(x) = (2 - x)(x + 2)(x + 1)$

29. $f(x) = x^4 - 4x^3 + 3x^2$

30. $f(x) = x^2(x - 2)^2$

31. $f(x) = (x^2 - x)(x^2 - 5x + 6)$

32. $f(x) = x^2(x^2 + 3x + 2)$

33. $f(x) = (x^2 - 1)(x^2 + 9)$

34. $f(x) = x^4 + 5x^2 - 6$

35. $f(x) = -x^4 + 2x^2 - 1$

36. $f(x) = x^4 - 6x^2 + 9$

37. $f(x) = x^4 + 3x^3$

38. $f(x) = x(x - 2)^3$

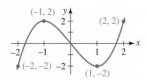

FIGURE 3.1.20 Graph for Problem 45

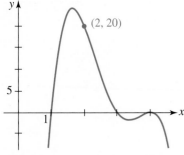

FIGURE 3.1.21 Graph for Problem 46

39. $f(x) = x^5 - 4x^3$

40. $f(x) = (x - 2)^5 - (x - 2)^3$

41. $f(x) = 3x(x + 1)^2(x - 1)^2$

42. $f(x) = (x + 1)^2(x - 1)^3$

43. $f(x) = -\frac{1}{2}x^2(x + 2)^3(x - 2)^2$

44. $f(x) = x(x + 1)^2(x - 2)(x - 3)$

45. The graph of $f(x) = x^3 - 3x$ is given in FIGURE 3.1.20.
   (a) Use the figure to obtain the graph of $g(x) = f(x) + 2$.
   (b) Using only the graph obtained in part (a) write an equation, in *factored* form, for $g(x)$. Then verify by multiplying out the factors that your equation for $g(x)$ is the same as $f(x) + 2 = x^3 - 3x + 2$.

46. Find a polynomial function f of lowest possible degree whose graph is consistent with the graph given in FIGURE 3.1.21.

47 Find the value of k such that $(2, 0)$ is an x-intercept for the graph of $f(x) = kx^5 - x^2 + 5x + 8$.

48. Find the values of k_1 and k_2 such that $(-1, 0)$ and $(1, 0)$ are x-intercepts for the graph of $f(x) = k_1x^4 - k_2x^3 + x - 4$.

49. Find the value of k such that $(0, 10)$ is the y-intercept for the graph of $f(x) = x^3 - 2x^2 + 14x - 3k$.

50. Consider the polynomial function $f(x) = (x - 2)^{n+1}(x + 5)$, where n is a positive integer. For what values of n does the graph of f touch, but not cross, the x-axis at $(2, 0)$?

51. Consider the polynomial function $f(x) = (x - 1)^{n+2}(x + 1)$, where n is a positive integer. For what values of n does the graph of f cross the x-axis at $(1, 0)$?

52. Consider the polynomial function $f(x) = (x - 5)^{2m}(x + 1)^{2n-1}$, where m and n are positive integers.
   (a) For what values of m does the graph of f cross the x-axis at $(5, 0)$?
   (b) For what values of n does the graph of f cross the x-axis at $(-1, 0)$?

Miscellaneous Calculus-Related Problems

53. Constructing a Box An open box can be made from a rectangular piece of cardboard by cutting a square of length x from each corner and bending up the sides. See FIGURE 3.1.22. If the cardboard measures 30 cm by 40 cm, show that the volume of the resulting box is given by

$$V(x) = x(30 - 2x)(40 - 2x).$$

Sketch the graph of $V(x)$ for $x > 0$. What is the domain of the function V?

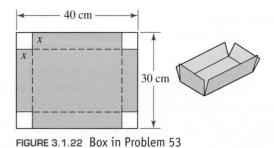

FIGURE 3.1.22 Box in Problem 53

54. Another Box In order to hold its shape, the box in Problem 53 will require tape or some other fastener at the corners. An open box that holds itself together can be made by cutting out a square of length x from each corner of a rectangular piece of cardboard, cutting on the solid line, and folding on the dashed lines, as shown in FIGURE 3.1.23. Find a polynomial function $V(x)$ that gives the volume of the resulting box if the original cardboard measures 30 cm by 40 cm. Sketch the graph of $V(x)$ for $x > 0$.

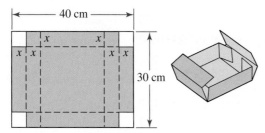

FIGURE 3.1.23 Box in Problem 54

55. **Making a Cup** A conical cup is made from a circular piece of paper of radius R by cutting out a circular sector and then joining the dashed edges as shown in FIGURE 3.1.24. Find a polynomial function $V(h)$ that gives the volume of the conical cup in terms of its height.
56. **Hourglass** Sand flows from the top half of the conical hourglass shown in FIGURE 3.1.25 to the bottom half at a constant rate. Find a polynomial function $V(h)$ that gives the volume of the bottom pile of sand in terms of the height of the sand. Assume that the top of the pile is level. [*Hint*: Use similar triangles.]

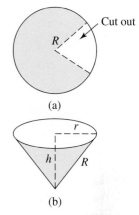

FIGURE 3.1.24 Cup in Problem 55

For Discussion

57. Examine Figure 3.1.5. Then discuss whether there can exist cubic polynomial functions that have no real zeros.
58. Suppose a polynomial function f has three zeros, -3, 2, and 4, and has the end behavior that its graph goes down to the left as $x \rightarrow -\infty$ and down to the right as $x \rightarrow \infty$. Discuss possible equations for f.

Calculator/Computer Problems

In Problems 59 and 60, use a graphing utility to examine the graph of the given polynomial function on the indicated intervals.

59. $f(x) = -(x - 8)(x + 10)^2$; $[-15, 15]$, $[-100, 100]$, $[-1000, 1000]$
60. $f(x) = (x - 5)^2(x + 5)^2$; $[-10, 10]$, $[-100, 100]$, $[-1000, 1000]$

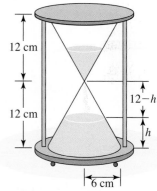

FIGURE 3.1.25 Hourglass in Problem 56

Division of Polynomial Functions

≡ **Introduction** If $p > 0$ and $s > 0$ are integers such that $p \geq s$, then p/s is called an **improper fraction**. By dividing p by s, we obtain unique numbers q and r that satisfy

$$\frac{p}{s} = q + \frac{r}{s} \qquad \text{or} \qquad p = sq + r, \tag{1}$$

where $0 \leq r < s$. The number p is called the **dividend**, s is the **divisor**, q is the **quotient**, and r is the **remainder**. For example, consider the improper fraction $\frac{1052}{23}$. Performing long division gives

$$
\begin{array}{r}
45 \quad \leftarrow \text{quotient} \\
23\overline{)1052} \quad \leftarrow \text{dividend} \\
\underline{92} \quad \leftarrow \text{subtract} \\
132 \\
\underline{115} \\
17. \quad \leftarrow \text{remainder}
\end{array}
$$

divisor $\rightarrow$ (2)

The result in (2) can be written as $\frac{1052}{23} = 45 + \frac{17}{23}$, where $\frac{17}{23}$ is a **proper fraction** since the numerator is less than the denominator; in other words, the fraction is less than 1. If we multiply this result by the divisor 23 we obtain the special way of writing the dividend p illustrated in the second equation in (1):

$$
1052 = 23 \cdot \underset{\underset{\text{divisor}}{\uparrow}}{\overset{\overset{\text{quotient} \quad \text{remainder}}{\downarrow \qquad \downarrow}}{45 + 17.}} \tag{3}
$$

☐ **Division of Polynomials** The method for dividing two polynomial functions $f(x)$ and $d(x)$ is similar to division of positive integers. If the degree of a polynomial $f(x)$ is greater than or equal to the degree of the polynomial $d(x)$, then $f(x)/d(x)$ is also called an **improper fraction**. A result analogous to (1) is called the **Division Algorithm** for polynomials.

THEOREM 3.2.1 Division Algorithm

Let $f(x)$ and $d(x) \neq 0$ be polynomials where the degree of $f(x)$ is greater than or equal to the degree of $d(x)$. Then there exist unique polynomials $q(x)$ and $r(x)$ such that

$$
\frac{f(x)}{d(x)} = q(x) + \frac{r(x)}{d(x)} \qquad \text{or} \qquad f(x) = d(x)q(x) + r(x), \tag{4}
$$

where $r(x)$ has degree less than the degree of $d(x)$.

The polynomial $f(x)$ is called the **dividend**, $d(x)$ the **divisor**, $q(x)$ the **quotient**, and $r(x)$ the **remainder**. Because $r(x)$ has degree less than the degree of $d(x)$, the rational expression $r(x)/d(x)$ is called a **proper fraction**.

Observe in (4) when $r(x) = 0$, then $f(x) = d(x)q(x)$, and so the divisor $d(x)$ is a factor of $f(x)$. In this case, we say that $f(x)$ is **divisible** by $d(x)$ or, in older terminology, $d(x)$ **divides evenly** into $f(x)$.

EXAMPLE 1   **Division of Two Polynomials**

Use long division to find the quotient $q(x)$ and remainder $r(x)$ when the polynomial $f(x) = 3x^3 - x^2 - 2x + 6$ is divided by the polynomial $d(x) = x^2 + 1$.

Solution By long division,

$$
\begin{array}{r}
3x \;-\; 1 \qquad \leftarrow \text{quotient} \\
x^2 + 1\overline{)3x^3 \;-\; x^2 \;-\; 2x \;+\; 6} \quad \leftarrow \text{dividend} \\
\underline{3x^3 \;+\; 0x^2 \;+\; 3x} \qquad \leftarrow \text{subtract} \\
-x^2 \;-\; 5x \;+\; 6 \\
\underline{-x^2 \;+\; 0x \;-\; 1} \\
-5x \;+\; 7 \qquad \leftarrow \text{remainder}
\end{array}
$$

(5)

The result of the division in (5) can be written

$$
\frac{3x^3 - x^2 - 2x + 6}{x^2 + 1} = \overbrace{3x - 1}^{q(x)} + \overbrace{\frac{-5x + 7}{x^2 + 1}}^{r(x)}.
$$

If we multiply both sides of the last equation by the divisor $x^2 + 1$, we get the second form given in (4):

$$
3x^3 - x^2 - 2x + 6 = (x^2 + 1)(3x - 1) + (-5x + 7). \qquad (6) \;\; \equiv
$$

If the divisor $d(x)$ is a linear polynomial $x - c$, it follows from the Division Algorithm that the degree of the remainder r is 0, that is to say, r is a constant. Thus (4) becomes

$$
f(x) = (x - c)q(x) + r. \qquad (7)
$$

When the number $x = c$ is substituted into (7), we discover an alternative way of evaluating a polynomial function:

$$
f(c) = (c - c)\overset{\underset{\downarrow}{0}}{q(c)} + r \quad \text{or} \quad f(c) = r.
$$

The foregoing result is called the **Remainder Theorem**.

THEOREM 3.2.2 Remainder Theorem

If a polynomial $f(x)$ is divided by a linear polynomial $x = c$, then the remainder r is the value of $f(x)$ at $x = c$, that is, $f(c) = r$.

EXAMPLE 2 **Finding the Remainder**

Use the Remainder Theorem to find r when $f(x) = 4x^3 - x^2 + 4$ is divided by $x - 2$.

Solution From the Remainder Theorem, the remainder r is the value of the function f evaluated at $x = 2$:

$$
r = f(2) = 4(2)^3 - (2)^2 + 4 = 32. \qquad (8) \;\; \equiv
$$

Example 2, where a remainder r is determined by calculating a function value $f(c)$, is more interesting than important. What *is* important is the reverse problem: Determine the function value $f(c)$ by finding the remainder r by division of f by $x - c$. The next two examples illustrate this concept.

EXAMPLE 3 **Evaluation by Division**

Use the Remainder Theorem to find $f(c)$ for $f(x) = x^5 - 4x^3 + 2x - 10$ when $c = -3$.

Solution The value $f(-3)$ is the remainder when $f(x) = x^5 - 4x^3 + 2x - 10$ is divided by $x - (-3) = x + 3$. For the purposes of long division we must account for the missing x^4 and x^2 terms by rewriting the dividend as

$$f(x) = x^5 + 0x^4 - 4x^3 + 0x^2 + 2x - 10.$$

Then,

$$
\begin{array}{r}
x^4 - 3x^3 + 5x^2 - 15x + 47 \\
x + 3 \overline{)\, x^5 + 0x^4 - 4x^3 + 0x^2 + 2x - 10} \\
\underline{x^5 + 3x^4} \\
-3x^4 - 4x^3 + 0x^2 + 2x - 10 \\
\underline{-3x^4 - 9x^3} \\
5x^3 + 0x^2 + 2x - 10 \\
\underline{5x^3 + 15x^2} \\
-15x^2 + 2x - 10 \\
\underline{-15x^2 - 45x} \\
47x - 10 \\
\underline{47x + 141} \\
-151
\end{array}
$$

(9)

The remainder r in the division is the value of the function f at $x = -3$, that is, $f(-3) = -151$. ≡

☐ **Synthetic Division** After working through Example 3 one could justifiably ask the question: Why would anyone want to calculate the value of a polynomial function f by division? The answer is: We would not bother do this were it not for **synthetic division**. Synthetic division is a shorthand method of dividing a polynomial $f(x)$ by a *linear* polynomial $x - c$; it does not require writing down the various powers of the variable x but only the coefficients of these powers in the dividend $f(x)$ (which must include all 0 coefficients). It is also a very efficient and quick way of evaluating $f(c)$, since the process utilizes only the arithmetic operations of multiplication and addition. No exponentiations such as 2^3 and 2^2 in (8) are involved. Here is the same division in (9) done synthetically:

For a review of synthetic division please ▶ see the *Student Resource Manual* that accompanies this text.

$$
\begin{array}{r|rrrrrr}
-3 & 1 & 0 & -4 & 0 & 2 & -10 \\
   &   & -3 & 9 & -15 & 45 & -141 \\
\hline
   & 1 & -3 & 5 & -15 & 47 & \boxed{-151} = r = f(-3)
\end{array}
$$

(10)

Recall that the bottom line of numbers in (10) are the coefficients of the various powers of x in the quotient $q(x)$ when $f(x) = x^5 - 4x^3 + 2x - 10$ is divided by $x + 3$. You should compare this with the quotient obtained by the long division in (9).

EXAMPLE 4 Using Synthetic Division to Evaluate a Function

Use the remainder theorem to find $f(c)$ for

$$f(x) = -3x^6 + 4x^5 + x^4 - 8x^3 - 6x^2 + 9$$

when $c = 2$.

Solution We will use synthetic division to find the remainder r in the division of f by $x - 2$. We begin by writing down all the coefficients in $f(x)$, including 0 as the coefficient of x. From

$$
\begin{array}{r|rrrrrrr}
2 & -3 & 4 & 1 & -8 & -6 & 0 & 9 \\
  &   & -6 & -4 & -6 & -28 & -68 & -136 \\
\hline
  & -3 & -2 & -3 & -14 & -34 & -68 & \boxed{-127} = r
\end{array}
$$

we see that $f(2) = -127$. ≡

CHAPTER 3 POLYNOMIAL AND RATIONAL FUNCTIONS

EXAMPLE 5 **Using Synthetic Division to Evaluate a Function**

Use synthetic division to evaluate $f(x) = x^3 - 7x^2 + 13x - 15$ at $x = 5$.

Solution From the synthetic division

$$
\begin{array}{r|rrrr}
5 & 1 & -7 & 13 & -15 \\
  &   & 5 & -10 & 15 \\
\hline
  & 1 & -2 & 3 & \boxed{0} = r
\end{array}
$$

we see that $f(5) = 0$.　　　　　　　　　　　　　　　　　　　　≡

The result in Example 5 that $f(5) = 0$ shows that 5 is a zero of the given function f. Moreover, we have found additionally that f is divisible by the linear polynomial $x - 5$, or put another way, $x - 5$ is a factor of f. The synthetic division shows that $f(x) = x^3 - 7x^2 + 13x - 15$ is equivalent to

$$f(x) = (x - 5)(x^2 - 2x + 3).$$

In the next section we will further explore the use of the Division Algorithm and the Remainder Theorem as a help in finding zeros and factors of a polynomial function.

3.2 **Exercises** Answers to selected odd-numbered problems begin on page ANS–10.

In Problems 1–10, use long division to find the quotient $q(x)$ and remainder $r(x)$ when the polynomial $f(x)$ is divided by the given polynomial $d(x)$. In each case write your answer in the form $f(x) = d(x)q(x) + r(x)$.

1. $f(x) = 8x^2 + 4x - 7;\quad d(x) = x^2$
2. $f(x) = x^2 + 2x - 3;\quad d(x) = x^2 + 1$
3. $f(x) = 5x^3 - 7x^2 + 4x + 1;\quad d(x) = x^2 + x - 1$
4. $f(x) = 14x^3 - 12x^2 + 6;\quad d(x) = x^2 - 1$
5. $f(x) = 2x^3 + 4x^2 - 3x + 5;\quad d(x) = (x + 2)^2$
6. $f(x) = x^3 + x^2 + x + 1;\quad d(x) = (2x + 1)^2$
7. $f(x) = 27x^3 + x - 2;\quad d(x) = 3x^2 - x$
8. $f(x) = x^4 + 8;\quad d(x) = x^3 + 2x - 1$
9. $f(x) = 6x^5 + 4x^4 + x^3;\quad d(x) = x^3 - 2$
10. $f(x) = 5x^6 - x^5 + 10x^4 + 3x^2 - 2x + 4;\quad d(x) = x^2 + x - 1$

In Problems 11–16, proceed as in Example 2 and use the Remainder Theorem to find r when $f(x)$ is divided by the given linear polynomial.

11. $f(x) = 2x^2 - 4x + 6;\quad x - 2$
12. $f(x) = 3x^2 + 7x - 1;\quad x + 3$
13. $f(x) = x^3 - 4x^2 + 5x + 2;\quad x - \frac{1}{2}$
14. $f(x) = 5x^3 + x^2 - 4x - 6;\quad x + 1$
15. $f(x) = x^4 - x^3 + 2x^2 + 3x - 5;\quad x - 3$
16. $f(x) = 2x^4 - 7x^2 + x - 1;\quad x + \frac{3}{2}$

In Problems 17–22, proceed as in Example 3 and use the Remainder Theorem to find $f(c)$ for the given value of c.

17. $f(x) = 4x^2 - 10x + 6; \quad c = 2$

18. $f(x) = 6x^2 + 4x - 2; \quad c = \frac{1}{4}$

19. $f(x) = x^3 + 3x^2 + 6x + 6; \quad c = -5$

20. $f(x) = 15x^3 + 17x^2 - 30; \quad c = \frac{1}{5}$

21. $f(x) = 3x^4 - 5x^2 + 20; \quad c = \frac{1}{2}$

22. $f(x) = 14x^4 - 60x^3 + 49x^2 - 21x + 19; \quad c = 1$

In Problems 23–32, use synthetic division to find the quotient $q(x)$ and remainder $r(x)$ when $f(x)$ is divided by the given linear polynomial.

23. $f(x) = 2x^2 - x + 5; \quad x - 2$

24. $f(x) = 4x^2 - 8x + 6; \quad x - \frac{1}{2}$

25. $f(x) = x^3 - x^2 + 2; \quad x + 3$

26. $f(x) = 4x^3 - 3x^2 + 2x + 4; \quad x - 7$

27. $f(x) = x^4 + 16; \quad x - 2$

28. $f(x) = 4x^4 + 3x^3 - x^2 - 5x - 6; \quad x + 3$

29. $f(x) = x^5 + 56x^2 - 4; \quad x + 4$

30. $f(x) = 2x^6 + 3x^3 - 4x^2 - 1; \quad x + 1$

31. $f(x) = x^3 - (2 + \sqrt{3})x^2 + 3\sqrt{3}x - 3; \quad x - \sqrt{3}$

32. $f(x) = x^8 - 3^8; \quad x - 3$

In Problems 33–38, use synthetic division and the Remainder Theorem to find $f(c)$ for the given value of c.

33. $f(x) = 4x^2 - 2x + 9; \quad c = -3$

34. $f(x) = 3x^4 - 5x^2 + 27; \quad c = \frac{1}{2}$

35. $f(x) = 14x^4 - 60x^3 + 49x^2 - 21x + 19; \quad c = 1$

36. $f(x) = 3x^5 + x^2 - 16; \quad c = -2$

37. $f(x) = 2x^6 - 3x^5 + x^4 - 2x + 1; \quad c = 4$

38. $f(x) = x^7 - 3x^5 + 2x^3 - x + 10; \quad c = 5$

In Problems 39 and 40, use long division to find a value of k such that $f(x)$ is divisible by $d(x)$.

39. $f(x) = x^4 + x^3 + 3x^2 + kx - 4; \quad d(x) = x^2 - 1$

40. $f(x) = x^5 - 3x^4 + 7x^3 + kx^2 + 9x - 5; \quad d(x) = x^2 - x + 1$

In Problems 41 and 42, use synthetic division to find a value of k such that $f(x)$ is divisible by $d(x)$.

41. $f(x) = kx^4 + 2x^2 + 9k; \quad d(x) = x - 1$

42. $f(x) = x^3 + kx^2 - 2kx + 4; \quad d(x) = x + 2$

43. Find a value of k such that the remainder in the division of $f(x) = 3x^2 - 4kx + 1$ by $d(x) = x + 3$ is $r = -20$.

44. When $f(x) = x^2 - 3x - 1$ is divided by $x - c$, the remainder is $r = 3$. Determine c.

3.3 Zeros and Factors of Polynomial Functions

≡ **Introduction** In Section 2.1 we saw that a zero of a function f is a number c for which $f(c) = 0$. A zero c of a function f can be a *real* or a *complex number*. Recall from algebra that a **complex number** is a number of the form

$$z = a + bi, \quad \text{where} \quad i^2 = -1,$$

and a and b are real numbers. The number a is called the **real part** of z and b is called the **imaginary part** of z. The symbol i is called the **imaginary unit** and it is common practice to write it as $i = \sqrt{-1}$. If $z = a + bi$ is a complex number, then $\bar{z} = a - bi$ is called its **conjugate**. Thus the simple polynomial function $f(x) = x^2 + 1$ has two complex zeros since the solutions of $x^2 + 1 = 0$ are $\pm\sqrt{-1}$, that is, i and $-i$.

◀ The arithmetic of complex numbers is reviewed in Appendix A.

In this section we explore the connection between the zeros of a polynomial function f, the operation of division, and the factors of f.

EXAMPLE 1   A Real Zero

Consider the polynomial function $f(x) = 2x^3 - 9x^2 + 6x - 1$. The real number $\frac{1}{2}$ is a zero of the function since

$$\begin{aligned}
f\left(\tfrac{1}{2}\right) &= 2\left(\tfrac{1}{2}\right)^3 - 9\left(\tfrac{1}{2}\right)^2 + 6\left(\tfrac{1}{2}\right) - 1 \\
&= 2\left(\tfrac{1}{8}\right) - \tfrac{9}{4} + 3 - 1 \\
&= \tfrac{1}{4} - \tfrac{9}{4} + \tfrac{8}{4} = 0.
\end{aligned}$$

≡

EXAMPLE 2   A Complex Zero

Consider the polynomial function $f(x) = x^3 - 5x^2 + 8x - 6$. The complex number $1 + i$ is a zero of the function. To verify this we use the binomial expansion of $(a + b)^3$ and the fact that $i^2 = -1$ and $i^3 = -i$:

◀ See (7) in Section 1.5.

$$\begin{aligned}
f(1 + i) &= (1 + i)^3 - 5(1 + i)^2 + 8(1 + i) - 6 \\
&= (1^3 + 3 \cdot 1^2 \cdot i + 3 \cdot 1 \cdot i^2 + i^3) - 5(1^2 + 2i + i^2) + 8(1 + i) - 6 \\
&= (-2 + 2i) - 5(2i) + (2 + 8i) \\
&= (-2 + 2) + (10 - 10)i = 0 + 0i = 0.
\end{aligned}$$

≡

□ **Factor Theorem** We can now relate the notion of a zero of a polynomial function f with division of polynomials. From the Remainder Theorem we know that when $f(x)$ is divided by the linear polynomial $x - c$ the remainder is $r = f(c)$. If c is a zero of f, then $f(c) = 0$ implies $r = 0$. From the form of the Division Algorithm given in (4) of Section 3.2 we can then write f as

$$f(x) = (x - c)q(x). \tag{1}$$

Thus, if c is a zero of a polynomial function f, then $x - c$ is a factor of $f(x)$. Conversely, if $x - c$ is a factor of $f(x)$, then f has the form given in (1). In this case, we see immediately that $f(c) = (c - c)q(c) = 0$. These results are summarized in the **Factor Theorem** given next.

THEOREM 3.3.1 Factor Theorem

A number c is a zero of a polynomial function f if and only if $x - c$ is a factor of $f(x)$.

Recall, if a polynomial function f is of degree n and $(x - c)^m$, $m \le n$, is a factor of $f(x)$, then c is said to be a **zero of multiplicity m**. When $m = 1$, c is a **simple zero**. Equivalently, we say that the number c is a **root of multiplicity m** of the equation $f(x) = 0$. We have already examined the graphical significance of repeated real zeros of a polynomial function f in Section 3.1. See Figure 3.1.7.

EXAMPLE 3 **Factors of a Polynomial**

Determine whether

(a) $x + 1$ is a factor of $f(x) = x^4 - 5x^2 + 6x - 1$,
(b) $x - 2$ is a factor of $f(x) = x^3 - 3x^2 + 4$.

Solution We use synthetic division to divide $f(x)$ by the given linear term.
(a) From the division

$$
\begin{array}{r|rrrrr}
-1 & 1 & 0 & -5 & 6 & -1 \\
   &   & -1 & 1 & 4 & -10 \\
\hline
   & 1 & -1 & -4 & 10 & \boxed{-11} = r = f(-1)
\end{array}
$$

we see that $f(-1) = -11$ and so -1 is not a zero of f. We conclude that $x - (-1) = x + 1$ is not a factor of $f(x)$.

(b) From the division

$$
\begin{array}{r|rrrr}
2 & 1 & -3 & 0 & 4 \\
  &   & 2 & -2 & -4 \\
\hline
  & 1 & -1 & -2 & \boxed{0} = r = f(2)
\end{array}
$$

we see that $f(2) = 0$. This means that 2 is a zero and that $x - 2$ is a factor of $f(x)$. From the division we see that the quotient is $q(x) = x^2 - x - 2$ and so $f(x) = (x - 2)(x^2 - x - 2)$. ≡

☐ **Number of Zeros** In Example 6 of Section 3.1 we graphed the polynomial function

$$f(x) = (x - 3)(x - 1)^2(x + 2)^3. \tag{2}$$

The number 3 is a simple zero of f; the number 1 is a zero of multiplicity 2; and -2 is a zero of multiplicity 3. Although the function f has three *distinct* zeros (different from one another), it is, nevertheless, standard practice to say that f has *six zeros* because we count the multiplicities of each zero. Hence for the function f in (2), the number of zeros is $1 + 2 + 3 = 6$. The question

How many zeros does a polynomial function f have?

is answered next.

THEOREM 3.3.2 Fundamental Theorem of Algebra

A polynomial function f of degree $n > 0$ has at least one zero.

The foregoing theorem, first proved by the German mathematician **Carl Friedrich Gauss** (1777–1855) in 1799, is considered one of the major milestones in the history of mathematics. At first reading this theorem does not appear to say much, but when combined with the Factor Theorem, the Fundamental Theorem of Algebra shows:

Every polynomial function f of degree n > 0 has exactly n zeros. (3)

Of course if a zero is repeated—say, it has multiplicity k—we count that zero k times. To prove (3), we know from the Fundamental Theorem of Algebra that f has a zero (call it c_1). By the Factor Theorem we can write

$$f(x) = (x - c_1)q_1(x),$$ (4)

where q_1 is a polynomial function of degree $n - 1$. If $n - 1 \neq 0$, then in like manner we know that q_1 must have a zero (call it c_2) and so (4) becomes

$$f(x) = (x - c_1)(x - c_2)q_2(x),$$

where q_2 is a polynomial function of degree $n - 2$. If $n - 2 \neq 0$, we continue and arrive at

$$f(x) = (x - c_1)(x - c_2)(x - c_3)q_3(x),$$ (5)

and so on. Eventually we arrive at a factorization of $f(x)$ with n linear factors and the last factor $q_n(x)$ of degree 0. In other words, $q_n(x) = a_n$, where a_n is a constant. We have arrived at the **complete factorization** of $f(x)$. Bear in mind that some or all the zeros $c_1, \ldots, c_n$ in (6) may be complex numbers $a + bi$, where $b \neq 0$.

THEOREM 3.3.3 Complete Factorization Theorem

Let $c_1, c_2, \ldots, c_n$ be the n (not necessarily distinct) zeros of the polynomial function of degree $n > 0$:

$$f(x) = a_n x^n + a_{n-1} x^{n-1} + \cdots + a_2 x^2 + a_1 x + a_0.$$

Then $f(x)$ can be written as a product of n linear factors

$$f(x) = a_n(x - c_1)(x - c_2) \cdots (x - c_n).$$ (6)

In the case of a second-degree, or quadratic, polynomial function $f(x) = ax^2 + bx + c$, where the coefficients a, b, and c are real numbers, the zeros c_1 and c_2 of f can be found using the quadratic formula:

$$c_1 = \frac{-b - \sqrt{b^2 - 4ac}}{2a} \quad \text{and} \quad c_2 = \frac{-b + \sqrt{b^2 - 4ac}}{2a}.$$ (7)

The results in (7) tell the whole story about the zeros of the quadratic function: the zeros are real and distinct when $b^2 - 4ac > 0$, real with multiplicity two when $b^2 - 4ac = 0$, and complex and distinct when $b^2 - 4ac < 0$. It follows from (6) that the complete factorization of a quadratic polynomial function is

$$f(x) = a(x - c_1)(x - c_2).$$ (8)

EXAMPLE 4   **Example 1 Revisited**

In Example 1 we demonstrated that $\frac{1}{2}$ is a zero of $f(x) = 2x^3 - 9x^2 + 6x - 1$. We now know that $x - \frac{1}{2}$ is a factor of $f(x)$ and that $f(x)$ has three zeros. The synthetic division

$$
\begin{array}{r|rrrr}
\frac{1}{2} & 2 & -9 & 6 & -1 \\
  &   & 1 & -4 & 1 \\
\hline
  & 2 & -8 & 2 & \boxed{0} = r
\end{array}
$$

again demonstrates that $\frac{1}{2}$ is a zero of $f(x)$ (the 0 remainder is the value of $f\left(\frac{1}{2}\right)$) and, additionally, gives us the quotient $q(x)$ obtained in the division of $f(x)$ by $x - \frac{1}{2}$, that is, $f(x) = \left(x - \frac{1}{2}\right)(2x^2 - 8x + 2)$. As shown in (8), we can now factor the quadratic quotient $q(x) = 2x^2 - 8x + 2$ by finding the roots of $2x^2 - 8x + 2 = 0$ by the quadratic formula:

$$\downarrow \sqrt{48} = \sqrt{16 \cdot 3} = \sqrt{16}\sqrt{3} = 4\sqrt{3}$$

$$x = \frac{-(-8) \pm \sqrt{(-8)^2 - 4(2)(2)}}{4} = \frac{8 \pm \sqrt{48}}{4} = \frac{8 \pm 4\sqrt{3}}{4}$$

$$= \frac{4(2 \pm \sqrt{3})}{4} = 2 \pm \sqrt{3}.$$

Thus the remaining zeros of $f(x)$ are the irrational numbers $2 + \sqrt{3}$ and $2 - \sqrt{3}$. With the identification of the leading coefficient as $a_3 = 2$, it follows from (6) that the complete factorization of $f(x)$ is then

$$f(x) = 2\left(x - \tfrac{1}{2}\right)\left(x - \left(2 + \sqrt{3}\right)\right)\left(x - \left(2 - \sqrt{3}\right)\right)$$
$$= 2\left(x - \tfrac{1}{2}\right)\left(x - 2 - \sqrt{3}\right)\left(x - 2 + \sqrt{3}\right). \qquad \equiv$$

EXAMPLE 5　　　　Using Synthetic Division

Find the complete factorization of

$$f(x) = x^4 - 12x^3 + 47x^2 - 62x + 26$$

given that 1 is a zero of f of multiplicity two.

Solution We know that $x - 1$ is a factor of $f(x)$, so by the division

$$
\begin{array}{r|rrrrr}
1\!\!\! & 1 & -12 & 47 & -62 & 26 \\
  &   & 1 & -11 & 36 & -26 \\
\hline
  & 1 & -11 & 36 & -26 & \boxed{0} = r \\
\end{array}
$$

we find $\qquad\qquad f(x) = (x - 1)(x^3 - 11x^2 + 36x - 26).$

Since 1 is a zero of multiplicity 2, $x - 1$ must also be a factor of the quotient $q(x) = x^3 - 11x^2 + 36x - 26$. By the division,

$$
\begin{array}{r|rrrr}
1\!\!\! & 1 & -11 & 36 & -26 \\
  &   & 1 & -10 & 26 \\
\hline
  & 1 & -10 & 26 & \boxed{0} = r \\
\end{array}
$$

we conclude that $q(x)$ can be written $q(x) = (x - 1)(x^2 - 10x + 26)$. Therefore,

$$f(x) = (x - 1)^2(x^2 - 10x + 26).$$

The remaining two zeros, found by solving $x^2 - 10x + 26 = 0$ by the quadratic formula, are the complex numbers $5 + i$ and $5 - i$. Since the leading coefficient is $a_4 = 1$ the complete factorization of $f(x)$ is

$$f(x) = (x - 1)^2(x - (5 + i))(x - (5 - i))$$
$$= (x - 1)^2(x - 5 - i)(x - 5 + i). \qquad \equiv$$

EXAMPLE 6　　　　Complete Linear Factorization

Find a polynomial function f of degree three, with zeros 1, -4, and 5 such that its graph possesses the y-intercept $(0, 5)$.

Solution Because we have three zeros $1, -4,$ and 5 we know $x - 1, x + 4,$ and $x - 5$ are factors of f. However, the function we seek is *not*

$$f(x) = (x - 1)(x + 4)(x - 5). \tag{9}$$

The reason for this is that any nonzero constant multiple of f is a different polynomial with the same zeros. Notice, too, that the function in (9) gives $f(0) = 20$, but we want $f(0) = 5$. Hence by (6) we must assume that f has the form

$$f(x) = a_3(x - 1)(x + 4)(x - 5), \tag{10}$$

where a_3 is some real constant. Using (10), $f(0) = 5$ gives

$$f(0) = a_3(0 - 1)(0 + 4)(0 - 5) = 20a_3 = 5$$

and so $a_3 = \frac{5}{20} = \frac{1}{4}$. The desired function is then

$$f(x) = \tfrac{1}{4}(x - 1)(x + 4)(x - 5). \qquad \equiv$$

We have seen in the introduction that complex zeros of $f(x) = x^2 + 1$ are i and $-i$. In Example 5 the complex zeros are $5 + i$ and $5 - i$. In each case the complex zeros of the polynomial function are conjugate pairs. In other words, one complex zero is the conjugate of the other. This is no coincidence; complex zeros of polynomials with *real* coefficients *always* appear in conjugate pairs. In order to prove this, we use the following results concerning conjugates.

If z_1 and z_2 are complex numbers, then

$$\overline{z_1 + z_2} = \overline{z_1} + \overline{z_2} \quad \text{and} \quad \overline{z_1^n} = \overline{z_1}^n. \tag{11}$$

See Problem 48 in Exercises 3.3.

THEOREM 3.3.4   Conjugate Zeros Theorem

Let $f(x)$ be a polynomial function of degree $n > 1$ with real coefficients. If z is a complex zero of $f(x)$, then the conjugate $\bar{z}$ is also a zero of $f(x)$.

PROOF: Let $f(x) = a_n x^n + a_{n-1} x^{n-1} + \cdots + a_2 x^2 + a_1 x + a_0$, where the coefficients $a_i, i = 0, 1, 2, \ldots, n$ are real numbers. By assumption $f(z) = 0$ so

$$a_n z^n + a_{n-1} z^{n-1} + \cdots + a_2 z^2 + a_1 z + a_0 = 0.$$

Taking the conjugate of both sides of this equation gives

$$\overline{a_n z^n + a_{n-1} z^{n-1} + \cdots + a_2 z^2 + a_1 z + a_0} = \overline{0}.$$

Now using (11) along with the fact that the conjugate of any real number is itself, we obtain

$$a_n \bar{z}^n + a_{n-1} \bar{z}^{n-1} + \cdots + a_2 \bar{z}^2 + a_1 \bar{z} + a_0 = 0.$$

This means $f(\bar{z}) = 0$ and, therefore, $\bar{z}$ is a zero of $f(x)$ whenever z is a zero. $\qquad \equiv$

■ EXAMPLE 7   Example 2 Revisited

In Example 2 we demonstrated that $1 + i$ is a complex zero of $f(x) = x^3 - 5x^2 + 8x - 6$. Since the coefficients of f are real numbers we conclude that another zero is the conjugate of $1 + i$, namely, $1 - i$. Thus we know two factors of $f(x)$, $x - (1 + i)$ and $x - (1 - i)$. Carrying out the multiplication, we find

$$(x - 1 - i)(x - 1 + i) = x^2 - 2x + 2.$$

Thus we can write

$$f(x) = (x - 1 - i)(x - 1 + i)q(x) = (x^2 - 2x + 2)q(x).$$

We determine $q(x)$ by performing the *long division* of $f(x)$ by $x^2 - 2x + 2$. (We can't do synthetic division because we are not dividing by a linear factor.) From

$$
\begin{array}{r}
x - 3 \phantom{{}-6} \\
x^2 - 2x + 2 \overline{\smash{)}\, x^3 - 5x^2 + 8x - 6} \\
\underline{x^3 - 2x^2 + 2x} \phantom{{}-6} \\
-3x^2 + 6x - 6 \\
\underline{-3x^2 + 6x - 6} \\
0
\end{array}
$$

we see that the complete factorization of $f(x)$ is

$$f(x) = (x - 1 - i)(x - 1 + i)(x - 3).$$

The three zeros of $f(x)$ are $1 + i$, $1 - i$, and 3. $\equiv$

3.3 **Exercises** Answers to selected odd-numbered problems begin on page ANS-10.

In Problems 1–6, determine whether the indicated real number is a zero of the given polynomial function f. If yes, find all other zeros and then give the complete factorization of $f(x)$.

1. 1; $f(x) = 4x^3 - 9x^2 + 6x - 1$ **2.** $\frac{1}{2}$; $f(x) = 2x^3 - x^2 + 32x - 16$
3. 5; $f(x) = x^3 - 6x^2 + 6x + 5$ **4.** 3; $f(x) = x^3 - 3x^2 + 4x - 12$
5. $-\frac{2}{3}$; $f(x) = 3x^3 - 10x^2 - 2x + 4$ **6.** -2; $f(x) = x^3 - 4x^2 - 2x + 20$

In Problems 7–12, verify that each of the indicated numbers are zeros of the given polynomial function f. Find all other zeros and then give the complete factorization of $f(x)$.

7. $-3, 5$; $f(x) = 4x^4 - 8x^3 - 61x^2 + 2x + 15$
8. $\frac{1}{4}, \frac{3}{2}$; $f(x) = 8x^4 - 30x^3 + 23x^2 + 8x - 3$
9. $1, -\frac{1}{3}$ (multiplicity 2); $f(x) = 9x^4 + 69x^3 - 29x^2 - 41x - 8$
10. $-\sqrt{5},\ \ \sqrt{5}$; $f(x) = 3x^4 + x^3 - 17x^2 - 5x + 10$
11. $1, 5$; $f(x) = x^5 - 3x^4 - 22x^3 + 74x^2 - 75x + 25$
12. $0, \frac{1}{2}$; $f(x) = 2x^5 - 17x^4 + 40x^3 - 16x^2$

In Problems 13–18, use synthetic division to determine whether the indicated linear polynomial is a factor of the given polynomial function f. If yes, find all other zeros and then give the complete factorization of $f(x)$.

13. $x - 5$; $f(x) = 2x^2 + 6x - 25$
14. $x + \frac{1}{2}$; $f(x) = 10x^2 - 27x + 11$
15. $x - 1$; $f(x) = x^3 + x - 2$
16. $x + \frac{1}{2}$; $f(x) = 2x^3 - x^2 + x + 1$
17. $x - \frac{1}{3}$; $f(x) = 3x^3 - 3x^2 + 8x - 2$
18. $x - 2$; $f(x) = x^3 - 6x^2 - 16x + 48$

In Problems 19–22, use division to show that the indicated polynomial is a factor of the given polynomial function f. Find all other zeros and then give the complete factorization of $f(x)$.

19. $(x - 1)(x - 2);$ $f(x) = x^4 - 3x^3 + 6x^2 - 12x + 8$
20. $x(3x - 1);$ $f(x) = 3x^4 - 7x^3 + 5x^2 - x$
21. $(x - 1)^2;$ $f(x) = 2x^4 + x^3 - 5x^2 - x + 3$
22. $(x + 3)^2;$ $f(x) = x^4 - 4x^3 - 22x^2 + 84x + 261$

In Problems 23–28, verify that the indicated complex number is a zero of the given polynomial function f. Proceed as in Example 7 to find all other zeros and then give the complete factorization of $f(x)$.

23. $2i;$ $f(x) = 3x^3 - 5x^2 + 12x - 20$
24. $\frac{1}{2}i;$ $f(x) = 12x^3 + 8x^2 + 3x + 2$
25. $-1 + i;$ $f(x) = 5x^3 + 12x^2 + 14x + 4$
26. $-i;$ $f(x) = 4x^4 - 8x^3 + 9x^2 - 8x + 5$
27. $1 + 2i;$ $f(x) = x^4 - 2x^3 - 4x^2 + 18x - 45$
28. $1 + i;$ $f(x) = 6x^4 - 11x^3 + 9x^2 + 4x - 2$

In Problems 29–34, find a polynomial function f with real coefficients of the indicated degree that possesses the given zeros.

29. degree 4; $2, 1, -3$ (multiplicity 2)   **30.** degree 5; $-4i, -\frac{1}{3}, \frac{1}{2}$ (multiplicity 2)
31. degree 5; $3 + i, 0$ (multiplicity 3)   **32.** degree 4; $5i, 2 - 3i$
33. degree 2; $1 - 6i$ **34.** degree 2; $4 + 3i$

In Problems 35–38, find the zeros of the given polynomial function f. State the multiplicity of each zero.

35. $f(x) = x(4x - 5)^2(2x - 1)^3$ **36.** $f(x) = x^4 + 6x^3 + 9x^2$
37. $f(x) = (9x^2 - 4)^2$ **38.** $f(x) = (x^2 + 25)(x^2 - 5x + 4)^2$

In Problems 39 and 40, find the value(s) of k such that the indicated number is a zero of $f(x)$. Then give the complete factorization of $f(x)$.

39. 3; $f(x) = 2x^3 - 2x^2 + k$ **40.** 1; $f(x) = x^3 + 5x^2 - k^2x + k$

In Problems 41 and 42, find a polynomial function f of the indicated degree whose graph is given in the figure.

41. degree 3

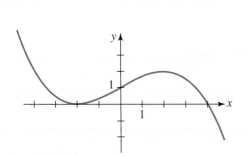

FIGURE 3.3.1 Graph for Problem 41

42. degree 5

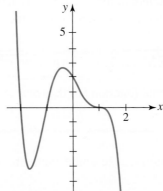

FIGURE 3.3.2 Graph for Problem 42

For Discussion

43. Discuss:
  (a) For what positive-integer values of n is $x - 1$ a factor of $f(x) = x^n - 1$?
  (b) For what positive-integer values of n is $x + 1$ a factor of $f(x) = x^n + 1$?

44. Suppose $f(x)$ is a polynomial function of degree three with real coefficients. Why can't $f(x)$ have three complex zeros? Put another way, why must at least one zero of a cubic polynomial function be a real number? Can you generalize this result?

45. What is the smallest degree that a polynomial function $f(x)$ with real coefficients can have such that $1 + i$ is a complex zero of multiplicity 2? Of multiplicity 3?

46. Let $z = a + bi$. Show that $z + \bar{z}$ and $z\bar{z}$ are real numbers.

47. Let $z = a + bi$. Use the results of Problem 46 to show that

$$f(x) = (x - z)(x - \bar{z})$$

is a polynomial function with real coefficients.

48. Let $z_1 = a + bi$ and $z_2 = c + di$. Show that

$$\overline{z_1 + z_2} = \bar{z}_1 + \bar{z}_2 \quad \text{and} \quad \overline{z_1^2} = \bar{z}_1^2.$$

3.4 Real Zeros of Polynomial Functions

≡ **Introduction** In the preceding section we saw that as a consequence of the Fundamental Theorem of Algebra, a polynomial function f of degree n has n zeros when the multiplicities of the zeros are counted. We also saw that a zero of a polynomial function could be either a real or a complex number. In this section we confine our attention to *real zeros* of polynomial functions with real coefficients.

☐ **Real Zeros** If a polynomial function f of degree $n > 0$ has m (not necessarily distinct) real zeros $c_1, c_2, \ldots, c_m$, then by the Factor Theorem each of the linear polynomials $x - c_1, x - c_2, \ldots, x - c_m$ are factors of $f(x)$. That is,

$$f(x) = (x - c_1)(x - c_2)\cdots(x - c_m)q(x),$$

where $q(x)$ is a polynomial. Thus n, the degree of f, must be greater than or possibly equal to m, the number of real zeros when each is counted according to its multiplicity. Using slightly different words, we restate the last sentence.

THEOREM 3.4.1 Number of Real Zeros

A polynomial function f of degree $n > 0$ has at most n real zeros (not necessarily distinct).

Let's summarize some facts about real zeros of a polynomial function f of degree n:

• f may not have any real zeros.

For example, the fourth degree polynomial function $f(x) = x^4 + 9$ has no real zeros, since there exists no real number x satisfying $x^4 + 9 = 0$ or $x^4 = -9$.

- f may have m real zeros where $m < n$.

For example, the third degree polynomial function $f(x) = (x - 1)(x^2 + 1)$ has one real zero.

- f may have n real zeros.

For example, by factoring the third degree polynomial function $f(x) = x^3 - x$ as $f(x) = x(x^2 - 1) = x(x + 1)(x - 1)$, we see that it has three real zeros.

- f has at least one real zero when its degree n is odd.

This is a consequence of the fact that complex zeros of a polynomial function f with real coefficients must appear in conjugate pairs. Thus if we write down an arbitrary cubic polynomial function such as $f(x) = x^3 + x + 1$, we know that f cannot have just one complex zero, nor can it have three complex zeros. Put another way, $f(x) = x^3 + x + 1$ either has exactly one real zero or it has exactly three real zeros.

- If the coefficients of $f(x)$ are positive and the constant term $a_0 \neq 0$, then any real zeros of f must be negative.

☐ **Finding Real Zeros** It is one thing to talk about the existence of real and complex zeros of a polynomial function; it is an entirely different problem to actually find these zeros. The problem of finding a *formula* that expresses the zeros of a general nth degree polynomial function f in terms of its coefficients perplexed mathematicians for centuries. We have seen in Sections 2.4 and 3.3 that in the case of a second-degree, or quadratic, polynomial function $f(x) = ax^2 + bx + c$, where the coefficients a, b, and c are real numbers, the zeros c_1 and c_2 of f can be found using the quadratic formula.

The problem of finding zeros of third-degree, or cubic, polynomial functions was solved in the sixteenth century through the pioneering work of the Italian mathematician **Niccolò Fontana** (1499–1557), also known as Tartaglia—"the stammerer." Around 1540 another Italian mathematician, **Lodovico Ferrari** (1522–1565) discovered an algebraic formula for determining the zeros for fourth degree, or quartic, polynomial functions. Since these formulas are complicated and difficult to use, they are seldom discussed in elementary courses.

For the next 284 years no one discovered any formulas for zeros for general polynomial functions of degrees five, six, ... For good reason! In 1824, at age 22, the Norwegian mathematician **Niels Henrik Abel** (1802–1829) proved it was impossible to find such formulas for the zeros of all general polynomials of degrees $n \geq 5$ in terms of their coefficients.

Niels Henrik Abel

☐ **Rational Zeros** Real zeros of a polynomial function are either rational or irrational numbers. A rational number is a number of the form p/s, where p and s are integers and $s \neq 0$. An irrational number is one that is not rational. For example, $\frac{1}{4}$ and -9 are rational numbers, but $\sqrt{2}$ and π are irrational, that is, neither $\sqrt{2}$ nor π can be written as a fraction p/s where p and s are integers. So how do we find real zeros for polynomial functions of degree $n > 2$? The bad news: For irrational real zeros, we *may* have to be content to use an accurate graph to "eyeball" their location on the x-axis and then use one of the many sophisticated methods for *approximating* the zero that have been developed over the years. The good news: We can always find the rational real zeros of *any* polynomial function with rational coefficients. We have already seen that synthetic division is a useful method for determining whether a given number c is a zero of a polynomial function $f(x)$. When the remainder in the division of $f(x)$ by $x - c$ is $r = 0$, we have found a zero of the polynomial function f, since $r = f(c) = 0$. For example, $\frac{2}{3}$ is a zero of $f(x) = 18x^3 - 15x^2 + 14x - 8$, since

$$
\begin{array}{r|rrrr}
\frac{2}{3} & 18 & -15 & 14 & -8 \\
& & 12 & -2 & 8 \\
\hline
& 18 & -3 & 12 & \boxed{0} = r.
\end{array}
$$

Hence by the Factor Theorem, both $x - \frac{2}{3}$ and the quotient $18x^2 - 3x + 12$ are factors of f and so we can write the polynomial function as the product

$$
\begin{aligned}
f(x) &= \left(x - \tfrac{2}{3}\right)(18x^2 - 3x + 12) \quad \leftarrow \text{factor 3 from the quadratic polynomial} \\
&= \left(x - \tfrac{2}{3}\right)(3)(6x^2 - x + 4) \\
&= (3x - 2)(6x^2 - x + 4).
\end{aligned} \tag{1}
$$

As discussed in the preceding section, if we can factor the polynomial to the point where the remaining factor is a quadratic polynomial, we can then find the remaining two zeros by the quadratic formula. For this example, the factorization in (1) is as far as we can go using real numbers since the zeros of the quadratic factor $6x^2 - x + 4$ are complex (verify). But the indicated multiplication in (1) illustrates something important about rational zeros. The leading coefficient 18 and the constant term -8 of $f(x)$ are obtained from the products

$$
\overbrace{(3x - 2)}\underbrace{(6x^2 - x + 4)}_{}.
$$

Thus we see that the denominator 3 of the rational zero $\frac{2}{3}$ is a *factor* of the leading coefficient 18 of $f(x) = 18x^3 - 15x^2 + 14x - 8$, and the numerator 2 of the rational zero is a factor of the constant term -8.

This example illustrates the following general principle for determining the rational zeros of a polynomial function. Read the following theorem carefully; the coefficients of f are not only real numbers—they must be *integers*.

THEOREM 3.4.2   Rational Zeros Theorem

Let p/s be a rational number in lowest terms and a zero of the polynomial function

$$
f(x) = a_n x^n + a_{n-1} x^{n-1} + \cdots + a_2 x^2 + a_1 x + a_0,
$$

where the coefficients $a_n, a_{n-1}, \ldots, a_2, a_1, a_0$ are integers with $a_n \neq 0$. Then p is an integer factor of the constant term a_0 and s is an integer factor of the leading coefficient a_n.

The Rational Zeros Theorem deserves to be read several times. Note that Theorem 3.4.2 *does not* assert that a polynomial function f with integer coefficients *must* have a rational zero; rather, it states that *if* a polynomial function f with integer coefficients has a rational zero p/s, then necessarily:

$$
\frac{p}{s}. \quad
\begin{array}{l}
\leftarrow \text{is an integer factor of } a_0 \\
\leftarrow \text{is an integer factor of } a_n
\end{array}
$$

By forming all possible quotients of each integer factor of a_0 to each integer factor of a_n, we can construct a list of *potential* rational zeros of f.

EXAMPLE 1 **Rational Zeros**

Find all rational zeros of $f(x) = 3x^4 - 10x^3 - 3x^2 + 8x - 2$.

Solution We identify the constant term $a_0 = -2$ and leading coefficient $a_4 = 3$, and then list all the integer factors of a_0 and a_4, respectively:

$$p: \pm 1, \pm 2,$$
$$s: \pm 1, \pm 3.$$

Now we form a list of all possible rational zeros p/s by dividing all the factors of p by ± 1 and then by ± 3:

$$\frac{p}{s}: \pm 1, \pm 2, \pm\tfrac{1}{3}, \pm\tfrac{2}{3}. \tag{2}$$

We know that the given fourth-degree polynomial function f has four zeros; if any of these zeros is a real number and is rational, then it must appear in the list (2).

To determine which, if any, of the numbers in (2) are zeros, we could use direct substitution into $f(x)$. Synthetic division, however, is usually a more efficient means of evaluating $f(x)$. We begin by testing -1:

$$\begin{array}{r|rrrrr} -1 & 3 & -10 & -3 & 8 & -2 \\ & & -3 & 13 & -10 & 2 \\ \hline & 3 & -13 & 10 & -2 & \boxed{0} = r. \end{array} \tag{3}$$

The zero remainder shows $r = f(-1) = 0$, and so -1 is a zero of f. Hence $x - (-1) = x + 1$ is a factor of f. Using the quotient found in (3) we can write

$$f(x) = (x + 1)(3x^3 - 13x^2 + 10x - 2). \tag{4}$$

From (4) we see that any other rational zero of f must be a zero of the quotient $3x^3 - 13x^2 + 10x - 2$. Since the latter polynomial is of lower degree, it will be easier to use synthetic division on it rather than on $f(x)$ to check the next rational zero. At this point in the process you should check to see whether the zero just found is a repeated zero. This is done by determining whether the found zero is also a zero of the quotient. A quick check, using synthetic division, shows that -1 is *not* a repeated zero of f since it is not a zero of $3x^3 - 13x^2 + 10x - 2$. So we move on and determine whether the number 1 is a rational zero of f. Indeed, it is *not* because the division

$$\begin{array}{r|rrrr} 1 & 3 & -13 & 10 & -2 \qquad \leftarrow \text{coefficients of the quotient in (3)} \\ & & 3 & -10 & 0 \\ \hline & 3 & -10 & 0 & \boxed{-2} = r \end{array} \tag{5}$$

shows that the remainder is $r = -2 \neq 0$. Checking $\tfrac{1}{3}$, we have

$$\begin{array}{r|rrrr} \tfrac{1}{3} & 3 & -13 & 10 & -2 \\ & & 1 & -4 & 2 \\ \hline & 3 & -12 & 6 & \boxed{0} = r. \end{array} \tag{6}$$

Thus $\tfrac{1}{3}$ is a zero. At this point we can stop using synthetic division since (6) indicates that the remaining factor of f is the quadratic polynomial $3x^2 - 12x + 6$. From the quadratic formula we find that the remaining real zeros are $2 + \sqrt{2}$ and $2 - \sqrt{2}$. Therefore the given polynomial function f has two rational zeros, -1 and $\tfrac{1}{3}$, and two irrational zeros, $2 + \sqrt{2}$ and $2 - \sqrt{2}$. ≡

If you have access to technology, your selection of rational numbers to test in Example 1 can be motivated by a graph of the function $f(x) = 3x^4 - 10x^3 - 3x^2 + 8x - 2$.

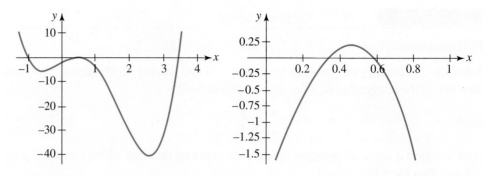

(a) Graph of f on the interval $[-1, 4]$ (b) Zoom-in of graph on the interval $[0, 1]$

FIGURE 3.4.1 Graph of function f in Example 1

With the aid of a graphing utility we obtain the graphs in FIGURE 3.4.1. In Figure 3.4.1(a) it would appear that f has at least three real zeros. But by "zooming-in" on the graph on the interval $[0, 1]$, Figure 3.4.1(b) reveals that f actually has four real zeros: one negative and three positive. Thus, once you have determined one negative rational zero of f you may disregard all other negative numbers as potential zeros.

▌EXAMPLE 2   Complete Factorization

Since the function f in Example 1 is of degree 4 and we have found four real zeros, we can give its complete factorization. Using the leading coefficient $a_4 = 3$, it follows from (6) of Section 3.3 that

$$f(x) = 3(x + 1)\left(x - \tfrac{1}{3}\right)\left(x - \left(2 - \sqrt{2}\right)\right)\left(x - \left(2 + \sqrt{2}\right)\right)$$
$$= 3(x + 1)\left(x - \tfrac{1}{3}\right)\left(x - 2 + \sqrt{2}\right)\left(x - 2 - \sqrt{2}\right). \qquad \equiv$$

▌EXAMPLE 3   Rational Zeros

Find all rational zeros of $f(x) = x^4 + 4x^3 + 5x^2 + 4x + 4$.

Solution In this case the constant term is $a_0 = 4$ and the leading coefficient is $a_4 = 1$. The integer factors of a_0 and a_4 are, respectively:

$$p: \pm 1, \pm 2, \pm 4,$$
$$s: \pm 1.$$

The list of all possible rational zeros p/s is:

$$\frac{p}{s}: \pm 1, \pm 2, \pm 4.$$

Since all the coefficients of f are positive, substituting a positive number from the foregoing list into $f(x)$ can never result in $f(x) = 0$. Thus the only numbers that are potential rational zeros are -1, -2, and -4. From the synthetic division

$$
\begin{array}{r|rrrrr}
-1 & 1 & 4 & 5 & 4 & 4 \\
   &   & -1 & -3 & -2 & -2 \\
\hline
   & 1 & 3 & 2 & 2 & \underline{2} = r
\end{array}
$$

we see that -1 is not a zero. However, from

$$
\begin{array}{r|rrrrr}
-2 & 1 & 4 & 5 & 4 & 4 \\
   &   & -2 & -4 & -2 & -4 \\
\hline
   & 1 & 2 & 1 & 2 & \boxed{0} = r
\end{array}
$$

we see -2 is a zero. We now test to see whether -2 is a repeated zero. Using the coefficients in the quotient,

$$
\begin{array}{r|rrrr}
-2 & 1 & 2 & 1 & 2 \\
   &   & -2 & 0 & -2 \\
\hline
   & 1 & 0 & 1 & \boxed{0} = r
\end{array}
$$

it follows that -2 is a zero of multiplicity 2. So far we have shown that

$$ f(x) = (x + 2)^2(x^2 + 1). $$

Since the zeros of $x^2 + 1$ are the complex conjugates i and $-i$, we can conclude that -2 is the only rational real zero of $f(x)$. ≡

EXAMPLE 4 No Rational Zeros

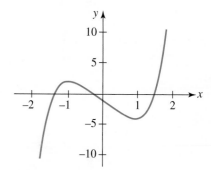

FIGURE 3.4.2 Graph of f in Example 4

Consider the polynomial function $f(x) = x^5 - 4x - 1$. The only possible rational zeros are -1 and 1, and it is easy to see that neither $f(-1)$ nor $f(1)$ are 0. Thus f has no rational zeros. Since f is of odd degree we know that it has at least one real zero, and so that zero must be an irrational number. With the aid of a graphing utility we obtain the graph in FIGURE 3.4.2. Note in the figure that the graph to the right of $x = 2$ cannot turn back down *and* the graph to the left of $x = -2$ cannot turn back up, so that the graph crosses the x-axis five times because that shape of the graph would be inconsistent with the end behavior of f. Thus we can conclude that the function f possesses three irrational real zeros and two complex conjugate zeros. The best we can do here is to approximate these zeros. Using a computer algebra system such as *Mathematica* we can approximate both the real and the complex zeros. We find these approximations to be $-1.34, -0.25, 1.47,$ $0.061 + 1.42i,$ and $0.061 - 1.42i$. ≡

Although the Rational Zeros Theorem requires that the coefficients of a polynomial function f be integers, in some circumstances we can apply the theorem to a polynomial function with some real *noninteger* coefficients. The next example illustrates the concept.

EXAMPLE 5 Noninteger Coefficients

Find the rational zeros of $f(x) = \frac{5}{6}x^4 - \frac{23}{12}x^3 + \frac{10}{3}x^2 - 3x - \frac{3}{4}$.

Solution By multiplying f by the least common denominator 12 of all the rational coefficients, we obtain a new function g with integer coefficients:

$$ g(x) = 10x^4 - 23x^3 + 40x^2 - 36x - 9. $$

In other words, $g(x) = 12f(x)$. If c is a zero of the function g, then c is also zero of f because $g(c) = 0 = 12f(c)$ implies $f(c) = 0$. After working through the numbers in the list of potential rational zeros

$$ \frac{p}{s}: \pm 1, \pm 3, \pm 9, \pm\tfrac{1}{2}, \pm\tfrac{3}{2}, \pm\tfrac{9}{2}, \pm\tfrac{1}{5}, \pm\tfrac{3}{5}, \pm\tfrac{9}{5}, \pm\tfrac{1}{10}, \pm\tfrac{3}{10}, \pm\tfrac{9}{10}, $$

we find that $-\frac{1}{5}$ and $\frac{3}{2}$ are zeros of g, and hence are rational zeros of f. ≡

In Problems 1–20, find all rational zeros of the given polynomial function f.

1. $f(x) = 5x^3 - 3x^2 + 8x + 4$

2. $f(x) = 2x^3 + 3x^2 - x + 2$

3. $f(x) = x^3 - 8x - 3$

4. $f(x) = 2x^3 - 7x^2 - 17x + 10$

5. $f(x) = 4x^4 - 7x^2 + 5x - 1$

6. $f(x) = 8x^4 - 2x^3 + 15x^2 - 4x - 2$

7. $f(x) = x^4 + 2x^3 + 10x^2 + 14x + 21$

8. $f(x) = 3x^4 + 5x^2 + 1$

9. $f(x) = 6x^4 - 5x^3 - 2x^2 - 8x + 3$

10. $f(x) = x^4 + 2x^3 - 2x^2 - 6x - 3$

11. $f(x) = x^4 + 6x^3 - 7x$

12. $f(x) = x^5 - 2x^2 - 12x$

13. $f(x) = x^5 + x^4 - 5x^3 + x^2 - 6x$

14. $f(x) = 128x^6 - 2$

15. $f(x) = \frac{1}{2}x^3 - \frac{9}{4}x^2 + \frac{17}{4}x - 3$

16. $f(x) = 0.2x^3 - x + 0.8$

17. $f(x) = 2.5x^3 + x^2 + 0.6x + 0.1$

18. $f(x) = \frac{3}{4}x^3 + \frac{9}{4}x^2 + \frac{5}{3}x + \frac{1}{3}$

19. $f(x) = 6x^4 + 2x^3 - \frac{11}{6}x^2 - \frac{1}{3}x + \frac{1}{6}$

20. $f(x) = x^4 + \frac{5}{2}x^3 + \frac{3}{2}x^2 - \frac{1}{2}x - \frac{1}{2}$

In Problems 21–30, find all real zeros of the given polynomial function f. Then factor $f(x)$ using only real numbers.

21. $f(x) = 8x^3 + 5x^2 - 11x + 3$

22. $f(x) = 6x^3 + 23x^2 + 3x - 14$

23. $f(x) = 10x^4 - 33x^3 + 66x - 40$

24. $f(x) = x^4 - 2x^3 - 23x^2 + 24x + 144$

25. $f(x) = x^5 + 4x^4 - 6x^3 - 24x^2 + 5x + 20$

26. $f(x) = 18x^5 + 75x^4 + 47x^3 - 52x^2 - 11x + 3$

27. $f(x) = 4x^5 - 8x^4 - 24x^3 + 40x^2 - 12x$

28. $f(x) = 6x^5 + 11x^4 - 3x^3 - 2x^2$

29. $f(x) = 16x^5 - 24x^4 + 25x^3 + 39x^2 - 23x + 3$

30. $f(x) = x^6 - 12x^4 + 48x^2 - 64$

In Problems 31–36, find all real solutions of the given equation.

31. $2x^3 + 3x^2 + 5x + 2 = 0$

32. $x^3 - 3x^2 = -4$

33. $2x^4 + 7x^3 - 8x^2 - 25x - 6 = 0$

34. $9x^4 + 21x^3 + 22x^2 + 2x - 4 = 0$

35. $x^5 - 2x^4 + 2x^3 - 4x^2 + 5x - 2 = 0$

36. $8x^4 - 6x^3 - 7x^2 + 6x - 1 = 0$

In Problems 37 and 38, find a polynomial function f of the indicated degree with integer coefficients that possesses the given rational zeros.

37. degree 4; $-4, \frac{1}{3}, 1, 3$

38. degree 5; $-2, -\frac{2}{3}, \frac{1}{2}, 1$ (multiplicity 2)

39. If $f(x) = 4x^3 - 11x^2 + 14x - 6$, then show the values $f(0)$ and $f(1)$ have different algebric signs. Explain why this information along with the fact that a polynomial function is a continous function (see page 138) enables us to conclude that f has a zero in the interval $[0, 1]$. Find the zero.

40. List, but do not test, all possible rational zeros of

$$f(x) = 24x^3 - 14x^2 + 36x + 105.$$

In Problems 41 and 42, find a cubic polynomial function f that satisfies the given conditions.

41. rational zeros 1 and 2, $f(0) = 1$ and $f(-1) = 4$

42. rational zero $\frac{1}{2}$, irrational zeros $1 + \sqrt{3}$ and $1 - \sqrt{3}$, coefficient of x is 2

Miscellaneous Calculus-Related Problems

43. Construction of a Box A box with no top is made from a square piece of cardboard by cutting square pieces from each corner and then folding up the sides. See FIGURE 3.4.3. The length of one side of the cardboard is 10 inches. Find the length of one side of the squares that were cut from the corners if the volume of the box is 48 in³.

44. Deflection of a Beam A cantilever beam 20 ft long with a load of 600 lb at its right end is deflected by an amount $d(x) = \frac{1}{16,000}(60x^2 - x^3)$, where d is measured in inches and x in feet. See FIGURE 3.4.4. Find x when the deflection is 0.1215 in. When the deflection is 1 in.

For Discussion

45. Discuss: What is the maximum number of times the graphs of the given polynomial functions can intersect?
  (a) $f(x) = a_3x^3 + a_2x^2 + a_1x + a_0$, $g(x) = b_2x^2 + b_1x + b_0$
  (b) $f(x) = x^3 + a_2x^2 + a_1x + a_0$, $g(x) = x^3 + b_2x^2 + b_1x + b_0$

46. Consider the polynomial function $f(x) = x^n + a_{n-1}x^{n-1} + \cdots + a_1x + a_0$, where the coefficients $a_{n-1}, \ldots, a_1, a_0$ are nonzero even integers. Discuss why -1 and 1 cannot be zeros of f.

47. If the leading coefficient of a polynomial function f with integer coefficients is 1, then what can be said about the possible real zeros of f?

48. If k is a prime number (a positive integer greater than 1 whose only positive integer factors are itself and 1) such that $k > 2$, then what are the possible rational zeros of $f(x) = 6x^4 - 9x^2 + k$?

49. (a) The real number $\sqrt{2}$ is a zero of the polynomial function $f(x) = x^2 - 2$. How does the discussion in this section prove that $\sqrt{2}$ is irrational?
  (b) Use the idea implied in part (a) to prove that the real number $1 + \sqrt{2}$ is irrational.

50. Without doing any work, explain why the polynomial function

$$f(x) = 4x^{10} + 9x^6 + 5x^4 + 13x^2 + 3$$

has no real zeros.

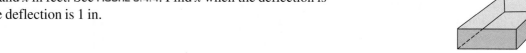

Cut out
Fold

(a)

(b)

FIGURE 3.4.3 Box in Problem 43

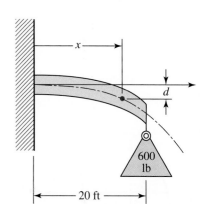

FIGURE 3.4.4 Cantilever beam in Problem 44

3.5 ■ Approximating Real Zeros

≡ **Introduction** A polynomial function f is a **continuous** function. Recall from Section 2.5, this means that the graph of $y = f(x)$ has no breaks, gaps, or holes in it. The following result is a direct consequence of continuity.

THEOREM 3.5.1 Intermediate Value Theorem

Suppose $y = f(x)$ is a continuous function on the closed interval $[a, b]$. If $f(a) \neq f(b)$ for $a < b$, and if N is any number between $f(a)$ and $f(b)$, then there exists a number c in the open interval (a, b) for which $f(c) = N$.

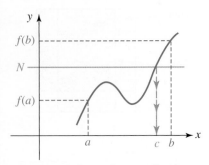

FIGURE 3.5.1 $f(x)$ takes on all values between $f(a)$ and $f(b)$

As we see in FIGURE 3.5.1, the Intermediate Value Theorem simply states that a continuous function $f(x)$ takes on all values between the numbers $f(a)$ and $f(b)$. In particular, if the function values $f(a)$ and $f(b)$ have opposite signs, then by identifying $N = 0$, we can say that there is at least one number in the open interval (a, b) for which $f(c) = 0$. In other words,

> *If either $f(a) > 0$, $f(b) < 0$ or $f(a) < 0$, $f(b) > 0$, then $f(x)$ has at least one zero c in the interval (a, b).* (1)

The plausibility of this conclusion is illustrated in FIGURE 3.5.2.

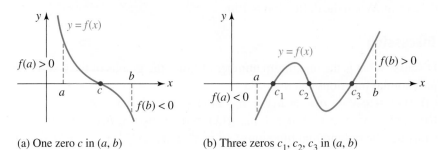

(a) One zero c in (a, b) (b) Three zeros c_1, c_2, c_3 in (a, b)

FIGURE 3.5.2 Locating zeros using the Intermediate Value Theorem

EXAMPLE 1   Using the Intermediate Value Theorem

Consider the polynomial function $f(x) = x^3 - 3x - 1$. From the data in the accompanying table we conclude from (1) that f has a real zero in each of the

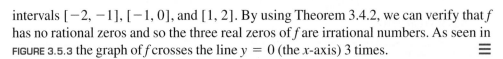

x	-2	-1	0	1	2
$f(x)$	-3	1	-1	-3	1

opposite signs

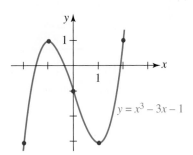

FIGURE 3.5.3 Graph of function in Example 1

intervals $[-2, -1]$, $[-1, 0]$, and $[1, 2]$. By using Theorem 3.4.2, we can verify that f has no rational zeros and so the three real zeros of f are irrational numbers. As seen in FIGURE 3.5.3 the graph of f crosses the line $y = 0$ (the x-axis) 3 times. ≡

In the next example we will obtain an approximation to one of the irrational zeros in Example 1 using a technique called the **bisection method**.

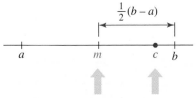

An approximation to c A zero of f

FIGURE 3.5.4 If $f(a)$ and $f(b)$ have opposite signs, then a zero c of f must lie either in $[a, m]$ or in $[m, b]$

☐ **Bisection Method** The basic idea of this method starts with the assumption that a function f is continuous and $f(a)$ and $f(b)$ have opposite signs. From this we know that there exists a number c in (a, b) for which $f(c) = 0$. Then the midpoint $m = (a + b)/2$ of the interval $[a, b]$ is an approximation to c. If $m = (a + b)/2$ is *not* a zero of f, then there is a zero c in an interval (either the open interval (a, m) or the open interval (m, b)) that is one-half the length of the original interval $[a, b]$. If, say, c lies in (m, b) as shown in FIGURE 3.5.4, we then divide this shorter interval in half: Either the new midpoint is a zero or the zero c lies in an interval that is one-fourth the length of the interval $[a, b]$. Continuing in this manner, we can locate the zero c of f in successively shorter intervals. We will then take the midpoints of these intervals as approximations to the zero c. Using this method, we see in Figure 3.5.4 that the error in an approximation to a zero in an interval is less than one-half the length of the interval.

We summarize the discussion as follows:

EXAMPLE 2   Using the Bisection Method

Find an approximation to the zero of $f(x) = x^3 - 3x - 1$ in the interval $[1, 2]$ that is accurate to three decimal places.

Solution Recall from Example 1 that $f(1) < 0$ and $f(2) > 0$. Now to obtain the desired accuracy, we must have the error less than 0.0005.* The first approximation to the zero in $[1, 2]$ is

$$m_1 = \frac{1 + 2}{2} = 1.5 \quad \text{with error} < \tfrac{1}{2}(2 - 1) = 0.5.$$

Since $f(1.5) = -2.15 < 0$, the zero lies in $[1.5, 2]$.
    The second approximation to the zero is

$$m_2 = \frac{1.5 + 2}{2} = 1.75 \quad \text{with error} < \tfrac{1}{2}(2 - 1.5) = 0.25.$$

Since $f(1.75) = -0.89065 < 0$, the zero lies in $[1.75, 2]$.

    The third approximation to the zero is

$$m_3 = \frac{1.75 + 2}{2} = 1.875 \quad \text{with error} < \tfrac{1}{2}(2 - 1.75) = 0.125.$$

Continuing in this manner we eventually find

$$m_{11} = 1.879395 \quad \text{with error} < 0.0005.$$

Thus the number 1.879 is an approximation to the zero of f in $[1, 2]$ that is accurate to three decimal places.   ≡

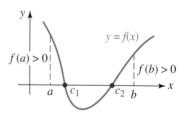

FIGURE 3.5.5 $f(a)$ and $f(b)$ are positive, yet there are two zeros in $[a, b]$

    In Example 2 we leave the approximation to the zeros of $f(x) = x^3 - 3x - 1$ in the intervals $[-2, -1]$ and $[-1, 0]$ as exercises.

---

* If we want an approximation that is accurate to two decimal places, we calculate the midpoints $m_i, i = 1, 2, \ldots, n$ until the error becomes less than 0.005.

Note of Caution ▶ If $f(a)$ and $f(b)$ have the same sign, the polynomial function f could still have one or more zeros in the interval $[a, b]$. See FIGURE 3.5.5 on page 163.

> **3.5** | Exercises Answers to selected odd-numbered problems begin on page ANS-10.

In Problems 1 and 2, find an approximation that is accurate to three decimal places to the zero of $f(x) = x^3 - 3x - 1$ in the given interval.

1. $[-2, -1]$ **2.** $[-1, 0]$

In Problems 3–6, use the bisection method to approximate to an accuracy of three decimal places the zero(s) indicated by the graph of the given function.

3. $f(x) = x^3 - x^2 + 4$

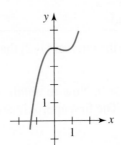

FIGURE 3.5.6 Graph for Problem 3

4. $f(x) = -x^3 - x + 11$

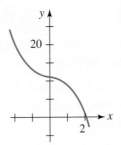

FIGURE 3.5.7 Graph for Problem 4

5. $f(x) = x^4 - 4x^3 + 10$

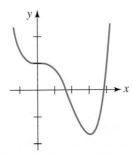

FIGURE 3.5.8 Graph for Problem 5

6. $f(x) = 3x^5 - 5x^3 + 1$

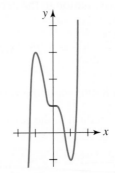

FIGURE 3.5.9 Graph for Problem 6

In Problems 7 and 8, use the bisection method to approximate to an accuracy of three decimal places the x-coordinates of the point(s) of intersection of the given graphs.

7.

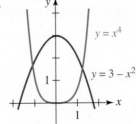

$y = x^4$

$y = 3 - x^2$

FIGURE 3.5.10 Graph for Problem 7

8.

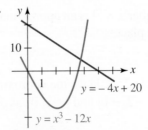

$y = -4x + 20$

$y = x^3 - 12x$

FIGURE 3.5.11 Graph for Problem 8

CHAPTER 3 POLYNOMIAL AND RATIONAL FUNCTIONS

Miscellaneous Applications

9. **Sinking Wooden Ball** A spherical wooden ball of radius r is placed in water. To determine the depth h to which the ball will sink, we equate the weight of the displaced water with the weight of the ball (Archimedes' principle):

$$\frac{\pi}{3}\rho_w h^2(3r - h) = \frac{4\pi}{3}\rho_b r^3,$$

where ρ_w and ρ_b are the densities of water and wood, respectively. See FIGURE 3.5.12. Suppose $\rho_b = 0.4\rho_w$ and $r = 2$ in. Use the bisection method to approximate to an accuracy of two decimal places the depth h to which a wooden ball will sink.

10. **Sag of a Cable** The length L of a cable between two vertical supports of a suspension bridge is given by

$$L = r + \frac{8}{3r}s^2 - \frac{32}{5r^3}s^4,$$

where r is the span of the supports and s is the sag of the cable between the supports. See FIGURE 3.5.13. If $r = 400$ ft and $L = 404$ ft, use the bisection method to approximate the sag s of the cable to an accuracy of two decimal places. [*Hint*: Consider the interval [20, 30].]

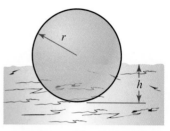

FIGURE 3.5.12 Floating ball in Problem 9

FIGURE 3.5.13 Suspension bridge in Problem 10

3.6 # Rational Functions

≡ **Introduction** Many functions are built up out of polynomial functions by means of arithmetic operations and function composition (see Section 2.6). In this section we construct a class of functions by forming the quotient of two polynomial functions.

DEFINITION 3.6.1 Rational Function

A **rational function** $y = f(x)$ is a function of the form

$$f(x) = \frac{P(x)}{Q(x)}, \qquad (1)$$

where P and Q are polynomial functions.

For example, the following functions are rational functions:

$$y = \frac{x}{x^2 + 5}, \qquad y = \frac{\overset{\text{polynomial}}{\overset{\downarrow}{x^3 - x + 7}}}{\underset{\uparrow}{\underset{\text{polynomial}}{x + 3}}}, \qquad y = \frac{1}{x}.$$

The function

$$y = \frac{\sqrt{x}}{x^2 - 1} \quad \leftarrow \text{not a polynomial}$$

is not a rational function. In (1) we cannot allow the denominator to be zero. So the **domain** of a rational function $f(x) = P(x)/Q(x)$ is the set of all real numbers *except* those numbers for which the denominator $Q(x)$ is zero. For example, the domain of the rational function $f(x) = (2x^3 - 1)/(x^2 - 9)$ is $\{x | x \neq -3, x \neq 3\}$ or $(-\infty, -3) \cup (-3, 3) \cup (3, \infty)$. It goes without saying that we also disallow the zero polynomial $Q(x) = 0$ as a denominator.

☐ **Graphs** Graphing a rational function f is a little more complicated than graphing a polynomial function because in addition to paying attention to

- intercepts,
- symmetry, and
- shifting/reflecting/stretching of known graphs,

you should also keep an eye on

- the domain of f, and
- the degrees of $P(x)$ and $Q(x)$.

The latter two topics are important in determining whether a graph of a rational function possesses *asymptotes*.

The y-intercept is the point $(0, f(0))$, provided the number 0 is in the domain of f. For example, the graph of the rational function $f(x) = (1 - x)/x$ does not cross the y-axis since $f(0)$ is not defined. If the polynomials $P(x)$ and $Q(x)$ have no common factors, then the x-intercepts of the graph of the rational function $f(x) = P(x)/Q(x)$ are the points whose x-coordinates are the real zeros of the numerator $P(x)$. In other words, the only way we can have $f(x) = P(x)/Q(x) = 0$ is to have $P(x) = 0$. The graph of a rational function f is symmetric with respect to the y-axis if $f(-x) = f(x)$, and symmetric with respect to the origin if $f(-x) = -f(x)$. Since it is easy to spot an even or an odd polynomial function (see page 135), here is an easy way to determine symmetry of the graph of a rational function. We again assume $P(x)$ and $Q(x)$ have no common factors.

- The quotient of two even functions is even. (2)
- The quotient of two odd functions is even. (3)
- The quotient of an even and an odd function is odd. (4)

See Problem 48 in Exercises 3.6.

We have already seen the graphs of two simple rational functions, $y = 1/x$ and $y = 1/x^2$, in Figures 2.2.1(e) and 2.2.1(f). You are encouraged to review those graphs at this time. Note that $P(x) = 1$ is an even function and $Q(x) = x$ is an odd function, so $y = 1/x$ is an odd function by (4). On the other hand, $P(x) = 1$ is an even function and $Q(x) = x^2$ is an even function, so $y = 1/x^2$ is an even function by (2).

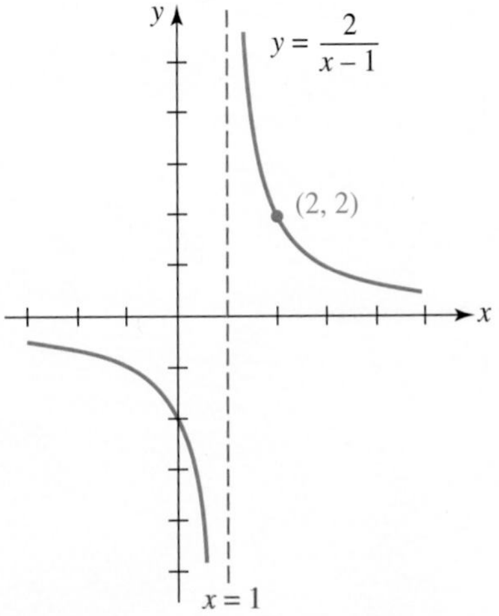

FIGURE 3.6.1 Graph of function in Example 1

EXAMPLE 1 Shifted Reciprocal Function

Graph the function $f(x) = \dfrac{2}{x - 1}$.

Solution The graph possesses no symmetry since $Q(x) = x - 1$ is neither even nor odd. Since $f(0) = -2$, the y-intercept is $(0, -2)$. Because $P(x) = 2$ is never 0, there are no x-intercepts. You might also recognize that the graph of this rational function is the graph of the reciprocal function $y = 1/x$ stretched vertically by a factor of 2 and shifted 1 unit to the right. The point $(1, 1)$ is on the graph of $y = 1/x$; in FIGURE 3.6.1, after the vertical stretch and horizontal shift, the corresponding point on the graph of $y = 2/(x-1)$ is $(2, 2)$.

The vertical line $x = 1$ and the horizontal line $y = 0$ (the equation of the x-axis) are of special importance for this graph.

The vertical dashed line $x = 1$ in Figure 3.6.1 is the y-axis in Figure 2.2.1(e) shifted 1 unit to the right. Although the number 1 is not in the domain of the given function, we can evaluate f at values of x that are *near* 1. For example, you should verify that

x	0.999	1.001
$f(x)$	-2000	2000

(5)

The table in (5) shows that for values of x close to 1, the corresponding function values $f(x)$ are large in absolute value. On the other hand, for values of x for which $|x|$ is large, the corresponding function values $f(x)$ are near 0. For example, you should verify that

x	-999	1001
$f(x)$	-0.002	0.002

(6)

Geometrically, as x approaches 1, the graph of the function approaches the vertical line $x = 1$, and as $|x|$ increases without bound the graph of the function approaches the horizontal line $y = 0$. ☰

☐ **Asymptotes** Recall from Section 1.5 that to indicate that x is approaching a number a, we use the arrow notation

- $x \rightarrow a^-$ to mean that x is approaching a from the *left*, that is, through numbers that are less than a;
- $x \rightarrow a^+$ to mean that x is approaching a from the *right*, that is, through numbers that are greater than a; and
- $x \rightarrow a$ to mean that x is approaching a from both the *left* and the *right*.

We also use the infinity symbols and the arrow notation

- $x \rightarrow -\infty$ to mean that x becomes *unbounded in the negative direction*, and
- $x \rightarrow \infty$ to mean that x becomes *unbounded in the positive direction*.

Similar interpretations are given to the symbols $f(x) \rightarrow -\infty$ and $f(x) \rightarrow \infty$. These notational devices are a convenient way of describing the behavior of a function either near a number $x = a$ or as x increases to the right or decreases to the left. Thus, in Example 1 it is apparent from (5) and Figure 3.6.1 that

$$f(x) \rightarrow -\infty \text{ as } x \rightarrow 1^- \qquad \text{and} \qquad f(x) \rightarrow \infty \text{ as } x \rightarrow 1^+.$$

In words, the notation in the preceding line signifies that the function values are decreasing without bound as x approaches 1 from the left, and the function values are increasing without bound as x approaches 1 from the right. From (6) and Figure 3.5.1 it should also be apparent that

$$f(x) \rightarrow 0 \text{ as } x \rightarrow -\infty \qquad \text{and} \qquad f(x) \rightarrow 0 \text{ as } x \rightarrow \infty.$$

In Figure 3.6.1, the vertical line whose equation is $x = 1$ is a called a **vertical asymptote** for the graph of f, and the horizontal line whose equation is $y = 0$ is called a **horizontal asymptote** for the graph of f.

In this section we will examine three types of asymptotes, which correspond to the three types of lines studied in Section 2.3: *vertical lines*, *horizontal lines*, and *slant* (or oblique) *lines*. The characteristic of any asymptote is that the graph of a function f must get close to, or approach, the line.

DEFINITION 3.6.2   Vertical Asymptote

A line $x = a$ is said to be a **vertical asymptote** for the graph of a function f if at least one of the following six statements is true:

$$f(x) \to -\infty \quad \text{as} \quad x \to a^-, \qquad f(x) \to \infty \quad \text{as} \quad x \to a^-,$$
$$f(x) \to -\infty \quad \text{as} \quad x \to a^+, \qquad f(x) \to \infty \quad \text{as} \quad x \to a^+, \qquad (7)$$
$$f(x) \to -\infty \quad \text{as} \quad x \to a, \qquad f(x) \to \infty \quad \text{as} \quad x \to a.$$

FIGURE 3.6.2 illustrates four of the possibilities listed in (7) of Definition 3.6.2 for the unbounded behavior of a function f near a vertical asymptote $x = a$. If the function exhibits the *same kind of unbounded behavior from both sides of $x = a$*, then we write either

$$f(x) \to \infty \qquad \text{as} \qquad x \to a, \qquad (8)$$

or
$$f(x) \to -\infty \qquad \text{as} \qquad x \to a. \qquad (9)$$

In Figure 3.6.2(d) we see that $f(x) \to \infty$ as $x \to a^-$ and $f(x) \to \infty$ as $x \to a^+$, and so we write $f(x) \to \infty$ as $x \to a$.

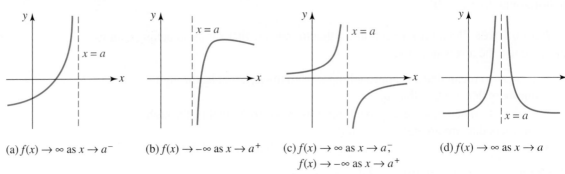

(a) $f(x) \to \infty$ as $x \to a^-$ (b) $f(x) \to -\infty$ as $x \to a^+$ (c) $f(x) \to \infty$ as $x \to a_-^-$
$f(x) \to -\infty$ as $x \to a^+$ (d) $f(x) \to \infty$ as $x \to a$

FIGURE 3.6.2 The line $x = a$ is a vertical asymptote

If $x = a$ is a vertical asymptote for the graph of a *rational function* $f(x) = P(x)/Q(x)$, then the function values $f(x)$ become unbounded as x approaches a from *both sides*, that is, from the right $(x \to a^+)$ *and* from the left $(x \to a^-)$. The graphs in Figures 3.6.2(c) and 3.6.2(d) (or the reflection of these graphs in the x-axis) are typical graphs of a rational function with a single vertical asymptote. As can be seen from these figures, a rational function with a vertical asymptote is a **discontinuous function**. There is an infinite break in each graph at $x = a$. As seen in Figures 3.6.2(c) and 3.6.2(d), a single vertical asymptote divides the xy-plane into two regions, and within each region there is a single piece or **branch** of the graph of the rational function f.

DEFINITION 3.6.3   Horizontal Asymptote

A line $y = c$ is said to be a **horizontal asymptote** for the graph of a function f if

$$f(x) \to c \text{ as } x \to -\infty \qquad \text{or} \qquad f(x) \to c \text{ as } x \to \infty. \qquad (10)$$

In FIGURE 3.6.3 we have illustrated some typical horizontal asymptotes. We note, in conjunction with Figure 3.6.3(d) that, in general, the graph of a function can have at most *two* horizontal asymptotes, but the graph of a *rational function* $f(x) = P(x)/Q(x)$ can have at most *one*. If the graph of a rational function f possesses a horizontal asymptote $y = c$, then as shown in Figure 3.6.3(c),

◀ Remember this.

$$f(x) \to c \text{ as } x \to -\infty \qquad \text{and} \qquad f(x) \to c \text{ as } x \to \infty.$$

The last line is a mathematical description of the end behavior of the graph of a rational function with a horizontal asymptote. Also, the graph of a function can *never* cross a vertical asymptote but, as suggested in Figure 3.6.3(a), a graph can cross a horizontal asymptote.

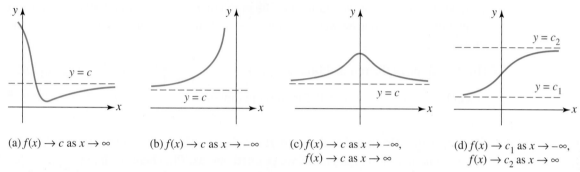

(a) $f(x) \to c$ as $x \to \infty$ (b) $f(x) \to c$ as $x \to -\infty$ (c) $f(x) \to c$ as $x \to -\infty$, $f(x) \to c$ as $x \to \infty$ (d) $f(x) \to c_1$ as $x \to -\infty$, $f(x) \to c_2$ as $x \to \infty$

FIGURE 3.6.3 The line $y = c$ is a horizontal asymptote in (a), (b), and (c)

DEFINITION 3.6.4   Slant Asymptote

A line $y = mx + b$, $m \neq 0$, is said to be a **slant asymptote** for the graph of a function f if

$$f(x) \to mx + b \text{ as } x \to -\infty$$

or

$$f(x) \to mx + b \text{ as } x \to \infty. \tag{11}$$

◀ A slant asymptote is also called an **oblique asymptote**.

The notation in (11) of Definition 3.6.4 means that the graph of f possesses a slant asymptote whenever the function values $f(x)$ become closer and closer to the values of y on the line $y = mx + b$ as x becomes large in absolute value. Another way of stating (11) is: A line $y = mx + b$ is a slant asymptote for the graph of f if the vertical distance $d(x)$ between points with the same x-coordinate on the two graphs satisfies

$$d(x) = f(x) - (mx + b) \to 0 \text{ as } x \to -\infty \text{ or as } x \to \infty.$$

See FIGURE 3.6.4. We note that if a graph of a rational function $f(x) = P(x)/Q(x)$ possesses a slant asymptote it can have vertical asymptotes, but the graph *cannot* have a horizontal asymptote.

On a practical level, vertical and horizontal asymptotes of the graph of a rational function f can be determined by inspection. So for the sake of discussion let us suppose that

$$f(x) = \frac{P(x)}{Q(x)} = \frac{a_n x^n + a_{n-1} x^{n-1} + \cdots + a_1 x + a_0}{b_m x^m + b_{m-1} x^{m-1} + \cdots + b_1 x + b_0}, \quad a_n \neq 0, b_m \neq 0, \tag{12}$$

represents a general rational function.

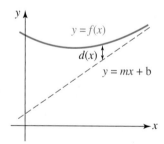

FIGURE 3.6.4   Slant asymptote is $y = mx + b$

☐ **Finding a Vertical Asymptote** Let us assume that the polynomial functions $P(x)$ and $Q(x)$ in (12) have no common factors. In that case:

- If a is a real number such that $Q(a) = 0$, then the line $x = a$ is a vertical asymptote for the graph of f.

Since $Q(x)$ is a polynomial function of degree m, it can have up to m real zeros, and so the graph of a rational function f can have up to m vertical asymptotes. If the graph of a rational function f has, say, k ($k \leq m$) vertical asymptotes, then the k vertical lines divide the xy-plane into $k + 1$ regions. Thus the graph of this rational function would have $k + 1$ branches.

EXAMPLE 2　　　Vertical Asymptotes

(a) Inspection of the rational function $f(x) = \dfrac{2x + 1}{x^2 - 4}$ shows that the denominator $Q(x) = x^2 - 4 = (x + 2)(x - 2) = 0$ at $x = -2$ and $x = 2$. These are equations of vertical asymptotes for the graph of f. The graph of f has three branches: one to the left of the line $x = -2$, one between the lines $x = -2$ and $x = 2$, and one to the right of the line $x = 2$.

(b) The graph of the rational function $f(x) = \dfrac{1}{x^2 + x + 4}$ has no vertical asymptotes, since $Q(x) = x^2 + x + 4 \neq 0$ for all real numbers. 　　　　　　　　　　　　≡

☐ **Finding a Horizontal Asymptote** When we discussed end behavior of a polynomial function $P(x)$ of degree n, we pointed out that $P(x)$ behaves like $y = a_n x^n$, that is, $P(x) \approx a_n x^n$, for values of x large in absolute value. As a consequence, we see from

lower powers of x are
irrelevant as $x \to \pm\infty$
↓

$$f(x) = \dfrac{a_n x^n \boxed{+\ a_{n-1}x^{n-1} + \cdots + a_1 x + a_0}}{b_m x^m \boxed{+\ b_{m-1}x^{m-1} + \cdots + b_1 x + b_0}}$$

that $f(x)$ behaves like $y = \dfrac{a_n}{b_m}x^{n-m}$ because $f(x) \approx \dfrac{a_n x^n}{b_m x^m} = \dfrac{a_n}{b_m}x^{n-m}$ for $x \to \pm\infty$.

Therefore:

0
↓

$$\text{If } n = m, \qquad f(x) \approx \dfrac{a_n}{b_m}x^{n-n} \to \dfrac{a_n}{b_m} \qquad \text{as } x \to \pm\infty. \tag{13}$$

negative
↓

$$\text{If } n < m, \qquad f(x) \approx \dfrac{a_n}{b_m}x^{n-m} = \dfrac{a_n}{b_m}\dfrac{1}{x^{m-n}} \to 0 \qquad \text{as } x \to \pm\infty. \tag{14}$$

positive
↓

$$\text{If } n > m, \qquad f(x) \approx \dfrac{a_n}{b_m}x^{n-m} \to \infty \qquad \text{as } x \to \pm\infty. \tag{15}$$

From (13), (14), and (15) we glean the following three facts about horizontal asymptotes for the graph of $f(x) = P(x)/Q(x)$:

- If degree of $P(x)$ = degree of $Q(x)$, then $y = a_n/b_m$ (the quotient of the leading coefficients) is a horizontal asymptote. (16)
- If degree of $P(x)$ < degree of $Q(x)$, then $y = 0$ is a horizontal asymptote. (17)
- If degree of $P(x)$ > degree of $Q(x)$, then the graph of f has *no* horizontal asymptote. (18)

CHAPTER 3 POLYNOMIAL AND RATIONAL FUNCTIONS

EXAMPLE 3   Horizontal Asymptotes

Determine whether the graph of each of the following rational functions possesses a horizontal asymptote.

(a) $f(x) = \dfrac{3x^2 + 4x - 1}{8x^2 + x}$ **(b)** $f(x) = \dfrac{4x^3 + 7x + 8}{2x^4 + 3x^2 - x + 6}$ **(c)** $f(x) = \dfrac{5x^3 + x^2 + 1}{2x + 3}$

Solution **(a)** Since the degree of the numerator $3x^2 + 4x - 1$ is the same as the degree of the denominator $8x^2 + x$ (both degrees are 2), we see from (13) that

$$f(x) \approx \frac{3}{8}x^{2-2} = \frac{3}{8} \quad \text{as} \quad x \to \pm\infty.$$

As summarized in (16), $y = \frac{3}{8}$ is a horizontal asymptote for the graph of f.

(b) Since the degree of the numerator $4x^3 + 7x + 8$ is 3 and the degree of the denominator $2x^4 + 3x^2 - x + 6$ is 4 (and $3 < 4$), we see from (14) that

$$f(x) \approx \frac{4}{2}x^{3-4} = \frac{2}{x} \to 0 \quad \text{as} \quad x \to \pm\infty.$$

As summarized in (17), $y = 0$ (the x-axis) is a horizontal asymptote for the graph of f.

(c) Since the degree of the numerator $5x^3 + x^2 - 1$ is 3 and the degree of the denominator $2x + 3$ is 1 (and $3 > 1$), we see from (15) that

$$f(x) \approx \frac{5}{2}x^{3-1} = \frac{5}{2}x^2 \to \infty \quad \text{as} \quad x \to \pm\infty.$$

As summarized in (18), the graph of f has no horizontal asymptote.   ≡

    In the graphing examples that follow we will assume again that $P(x)$ and $Q(x)$ in (12) have no common factors.

EXAMPLE 4   Graph of a Rational Function

Graph the function $f(x) = \dfrac{3 - x}{x + 2}$.

Solution Here are some things we look at to sketch the graph of f.

Symmetry: No symmetry. $P(x) = 3 - x$ and $Q(x) = x + 2$ are neither even nor odd.

Intercepts: $f(0) = \frac{3}{2}$ and so the y-intercept is $\left(0, \frac{3}{2}\right)$. Setting $P(x) = 0$ or $3 - x = 0$ implies 3 is a zero of P. The single x-intercept is $(3, 0)$.

Vertical Asymptotes: Setting $Q(x) = 0$ or $x + 2 = 0$ gives $x = -2$. The line $x = -2$ is a vertical asymptote.

Branches: Because there is only a single vertical asymptote, the graph of f consists of two distinct branches, one to the left of $x = -2$ and one to the right of $x = -2$.

Horizontal Asymptote: The degree of P and the degree of Q are the same (namely, 1), and so the graph of f has a horizontal asymptote. By rewriting f as

$f(x) = \dfrac{-x + 3}{x + 2}$ we see that the ratio of leading coefficients is $-1/1 = -1$. From (16) we see that the line $y = -1$ is a horizontal asymptote.

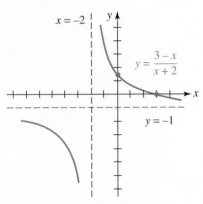

$x = -2$

$y = \dfrac{3 - x}{x + 2}$

$y = -1$

FIGURE 3.6.5 Graph of function in Example 4

The Graph: We draw the vertical and horizontal asymptotes using dashed lines. The right branch of the graph of f is drawn through the intercepts $\left(0, \frac{3}{2}\right)$ and $(3, 0)$ in such a manner that it approaches both asymptotes. The left branch is drawn *below* the horizontal asymptote $y = -1$. Were we to draw this branch above the horizontal asymptote it would have to be near the horizontal asymptote from above and near the vertical asymptote from the left. In order to do this the branch of the graph would have to cross the x-axis, but since there are no more x-intercepts this is impossible. See **FIGURE 3.6.5**. ≡

EXAMPLE 5   Example 4 Using Transformations

Long division and rigid transformations can sometimes be aids in graphing rational functions. Note that if we carry out the long division for the function f in Example 4, we see that

$$f(x) = \frac{3 - x}{x + 2} \quad \text{is the same as} \quad f(x) = -1 + \frac{5}{x + 2}.$$

Thus, starting with the graph of $y = 1/x$, we stretch it vertically by a factor of 5. Next, shift the graph of $y = 5/x$ two units to the left. Finally, shift $y = 5/(x + 2)$ one unit vertically downward. You should verify that the net result is the graph in Figure 3.6.5. ≡

EXAMPLE 6   Graph of a Rational Function

Graph the function $f(x) = \dfrac{x}{1 - x^2}$.

Solution *Symmetry*: Since $P(x) = x$ is odd and $Q(x) = 1 - x^2$ is even, the quotient $P(x)/Q(x)$ is odd. The graph of f is symmetric with respect to the origin.

Intercepts: $f(0) = 0$, and so the y-intercept is $(0, 0)$. Setting $P(x) = x = 0$ gives $x = 0$. Thus the only intercept is $(0, 0)$.

Vertical Asymptotes: Setting $Q(x) = 0$ or $1 - x^2 = 0 = 0$ gives $x = -1$ and $x = 1$. The lines $x = -1$ and $x = 1$ are vertical asymptotes.

Branches: Because there are two vertical asymptotes, the graph of f consists of three distinct branches, one to the left of the line $x = -1$, one between the lines $x = -1$ and $x = 1$, and one to the right of the line $x = 1$.

Horizontal Asymptote: Since the degree of the numerator x is 1 and the degree of the denominator $1 - x^2$ is 2 (and $1 < 2$), it follows from (14) and (17) that $y = 0$ is a horizontal asymptote for the graph of f.

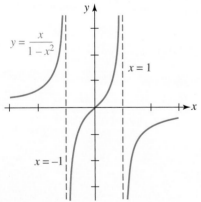

$y = \dfrac{x}{1 - x^2}$

$x = 1$

$x = -1$

FIGURE 3.6.6 Graph of function in Example 6

The Graph: We can plot the graph for $x \geq 0$ and then use symmetry to obtain the remaining part of the graph of $x < 0$. We begin by drawing the vertical asymptotes using dashed lines. The half-branch of the graph of f on the interval $[0, 1)$ is drawn starting at $(0, 0)$. The function f must then increase because $P(x) = x > 0$, and $Q(x) = 1 - x^2 > 0$ indicates that $f(x) > 0$ for $0 < x < 1$. This implies that near the vertical asymptote $x = 1, f(x) \to \infty$ as $x \to 1^-$. The branch of the graph for $x > 1$ is drawn below the horizontal asymptote $y = 0$, since $P(x) = x > 0$ and $Q(x) = 1 - x^2 < 0$ imply $f(x) < 0$. Thus $f(x) \to -\infty$ as $x \to 1^+$ and $f(x) \to 0$ as $x \to \infty$. The remainder of the graph for $x < 0$ is obtained by reflecting the graph for $x > 0$ through the origin. See **FIGURE 3.6.6**. ≡

CHAPTER 3 POLYNOMIAL AND RATIONAL FUNCTIONS

EXAMPLE 7　　　Graph of a Rational Function

Graph the function $f(x) = \dfrac{x}{1 + x^2}$.

Solution The given function f is similar to the function in Example 6 in that f is an odd function, $(0, 0)$ is the only intercept of its graph, and its graph has the horizontal asymptote $y = 0$. However, note that since $1 + x^2 > 0$ for all real numbers, there are no vertical asymptotes. Thus there are no branches; the graph is one continuous curve. For $x \geq 0$, the graph passes through $(0, 0)$ and then must increase since $f(x) > 0$ for $x > 0$. Also, f must attain a relative maximum and then decrease in order to satisfy the condition $f(x) \to 0$ as $x \to \infty$. As mentioned in Section 3.1, the exact value of this relative maximum can be obtained through calculus techniques. Finally, we reflect the portion of the graph for $x > 0$ through the origin. The graph must look something like that shown in **FIGURE 3.6.7**. ≡

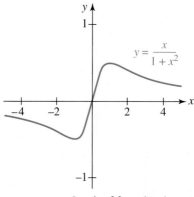

FIGURE 3.6.7 Graph of function in Example 7

☐ **Finding a Slant Asymptote** Let us again assume that the polynomials $P(x)$ and $Q(x)$ in (12) have no common factors. In that case we can recognize the existence of a slant asymptote in the following manner:

- If the degree of $P(x)$ is precisely one greater than the degree of $Q(x)$, that is, if the degree of $Q(x)$ is m and the degree of $P(x)$ is $m + 1$, then the graph of f possesses a slant asymptote.

We find the slant asymptote by division. Using long division to divide $P(x)$ by $Q(x)$ yields a quotient that is a linear polynomial $mx + b$ and a polynomial remainder $R(x)$:

$$f(x) = \frac{P(x)}{Q(x)} = mx + b + \frac{R(x)}{Q(x)}. \qquad (19)$$

where $mx + b$ is the quotient and $R(x)$ is the remainder.

Because the degree of $R(x)$ must be less than the degree of the divisor $Q(x)$, we have $R(x)/Q(x) \to 0$ as $x \to -\infty$ and as $x \to \infty$, and consequently

$$f(x) \to mx + b \text{ as } x \to -\infty \qquad \text{and} \qquad f(x) \to mx + b \text{ as } x \to \infty.$$

In other words, an equation of the slant asymptote is $y = mx + b$, where $mx + b$ is the quotient in (19).

If the denominator $Q(x)$ is a *linear* polynomial, we can then use synthetic division to carry out the long division.

EXAMPLE 8　　　Graph with a Slant Asymptote

Graph the function $f(x) = \dfrac{x^2 - x - 6}{x - 5}$.

Solution *Symmetry*: No symmetry. $P(x) = x^2 - x - 6$ and $Q(x) = x - 5$ are neither even nor odd.

Intercepts: $f(0) = \frac{6}{5}$, and so the y-intercept is $\left(0, \frac{6}{5}\right)$. Setting $P(x) = 0$ or $x^2 - x - 6 = 0$ or $(x + 2)(x - 3) = 0$ shows that -2 and 3 are zeros of $P(x)$. The x-intercepts are $(-2, 0)$ and $(3, 0)$.

Vertical Asymptotes: Setting $Q(x) = 0$ or $x - 5 = 0$ gives $x = 5$. The line $x = 5$ is a vertical asymptote.

Branches: The graph of f consists of two branches, one to the left of $x = 5$ and one to the right of $x = 5$.

Horizontal Asymptote: None.

Slant Asymptote: Since the degree of $P(x) = x^2 - x - 6$ (which is 2) is exactly one greater than the degree of $Q(x) = x - 5$ (which is 1), the graph of $f(x)$ has a slant asymptote. To find it, we divide $P(x)$ by $Q(x)$. Because $Q(x)$ is a linear polynomial we can use synthetic division:

$$\begin{array}{r|rrr} 5 & 1 & -1 & -6 \\ & & 5 & 20 \\ \hline & 1 & 4 & \boxed{14.} \end{array}$$

Recall that the latter notation means that

$$y = x + 4 \text{ is the slant asymptote}$$
$$\downarrow$$
$$\frac{x^2 - x - 6}{x - 5} = x + 4 + \frac{14}{x - 5}.$$

Note again that $14/(x - 5) \to 0$ as $x \to \pm\infty$. Hence the line $y = x + 4$ is a slant asymptote.

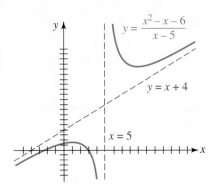

FIGURE 3.6.8 Graph of function in Example 8

The Graph: Using the foregoing information we obtain the graph in FIGURE 3.6.8. The asymptotes are the dashed lines in the figure. ≡

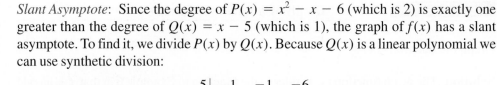

EXAMPLE 9　　　Graph with a Slant Asymptote

By inspection it should be apparent that the graph of the rational function $f(x) = \dfrac{x^3 - 8x + 12}{x^2 + 1}$ possesses a slant asymptote but no vertical asymptotes. Since the denominator is a quadratic polynomial we resort to long division to obtain

$$\frac{x^3 - 8x + 12}{x^2 + 1} = x + \frac{-9x + 12}{x^2 + 1}.$$

The slant asymptote is the line $y = x$. The graph has no symmetry. The y-intercept is $(0, 12)$. The lack of vertical asymptotes indicates that the function f is continuous; its graph consists of an unbroken curve. Because the numerator is a polynomial of odd degree, we know that it has at least one real zero. Since $x^3 - 8x + 12 = 0$ has no rational roots, we use approximation or graphical techniques to show that the equation possesses only one real irrational root. Thus the x-intercept is approximately $(-3.4, 0)$. The graph of f is given in FIGURE 3.6.9. Notice in the figure that the graph of f crosses the slant asymptote. ≡

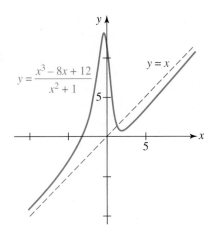

FIGURE 3.6.9 Graph of function in Example 9

☐ **Graph with a Hole** We assumed throughout the preceding discussion of asymptotes of rational functions that the polynomial functions $P(x)$ and $Q(x)$ in (1) have no common factors. We now know that if a is a real number such that $Q(a) = 0$, and $P(x)$ and $Q(x)$ have no common factors, then the line $x = a$ is a vertical asymptote for the graph of f. Because Q is a polynomial function, it follows from the Factor Theorem that $Q(x) = (x - a)q(x)$. The assumption that the numerator P and denominator Q have no common factors tells us that $x - a$ is not a factor of P and so $P(a) \neq 0$. When $P(a) = 0$ *and* $Q(a) = 0$, then $x = a$ *may not* be a vertical asymptote. For example, when a is a *simple zero* of both P and Q, then $x = a$ is *not* a vertical asymptote for the graph of $f(x) = P(x)/Q(x)$. To see this, we know from the Factor Theorem that if $P(a) = 0$ and $Q(a) = 0$, then $x - a$ is a common factor of P and Q:

$$P(x) = (x - a)p(x) \qquad \text{and} \qquad Q(x) = (x - a)q(x),$$

where p and q are polynomials such that $p(a) \neq 0$ and $q(a) \neq 0$. After canceling

$$f(x) = \frac{P(x)}{Q(x)} = \frac{(x - a)p(x)}{(x - a)q(x)} = \frac{p(x)}{q(x)}, \quad x \neq a,$$

we see that $f(x)$ is undefined at a, but the function values $f(x)$ do not become unbounded as $x \to a^-$ or as $x \to a^+$ because $q(x)$ is not approaching 0. As an example, we saw in Section 2.5 that the graph of the rational function

$$f(x) = \frac{x^2 - 4}{x - 2} = \frac{(x - 2)(x + 2)}{x - 2} = x + 2, \quad x \neq 2,$$

is basically a straight line. But since $f(2)$ is undefined there is no point $(2, f(2))$ on the line. Instead, there is a **hole** in the graph at the point $(2, 4)$. See Figure 2.5.5(a) on page 87.

EXAMPLE 10　　Graph with a Hole

Graph the function $f(x) = \dfrac{x^2 - 2x - 3}{x^2 - 1}$.

Solution Although $x^2 - 1 = 0$ for $x = -1$ and $x = 1$, only $x = 1$ is a vertical asymptote. Note that the numerator $P(x)$ and denominator $Q(x)$ have the common factor $x + 1$, which we cancel provided $x \neq -1$:

equality is true for $x \neq -1$
↓

$$f(x) = \frac{(x + 1)(x - 3)}{(x + 1)(x - 1)} = \frac{x - 3}{x - 1}. \tag{20}$$

Thus we see from (20) that there is no infinite break in the graph at $x = -1$. We graph $y = \dfrac{x - 3}{x - 1}, x \neq -1$, by observing that the y-intercept is $(0, 3)$, an x-intercept is $(3, 0)$, a vertical asymptote is $x = 1$, and a horizontal asymptote is $y = 1$. The graph of this function has two branches, but the branch to the left of the vertical asymptote $x = 1$ has a hole in it corresponding to the point $(-1, 2)$. See FIGURE 3.6.10. ≡

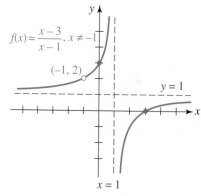

FIGURE 3.6.10 Graph of function in Example 10

NOTES FROM THE CLASSROOM

When asked whether they have ever heard the statement "*An asymptote is a line that the graph approaches but does not cross,*" a surprising number of students will raise their hands. First, let's make it clear that the statement is false; a graph *can* cross a horizontal asymptote and *can* cross a slant asymptote. A graph can never cross a *vertical* asymptote $x = a$, since the function is inherently undefined at $x = a$. We can even find the points where a graph crosses a horizontal or slant asymptote. For example, the rational function $f(x) = \dfrac{x^2 + 2x}{x^2 - 1}$ has the horizontal asymptote $y = 1$. Determining whether the graph of f crosses the horizontal line $y = 1$ is equivalent to asking whether $y = 1$ is in the range of the function f. Setting $f(x)$ equal to 1, that is,

$$\frac{x^2 + 2x}{x^2 - 1} = 1$$

implies $\quad\quad x^2 + 2x = x^2 - 1 \quad$ and $\quad x = -\frac{1}{2}$.

Since $x = -\frac{1}{2}$ is in the domain of f, the graph of f crosses the horizontal asymptote at $\left(-\frac{1}{2}, f\left(-\frac{1}{2}\right)\right) = \left(-\frac{1}{2}, 1\right)$. Observe in Example 9 we can find the point where the slant asymptote crosses the graph of $y = x$ by solving $f(x) = x$. You should verify that the point of intersection is $\left(\frac{4}{3}, \frac{4}{3}\right)$. See Problems 31–36 in Exercises 3.6.

In Problems 1 and 2, use a calculator to fill out the given table for the rational function $f(x) = \dfrac{2x}{x - 3}$.

1. $x = 3$ is a vertical asymptote for the graph of f

x	3.1	3.01	3.001	3.0001	3.00001
$f(x)$					
x	2.9	2.99	2.999	2.9999	2.99999
$f(x)$					

2. $y = 2$ is a horizontal asymptote for the graph of f

x	10	100	1000	10,000	100,000
$f(x)$					
x	-10	-100	-1000	$-10,000$	$-100,000$
$f(x)$					

In Problems 3–22, find the vertical and horizontal asymptotes for the graph of the given rational function. Find x- and y-intercepts of the graph. Sketch the graph of f.

3. $f(x) = \dfrac{1}{x - 2}$

4. $f(x) = \dfrac{4}{x + 3}$

5. $f(x) = \dfrac{x}{x + 1}$

6. $f(x) = \dfrac{x}{2x - 5}$

7. $f(x) = \dfrac{4x - 9}{2x + 3}$

8. $f(x) = \dfrac{2x + 4}{x - 2}$

9. $f(x) = \dfrac{1 - x}{x + 1}$

10. $f(x) = \dfrac{2x - 3}{x}$

11. $f(x) = \dfrac{1}{(x - 1)^2}$

12. $f(x) = \dfrac{4}{(x + 2)^3}$

13. $f(x) = \dfrac{1}{x^3}$

14. $f(x) = \dfrac{8}{x^4}$

15. $f(x) = \dfrac{x}{x^2 - 1}$

16. $f(x) = \dfrac{x^2}{x^2 - 4}$

17. $f(x) = \dfrac{1}{x(x - 2)}$

18. $f(x) = \dfrac{1}{x^2 - 2x - 8}$

19. $f(x) = \dfrac{1 - x^2}{x^2}$

20. $f(x) = \dfrac{16}{x^2 + 4}$

21. $f(x) = \dfrac{-2x^2 + 8}{(x - 1)^2}$

22. $f(x) = \dfrac{x(x - 5)}{x^2 - 9}$

In Problems 23–30, find the vertical and slant asymptotes for the graph of the given rational function. Find x- and y-intercepts of the graph. Sketch the graph f.

23. $f(x) = \dfrac{x^2 - 9}{x}$

24. $f(x) = \dfrac{x^2 - 3x - 10}{x}$

25. $f(x) = \dfrac{x^2}{x + 2}$

26. $f(x) = \dfrac{x^2 - 2x}{x + 2}$

27. $f(x) = \dfrac{x^2 - 2x - 3}{x - 1}$

28. $f(x) = \dfrac{-(x - 1)^2}{x + 2}$

29. $f(x) = \dfrac{x^3 - 8}{x^2 - x}$

30. $f(x) = \dfrac{5x(x + 1)(x - 4)}{x^2 + 1}$

In Problems 31–34, find the point where the graph of f crosses its horizontal asymptote. Sketch the graph of f.

31. $f(x) = \dfrac{x - 3}{x^2 + 3}$

32. $f(x) = \dfrac{(x - 3)^2}{x^2 - 5x}$

33. $f(x) = \dfrac{4x(x - 2)}{(x - 3)(x + 4)}$

34. $f(x) = \dfrac{2x^2}{x^2 + x + 1}$

In Problems 35 and 36, find the point where the graph of f crosses its slant asymptote. Use a graphing utility to obtain the graph of f and the slant asymptote in the same coordinate plane.

35. $f(x) = \dfrac{x^3 - 3x^2 + 2x}{x^2 + 1}$

36. $f(x) = \dfrac{x^3 + 2x - 4}{x^2}$

In Problems 37–40, find a rational function that satisfies the given conditions. There is no unique answer.

37. vertical asymptote: $x = 2$
horizontal asymptote: $y = 1$
x-intercept: $(5, 0)$

38. vertical asymptote: $x = 1$
horizontal asymptote: $y = -2$
y-intercept: $(0, -1)$

39. vertical asymptotes: $x = -1, x = 2$
horizontal asymptote: $y = 3$
x-intercept: $(3, 0)$

40. vertical asymptote: $x = 4$
slant asymptote: $y = x + 2$

In Problems 41–44, find the asymptotes and any holes in the graph of the given rational function. Find x- and y-intercepts of the graph. Sketch the graph f.

41. $f(x) = \dfrac{x^2 - 1}{x - 1}$

42. $f(x) = \dfrac{x - 1}{x^2 - 1}$

43. $f(x) = \dfrac{x + 1}{x(x^2 + 4x + 3)}$

44. $f(x) = \dfrac{x^3 + 8}{x + 2}$

Miscellaneous Applications

45. Parallel Resistors A 5-ohm resistor and a variable resistor are placed in parallel as shown in **FIGURE 3.6.11**. The resulting resistance R (in ohms) is related to the resistance r (in ohms) of the variable resistor by the equation

$$R = \frac{5r}{5 + r}.$$

Sketch the graph of R as a function of r for $r > 0$. What is the resulting resistance R as r becomes very large?

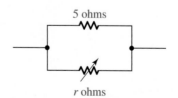

5 ohms

r ohms

FIGURE 3.6.11 Parallel resistors in Problem 45

46. Power The electrical power P produced by a certain source is given by

$$P = \frac{E^2 r}{R^2 + 2Rr + r^2},$$

where E is the voltage of the source, R is the resistance of the source, and r is the resistance in the circuit. Sketch the graph of P as a function of r using the values $E = 5$ volts and $R = 1$ ohm.

47. Illumination Intensity The intensity of illumination from a light source at any point is directly proportional to the strength of the source and inversely proportional to the square of the distance from the source. Given two sources of strengths 16 units and 2 units that are 100 cm apart, as shown in FIGURE 3.6.12, the intensity I at any point P between them is given by

$$I(x) = \frac{16}{x^2} + \frac{2}{(100 - x)^2},$$

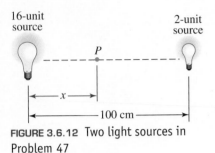

16-unit source

2-unit source

P

x

100 cm

FIGURE 3.6.12 Two light sources in Problem 47

where x is the distance from the 16-unit source. Sketch the graph of $I(x)$ on the interval $(0, 100)$. Describe the behavior of $I(x)$ as $x \to 0^+$. As $x \to 100^-$.

For Discussion

48. Suppose $f(x) = P(x)/Q(x)$. Prove the symmetry rules (2), (3), and (4) for rational functions. Assume $P(x)$ and $Q(x)$ have no common factors.

49. Construct a rational function $f(x) = P(x)/Q(x)$ whose graph crosses its slant asymptote twice.

50. If you have studied Section 1.5, then discuss how topics in this section and Section 3.2 can be used to help find the limit: $\lim\limits_{x \to 1} \dfrac{x^5 - 1}{x - 1}$.

3.7 **The Area Problem**

∫**Calculus** ≡ **Introduction** As we saw in Section 2.9, the fundamental
PREVIEW motivating problem of differential calculus, *Find a tangent line to the graph of a function f*, is answered by the notion of the **derivative** of the function. Differential calculus is the study of the properties and applications of the derivative of a function $y = f(x)$. Integral calculus, on the other hand, is the study of the properties and the applications of the **definite integral** of a function $y = f(x)$. As mentioned in Section 2.9, the historical problem that leads to the concept of the definite integral is, *Find the area under the graph of a function f.* We examine the area problem in this section.

☐ **Area Under a Graph** Throughout the discussion that follows we will assume that $y = f(x)$ is a function that is continuous and nonnegative on an interval $[a, b]$. Recall that the concept of continuity has been mentioned several times in previous sections; in this case, the graph of f has no breaks, gaps, or holes in it anywhere on the interval

$[a, b]$. The requirement that f be nonnegative, that is, $f(x) \geq 0$ for all x in $[a, b]$, means that no portion of its graph on the interval is below the x-axis. Specifically, then,

> By the **area under a graph** we mean the area A of the region in the plane bounded by the graph of f, the lines $x = a$ and $x = b$, and the x-axis.

See **FIGURE 3.7.1**.

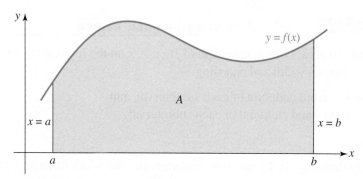

FIGURE 3.7.1 Area A under a graph

To get to the answer of the question, What is the *exact* value of A? we begin with a method for systematically *approximating* A. The basic idea is simply this: build rectangles across the interval $[a, b]$ and use the sum of the areas of the rectangles as an approximation for A.

☐ **Approximating the Area** One possible systematic procedure for approximating the value of the area A under a graph is summarized next.

(*i*) Subdivide the interval $[a, b]$ into n subintervals $[x_{k-1}, x_k]$, where

$$a = x_0 < x_1 < x_2 < \cdots < x_{n-1} < x_n = b,$$

so that each subinterval has the same width $\Delta x = \dfrac{b - a}{n}$. This is called a **regular partition** of the interval $[a, b]$.

(*ii*) Choose a number x_k^* in each of the n subintervals $[x_{k-1}, x_k]$ and form the n products $f(x_k^*) \, \Delta x$. Since the area of a rectangle is *length* × *width*, $f(x_k^*) \, \Delta x$ is the area of the rectangle of length $f(x_k^*)$ and width Δx built up on the kth subinterval $[x_{k-1}, x_k]$. The n numbers $x_1^*, x_2^*, x_3^*, \ldots, x_n^*$ are called **sample points**. See **FIGURE 3.7.2**.

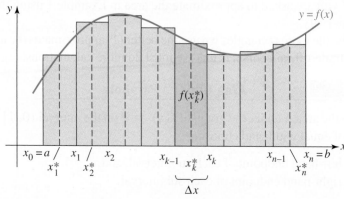

FIGURE 3.7.2 n rectangles of width Δx and length $f(x_k^*)$

(*iii*) The sum of the areas of the n rectangles represents an approximation to the value of the area,

$$A \approx f(x_1^*)\Delta x + f(x_2^*)\Delta x + f(x_3^*)\Delta x + \cdots + f(x_n^*)\Delta x. \tag{1}$$

To simplify the hand calculations, the sample points x_k^*, $k = 1, 2, \ldots, n$, are generally chosen to be either the left-hand endpoint or the right-hand endpoint of each subinterval $[x_{k-1}, x_k]$.

■ EXAMPLE 1   Area of a Triangular Region

Approximate the area A under the graph of $f(x) = x$ on the interval $[0, 1]$ using four subintervals of equal width and choosing

(a) x_k^* as the left-hand endpoint of each subinterval, and
(b) x_k^* as the right-hand endpoint of each subinterval.

See **FIGURE 3.7.3**.

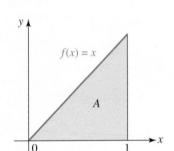

FIGURE 3.7.3 Area A in Example 1

Solution By dividing $[0, 1]$ into four subintervals, the width of each subinterval is $\Delta x = \frac{1-0}{4} = \frac{1}{4}$.
(a) If x_k^* is the left-hand endpoint of each of the four subintervals, then $x_1^* = 0$, $x_2^* = \frac{1}{4}, x_3^* = \frac{2}{4} = \frac{1}{2}$, and $x_4^* = \frac{3}{4}$. See **FIGURE 3.7.4(a)**. We have from (1),

$$A \approx f(0)\tfrac{1}{4} + f\left(\tfrac{1}{4}\right)\tfrac{1}{4} + f\left(\tfrac{1}{2}\right)\tfrac{1}{4} + f\left(\tfrac{3}{4}\right)\tfrac{1}{4}$$
$$= 0 \cdot \tfrac{1}{4} + \tfrac{1}{4} \cdot \tfrac{1}{4} + \tfrac{1}{2} \cdot \tfrac{1}{4} + \tfrac{3}{4} \cdot \tfrac{1}{4}$$
$$= \tfrac{3}{8} = 0.375.$$

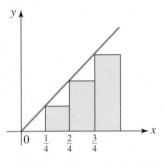

(a) Using left-hand endpoints

(b) If x_k^* is the right-hand endpoint of each of the four subintervals, then $x_1^* = \frac{1}{4}$, $x_2^* = \frac{2}{4} = \frac{1}{2}, x_3^* = \frac{3}{4}$, and $x_4^* = \frac{4}{4} = 1$. See Figure 3.7.4(b). We have from (1),

$$A \approx f\left(\tfrac{1}{4}\right)\tfrac{1}{4} + f\left(\tfrac{1}{2}\right)\tfrac{1}{4} + f\left(\tfrac{3}{4}\right)\tfrac{1}{4} + f(1)\tfrac{1}{4}$$
$$= \tfrac{1}{4} \cdot \tfrac{1}{4} + \tfrac{1}{2} \cdot \tfrac{1}{4} + \tfrac{3}{4} \cdot \tfrac{1}{4} + 1 \cdot \tfrac{1}{4}$$
$$= \tfrac{5}{8} = 0.625. \qquad \equiv$$

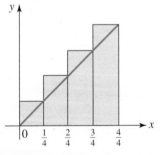

(b) Using right-hand endpoints
FIGURE 3.7.4 Approximating the area A in Example 1

As can be seen in Figures 3.7.4(a) and 3.7.4(b), the value obtained in part (a) of Example 1 underestimates the area A, whereas the value in part (b) overestimates A, that is, $0.375 < A < 0.625$. We can compare these approximations with the actual area. Since the area under the graph of $f(x) = x$ on the interval $[0, 1]$ is the area of a right triangle of base $= 1$ and height $= 1$, the exact area is $A = \frac{1}{2} \cdot$ base $\cdot$ height $= \frac{1}{2} \cdot 1 \cdot 1 = \frac{1}{2} = 0.5$.

There is no special reason that we chose the sample points x_k^*, $k = 1, 2, \ldots, n$, to be the left-hand and then the right-hand endpoints of the subintervals $[x_{k-1}, x_k]$, other than *convenience*. We could pick x_k^* randomly in each subinterval. In Problem 3 of Exercises 3.7 you are asked to approximate the area in Example 1 using the midpoint of each subinterval.

Intuitively, the more rectangles we use the better (1) approximates the area A under a graph. The trade-off, of course, is that we must do more calculations.

■ EXAMPLE 2   Area Under a Parabola

Approximate the area A under the graph of $f(x) = x^2$ on the interval $[0, 1]$ using eight subintervals of equal width and choosing

(a) x_k^* as the left-hand endpoint of each subinterval, and
(b) x_k^* as the right-hand endpoint of each subinterval.

See **FIGURE 3.7.5**.

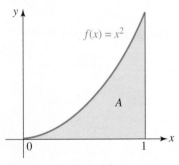

FIGURE 3.7.5 Area A in Example 2

Solution By dividing $[0, 1]$ into eight subintervals, the width of each subinterval is $\Delta x = \frac{1-0}{8} = \frac{1}{8}$.

(a) If x_k^* is the left-hand endpoint of each of the four subintervals, then $x_1^* = 0$, $x_2^* = \frac{1}{8}$, $x_3^* = \frac{2}{8} = \frac{1}{4}$, $x_4^* = \frac{3}{8}$, $x_5^* = \frac{4}{8} = \frac{1}{2}$, $x_6^* = \frac{5}{8}$, $x_7^* = \frac{6}{8} = \frac{3}{4}$, $x_8^* = \frac{7}{8}$. See **FIGURE 3.7.6(a)**. We have from (1),

$$A \approx f(0) \cdot \tfrac{1}{8} + f\left(\tfrac{1}{8}\right) \cdot \tfrac{1}{8} + f\left(\tfrac{1}{4}\right) \cdot \tfrac{1}{8} + f\left(\tfrac{3}{8}\right) \cdot \tfrac{1}{8} + f\left(\tfrac{1}{2}\right) \cdot \tfrac{1}{8} + f\left(\tfrac{5}{8}\right) \cdot \tfrac{1}{8} + f\left(\tfrac{3}{4}\right) \cdot \tfrac{1}{8} + f\left(\tfrac{7}{8}\right) \cdot \tfrac{1}{8}$$

$$= 0 \cdot \tfrac{1}{8} + \tfrac{1}{64} \cdot \tfrac{1}{8} + \tfrac{1}{16} \cdot \tfrac{1}{8} + \tfrac{9}{64} \cdot \tfrac{1}{8} + \tfrac{1}{4} \cdot \tfrac{1}{8} + \tfrac{25}{64} \cdot \tfrac{1}{8} + \tfrac{9}{16} \cdot \tfrac{1}{8} + \tfrac{49}{64} \cdot \tfrac{1}{8}$$

$$= \tfrac{35}{128} = 0.2734375.$$

(b) If x_k^* is the right-hand endpoint of each of the four subintervals, then $x_1^* = \frac{1}{8}$, $x_2^* = \frac{2}{8} = \frac{1}{4}$, $x_3^* = \frac{3}{8}$, $x_4^* = \frac{4}{8} = \frac{1}{2}$, $x_5^* = \frac{5}{8}$, $x_6^* = \frac{6}{8} = \frac{3}{4}$, $x_7^* = \frac{7}{8}$, $x_8^* = \frac{8}{8} = 1$. See Figure 3.7.6(b). We have from (1),

$$A \approx f\left(\tfrac{1}{8}\right) \cdot \tfrac{1}{8} + f\left(\tfrac{1}{4}\right) \cdot \tfrac{1}{8} + f\left(\tfrac{3}{8}\right) \cdot \tfrac{1}{8} + f\left(\tfrac{1}{2}\right) \cdot \tfrac{1}{8} + f\left(\tfrac{5}{8}\right) \cdot \tfrac{1}{8} + f\left(\tfrac{3}{4}\right) \cdot \tfrac{1}{8} + f\left(\tfrac{7}{8}\right) \cdot \tfrac{1}{8} + f(1) \cdot \tfrac{1}{8}$$

$$= \tfrac{1}{64} \cdot \tfrac{1}{8} + \tfrac{1}{16} \cdot \tfrac{1}{8} + \tfrac{9}{64} \cdot \tfrac{1}{8} + \tfrac{1}{4} \cdot \tfrac{1}{8} + \tfrac{25}{64} \cdot \tfrac{1}{8} + \tfrac{9}{16} \cdot \tfrac{1}{8} + \tfrac{49}{64} \cdot \tfrac{1}{8} + 1 \cdot \tfrac{1}{8}$$

$$= \tfrac{51}{128} = 0.3984375. \qquad\qquad \equiv$$

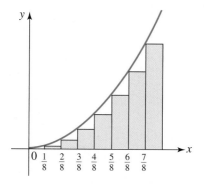

(a) Using left-hand endpoints

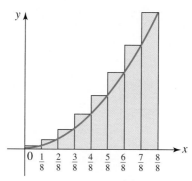

(b) Using right-hand endpoints

FIGURE 3.7.6 Approximating the area A in Example 2

From Figure 3.7.6(a) we see that the area of the seven rectangles underestimates A in Example 2, whereas the eight rectangles in Figure 3.7.6(b) overestimates A. From the calculations in Example 2 we can write $0.2734375 < A < 0.3984375$. But an observation is in order at this point. Don't assume that by using left-hand endpoints followed by the right-hand endpoints of the subintervals for x_k^* that we *always* get, in turn, a lower estimate followed by an upper estimate of the area A under the graph of f on $[a, b]$. This occurred in Examples 1 and 2 simply because, in both cases, the function f was increasing on the interval $[0, 1]$.

☐ **Summation Notation** Writing out sums such as (1) can become very tedious. To facilitate the discussion of the area problem, a special notation for summation is used in calculus. Suppose a_k denotes a real number that depends on an integer k. The sum of n such real numbers a_k, $a_1 + a_2 + a_3 + \cdots + a_n$, is denoted by the symbol $\sum_{k=1}^{n} a_k$, that is,

$$\sum_{k=1}^{n} a_k = a_1 + a_2 + a_3 + \cdots + a_n. \qquad (2)$$

Since Σ is the capital Greek letter sigma, (2) is called **sigma notation** or, more commonly, **summation notation**. The integer k is called the **index of summation** and takes on consecutive integer values starting with $k = 1$ and ending with $k = n$. For example, the sum of the first 100 squared positive integers,

$$1^2 + 2^2 + 3^2 + 4^2 + \cdots + 98^2 + 99^2 + 100^2,$$

can be written compactly as

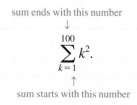

$$\text{sum ends with this number}$$
$$\downarrow$$
$$\sum_{k=1}^{100} k^2.$$
$$\uparrow$$
$$\text{sum starts with this number}$$

Using summation notation, the sum of the areas in (1) can be written as

$$A \approx \sum_{k=1}^{n} f(x_k^*) \Delta x.$$

☐ **Area** It should seem believable that we can reduce the error inherent in the method of approximating an area A under a graph by summing areas of rectangles by using more and more rectangles ($n \to \infty$) of decreasing width $\left(\Delta x = \dfrac{b-a}{n} \to 0 \right)$. Thus the 32 rectangles in FIGURE 3.7.7 should give us a better approximation to area A in Figure 3.7.1 than the eight rectangles shown in Figure 3.7.2. Indeed that is the case. It can be proved that when f is continuous on $[a, b]$ and $f(x) \geq 0$ for all x in the interval, the area A under the graph of the function $y = f(x)$ on the interval is given by the limit

$$A = \lim_{n \to \infty} \sum_{k=1}^{n} f(x_k^*) \Delta x. \tag{3}$$

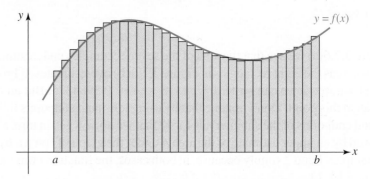

FIGURE 3.7.7 Using more rectangles improves the approximation to area A

The limit (3) exists regardless of how the sample points $x_1^*, x_2^*, x_3^*, \ldots, x_n^*$ are chosen in the subintervals $[x_0, x_1], [x_1, x_2], [x_2, x_3], \ldots, [x_{n-1}, x_n]$. Thus in (3), each sample point x_k^* could always be chosen, say, to be the right-hand endpoint of each subinterval. Since we are in no position to deal, in general terms, with limits of the kind given in (3), we leave that aspect of the area problem to a course in calculus. But if you are willing to put in the time to work Problems 15–22, then Problems 23 and 24 will give you a small taste of what is involved in computing area A by the limiting process given in (3).

We said at the start that the area problem is the motivating problem for the definite integral. You ask: So what is a definite integral? It is now just a small jump from (3) to the concept of the definite integral.

DEFINITION 3.7.1   Definite Integral

Let the function f be continuous on $[a, b]$. The **definite integral** of f from $x = a$ to $x = b$, denoted by $\int_a^b f(x)\,dx$, is

$$\int_a^b f(x)\,dx = \lim_{n \to \infty} \sum_{k=1}^{n} f(x_k^*) \Delta x. \tag{4}$$

The integral symbol $\int$ in (4), as used by **Wilhelm Gottfried Leibniz** (1646–1716) who is considered the co-inventor of calculus, along with **Isaac Newton** (1643–1727), is simply an elongated S for the word "sum."

NOTES FROM THE CLASSROOM

(*i*) If read quickly, you might conclude that formula (4) is the same as (3). In a way this is correct; however, (4) is a more general concept (notice that we are not requiring f to be nonnegative on the interval $[a, b]$). Thus, a *definite integral need not be area*. Also, in its most general setting, even the conditions of continuity of f and the use of a regular partition are dropped in the definition of the definite integral. What, then, is a definite integral? For now, accept the fact that a definite integral is simply a real number that can be negative, zero, or positive. When the conditions of continuity and nonnegativity are imposed on $y = f(x)$ on the interval $[a, b]$, then the area under the graph is $A = \int_a^b f(x)\, dx$. Also, you should be aware that the interpretations of derivative and the definite integral are much broader than just slopes of tangent lines and areas under graphs. As you progress through courses in mathematics, sciences, and engineering you will see many diverse applications of the derivative and the definite integral.

(*ii*) In this chapter we worked principally with polynomial functions. Polynomial functions are the fundamental building blocks of a class known as **algebraic functions**. In Section 3.6 we saw that a rational function is the quotient of two polynomial functions. In general, an algebraic function f involves a finite number of additions, subtractions, multiplications, divisions, and roots of polynomial functions. Thus

$$y = 2x^2 - 5x, \quad y = \sqrt[3]{x^2}, \quad y = x^4 + \sqrt{x^2 + 5}, \text{ and } y = \frac{\sqrt{x}}{x^3 - 2x^2 + 7}$$

are algebraic functions. Indeed, all the functions in Chapters 2 and 3 are algebraic functions. Starting with the next chapter we consider functions that belong to a different class known as **transcendental functions**. A transcendental function f is defined to be one that is *not algebraic*. The six trigonometric functions (Chapter 4) and the exponential and logarithmic functions (Chapter 6) are examples of transcendental functions.

3.7 | Exercises Answers to selected odd-numbered problems begin on page ANS–11.

In Problems 1–4, the function f and the interval are given in Example 1.

1. Approximate the area A, this time using eight subintervals of equal width and choosing x_k^* as the left-hand endpoint of each subinterval. Draw the eight rectangles.

2. Approximate the area A, this time using eight subintervals of equal width and choosing x_k^* as the right-hand endpoint of each subinterval. Draw the eight rectangles.

3. Approximate the area A, this time using four subintervals of equal width but choosing x_k^* as the midpoint of each subinterval. Draw the four rectangles.

4. Compare the approximation obtained in Problem 3 of the exact area $A = 0.5$. Explain why your answer in Problem 3 is not surprising.

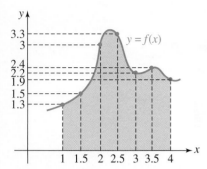

FIGURE 3.7.8 Graph for Problem 11

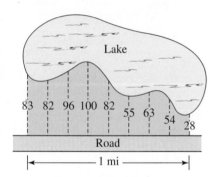

FIGURE 3.7.9 Graph for Problem 12

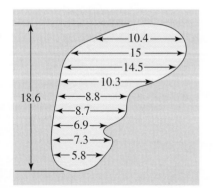

FIGURE 3.7.10 Land in Problem 13

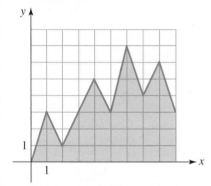

FIGURE 3.7.11 Fish pond in Problem 14

5. Approximate the area under the graph of $f(x) = x + 2$ on the interval $[-1, 2]$ using six subintervals of equal width and choosing:
   (a) x_k^* as the left-hand endpoint of each subinterval, and
   (b) x_k^* as the right-hand endpoint of each subinterval.
6. Repeat Problem 5 using twelve subintervals of equal width.
7. Approximate the area under the graph of $f(x) = -x^2 + 5x$ on the interval $[0, 5]$ using five subintervals of equal width and choosing:
   (a) x_k^* as the left-hand endpoint of each subinterval, and
   (b) x_k^* as the right-hand endpoint of each subinterval.
8. Repeat Problem 7 using ten subintervals of equal width.
9. Approximate the area under the graph of $f(x) = -x^2 + 5x$ on the interval $[0, 5]$ using five subintervals of equal width and choosing x_k^* as the midpoints of each subinterval.
10. Approximate the area under the graph of $f(x) = -x^3 + 2x^2$ on the interval $[0, 2]$ using ten subintervals of equal width and choosing x_k^* as the right-hand endpoint of each subinterval.
11. Find two different approximations for the area A under the graph $y = f(x)$ on the interval $[1, 4]$ shown in **FIGURE 3.7.8**.
12. Find two different approximations for the area A under the graph $y = f(x)$ on the interval shown in **FIGURE 3.7.9**.

Miscellaneous Applications

13. Lakefront Property Suppose a realtor wants to find the area of an irregularly shaped piece of land that is bounded between a 1 mile-long segment of a straight road and the shore of a lake. Measurements (in feet) of the perpendicular distances from the road to the lake are taken at equally spaced intervals along the road as shown in **FIGURE 3.7.10**. Find two different approximations of the area of the land. Express your answer in acres using the fact that 1 acre = 43,560 ft^2.
14. For the Fish The large irregularly shaped fish pond shown in **FIGURE 3.7.11** is filled with water to a uniform depth of 4 ft. Find an approximation to the number of gallons of water in the pond. Measurements are in feet and the vertical spacing between the horizontal measurements is 1.86 ft. There are 7.48 gallons in 1 cubic foot of water. [*Hint*: The volume of water is the area of the surface $\times$ depth.]

For Discussion

If c denotes a constant—that is, independent of the summation index k—then $\sum_{k=1}^{n} c$ means $c + c + c + \cdots + c$. Since there are n c's in this sum, we have

$$\sum_{k=1}^{n} c = nc. \tag{5}$$

In Problems 15 and 16, use (5) to find the numerical value of the given sum.

15. $\displaystyle\sum_{k=1}^{75} 6$ **16.** $\displaystyle\sum_{k=1}^{25} 10$

The sum of the first n positive integers can be written $\sum_{k=1}^{n} k$. If this sum is denoted by S, then

$$S = 1 + 2 + 3 + \cdots + (n - 1) + n \tag{6}$$

can also be written as

$$S = n + (n - 1) + \cdots + 3 + 2 + 1. \tag{7}$$

If we add (6) and (7), then

$$2S = \underbrace{(n + 1) + (n + 1) + (n + 1) + \cdots + (n + 1)}_{n \text{ terms of } n + 1} = n(n + 1).$$

Solving for S gives $S = \frac{1}{2}n(n + 1)$, or

$$\sum_{k=1}^{n} k = \frac{1}{2}n(n + 1). \tag{8}$$

In Problems 17 and 18, use (8) to find the numerical value of the given sum.

17. $\displaystyle\sum_{k=1}^{50} k$ **18.** $\displaystyle\sum_{k=1}^{1000} k$

Here are two properties of summation notation:

$$\sum_{k=1}^{n} ca_k = c \sum_{k=1}^{n} a_k, \quad c \text{ a constant}, \tag{9}$$

$$\sum_{k=1}^{n} (a_k \pm b_k) = \sum_{k=1}^{n} a_k \pm \sum_{k=1}^{n} b_k. \tag{10}$$

In Problems 19–22, use (5) and (8)–(10) to find the numerical value of the given sum.

19. $\displaystyle\sum_{k=1}^{20} 2k$ **20.** $\displaystyle\sum_{k=1}^{15} (-6k)$

21. $\displaystyle\sum_{k=1}^{10} (4k + 5)$ **22.** $\displaystyle\sum_{k=1}^{20} (4k - 3)$

In Problems 23 and 24, use the results in (5) and (8)–(10) and the limit definition of area given in (3) to find the exact value of the area A. In each case, partition the given interval into n subintervals of width $\Delta x = (b - a)/n$ and use x_k^* as the right-hand end-point of each subinterval.

23. A is the area under the graph of $f(x) = 2x + 1$ on the interval $[0, 4]$
24. A is the area under the graph of $f(x) = -3x + 12$ on the interval $[1, 3]$

25. Consider the trapezoid given in FIGURE 3.7.12.
    (a) Discuss how the area A can be approximated using (1) of this section.
    (b) Using well-known area formulas, find a formula that expresses A in terms of h_1, h_2, and b.

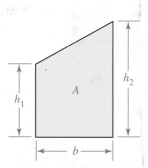

FIGURE 3.7.12 Trapezoid in Problem 25

A. Fill in the Blanks

In Problems 1–20, fill in the blanks.

1. The graph of the polynomial function $f(x) = x^3(x - 1)^2(x - 5)$ is tangent to the x-axis at _____ and passes through the x-axis at _____.

2. A third-degree polynomial function with zeros 1 and $3i$ is _____.

3. The end behavior of the graph of $f(x) = x^2(x + 3)(x - 5)$ resembles the graph of the power function $f(x) =$ _____.

4. The polynomial function $f(x) = x^4 - 3x^3 + 17x^2 - 2x + 2$ has _____ (how many) possible rational zeros.

5. For $f(x) = kx^2(x - 2)(x - 3), f(-1) = 8$ if $k =$ _____.

6. The y-intercept of the graph of the rational function $f(x) = \dfrac{2x + 8}{x^2 - 5x + 4}$ is _____.

7. The vertical asymptotes for the graph of the rational function
$f(x) = \dfrac{2x + 8}{x^2 - 5x + 4}$ are _____.

8. The x-intercepts of the graph of the rational function $f(x) = \dfrac{x^3 - x}{4 - 2x^3}$ are _____.

9. The horizontal asymptote for the graph of the rational function $f(x) = \dfrac{x^3 - x}{4 - 2x^3}$ is _____.

10. A rational function whose graph has the horizontal asymptote $y = 1$ and x-intercept $(3, 0)$ is _____.

11. The graph of the rational function $f(x) = \dfrac{x^n}{x^3 + 1}$, where n is a nonnegative integer, has the horizontal asymptote $y = 0$ when $n =$ _____.

12. The graph of the polynomial function $f(x) = 3x^5 - 4x^2 + 5x - 2$ has at most _____ turning points.

13. If i is a zero of $f(x) = x^4 + 2x^3 + 3x^2 + 2x + 2$ then three other zeros are _____.

14. The rational function $f(x) = \dfrac{x^3}{x^2 + 1}$ has _____ (how many) asymptotes.

15. Suppose that when a polynomial function f of degree 5 is divided by $x - 3$ we get $\dfrac{f(x)}{x - 3} = q(x) + \dfrac{7}{x - 3}$. The degree of the quotient $q(x)$ is _____.

16. If f is the polynomial function in Problem 15, then $f(3) =$ _____.

17. A rational function f can be written $f(x) = 2x + 4 + \dfrac{x - 3}{x^2 - x}$. The asymptotes for the graph of f are _____.

18. If $f(x) = \dfrac{x + 10}{2 - x}$, then $f(x) \rightarrow$ _____ as $x \rightarrow 2^+$.

19. The graph of the rational function $f(x) = \dfrac{x^3}{x^2 - 9}$ is symmetric with respect to _____.

20. If $f(-2) > 1$ and $f(3) < -1$, then the polynomial function f has at least one zero in the interval _____.

B. True/False

In Problems 1–20, answer true or false.

1. $f(x) = 2x^3 - 8x^{-2} + 5$ is not a polynomial function. _____

2. $f(x) = x + \dfrac{1}{x}$ is a rational function. _____

3. The graph of a polynomial function f can have no holes in it. _____

4. A polynomial function f of degree 4 has exactly four real zeros. _____

5. When a polynomial of degree greater than one is divided by $x - 1$, the remainder is always a constant. _____

6. If the coefficients a, b, c, and d of the polynomial function $f(x) = ax^3 + bx^2 + cx + d$ are positive integers, then f has no positive real zeros. _____

7. The polynomial equation $2x^7 = 1 - x$ has a solution in the interval $[0, 1]$. _____

8. The graph of the rational function $f(x) = (x^2 + 1)/x$ has a slant asymptote. _____

9. The graph of the polynomial function $f(x) = 4x^6 + 3x^2$ is symmetric with respect to the y-axis. _____

10. The graph of a polynomial function f that is an odd function, must pass through the origin. _____

11. An asymptote is a line that the graph of a function approaches but never crosses. _____

12. The point $\left(\frac{1}{3}, \frac{7}{4}\right)$ is on the graph of $f(x) = \dfrac{2x + 4}{3 - x}$. _____

13. The graph of a rational function $f(x) = P(x)/Q(x)$ has a slant asymptote when the degree of P is greater than the degree of Q. _____

14. If $3 - 4i$ is a zero of a polynomial function $f(x)$ with real coefficients, then $3 + 4i$ is also a zero of f. _____

15. A polynomial function f must have at least one rational zero. _____

16. The graph of $f(x) = x^4 + 5x^2 + 2$ does not cross the x-axis. _____

17. If $(-1, 6)$ and $(4, -2)$ are two points on the graph of a polynomial function f, then f has at least one zero in the open interval $(-1, 4)$. _____

18. If the end behavior of a polynomial function f is that its graph goes up for large values of $|x|$, then the degree of f must be an even positive integer. _____

19. The graph of a rational function f can have at most one horizontal asymptote. _____

20. If a polynomial function f is an odd function and 3 is a zero of f, then -3 is also a zero of f. _____

C. Review Exercises

In Problems 1 and 2, use long division to divide $f(x)$ by $d(x)$.

1. $f(x) = 6x^5 - 4x^3 + 2x^2 + 4$, $\quad d(x) = 2x^2 - 1$
2. $f(x) = 15x^4 - 2x^3 + 8x + 6$, $\quad d(x) = 5x^3 + x + 2$

In Problems 3 and 4, use synthetic division to divide $f(x)$ by $d(x)$.

3. $f(x) = 7x^4 - 6x^2 + 9x + 3$, $\quad d(x) = x - 2$
4. $f(x) = 4x^3 + 7x^2 - 8x$, $\quad d(x) = x + 1$

5. Without actually performing the division, determine the remainder when $f(x) = 5x^3 - 4x^2 + 6x - 9$ is divided by $d(x) = x + 3$.

6. Use synthetic division and the Remainder Theorem to find $f(c)$ for

$$f(x) = x^6 - 3x^5 + 2x^4 + 3x^3 - x^2 + 5x - 1$$

when $c = 2$.

7. Determine the values of the positive integer n such that $f(x) = x^n + c^n$ is divisible by $d(x) = x + c$.

8. Suppose that

$$f(x) = 36x^{98} - 40x^{25} + 18x^{14} - 3x^7 + 40x^4 + 5x^2 - x + 2$$

is divided by $d(x) = x - 1$. What is the remainder?

9. List, but do not test, all possible rational zeros of

$$f(x) = 8x^4 + 19x^3 + 31x^2 + 38x - 15.$$

10. Find the complete factorization of $f(x) = 12x^3 + 16x^2 + 7x + 1$.

In Problems 11 and 12, verify that each of the indicated numbers is a zero of the given polynomial function $f(x)$. Find all other zeros and then give the complete factorization of $f(x)$.

11. 2; $f(x) = (x - 3)^3 + 1$ **12.** -1; $f(x) = (x + 2)^4 - 1$

In Problems 13–16, find the real value of k so that the given condition is satisfied.

13. the remainder in the division of $f(x) = x^4 - 3x^3 - x^2 + kx - 1$ by $g(x) = x - 4$ is $r = 5$

14. $x + \frac{1}{2}$ is a factor of $f(x) = 8x^2 - 4kx + 9$

15. $x - k$ is a factor of $f(x) = 2x^3 + x^2 + 2x - 12$

16. the graph of $f(x) = \dfrac{x - k}{x^2 + 5x + 6}$ has a hole at $x = k$

In Problems 17 and 18, find a polynomial function f of indicated degree whose graph is given in the figure.

17. fifth degree

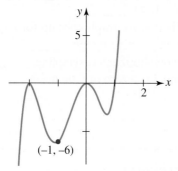

FIGURE 3.R.1 Graph for Problem 17

18. sixth degree

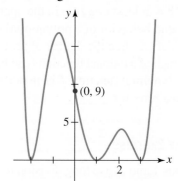

FIGURE 3.R.2 Graph for Problem 18

In Problems 19 and 20, find a rational function f whose graph is given in the figure.

19.

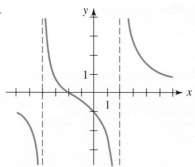

FIGURE 3.R.3 Graph for Problem 19

20.

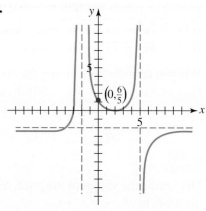

FIGURE 3.R.4 Graph for Problem 20

In Problems 21–30, match the given rational function with one of the graphs (a)–(j).

(a)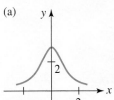

FIGURE 3.R.5

(b)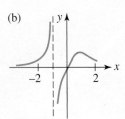

FIGURE 3.R.6

(c)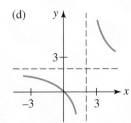

FIGURE 3.R.7

(d)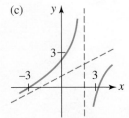

FIGURE 3.R.8

(e)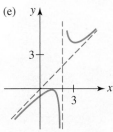

FIGURE 3.R.9

(f)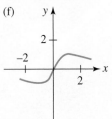

FIGURE 3.R.10

(g)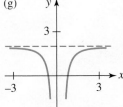

FIGURE 3.R.11

(h)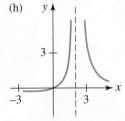

FIGURE 3.R.12

(i)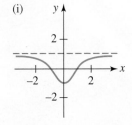

FIGURE 3.R.13

(j)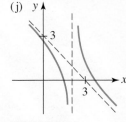

FIGURE 3.R.14

21. $f(x) = \dfrac{2x}{x^2 + 1}$

22. $f(x) = \dfrac{x^2 - 1}{x^2 + 1}$

23. $f(x) = \dfrac{2x}{x - 2}$

24. $f(x) = 2 - \dfrac{1}{x^2}$

25. $f(x) = \dfrac{x}{(x - 2)^2}$

26. $f(x) = \dfrac{(x - 1)^2}{x - 2}$

27. $f(x) = \dfrac{x^2 - 10}{2x - 4}$

28. $f(x) = \dfrac{-x^2 + 5x - 5}{x - 2}$

29. $f(x) = \dfrac{2x}{x^3 + 1}$

30. $f(x) = \dfrac{3}{x^2 + 1}$

In Problems 31 and 32, find the asymptotes for the graph of the given rational function. Find x- and y-intercepts of the graph. Sketch the graph of f.

31. $f(x) = \dfrac{x + 2}{x^2 + 2x - 8}$

32. $f(x) = \dfrac{-x^3 + 2x^2 + 9}{x^2}$

4 Trigonometric Functions

Chapter Outline

4.1 Angles and Their Measurement

≡ **Introduction** We begin our study of trigonometry by discussing angles and two methods of measuring them: degrees and radians. As we will see it is the radian measure of an angle that enables us to define trigonometric functions on sets of real numbers.

□ **Angles** An **angle** is formed by two half-rays, or half-lines, which have a common endpoint, called the **vertex**. We designate one ray the **initial side** of the angle and the other the **terminal side**. It is useful to consider the angle as having been formed by a rotation from the initial side to the terminal side as shown in FIGURE 4.1.1(a). An angle is said to be in **standard position** if its vertex is placed at the origin of a rectangular coordinate system with its initial side coinciding with the positive *x*-axis, as shown in Figure 4.1.1(b).

□ **Degree Measure** The **degree measure** of an angle is based on the assignment of 360 degrees (written 360°) to the angle formed by one complete counterclockwise rotation, as shown in FIGURE 4.1.2. Other angles are then measured in terms of a 360° angle, with a 1° angle being formed by $\frac{1}{360}$ of a complete rotation. If the rotation is counterclockwise, the measure will be *positive*; if clockwise, the measure is *negative*. For example, the angle in FIGURE 4.1.3(a) obtained by one-fourth of a complete counterclockwise rotation will be

$$\tfrac{1}{4}(360°) = 90°.$$

Shown in Figure 4.1.3(b) is the angle formed by three-fourths of a complete clockwise rotation. This angle has measure

$$\tfrac{3}{4}(-360°) = -270°.$$

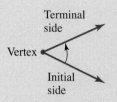
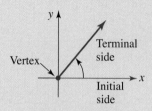

(a) Two half-rays

(b) Standard position

FIGURE 4.1.1 Initial and terminal sides of an angle

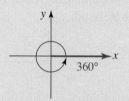

FIGURE 4.1.2 Angle of 360 degrees

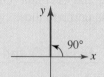

(a) 90° angle

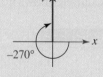

(b) –270° angle

FIGURE 4.1.3 Positive measure in (a); negative measure in (b)

□ **Coterminal Angles** Comparison of Figure 4.1.3(a) with Figure 4.1.3(b) shows that the terminal side of a 90° angle coincides with the terminal side of a −270° angle. When two angles in standard position have the same terminal sides we say they are **coterminal**. For example, the angles θ, θ + 360°, and θ − 360° shown in FIGURE 4.1.4 are coterminal. In fact, the addition of any integer multiple of 360° to a given angle results in a coterminal angle. Conversely, any two coterminal angles have degree measures that differ by an integer multiple of 360°.

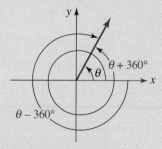

FIGURE 4.1.4 Three coterminal angles

EXAMPLE 1	**Angles and Coterminal Angles**

For a 960° angle:
(a) Locate the terminal side and sketch the angle.
(b) Find a coterminal angle between 0° and 360°.
(c) Find a coterminal angle between −360° and 0°.

Solution **(a)** We first determine how many full rotations are made in forming this angle. Dividing 960 by 360 we obtain a quotient of 2 and a remainder of 240. Equivalently we can write

$$960 = 2(360) + 240.$$

Thus, this angle is formed by making two counterclockwise rotations before completing $\frac{240}{360} = \frac{2}{3}$ of another rotation. As illustrated in FIGURE 4.1.5(a), the terminal side of $960°$ lies in the third quadrant.

(b) Figure 4.1.5(b) shows that the angle $240°$ is coterminal with a $960°$ angle.

(c) Figure 4.1.5(c) shows that the angle $-120°$ is coterminal with a $960°$ angle.

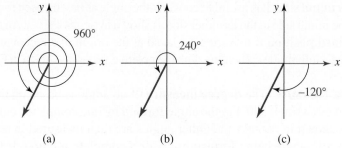

(a)   (b)   (c)

FIGURE 4.1.5 Angles in (b) and (c) are coterminal with the angle in (a) ≡

☐ **Minutes and Seconds** With calculators it is convenient to represent fractions of degrees by decimals, such as $42.23°$. Traditionally, however, fractions of degrees were expressed in **minutes** and **seconds**, where

$$1° = 60 \text{ minutes (written } 60'\text{)}* \qquad (1)$$

and

$$1' = 60 \text{ seconds (written } 60''\text{).} \qquad (2)$$

For example, an angle of 7 degrees, 30 minutes, and 5 seconds is expressed as $7°30'5''$. Some calculators have a special DMS key for converting an angle given in decimal degrees to Degrees, Minutes, and Seconds (DMS notation), and vice versa. The following example shows how to perform these conversions by hand.

▌EXAMPLE 2   Using (1) and (2)

Convert:

(a) $86.23°$ to degrees, minutes, and seconds,

(b) $17°47'13''$ to decimal notation.

Solution In each case we will use (1) and (2).

(a) Since $0.23°$ represents $\frac{23}{100}$ of $1°$ and $1° = 60'$, we have

$$86.23° = 86° + 0.23°$$
$$= 86° + (0.23)(60')$$
$$= 86° + 13.8'.$$

Now $13.8' = 13' + 0.8'$, so we must convert $0.8'$ to seconds. Since $0.8'$ represents $\frac{8}{10}$ of $1'$ and $1' = 60''$, we have

$$86° + 13' + 0.8' = 86° + 13' + (0.8)(60'')$$
$$= 86° + 13' + 48''.$$

Hence, $86.23° = 86°13'48''$.

---

*The use of the number 60 as a base dates back to the Babylonians. Another example of the use of this base in our culture in the measurement of time (1 hour = 60 minutes and 1 minute = 60 seconds).

(b) Since $1° = 60'$, it follows that $1' = \left(\frac{1}{60}\right)°$. Similarly, $1'' = \left(\frac{1}{60}\right)' = \left(\frac{1}{3600}\right)°$. Thus we have

$$17°47'13'' = 17° + 47' + 13''$$
$$= 17° + 47\left(\frac{1}{60}\right)° + 13\left(\frac{1}{3600}\right)°$$
$$\approx 17° + 0.7833° + 0.0036°.$$

Thus we see that $17°47'13'' \approx 17.7869°$.

≡

☐ **Radian Measure** Another measure for angles is **radian measure**, which is generally used in almost all applications of trigonometry that involve calculus. The radian measure of an angle θ is based on the length of an arc on a circle. If we place the vertex of the angle θ at the center of a circle of radius r, then θ is called a **central angle**. As we know, an angle θ in standard position can be viewed as having been formed by the initial side rotating from the positive x-axis to the terminal side. The region formed by the initial and terminal sides with a central angle θ is called a **sector** of a circle. As shown in FIGURE 4.1.6(a), if the initial side of θ traverses a distance s along the circumference of the circle, then the **radian measure of θ** is defined by

$$\theta = \frac{s}{r}. \tag{3}$$

A slice of pizza is an example of a circular sector

In the case when the terminal side of θ traverses an arc of length s along the circumference of the circle equal to the radius r of the circle, then we see from (3) that the measure of the angle θ is **1 radian**. See Figure 4.1.6(b).

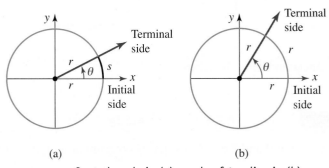

(a)   (b)

FIGURE 4.1.6 Central angle in (a); angle of 1 radian in (b)

The definition given in (3) does not depend on the size of the circle. To see this, all we need do is to draw another circle centered at the vertex of θ of radius r' and subtended arc length s'. See FIGURE 4.1.7. Because the two circular sectors are similar the ratios s/r and s'/r' are equal. Therefore, regardless of which circle we use, we obtain the same radian measure for θ.

In equation (3) any convenient unit of length may be used for s and r, but the same unit must be used for *both* s and r. Thus,

$$\theta\,(\text{in radians}) = \frac{s\,(\text{units of length})}{r\,(\text{units of length})}$$

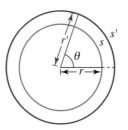

FIGURE 4.1.7 Concentric circles

appears to be a "dimensionless" quantity. For example, if $s = 4$ in. and $r = 2$ in., then the radian measure of the angle is

$$\theta = \frac{4 \text{ in.}}{2 \text{ in.}} = 2,$$

where 2 is simply a real number. This is the reason why sometimes the word *radians* is omitted when an angle is measured in radians. We will come back to this idea in Section 4.2.

One complete rotation of the initial side of θ will traverse an arc equal in length to the circumference of the circle $2\pi r$. It follows from (3) that

$$\text{one rotation} = \frac{s}{r} = \frac{2\pi r}{r} = 2\pi \text{ radians.}$$

We have the same convention as before: An angle formed by a counterclockwise rotation is considered positive, whereas an angle formed by a clockwise rotation is negative. In FIGURE 4.1.8 we illustrate angles in standard position of $\pi/2$, $-\pi/2$, π, and 3π radians, respectively. Note that the angle of $\pi/2$ radians shown in 4.1.8(a) is obtained by one-fourth of a complete counterclockwise rotation; that is

$$\frac{1}{4}(2\pi \text{ radians}) = \frac{\pi}{2} \text{ radians.}$$

The angle shown in Figure 4.1.8(b), obtained by one-fourth of a complete clockwise rotation, is $-\pi/2$ radians. The angle shown in Figure 4.1.8(c) is coterminal with the angle shown in Figure 4.1.8(d). In general, the addition of any integer multiple of 2π radians to an angle measured in radians results in a coterminal angle. Conversely, any two coterminal angles measured in radians will differ by an integer multiple of 2π.

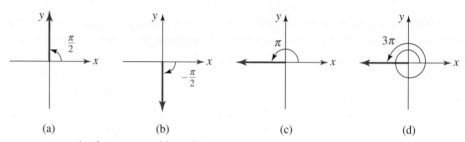

(a)　　　　　　(b)　　　　　　(c)　　　　　　(d)

FIGURE 4.1.8 Angles measured in radians

EXAMPLE 3　　　A Coterminal Angle

Find an angle between 0 and 2π radians that is coterminal with $\theta = 11\pi/4$ radians. Sketch the angle.

Solution Since $2\pi < 11\pi/4 < 3\pi$, we subtract the equivalent of one rotation, or 2π radians, to obtain

$$\frac{11\pi}{4} - 2\pi = \frac{11\pi}{4} - \frac{8\pi}{4} = \frac{3\pi}{4}.$$

Alternatively, we can proceed as in part (a) of Example 1 and divide: $11\pi/4 = 2\pi + 3\pi/4$. Thus, an angle of $3\pi/4$ radians is coterminal with θ, as illustrated in FIGURE 4.1.9. ≡

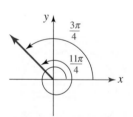

FIGURE 4.1.9 Coterminal angles in Example 3

□ **Conversion Formulas** Although many scientific calculators have keys that convert between degree and radian measure, there is an easy way to remember the relationship between the two measures. Since the circumference of a unit circle is 2π, one complete rotation has measure 2π radians as well as 360°. It follows that $360° = 2\pi$ radians or

$$180° = \pi \text{ radians.} \tag{4}$$

If we interpret (4) as $180(1°) = \pi(1 \text{ radian})$, then we obtain the following two formulas for converting between degree and radian measure.

Conversion between Degrees and Radians

$$1° = \frac{\pi}{180} \text{ radian} \tag{5}$$

$$1 \text{ radian} = \left(\frac{180}{\pi}\right)° \tag{6}$$

Using a calculator to carry out the divisions in (5) and (6), we find that

$$1° \approx 0.0174533 \text{ radian} \quad \text{and} \quad 1 \text{ radian} \approx 57.29578°.$$

EXAMPLE 4 — Conversion Between Degrees and Radians

Convert:

(a) $20°$ to radians (b) $7\pi/6$ radians to degrees (c) 2 radians to degrees.

Solution (a) To convert from degrees to radians we use (5):

$$20° = 20(1°) = 20 \cdot \left(\frac{\pi}{180} \text{ radian}\right) = \frac{\pi}{9} \text{ radian}.$$

(b) To convert from radians to degrees we use (6):

$$\frac{7\pi}{6} \text{ radians} = \frac{7\pi}{6} \cdot (1 \text{ radian}) = \frac{7\pi}{6}\left(\frac{180}{\pi}\right)° = 210°.$$

(c) We again use (6):

approximate answer
rounded to two
decimal places

$$2 \text{ radians} = 2 \cdot (1 \text{ radian}) = 2 \cdot \left(\frac{180}{\pi}\right)° = \left(\frac{360}{\pi}\right)° \approx \overbrace{114.59°}.$$ ≡

Table 4.1.1 provides the radian and degree measure of the most commonly used angles.

TABLE 4.1.1

Degrees	0	30	45	60	90	180
Radians	0	$\frac{\pi}{6}$	$\frac{\pi}{4}$	$\frac{\pi}{3}$	$\frac{\pi}{2}$	π

☐ **Terminology** You may recall from geometry that a 90° angle is called a **right angle** and a 180° angle is called a **straight angle**. In radian measure, $\pi/2$ is a right angle and π is a straight angle. An **acute angle** has measure between 0° and 90° (or between 0 and $\pi/2$ radians); and an **obtuse angle** has measure between 90° and 180° (or between $\pi/2$ and π radians). Two acute angles are said to be **complementary** if their sum is 90° (or $\pi/2$ radians). Two positive angles are **supplementary** if their sum is 180° (or π radians). The angle 180° (or π radians) is a **straight angle**. An angle whose terminal side coincides with a coordinate axis is called a **quadrantal angle**. For example, 90° (or $\pi/2$ radians) is a quadrantal angle. A triangle that contains a right angle is called a **right triangle**. The lengths a, b, and c of the sides of a right triangle satisfy the Pythagorean theorem $a^2 + b^2 = c^2$, where c is the length of the side opposite the right angle (the hypotenuse).

EXAMPLE 5   Complementary and Supplementary Angles

(a) Find the angle that is complementary to $\theta = 74.23°$.

(b) Find the angle that is supplementary to $\phi = \pi/3$ radians.

Solution (a) Since two angles are complementary if their sum is 90°, we find the angle that is complementary to $\theta = 74.23°$ is

$$90° - \theta = 90° - 74.23° = 15.77°.$$

(b) Since two angles are supplementary if their sum is π radians, we find the angle that is supplementary to $\phi = \pi/3$ radians is

$$\pi - \phi = \pi - \frac{\pi}{3} = \frac{3\pi}{3} - \frac{\pi}{3} = \frac{2\pi}{3} \quad \text{radians.} \qquad \equiv$$

☐ **Arc Length** In many applications it is necessary to find the length s of the arc subtended by a central angle θ in a circle of radius r. See **FIGURE 4.1.10**. From the definition of radian measure given in (3),

$$\theta \text{ (in radians)} = \frac{s}{r}.$$

By multiplying both sides of the last equation by r we obtain the **arc length formula** $s = r\theta$. We summarize the result.

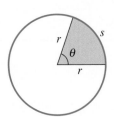

FIGURE 4.1.10 Length of arc s determined by a central angle θ

THEOREM 4.1.1 Arc Length Formula

For a circle of radius r, a central angle of θ radians subtends an **arc of length**

$$s = r\theta \qquad (7)$$

EXAMPLE 6   Finding Arc Length

Find the arc length subtended by a central angle of (a) 2 radians in a circle of radius 6 inches, (b) 30° in a circle of radius 12 feet.

Solution (a) From the arc length formula (7) with $\theta = 2$ radians and $r = 6$ inches, we have $s = r\theta = 2 \cdot 6 = 12$. So the arc length is 12 inches.

Students often apply the arc length formula ▶ incorrectly by using degree measure. Remember $s = r\theta$ is valid only if θ is measured in radians.

(b) We must first express 30° in radians. Recall that $30° = \pi/6$ radians. Then from the arc length formula (7) we have $s = r\theta = (12)(\pi/6) = 2\pi$. So the arc length is $2\pi \approx 6.28$ feet. $\qquad \equiv$

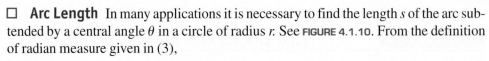

4.1   Exercises   Answers to selected odd-numbered problems begin on page ANS-12.

In Problems 1–16, draw the given angle in standard position. Bear in mind that the lack of a degree symbol (°) in an angular measurement indicates that the angle is measured in radians.

1. 60°   **2.** $-120°$ **3.** 135°   **4.** 150°

5. 1140°   **6.** $-315°$ **7.** $-240°$ **8.** $-210°$

9. $\dfrac{\pi}{3}$ **10.** $\dfrac{5\pi}{4}$ **11.** $\dfrac{7\pi}{6}$ **12.** $-\dfrac{2\pi}{3}$

13. $-\dfrac{\pi}{6}$ **14.** -3π **15.** 3   **16.** 4

In Problems 17–20, express the given angle in decimal notation.

17. $10°39'17''$ **18.** $143°7'2''$ **19.** $5°10'$ **20.** $10°25'$

In Problems 21–24, express the given angle in terms of degrees, minutes, and seconds.

21. $210.78°$ **22.** $15.45°$ **23.** $30.81°$ **24.** $110.5°$

In Problems 25–32, convert from degrees to radians.

25. $10°$ **26.** $15°$ **27.** $45°$ **28.** $215°$

29. $270°$ **30.** $-120°$ **31.** $-230°$ **32.** $540°$

In Problems 33–40, convert from radians to degrees.

33. $\dfrac{2\pi}{9}$ **34.** $\dfrac{11\pi}{6}$ **35.** $\dfrac{2\pi}{3}$ **36.** $\dfrac{5\pi}{12}$

37. $\dfrac{5\pi}{4}$ **38.** 7π **39.** 3.1   **40.** 12

In Problems 41–44, find the angle between $0°$ and $360°$ that is coterminal with the given angle.

41. $875°$ **42.** $400°$ **43.** $-610°$ **44.** $-150°$

45. Find the angle between $-360°$ and $0°$ that is coterminal with the angle in Problem 41.

46. Find the angle between $-360°$ and $0°$ that is coterminal with the angle in Problem 43.

In Problems 47–52, find the angle between 0 and 2π that is coterminal with the given angle.

47. $-\dfrac{9\pi}{4}$ **48.** $\dfrac{17\pi}{2}$ **49.** 5.3π

50. $-\dfrac{9\pi}{5}$ **51.** -4 **52.** 7.5

53. Find the angle between -2π and 0 radians that is coterminal with the angle in Problem 47.

54. Find the angle between -2π and 0 radians that is coterminal with the angle in Problem 49.

In Problems 55–62, find an angle that is **(a)** complementary and **(b)** supplementary to the given angle, or state why no such angle can be found.

55. $48.25°$ **56.** $93°$ **57.** $98.4°$ **58.** $63.08°$

59. $\dfrac{\pi}{4}$ **60.** $\dfrac{\pi}{6}$ **61.** $\dfrac{2\pi}{3}$ **62.** $\dfrac{5\pi}{6}$

Planet Mercury in
Problem 68

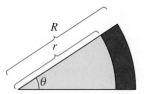

FIGURE 4.1.11 Circular band in
Problem 74

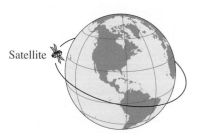

FIGURE 4.1.12 Satellite in
Problem 75

63. Find both the degree and the radian measures of the angle formed by **(a)** three-fifths of a counterclockwise rotation and **(b)** five and one-eighth clockwise rotations.

64. Find both the degree and the radian measures of the obtuse angle formed by the hands of a clock **(a)** at 8:00, **(b)** at 1:00, and **(c)** at 7:30.

65. Find both the degree and the radian measures of the angle through which the hour hand on a clock rotates in 2 hours.

66. Answer the question in Problem 65 for the minute hand.

67. The Earth rotates on its axis once every 24 hours. How long does it take the Earth to rotate through an angle of **(a)** 240° and **(b)** $\pi/6$ radians?

68. The planet Mercury completes one rotation on its axis every 59 days. Through what angle (measured in degrees) does it rotate in **(a)** 1 day, **(b)** 1 hour, and **(c)** 1 minute?

69. Find the arc length subtended by a central angle of 3 radians in a circle of **(a)** radius 3 and **(b)** radius 5.

70. Find the arc length subtended by a central angle of 30° in a circle of **(a)** radius 2 and **(b)** radius 4.

71. Find the measure of a central angle θ in a circle of radius 5 if θ subtends an arc of length 7.5. Give θ in **(a)** radians and **(b)** degrees.

72. Find the measure of a central angle θ in a circle of radius 1 if θ subtends an arc of length $\pi/6$. Give θ in **(a)** radians and **(b)** degrees.

73. Show that the area A of a sector formed by a central angle of θ radians in a circle of radius r is given by $A = \frac{1}{2}r^2\theta$. See Figure 4.1.10. [*Hint*: Use the proportionality property from geometry that the ratio of the area A of a circular sector to the total area πr^2 of the circle equals the ratio of the central angle θ to one complete revolution 2π.]

74. What is the area of the red shaded circular band shown in FIGURE 4.1.11 if θ is measured **(a)** in radians and **(b)** in degrees? [*Hint*: Use the result of Problem 73.]

Miscellaneous Applications

75. Angular and Linear Speed If we divide (7) by time t we get the relationship $v = r\omega$, where $v = s/t$ is called the **linear speed** of a point on the circumference of a circle and $\omega = \theta/t$ is called the **angular speed** of the point. A communications satellite is placed in a circular geosynchronous orbit 35,786 km above the surface of the Earth. The time it takes the satellite to make one full revolution around the Earth is 23 hours, 56 minutes, 4 seconds and the radius of the Earth is 6378 km. See FIGURE 4.1.12.

(a) What is the angular speed of the satellite in rad/s?
(b) What is the linear speed of the satellite in km/s?

76. Pendulum Clock A clock pendulum is 1.3 m long and swings back and forth along a 15-cm arc. Find **(a)** the central angle and **(b)** the area of the sector through which the pendulum sweeps in one swing. [*Hint*: To answer part (b), use the result of Problem 73.]

77. Sailing at Sea A nautical mile is defined as the arc length subtended on the surface of the Earth by an angle of measure 1 minute. If the diameter of the Earth is 7927 miles, find how many statute (land) miles there are in a nautical mile.

78. Circumference of the Earth Around 230 B.C.E. **Eratosthenes** calculated the circumference of the Earth from the following observations. At noon on the longest day of the year, the Sun was directly overhead in Syene, while it was inclined 7.2° from the vertical in Alexandria. He believed the two cities to be on the same longitudinal line and assumed that the rays of the Sun are parallel. Thus he concluded that the arc from Syene to Alexandria was subtended by a central angle of 7.2° at the center of the Earth. See **FIGURE 4.1.13**. At that time the distance from Syene to Alexandria was measured as 5000 stades. If one stade = 559 feet, find the circumference of the Earth in **(a)** stades and **(b)** miles. Show that Eratosthenes' data gives a result that is within 7% of the correct value if the polar diameter of the Earth is 7900 miles (to the nearest mile).

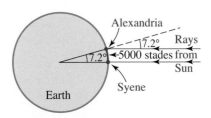

FIGURE 4.1.13 Earth in Problem 78

79. **Circular Motion of a Yo-Yo** A yo-yo is whirled around in a circle at the end of its 100-cm string.
    (a) If it makes six revolutions in 4 seconds, find its rate of turning, or angular speed, in radians per second.
    (b) Find the speed at which the yo-yo travels in centimeters per second; that is its linear speed.

80. **More Yo-Yos** If there is a knot in the yo-yo string described in Problem 79 at a point 40 cm from the yo-yo, find **(a)** the angular speed of the knot and **(b)** the linear speed.

81. **Circular Motion of a Tire** An automobile with 26-in. diameter tires is traveling at a rate of 55 mi/h.
    (a) Find the number of revolutions per minute that its tires are making.
    (b) Find the angular speed of its tires in radians per minute.

82. **Diameter of the Moon** The average distance from the Earth to the Moon as given by NASA is 238,855 miles. If the angle subtended by the Moon at the eye of an observer on Earth is 0.52°, then what is the approximate diameter of the Moon? **FIGURE 4.1.14** is not to scale.

Yo-yo in
Problems 79 and 80

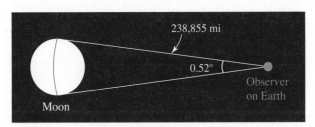

FIGURE 4.1.14 The curved red arc represents the approximate diameter of the Moon

4.2 The Sine and Cosine Functions

≡ **Introduction** Originally, the trigonometric functions were defined using angles in right triangles. A more modern approach, and one that is used in calculus, is to define the trigonometric functions on sets of real numbers. As we will see, the radian measure for angles is key in making these definitions.

☐ **Trigonometric Functions** For each real number t there corresponds an angle of t radians in standard position. As shown in **FIGURE 4.2.1** we denote the point of intersection of the terminal side of the angle t with the **unit circle** by $P(t)$. The x and y coordinates of this point give us the values of the six basic trigonometric functions. The y-coordinate of $P(t)$ is called the **sine of t**, while the x-coordinate of $P(t)$ is called the **cosine of t**.

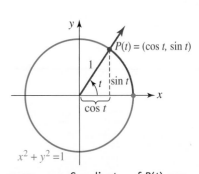

FIGURE 4.2.1 Coordinates of $P(t)$ are $(\cos t, \sin t)$

DEFINITION 4.2.1 Sine and Cosine Functions

Let t be any real number and $P(t) = (x, y)$ be the point of intersection of the unit circle with the terminal side of the angle of t radians in standard position. Then, the **sine of t**, denoted $\sin t$, and the **cosine of t**, denoted $\cos t$, are

$$\sin t = y \tag{1}$$

and

$$\cos t = x \tag{2}$$

Since to each real number t there corresponds a unique point $P(t) = (\cos t, \sin t)$, we have just defined two functions – the sine function and the cosine function – each with domain the set R of real numbers. Four additional trigonometric functions are defined in terms of the coordinates of $P(t) = (x, y)$.

DEFINITION 4.2.2 Tangent, Cotangent, Secant, and Cosecant Functions

---

The **tangent, cotangent, secant,** and **cosecant functions** of the real number t are

$$\tan t = \frac{y}{x}, \quad x \neq 0 \tag{3}$$

$$\cot t = \frac{x}{y}, \quad y \neq 0 \tag{4}$$

$$\sec t = \frac{1}{x}, \quad x \neq 0 \tag{5}$$

and

$$\csc t = \frac{1}{y}, \quad y \neq 0 \tag{6}$$

Using $\sin t = y$ and $\cos t = x$ in (3)–(6) of Definition 4.2.2 we obtain the important identities:

$$\tan t = \frac{\sin t}{\cos t} \qquad \cot t = \frac{\cos t}{\sin t} \tag{7}$$

$$\sec t = \frac{1}{\cos t} \qquad \csc t = \frac{1}{\sin t} \tag{8}$$

Because of the role played by the unit circle in Definitions 4.2.1 and 4.2.2, the six trigonometric functions are referred to as the **circular functions**.

For the remainder of this section and the next we are going to examine the sine and cosine functions in detail. We will come back to the tangent, cotangent, secant, and cosecant functions in Section 4.4.

A number of properties of the sine and cosine functions follow from the fact that $P(t) = (\cos t, \sin t)$ lies on the unit circle. For instance, the coordinates of $P(t)$ must satisfy the equation of the circle:

$$x^2 + y^2 = 1. \tag{9}$$

Substituting $x = \cos t$ and $y = \sin t$ gives an important relationship between the sine and the cosine called the **Pythagorean identity**:

$$(\cos t)^2 + (\sin t)^2 = 1.$$

From now on we will follow two standard practices in writing this identity: $(\cos t)^2$ and $(\sin t)^2$ will be written as $\cos^2 t$ and $\sin^2 t$, respectively, and the $\sin^2 t$ term will be written first.

THEOREM 4.2.1 Pythagorean Identity

For all real numbers t,

$$\sin^2 t + \cos^2 t = 1 \qquad\qquad (10)$$

Again, if $P(x, y)$ denotes a point on the unit circle (9), it follows that the coordinates of P must satisfy the inequalities $-1 \le x \le 1$ and $-1 \le y \le 1$. Because $x = \cos t$ and $y = \sin t$ we have the following bounds on the values of the sine and cosine functions.

THEOREM 4.2.2 Bounds on the Values of Sine and Cosine

For all real numbers t,

$$-1 \le \sin t \le 1 \qquad \text{and} \qquad -1 \le \cos t \le 1$$

The foregoing inequalities can also be expressed as $|\sin t| \le 1$ and $|\cos t| \le 1$. Thus, for example, there is no real number t such that $\sin t = \frac{3}{2}$.

□ **Domain and Range** From the preceding observations we have the sine and cosine functions $f(t) = \sin t$ and $g(t) = \cos t$ each with **domain** the set R of real numbers and **range** the interval $[-1, 1]$.

□ **Signs of the Circular Functions** The signs of the function values $\sin t$ and $\cos t$ are determined by the quadrant in which the point $P(t)$ lies, and conversely. For example, if $\sin t$ and $\cos t$ are both negative, then the point $P(t)$ and terminal side of the corresponding angle of t radians must lie in quadrant III. **FIGURE 4.2.2** displays the signs of the cosine and sine functions in each of the four quadrants.

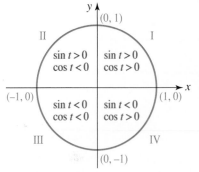

FIGURE 4.2.2 Algebraic signs of $\sin t$ and $\cos t$ in the four quadrants

---

EXAMPLE 1 Using the Pythagorean Identity

Given that $\cos t = \frac{1}{3}$ and that $P(t)$ is a point in the fourth quadrant, find $\sin t$.

Solution Substitution of $\cos t = \frac{1}{3}$ into the Pythagorean identity (10) gives $\sin^2 t + \left(\frac{1}{3}\right)^2 = 1$ or $\sin^2 t = \frac{8}{9}$. Since $\sin t$ is the y-coordinate of $P(t)$, a point in the fourth quadrant, we must take the negative square root for $\sin t$:

$$\sin t = -\sqrt{\frac{8}{9}} = -\frac{2\sqrt{2}}{3}. \qquad\qquad \equiv$$

EXAMPLE 2 Sine and Cosine of a Real Number

Use a calculator to approximate $\sin 3$ and $\cos 3$ and give a geometric interpretation of these values.

Solution From a calculator set in *radian mode*, we obtain $\cos 3 \approx -0.9899925$ and $\sin 3 \approx 0.1411200$. These values represent the x- and y-coordinates, respectively, of the point of intersection of the terminal side of the angle of 3 radians in standard position

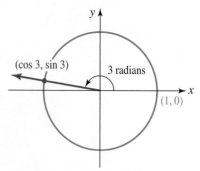

FIGURE 4.2.3 The point $P(3)$

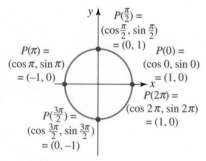

FIGURE 4.2.4 Sine and cosine values for quadrantal angles

with the unit circle. As shown in FIGURE 4.2.3, this point lies in the second quadrant because $\pi/2 < 3 < \pi$. This would also be expected in view of Figure 4.2.2 since $\cos 3$, the x-coordinate, is *negative* and $\sin 3$, the y-coordinate, is *positive*. ≡

☐ **Values Corresponding to Unit Circle Intercepts** As shown in FIGURE 4.2.4, the x- and y-intercepts of the unit circle give us the values of the sine and cosine functions for the real numbers corresponding to **quadrantal angles** listed next.

Values of the Sine and Cosine

For $t = 0$: $\sin 0 = 0$ and $\cos 0 = 1$

For $t = \dfrac{\pi}{2}$: $\sin \dfrac{\pi}{2} = 1$ and $\cos \dfrac{\pi}{2} = 0$

For $t = \pi$: $\sin \pi = 0$ and $\cos \pi = -1$

For $t = \dfrac{3\pi}{2}$: $\sin \dfrac{3\pi}{2} = -1$ and $\cos \dfrac{3\pi}{2} = 0$

☐ **Periodicity** In Section 4.1 we saw that for any real number t, the angles of t radians and $t \pm 2\pi$ radians are coterminal. Thus they determine the same point (x, y) on the unit circle. Therefore

$$\sin t = \sin(t \pm 2\pi) \quad \text{and} \quad \cos t = \cos(t \pm 2\pi). \tag{11}$$

In other words, the sine and cosine functions repeat their values every 2π units. It also follows that for any integer n:

$$\sin(t + 2n\pi) = \sin t \quad \text{and} \quad \cos(t + 2n\pi) = \cos t. \tag{12}$$

DEFINITION 4.2.3   Periodic Functions

A nonconstant function f is said to be **periodic** if there is a positive number p such that

$$f(t) = f(t + p) \tag{13}$$

for every t in the domain of f. If p is the smallest positive number for which (13) is true, then p is called the **period** of the function f.

The equations in (11) imply that the sine and the cosine functions are periodic with period $p \leq 2\pi$. To see that the period of $\sin t$ is actually 2π, we observe that there is only one point on the unit circle with y-coordinate 1, namely, $P(\pi/2) = (\cos(\pi/2), \sin(\pi/2)) = (0, 1)$. Therefore,

$$\sin t = 1 \quad \text{only for} \quad t = \frac{\pi}{2}, \frac{\pi}{2} \pm 2\pi, \frac{\pi}{2} \pm 4\pi,$$

and so on. Thus the smallest possible positive value of p is 2π.

THEOREM 4.2.3   Period of the Sine and Cosine

The sine and cosine functions are periodic with period **2π**. Therefore,

$$\sin(t + 2\pi) = \sin t \quad \text{and} \quad \cos(t + 2\pi) = \cos t \tag{14}$$

for every real number t.

☐ **Even–Odd Properties** The symmetry of the unit circle endows the circular functions with several additional properties. For any real number t, the points $P(t)$ and $P(-t)$ on the unit circle are located on the terminal side of an angle of t and $-t$ radians, respectively. These two points will always be symmetric with respect to the x-axis. FIGURE 4.2.5 illustrates the situation for a point $P(t)$ lying in the first quadrant: The x-coordinates of the two points are identical; however, the y-coordinates have equal magnitudes but opposite signs. The same symmetries will hold regardless of which quadrant contains $P(t)$. Thus, for any real number t, $\cos(-t) = \cos t$ and $\sin(-t) = -\sin t$. Applying the definitions of **even** and **odd functions** from Section 2.2 we have the following result.

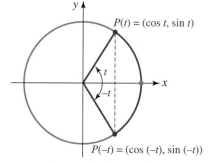

FIGURE 4.2.5 Coordinates of $P(t)$ and $P(-t)$

THEOREM 4.2.4 Even and Odd Functions

The cosine function is **even** and the sine function is **odd**. That is, for every real number t,

$$\cos(-t) = \cos t \quad \text{and} \quad \sin(-t) = -\sin t \qquad (15)$$

The following additional properties of the sine and cosine functions can be verified by considering the symmetries of appropriately chosen points on the unit circle.

THEOREM 4.2.5 Additional Properties

$$\cos\left(\frac{\pi}{2} - t\right) = \sin t \quad \text{and} \quad \sin\left(\frac{\pi}{2} - t\right) = \cos t \qquad (16)$$

$$\cos(t + \pi) = -\cos t \quad \text{and} \quad \sin(t + \pi) = -\sin t \qquad (17)$$

$$\cos(\pi - t) = -\cos t \quad \text{and} \quad \sin(\pi - t) = \sin t \qquad (18)$$

For example, to justify the properties in (16) of Theorem 4.2.5 for $0 < t < \pi/2$, consider FIGURE 4.2.6. Since the points $P(t)$ and $P(\pi/2 - t)$ are symmetric with respect to the line $y = x$, we can obtain the coordinates of $P(\pi/2 - t)$, by interchanging the coordinates of $P(t)$. Thus,

$$\cos t = \sin\left(\frac{\pi}{2} - t\right) \quad \text{and} \quad \sin t = \cos\left(\frac{\pi}{2} - t\right).$$

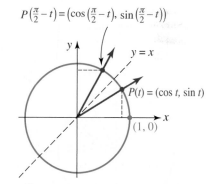

FIGURE 4.2.6 Geometric justification of (16) in Theorem 4.2.5

The special properties of the sine and cosine functions in Theorem 4.2.5 will become quite useful as soon as we determine additional values for $\sin t$ and $\cos t$ in the interval $[0, 2\pi)$. Using results from plane geometry we will now find the values of the sine and cosine functions for $t = \pi/6, t = \pi/4$, and $t = \pi/3$.

☐ **Finding $\sin(\pi/4)$ and $\cos(\pi/4)$** We draw an angle of $\pi/4$ radians (45°) in standard position and locate and label $P(\pi/4) = (\cos(\pi/4), \sin(\pi/4))$ on the unit circle. As shown in FIGURE 4.2.7, we form a right triangle by dropping a perpendicular from $P(\pi/4)$ to the x-axis. Since the sum of the angles in any triangle is π radians (180°), the third angle of this triangle is also $\pi/4$ radians, hence the triangle is isosceles. Therefore

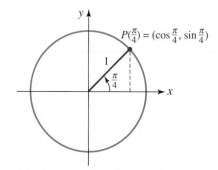

FIGURE 4.2.7 The point $P(\pi/4)$

the coordinates of $P(\pi/4)$ are equal; that is, $\cos(\pi/4) = \sin(\pi/4)$. It follows from the Pythagorean identity (10)

$$\sin^2\frac{\pi}{4} + \cos^2\frac{\pi}{4} = 1 \qquad \text{that} \qquad 2\cos^2\frac{\pi}{4} = 1.$$

Dividing by 2 and taking the square root, we obtain $\cos(\pi/4) = \pm\sqrt{2}/2$. Since $P(\pi/4)$ lies in the first quadrant, both coordinates must be positive. So we have found the (equal) coordinates of $P(\pi/4)$:

$$\cos\frac{\pi}{4} = \frac{\sqrt{2}}{2} \qquad \text{and} \qquad \sin\frac{\pi}{4} = \frac{\sqrt{2}}{2}.$$

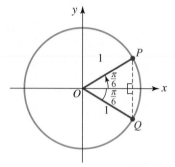

FIGURE 4.2.8 The point $P(\pi/6)$

☐ **Finding $\sin(\pi/6)$ and $\cos(\pi/6)$** We construct two angles of $\pi/6$ radians ($30°$) in the first and fourth quadrants, as shown in **FIGURE 4.2.8**, and label the points of intersection with the unit circle $P(\pi/6)$ and Q, respectively. By drawing perpendicular line segments from P and Q to the x-axis, we obtain two *congruent* right triangles because each triangle has a hypotenuse of length 1 and angles of $30°$, $60°$, and $90°$. Since the $90°$ angles form a straight angle, these two right triangles form an *equilateral* triangle $\triangle POQ$ with sides of length 1. Since $\sin(\pi/6)$ is equal to half of the vertical side of $\triangle POQ$, we have

$$\sin\frac{\pi}{6} = \frac{1}{2}.$$

From this result and the Pythagorean identity (10) we find the value of $\cos(\pi/6)$:

$$\left(\frac{1}{2}\right)^2 + \cos^2\frac{\pi}{6} = 1 \qquad \text{implies} \qquad \cos^2\frac{\pi}{6} = \frac{3}{4}$$

We take the positive square root here
because $P(\pi/6)$ lies in the first quadrant. ▶   or

$$\cos\frac{\pi}{6} = \frac{\sqrt{3}}{2}.$$

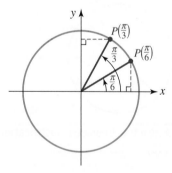

FIGURE 4.2.9 The point $P(\pi/3)$

☐ **Finding $\sin(\pi/3)$ and $\cos(\pi/3)$** We draw angles of $\pi/6$ and $\pi/3$ in standard position and locate and label the points $P(\pi/6)$ and $P(\pi/3)$, as shown in **FIGURE 4.2.9**. We then construct two congruent $30°$-$60°$-$90°$ triangles by dropping perpendiculars to the x- and y-axes, respectively. It follows from the congruence of these triangles that

$$\cos\frac{\pi}{3} = \sin\frac{\pi}{6} = \frac{1}{2} \qquad \text{and} \qquad \sin\frac{\pi}{3} = \cos\frac{\pi}{6} = \frac{\sqrt{3}}{2}.$$

The foregoing results also follow from (16) of Theorem 4.2.5 with $t = \pi/6$.

We summarize the values of the sine and cosine functions corresponding to the basic fractional multiples of π that we have determined so far.

Values of the Sine and Cosine (Continued)		
For $t = \dfrac{\pi}{6}$: $\sin\dfrac{\pi}{6} = \dfrac{1}{2}$	and	$\cos\dfrac{\pi}{6} = \dfrac{\sqrt{3}}{2}$
For $t = \dfrac{\pi}{4}$: $\sin\dfrac{\pi}{4} = \dfrac{\sqrt{2}}{2}$	and	$\cos\dfrac{\pi}{4} = \dfrac{\sqrt{2}}{2}$
For $t = \dfrac{\pi}{3}$: $\sin\dfrac{\pi}{3} = \dfrac{\sqrt{3}}{2}$	and	$\cos\dfrac{\pi}{3} = \dfrac{1}{2}$

CHAPTER 4 TRIGONOMETRIC FUNCTIONS

□ **Reference Angle** As we noted at the beginning of this section, for each real number t there is a unique angle of t radians in standard position that determines the point $P(t)$, with coordinates $(\cos t, \sin t)$, on the unit circle. As shown in FIGURE 4.2.10, the terminal side of any angle of t radians (with $P(t)$ not on an axis) will form an acute angle with the x-axis. We can then locate an angle of t' radians in the first quadrant that is congruent to this acute angle. The angle of t' radians is called the **reference angle** for t. Because of the symmetry of the unit circle, the coordinates of $P(t')$ will be equal *in absolute value* to the respective coordinates of $P(t)$. Hence

$$\sin t = \pm \sin t' \quad \text{and} \quad \cos t = \pm \cos t'.$$

As the following examples will show, reference angles can be used to find the trigonometric function values of any integer multiple of $\pi/6$, $\pi/4$, and $\pi/3$.

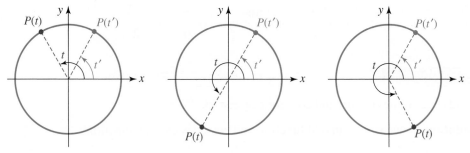

FIGURE 4.2.10 Reference angle t' is an acute angle

EXAMPLE 3   Using Reference Angles

Find exact values of $\sin t$ and $\cos t$ for the given real number t.

(a) $t = 5\pi/3$ **(b)** $t = -3\pi/4$

Solution In each part we begin by finding the reference angle corresponding to the given value of t.

(a) From FIGURE 4.2.11 we find that an angle of $t = 5\pi/3$ radians determines a point $P(5\pi/3)$ in the fourth quadrant and has the reference angle $t' = \pi/3$ radians. After adjusting the signs of the coordinates of $P(\pi/3) = (1/2, \sqrt{3}/2)$ to obtain the fourth quadrant point $P(5\pi/3) = (1/2, -\sqrt{3}/2)$, we find that

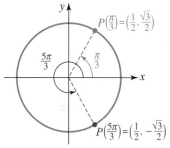

FIGURE 4.2.11 Reference angle in part (a) of Example 3

$$\sin\frac{5\pi}{3} = -\sin\frac{\pi}{3} = -\frac{\sqrt{3}}{2} \quad \text{and} \quad \cos\frac{5\pi}{3} = \cos\frac{\pi}{3} = \frac{1}{2}.$$

(with "reference angle" labeled above the $-\sin\frac{\pi}{3}$ and $\cos\frac{\pi}{3}$ terms)

(b) The point $P(-3\pi/4)$ lies in the third quadrant and has reference angle $\pi/4$, as shown in FIGURE 4.2.12. Therefore,

$$\sin\left(-\frac{3\pi}{4}\right) = -\sin\frac{\pi}{4} = -\frac{\sqrt{2}}{2} \quad \text{and} \quad \cos\left(-\frac{3\pi}{4}\right) = -\cos\frac{\pi}{4} = -\frac{\sqrt{2}}{2}. \quad \equiv$$

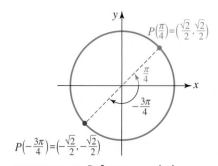

FIGURE 4.2.12 Reference angle in part (b) of Example 3

Sometimes, in order to find the trigonometric values of multiples of our basic fractions of π we must use periodicity or the even-odd function properties in addition to reference numbers.

Find exact values of $\sin t$ and $\cos t$ for $t = 29\pi/6$.

Solution Since $29\pi/6$ is greater than 2π, we rewrite $29\pi/6$ as an integer multiple of 2π plus a number less than 2π:

$$\frac{29\pi}{6} = 4\pi + \frac{5\pi}{6} = 2(2\pi) + \frac{5\pi}{6}.$$

From the periodicity equations (12) with $n = 2$ and $t = 5\pi/6$ we know that $\sin(29\pi/6) = \sin(5\pi/6)$ and $\cos(29\pi/6) = \cos(5\pi/6)$. Next we see from FIGURE 4.2.13 that the reference angle for $5\pi/6$ is $\pi/6$. Since $P(5\pi/6)$ is a second quadrant point, we have

$$\sin\frac{29\pi}{6} = \sin\frac{5\pi}{6} = \sin\frac{\pi}{6} = \frac{1}{2}$$

and

$$\cos\frac{29\pi}{6} = \cos\frac{5\pi}{6} = -\cos\frac{\pi}{6} = -\frac{\sqrt{3}}{2}. \qquad \equiv$$

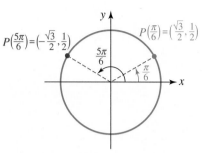

FIGURE 4.2.13 Reference angle in Example 4

$P\left(\frac{5\pi}{6}\right) = \left(-\frac{\sqrt{3}}{2}, \frac{1}{2}\right)$ $P\left(\frac{\pi}{6}\right) = \left(\frac{\sqrt{3}}{2}, \frac{1}{2}\right)$

Find exact values of $\sin t$ and $\cos t$ for $t = -\pi/6$.

Solution Since sine is an odd function and cosine is an even function,

See Theorem 4.2.4. ▶

$$\sin\left(-\frac{\pi}{6}\right) = -\sin\left(\frac{\pi}{6}\right) = -\frac{1}{2} \qquad \text{and} \qquad \cos\left(-\frac{\pi}{6}\right) = \cos\left(\frac{\pi}{6}\right) = \frac{\sqrt{3}}{2}.$$

This problem could also have been solved by using a reference angle. $\qquad \equiv$

□ **Trigonometric Functions of Angles** In this section we have defined sine and cosine functions of the real number t by using the coordinates of a point $P(t)$ on the unit circle. It is now possible to define the **trigonometric functions of any angle θ**. For any angle θ, we simply let

$$\sin\theta = \sin t \qquad \text{and} \qquad \cos\theta = \cos t,$$

where the real number t is the radian measure of θ. As mentioned in Section 4.1, it is common to omit the word radians when measuring an angle. So we write $\sin(\pi/6)$ for both the sine of the real number $\pi/6$ and for the sine of the angle of $\pi/6$ radians. Furthermore, since the values of the trigonometric functions are determined by the coordinates of the point $P(t)$ on the unit circle, it really does not matter whether θ is measured in radians or in degrees. For example, regardless of whether we are given $\theta = \pi/6$ radians or $\theta = 30°$, the point on the unit circle corresponding to this angle in standard position is $\left(\sqrt{3}/2, 1/2\right)$. Thus,

$$\sin\frac{\pi}{6} = \sin 30° = \frac{1}{2} \qquad \text{and} \qquad \cos\frac{\pi}{6} = \cos 30° = \frac{\sqrt{3}}{2}.$$

4.2   Exercises   Answers to selected odd-numbered problems begin on page ANS–12.

1. Given that $\cos t = -\frac{2}{5}$ and that $P(t)$ is a point in the second quadrant, find $\sin t$.
2. Given that $\sin t = \frac{1}{4}$ and that $P(t)$ is a point in the second quadrant, find $\cos t$.
3. Given that $\sin t = -\frac{2}{3}$ and that $P(t)$ is a point in the third quadrant, find $\cos t$.

4. Given that $\cos t = \frac{3}{4}$ and that $P(t)$ is a point in the fourth quadrant, find $\sin t$.
5. If $\sin t = -\frac{2}{7}$, find all possible values of $\cos t$.
6. If $\cos t = \frac{3}{10}$, find all possible values of $\sin t$.
7. If $\cos t = -0.2$, find all possible values of $\sin t$.
8. If $\sin t = 0.4$, find all possible values of $\cos t$.
9. If $2\sin t - \cos t = 0$, find all possible values of $\sin t$ and $\cos t$.
10. If $3\sin t - 2\cos t = 0$, find all possible values of $\sin t$ and $\cos t$.

In Problems 11–14, find the exact value of **(a)** $\sin t$ and **(b)** $\cos t$ for the given value of t. Do not use a calculator.

11. $t = -\pi/2$ **12.** $t = 3\pi$
13. $t = 8\pi$ **14.** $t = -3\pi/2$

In Problems 15–26, for the given value of t determine the reference angle t' and the exact values of $\sin t$ and $\cos t$. Do not use a calculator.

15. $t = 2\pi/3$ **16.** $t = 4\pi/3$ **17.** $t = 5\pi/4$
18. $t = 3\pi/4$ **19.** $t = 11\pi/6$ **20.** $t = 7\pi/6$
21. $t = -\pi/4$ **22.** $t = -7\pi/4$ **23.** $t = -5\pi/6$
24. $t = -11\pi/6$ **25.** $t = -5\pi/3$ **26.** $t = -2\pi/3$

In Problems 27–32, find the given trigonometric function value. Do not use a calculator.

27. $\sin(-11\pi/3)$ **28.** $\cos(17\pi/6)$ **29.** $\cos(-7\pi/4)$
30. $\sin(-19\pi/2)$ **31.** $\cos 5\pi$ **32.** $\sin(23\pi/3)$

In Problems 33–38, justify the given statement with one of the properties of the trigonometric functions.

33. $\sin \pi = \sin 3\pi$ **34.** $\cos(\pi/4) = \sin(\pi/4)$
35. $\sin(-3 - \pi) = -\sin(3 + \pi)$ **36.** $\cos 16.8\pi = \cos 14.8\pi$
37. $\cos 0.43 = \cos(-0.43)$ **38.** $\sin(2\pi/3) = \sin(\pi/3)$

In Problems 39–46, find the given trigonometric function value. Do not use a calculator.

39. $\sin 135°$ **40.** $\cos 150°$
41. $\cos 210°$ **42.** $\sin 270°$
43. $\cos 330°$ **44.** $\sin(-180°)$
45. $\sin(-60°)$ **46.** $\cos(-300°)$

In Problems 47–50, find all angles t, where $0 \leq t < 2\pi$, that satisfy the given condition.

47. $\sin t = 0$ **48.** $\cos t = -1$
49. $\cos t = \sqrt{2}/2$ **50.** $\sin t = \frac{1}{2}$

In Problems 51–54, find all angles θ, where $0° \leq \theta < 360°$, that satisfy the given condition.

51. $\cos \theta = \sqrt{3}/2$ **52.** $\sin \theta = -\frac{1}{2}$
53. $\sin \theta = -\sqrt{2}/2$ **54.** $\cos \theta = 1$

Miscellaneous Applications

55. Free Throw Under certain conditions the maximum height y attained by a basketball released from a height h at an angle α measured from the horizontal with

Free throw

an initial velocity v_0 is given by $y = h + (v_0^2 \sin^2 \alpha)/2g$, where g is the acceleration due to gravity. Compute the maximum height reached by a free throw if $h = 2.15$ m, $v_0 = 8$ m/s, $\alpha = 64.47°$, and $g = 9.81$ m/s^2.

56. **Putting the Shot** The range of a shot put released from a height h above the ground with an initial velocity v_0 at an angle α to the horizontal can be approximated by

$$R = \frac{v_0 \cos \alpha}{g} \left(v_0 \sin \alpha + \sqrt{v_0^2 \sin^2 \alpha + 2gh} \right),$$

where g is the acceleration due to gravity. If $v_0 = 13.7$ m/s, $\alpha = 40°$, and $g = 9.81$ m/s^2, compare the ranges achieved for the release heights **(a)** $h = 2.0$ m and **(b)** $h = 2.4$ m. **(c)** Explain why an increase in h yields an increase in R if the other parameters are held fixed. **(d)** What does this imply about the advantage that height gives a shot-putter?

57. **Acceleration Due to Gravity** Because of its rotation, the Earth bulges at the equator and is flattened at the poles. As a result, the acceleration due to gravity is not a constant 980 cm/s^2, but varies with latitude θ. Satellite studies have shown that the acceleration due to gravity g_{sat} is approximated by the function

$$g_{sat} = 978.0309 + 5.18552 \sin^2 \theta - 0.00570 \sin^2 2\theta.$$

(a) Find g_{sat} at the equator $(\theta = 0°)$, **(b)** at the North Pole, and **(c)** at 45° north latitude.

For Discussion

58. Discuss how it is possible to determine without a calculator that the point $P(6) = (\cos 6, \sin 6)$ lies in the fourth quadrant.
59. Discuss how it is possible to determine without the aid of a calculator that both $\sin 4$ and $\cos 4$ are negative.
60. Is there a real number t satisfying $3 \sin t = 5$? Explain why or why not.
61. Is there an angle θ satisfying $\cos \theta = -2$? Explain why or why not.
62. Suppose f is a periodic function with period p. Show that $F(x) = f(ax)$, $a > 0$, is periodic with period p/a.

4.3 Graphs of Sine and Cosine Functions

☰ **Introduction** One way to further your understanding of the trigonometric functions is to examine their graphs. In this section we consider the graphs of the sine and cosine functions.

☐ **Graphs of Sine and Cosine** In Section 4.2 we saw that the domain of the sine function $f(t) = \sin t$ is the set of real numbers $(-\infty, \infty)$ and the interval $[-1, 1]$ is its range. Since the sine function has period 2π, we begin by sketching its graph on the interval $[0, 2\pi]$. We obtain a rough sketch of the graph given in FIGURE 4.3.1(b) by considering various positions of the point $P(t)$ on the unit circle, as shown in Figure 4.3.1(a). As t varies from 0 to $\pi/2$, the value $\sin t$ increases from 0 to its maximum value 1. But as t varies from $\pi/2$ to $3\pi/2$, the value $\sin t$ decreases from 1 to its minimum value -1. We note that $\sin t$ changes from positive to negative at $t = \pi$. For t between $3\pi/2$ and 2π, we see that the corresponding values of $\sin t$ increase

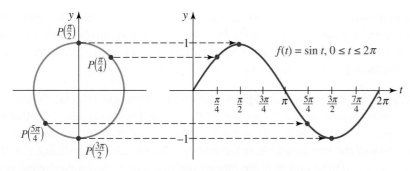

(a) Unit circle   (b) One cycle of sine graph

FIGURE 4.3.1 Points $P(t)$ on a circle corresponding to points on the graph

from -1 to 0. The graph of *any* periodic function over an interval of length equal to its period is said to be one **cycle of its graph**. In the case of the sine function, the graph over the interval $[0, 2\pi]$ in Figure 4.3.1(b) is one cycle of the graph of $f(t) = \sin t$.

◀ Note: Change of symbols

From this point on we will revert to the traditional symbols x and y when graphing trigonometric functions. Thus, $f(t) = \sin t$ will either be written $f(x) = \sin x$ or simply $y = \sin x$.

The graph of a periodic function is easily obtained by repeatedly drawing one cycle of its graph. In other words, the graph of $y = \sin x$ on, say, the intervals $[-2\pi, 0]$ and $[2\pi, 4\pi]$ is the same as that given in Figure 4.3.1(b). Recall from Theorem 4.2.4 of Section 4.2 that the sine function is an odd function since $f(-x) = \sin(-x) = -\sin x = -f(x)$. In other words, if (x, y) is a point on the graph of f, then so is $(-x, -y)$. Thus, from Theorem 2.2.1 it follows that the graph of $y = \sin x$ shown in **FIGURE 4.3.2** is symmetric with respect to the origin.

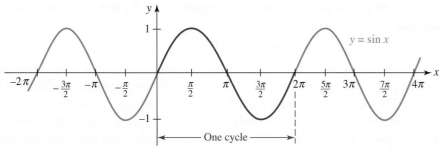

FIGURE 4.3.2 Graph of $y = \sin x$

By working again with the unit circle we can obtain one cycle of the graph of the cosine function $g(x) = \cos x$ on the interval $[0, 2\pi]$. In contrast to the graph of $f(x) = \sin x$ where $f(0) = f(2\pi) = 0$, for the cosine function we have $g(0) = g(2\pi) = 1$. **FIGURE 4.3.3** shows

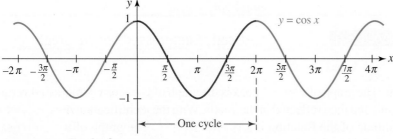

FIGURE 4.3.3 Graph of $y = \cos x$

one cycle (in red) of $y = \cos x$ on $[0, 2\pi]$, along with the extension of that cycle (in blue) to the adjacent intervals $[-2\pi, 0]$ and $[2\pi, 4\pi]$. We see from this figure that the graph of the cosine function is symmetric with respect to the y-axis. This is a consequence of g being an even function: $g(-x) = \cos(-x) = \cos x = g(x)$.

☐ **Intercepts** In this and subsequent courses in mathematics it is important that you know the x-coordinates of the x-intercepts of the sine and cosine graphs—in other words, the zeros of $f(x) = \sin x$ and $g(x) = \cos x$. From the sine graph in Figure 4.3.2, we see that the zeros of the sine function, or the numbers for which $\sin x = 0$, are $x = 0, \pm\pi,$ $\pm 2\pi, \pm 3\pi, \ldots$. These numbers are integer multiples of π. From the cosine graph in Figure 4.3.3, we see that $\cos x = 0$ when $x = \pm\pi/2, \pm 3\pi/2, \pm 5\pi/2, \ldots$. These numbers are odd-integer multiples of $\pi/2$.

If n represents an integer, then $2n + 1$ is an odd integer. Therefore the zeros of $f(x) = \sin x$ and $g(x) = \cos x$ can be written in a compact form.

Zeros of the Sine and Cosine Functions

$$\sin x = 0 \quad \text{for} \quad x = n\pi, n \text{ an integer} \tag{1}$$

$$\cos x = 0 \quad \text{for} \quad x = (2n + 1)\frac{\pi}{2}, n \text{ an integer} \tag{2}$$

Using the distributive law, the result in (2) is often written as $x = \pi/2 + n\pi$.

As we did in Chapters 2 and 3, we can obtain variations of the basic sine and cosine graphs through rigid and nonrigid transformations. For the remainder of the discussion we will consider graphs of functions of the form

$$y = A\sin(Bx + C) + D \quad \text{or} \quad y = A\cos(Bx + C) + D, \tag{3}$$

where A, B, C, and D are real constants.

☐ **Graphs of $y = A \sin x + D$ and $y = A \cos x + D$** We begin by considering the special cases of (3):

$$y = A\sin x \quad \text{and} \quad y = A\cos x.$$

For $A > 0$, graphs of these functions are either a vertical stretch or a vertical compression of the graphs of $y = \sin x$ or $y = \cos x$. For $A < 0$, the graphs are also reflected in the x-axis. For example, as FIGURE 4.3.4 shows, we obtain the graph of $y = 2\sin x$ by stretching the graph of $y = \sin x$ vertically by a factor of 2. Note that the maximum and minimum values of $y = 2\sin x$ occur at the same x-values as the maximum and minimum values of $y = \sin x$. In general, the maximum distance from any point on the graph of $y = A\sin x$ or $y = A\cos x$ to the x-axis is $|A|$. The number $|A|$ is called the **amplitude** of the functions or of their graphs. The amplitude of the basic functions $y = \sin x$ and $y = \cos x$ is $|A| = 1$. In general, if a periodic function f is continuous, then over a closed interval of length equal to its period, f has both a maximum value M and a minimum value m. The amplitude is defined by

$$\text{amplitude} = \tfrac{1}{2}[M - m]. \tag{4}$$

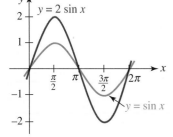

FIGURE 4.3.4 Vertical stretch of $y = \sin x$

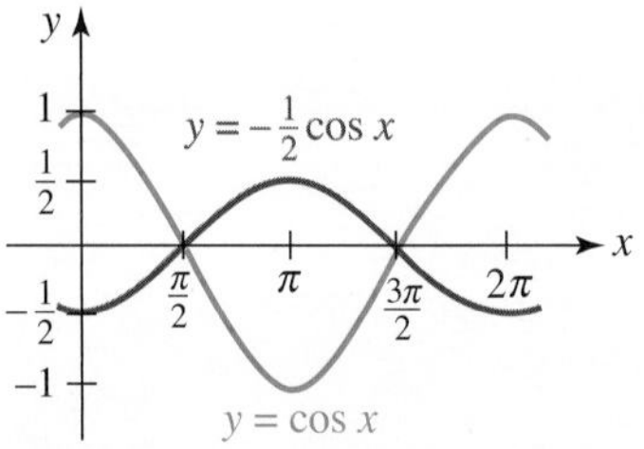

FIGURE 4.3.5 Graph of function in Example 1

▌ EXAMPLE 1   Vertically Compressed Cosine Graph

Graph $y = -\tfrac{1}{2}\cos x$.

Solution The graph of $y = -\tfrac{1}{2}\cos x$ is the graph of $y = \cos x$ compressed vertically by a factor of $\tfrac{1}{2}$ and then reflected in the x-axis. With the identification $A = -\tfrac{1}{2}$, we see that the amplitude of the function is $|A| = |-\tfrac{1}{2}| = \tfrac{1}{2}$. The graph of $y = -\tfrac{1}{2}\cos x$ on the interval $[0, 2\pi]$ is shown in red in FIGURE 4.3.5. ≡

CHAPTER 4 TRIGONOMETRIC FUNCTIONS

The graphs of

$$y = A\sin x + D \qquad \text{and} \qquad y = A\cos x + D$$

are the graphs of $y = A\sin x$ and $y = A\cos x$ shifted vertically, up for $D > 0$ and down for $D < 0$. For example, the graph of $y = 1 + 2\sin x$ is the graph of $y = 2\sin x$ (Figure 4.3.4) shifted up 1 unit. The amplitude of the graph of either $y = A\sin x + D$ or $y = A\cos x + D$ is still $|A|$. Note that in FIGURE 4.3.6, the maximum of $y = 1 + 2\sin x$ is $y = 3$ at $x = \pi/2$ and the minimum is $y = -1$ at $x = 3\pi/2$. From (4), the amplitude of $y = 1 + 2\sin x$ is then $\frac{1}{2}[3 - (-1)] = 2$.

By interpreting x as a placeholder in (1) and (2), we can find the x-coordinates of the x-intercepts of the graphs of sine and cosine functions of the form $y = A\sin Bx$ and $y = A\cos Bx$. (We consider this next.) For example, to solve $\sin 2x = 0$, we have from (1)

$$2x = n\pi \qquad \text{so that} \qquad x = \tfrac{1}{2}n\pi, n = 0, \pm 1, \pm 2, \dots,$$

that is, $x = 0, \pm\frac{1}{2}\pi, \pm\frac{2}{2}\pi = \pi, \pm\frac{3}{2}\pi, \pm\frac{4}{2}\pi = 2\pi$, and so on. See FIGURE 4.3.7.

□ **Graphs of $y = A\sin Bx$ and $y = A\cos Bx$** We now consider the graph of $y = \sin Bx$ for $B > 0$. The function has amplitude 1 since $A = 1$. Because the period of $y = \sin x$ is 2π, a cycle of the graph of $y = \sin Bx$ begins at $x = 0$ and will start to repeat its values when $Bx = 2\pi$. In other words, a cycle of the function $y = \sin Bx$ is completed on the interval defined by $0 \le Bx \le 2\pi$. Dividing the last inequality by B shows that the **period** of the function $y = \sin Bx$ is $2\pi/B$ and that the graph over the interval $[0, 2\pi/B]$ is one **cycle** of its graph. For example, the period of $y = \sin 2x$ is $2\pi/2 = \pi$, and therefore one cycle of the graph is completed on the interval $[0, \pi]$. Figure 4.3.7 shows that two cycles of the graph of $y = \sin 2x$ (in red and blue) are completed on the interval $[0, 2\pi]$, whereas the graph of $y = \sin x$ (in green) has completed only one cycle. In terms of transformations, we can characterize the cycle of $y = \sin 2x$ on $[0, \pi]$ as a **horizontal compression** of the cycle of $y = \sin x$ on $[0, 2\pi]$.

◀ Be careful here: $\sin 2x \ne 2\sin x$

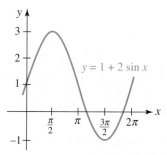

FIGURE 4.3.6 Graph of $y = 2\sin x$ shifted up 1 unit

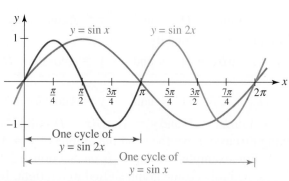

FIGURE 4.3.7 Comparison of the graphs of $y = \sin x$ and $y = \sin 2x$

In summary, the graphs of

$$y = A\sin Bx \qquad \text{and} \qquad y = A\cos Bx$$

for $B > 0$ each have amplitude $|A|$ and period $2\pi/B$.

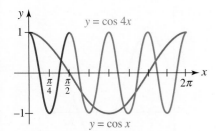

FIGURE 4.3.8 Graph of function in Example 2

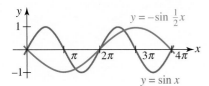

FIGURE 4.3.9 Graph of function in Example 3

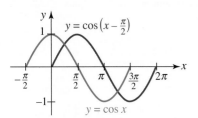

FIGURE 4.3.10 Horizontally shifted cosine graph

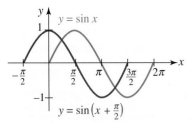

FIGURE 4.3.11 Horizontally shifted sine graph

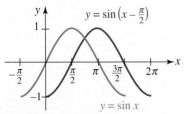

FIGURE 4.3.12 Horizontally shifted sine graph

EXAMPLE 2 **Horizontally Compressed Cosine Graph**

Find the period of $y = \cos 4x$ and graph the function.

Solution Since $B = 4$, we see that the period of $y = \cos 4x$ is $2\pi/4 = \pi/2$. We conclude that the graph of $y = \cos 4x$ is the graph of $y = \cos x$ compressed horizontally. To graph the function, we draw one cycle of the cosine graph with amplitude 1 on the interval $[0, \pi/2]$ and then use periodicity to extend the graph. FIGURE 4.3.8 shows four complete cycles of $y = \cos 4x$ (the basic cycle in red and the extended graph in blue) and one cycle of $y = \cos x$ (in green) on $[0, 2\pi]$. Notice that $y = \cos 4x$ attains its minimum at $x = \pi/4$ since $\cos 4(\pi/4) = \cos \pi = -1$, and attains its maximum at $x = \pi/2$ since $\cos 4(\pi/2) = \cos 2\pi = 1$. ≡

If $B < 0$ in either $y = A\sin Bx$ or $y = A\cos Bx$, we can use the even–odd properties (see (8) of Section 4.2) to rewrite the function with positive B. This is illustrated in the next example.

EXAMPLE 3 **Horizontally Stretched Sine Graph**

Find the amplitude and period of $y = \sin\left(-\frac{1}{2}x\right)$. Graph the function.

Solution Since we require $B > 0$, we use $\sin(-x) = -\sin x$ to rewrite the function as

$$y = \sin\left(-\frac{1}{2}x\right) = -\sin\frac{1}{2}x.$$

With the identification $A = -1$, the amplitude is seen to be $|A| = |-1| = 1$. Now with $B = \frac{1}{2}$, we find that the period is $2\pi/\frac{1}{2} = 4\pi$. Hence we can interpret the cycle of $y = -\sin\frac{1}{2}x$ on $[0, 4\pi]$ as a horizontal stretch and a reflection (in the x-axis because $A < 0$) of the cycle of $y = \sin x$ on $[0, 2\pi]$. FIGURE 4.3.9 shows that on the interval $[0, 4\pi]$, the graph of $y = -\sin\frac{1}{2}x$ (in blue) completes one cycle, whereas the graph of $y = \sin x$ (in green) completes two cycles. ≡

☐ **Graphs of $y = A\sin(Bx + C)$ and $y = A\cos(Bx + C)$** We have seen that the basic graphs of $y = \sin x$ and $y = \cos x$ can be stretched or compressed vertically

$$y = A\sin x \qquad \text{and} \qquad y = A\cos x,$$

shifted vertically

$$y = A\sin x + D \qquad \text{and} \qquad y = A\cos x + D,$$

and stretched or compressed horizontally

$$y = A\sin Bx + D \qquad \text{and} \qquad y = A\cos Bx + D.$$

The graphs of

$$y = A\sin(Bx + C) + D \qquad \text{and} \qquad y = A\cos(Bx + C) + D$$

are the graphs of $y = A\sin Bx + D$ and $y = A\cos Bx + D$ shifted horizontally.

In the remaining discussion we focus on the graphs of $y = A\sin(Bx + C)$ and $y = A\cos(Bx + C)$. For example, we know from Section 2.2 that the graph of $y = \cos(x - \pi/2)$ is the basic cosine graph shifted to the right. In FIGURE 4.3.10, the graph of $y = \cos(x - \pi/2)$ (in red) on the interval $[0, 2\pi]$ is one cycle of $y = \cos x$ on the interval $[-\pi/2, 3\pi/2]$ (in blue) shifted horizontally $\pi/2$ units to the right. Similarly, the graphs of $y = \sin(x + \pi/2)$ and $y = \sin(x - \pi/2)$ are the basic sine graphs shifted $\pi/2$ units to the left and to the right, respectively. See FIGURES 4.3.11 and 4.3.12.

By comparing the red graphs in Figures 4.3.10–4.3.12 with the graphs in Figures 4.3.2 and 4.3.3 we see that

- the cosine graph shifted $\pi/2$ units to the right is the sine graph,
- the sine graph shifted $\pi/2$ units to the left is the cosine graph, and

- the sine graph shifted $\pi/2$ units to the right is the cosine graph reflected in the x-axis.

In other words, we have graphically verified the identities

$$\cos\left(x - \frac{\pi}{2}\right) = \sin x, \quad \sin\left(x + \frac{\pi}{2}\right) = \cos x, \quad \text{and} \quad \sin\left(x - \frac{\pi}{2}\right) = -\cos x. \quad (5)$$

We now consider the graph of $y = A\sin(Bx + C)$ for $B > 0$. Since the values of $\sin(Bx + C)$ range from -1 to 1, it follows that $A\sin(Bx + C)$ varies between $-A$ and A. That is, the **amplitude** of $y = A\sin(Bx + C)$ is $|A|$. Also, as $Bx + C$ varies from 0 to 2π, the graph will complete one cycle. By solving $Bx + C = 0$ and $Bx + C = 2\pi$, we find that one cycle is completed as x varies from $-C/B$ to $(2\pi - C)/B$. Therefore, the function $y = A\sin(Bx + C)$ has the **period**

$$\frac{2\pi - C}{B} - \left(-\frac{C}{B}\right) = \frac{2\pi}{B}.$$

Moreover, if $f(x) = A\sin Bx$, then

$$f\left(x + \frac{C}{B}\right) = A\sin B\left(x + \frac{C}{B}\right) = A\sin(Bx + C). \quad (6)$$

The result in (6) shows that the graph of $y = A\sin(Bx + C)$ can be obtained by shifting the graph of $f(x) = A\sin Bx$ horizontally a distance $|C|/B$. If $C < 0$ the shift is to the right, whereas if $C > 0$ the shift is to the left. The number $|C|/B$ is called the **phase shift** of the graph of $y = A\sin(Bx + C)$.

EXAMPLE 4   Equation of a Shifted Cosine Graph

The graph of $y = 10\cos 4x$ is shifted $\pi/12$ units to the right. Find its equation.

Solution By writing $f(x) = 10\cos 4x$ and using (6), we find

$$f\left(x - \frac{\pi}{12}\right) = 10\cos 4\left(x - \frac{\pi}{12}\right) \quad \text{or} \quad y = 10\cos\left(4x - \frac{\pi}{3}\right).$$

In the last equation we would identify $C = -\pi/3$. The phase shift is $\pi/12$. ≡

As a practical matter the phase shift of $y = A\sin(Bx + C)$ can be obtained by factoring the number B from $Bx + C$:

◀ Note

$$y = A\sin(Bx + C) = A\sin B\left(x + \frac{C}{B}\right).$$

For convenience we summarize the preceding information.

Shifted Sine and Cosine Graphs

The graphs of

$$y = A\sin(Bx + C) \quad \text{and} \quad y = A\cos(Bx + C),$$

$B > 0$, are, respectively, the graphs of $y = A\sin Bx$ and $y = A\cos Bx$ shifted horizontally by $|C|/B$. The shift is to the right if $C < 0$ and to the left if $C > 0$. The number $|C|/B$ is called the **phase shift**. The **amplitude** of each graph is $|A|$ and the **period** of each graph is $2\pi/B$.

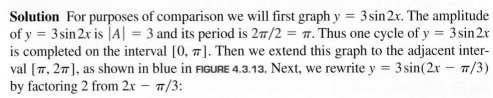

EXAMPLE 5   Horizontally Shifted Sine Graph

Graph $y = 3\sin(2x - \pi/3)$.

Solution For purposes of comparison we will first graph $y = 3\sin 2x$. The amplitude of $y = 3\sin 2x$ is $|A| = 3$ and its period is $2\pi/2 = \pi$. Thus one cycle of $y = 3\sin 2x$ is completed on the interval $[0, \pi]$. Then we extend this graph to the adjacent interval $[\pi, 2\pi]$, as shown in blue in FIGURE 4.3.13. Next, we rewrite $y = 3\sin(2x - \pi/3)$ by factoring 2 from $2x - \pi/3$:

$$y = 3\sin\left(2x - \frac{\pi}{3}\right) = 3\sin 2\left(x - \frac{\pi}{6}\right).$$

From the last form we see that the phase shift is $\pi/6$. The graph of the given function, shown in red in Figure 4.3.13, is obtained by shifting the graph of $y = 3\sin 2x$ to the right $\pi/6$ units. Remember, this means that if (x, y) is a point on the blue graph, then $(x + \pi/6, y)$ is the corresponding point on the red graph. For example, $x = 0$ and $x = \pi$ are the x-coordinates of two x-intercepts of the blue graph. Thus $x = 0 + \pi/6 = \pi/6$ and $x = \pi + \pi/6 = 7\pi/6$ are x-coordinates of the x-intercepts of the red or shifted graph. These numbers are indicated by the arrows in Figure 4.3.13.   ☰

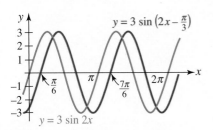

FIGURE 4.3.13 Graph of function in Example 5

EXAMPLE 6   Horizontally Shifted Graphs

Determine the amplitude, period, phase shift, and direction of horizontal shift for each of the following functions.

(a) $y = 15\cos\left(5x - \dfrac{3\pi}{2}\right)$ **(b)** $y = -8\sin\left(2x + \dfrac{\pi}{4}\right)$

Solution (a) We first make the identifications $A = 15$, $B = 5$, and $C = -3\pi/2$. Thus the amplitude is $|A| = 15$ and the period is $2\pi/B = 2\pi/5$. The phase shift can be computed either by $(|{-3\pi}|/2)/5 = 3\pi/10$ or by rewriting the function as

$$y = 15\cos 5\left(x - \frac{3\pi}{10}\right).$$

The last form indicates that the graph of $y = 15\cos(5x - 3\pi/2)$ is the graph of $y = 15\cos 5x$ shifted $3\pi/10$ units to the right.

(b) Since $A = -8$, the amplitude is $|A| = |-8| = 8$. With $B = 2$ the period is $2\pi/2 = \pi$. By factoring 2 from $2x + \pi/4$, we see from

$$y = -8\sin\left(2x + \frac{\pi}{4}\right) = -8\sin 2\left(x + \frac{\pi}{8}\right)$$

that the phase shift is $\pi/8$. The graph of $y = -8\sin(2x + \pi/4)$ is the graph of $y = -8\sin 2x$ shifted $\pi/8$ units to the left.   ☰

EXAMPLE 7   Horizontally Shifted Cosine Graph

Graph $y = 2\cos(\pi x + \pi)$.

Solution The amplitude of $y = 2\cos \pi x$ is $|A| = 2$ and the period is $2\pi/\pi = 2$. Thus one cycle of $y = 2\cos \pi x$ is completed on the interval $[0, 2]$. In FIGURE 4.3.14, two cycles of the graph of $y = 2\cos \pi x$ (in blue) are shown. The x-intercepts of this graph correspond to the values of x for which $\cos \pi x = 0$. By (2), this implies $\pi x = (2n + 1)\pi/2$ or $x = (2n + 1)/2$, n an integer. In other words, for $n = 0, -1, 1, -2$,

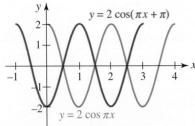

FIGURE 4.3.14 Graph of function in Example 7

CHAPTER 4 TRIGONOMETRIC FUNCTIONS

2, −3, . . . , we get $x = \pm\frac{1}{2}, \pm\frac{3}{2}, \pm\frac{5}{2}$, and so on. Now by rewriting the given function as

$$y = 2\cos\pi(x + 1)$$

we see the phase shift is 1. The graph of $y = 2\cos(\pi x + \pi)$ (in red) in Figure 4.3.14 is obtained by shifting the graph of $y = 2\cos\pi x$ to the left 1 unit. This means that the x-intercepts are the same for both graphs. ≡

EXAMPLE 8   Alternating Current

The current I (in amperes) in a wire of an alternating-current circuit is given by $I(t) = 30\sin 120\pi t$, where t is time measured in seconds. Sketch one cycle of the graph. What is the maximum value of the current?

Solution The graph has amplitude 30 and period $2\pi/120\pi = \frac{1}{60}$. Therefore, we sketch one cycle of the basic sine curve on the interval $\left[0, \frac{1}{60}\right]$, as shown in FIGURE 4.3.15. From the figure it is evident that the maximum value of the current is $I = 30$ amperes and occurs at $t = \frac{1}{240}$ since

$$I\left(\frac{1}{240}\right) = 30\sin\left(120\pi \cdot \frac{1}{240}\right) = 30\sin\frac{\pi}{2} = 30.$$ ≡

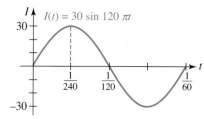

FIGURE 4.3.15 Graph of current in Example 8

4.3   Exercises   Answers to selected odd-numbered problems begin on page ANS–12.

In Problems 1–6, use the techniques of shifting, stretching, compressing, and reflecting to sketch at least one cycle of the graph of the given function.

1. $y = \frac{1}{2} + \cos x$ **2.** $y = -1 + \cos x$ **3.** $y = 2 - \sin x$
4. $y = 3 + 3\sin x$ **5.** $y = -2 + 4\cos x$ **6.** $y = 1 - 2\sin x$

In Problems 7–10, the given figure shows one cycle of a sine or cosine graph. From the figure, determine A and D and write an equation of the form $y = A\sin x + D$ or $y = A\cos x + D$ for the graph.

7.

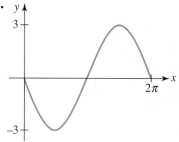

FIGURE 4.3.16 Graph for Problem 7

8.

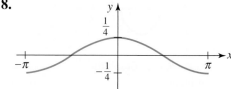

FIGURE 4.3.17 Graph for Problem 8

9.

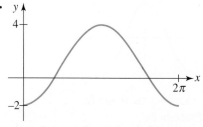

FIGURE 4.3.18 Graph for Problem 9

10.

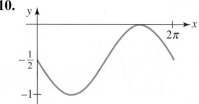

FIGURE 4.3.19 Graph for Problem 10

In Problems 11–16, use (1) and (2) of this section to find the x-intercepts for the graph of the given function. Do not graph.

11. $y = \sin \pi x$

12. $y = -\cos 2x$

13. $y = 10\cos\dfrac{x}{2}$

14. $y = 3\sin(-5x)$

15. $y = \sin\left(x - \dfrac{\pi}{4}\right)$

16. $y = \cos(2x - \pi)$

In Problems 17 and 18, find the x-intercepts of the graph of the given function on the interval $[0, 2\pi]$. Then find all intercepts using periodicity.

17. $y = -1 + \sin x$

18. $y = 1 - 2\cos x$

In Problems 19–24, the given figure shows one cycle of a sine or cosine graph. From the figure, determine A and B and write an equation of the form $y = A\sin Bx$ or $y = A\cos Bx$ for the graph.

19.

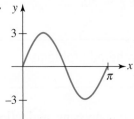

FIGURE 4.3.20 Graph for Problem 19

20.

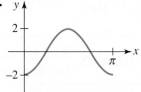

FIGURE 4.3.21 Graph for Problem 20

21.

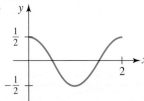

FIGURE 4.3.22 Graph for Problem 21

22.

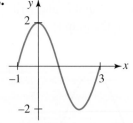

FIGURE 4.3.23 Graph for Problem 22

23.

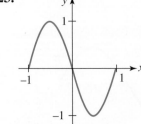

FIGURE 4.3.24 Graph for Problem 23

24.

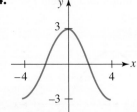

FIGURE 4.3.25 Graph for Problem 24

In Problems 25–32, find the amplitude and period of the given function. Sketch at least one cycle of the graph.

25. $y = 4\sin \pi x$

26. $y = -5\sin \dfrac{x}{2}$

27. $y = -3\cos 2\pi x$

28. $y = \dfrac{5}{2}\cos 4x$

29. $y = 2 - 4\sin x$

30. $y = 2 - 2\sin \pi x$

31. $y = 1 + \cos \dfrac{2x}{3}$

32. $y = -1 + \sin \dfrac{\pi x}{2}$

In Problems 33–42, find the amplitude, period, and phase shift of the given function. Sketch at least one cycle of the graph.

33. $y = \sin\left(x - \dfrac{\pi}{6}\right)$

34. $y = \sin\left(3x - \dfrac{\pi}{4}\right)$

35. $y = \cos\left(x + \dfrac{\pi}{4}\right)$

36. $y = -2\cos\left(2x - \dfrac{\pi}{6}\right)$

37. $y = 4\cos\left(2x - \dfrac{3\pi}{2}\right)$

38. $y = 3\sin\left(2x + \dfrac{\pi}{4}\right)$

39. $y = 3\sin\left(\dfrac{x}{2} - \dfrac{\pi}{3}\right)$

40. $y = -\cos\left(\dfrac{x}{2} - \pi\right)$

41. $y = -4\sin\left(\dfrac{\pi}{3}x - \dfrac{\pi}{3}\right)$

42. $y = 2\cos\left(-2\pi x - \dfrac{4\pi}{3}\right)$

In Problems 43 and 44, write an equation of the function whose graph is described in words.

43. The graph of $y = \cos x$ is vertically stretched up by a factor of 3 and shifted down by 5 units. One cycle of $y = \cos x$ on $[0, 2\pi]$ is compressed to $[0, \pi/3]$ and then the compressed cycle is shifted horizontally $\pi/4$ units to the left.

44. One cycle of $y = \sin x$ on $[0, 2\pi]$ is stretched to $[0, 8\pi]$ and then the stretched cycle is shifted horizontally $\pi/12$ units to the right. The graph is also compressed vertically by a factor of $\frac{3}{4}$ and then reflected in the x-axis.

In Problems 45–48, find horizontally shifted sine and cosine functions so that each function satisfies the given conditions. Graph the functions.

45. Amplitude 3, period $2\pi/3$, shifted by $\pi/3$ units to the right
46. Amplitude $\frac{2}{3}$, period π, shifted by $\pi/4$ units to the left
47. Amplitude 0.7, period 0.5, shifted by 4 units to the right
48. Amplitude $\frac{5}{4}$, period 4, shifted by $1/2\pi$ units to the left

In Problems 49 and 50, graphically verify the given identity.

49. $\cos(x + \pi) = -\cos x$

50. $\sin(x + \pi) = -\sin x$

Miscellaneous Applications

51. Pendulum The angular displacement θ of a pendulum from the vertical at time t seconds is given by $\theta(t) = \theta_0 \cos \omega t$, where θ_0 is the initial displacement at $t = 0$ seconds. See **FIGURE 4.3.26**. For $\omega = 2$ rad/s and $\theta_0 = \pi/10$, sketch two cycles of the resulting function.

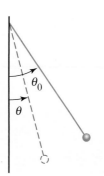

FIGURE 4.3.26 Pendulum in Problem 51

52. Current In a certain kind of electrical circuit, the current I measured in amperes at time t seconds is given by

$$I(t) = 10\cos\left(120\pi t + \frac{\pi}{3}\right).$$

Sketch two cycles of the graph of I as a function of time t.

53. Depth of Water The depth d of water at the entrance to a small harbor at time t is modeled by a function of the form

$$d(t) = A\sin B\left(t - \frac{\pi}{2}\right) + C,$$

where A is one-half the difference between the high- and low-tide depths; $2\pi/B$, $B > 0$, is the tidal period; and C is the average depth. Assume that the tidal period is 12 hours, the depth at high tide is 18 feet, and the depth at low tide is 6 feet. Sketch two cycles of the graph of d.

54. Fahrenheit Temperature Suppose that

$$T(t) = 50 + 10\sin\frac{\pi}{12}(t - 8),$$

$0 \le t \le 24$, is a mathematical model of the Fahrenheit temperature at t hours after midnight on a certain day of the week.
(a) What is the temperature at 8 A.M.?
(b) At what time(s) does $T(t) = 60$?
(c) Sketch the graph of T.
(d) Find the maximum and minimum temperatures and the times at which they occur.

Calculator Problems

In Problems 55–58, use a calculator to investigate whether the given function is periodic.

55. $f(x) = \sin\left(\dfrac{1}{x}\right)$

56. $f(x) = \dfrac{1}{\sin 2x}$

57. $f(x) = 1 + (\cos x)^2$

58. $f(x) = x\sin x$

For Discussion

In Problems 59 and 60, find the period of the given function.

59. $f(x) = \sin\frac{1}{2}x \sin 2x$

60. $f(x) = \sin\frac{3}{2}x + \cos\frac{5}{2}x$

In Problems 61 and 62, discuss and then sketch the graph of the given function.

61. $f(x) = |\sin x|$

62. $f(x) = |\cos x|$

Other Trigonometric Functions

≡ **Introduction** Recall that the remaining four trigonometric functions are the **tangent**, **cotangent**, **secant**, and **cosecant functions** and are denoted, in turn, as $\tan x$, $\cot x$, $\sec x$, and $\csc x$. We saw in Section 4.2 that by using (1) and (2) of Definition 4.2.1 in (3)–(6) of Definition 4.2.2 we can express these four new functions in terms of $\sin x$ and $\cos x$:

$$\tan x = \frac{\sin x}{\cos x} \qquad \cot x = \frac{\cos x}{\sin x} \tag{1}$$

$$\sec x = \frac{1}{\cos x} \qquad \csc x = \frac{1}{\sin x} \tag{2}$$

☐ **Domain and Range** Because the functions in (1) and (2) are quotients, we know from Definition 2.6.1 that the **domain** of each function consists of the set of real numbers *except* those numbers for which the denominator is zero. We have seen in (2) of Section 4.3 that $\cos x = 0$ for $x = (2n + 1)\pi/2$, $n = 0, \pm 1, \pm 2, \ldots$, and so

- the domain of $\tan x$ and of $\sec x$ is $\{x \,|\, x \ne (2n + 1)\pi/2, n = 0, \pm 1, \pm 2, \ldots\}$.

Similarly, from (1) of Section 4.3, $\sin x = 0$ for $x = n\pi$, $n = 0, \pm 1, \pm 2, \ldots$, and so it follows that

- the domain of $\cot x$ and of $\csc x$ is $\{x \,|\, x \ne n\pi, n = 0, \pm 1, \pm 2, \ldots\}$.

We know that the values of the sine and cosine are bounded, that is, $|\sin x| \le 1$ and $|\cos x| \le 1$. From these last inequalities we have

$$|\sec x| = \left|\frac{1}{\cos x}\right| = \frac{1}{|\cos x|} \ge 1 \tag{3}$$

and

$$|\csc x| = \left|\frac{1}{\sin x}\right| = \frac{1}{|\sin x|} \ge 1. \tag{4}$$

Recall that an inequality such as (3) means that $\sec x \ge 1$ or $\sec x \le -1$. Hence the range of the secant function is $(-\infty, -1] \cup [1, \infty)$. The inequality in (4) implies that the cosecant function has the same range $(-\infty, -1] \cup [1, \infty)$. When we consider the graphs of the tangent and cotangent functions we will see that they have the same range: $(-\infty, \infty)$.

If we interpret x as an angle, then FIGURE 4.4.1 illustrates the algebraic signs of the tangent, cotangent, secant, and cosecant functions in each of the four quadrants. This is easily verified using the signs of the sine and cosine functions displayed in Figure 4.2.2.

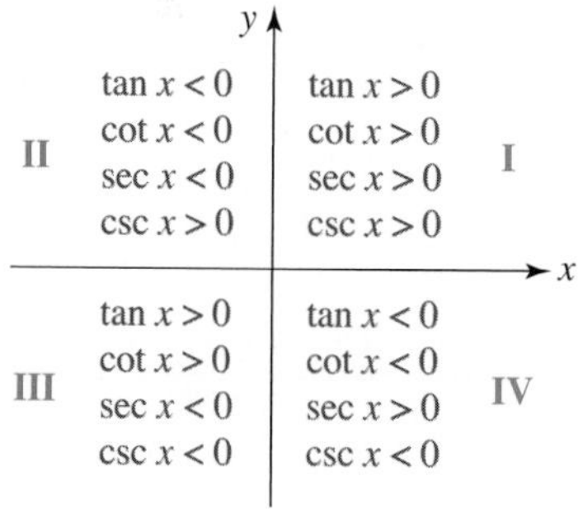

FIGURE 4.4.1 Signs of $\tan x$, $\cot x$, $\sec x$, and $\csc x$, in the four quadrants

---

EXAMPLE 1   **Example 5 of Section 4.2 Revisited**

Find $\tan x$, $\cot x$, $\sec x$, and $\csc x$ for $x = -\pi/6$.

Solution In Example 5 of Section 4.2 we saw that

$$\sin\left(-\frac{\pi}{6}\right) = -\sin\frac{\pi}{6} = -\frac{1}{2} \qquad \text{and} \qquad \cos\left(-\frac{\pi}{6}\right) = \cos\frac{\pi}{6} = \frac{\sqrt{3}}{2}.$$

Therefore, by the definitions in (1) and (2):

$$\tan\left(-\frac{\pi}{6}\right) = \frac{-1/2}{\sqrt{3}/2} = -\frac{1}{\sqrt{3}}, \quad \cot\left(-\frac{\pi}{6}\right) = \frac{\sqrt{3}/2}{-1/2} = -\sqrt{3}, \quad \leftarrow \begin{cases} \text{We could also use} \\ \cot x = 1/\tan x \end{cases}$$

$$\sec\left(-\frac{\pi}{6}\right) = \frac{1}{\sqrt{3}/2} = \frac{2}{\sqrt{3}}, \quad \csc\left(-\frac{\pi}{6}\right) = \frac{1}{-1/2} = -2. \qquad \equiv$$

TABLE 4.4.1

x	0	$\frac{\pi}{6}$	$\frac{\pi}{4}$	$\frac{\pi}{3}$	$\frac{\pi}{2}$
$\tan x$	0	$\frac{1}{\sqrt{3}}$	1	$\sqrt{3}$	$-$
$\cot x$	$-$	$\sqrt{3}$	1	$\frac{1}{\sqrt{3}}$	0
$\sec x$	1	$\frac{2}{\sqrt{3}}$	$\sqrt{2}$	2	$-$
$\csc x$	$-$	2	$\sqrt{2}$	$\frac{2}{\sqrt{3}}$	1

Table 4.4.1 summarizes some important values of the tangent, cotangent, secant, and cosecant and was constructed using values of the sine and cosine from Section 4.2. A dash in the table indicates that the trigonometric function is not defined at that particular value of x.

□ **Identities** The tangent is related to the secant by a useful identity. If we divide the Pythagorean identity

$$\sin^2 x + \cos^2 x = 1 \qquad (5)$$

by $\cos^2 x$, we see that

$$\frac{\sin^2 x}{\cos^2 x} + \frac{\cos^2 x}{\cos^2 x} = \frac{1}{\cos^2 x}. \qquad (6)$$

Similarly, dividing (5) by $\sin^2 x$ yields an identity relating the cotangent with the cosecant:

$$\frac{\sin^2 x}{\sin^2 x} + \frac{\cos^2 x}{\sin^2 x} = \frac{1}{\sin^2 x}. \qquad (7)$$

Using the laws of exponents,

$$\frac{\sin^2 x}{\cos^2 x} = \left(\frac{\sin x}{\cos x}\right)^2 = \tan^2 x, \qquad \frac{1}{\cos^2 x} = \left(\frac{1}{\cos x}\right)^2 = \sec^2 x,$$

$$\frac{\cos^2 x}{\sin^2 x} = \left(\frac{\cos x}{\sin x}\right)^2 = \cot^2 x, \qquad \frac{1}{\sin^2 x} = \left(\frac{1}{\sin x}\right)^2 = \csc^2 x,$$

we see that (6) and (7) can be written in a simpler manner:

$$1 + \tan^2 x = \sec^2 x$$
$$1 + \cot^2 x = \csc^2 x.$$

Since last two identities are direct consequences of $\sin^2 x + \cos^2 x = 1$ they too are called **Pythagorean identities**.

Finally, note that the tangent and cotangent function are related by the **reciprocal identity**

$$\cot x = \frac{\cos x}{\sin x} = \frac{1}{\dfrac{\sin x}{\cos x}} = \frac{1}{\tan x}.$$

□ **Summary** For future reference, especially for the work in the next section, we pause here to summarize a small collection of identities that are so basic to the study of trigonometry that they are known collectively as the **fundamental identities**. You should firmly commit these identities to memory.

CHAPTER 4 TRIGONOMETRIC FUNCTIONS

Fundamental Trigonometric Identities

Pythagorean identities:

$$\sin^2 x + \cos^2 x = 1 \tag{8}$$
$$1 + \tan^2 x = \sec^2 x \tag{9}$$
$$1 + \cot^2 x = \csc^2 x \tag{10}$$

Quotient identities:

$$\tan x = \frac{\sin x}{\cos x} \qquad \cot x = \frac{\cos x}{\sin x} \tag{11}$$

Reciprocal identities:

$$\sec x = \frac{1}{\cos x} \qquad \csc x = \frac{1}{\sin x} \qquad \cot x = \frac{1}{\tan x} \tag{12}$$

EXAMPLE 2   Using a Pythagorean Identity

Given that $\csc x = -5$ and $3\pi/2 < x < 2\pi$, determine the values of $\tan x$ and $\cot x$.

Solution We first compute $\cot x$. It follows from (10) that

$$\cot^2 x = \csc^2 x - 1.$$

For $3\pi/2 < x < 2\pi$, we see from Figure 4.4.1 that $\cot x$ must be negative and so we take the negative square root:

$$\cot x = -\sqrt{\csc^2 x - 1} = -\sqrt{(-5)^2 - 1} = -\sqrt{24} = -2\sqrt{6}.$$

Using $\cot x = 1/\tan x$, we have

$$\tan x = \frac{1}{\cot x} = \overset{\text{rationalizing the denominator}}{\underset{\downarrow}{\frac{1}{-2\sqrt{6}}}} = -\frac{\sqrt{6}}{12}. \qquad\qquad \equiv$$

In Example 2, given the information $\csc x = -5$ and $3\pi/2 < x < 2\pi$, we could easily find the values of the remaining five trigonometric functions. One way of proceeding would be to use $\csc x = 1/\sin x$ to find $\sin x = 1/\csc x = -\frac{1}{5}$. Then we use $\sin^2 x + \cos^2 x = 1$ to find $\cos x$. After we have found $\cos x$, the remaining three trigonometric functions can be obtained from (1) and (2).

☐ **Periodicity** Because the sine and cosine functions are 2π periodic, each of the functions in (1) and (2) have a period 2π. But from (17) of Theorem 4.2.5 we have

$$\tan(x + \pi) = \frac{\sin(x + \pi)}{\cos(x + \pi)} = \frac{-\sin x}{-\cos x} = \tan x. \tag{13}$$

◀ Also see Problems 49 and 50 in Exercises 4.3.

Thus (13) implies that $\tan x$ and $\cot x$ are periodic with a period $p \le \pi$. In the case of the tangent function, $\tan x = 0$ only if $\sin x = 0$, that is, only if $x = 0, \pm\pi, \pm 2\pi$, and so on. Therefore, the smallest positive number p for which $\tan(x + p) = \tan x$ is $p = \pi$. The cotangent function has the same period since it is the reciprocal of the tangent function.

THEOREM 4.4.1 Period of the Tangent and Cotangent

The tangent and cotangent functions are periodic with **period π**. Therefore,

$$\tan(x + \pi) = \tan x \qquad \text{and} \qquad \cot(x + \pi) = \cot x \qquad (14)$$

for every real number x for which the functions are defined.

THEOREM 4.4.2 Period of the Secant and Cosecant

The secant and cosecant functions are periodic with **period 2π**. Therefore,

$$\sec(x + 2\pi) = \sec x \qquad \text{and} \qquad \csc(x + 2\pi) = \csc x \qquad (15)$$

for every real number x for which the functions are defined.

☐ **Even-Odd Properties** Because the cosine function is even and the sine function is odd, each of the remaining four trigonometric functions is either even or odd.

THEOREM 4.4.3 Even and Odd Functions

The tangent, cotangent, and cosecant functions are **odd functions**, whereas the secant function is an **even function**. That is,

$$\tan(-x) = -\tan x \qquad \text{and} \qquad \cot(-x) = -\cot x \qquad (16)$$
$$\sec(-x) = \sec x \qquad \text{and} \qquad \csc(-x) = -\csc x \qquad (17)$$

for every real number x for which the functions are defined,

PROOF: We prove the first entries in (16) and (17). Because $\cos(-x) = \cos x$ and $\sin(-x) = -\sin x$,

$$\tan(-x) = \frac{\sin(-x)}{\cos(-x)} = \frac{-\sin x}{\cos x} = -\frac{\sin x}{\cos x} = -\tan x,$$
$$\sec(-x) = \frac{1}{\cos(-x)} = \frac{1}{\cos x} = \sec x$$

This is a good time to review (7) of Section 3.6. ▶ proves, in turn, that $\tan x$ is an odd function and $\sec x$ is an even function. ≡

☐ **Graphs of $y = \tan x$ and $y = \cot x$** The numbers that make the denominators of $\tan x$, $\cot x$, $\sec x$, and $\csc x$ equal to zero correspond to vertical asymptotes of their graphs. For example, we encourage you to verify, using a calculator, that

$$\tan x \to -\infty \text{ as } x \to -\frac{\pi}{2}^{+} \qquad \text{and} \qquad \tan x \to \infty \text{ as } x \to \frac{\pi}{2}^{-}.$$

In other words, $x = -\pi/2$ and $x = \pi/2$ are vertical asymptotes. The graph of $y = \tan x$ on the interval $(-\pi/2, \pi/2)$ given in FIGURE 4.4.2 is one **cycle** of the graph of $y = \tan x$. Using periodicity we extend the cycle in Figure 4.4.2 to adjacent intervals of length π, as shown in FIGURE 4.4.3. The x-intercepts of the graph of the tangent function are $(0, 0)$, $(\pm\pi, 0)$, $(\pm 2\pi, 0)$, ..., and the vertical asymptotes of the graph are $x = \pm\pi/2$, $\pm 3\pi/2$, $\pm 5\pi/2$, ...

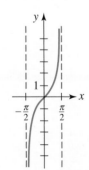

FIGURE 4.4.2 One cycle of the graph of $y = \tan x$

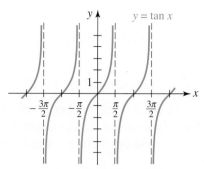

FIGURE 4.4.3 Graph of $y = \tan x$

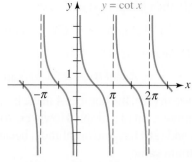

FIGURE 4.4.4 Graph of $y = \cot x$

The graph of $y = \cot x$ is similar to the graph of the tangent function and is given in FIGURE 4.4.4. In this case, the graph of $y = \cot x$ on the interval $(0, \pi)$ is one **cycle** of the graph of $y = \cot x$. The x-intercepts of the graph of the cotangent function are $(\pm \pi/2, 0)$, $(\pm 3\pi/2, 0)$, $(\pm 5\pi/2, 0)$, ..., and the vertical asymptotes of the graph are $x = 0$, $\pm \pi$, $\pm 2\pi$, $\pm 3\pi$, ... Because $y = \tan x$ and $y = \cot x$ are odd functions, their graphs are symmetric with respect to the origin.

☐ **Graphs of $y = \sec x$ and $y = \csc x$** For both $y = \sec x$ and $y = \csc x$ we know that $|y| \geq 1$, and so no portion of their graphs can appear in the horizontal strip $-1 < y < 1$ of the Cartesian plane. Hence the graphs of $y = \sec x$ and $y = \csc x$ have no x-intercepts. Both $y = \sec x$ and $y = \csc x$ have period 2π. The vertical asymptotes for the graph of $y = \sec x$ are the same as $y = \tan x$, namely, $x = \pm \pi/2$, $\pm 3\pi/2$, $\pm 5\pi/2$, ... Because $y = \sec x$ is an even function, its graph is symmetric with respect to the y-axis. On the other hand, the vertical asymptotes for the graph of $y = \csc x$ are the same as $y = \cot x$, namely, $x = 0$, $\pm \pi$, $\pm 2\pi$, $\pm 3\pi$, ... Because $y = \csc x$ is an odd function, its graph is symmetric with respect to the origin. One cycle of the graph of $y = \sec x$ on $[0, 2\pi]$ is extended to the interval $[-2\pi, 0]$ by periodicity (or y-axis symmetry) in FIGURE 4.4.5. Similarly, in FIGURE 4.4.6 we extend one cycle of $y = \csc x$ on $(0, 2\pi)$ to the interval $(-2\pi, 0)$ by periodicity (or origin symmetry).

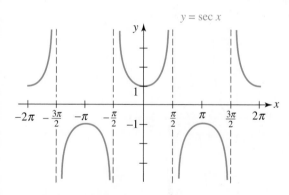

FIGURE 4.4.5 Graph of $y = \sec x$

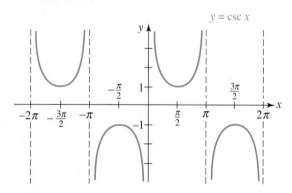

FIGURE 4.4.6 Graph of $y = \csc x$

☐ **Transformations and Graphs** Similar to the sine and cosine graphs, rigid and nonrigid transformations can be applied to the graphs of $y = \tan x$, $y = \cot x$, $y = \sec x$, and $y = \csc x$. For example, a function such as $y = A\tan(Bx + C) + D$ can be analyzed in the following manner:

vertical stretch/compression/reflection   vertical shift
↓   ↓

$$y = A\tan(Bx + C) + D \qquad (18)$$

↑   ↑
horizontal stretch/compression by changing period   horizontal shift

If $B > 0$, then the period of
$$y = A\tan(Bx + C) \qquad \text{and} \qquad y = A\cot(Bx + C) \text{ is } \pi/B, \qquad (19)$$
whereas the period of
$$y = A\sec(Bx + C) \qquad \text{and} \qquad y = A\csc(Bx + C) \text{ is } 2\pi/B. \qquad (20)$$

As we see in (18), the number A in each case can be interpreted as either a vertical stretch or a compression of a graph. However, you should be aware of the fact that the functions in (19) and (20) have no amplitude, because none of the functions has a maximum *and* a minimum value.

Of the six trigonometric functions, only the ▶ sine and cosine functions have an amplitude.

▌EXAMPLE 3   Comparison of Graphs

Find the period, x-intercepts, and vertical asymptotes for the graph of $y = \tan 2x$. Graph the function on $[0, \pi]$.

Solution With the identification $B = 2$, we see from (19) that the period is $\pi/2$. Since $\tan 2x = \sin 2x/\cos 2x$, the x-intercepts of the graph occur at the zeros of $\sin 2x$. From (1) of Section 4.3, $\sin 2x = 0$ for
$$2x = n\pi \qquad \text{so that} \qquad x = \tfrac{1}{2}n\pi, n = 0, \pm 1, \pm 2, \dots.$$
That is, $x = 0, \pm\pi/2, \pm 2\pi/2 = \pi, \pm 3\pi/2, \pm 4\pi/2 = 2\pi$, and so on. The x-intercepts are $(0, 0), (\pm\pi/2, 0), (\pm\pi, 0), (\pm 3\pi/2, 0), \dots$. The vertical asymptotes of the graph occur at zeros of $\cos 2x$. From (2) of Section 4.3, the numbers for which $\cos 2x = 0$ are found in the following manner:
$$2x = (2n + 1)\frac{\pi}{2} \qquad \text{so that} \qquad x = (2n + 1)\frac{\pi}{4}, n = 0, \pm 1, \pm 2, \dots.$$
That is, the vertical asymptotes are $x = \pm\pi/4, \pm 3\pi/4, \pm 5\pi/4, \dots$. On the interval $[0, \pi]$, the graph of $y = \tan 2x$ has three intercepts $(0, 0), (\pi/2, 0)$, and $(\pi, 0)$ and two vertical asymptotes $x = \pi/4$ and $x = 3\pi/4$. In **FIGURE 4.4.7**, we have compared the graphs of $y = \tan x$ and $y = \tan 2x$ on the interval. The graph of $y = \tan 2x$ is a horizontal compression of the graph of $y = \tan x$.

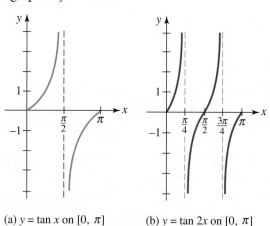

(a) $y = \tan x$ on $[0, \pi]$ (b) $y = \tan 2x$ on $[0, \pi]$

FIGURE 4.4.7 Graph of functions in Example 3

▌EXAMPLE 4   Comparison of Graphs

Compare one cycle of the graphs of $y = \tan x$ and $y = \tan(x - \pi/4)$.

Solution The graph of $y = \tan(x - \pi/4)$ is the graph of $y = \tan x$ shifted horizontally $\pi/4$ units to the right. The intercept $(0, 0)$ for the graph of $y = \tan x$ is shifted to $(\pi/4, 0)$ on the graph of $y = \tan(x - \pi/4)$. The vertical asymptotes $x = -\pi/2$ and

$x = \pi/2$ for the graph of $y = \tan x$ are shifted to $x = -\pi/4$ and $x = 3\pi/4$ for the graph of $y = \tan(x - \pi/4)$. In FIGURES 4.4.8(a) and 4.4.8(b) we see, respectively, that a cycle of the graph of $y = \tan x$ on the interval $(-\pi/2, \pi/2)$ is shifted to the right to yield a cycle of the graph of $y = \tan(x - \pi/4)$ on the interval $(-\pi/4, 3\pi/4)$.

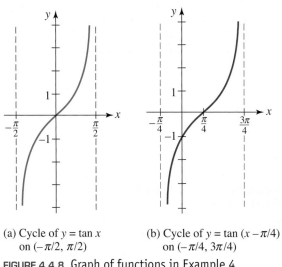

(a) Cycle of $y = \tan x$ (b) Cycle of $y = \tan(x - \pi/4)$
on $(-\pi/2, \pi/2)$ on $(-\pi/4, 3\pi/4)$

FIGURE 4.4.8 Graph of functions in Example 4

As we did in the analysis of the graphs of $y = A\sin(Bx + C)$ and $y = A\cos(Bx + C)$, we can determine the amount of horizontal shift for graphs of functions such as $y = A\tan(Bx + C)$ and $y = A\sec(Bx + C)$ by factoring the number $B > 0$ from $Bx + C$.

EXAMPLE 5   Two Shifts and Two Compressions

Graph $y = 2 - \frac{1}{2}\sec(3x - \pi/2)$.

Solution Let's break down the analysis of the graph into four parts, namely, by transformations.

(*i*) One cycle of the graph of $y = \sec x$ occurs on $[0, 2\pi]$. Since the period of $y = \sec 3x$ is $2\pi/3$, one cycle of its graph occurs on the interval $[0, 2\pi/3]$. In other words, the graph of $y = \sec 3x$ is a horizontal compression of the graph of $y = \sec x$. Since $\sec 3x = 1/\cos 3x$, the vertical asymptotes occur at the zeros of $\cos 3x$. Using (2) of Section 4.3, we find

$$3x = (2n + 1)\frac{\pi}{2} \quad \text{or} \quad x = (2n + 1)\frac{\pi}{6}, n = 0, \pm 1, \pm 2, \ldots.$$

FIGURE 4.4.9(a) shows two cycles of the graph $y = \sec 3x$; one cycle on $[-2\pi/3, 0]$ and another on $[0, 2\pi/3]$. Within those intervals the vertical asymptotes are $x = -\pi/2$, $x = -\pi/6$, $x = \pi/6$, and $x = \pi/2$.

(*ii*) The graph of $y = -\frac{1}{2}\sec 3x$ is the graph of $y = \sec 3x$ compressed vertically by a factor of $\frac{1}{2}$ and then reflected in the x-axis. See Figure 4.4.9(b).

(*iii*) By factoring 3 from $3x - \pi/2$, we see from

$$y = -\frac{1}{2}\sec\left(3x - \frac{\pi}{2}\right) = -\frac{1}{2}\sec 3\left(x - \frac{\pi}{6}\right)$$

that the graph of $y = -\frac{1}{2}\sec(3x - \pi/2)$ is the graph of $y = -\frac{1}{2}\sec 3x$ shifted $\pi/6$ units to the right. By shifting the two intervals $[-2\pi/3, 0]$ and $[0, 2\pi/3]$ in Figure 4.4.9(b)

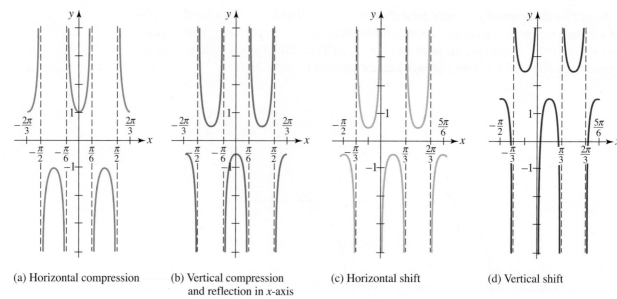

(a) Horizontal compression   (b) Vertical compression   (c) Horizontal shift   (d) Vertical shift
and reflection in *x*-axis

FIGURE 4.4.9 Graph of function in Example 5

to the right $\pi/6$ units, we see in Figure 4.4.9(c) two cycles of $y = -\frac{1}{2}\sec(3x - \pi/2)$ on the intervals $[-\pi/2, \pi/6]$ and $[\pi/6, 5\pi/6]$. The vertical asymptotes $x = -\pi/2$, $x = -\pi/6, x = \pi/6$, and $x = \pi/2$ shown in Figure 4.4.9(b) are shifted to $x = -\pi/3$, $x = 0, x = \pi/3$, and $x = 2\pi/3$. Observe that the *y*-intercept $\left(0, -\frac{1}{2}\right)$ in Figure 4.4.9(b) is now moved to $\left(\pi/6, -\frac{1}{2}\right)$ in Figure 4.4.9(c).

(*iv*) Finally, we obtain the graph $y = 2 - \frac{1}{2}\sec(3x - \pi/2)$ in Figure 4.4.9(d) by shifting the graph of $y = -\frac{1}{2}\sec(3x - \pi/2)$ in Figure 4.4.9(c) two units upward. ≡

4.4   Exercises   Answers to selected odd-numbered problems begin on page ANS-13.

In Problems 1 and 2, complete the given table.

1.

x	$\frac{2\pi}{3}$	$\frac{3\pi}{4}$	$\frac{5\pi}{6}$	π	$\frac{7\pi}{6}$	$\frac{5\pi}{4}$	$\frac{4\pi}{3}$	$\frac{3\pi}{2}$	$\frac{5\pi}{3}$	$\frac{7\pi}{4}$	$\frac{11\pi}{6}$	2π
$\tan x$												
$\cot x$												

2.

x	$\frac{2\pi}{3}$	$\frac{3\pi}{4}$	$\frac{5\pi}{6}$	π	$\frac{7\pi}{6}$	$\frac{5\pi}{4}$	$\frac{4\pi}{3}$	$\frac{3\pi}{2}$	$\frac{5\pi}{3}$	$\frac{7\pi}{4}$	$\frac{11\pi}{6}$	2π
$\sec x$												
$\csc x$												

In Problems 3–18, find the indicated value without the use of a calculator.

3. $\cot\dfrac{13\pi}{6}$ **4.** $\csc\left(-\dfrac{3\pi}{2}\right)$ **5.** $\tan\dfrac{9\pi}{2}$ **6.** $\sec 7\pi$

7. $\csc\left(-\dfrac{\pi}{3}\right)$ **8.** $\cot\left(-\dfrac{13\pi}{3}\right)$ **9.** $\tan\dfrac{23\pi}{4}$ **10.** $\tan\left(-\dfrac{5\pi}{6}\right)$

11. $\sec\dfrac{10\pi}{3}$ **12.** $\cot\dfrac{17\pi}{6}$ **13.** $\csc 5\pi$ **14.** $\sec\dfrac{29\pi}{4}$

15. $\sec(-120°)$ **16.** $\tan 405°$ **17.** $\csc 495°$ **18.** $\cot(-720°)$

In Problems 19–26, use the given information to find the values of the remaining five trigonometric functions.

19. $\tan x = -2$, $\pi/2 < x < \pi$ **20.** $\cot x = \frac{1}{2}$, $\pi < x < 3\pi/2$

21. $\csc x = \frac{4}{3}$, $0 < x < \pi/2$ **22.** $\sec x = -5$, $\pi/2 < x < \pi$

23. $\sin x = \frac{1}{3}$, $\pi/2 < x < \pi$ **24.** $\cos x = -1/\sqrt{5}$, $\pi < x < 3\pi/2$

25. $\cos x = \frac{12}{13}$, $3\pi/2 < x < 2\pi$ **26.** $\sin x = \frac{4}{5}$, $0 < x < \pi/2$

27. If $3\cos x = \sin x$, find all values of $\tan x$, $\cot x$, $\sec x$, and $\csc x$.

28. If $\csc x = \sec x$, find all values of $\tan x$, $\cot x$, $\sin x$, and $\cos x$.

In Problems 29–36, find the period, x-intercepts, and the vertical asymptotes of the given function. Sketch at least one cycle of the graph.

29. $y = \tan \pi x$ **30.** $y = \tan\dfrac{x}{2}$

31. $y = \cot 2x$ **32.** $y = -\cot\dfrac{\pi x}{3}$

33. $y = \tan\left(\dfrac{x}{2} - \dfrac{\pi}{4}\right)$ **34.** $y = \dfrac{1}{4}\cot\left(x - \dfrac{\pi}{2}\right)$

35. $y = -1 + \cot \pi x$ **36.** $y = \tan\left(x + \dfrac{5\pi}{6}\right)$

In Problems 37–44, find the period and the vertical asymptotes of the given function. Sketch at least one cycle of the graph.

37. $y = -\sec x$ **38.** $y = 2\sec\dfrac{\pi x}{2}$

39. $y = 3\csc \pi x$ **40.** $y = -2\csc\dfrac{x}{3}$

41. $y = \sec\left(3x - \dfrac{\pi}{2}\right)$ **42.** $y = \csc(4x + \pi)$

43. $y = 3 + \csc\left(2x + \dfrac{\pi}{2}\right)$ **44.** $y = -1 + \sec(x - 2\pi)$

In Problems 45 and 46, use the graphs of $y = \tan x$ and $y = \sec x$ to find numbers A and C for which the given equality is true.

45. $\cot x = A\tan(x + C)$ **46.** $\csc x = A\sec(x + C)$

Calculator Problems

47. Using a calculator in radian mode, compare the values of $\tan 1.57$ and $\tan 1.58$. Explain the difference in these values.

48. Using a calculator in radian mode, compare the values of $\cot 3.14$ and $\cot 3.15$.

For Discussion

49. Can $9\csc x = 1$ for any real number x?

50. Can $7 + 10\sec x = 0$ for any real number x?

51. For which real numbers x is **(a)** $\sin x \leq \csc x$? **(b)** $\sin x < \csc x$?

52. For which real numbers x is **(a)** $\sec x \leq \cos x$? **(b)** $\sec x < \cos x$?

53. Discuss and then sketch the graphs of $y = |\sec x|$ and $y = |\csc x|$.

4.5 Verifying Trigonometric Identities

≡ **Introduction** There are *many* identities involving trigonometric functions; in this section we will illustrate how to verify some of them.

A **trigonometric identity** is an equation or formula involving only trigonometric functions that is valid for all angles measured in degrees or radians or for real numbers for which both sides of the equality are defined. To verify a trigonometric identity we use

- the fundamental trigonometric identities, ← (8)−(12) in Section 4.4

- the even-odd properties, ← $\begin{cases} (15) \text{ in Section 4.2 and} \\ (16) \text{ and } (17) \text{ in Section 4.4} \end{cases}$

- and basic arithmetic and algebraic operations.

For example,

$$\frac{\sin x}{\tan x} = \cos x \tag{1}$$

is an identity for all real numbers for which $\tan x$ is defined and $\tan x \neq 0$. To verify the identity we start on one side of the equation and deduce through valid manipulations the equivalence with the other side. In the case of (1) we will start on the left-hand side:

$$\frac{\sin x}{\tan x} = \frac{\sin x}{\dfrac{\sin x}{\cos x}} \qquad \leftarrow \begin{cases} \text{quotient identity;} \\ (11) \text{ of Section 4.4} \end{cases}$$

$$= \sin x \left(\frac{\cos x}{\sin x} \right) \qquad \leftarrow \text{division; invert and multiply}$$

$$= \cos x. \qquad \leftarrow \text{cancellation of } \sin x$$

EXAMPLE 1   Using a Reciprocal Identity

Write $\sin x \sec x$ as a single trigonometric function.

Solution Using the reciprocal identity $\sec x = 1/\cos x$, we find

$$\sin x \sec x = \sin x \frac{1}{\cos x} = \frac{\sin x}{\cos x} = \tan x. \qquad \equiv$$

EXAMPLE 2   Simplification

Simplify $(1 + \sin x)(1 + \sin(-x))$.

Solution First we recall that the sine is an odd function, that is, $\sin(-x) = -\sin x$. Then from algebra we know $(a + b)(a - b) = a^2 - b^2$. With the identifications $a = 1$ and $b = \sin x$ we can write

$$(1 + \sin x)(1 + \sin(-x)) = (1 + \sin x)(1 - \sin x) = 1 - \sin^2 x.$$

Finally, the Pythagorean identity $\sin^2 x + \cos^2 x = 1$ implies $1 - \sin^2 x = \cos^2 x$. Therefore,

$$(1 + \sin x)(1 + \sin(-x)) = \cos^2 x. \qquad \equiv$$

EXAMPLE 3   Verification

Verify the identity $\sec^2 t + \csc^2 t = \sec^2 t \csc^2 t$.

Solution We begin with the left-hand side of the equation:

$$
\begin{aligned}
\sec^2 t + \csc^2 t &= \frac{1}{\cos^2 t} + \frac{1}{\sin^2 t} &&\leftarrow \begin{cases} \text{reciprocal identities;} \\ (12) \text{ of Section 4.4} \end{cases} \\[2mm]
&= \frac{1}{\cos^2 t}\frac{\sin^2 t}{\sin^2 t} + \frac{1}{\sin^2 t}\frac{\cos^2 t}{\cos^2 t} &&\leftarrow \text{common denominator} \\[2mm]
&= \frac{\sin^2 t + \cos^2 t}{\cos^2 t \sin^2 t} &&\leftarrow \text{adding fractions} \\[2mm]
&= \frac{1}{\cos^2 t \sin^2 t} &&\leftarrow \begin{cases} \text{Pythagorean identity;} \\ (8) \text{ of Section 4.4} \end{cases} \\[2mm]
&= \left(\frac{1}{\cos t}\right)^2 \left(\frac{1}{\sin t}\right)^2 &&\leftarrow \begin{cases} \text{algebra: laws of exponents} \\ \text{and multiplication of fractions} \end{cases} \\[2mm]
&= (\sec t)^2 (\csc t)^2 &&\leftarrow \begin{cases} \text{reciprocal identities;} \\ (12) \text{ of Section 4.4} \end{cases} \\[2mm]
&= \sec^2 t \csc^2 t. &&\qquad\qquad\qquad \equiv
\end{aligned}
$$

Implicit in Example 3 is the assumption that the identity is valid only for those values of t for which both sides of the identity are defined. In Example 3, for t a real number, we must require that $t \ne n\pi$ and $t \ne (2n + 1)\pi/2$, where n is an integer. In the remaining examples, we will not mention the restrictions on the variable.

□ **Suggestions** In order to verify a trigonometric identity, we are required to show that the given expressions are equivalent. In the preceding example, we worked with the expression on the left-hand side of the equation to show that that side was equivalent to the other. Starting on one side of an equation and deducing the other side is standard practice in verifying trigonometric identities. But often we can perform correct work on one side of an equation and reach a point where that side is simplified but appears to be not identical to the other side. This does not mean the identity is false; sometimes we must reduce each side of an equation *separately* to the same expression. However, the same algebraic operations should not be performed on both sides of the equation *simultaneously*. In other words, do not treat a trigonometric equation as an identity until after you have proven that it is really true. Although there is no general method for demonstrating that a trigonometric equation is an identity, we list below a few suggestions that may be useful.

Suggestions for Verifying Identities

    (*i*)   Simplify the more complicated side of the equation first.
    (*ii*)   Find least common denominators for sums or differences of fractions.
    (*iii*)   If the two preceding suggestions fail, then express all trigonometric functions in terms of sines and cosines and try to simplify.

EXAMPLE 4   Verification

Verify the identity

$$\sin\theta\cos\theta = \frac{1}{\tan\theta + \cot\theta}.$$

Solution In this example we show that the right-hand side of the equation is equivalent to the left-hand side. You should be able to supply a justification for each step of the solution.

$$\frac{1}{\tan\theta + \cot\theta} = \frac{1}{\dfrac{\sin\theta}{\cos\theta} + \dfrac{\cos\theta}{\sin\theta}}$$

$$= \frac{1}{\dfrac{\sin\theta}{\cos\theta}\dfrac{\sin\theta}{\sin\theta} + \dfrac{\cos\theta}{\sin\theta}\dfrac{\cos\theta}{\cos\theta}}$$

$$= \frac{1}{\dfrac{\sin^2\theta + \cos^2\theta}{\sin\theta\cos\theta}}$$

$$= \frac{\sin\theta\cos\theta}{\sin^2\theta + \cos^2\theta}$$

$$= \sin\theta\cos\theta.$$

$\equiv$

EXAMPLE 5   Verification

Verify the identity

$$\sin x + \sin x \cot^2 x = \cos x \csc x \sec x.$$

Solution Because the left-hand side of the equation looks a bit more complicated than the right-hand side, we begin there:

$$\sin x + \sin x \cot^2 x = \sin x(1 + \cot^2 x) \quad \leftarrow \text{factor out } \sin x$$

$$= \sin x(\csc^2 x) \quad \leftarrow \text{Pythagorean identity}$$

$$= \frac{1}{\csc x}\csc^2 x \quad \leftarrow \text{reciprocal identity}$$

$$= \csc x.$$

Because we have arrived at such a simple expression, we now try to reduce the right-hand side to the same quantity:

$$\cos x \csc x \sec x = \cos x \csc x \frac{1}{\cos x} \quad \leftarrow \text{reciprocal identity}$$

$$= \csc x. \quad \leftarrow \text{cancellation of } \cos x$$

Since both sides of the original equation are equivalent to $\csc x$, they are equivalent to each other. Therefore, the equation is an identity.

$\equiv$

EXAMPLE 6   Verification

Verify the identity

$$\frac{\sin x}{1 + \cos x} = \csc x - \cot x.$$

Solution If we multiply numerator and denominator of the left-hand side of the equation by the conjugate factor of the denominator we can use a Pythagorean identity:

$$\frac{\sin x}{1 + \cos x} \frac{1 - \cos x}{1 - \cos x} = \frac{\sin x(1 - \cos x)}{1 - \cos^2 x}$$

$$= \frac{\sin x(1 - \cos x)}{\sin^2 x}$$

$$= \frac{1 - \cos x}{\sin x}$$

$$= \frac{1}{\sin x} - \frac{\cos x}{\sin x} \quad \leftarrow \left\{ \begin{array}{l} \text{now use reciprocal} \\ \text{and quotient identities} \end{array} \right.$$

$$= \csc x - \cot x. \qquad \equiv$$

☐ **Trigonometric Substitutions** In a calculus course it is often useful to make use of a **trigonometric substitution** using a sine, cosine, or tangent function to change the form of an algebraic expression involving a radical into a trigonometric expression with no radical. Generally, this is done using one of the Pythagorean identities. The following example illustrates the technique.

EXAMPLE 7 **Rewriting a Radical**
___

Rewrite $\sqrt{a^2 - x^2}$ as a trigonometric expression without a radical by means of the substitution $x = a\sin\theta$, $a > 0$ and $-\pi/2 \le \theta \le \pi/2$.

Solution If $x = a\sin\theta$, then

$$\sqrt{a^2 - x^2} = \sqrt{a^2 - (a\sin\theta)^2}$$

$$= \sqrt{a^2 - a^2\sin^2\theta}$$

$$= \sqrt{a^2(1 - \sin^2\theta)} \quad \leftarrow \text{now use (8) of Section 4.4}$$

$$= \sqrt{a^2\cos^2\theta}.$$

The restriction of the variable θ enables us to take the foregoing square root without recourse to absolute values. In other words, since $a > 0$ and $\cos\theta \ge 0$ for $-\pi/2 \le \theta \le \pi/2$, the original radical is the same as

$$\sqrt{a^2 - x^2} = \sqrt{a^2\cos^2\theta} = a\cos\theta. \qquad \equiv$$

4.5 | Exercises Answers to selected odd-numbered problems begin on page ANS–14.

In Problems 1–10, use the fundamental identities and the even-odd identities to simplify each expression.

1. $\sec t \cos t$

2. $\tan \alpha \cos \alpha$

3. $\dfrac{\sin \theta}{\csc \theta} + \dfrac{\cos \theta}{\sec \theta}$

4. $\dfrac{\csc^2 x - 1}{\cot x}$

5. $\tan^2 t - \sec^2 t$

6. $1 + \tan^2(-\theta)$

7. $\sin(-t) + \sin t$

8. $\cos^2 t + \dfrac{1}{\csc^2 t}$

9. $\sec(-x)\cos x$

10. $1 + \dfrac{\cot\beta}{\tan\beta}$

In Problems 11–22, reduce the given expression to a single trigonometric function.

11. $\dfrac{\sin t + \sin t \cos t}{1 + \cos t}$

12. $\cos x + \cos x \tan^2 x$

13. $\dfrac{\sec^2 \alpha - 1}{\tan\alpha}$

14. $\dfrac{\tan t + \cot t}{\csc t}$

15. $\sin x + \cos x \cot x$

16. $\sin\theta \tan\theta \csc^2\theta - \sin\theta \tan\theta$

17. $\dfrac{\sec^2\alpha}{\cos\alpha + \cos\alpha \tan^2\alpha}$

18. $\dfrac{\sin^2\theta \cos\theta + \cos^3\theta - \cos\theta + \sin\theta}{\cos\theta}$

19. $\sin t \cos t \tan t \sec t \cot t$

20. $\dfrac{\sin\alpha \tan\alpha}{\csc\alpha} + \dfrac{\sin\alpha}{\sec\alpha}$

21. $\dfrac{1}{1 + \sin t} + \dfrac{1}{1 - \sin t}$

22. $(\sin^2 x - 1)(\cot^2 x + 1)$

In Problems 23–64, verify the given identity.

23. $\dfrac{\sin t}{\csc t} = 1 - \dfrac{\cos t}{\sec t}$

24. $\dfrac{1 + \sin x}{\cos x} = \sec x + \tan x$

25. $1 - \cos^4\theta = (2 - \sin^2\theta)\sin^2\theta$

26. $\dfrac{1 + \tan t}{\tan t} = \cot t + \sec^2 t - \tan^2 t$

27. $1 - 2\sin^2 t = 2\cos^2 t - 1$

28. $\tan^2\beta - \sin^2\beta = \tan^2\beta \sin^2\beta$

29. $\dfrac{\sec z - \csc z}{\sec z + \csc z} = \dfrac{\tan z - 1}{\tan z + 1}$

30. $\dfrac{\sin t + \tan t}{1 + \cos t} = \tan t$

31. $\dfrac{\sec^4 t - \tan^4 t}{1 + 2\tan^2 t} = 1$

32. $\dfrac{1 + \sin t}{\cos t} + \dfrac{\cos t}{1 + \sin t} = 2\sec t$

33. $\sin^2 x \cot^2 x + \cos^2 x \tan^2 x = 1$

34. $\dfrac{\sin\alpha + \tan\alpha}{\cot\alpha + \csc\alpha} = \sin^2\alpha \sec\alpha$

35. $\sec t - \dfrac{\cos t}{1 + \sin t} = \tan t$

36. $\dfrac{1}{\sec t - \tan t} = \sec t + \tan t$

37. $\dfrac{\tan^2\beta}{1 + \cos\beta} = \dfrac{\sec\beta - 1}{\cos\beta}$

38. $\dfrac{\tan^2 t - 1}{\sin t + \cos t} = \dfrac{\sin t - \cos t}{\cos^2 t}$

39. $(\csc t - \cot t)^2 = \dfrac{1 - \cos t}{1 + \cos t}$

40. $\cos\theta - \sin\theta + \csc\theta = \dfrac{\sin\theta + \cos\theta}{\tan\theta}$

41. $1 + \dfrac{1}{\cos x} = \dfrac{\tan^2 x}{\sec x - 1}$

42. $\dfrac{\tan t + \cot t}{\cos^2 t} - \sin t \sec^3 t = \sec t \csc t$

43. $\dfrac{\cot t - \tan t}{\cot t + \tan t} = 1 - 2\sin^2 t$

44. $\dfrac{1 + \sec t}{\sin t + \tan t} = \csc t$

45. $\cos(-t)\csc(-t) = -\cot t$

46. $\dfrac{\tan(-t)}{\sin(-t)} = \sec t$

47. $\sqrt{\dfrac{1 + \sin\theta}{1 - \sin\theta}} = \dfrac{1 + \sin\theta}{|\cos\theta|}$

48. $\sqrt{\dfrac{1 + \cos\alpha}{1 - \cos\alpha}} = \dfrac{|\sin\alpha|}{1 - \cos\alpha}$

49. $\left(\dfrac{\sin^2\theta}{\cot^4\theta}\right)^4 \cdot \left(\dfrac{\csc\theta}{\tan^2\theta}\right)^8 = 1$

50. $\dfrac{\cos^3 x + \sin^3 x}{\cos x + \sin x} = 1 - \cos x \sin x$

51. $(\tan^2 t + 1)(\cos^2 t - 1) = 1 - \sec^2 t$

52. $\dfrac{1}{1 - \cos\alpha} + \dfrac{1}{1 + \cos\alpha} = 2\csc^2\alpha$

53. $\dfrac{1 + \cos\phi}{\sin\phi} = \dfrac{\sin\phi}{1 - \cos\phi}$

54. $\dfrac{1 - \cos\alpha}{1 + \cos\alpha} = \dfrac{\sec\alpha - 1}{\sec\alpha + 1}$

55. $(1 - \tan\beta)^2(1 + \tan\beta)^2 + 4\tan^2\beta = \sec^4\beta$

56. $\dfrac{\cos(-t)}{1 + \tan(-t)} - \dfrac{\sin(-t)}{1 + \cot(-t)} = \sin t + \cos t$

57. $\dfrac{\sin\theta}{1 - \cot\theta} + \dfrac{\cos\theta}{1 - \tan\theta} = \cos\theta + \sin\theta$

58. $\sin^6 t + \cos^6 t = 1 - 3\sin^2 t\cos^2 t$

59. $\csc^4 t - \csc^2 t = \cot^4 t + \cot^2 t$

60. $\dfrac{\tan x - \cot x}{\sin x \cos x} = \sec^2 x - \csc^2 x$

61. $\dfrac{\cos t}{1 - \sin t} = \sec t + \tan t$

62. $\dfrac{\sin x + \cos x}{\cos x} = 1 + \tan x$

63. $\dfrac{\tan\alpha - \tan\beta}{1 + \tan\alpha\tan\beta} = \dfrac{\cot\beta - \cot\alpha}{\cot\alpha\cot\beta + 1}$

64. $\dfrac{\sin\alpha\cos\beta + \cos\alpha\sin\beta}{\cos\alpha\cos\beta - \sin\alpha\sin\beta} = \dfrac{\tan\alpha + \tan\beta}{1 - \tan\alpha\tan\beta}$

In Problems 65–74, rewrite the given expression as a trigonometric expression without radicals by making the indicated substitution. Assume that $a > 0$.

65. $\sqrt{a^2 - x^2}$, $x = a\cos\theta$, $0 \le \theta \le \pi$

66. $\sqrt{a^2 + x^2}$, $x = a\tan\theta$, $-\pi/2 < \theta < \pi/2$

67. $\sqrt{x^2 - a^2}$, $x = a\sec\theta$, $0 \le \theta < \pi/2$

68. $\sqrt{16 - 25x^2}$, $x = \tfrac{4}{5}\sin\theta$, $-\pi/2 \le \theta < \pi/2$

69. $\dfrac{x}{\sqrt{9 - x^2}}$, $x = 3\sin\theta$, $-\pi/2 < \theta < \pi/2$

70. $x^2\sqrt{x^2 - 4}$, $x = 2\sec\theta$, $0 \le \theta < \pi/2$

71. $\dfrac{\sqrt{x^2 - 3}}{x^2}$, $x = \sqrt{3}\sec\theta$, $0 \le \theta < \pi/2$

72. $(36 + x^2)^{3/2}$, $x = 6\tan\theta$, $-\pi/2 < \theta < \pi/2$

73. $\dfrac{1}{\sqrt{7 + x^2}}$, $x = \sqrt{7}\tan\theta$, $-\pi/2 < \theta < \pi/2$

74. $\dfrac{\sqrt{5 - x^2}}{x}$, $x = \sqrt{5}\cos\theta$, $0 \le \theta \le \pi$

For Discussion

In Problems 75 and 76, show that the given expression is not an identity.

75. $\sin t = \sqrt{1 - \cos^2 t}$

76. $(\sin x + \cos x)^2 = \sin^2 x + \cos^2 x$

4.6 | Sum and Difference Formulas

≡ **Introduction** In this section we continue our examination of trigonometric identities. But this time we are going to develop only those identities that are of particular importance in courses in mathematics and science.

☐ **Sum and Difference Formulas** The **sum** and **difference formulas** for the cosine and sine functions are identities that reduce $\cos(x_1 + x_2)$, $\cos(x_1 - x_2)$, $\sin(x_1 + x_2)$, and $\sin(x_1 - x_2)$ to expressions that involve $\cos x_1$, $\cos x_2$, $\sin x_1$, and $\sin x_2$. We will derive the formula for $\cos(x_1 - x_2)$ first, and then we will use that result to obtain the others.

For convenience, let us suppose that x_1 and x_2 represent angles measured in radians. As shown in **FIGURE 4.6.1(a)**, let d denote the distance between $P(x_1)$ and $P(x_2)$. If we place the angle $x_1 - x_2$ in standard position as shown in Figure 4.6.1(b), then d is also the distance between $P(x_1 - x_2)$ and $P(0)$. Equating the squares of these distances gives

$$(\cos x_1 - \cos x_2)^2 + (\sin x_1 - \sin x_2)^2 = (\cos(x_1 - x_2) - 1)^2 + \sin^2(x_1 - x_2)$$

or

$$\cos^2 x_1 - 2\cos x_1 \cos x_2 + \cos^2 x_2 + \sin^2 x_1 - 2\sin x_1 \sin x_2 + \sin^2 x_2$$
$$= \cos^2(x_1 - x_2) - 2\cos(x_1 - x_2) + 1 + \sin^2(x_1 - x_2).$$

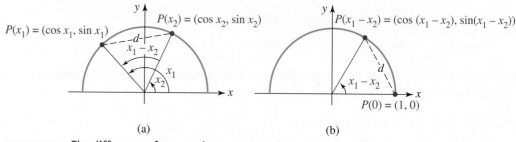

(a)   (b)

FIGURE 4.6.1 The difference of two angles

In view of (8), of Section 4.4,

$$\cos^2 x_1 + \sin^2 x_1 = 1, \quad \cos^2 x_2 + \sin^2 x_2 = 1, \quad \cos^2(x_1 - x_2) + \sin^2(x_1 - x_2) = 1,$$

and so the preceding equation simplifies to

$$\cos(x_1 - x_2) = \cos x_1 \cos x_2 + \sin x_1 \sin x_2.$$

This last result can be put to work immediately to find the cosine of the sum of two angles. Since $x_1 + x_2$ can be rewritten as the difference $x_1 - (-x_2)$,

$$\cos(x_1 + x_2) = \cos(x_1 - (-x_2))$$
$$= \cos x_1 \cos(-x_2) + \sin x_1 \sin(-x_2).$$

By the even–odd identities, $\cos(-x_2) = \cos x_2$ and $\sin(-x_2) = -\sin x_2$, it follows that the last line is the same as

$$\cos(x_1 + x_2) = \cos x_1 \cos x_2 - \sin x_1 \sin x_2.$$

The two results just obtained are summarized in the next theorem.

THEOREM 4.6.1 Sum and Difference Formulas for the Cosine

For all real numbers x_1 and x_2,

$$\cos(x_1 + x_2) = \cos x_1 \cos x_2 - \sin x_1 \sin x_2 \tag{1}$$
$$\cos(x_1 - x_2) = \cos x_1 \cos x_2 + \sin x_1 \sin x_2 \tag{2}$$

EXAMPLE 1   **Cosine of a Sum**

Evaluate $\cos(7\pi/12)$.

Solution We have no way of evaluating $\cos(7\pi/12)$ directly. However, observe that

$$\frac{7\pi}{12} \text{ radians} = 105° = 60° + 45° = \frac{\pi}{3} + \frac{\pi}{4}.$$

Because $7\pi/12$ radians is a second-quadrant angle, we know that the value of $\cos(7\pi/12)$ is negative. Proceeding, the sum formula (1) of the Theorem 4.6.1 gives

$$\cos\frac{7\pi}{12} = \cos\left(\frac{\pi}{3} + \frac{\pi}{4}\right) = \cos\frac{\pi}{3}\cos\frac{\pi}{4} - \sin\frac{\pi}{3}\sin\frac{\pi}{4}$$

$$= \frac{1}{2}\frac{\sqrt{2}}{2} - \frac{\sqrt{3}}{2}\frac{\sqrt{2}}{2} = \frac{\sqrt{2}}{4}(1 - \sqrt{3}).$$

Using $\sqrt{2}\sqrt{3} = \sqrt{6}$, this result can also be written as $\cos(7\pi/12) = (\sqrt{2} - \sqrt{6})/4$. Since $\sqrt{6} > \sqrt{2}$, we see that $\cos(7\pi/12) < 0$, as expected.   ≡

To obtain the corresponding sum/difference identities for the sine function we will make use of two identities:

$$\cos\left(x - \frac{\pi}{2}\right) = \sin x \quad \text{and} \quad \sin\left(x - \frac{\pi}{2}\right) = -\cos x. \qquad (3) \qquad ◀ \text{See (5) in Section 4.3.}$$

These identities were first discovered in Section 4.3 by shifting the graphs of the cosine and sine. However, both results in (3) can now be proved using (2):

$$\cos\left(x - \frac{\pi}{2}\right) = \cos x\underbrace{\cos\frac{\pi}{2}}_{\text{zero}} + \sin x\sin\frac{\pi}{2} = \cos x \cdot 0 + \sin x \cdot 1 = \sin x,$$

$$\cos x = \cos\left(\frac{\pi}{2} - \frac{\pi}{2} + x\right) = \cos\left(\frac{\pi}{2} - \left(\frac{\pi}{2} - x\right)\right) = \sin\left(\frac{\pi}{2} - x\right) = -\sin\left(x - \frac{\pi}{2}\right).$$

by (2) of Theorem 4.6.1

Now from the first equation in (3), the sine of the sum $x_1 + x_2$ can be written

$$\sin(x_1 + x_2) = \cos\left((x_1 + x_2) - \frac{\pi}{2}\right)$$

$$= \cos\left(x_1 + \left(x_2 - \frac{\pi}{2}\right)\right)$$

$$= \cos x_1\cos\left(x_2 - \frac{\pi}{2}\right) - \sin x_1\sin\left(x_2 - \frac{\pi}{2}\right) \quad ← \text{by (1) of Theorem 4.6.1}$$

$$= \cos x_1\sin x_2 - \sin x_1(-\cos x_2). \qquad ← \text{by (3)}$$

The last line is traditionally written as

$$\sin(x_1 + x_2) = \sin x_1\cos x_2 + \cos x_1\sin x_2.$$

To obtain the sine of the difference $x_1 - x_2$, we use again $\cos(-x_2) = \cos x_2$ and $\sin(-x_2) = -\sin x_2$:

$$\sin(x_1 - x_2) = \sin(x_1 + (-x_2)) = \sin x_1\cos(-x_2) + \cos x_1\sin(-x_2)$$

$$= \sin x_1\cos x_2 - \cos x_1\sin x_2.$$

> **THEOREM 4.6.2** Sum and Difference Formulas for the Sine

For all real numbers x_1 and x_2,

$$\sin(x_1 + x_2) = \sin x_1 \cos x_2 + \cos x_1 \sin x_2 \qquad (4)$$
$$\sin(x_1 - x_2) = \sin x_1 \cos x_2 - \cos x_1 \sin x_2 \qquad (5)$$

EXAMPLE 2 **Sine of a Sum**

Evaluate $\sin(7\pi/12)$.

Solution We proceed as in Example 1, except we use the sum formula (4) of Theorem 4.6.2:

$$\sin\frac{7\pi}{12} = \sin\left(\frac{\pi}{3} + \frac{\pi}{4}\right) = \sin\frac{\pi}{3}\cos\frac{\pi}{4} + \cos\frac{\pi}{3}\sin\frac{\pi}{4}$$

$$= \frac{\sqrt{3}}{2}\frac{\sqrt{2}}{2} + \frac{1}{2}\frac{\sqrt{2}}{2} = \frac{\sqrt{2}}{4}(1 + \sqrt{3}).$$

As in Example 1, the result can be rewritten as $\sin(7\pi/12) = (\sqrt{2} + \sqrt{6})/4$. ☰

Since we know the value of $\cos(7\pi/12)$ from Example 1, we can also compute the value of $\sin(7\pi/12)$ using the Pythagorean identity (8) of Section 4.4:

$$\sin^2\frac{7\pi}{12} + \cos^2\frac{7\pi}{12} = 1.$$

We solve for $\sin(7\pi/12)$ and take the positive square root:

$$\sin\frac{7\pi}{12} = \sqrt{1 - \cos^2\frac{7\pi}{12}} = \sqrt{1 - \left[\frac{\sqrt{2}}{4}(1 - \sqrt{3})\right]^2}$$

$$= \sqrt{\frac{4 + 2\sqrt{3}}{8}} = \frac{\sqrt{2 + \sqrt{3}}}{2}. \qquad (6)$$

Although the number in (6) does not look like the result obtained in Example 2, the values are the same. See Problem 72 in Exercises 4.6.

There are sum and difference formulas for the tangent function as well. We can derive the sum formula using the sum formulas for the sine and cosine as follows:

$$\tan(x_1 + x_2) = \frac{\sin(x_1 + x_2)}{\cos(x_1 + x_2)} = \frac{\sin x_1 \cos x_2 + \cos x_1 \sin x_2}{\cos x_1 \cos x_2 - \sin x_1 \sin x_2}. \qquad (7)$$

We now divide the numerator and denominator of (7) by $\cos x_1 \cos x_2$ (assuming that x_1 and x_2 are such that $\cos x_1 \cos x_2 \neq 0$),

$$\tan(x_1 + x_2) = \frac{\dfrac{\sin x_1}{\cos x_1}\dfrac{\cos x_2}{\cos x_2} + \dfrac{\cos x_1}{\cos x_1}\dfrac{\sin x_2}{\cos x_2}}{\dfrac{\cos x_1}{\cos x_1}\dfrac{\cos x_2}{\cos x_2} - \dfrac{\sin x_1}{\cos x_1}\dfrac{\sin x_2}{\cos x_2}} = \frac{\tan x_1 + \tan x_2}{1 - \tan x_1 \tan x_2}. \qquad (8)$$

The derivation of the difference formula for $\tan(x_1 - x_2)$ is obtained in a similar manner. We summarize the two results.

THEOREM 4.6.3 Sum and Difference Formulas for the Tangent

For real numbers x_1 and x_2 for which the functions are defined,

$$\tan(x_1 + x_2) = \frac{\tan x_1 + \tan x_2}{1 - \tan x_1 \tan x_2} \qquad (9)$$

$$\tan(x_1 - x_2) = \frac{\tan x_1 - \tan x_2}{1 + \tan x_1 \tan x_2} \qquad (10)$$

EXAMPLE 3   Tangent of a Difference

Evaluate $\tan(\pi/12)$.

Solution If we think of $\pi/12$ as an angle in radians, then

$$\frac{\pi}{12} \text{ radians} = 15° = 45° - 30° = \frac{\pi}{4} - \frac{\pi}{6} \text{ radians.}$$

It follows from formula (10) of Theorem 4.6.3:

$$\tan\frac{\pi}{12} = \tan\left(\frac{\pi}{4} - \frac{\pi}{6}\right) = \frac{\tan\dfrac{\pi}{4} - \tan\dfrac{\pi}{6}}{1 + \tan\dfrac{\pi}{4}\tan\dfrac{\pi}{6}}$$

$$= \frac{1 - \frac{1}{\sqrt{3}}}{1 + 1 \cdot \frac{1}{\sqrt{3}}} = \frac{\sqrt{3} - 1}{\sqrt{3} + 1}$$

$$= \frac{\sqrt{3} - 1}{\sqrt{3} + 1} \cdot \frac{\sqrt{3} - 1}{\sqrt{3} - 1} \qquad \leftarrow \text{rationalizing the denominator}$$

$$= \frac{\left(\sqrt{3} - 1\right)^2}{2} = \frac{4 - 2\sqrt{3}}{2} = 2 - \sqrt{3}. \qquad \equiv$$

◀ You should rework this example using
$\pi/12 = \pi/3 - \pi/4$
to see that the result is the same.

Strictly speaking, we really do not need the identities for $\tan(x_1 \pm x_2)$, since we can always compute $\sin(x_1 \pm x_2)$ and $\cos(x_1 \pm x_2)$ using (1), (2), (4), (5) and then proceed as in (7), that is, form the quotient $\sin(x_1 \pm x_2)/\cos(x_1 \pm x_2)$.

☐ **Double-Angle Formulas** Many useful trigonometric formulas can be derived from the sum and difference formulas. The **double-angle formulas** for the cosine and sine functions express the cosine and sine of $2x$ in terms of the cosine and sine of x.

If we set $x_1 = x_2 = x$ in (1) and use $\cos(x + x) = \cos 2x$, then

$$\cos 2x = \cos x \cos x - \sin x \sin x = \cos^2 x - \sin^2 x.$$

Similarly, by setting $x_1 = x_2 = x$ in (4) and using $\sin(x + x) = \sin 2x$, then

these two terms are equal
↓   ↓

$$\sin 2x = \sin x \cos x + \cos x \sin x = 2 \sin x \cos x.$$

We summarize the last two results along with the double-angle formula for the tangent function.

THEOREM 4.6.4 Double-Angle Formulas

For any real number x,

$$\cos 2x = \cos^2 x - \sin^2 x \qquad (11)$$
$$\sin 2x = 2\sin x \cos x \qquad (12)$$
$$\tan 2x = \frac{2\tan x}{1 - \tan^2 x} \qquad (13)$$

EXAMPLE 4   Using the Double-Angle Formulas

If $\sin x = -\frac{1}{4}$ and $\pi < x < 3\pi/2$, find the exact values of $\cos 2x$ and $\sin 2x$.

Solution First, we compute $\cos x$ using $\sin^2 x + \cos^2 x = 1$. Since $\pi < x < 3\pi/2$, $\cos x < 0$, and so we choose the negative square root:

$$\cos x = -\sqrt{1 - \sin^2 x} = -\sqrt{1 - \left(-\frac{1}{4}\right)^2} = -\frac{\sqrt{15}}{4}.$$

From the double-angle formula (11) of Theorem 4.6.4,

$$\cos 2x = \cos^2 x - \sin^2 x$$
$$= \left(-\frac{\sqrt{15}}{4}\right)^2 - \left(-\frac{1}{4}\right)^2$$
$$= \frac{15}{16} - \frac{1}{16} = \frac{14}{16} = \frac{7}{8}.$$

Finally, from the double-angle formula (12),

$$\sin 2x = 2\sin x \cos x = 2\left(-\frac{1}{4}\right)\left(-\frac{\sqrt{15}}{4}\right) = \frac{\sqrt{15}}{8}. \qquad \equiv$$

The formula in (11) has two useful alternative forms. By (8) of Section 4.4, we know that $\sin^2 x = 1 - \cos^2 x$. Substituting the last expression into (11) yields $\cos 2x = \cos^2 x - (1 - \cos^2 x)$ or

$$\cos 2x = 2\cos^2 x - 1. \qquad (14)$$

On the other hand, if we substitute $\cos^2 x = 1 - \sin^2 x$ into (11) we get

$$\cos 2x = 1 - 2\sin^2 x. \qquad (15)$$

□ **Half-Angle Formulas** The alternative forms (14) and (15) of the double-angle formula (11) are the source of two **half-angle formulas**. Solving (14) and (15) for $\cos^2 x$ and $\sin^2 x$ gives, respectively,

$$\cos^2 x = \frac{1}{2}(1 + \cos 2x) \qquad \text{and} \qquad \sin^2 x = \frac{1}{2}(1 - \cos 2x). \qquad (16)$$

By replacing the symbol x in (16) by $x/2$ and using $2(x/2) = x$, we obtain the half-angle formulas for the cosine and sine functions.

THEOREM 4.6.5 Half-Angle Formulas

For any real number x,

$$\cos^2\frac{x}{2} = \frac{1}{2}(1 + \cos x) \tag{17}$$

$$\sin^2\frac{x}{2} = \frac{1}{2}(1 - \cos x) \tag{18}$$

$$\tan^2\frac{x}{2} = \frac{1 - \cos x}{1 + \cos x} \tag{19}$$

EXAMPLE 5 **Using the Half-Angle Formulas**

Find the exact values of $\cos(5\pi/8)$ and $\sin(5\pi/8)$.

Solution If we let $x = 5\pi/4$, then $x/2 = 5\pi/8$ and formulas (17) and (18) of Theorem 4.6.5 yield, respectively,

$$\cos^2\left(\frac{5\pi}{8}\right) = \frac{1}{2}\left(1 + \cos\frac{5\pi}{4}\right) = \frac{1}{2}\left[1 + \left(-\frac{\sqrt{2}}{2}\right)\right] = \frac{2 - \sqrt{2}}{4},$$

and $\quad \sin^2\left(\frac{5\pi}{8}\right) = \frac{1}{2}\left(1 - \cos\frac{5\pi}{4}\right) = \frac{1}{2}\left[1 - \left(-\frac{\sqrt{2}}{2}\right)\right] = \frac{2 + \sqrt{2}}{4}.$

Because $5\pi/8$ radians is a second-quadrant angle, $\cos(5\pi/8) < 0$ and $\sin(5\pi/8) > 0$. Therefore, we take the negative square root for the value of the cosine,

$$\cos\left(\frac{5\pi}{8}\right) = -\sqrt{\frac{2 - \sqrt{2}}{4}} = -\frac{\sqrt{2 - \sqrt{2}}}{2},$$

and the positive square root for the value of the sine,

$$\sin\left(\frac{5\pi}{8}\right) = \sqrt{\frac{2 + \sqrt{2}}{4}} = \frac{\sqrt{2 + \sqrt{2}}}{2}. \qquad \equiv$$

If want the exact value of, say, $\tan(5\pi/8)$ we can use the results of Example 5 or formula (19) with $x = 5\pi/4$. Either way, the result is the same

$$\tan\frac{5\pi}{8} = -\frac{\sqrt{2 + \sqrt{2}}}{\sqrt{2 - \sqrt{2}}} = -\frac{\sqrt{2 + \sqrt{2}}}{\sqrt{2 - \sqrt{2}}} = -1 - \sqrt{2}. \quad \leftarrow \left\{\begin{array}{l}\text{rationalizing} \\ \text{the denominator}\end{array}\right.$$

NOTES FROM THE CLASSROOM

(*i*) Should you memorize all the identities presented in this section? You should consult your instructor about this, but in the opinion of the authors, you should at the very least memorize formulas (1), (2), (4), (5), (11) and (12).

(*ii*) When you enroll in a calculus course, check the title of your text. If it has the words *Early Transcendentals* in its title, then your knowledge of the graphs and properties of the trigonometric functions will come into play almost immediately.

(*iii*) As discussed in Sections 2.9 and 3.7, the principal topics of study in calculus are *derivatives and integrals* of functions. The sum identities (1) and (4) are used to find the derivatives of $\sin x$ and $\cos x$. See Section 4.11. Identities are especially useful in integral calculus. Replacing a radical by a trigonometric function as illustrated in Example 7 of Section 4.5 is a standard technique for evaluating some types of integrals. Also, to evaluate integrals of $\cos^2 x$ and $\sin^2 x$ you would use the half-angle formulas in the form given in (16):

$$\cos^2 x = \frac{1}{2}(1 + \cos 2x) \qquad \text{and} \qquad \sin^2 x = \frac{1}{2}(1 - \cos 2x).$$

At some point in your study of integral calculus you may be required to evaluate integrals of products such as

$$\sin 2x \sin 5x \qquad \text{and} \qquad \sin 10x \cos 4x.$$

One way of doing this is to use the sum/difference formulas to devise an identity that converts these products into either a sum of sines or a sum of cosines. This will be the topic of discussion in the next section.

4.6 | Exercises Answers to selected odd-numbered problems begin on page ANS–14.

In Problems 1–22, use a sum or difference formula to find the exact value of the given trigonometric function. Do not use a calculator.

1. $\cos\dfrac{\pi}{12}$
2. $\sin\dfrac{\pi}{12}$
3. $\sin 75°$
4. $\cos 75°$
5. $\sin\dfrac{7\pi}{12}$
6. $\cos\dfrac{11\pi}{12}$
7. $\tan\dfrac{5\pi}{12}$
8. $\cos\left(-\dfrac{5\pi}{12}\right)$
9. $\sin\left(-\dfrac{\pi}{12}\right)$
10. $\tan\dfrac{11\pi}{12}$
11. $\sin\dfrac{11\pi}{12}$
12. $\tan\dfrac{7\pi}{12}$
13. $\cos 165°$
14. $\sin 165°$
15. $\tan 165°$
16. $\cos 195°$
17. $\sin 195°$
18. $\tan 195°$
19. $\cos 345°$
20. $\sin 345°$
21. $\cos\dfrac{13\pi}{12}$
22. $\tan\dfrac{17\pi}{12}$

In Problems 23–28, use a double-angle formula to write the given expression as a single trigonometric function of twice the angle.

23. $2\cos\beta\sin\beta$
24. $\cos^2 2t - \sin^2 2t$
25. $1 - 2\sin^2\dfrac{\pi}{5}$
26. $2\cos^2\left(\dfrac{19}{2}x\right) - 1$
27. $\dfrac{\tan 3t}{1 - \tan^2 3t}$
28. $2\sin\dfrac{y}{2}\cos\dfrac{y}{2}$

In Problems 29–34, use the given information to find (a) $\cos 2x$, (b) $\sin 2x$, and (c) $\tan 2x$.

29. $\sin x = \sqrt{2}/3, \quad \pi/2 < x < \pi$
30. $\cos x = \sqrt{3}/5, \quad 3\pi/2 < x < 2\pi$
31. $\tan x = \frac{1}{2}, \quad \pi < x < 3\pi/2$
32. $\csc x = -3, \quad \pi < x < 3\pi/2$
33. $\sec x = -\frac{13}{5}, \quad \pi/2 < x < \pi$
34. $\cot x = \frac{4}{3}, \quad 0 < x < \pi/2$

In Problems 35–44, use a half-angle formula to find the exact value of the given trigonometric function. Do not use a calculator.

35. $\cos(\pi/12)$
36. $\sin(\pi/8)$
37. $\sin(3\pi/8)$
38. $\tan(\pi/12)$
39. $\cos 67.5°$
40. $\sin 15°$
41. $\tan 105°$
42. $\cot 157.5°$
43. $\csc(13\pi/12)$
44. $\sec(-3\pi/8)$

In Problems 45–50, use the given information to find (a) $\cos(x/2)$, (b) $\sin(x/2)$, and (c) $\tan(x/2)$.

45. $\sin x = \frac{12}{13}, \pi/2 < x < \pi$
46. $\cos x = \frac{4}{5}, 3\pi/2 < x < 2\pi$
47. $\tan x = 2, \pi < x < 3\pi/2$
48. $\csc x = 9, 0 < x < \pi/2$
49. $\sec x = \frac{3}{2}, 0 < x < 90°$
50. $\cot x = -\frac{1}{4}, 90° < x < 180°$

In Problems 51–60, verify the given identity.

51. $\sin 4x = 4\cos x(\sin x - 2\sin^3 x)$
52. $\cos 3x = 4\cos^3 x - 3\cos x$
53. $(\sin x + \cos x)^2 = 1 + \sin 2x$
54. $\cos 2x = \cos^4 x - \sin^4 x$
55. $\cot 2x = \frac{1}{2}(\cot x - \tan x)$
56. $\sec 2x = \dfrac{1}{2\cos^2 x - 1}$
57. $\dfrac{2\tan x}{1 + \tan^2 x} = \sin 2x$
58. $\dfrac{\cot x - \tan x}{\cot x + \tan x} = \cos 2x$
59. $\tan\dfrac{x}{2} = \dfrac{1 - \cos x}{\sin x}$
60. $\tan\dfrac{x}{2} = \dfrac{\sin x}{1 + \cos x}$

In Problems 61–64, rewrite the given function as a single trigonometric function involving no products or squares. Give the amplitude and period of the function.

61. $y = 4\cos^2 x - 2$
62. $y = \sin(x/2)\cos(x/2)$
63. $y = 2\sin 2x \cos 2x$
64. $y = 5\cos^2 4x - 5\sin^2 4x$

65. If $P(x_1)$ and $P(x_2)$ are points in quadrant II on the terminal side of the angles x_1 and x_2, respectively, with $\cos x_1 = -\frac{1}{3}$ and $\sin x_2 = \frac{2}{3}$, find (a) $\sin(x_1 + x_2)$, (b) $\cos(x_1 + x_2)$, (c) $\sin(x_1 - x_2)$, and (d) $\cos(x_1 - x_2)$.
66. If x_1 is a quadrant II angle, x_2 is a quadrant III angle, $\sin x_1 = \frac{8}{17}$, and $\tan x_2 = \frac{3}{4}$, find (a) $\sin(x_1 + x_2)$, (b) $\sin(x_1 - x_2)$, (c) $\cos(x_1 + x_2)$, and (d) $\cos(x_1 - x_2)$.

Miscellaneous Applications

67. Mach Number The ratio of the speed of an airplane to the speed of sound is called the Mach number M of the plane. If $M > 1$, the plane makes sound waves that form a (moving) cone, as shown in FIGURE 4.6.2. A sonic boom is heard at the intersection of the cone with the ground. If the vertex angle of the cone is θ, then

$$\sin\frac{\theta}{2} = \frac{1}{M}.$$

If $\theta = \pi/6$, find the exact value of the Mach number.
68. Cardiovascular Branching A mathematical model for blood flow in a large blood vessel predicts that the optimal values of the angles θ_1 and θ_2, which

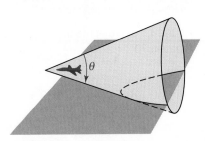

FIGURE 4.6.2 Airplane in Problem 67

represent the (positive) angles of the smaller daughter branches (vessels) with respect to the axis of the parent branch, are given by

$$\cos\theta_1 = \frac{A_0^2 + A_1^2 - A_2^2}{2A_0A_1} \quad \text{and} \quad \cos\theta_2 = \frac{A_0^2 - A_1^2 + A_2^2}{2A_0A_2},$$

where A_0 is the cross-sectional area of the parent branch and A_1 and A_2 are the cross-sectional areas of the daughter branches. See FIGURE 4.6.3. Let $\psi = \theta_1 + \theta_2$ be the junction angle, as shown in the figure.

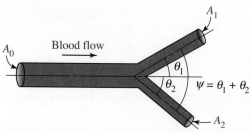

FIGURE 4.6.3 Branching of a large blood vessel in Problem 68

(a) Show that

$$\cos\psi = \frac{A_0^2 - A_1^2 - A_2^2}{2A_1A_2}.$$

(b) Show that for the optimal values of θ_1 and θ_2, the cross-sectional area of the daughter branches, $A_1 + A_2$, is greater than or equal to that of the parent branch. Therefore, the blood must slow down in the daughter branches.

69. **Area of a Triangle** Show that the area of an isosceles triangle with equal sides of length x is

$$A = \tfrac{1}{2}x^2\sin\theta,$$

where θ is the angle formed by the two equal sides. See FIGURE 4.6.4. [*Hint*: Consider $\theta/2$ as shown in the figure.]

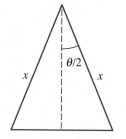

FIGURE 4.6.4 Isosceles triangle in Problem 69

70. **Putting the Shot** If a projectile, such as a shot put, is released from a height h, upward at an angle ϕ with velocity v_0, the range R at which it strikes the ground is given by

$$R = \frac{v_0^2\cos\phi}{g}\left(\sin\phi + \sqrt{\sin^2\phi + (2gh/v_0^2)}\right),$$

where g is the acceleration due to gravity. See FIGURE 4.6.5. It can be shown that the maximum range $R_{\max}$ is achieved if the angle ϕ satisfies the equation

$$\cos 2\phi = \frac{gh}{v_0^2 + gh}.$$

Show that
$$R_{\max} = \frac{v_0\sqrt{v_0^2 + 2gh}}{g},$$

by using the expressions for R and $\cos 2\phi$ and the half-angle formulas for the sine and the cosine with $t = 2\phi$.

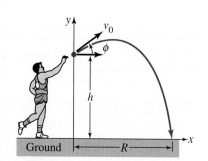

FIGURE 4.6.5 Projectile in Problem 70

For Discussion

71. Discuss: Why would you expect to get an error message from your calculator when you try to evaluate

$$\frac{\tan 35° + \tan 55°}{1 - \tan 35°\tan 55°}?$$

72. In Example 2 we showed that $\sin(7\pi/12) = \frac{1}{4}(\sqrt{2} + \sqrt{6})$. Following the example, we then showed that $\sin(7\pi/12) = \frac{1}{2}\sqrt{2 + \sqrt{3}}$. Demonstrate that these two answers are equivalent.

73. Discuss: How would you find a formula that expresses $\sin 3\theta$ in terms of $\sin\theta$? Carry out your ideas.

74. In Problem 65, in what quadrants do $P(x_1 + x_2)$ and $P(x_1 - x_2)$ lie?

75. In Problem 66, in which quadrant does the terminal side of $x_1 + x_2$ lie? The terminal side of $x_1 - x_2$?

4.7 Product-to-Sum and Sum-to-Product Formulas

≡ **Introduction** There are instances, especially in integral calculus, where it is necessary to convert a product of sine and cosine functions to a sum of these functions. Moreover, in solving trigonometric equations we may find it convenient to convert a sum of sine and cosine functions into a product of these functions. In the discussion that follows we establish trigonometric identities or formulas that do the job.

☐ **Reduction of a Product to a Sum** The **product-to-sum formulas** given in the next theorem are direct consequences of the sum and difference formulas for the cosine and sine functions in Section 4.6.

THEOREM 4.7.1   Product-to-Sum Formulas

For all real number x_1 and x_2,

$$\sin x_1 \sin x_2 = \tfrac{1}{2}[\cos(x_1 - x_2) - \cos(x_1 + x_2)] \qquad (1)$$
$$\cos x_1 \cos x_2 = \tfrac{1}{2}[\cos(x_1 - x_2) + \cos(x_1 + x_2)] \qquad (2)$$
$$\sin x_1 \cos x_2 = \tfrac{1}{2}[\sin(x_1 + x_2) + \sin(x_1 - x_2)] \qquad (3)$$

PROOF: To prove (1), we use (1) and (2) of Theorem 4.6.1:

$$\cos(x_1 - x_2) = \cos x_1 \cos x_2 + \sin x_1 \sin x_2 \qquad (4)$$
$$\cos(x_1 + x_2) = \cos x_1 \cos x_2 - \sin x_1 \sin x_2. \qquad (5)$$

Subtracting (5) from (4) yields

$$\cos(x_1 - x_2) - \cos(x_1 + x_2) = 2\sin x_1 \sin x_2.$$

And so,

$$\sin x_1 \sin x_2 = \tfrac{1}{2}[\cos(x_1 - x_2) - \cos(x_1 + x_2)]$$

which is (1). Similarly, by adding (4) and (5) we get

$$\cos(x_1 - x_2) + \cos(x_1 + x_2) = 2\cos x_1 \cos x_2,$$

which in turn, yields formula (2):

$$\cos x_1 \cos x_2 = \tfrac{1}{2}[\cos(x_1 - x_2) + \cos(x_1 + x_2)].$$

Formula (3) follows analogously by adding the sum and difference formulas for the sine, (4) and (5) in Theorem 4.6.2 of Section 4.6.   ≡

Although we do not feel that it is necessary to memorize (1)–(3) in Theorem 4.7.1, you should listen to what your instructor requires. By remembering the *procedure* just illustrated in the proof of Theorem 4.7.1 each of these formulas can be derived on the spot.

■ EXAMPLE 1   Using (2) of Theorem 4.7.1

Use a product-to-sum formula to rewrite the product $\cos 2\theta \cos 3\theta$ as a sum.

Solution From formula (2) of Theorem 4.7.1 with the identifications $x_1 = 2\theta$ and $x_2 = 3\theta$, we obtain

$$
\begin{aligned}
\cos 2\theta \cos 3\theta &= \tfrac{1}{2}[\cos(2\theta - 3\theta) + \cos(2\theta + 3\theta)] \\
&= \tfrac{1}{2}[\cos(-\theta) + \cos 5\theta] \qquad \leftarrow \cos(-\theta) = \cos\theta \\
&= \tfrac{1}{2}[\cos\theta + \cos 5\theta].
\end{aligned}
$$

$\equiv$

■ EXAMPLE 2   Using (3) of Theorem 4.7.1

Use a product-to-sum formula to find the exact value of the product $\sin 45° \cos 15°$.

Solution Using formula (3) of Theorem 4.7.1 with $x_1 = 45°$ and $x_2 = 15°$, we have

$$
\begin{aligned}
\sin 45° \cos 15° &= \tfrac{1}{2}[\sin(45° + 15°) + \sin(45° - 15°)] \\
&= \tfrac{1}{2}[\sin 60° + \sin 30°].
\end{aligned}
$$

Because $\sin 60° = \tfrac{1}{2}\sqrt{3}$ and $\sin 30° = \tfrac{1}{2}$ we observe that the exact value of the given product is

$$
\sin 45° \cos 15° = \tfrac{1}{2}[\sin 60° + \sin 30°] = \tfrac{1}{2}\left(\tfrac{1}{2}\sqrt{3} + \tfrac{1}{2}\right) = \tfrac{1}{4}\left(\sqrt{3} + 1\right).
$$

$\equiv$

□ **Reduction of a Sum to a Product** The results in Theorem 4.7.1 can now be used to derive the **sum-to-product formulas**.

THEOREM 4.7.2 Sum-to-Product Formulas

For all real number x_1 and x_2,

$$
\sin x_1 + \sin x_2 = 2\sin\frac{x_1 + x_2}{2}\cos\frac{x_1 - x_2}{2} \tag{6}
$$

$$
\sin x_1 - \sin x_2 = 2\cos\frac{x_1 + x_2}{2}\sin\frac{x_1 - x_2}{2} \tag{7}
$$

$$
\cos x_1 + \cos x_2 = 2\cos\frac{x_1 + x_2}{2}\cos\frac{x_1 - x_2}{2} \tag{8}
$$

$$
\cos x_1 - \cos x_2 = -2\sin\frac{x_1 + x_2}{2}\sin\frac{x_1 - x_2}{2} \tag{9}
$$

PROOF: By replacing, in turn, the symbols x_1 and x_2 in (1) of Theorem 4.7.1 by

$$
\frac{x_1 + x_2}{2} \qquad \text{and} \qquad \frac{x_1 - x_2}{2}, \tag{10}
$$

we get

$$\sin\frac{x_1 + x_2}{2}\sin\frac{x_1 - x_2}{2} = \frac{1}{2}\left[\cos\left(\frac{x_1 + x_2}{2} - \frac{x_1 - x_2}{2}\right) - \cos\left(\frac{x_1 + x_2}{2} + \frac{x_1 - x_2}{2}\right)\right]$$
$$= \frac{1}{2}[\cos x_2 - \cos x_1].$$

Multiplying the last expression by -2 yields

$$-2\sin\frac{x_1 + x_2}{2}\sin\frac{x_1 - x_2}{2} = \cos x_1 - \cos x_2,$$

which is the formula (9). In a similar manner, each of the remaining product-to-sum formulas together with the substitutions in (10) yields one of the sum-to-product formulas. ≡

EXAMPLE 3   Using (9) of Theorem 4.7.2

Use a sum-to-product formula to rewrite the sum $\cos t - \cos 5t$ as a product.

Solution We use formula (9) of Theorem 4.7.2 with $x_1 = t$ and $x_2 = 5t$:

$$\cos t - \cos 5t = -2\sin\left(\frac{t + 5t}{2}\right)\sin\left(\frac{t - 5t}{2}\right)$$
$$= -2\sin 3t \sin(-2t) \qquad \leftarrow \sin(-2t) = -\sin 2t$$
$$= 2\sin 3t \sin 2t$$ ≡

EXAMPLE 4   Using (6) of Theorem 4.7.2

Use a sum-to-product formula to find the exact value of the sum $\sin 75° + \sin 15°$.

Solution In this case we use formula (6) of Theorem 4.7.2 with $x_1 = 75°$ and $x_2 = 15°$:

$$\sin 75° + \sin 15° = 2\sin\left(\frac{75° + 15°}{2}\right)\cos\left(\frac{75° - 15°}{2}\right)$$
$$= 2\sin 45° \cos 30°.$$

Because $\sin 45° = \frac{1}{2}\sqrt{2}$ and $\cos 30° = \frac{1}{2}\sqrt{3}$ the exact value of the given sum

$$\sin 75° + \sin 15° = 2\sin 45° \cos 30° = 2\left(\tfrac{1}{2}\sqrt{2}\right)\left(\tfrac{1}{2}\sqrt{3}\right) = \tfrac{1}{2}\sqrt{6}.$$ ≡

| 4.7 | Exercises | Answers to selected odd-numbered problems begin on page ANS–14. |

In Problems 1–12, use a product-to-sum formula in Theorem 4.7.1 to write the given product as a sum of cosines or a sum of sines.

1. $\cos 4\theta \cos 3\theta$

2. $\sin\dfrac{3t}{2}\cos\dfrac{t}{2}$

3. $\sin 2x \sin 5x$

4. $\sin 10x \cos 4x$

5. $\cos\dfrac{4x}{3}\cos\dfrac{x}{3}$

6. $-\sin t \sin 2t$

7. $\sin 8x \cos 12x$

8. $\sin \pi\theta \cos 7\pi\theta$

9. $2\cos 3\beta \sin \beta$

10. $6\sin \alpha \sin 4\alpha$

11. $2\sin\left(x + \dfrac{\pi}{4}\right)\sin\left(x - \dfrac{\pi}{4}\right)$

12. $2\sin\left(t + \dfrac{\pi}{2}\right)\cos\left(t - \dfrac{\pi}{2}\right)$

In Problems 13–18, use a product-to-sum formula to find the exact value of the expression. Do not use a calculator.

13. $\cos\dfrac{5\pi}{12}\sin\dfrac{\pi}{12}$

14. $\sin\dfrac{5\pi}{8}\cos\dfrac{\pi}{8}$

15. $\sin 75° \sin 15°$

16. $\cos 15° \cos 45°$

17. $\sin 97.5° \sin 52.5°$

18. $\sin 105° \cos 195°$

In Problems 19–30, use a sum-to-product-formula in Theorem 4.7.2 to write the given sum as a product of cosines, a product of sines, or a product of a sine and a cosine.

19. $\sin y - \sin 5y$

20. $\cos 3\theta - \cos\theta$

21. $\cos\dfrac{9x}{2} - \cos\dfrac{x}{2}$

22. $\sin\dfrac{x}{2} - \sin\dfrac{3x}{2}$

23. $\cos 2x + \cos 6x$

24. $\sin 5t + \sin 3t$

25. $\sin\omega_1 t + \sin\omega_2 t$

26. $\frac{1}{2}(\cos 2\alpha + \cos 2\beta)$

27. $-\frac{1}{2}\cos t + \frac{1}{2}\cos 5t$

28. $\sin(\theta + \pi) + \sin(\theta - \pi)$

29. $\sin\left(t + \dfrac{\pi}{2}\right) + \sin\left(t - \dfrac{\pi}{2}\right)$

30. $\cos\left(t + \dfrac{\pi}{2}\right) - \cos\left(t - \dfrac{\pi}{2}\right)$

In Problems 31–36, use a sum-to-product-formula to find the exact value of the expression. Do not use a calculator.

31. $\sqrt{2}\sin\dfrac{13\pi}{12} + \sqrt{2}\sin\dfrac{5\pi}{12}$

32. $\sin\dfrac{\pi}{12} - \sin\dfrac{5\pi}{12}$

33. $\cos 105° - \cos 15°$

34. $\cos 15° + \cos 75°$

35. $\sin 195° + \sin 105°$

36. $2\cos 195° - 2\cos 105°$

Miscellaneous Applications

37. Sound Wave A note produced by a certain musical instrument results in a sound wave described by

$$f(t) = 0.03\sin 500\pi t + 0.03\sin 1000\pi t,$$

where $f(t)$ is the difference between atmospheric pressure and air pressure in dynes per square centimeter at the eardrum after t seconds. Express f as the product of a sine and a cosine function.

38. Beats If two piano wires struck by the same key are slightly out of tune, the difference between the atmospheric pressure and air pressure at the eardrum can be represented by the function

$$f(t) = a\cos 2\pi b_1 t + a\cos 2\pi b_2 t,$$

where the value of the constant b_1 is close to the value of constant b_2. The variations in loudness that occur are called **beats**. See FIGURE 4.7.1. The two strings can be tuned to the same frequency by tightening one of them while sounding both until the beats disappear.

(a) Use a sum formula to write $f(t)$ as a product.

(b) Show that $f(t)$ can be considered a cosine function with period $2/(b_1 + b_2)$ and variable amplitude $2a\cos\pi(b_1 - b_2)t$.

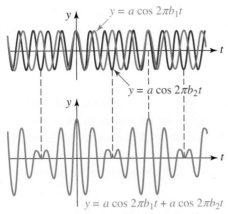

$y = a \cos 2\pi b_1 t$

$y = a \cos 2\pi b_2 t$

$y = a \cos 2\pi b_1 t + a \cos 2\pi b_2 t$

FIGURE 4.7.1 Graph for Problem 38

(c) Use a graphing utility to obtain the graph of f in the case $2\pi b_1 = 5$, $2\pi b_2 = 4$, and $a = \frac{1}{2}$.

39. Alternating Current The term $\sin \omega t \sin(\omega t + \phi)$ is encountered in the derivation of an expression for the power in an alternating-current circuit. Show that this term can be written as $\frac{1}{2}[\cos \phi - \cos(2\omega t + \phi)]$.

For Discussion

40. If $x_1 + x_2 + x_3 = \pi$, then show that
$$\sin 2x_1 + \sin 2x_2 + \sin 2x_3 = 4\sin x_1 \sin x_2 \sin x_3.$$

41. Write as a product of cosines: $1 + \cos 2t + \cos 4t + \cos 6t$.

42. Simplify: $2\cos 2t \cos t - \cos 3t$.

4.8　Inverse Trigonometric Functions

☰ **Introduction** Although we can find the values of the trigonometric functions of real numbers or angles, in many applications we must do the reverse: Given the value of a trigonometric function, find a corresponding angle or number. This suggests we consider inverse trigonometric functions. Before we define the inverse trigonometric functions, let's recall from Section 2.7 some of the properties of a one-to-one function f and its inverse f^{-1}.

◀ Recall, a function f is one-to-one if every y in its range corresponds to exactly one x in its domain.

☐ **Properties of Inverse Functions** If $y = f(x)$ is a one-to-one function, then there is a unique inverse function f^{-1} with the following properties:

Properties of Inverse Functions

- The domain of f^{-1} = range of f.
- The range of f^{-1} = domain of f.
- $y = f(x)$ is equivalent to $x = f^{-1}(y)$.
- The graphs of f and f^{-1} are reflections in the line $y = x$.
- $f(f^{-1}(x)) = x$ for x in the domain of f^{-1}.
- $f^{-1}(f(x)) = x$ for x in the domain of f.

Inspection of the graphs of the various trigonometric functions clearly shows that *none* of these functions are one-to-one. In Section 2.7 we discussed the fact that if a function f

See Example 8 in Section 2.7. ▶

is not one-to-one, it may be possible to restrict the function to a portion of its domain where it is one-to-one. Then we can define an inverse for f on that restricted domain. Normally, when we restrict the domain, we make sure to preserve the entire range of the original function.

☐ **Arcsine Function** From FIGURE 4.8.1(a) we see that the function $y = \sin x$ on the closed interval $[-\pi/2, \pi/2]$ takes on all values in its range $[-1, 1]$. Notice that any horizontal line drawn to intersect the red portion of the graph can do so at most once. Thus the sine function on this restricted domain is one-to-one and has an inverse. There are two commonly used notations to denote the inverse of the function shown in Figure 4.8.1(b):

$$\arcsin x \qquad \text{or} \qquad \sin^{-1}x,$$

and are read **arcsine of x** and **inverse sine of x**, respectively.

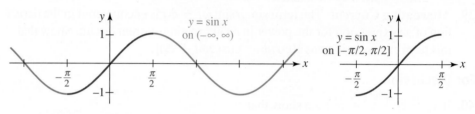

(a) Not a one-to-one function　　　　　　　　　　(b) A one-to-one function

FIGURE 4.8.1 Restricting the domain of $y = \sin x$ to produce a one-to-one function

In FIGURE 4.8.2(a) we have reflected the portion of the graph of $y = \sin x$ on the interval $[-\pi/2, \pi/2]$ (the red graph in Figure 4.8.1(b)) about the line $y = x$ to obtain the graph of $y = \arcsin x$ (in blue). For clarity, we have reproduced this blue graph in Figure 4.8.2(b). As this curve shows, the domain of the arcsine function is $[-1, 1]$ and the range is $[-\pi/2, \pi/2]$.

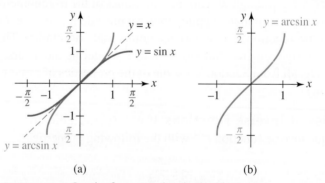

(a)　　　　　　　　(b)

FIGURE 4.8.2 Graph of $y = \arcsin x$ is the blue curve

DEFINITION 4.8.1　Arcsine Function

The **arcsine function**, or **inverse sine function**, is defined by

$$y = \arcsin x \qquad \text{if and only if} \qquad x = \sin y, \qquad (1)$$

where $-1 \le x \le 1$ and $-\pi/2 \le y \le \pi/2$.

In other words:

> *The arcsine of the number x is that number y (or radian-measured angle) between $-\pi/2$ and $\pi/2$ whose sine is x.*

When using the notation $\sin^{-1}x$ it is important to realize that "-1" is not an exponent; rather, it denotes an inverse function. The notation $\arcsin x$ has an advantage over the notation $\sin^{-1}x$ in that there is no "-1" and hence no potential for misinterpretation; moreover, the prefix "arc" refers to an angle—*the* angle whose sine is x. But since $y = \arcsin x$ and $y = \sin^{-1}x$ are used interchangeably in calculus and in applications, we will continue to alternate their use so that you become comfortable with both notations.

◀ **Note of Caution:**
$$(\sin x)^{-1} = \frac{1}{\sin x} \neq \sin^{-1}x$$

◼ EXAMPLE 1   Evaluating the Inverse Sine Function

Find **(a)** $\arcsin\frac{1}{2}$, **(b)** $\sin^{-1}\left(-\frac{1}{2}\right)$, and **(c)** $\sin^{-1}(-1)$.

Solution (a) If we let $y = \arcsin\frac{1}{2}$, then by (1) we must find the number y (or radian-measured angle) that satisfies $\sin y = \frac{1}{2}$ *and* $-\pi/2 \le y \le \pi/2$. Since $\sin(\pi/6) = \frac{1}{2}$ and $\pi/6$ satisfies the inequality $-\pi/2 \le y \le \pi/2$ it follows that $y = \pi/6$.

(b) If we let $y = \sin^{-1}\left(-\frac{1}{2}\right)$, then $\sin y = -\frac{1}{2}$. Since we must choose y such that $-\pi/2 \le y \le \pi/2$, we find that $y = -\pi/6$.

(c) Letting $y = \sin^{-1}(-1)$, we have that $\sin y = -1$ and $-\pi/2 \le y \le \pi/2$.

Hence $y = -\pi/2$.   ≡

In parts (b) and (c) of Example 1 we were careful to choose y so that $-\pi/2 \le y \le \pi/2$. For example, it is a common error to think that because $\sin(3\pi/2) = -1$, then necessarily $\sin^{-1}(-1)$ can be taken to be $3\pi/2$. Remember: If $y = \sin^{-1}x$, then y is subject to the restriction $-\pi/2 \le y \le \pi/2$ and $3\pi/2$ does not satisfy this inequality.

◀ Read this paragraph several times.

◼ EXAMPLE 2   Evaluating a Composition

Without using a calculator, find $\tan\left(\sin^{-1}\frac{1}{4}\right)$.

Solution We must find the tangent of the angle of t radians with sine equal to $\frac{1}{4}$, that is, $\tan t$, where $t = \sin^{-1}\frac{1}{4}$. The angle t is shown in FIGURE 4.8.3. Since

$$\tan t = \frac{\sin t}{\cos t} = \frac{\frac{1}{4}}{\cos t},$$

we want to determine the value of $\cos t$. From Figure 4.8.3 and the Pythagorean identity $\sin^2 t + \cos^2 t = 1$, we see that

$$\left(\frac{1}{4}\right)^2 + \cos^2 t = 1 \quad \text{or} \quad \cos t = \frac{\sqrt{15}}{4}.$$

Hence we have

$$\tan t = \frac{1/4}{\sqrt{15}/4} = \frac{1}{\sqrt{15}} = \frac{\sqrt{15}}{15},$$

and so

$$\tan\left(\sin^{-1}\frac{1}{4}\right) = \tan t = \frac{\sqrt{15}}{15}. \qquad ≡$$

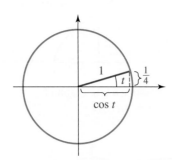

FIGURE 4.8.3 The angle $t = \sin^{-1}\frac{1}{4}$ in Example 2

☐ **Arccosine Function** If we restrict the domain of the cosine function to the closed interval $[0, \pi]$, the resulting function is one-to-one and thus has an inverse. We denote this inverse by

$$\arccos x \qquad \text{or} \qquad \cos^{-1}x,$$

which gives us the following definition.

DEFINITION 4.8.2 Arccosine Function

The **arccosine function**, or **inverse cosine function**, is defined by

$$y = \arccos x \qquad \text{if and only if} \qquad x = \cos y, \qquad (2)$$

where $-1 \leq x \leq 1$ and $0 \leq y \leq \pi$.

The graphs shown in FIGURE 4.8.4 illustrate how the function $y = \cos x$ restricted to the interval $[0, \pi]$ becomes a one-to-one function. The inverse of the function shown in Figure 7.5.4(b) is $y = \arccos x$.

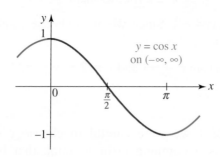

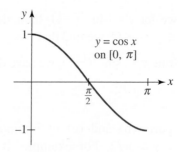

(a) Not a one-to-one function   (b) A one-to-one function

FIGURE 4.8.4 Restricting the domain of $y = \cos x$ to produce a one-to-one function

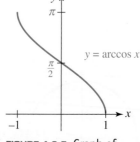

FIGURE 4.8.5 Graph of $y = \arccos x$

By reflecting the graph of the one-to-one function in Figure 4.8.4(b) in the line $y = x$ we obtain the graph of $y = \arccos x$ shown in FIGURE 4.8.5.

Note that the figure clearly shows that the domain and range of $y = \arccos x$ are $[-1, 1]$ and $[0, \pi]$, respectively.

EXAMPLE 3   **Evaluating the Inverse Cosine Function**

Find **(a)** $\arccos\left(\sqrt{2}/2\right)$ **(b)** $\cos^{-1}(-\sqrt{3}/2)$.

Solution (a) If we let $y = \arccos\left(\sqrt{2}/2\right)$, then $\cos y = \sqrt{2}/2$ and $0 \leq y \leq \pi$. Thus $y = \pi/4$.
(b) Letting $y = \cos^{-1}(-\sqrt{3}/2)$, we have that $\cos y = -\sqrt{3}/2$, and we must find y such that $0 \leq y \leq \pi$. Therefore, $y = 5\pi/6$ since $\cos(5\pi/6) = -\sqrt{3}/2$. ≡

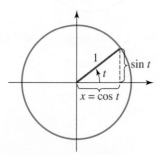

FIGURE 4.8.6 The angle $t = \cos^{-1}x$ in Example 4

EXAMPLE 4   **Evaluating the Compositions of Functions**

Write $\sin(\cos^{-1}x)$ as an algebraic expression in x.

Solution In FIGURE 4.8.6 we have constructed an angle of t radians with cosine equal to x. Then $t = \cos^{-1}x$, or $x = \cos t$, where $0 \leq t \leq \pi$. Now to find $\sin(\cos^{-1}x) = \sin t$, we use the identity $\sin^2 t + \cos^2 t = 1$. Thus

CHAPTER 4 TRIGONOMETRIC FUNCTIONS

$$\sin^2 t + x^2 = 1$$
$$\sin^2 t = 1 - x^2$$
$$\sin t = \sqrt{1 - x^2}$$
$$\sin(\cos^{-1} x) = \sqrt{1 - x^2}.$$

We use the positive square root of $1 - x^2$, since the range of $\cos^{-1} x$ is $[0, \pi]$, and the sine of an angle t in the first or second quadrant is positive. ≡

☐ **Arctangent Function** If we restrict the domain of $\tan x$ to the open interval $(-\pi/2, \pi/2)$, then the resulting function is one-to-one and thus has an inverse. This inverse is denoted by

$$\arctan x \qquad \text{or} \qquad \tan^{-1} x.$$

DEFINITION 4.8.3 Arctangent Function

The **arctangent**, or **inverse tangent**, function is defined by

$$y = \arctan x \qquad \text{if and only if} \qquad x = \tan y, \tag{3}$$

where $-\infty < x < \infty$ and $-\pi/2 < y < \pi/2$.

The graphs shown in FIGURE 4.8.7 illustrate how the function $y = \tan x$ restricted to the open interval $(-\pi/2, \pi/2)$ becomes a one-to-one function.

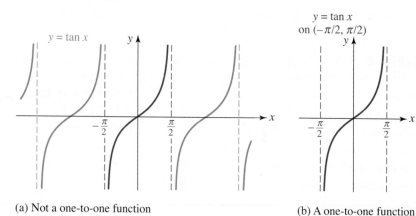

(a) Not a one-to-one function   (b) A one-to-one function

FIGURE 4.8.7 Restricting the domain of $y = \tan x$ to produce a one-to-one function

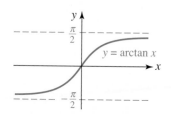

FIGURE 4.8.8 Graph of $y = \arctan x$

By reflecting the graph of the one-to-one function in Figure 4.8.7(b) in the line $y = x$ we obtain the graph of $y = \arctan x$ shown in FIGURE 4.8.8. We see in the figure that the domain and range of $y = \arctan x$ are, in turn, the intervals $(-\infty, \infty)$ and $(-\pi/2, \pi/2)$.

 **EXAMPLE 5**   **Evaluating the Inverse Tangent Function**

Find $\tan^{-1}(-1)$.

Solution If $\tan^{-1}(-1) = y$, then $\tan y = -1$, where $-\pi/2 < y < \pi/2$. It follows that $\tan^{-1}(-1) = y = -\pi/4$. ≡

EXAMPLE 6   Evaluating the Compositions of Functions

Without using a calculator, find $\sin\left(\arctan\left(-\frac{5}{3}\right)\right)$.

Solution If we let $t = \arctan\left(-\frac{5}{3}\right)$, then $\tan t = -\frac{5}{3}$. The Pythagorean identity $1 + \tan^2 t = \sec^2 t$ can be used to find $\sec t$:

$$1 + \left(-\frac{5}{3}\right)^2 = \sec^2 t$$

$$\sec t = \sqrt{1 + \frac{25}{9}} = \sqrt{\frac{34}{9}} = \frac{\sqrt{34}}{3}.$$

In the preceding line we take the positive square root because $t = \arctan\left(-\frac{5}{3}\right)$ is in the interval $(-\pi/2, \pi/2)$ (the range of the arctangent function) and the secant of an angle t in the first or fourth quadrant is positive. Also, from $\sec t = \sqrt{34}/3$ we find the value of $\cos t$ from the reciprocal identity:

$$\cos t = \frac{1}{\sec t} = \frac{1}{\sqrt{34}/3} = \frac{3}{\sqrt{34}}.$$

Finally, we can use the identity $\tan t = \sin t / \cos t$ in the form $\sin t = \tan t \cos t$ to compute $\sin\left(\arctan\left(-\frac{5}{3}\right)\right)$. It follows that

$$\sin t = \tan t \cos t = \left(-\frac{5}{3}\right)\left(\frac{3}{\sqrt{34}}\right) = -\frac{5}{\sqrt{34}}.$$

☐ **Properties of the Inverses** Recall from Section 2.7 that $f^{-1}(f(x)) = x$ and $f(f^{-1}(x)) = x$ hold for any function f and its inverse under suitable restrictions on x. Thus for the inverse trigonometric functions, we have the following properties.

THEOREM 4.8.1   *Properties of Inverse Trigonometric Functions*

(*i*) $\arcsin(\sin x) = \sin^{-1}(\sin x) = x$	if	$-\pi/2 \le x \le \pi/2$
(*ii*) $\sin(\arcsin x) = \sin(\sin^{-1} x) = x$	if	$-1 \le x \le 1$
(*iii*) $\arccos(\cos x) = \cos^{-1}(\cos x) = x$	if	$0 \le x \le \pi$
(*iv*) $\cos(\arccos x) = \cos(\cos^{-1} x) = x$	if	$-1 \le x \le 1$
(*v*) $\arctan(\tan x) = \tan^{-1}(\tan x) = x$	if	$-\pi/2 < x < \pi/2$
(*vi*) $\tan(\arctan x) = \tan(\tan^{-1} x) = x$	if	$-\infty < x < \infty$

EXAMPLE 7   Using the Inverse Properties

Without using a calculator, evaluate:

(a) $\sin^{-1}\left(\sin\dfrac{\pi}{12}\right)$ **(b)** $\cos\left(\cos^{-1}\dfrac{1}{3}\right)$ **(c)** $\tan^{-1}\left(\tan\dfrac{3\pi}{4}\right)$.

Solution In each case we use the properties of the inverse trigonometric functions given in Theorem 4.8.1.

(a) Because $\pi/12$ satisfies $-\pi/2 \le x \le \pi/2$ it follows from property (*i*) that

$$\sin^{-1}\left(\sin\frac{\pi}{12}\right) = \frac{\pi}{12}.$$

(b) By property (*iv*), $\cos\left(\cos^{-1}\frac{1}{3}\right) = \frac{1}{3}$.

(c) In this case we *cannot* apply property (v), since $3\pi/4$ is not in the interval $(-\pi/2, \pi/2)$. If we first evaluate $\tan(3\pi/4) = -1$, then we have

$$\tan^{-1}\left(\tan\frac{3\pi}{4}\right) = \overset{\text{see Example 5}}{\tan^{-1}(-1)} = -\frac{\pi}{4}. \qquad \equiv$$

In the next section we illustrate how inverse trigonometric functions can be used to solve trigonometric equations.

☐ **Postscript—The Other Inverse Trig Functions** The functions $\cot x$, $\sec x$, and $\csc x$ also have inverses when their domains are suitably restricted. See Problems 49–51 in Exercises 4.8. Because these functions are not used as often as arctan, arccos, and arcsin, most scientific calculators do not have keys for them. However, any calculator that computes arcsin, arccos, and arctan can be used to obtain values for **arccsc**, **arcsec**, and **arccot**. Unlike the fact that $\sec x = 1/\cos x$, we note that $\sec^{-1}x \neq 1/\cos^{-1}x$; rather, $\sec^{-1}x = \cos^{-1}(1/x)$ for $|x| \geq 1$. Similar relationships hold for $\csc^{-1}x$ and $\cot^{-1}x$. See Problems 56–58 in Exercises 4.8.

4.8 Exercises   Answers to selected odd-numbered problems begin on page ANS-14.

In Problems 1–14, find the exact value of the given trigonometric expression. Do not use a calculator.

1. $\sin^{-1}0$

2. $\tan^{-1}\sqrt{3}$

3. $\arccos(-1)$

4. $\arcsin\dfrac{\sqrt{3}}{2}$

5. $\arccos\frac{1}{2}$

6. $\arctan\left(-\sqrt{3}\right)$

7. $\sin^{-1}\left(-\dfrac{\sqrt{3}}{2}\right)$

8. $\cos^{-1}\dfrac{\sqrt{3}}{2}$

9. $\tan^{-1}1$

10. $\sin^{-1}\dfrac{\sqrt{2}}{2}$

11. $\arctan\left(-\dfrac{\sqrt{3}}{3}\right)$

12. $\arccos\left(-\frac{1}{2}\right)$

13. $\sin^{-1}\left(-\dfrac{\sqrt{2}}{2}\right)$

14. $\arctan 0$

In Problems 15–32, find the exact value of the given trigonometric expression. Do not use a calculator.

15. $\sin\left(\cos^{-1}\frac{3}{5}\right)$

16. $\cos\left(\sin^{-1}\frac{1}{3}\right)$

17. $\tan\left(\arccos\left(-\frac{2}{3}\right)\right)$

18. $\sin\left(\arctan\frac{1}{4}\right)$

19. $\cos(\arctan(-2))$

20. $\tan\left(\sin^{-1}\left(-\frac{1}{6}\right)\right)$

21. $\csc\left(\sin^{-1}\frac{3}{5}\right)$

22. $\sec(\tan^{-1}4)$

23. $\sin\left(\sin^{-1}\frac{1}{5}\right)$

24. $\cos\left(\cos^{-1}\left(-\frac{4}{5}\right)\right)$

25. $\tan(\tan^{-1}1.2)$

26. $\sin(\arcsin 0.75)$

27. $\arcsin\left(\sin\dfrac{\pi}{16}\right)$

28. $\arccos\left(\cos\dfrac{2\pi}{3}\right)$

29. $\tan^{-1}(\tan\pi)$

30. $\sin^{-1}\left(\sin\dfrac{5\pi}{6}\right)$

31. $\cos^{-1}\left(\cos\left(-\dfrac{\pi}{4}\right)\right)$

32. $\arctan\left(\tan\dfrac{\pi}{7}\right)$

In Problems 33–40, write the given expression as an algebraic expression in x.

33. $\sin(\tan^{-1}x)$

34. $\cos(\tan^{-1}x)$

35. $\tan(\arcsin x)$

36. $\sec(\arccos x)$

37. $\cot(\sin^{-1}x)$

38. $\cos(\sin^{-1}x)$

39. $\csc(\arctan x)$

40. $\tan(\arccos x)$

In Problems 41–48, sketch the graph of the given function.

41. $y = \arctan|x|$

42. $y = \dfrac{\pi}{2} - \arctan x$

43. $y = |\arcsin x|$

44. $y = \sin^{-1}(x + 1)$

45. $y = 2\cos^{-1}x$

46. $y = \cos^{-1}2x$

47. $y = \arccos(x - 1)$

48. $y = \cos(\arcsin x)$

49. The **arccotangent** function can be defined by $y = \operatorname{arccot}x$ (or $y = \cot^{-1}x$) if and only if $x = \cot y$, where $0 < y < \pi$. Graph $y = \operatorname{arccot}x$, and give the domain and the range of this function.

50. The **arccosecant** function can be defined by $y = \operatorname{arccsc}x$ (or $y = \csc^{-1}x$) if and only if $x = \csc y$, where $-\pi/2 \le y \le \pi/2$ and $y \ne 0$. Graph $y = \operatorname{arccsc}x$, and give the domain and the range of this function.

51. One definition of the **arcsecant** function is $y = \operatorname{arcsec}x$ (or $y = \sec^{-1}x$) if and only if $x = \sec y$, where $0 \le y \le \pi$ and $y \ne \pi/2$. (See Problem 52 for an alternative definition.) Graph $y = \operatorname{arcsec}x$, and give the domain and the range of this function.

52. An alternative definition of the arcsecant function can be made by restricting the domain of the secant function to $[0, \pi/2) \cup [\pi, 3\pi/2)$. Under this restriction, define the arcsecant function. Graph $y = \operatorname{arcsec}x$, and give the domain and the range of this function.

53. Using the definition of the arccotangent function from Problem 49, for what values of x is it true that **(a)** $\cot(\operatorname{arccot}x) = x$ and **(b)** $\operatorname{arccot}(\cot x) = x$?

54. Using the definition of the arccosecant function from Problem 50, for what values of x is it true that **(a)** $\csc(\operatorname{arccsc}x) = x$ and **(b)** $\operatorname{arccsc}(\csc x) = x$?

55. Using the definition of the arcsecant function from Problem 51, for what values of x is it true that **(a)** $\sec(\operatorname{arcsec}x) = x$ and **(b)** $\operatorname{arcsec}(\sec x) = x$?

56. Verify that $\operatorname{arccot}x = \dfrac{\pi}{2} - \arctan x$, for all real numbers x.

57. Verify that $\operatorname{arccsc}x = \arcsin(1/x)$ for $|x| \ge 1$.

58. Verify that $\operatorname{arcsec}x = \arccos(1/x)$ for $|x| \ge 1$.

In Problems 59–64, use the results of Problems 56–58 and a calculator to find the indicated value.

59. $\cot^{-1}0.75$

60. $\csc^{-1}(-1.3)$

61. $\operatorname{arccsc}(-1.5)$

62. $\operatorname{arccot}(-0.3)$

63. $\operatorname{arcsec}(-1.2)$

64. $\sec^{-1}2.5$

Miscellaneous Applications

65. **Projectile Motion** The departure angle θ for a bullet to hit a target at a distance R (assuming that the target and the gun are at the same height) satisfies

$$R = \frac{v_0^2 \sin 2\theta}{g},$$

where v_0 is the muzzle velocity and g is the acceleration due to gravity. If the target is 800 ft from the gun and the muzzle velocity is 200 ft/s, find the departure angle. Use $g = 32$ ft/s². [*Hint*: There are two solutions.]

66. **Olympic Sports** For the Olympic event, the hammer throw, it can be shown that the maximum distance is achieved for the release angle θ (measured from the horizontal) that satisfies

$$\cos 2\theta = \frac{gh}{v_0^2 + gh},$$

where h is the height of the hammer above the ground at release, v_0 is the initial velocity, and g is the acceleration due to gravity. For $v_0 = 13.7$ m/s and $h = 2.25$ m, find the optimal release angle. Use $g = 9.81$ m/s².

67. **Highway Design** In the design of highways and railroads, curves are banked to provide centripetal force for safety. The optimal banking angle θ is given by $\tan \theta = v^2/Rg$, where v is the speed of the vehicle, R is the radius of the curve, and g is the acceleration due to gravity. See FIGURE 4.8.9. As the formula indicates, for a given radius there is no one correct angle for all speeds. Consequently, curves are banked for the average speed of the traffic over them. Find the correct banking angle for a curve of radius 600 ft on a country road where speeds average 30 mi/h. Use $g = 32$ ft/s². [*Hint*: Use consistent units.]

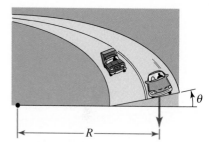

FIGURE 4.8.9 Banked curve in Problem 67

68. **Highway Design—Continued** If μ is the coefficient of friction between the car and the road, then the maximum velocity v_m that a car can travel around a curve without slipping is given by $v_m^2 = gR \tan(\theta + \tan^{-1}\mu)$, where θ is the banking angle of the curve. Find v_m for the country road in Problem 67 if $\mu = 0.26$.

69. **Geology** Viewed from the side, a volcanic cinder cone usually looks like an isosceles trapezoid. See FIGURE 4.8.10. Studies of cinder cones less than 50,000 years old indicate that cone height H_{co} and crater width W_{cr} are related to the cone width W_{co} by the equations $H_{co} = 0.18W_{co}$ and $W_{cr} = 0.40W_{co}$. If $W_{co} = 1.00$, use these equations to determine the base angle ϕ of the trapezoid in Figure 4.8.10.

Volcanic cone

For Discussion

70. Using a calculator set in radian mode, evaluate arctan (tan 1.8), arccos (cos 1.8), and arcsin (sin 1.8). Explain the results.

71. Using a calculator set in radian mode, evaluate $\tan^{-1}(\tan(-1))$, $\cos^{-1}(\cos(-1))$, and $\sin^{-1}(\sin(-1))$. Explain the results.

72. In Section 4.3 we saw that the graphs of $y = \sin x$ and $y = \cos x$ are related by shifting and reflecting. Justify the identity

$$\arcsin x + \arccos x = \frac{\pi}{2},$$

for all x in $[-1, 1]$, by finding a similar relationship between the graphs of $y = \arcsin x$ and $y = \arccos x$.

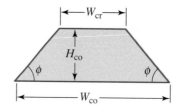

FIGURE 4.8.10 Volcanic cinder cone in Problem 69

73. With a calculator set in radian mode determine which of the following inverse trigonometric evaluations result in an error message: **(a)** $\sin^{-1}(-2)$, **(b)** $\cos^{-1}(-2)$, **(c)** $\tan^{-1}(-2)$. Explain.

74. Discuss: Can any periodic function be one-to-one?

75. Show that $\arcsin \frac{3}{5} + \arcsin \frac{5}{13} = \arcsin \frac{56}{65}$. [*Hint*: See (4) of Section 4.6.]

4.9 Trigonometric Equations

≡ **Introduction** In Section 4.5, 4.6, and 4.7 we considered **trigonometric identities**, which are equations involving trigonometric functions that are satisfied by all values of the variable for which both sides of the equality are defined. In this section we examine **conditional trigonometric equations**, that is, equations that are true for only certain values of the variable. We discuss techniques for finding those values of the variable (if any) that satisfy the equation.

We begin by considering the problem of finding all real numbers x that satisfy $\sin x = \sqrt{2}/2$. Interpreted as the x-coordinates of the points of intersection of the graphs of $y = \sin x$ and $y = \sqrt{2}/2$, **FIGURE 4.9.1** shows that there exists infinitely many solutions of the equation $\sin x = \sqrt{2}/2$:

$$\ldots, -\frac{7\pi}{4}, \frac{\pi}{4}, \frac{9\pi}{4}, \frac{17\pi}{4}, \ldots \tag{1}$$

$$\ldots, -\frac{5\pi}{4}, \frac{3\pi}{4}, \frac{11\pi}{4}, \frac{19\pi}{4}, \ldots \tag{2}$$

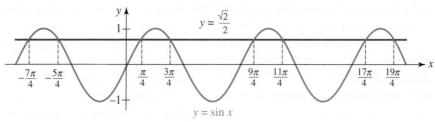

FIGURE 4.9.1 Graphs of $y = \sin x$ and $y = \frac{\sqrt{2}}{2}$

Note that in each of the lists (1) and (2), two successive solutions differ by $2\pi = 8\pi/4$. This is a consequence of the periodicity of the sine function. It is common for trigonometric equations to have an infinite number of solutions because of the periodicity of the trigonometric functions. In general, to obtain solutions of an equation such as $\sin x = \sqrt{2}/2$, it is more convenient to use a unit circle and reference angles rather than a graph of the trigonometric function. We illustrate this approach in the following example.

EXAMPLE 1   Using the Unit Circle

Find all real numbers x satisfying $\sin x = \sqrt{2}/2$.

Solution If $\sin x = \sqrt{2}/2$, the reference angle for x is $\pi/4$ radian. Since the value of $\sin x$ is positive, the terminal side of the angle x lies in either the first or second quadrant. Thus, as shown in **FIGURE 4.9.2**, the only solutions between 0 and 2π are

$$x = \frac{\pi}{4} \quad \text{and} \quad x = \frac{3\pi}{4}.$$

Since the sine function is periodic with period 2π, all of the remaining solutions can be obtained by adding integer multiples of 2π to these solutions. The two solutions are

$$x = \frac{\pi}{4} + 2n\pi \quad \text{and} \quad x = \frac{3\pi}{4} + 2n\pi, \tag{3}$$

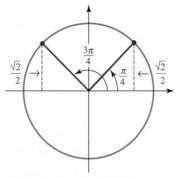

FIGURE 4.9.2 Unit circle in Example 1

CHAPTER 4 TRIGONOMETRIC FUNCTIONS

where n is an integer The numbers that you see in (1) and (2) correspond, respectively, to letting $n = -1$, $n = 0$, $n = 1$, and $n = 2$ in the first and second formulas in (3). ≡

When we are faced with a more complicated equation, such as

$$4\sin^2 x - 8\sin x + 3 = 0,$$

the basic approach is to solve for a single trigonometric function (in this case, it would be $\sin x$) by using methods similar to those for solving algebraic equations.

EXAMPLE 2   Solving a Trigonometric Equation by Factoring

Find all solutions of $4\sin^2 x - 8\sin x + 3 = 0$.

Solution We first observe that this is a quadratic equation in $\sin x$, and that it factors as

$$(2\sin x - 3)(2\sin x - 1) = 0.$$

This implies that either

$$\sin x = \frac{3}{2} \quad \text{or} \quad \sin x = \frac{1}{2}.$$

The first equation has no solution since $|\sin x| \le 1$. As we see in FIGURE 4.9.3 the two angles between 0 and 2π for which $\sin x$ equals $\frac{1}{2}$ are

$$x = \frac{\pi}{6} \quad \text{and} \quad x = \frac{5\pi}{6}.$$

Therefore, by the periodicity of the sine function, the solutions are

$$x = \frac{\pi}{6} + 2n\pi \quad \text{and} \quad x = \frac{5\pi}{6} + 2n\pi,$$

where n is an integer.   ≡

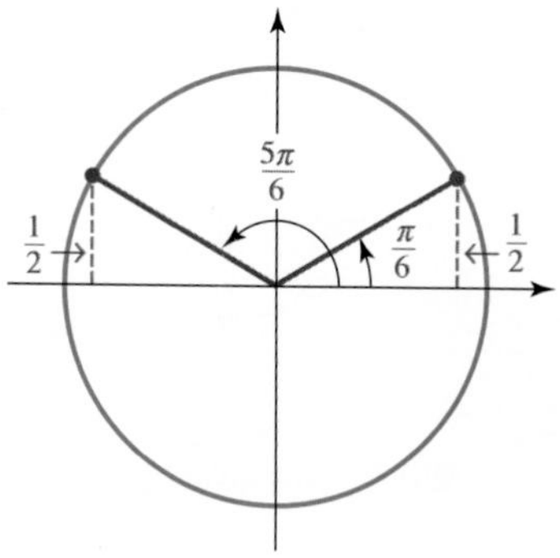

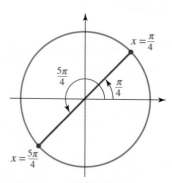

FIGURE 4.9.3 Unit circle in Example 2

EXAMPLE 3   Checking for Lost Solutions

Find all solutions of

$$\sin x = \cos x. \tag{4}$$

Solution In order to work with a single trigonometric function, we divide both sides of the equation by $\cos x$ to obtain

$$\tan x = 1. \tag{5}$$

Equation (5) is equivalent to (4) *provided* that $\cos x \ne 0$. We observe that if $\cos x = 0$, then as we have seen in Section 4.3, $x = (2n + 1)\pi/2 = \pi/2 + n\pi$, for n an integer. By the sum formula for the sine,

◀ $\cos 0 = 1$, $\cos \pi = -1$, $\cos 2\pi = 1$, $\cos 3\pi = -1$, and so on. In general, $\cos n\pi = (-1)^n$, where n is an integer.

$$\sin\!\left(\frac{\pi}{2} + n\pi\right) = \overset{\underset{\text{see (4) of Section 4.6}}{\downarrow}}{\sin\frac{\pi}{2}}\overset{\underset{(-1)^n}{\downarrow}}{\cos n\pi} + \cos\frac{\pi}{2}\overset{\underset{0}{\downarrow}}{\sin n\pi} = (-1)^n \ne 0,$$

we see that these values of x do not satisfy the original equation. Thus we will find *all* the solutions to (4) by solving equation (5).

Now $\tan x = 1$ implies that the reference angle for x is $\pi/4$ radian. Since $\tan x = 1 > 0$, the terminal side of the angle of x radians can lie either in the first or in the third quadrant, as shown in FIGURE 4.9.4. Thus the solutions are

$$x = \frac{\pi}{4} + 2n\pi \quad \text{and} \quad x = \frac{5\pi}{4} + 2n\pi,$$

FIGURE 4.9.4 Unit circle in Example 3

where n is an integer. We can see from Figure 4.9.4 that these two sets of numbers can be written more compactly as

$$x = \frac{\pi}{4} + n\pi,$$

where n is an integer.

This also follows from the fact that $\tan x$ is ▶ π-periodic.

□ **Losing Solutions** When solving an equation, if you divide by an expression containing a variable, you may lose some solutions of the original equation. For example, in algebra a common mistake in solving equations such as $x^2 = x$ is to divide by x to obtain $x = 1$. But by writing $x^2 = x$ as $x^2 - x = 0$ or $x(x - 1) = 0$ we see that in fact $x = 0$ or $x = 1$. To prevent the loss of a solution you must determine the values that make the expression zero and check to see whether they are solutions of the original equation. Note that in Example 3, when we divided by $\cos x$, we took care to check that no solutions were lost.

Whenever possible, it is preferable to avoid dividing by a variable expression. As illustrated with the algebraic equation $x^2 = x$, this can frequently be accomplished by collecting all nonzero terms on one side of the equation and then factoring (something we could not do in Example 3). Example 4 illustrates this technique.

▮EXAMPLE 4   Solving a Trigonometric Equation by Factoring

Solve
$$2\sin x\cos^2 x = -\frac{\sqrt{3}}{2}\cos x. \tag{6}$$

Solution To avoid dividing by $\cos x$, we write the equation as

$$2\sin x\cos^2 x + \frac{\sqrt{3}}{2}\cos x = 0$$

and factor:
$$\cos x\left(2\sin x\cos x + \frac{\sqrt{3}}{2}\right) = 0.$$

Thus either

$$\cos x = 0 \qquad \text{or} \qquad 2\sin x\cos x + \frac{\sqrt{3}}{2} = 0.$$

Since the cosine is zero for all odd multiples of $\pi/2$, the solutions of $\cos x = 0$ are

$$x = (2n + 1)\frac{\pi}{2} = \frac{\pi}{2} + n\pi,$$

where n is an integer.

In the second equation we replace $2\sin x\cos x$ by $\sin 2x$ from the double-angle formula for the sine function to obtain an equation with a single trigonometric function:

See (12) in Section 4.6. ▶

$$\sin 2x + \frac{\sqrt{3}}{2} = 0 \qquad \text{or} \qquad \sin 2x = -\frac{\sqrt{3}}{2}.$$

Thus the reference angle for $2x$ is $\pi/3$. Since the sine is negative, the angle $2x$ must be in either the third quadrant or the fourth quadrant. As **FIGURE 4.9.5** illustrates, either

$$2x = \frac{4\pi}{3} + 2n\pi \qquad \text{or} \qquad 2x = \frac{5\pi}{3} + 2n\pi.$$

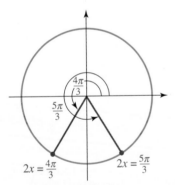

FIGURE 4.9.5 Unit circle in Example 4

Dividing by 2 gives

$$x = \frac{2\pi}{3} + n\pi \quad \text{or} \quad x = \frac{5\pi}{6} + n\pi.$$

Therefore, all solutions of (6) are

$$x = \frac{\pi}{2} + n\pi, \quad x = \frac{2\pi}{3} + n\pi, \quad \text{and} \quad x = \frac{5\pi}{6} + n\pi,$$

where n is an integer.   ≡

In Example 4 had we simplified the equation by dividing by $\cos x$ and not checked to see whether the values of x for which $\cos x = 0$ satisfied equation (6), we would have lost the solutions $x = \pi/2 + n\pi$, where n is an integer.

■ EXAMPLE 5 Using a Trigonometric Identity

Solve $3\cos^2 x - \cos 2x = 1$.

Solution We observe that the given equation involves both the cosine of x and the cosine of $2x$. Consequently, we use the double-angle formula for the cosine in the form

$$\cos 2x = 2\cos^2 x - 1 \qquad \leftarrow \text{See (14) of Section 4.6}$$

to replace the equation by an equivalent equation that involves $\cos x$ only. We find that

$$3\cos^2 x - (2\cos^2 x - 1) = 1 \quad \text{becomes} \quad \cos^2 x = 0.$$

Therefore, $\cos x = 0$, and the solutions are

$$x = (2n + 1)\frac{\pi}{2} = \frac{\pi}{2} + n\pi,$$

where n is an integer.   ≡

We are often interested in finding roots of an equation only in a specified interval.

■ EXAMPLE 6 Using a Trigonometric Identity

Find all solutions of the equation $\cos t - \cos 5t = 0$ in the interval $[0, 2\pi)$.

Solution In this case it is helpful to use a sum-to-product formula. In Example 3 of Section 4.7 we saw that by identifying $x_1 = t$ and $x_2 = 5t$, (9) of Theorem 4.7.2 gives

$$\overset{\displaystyle \sin(-2t) = -\sin 2t}{\downarrow}$$

$$\cos t - \cos 5t = -2\sin\frac{t + 5t}{2}\sin\frac{t - 5t}{2} = -2\sin 3t\sin(-2t) = 2\sin 3t\sin 2t$$

Replacing the sum $\cos t - \cos 5t$ in the given equation by the product $2\sin 3t\sin 2t$ gives the equivalent equation $2\sin 3t\sin 2t = 0$, or

$$\sin 3t\sin 2t = 0.$$

The last equation is satisfied if either $\sin 3t = 0$ or $\sin 2t = 0$. Then from (1) in Section 4.3 we see that $\sin 3t = 0$ implies

$$3t = n\pi, \, n = 0, 1, 2, 3, \ldots \tag{7}$$

whereas $\sin 2t = 0$ implies

$$2t = n\pi, \, n = 1, 2, 3, \ldots \tag{8}$$

If you think in terms of angles measured in radians and the unit circle, then the only angles satisfying the condition that t be in the interval $[0, 2\pi)$ correspond to $n = 1, 2, 3, 4, 5$ in (7) and $n = 1, 2, 3$ in (8):

$$t = 0, \pi/3, 2\pi/3, \pi, 4\pi/3, 5\pi/3, \qquad (9)$$

or
$$t = \pi/2, \pi, 3\pi/2. \qquad (10)$$

The solution set of the original equation is then the union of the two sets defined by the numbers in (9) and (10), that is

$$\{0, \pi/3, \pi/2, 2\pi/3, \pi, 4\pi/3, 3\pi/2, 5\pi/3\}. \qquad \equiv$$

So far in this section we have viewed the variable in the trigonometric equation as representing either a real number or an angle measured in radians. If the variable represents an angle measured in degrees, the technique for solving is the same.

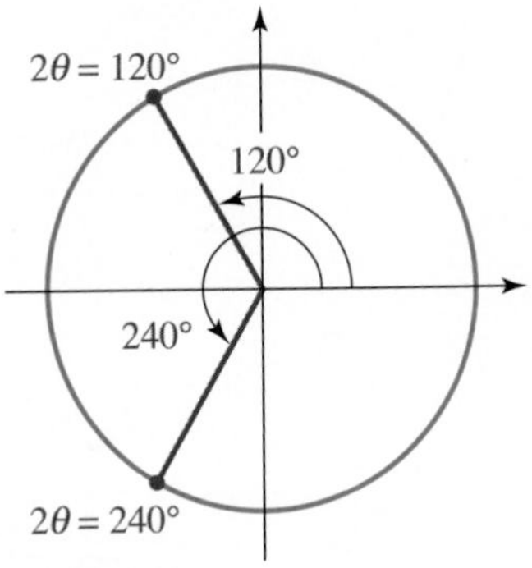

FIGURE 4.9.6 Unit circle in Example 7

EXAMPLE 7 **Equation When the Angle Is in Degrees**

Solve $\cos 2\theta = -\frac{1}{2}$, where θ is an angle measured in degrees.

Solution Since $\cos 2\theta = -\frac{1}{2}$, the reference angle for 2θ is $60°$ and the angle 2θ must be in either the second or the third quadrant. FIGURE 4.9.6 illustrates that either $2\theta = 120°$ or $2\theta = 240°$. Any angle that is coterminal with one of these angles will also satisfy $\cos 2\theta = -\frac{1}{2}$. These angles are obtained by adding any integer multiple of $360°$ to $120°$ or to $240°$:

$$2\theta = 120° + 360°n \qquad \text{or} \qquad 2\theta = 240° + 360°n,$$

where n is an integer. Dividing by 2 the last line yields the two solutions

$$\theta = 60° + 180°n \qquad \text{and} \qquad \theta = 120° + 180°n. \qquad \equiv$$

☐ **Extraneous Solutions** The next example shows that by squaring a trigonometric equation we may introduce extraneous solutions. In other words, the resulting equation after squaring may *not* be equivalent to the original.

EXAMPLE 8 **Extraneous Roots**

Find all solutions of $1 + \tan \alpha = \sec \alpha$, where α is an angle measured in degrees.

Solution The equation does not factor, but we see that if we square both sides, we can use a fundamental identity to obtain an equation involving a single trigonometric function:

$$\begin{aligned}
(1 + \tan \alpha)^2 &= (\sec \alpha)^2 \\
1 + 2\tan \alpha + \tan^2\alpha &= \sec^2\alpha \qquad \leftarrow \text{now use (9) of Section 4.4} \\
1 + 2\tan \alpha + \tan^2\alpha &= 1 + \tan^2\alpha \\
2\tan \alpha &= 0 \\
\tan \alpha &= 0.
\end{aligned}$$

The values of α for $0° \le \alpha < 360°$ at which $\tan \alpha = 0$ are

$$\alpha = 0° \qquad \text{and} \qquad \alpha = 180°.$$

Since we squared each side of the original equation, we may have introduced extraneous solutions. Therefore, it is important that we check all solutions in the original equation. Substituting $\alpha = 0°$ into $1 + \tan\alpha = \sec\alpha$, we obtain the *true* statement $1 + 0 = 1$. But after substituting $\alpha = 180°$, we obtain the *false* statement $1 + 0 = -1$. Therefore, $180°$ is an extraneous solution and $\alpha = 0°$ is the only solution satisfying $0° \le \alpha < 360°$. Thus, all the solutions of the equation are given by

$$\alpha = 0° + 360°n = 360°n,$$

where n is an integer. For $n \ne 0$, these are the angles that are coterminal with $0°$. ≡

Recall from Section 2.1 that to find the x-intercepts of the graph of a function $y = f(x)$ we find the zeros of f, that is, we must solve the equation $f(x) = 0$. The following example makes use of this fact.

EXAMPLE 9   Intercepts of a Graph

Find the first three x-intercepts of the graph of $f(x) = \sin 2x \cos x$ on the positive x-axis.

Solution We must solve $f(x) = 0$, that is, $\sin 2x \cos x = 0$. It follows that either $\sin 2x = 0$ or $\cos x = 0$.

From $\sin 2x = 0$, we obtain $2x = n\pi$, where n is an integer, or $x = n\pi/2$, where n is an integer. From $\cos x = 0$, we find $x = \pi/2 + n\pi$, where n is an integer. Then for $n = 2$, $x = n\pi/2$ gives $x = \pi$, whereas for $n = 0$ and $n = 1$, $x = \pi/2 + n\pi$ gives $x = \pi/2$ and $x = 3\pi/2$, respectively. Thus the first three x-intercepts on the positive x-axis are $(\pi/2, 0)$, $(\pi, 0)$, and $(3\pi/2, 0)$. ≡

☐ **Using Inverse Functions** So far all of the trigonometric equations have had solutions that were related by reference angles to the special angles 0, $\pi/6$, $\pi/4$, $\pi/3$, or $\pi/2$. If this is not the case, we will see in the next example how to use inverse trigonometric functions and a calculator to find solutions.

EXAMPLE 10   Solving Equations Using Inverse Functions

Find the solutions of $4\cos^2 x - 3\cos x - 2 = 0$ in the interval $[0, \pi]$.

Solution We recognize that this is a quadratic equation in $\cos x$. Since the left-hand side of the equation does not readily factor, we apply the quadratic formula to obtain

$$\cos x = \frac{3 \pm \sqrt{41}}{8}.$$

At this point we can discard the value $(3 + \sqrt{41})/8 \approx 1.18$, because $\cos x$ cannot be greater than 1. We then use the inverse cosine function (and the aid of a calculator) to solve the remaining equation:

$$\cos x = \frac{3 - \sqrt{41}}{8} \quad \text{which implies} \quad x = \cos^{-1}\left(\frac{3 - \sqrt{41}}{8}\right) \approx 2.01. ≡$$

Of course in Example 10, had we attempted to compute $\cos^{-1}[(3 + \sqrt{41})/8]$ with a calculator, we would have received an error message.

In Problems 1–6, find all solutions of the given trigonometric equation if x represents an angle measured in radians.

1. $\sin x = \sqrt{3}/2$
2. $\cos x = -\sqrt{2}/2$
3. $\sec x = \sqrt{2}$
4. $\tan x = -1$
5. $\cot x = -\sqrt{3}$
6. $\csc x = 2$

In Problems 7–12, find all solutions of the given trigonometric equation if x represents a real number.

7. $\cos x = -1$
8. $2\sin x = -1$
9. $\tan x = 0$
10. $\sqrt{3}\sec x = 2$
11. $-\csc x = 1$
12. $\sqrt{3}\cot x = 1$

In Problems 13–18, find all solutions of the given trigonometric equation if θ represents an angle measured in degrees.

13. $\csc\theta = 2\sqrt{3}/3$
14. $2\sin\theta = \sqrt{2}$
15. $1 + \cot\theta = 0$
16. $\sqrt{3}\sin\theta = \cos\theta$
17. $\sec\theta = -2$
18. $2\cos\theta + \sqrt{2} = 0$

In Problems 19–46, find all solutions of the given trigonometric equation if x is a real number and θ is an angle measured in degrees.

19. $\cos^2 x - 1 = 0$
20. $2\sin^2 x - 3\sin x + 1 = 0$
21. $3\sec^2 x = \sec x$
22. $\tan^2 x + (\sqrt{3} - 1)\tan x - \sqrt{3} = 0$
23. $2\cos^2\theta - 3\cos\theta - 2 = 0$
24. $2\sin^2\theta - \sin\theta - 1 = 0$
25. $\cot^2\theta + \cot\theta = 0$
26. $2\sin^2\theta + (2 - \sqrt{3})\sin\theta - \sqrt{3} = 0$
27. $\cos 2x = -1$
28. $\sec 2x = 2$
29. $2\sin 3\theta = 1$
30. $\tan 4\theta = -1$
31. $\cot (x/2) = 1$
32. $\csc (\theta/3) = -1$
33. $\sin 2x + \sin x = 0$
34. $\cos 2x + \sin^2 x = 1$
35. $\cos 2\theta = \sin\theta$
36. $\sin 2\theta + 2\sin\theta - 2\cos\theta = 2$
37. $\sin^4 x - 2\sin^2 x + 1 = 0$
38. $\tan^4\theta - 2\sec^2\theta + 3 = 0$
39. $\sec x \sin^2 x = \tan x$
40. $\dfrac{1 + \cos\theta}{\cos\theta} = 2$
41. $\sin\theta + \cos\theta = 1$
42. $\sin x + \cos x = 0$
43. $\sqrt{\dfrac{1 + 2\sin x}{2}} = 1$
44. $\sin x + \sqrt{\sin x} = 0$
45. $\cos\theta - \sqrt{\cos\theta} = 0$
46. $\cos\theta\sqrt{1 + \tan^2\theta} = 1$

In Problems 47–52, use a sum-to-product formula (as in Example 6) to solve the given equation on the indicated interval.

47. $\sin 6t - \sin 4t = 0$, $[0, 2\pi)$
48. $\cos 2t + \cos 3t = 0$, $[0, 2\pi)$
49. $\cos\theta - \cos 4\theta = 0$, $[-\pi, \pi)$
50. $\sin 5\alpha + \sin 3\alpha = 0$, $[0, 2\pi)$
51. $\sin 7x - \sin x - 2\sin 3x = 0$, $(-\pi, \pi)$
52. $\sin x + \cos 2x - \sin 3x = 0$, $[0, 3\pi)$

In Problems 53–60, find the first three x-intercepts of the graph of the given function on the positive x-axis.

53. $f(x) = -5\sin(3x + \pi)$

54. $f(x) = 2\cos\left(x + \dfrac{\pi}{4}\right)$

55. $f(x) = 2 - \sec\dfrac{\pi}{2}x$

56. $f(x) = 1 + \cos\pi x$

57. $f(x) = \sin x + \tan x$

58. $f(x) = 1 - 2\cos\left(x + \dfrac{\pi}{3}\right)$

59. $f(x) = \sin x - \sin 2x$

60. $f(x) = \cos x + \cos 3x$
[*Hint*: Write $3x = x + 2x$.]

In Problems 61–64, by graphing determine whether the given equation has any solutions.

61. $\tan x = x$ [*Hint*: Graph $y = \tan x$ and $y = x$ on the same set of axes.]
62. $\sin x = x$
63. $\cot x - x = 0$

64. $\cos x + x + 1 = 0$

In Problems 65–70, using a inverse trigonometric function find the solutions of the given equation in the indicated interval. Round your answers to two decimal places.

65. $20\cos^2 x + \cos x - 1 = 0,\quad [0, \pi]$
66. $3\sin^2 x - 8\sin x + 4 = 0,\quad [-\pi/2, \pi/2]$
67. $\tan^2 x + \tan x - 1 = 0,\quad (-\pi/2, \pi/2)$
68. $3\sin 2x + \cos x = 0,\quad [-\pi/2, \pi/2]$
69. $5\cos^3 x - 3\cos^2 x - \cos x = 0,\quad [0, \pi]$
70. $\tan^4 x - 3\tan^2 x + 1 = 0,\quad (-\pi/2, \pi/2)$

Miscellaneous Applications

71. Isosceles Triangle From Problem 69 in Exercises 4.6, the area of the isosceles triangle with vertex angle θ as shown in Figure 4.6.4 is given by $A = \frac{1}{2}x^2\sin\theta$. If the length x is 4, what value of θ will give a triangle with area 4?

72. Circular Motion An object travels in a circular path centered at the origin with constant angular speed. The y-coordinate of the object at any time t seconds is given by $y = 8\cos(\pi t - \pi/12)$. At what time(s) does the object cross the x-axis?

73. Mach Number Use Problem 67 in Exercises 4.6 to find the vertex angle of the cone of sound waves made by an airplane flying at Mach 2.

74. Alternating Current An electric generator produces a 60-cycle alternating current given by $I(t) = 30\sin 120\pi\left(t - \frac{7}{36}\right)$, where $I(t)$ is the current in amperes at t seconds. Find the smallest positive value of t for which the current is 15 amperes.

75. Electrical Circuits If the voltage given by $V = V_0\sin(\omega t + \alpha)$ is impressed on a series circuit, an alternating current is produced. If $V_0 = 110$ volts, $\omega = 120\pi$ radians per second, and $\alpha = -\pi/6$, when is the voltage equal to zero?

76. Refraction of Light Consider a ray of light passing from one medium (such as air) into another medium (such as a crystal). Let ϕ be the angle of incidence and θ the angle of refraction. As shown in FIGURE 4.9.7, these angles are measured from a vertical line. According to **Snell's law**, there is a constant c that depends on the two mediums, such that $\dfrac{\sin\phi}{\sin\theta} = c$. Assume that for light passing from air into a crystal, $c = 1.437$. Find ϕ and θ such that the angle of incidence is twice the angle of refraction.

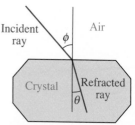

FIGURE 4.9.7 Light rays in Problem 76

71. Snow Cover On the basis of data collected from 1966 to 1980, the extent of snow cover S in the northern hemisphere, measured in millions of square kilometers, can be modeled by the function

$$S(w) = 25 + 21\cos\frac{\pi}{26}(w - 5),$$

where w is the number of weeks past January 1.

   (a) How much snow cover does this formula predict for April Fool's Day? (Round w to the nearest integer.)
   (b) In which week does the formula predict the least amount of snow cover?
   (c) What month does this fall in?

4.10 Simple Harmonic Motion

≡ **Introduction** Many physical objects vibrate or oscillate in a regular manner, repeatedly moving back and forth over a definite time interval. Some examples are clock pendulums, a mass on a spring, sound waves, strings on a guitar when plucked, the human heart, tides, and alternating current. In this section we will focus on mathematical models of the undamped oscillatory motion of a mass on a spring.

Before proceeding with the main discussion we need to discuss the graph of the sum of constant multiples of $\cos Bx$ and $\sin Bx$, that is, $y = c_1 \cos Bx + c_2 \sin Bx$, where c_1 and c_2 are constants.

☐ **Addition of Two Sinusoidal Functions** In Section 4.3 we examined the graphs of horizontally shifted sine and cosine graphs. It turns out that any linear combination of a sine function and a cosine function of the form

$$y = c_1 \cos Bx + c_2 \sin Bx, \tag{1}$$

where c_1 and c_2 are constants, can be expressed either as a shifted sine function $y = A\sin(Bx + \phi)$, $B > 0$, or as a shifted cosine function $y = A\cos(Bx + \phi)$. Note that in (1) the sine and cosine functions have the same period $2\pi/B$.

| EXAMPLE 1 | Addition of a Sine and a Cosine |

Graph the function $y = \cos 2x - \sqrt{3}\sin 2x$.

Solution Using a graphing utility we have shown in FIGURE 4.10.1 four cycles of the graphs of $y = \cos 2x$ (in red) and $y = -\sqrt{3}\sin 2x$ (in green). It is apparent in FIGURE 4.10.2 that the period of the sum of these two functions is π, the common period of $\cos 2x$ and $\sin 2x$. Also apparent is that the blue graph is a horizontally shifted sine (or cosine) function. Although Figure 4.10.2 suggests that the amplitude of the function $y = \cos 2x - \sqrt{3}\sin 2x$ is 2, the exact phase shift of the graph is certainly is *not* apparent.

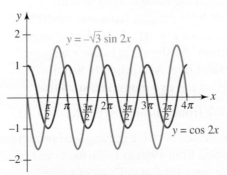

FIGURE 4.10.1 Superimposed graphs of $y = \cos 2x$ and $y = -\sqrt{3}\sin 2x$ in Example 1

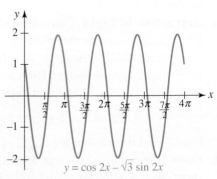

FIGURE 4.10.2 Graph of the sum $y = \cos 2x - \sqrt{3}\sin 2x$ in Example 1

□ **Reduction to a Sine Function** We examine only the reduction of (1) to the form $y = A\sin(Bx + \phi)$, $B > 0$.

◄ The sine form $y = A\sin(Bx + \phi)$ is slightly easier to use than the cosine form $y = A\cos(Bx + \phi)$.

THEOREM 4.10.1 Reduction of (1) to (2)

For real numbers c_1, c_2, B, and x,

$$c_1\cos Bx + c_2\sin Bx = A\sin(Bx + \phi), \tag{2}$$

where A and ϕ are defined by

$$A = \sqrt{c_1^2 + c_2^2}, \tag{3}$$

and

$$\left. \begin{array}{l} \sin\phi = \dfrac{c_1}{A} \\[2mm] \cos\phi = \dfrac{c_2}{A} \end{array} \right\} \tan\phi = \dfrac{c_1}{c_2}. \tag{4}$$

PROOF: To prove (2), we use the sum formula (4) of Section 4.6:

$$\begin{aligned} A\sin(Bx + \phi) &= A\sin Bx\cos\phi + A\cos Bx\sin\phi \\ &= (A\sin\phi)\cos Bx + (A\cos\phi)\sin Bx \\ &= c_1\cos Bx + c_2\sin Bx \end{aligned}$$

and identify $A\sin\phi = c_1$, $A\cos\phi = c_2$. Thus, $\sin\phi = c_1/A = c_1/\sqrt{c_1^2 + c_2^2}$ and $\cos\phi = c_2/A = c_2/\sqrt{c_1^2 + c_2^2}$. ≡

EXAMPLE 2   **Example 1 Revisited**

Express $y = \cos 2x - \sqrt{3}\sin 2x$ as a single sine function.

Solution With the identifications $c_1 = 1$, $c_2 = -\sqrt{3}$, and $B = 2$, we have from (3) and (4),

$$A = \sqrt{c_1^2 + c_2^2} = \sqrt{1^2 + (-\sqrt{3})^2} = \sqrt{4} = 2,$$

$$\left. \begin{array}{l} \sin\phi = \dfrac{1}{2} \\[2mm] \cos\phi = -\dfrac{\sqrt{3}}{2} \end{array} \right\} \tan\phi = -\dfrac{1}{\sqrt{3}}.$$

Although $\tan\phi = -1/\sqrt{3}$ we cannot blindly assume that $\phi = \tan^{-1}(-1/\sqrt{3})$. The angle we take for ϕ must be consistent with the algebraic signs of $\sin\phi$ and $\cos\phi$. Because $\sin\phi > 0$ and $\cos\phi < 0$ the terminal side of the angle ϕ lies in the second quadrant. But since the range of the inverse tangent function is the interval $(-\pi/2, \pi/2)$, $\tan^{-1}(-1/\sqrt{3}) = -\pi/6$ is a fourth-quadrant angle. The correct angle is found by using the reference angle $\pi/6$ for $\tan^{-1}(-1/\sqrt{3})$ to find the second-quadrant angle

$$\phi = \pi - \frac{\pi}{6} = \frac{5\pi}{6} \text{ radians.}$$

Therefore $y = \cos 2x - \sqrt{3}\sin 2x$ can be rewritten as

$$y = 2\sin\left(2x + \frac{5\pi}{6}\right) \quad \text{or} \quad y = 2\sin 2\left(x + \frac{5\pi}{12}\right).$$

Hence the graph of $y = \cos 2x - \sqrt{3}\sin 2x$ is the graph of $y = 2\sin 2x$, which has amplitude 2, period $2\pi/2 = \pi$, and is shifted $5\pi/12$ units to the left. ≡

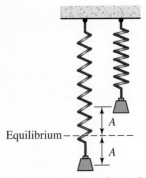

Equilibrium

FIGURE 4.10.3 An undamped spring/mass system exhibits simple harmonic motion

□ **Simple Harmonic Motion** Consider the motion of a mass on a spring as shown in FIGURE 4.10.3. In the absence of frictional or damping forces, a mathematical model for the displacement (or directed distance) of the mass measured from a position called the **equilibrium position** is given by the function

$$y(t) = y_0\cos \omega t + \frac{v_0}{\omega}\sin \omega t. \tag{5}$$

Oscillatory motion modeled by the function (5) is said to be **simple harmonic motion**. More precisely, we have the following definition.

DEFINITION 4.10.1 Simple Harmonic Motion

A point moving on a coordinate line whose position at time t is given by

$$y(t) = A\sin(\omega t + \phi) \quad \text{or} \quad y(t) = A\cos(\omega t + \phi), \tag{6}$$

where A, $\omega > 0$, and ϕ are constants, is said to exhibit **simple harmonic motion**.

Special cases of the trigonometric functions in (6) are $y(t) = A\sin \omega t$, $y(t) = A\cos \omega t$, and $y(t) = c_1\cos \omega t + c_2\sin \omega t$.

□ **Terminology** The function (5) is said to be the **equation of motion** of the mass. Also, in (5), $\omega = \sqrt{k/m}$, where k is the **spring constant** (an indicator of the stiffness of the spring), m is the **mass** attached to the spring (measured in slugs or kilograms), y_0 is the **initial displacement** of the mass (measured above or below the equilibrium position), v_0 is the **initial velocity** of the mass, t is **time** measured in seconds, and the **period** p of motion is $p = 2\pi/\omega$ seconds. The number $f = 1/p = 1/(2\pi/\omega) = \omega/2\pi$ is called the **frequency** of motion. The frequency indicates the number of cycles completed by the graph per unit time. For example, if the period of (5) is, say, $p = 2$ seconds, then we know that one cycle of the function is complete in 2 seconds. The frequency $f = 1/p = \frac{1}{2}$ means one-half of a cycle is complete in 1 second.

In the study of simple harmonic motion it is convenient to recast the equation of motion (5) as a single expression involving only the sine function:

$$y(t) = A\sin(\omega t + \phi). \tag{7}$$

The reduction of (5) to the sine function (7) can be done in exactly the same manner as illustrated in Example 2. In this situation we make the following identifications between (2) and (5):

$$c_1 = y_0, \quad c_2 = v_0/\omega, \quad A = \sqrt{c_1^2 + c_2^2}, \quad \text{and} \quad B = \omega.$$

EXAMPLE 3　　　　　**Equation of Motion**

(a) Find the equation of simple harmonic motion (5) for a spring mass system if $m = \frac{1}{16}$ slug, $y_0 = -\frac{2}{3}$ ft, $k = 4$ lb/ft, and $v_0 = \frac{4}{3}$ ft/s.
(b) Find the period and frequency of motion.

Solution (a) We begin with the simple harmonic motion equation (5). Since $k/m = 4/(\frac{1}{16}) = 64$, $\omega = \sqrt{k/m} = 8$, and $v_0/\omega = (\frac{4}{3})/8 = \frac{1}{6}$, therefore (5) becomes

$$y(t) = -\frac{2}{3}\cos 8t + \frac{1}{6}\sin 8t. \tag{8}$$

(b) The period of motion is $2\pi/8 = \pi/4$ second; the frequency is $4/\pi \approx 1.27$ cycles per second.

EXAMPLE 4   **Example 3 Continued**

Express the equation of motion (8) as a single sine function (7).

Solution With $c_1 = -\frac{2}{3}$, $c_2 = \frac{1}{6}$, we find the amplitude of motion is

$$A = \sqrt{\left(-\tfrac{2}{3}\right)^2 + \left(\tfrac{1}{6}\right)^2} = \tfrac{1}{6}\sqrt{17} \text{ ft.}$$

Then from

$$\left.\begin{array}{l} \sin\phi = -\tfrac{2}{3} \big/ \tfrac{\sqrt{17}}{6} < 0 \\[4pt] \cos\phi = \tfrac{1}{6} \big/ \tfrac{\sqrt{17}}{6} > 0 \end{array}\right\} \tan\phi = -4$$

we can see from algebraic signs $\sin\phi < 0$ and $\cos\phi > 0$ that the terminal side of the angle ϕ lies in the fourth quadrant. Hence the correct value of ϕ is $\tan^{-1}(-4) \approx -1.3258$. The equation of motion is then $y(t) = \frac{1}{6}\sqrt{17}\sin(8t - 1.3258)$. As shown in FIGURE 4.10.4, the amplitude of motion is $A = \sqrt{17}/6 \approx 0.6872$. Since we are assuming that that is no resistance to the motion, once the spring/mass system is set in motion the model indicates it stays in motion bouncing back and forth between its maximum displacement $\sqrt{17}/6$ feet above the equilibrium position and a minimum of $-\sqrt{17}/6$ feet below the equilibrium position.

$$\equiv$$

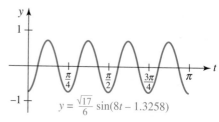

$$y = \frac{\sqrt{17}}{6}\sin(8t - 1.3258)$$

FIGURE 4.10.4 Graph of the equation of motion in Example 4

Only in the two cases, $c_1 > 0$, $c_2 > 0$ or $c_1 < 0$, $c_2 > 0$, can we use $\tan\phi$ in (4) to write $\phi = \tan^{-1}(c_1/c_2)$. (Why?) Correspondingly, ϕ is a first or a fourth-quadrant angle.

4.10 **Exercises** Answers to selected odd-numbered problems begin on page ANS-15.

In Problems 1–6, proceed as in Example 2 and reduce the given trigonometric expression to the form $y = A\sin(Bx + \phi)$. Sketch the graph and give the amplitude, the period, and the phase shift.

1. $y = \cos\pi x - \sin\pi x$

2. $y = \sin\dfrac{\pi}{2}x - \sqrt{3}\cos\dfrac{\pi}{2}x$

3. $y = \sqrt{3}\sin 2x + \cos 2x$

4. $y = \sqrt{3}\cos 4x - \sin 4x$

5. $y = \dfrac{\sqrt{2}}{2}(-\sin x - \cos x)$

6. $y = \sin x + \cos x$

In Problems 7–10, proceed as in Examples 3 and 4 and use the given information to express the equation of simple harmonic motion (5) for a spring/mass system in the trigonometric form (7). Give the amplitude, period, and frequency of motion.

7. $m = \frac{1}{4}$ slug, $y_0 = \frac{1}{2}$ ft, $k = 1$ lb/ft, and $v_0 = \frac{3}{2}$ ft/s

8. $m = 1.6$ slug, $y_0 = -\frac{1}{3}$ ft, $k = 40$ lb/ft, and $v_0 = -\frac{5}{4}$ ft/s

9. $m = 1$ slug, $y_0 = -1$ ft, $k = 16$ lb/ft, and $v_0 = -2$ ft/s

10. $m = 2$ slug, $y_0 = -\frac{2}{3}$ ft, $k = 200$ lb/ft, and $v_0 = 5$ ft/s

11. The equation of simple harmonic motion of a spring/mass system is $y(t) = \frac{5}{2}\sin(2t - \pi/3)$. Determine the initial displacement y_0 and initial velocity v_0 of the mass. [*Hint*: Use (5).]

12. Use the equation of simple harmonic motion of the spring/mass system given in Problem 11 to find the times for which the mass passes through the equilibrium position $y = 0$.

13. Find the first three x-intercepts of the function $y = \cos 2x - \sqrt{3}\sin 2x$ on the positive x-axis. See Figure 4.10.2.

Miscellaneous Applications

14. Electrical Circuits Under certain conditions the current $I(t)$ in an electrical circuit at time t is given by $I(t) = I_0[\sin(\omega t + \theta)\cos\phi + \cos(\omega t + \theta)\sin\phi]$. Express $I(t)$ as a single sine function of the form given in (7). [*Hint*: Review the sum formula in (4) of Theorem 4.6.2.]

4.11 The Limit Concept Revisited

∫**Calculus** **PREVIEW**

≡ **Introduction** As we saw in Section 2.9, the fundamental motivating problem of differential calculus, *find a tangent line to the graph of the function*, is answered by the concept of a *limit*. In that section we purposely kept the discussion about limits at an intuitive level; our emphasis was on reviewing the appropriate algebra, such as factoring and rationalization, necessary to be able to compute a limit analytically. In the study of the calculus of the trigonometric functions you will, of course, be expected to compute limits involving trigonometric functions. As the examples in this section will illustrate, computation of trigonometric limits entail both algebraic manipulations and knowledge of basic trigonometric identities.

We begin with a fundamental limit result for the sine function.

☐ **An Important Trigonometric Limit** To do the calculus of the trigonometric functions, $\sin x$, $\cos x$, $\tan x$, and so on, it is important to realize that the variable x is a real number or an angle x measured in radians. With that in mind, consider the numerical values of $(\sin x)/x$ as x approaches 0 from the right $(x \to 0^+)$ given in the table that follows.

$x \to 0^+$	0.1	0.01	0.001	0.0001
$\dfrac{\sin x}{x}$	0.99833416	0.99998333	0.99999983	0.99999999

It is easy to see that the same results given in the table hold as $x \to 0^-$. Because $\sin x$ is an odd function, for $x > 0$ and $-x < 0$ we have $\sin(-x) = -\sin x$ and as a consequence $\dfrac{\sin(-x)}{-x} = \dfrac{\sin x}{x}$. In other words, when the value of x is small in absolute value

$$\frac{\sin x}{x} \approx 1.$$

While numerical calculations such as this do not constitute a proof, they do suggest that $\dfrac{\sin x}{x} \to 1$ as $x \to 0$. Using the limit symbol, we have motivated the following result

$$\lim_{x \to 0} \frac{\sin x}{x} = 1. \tag{1}$$

See Problem 26 in Exercises 5.1 for a guided tour through the basic steps of a proof of (1) that is usually presented in calculus.

Important ▶   In this discussion we make the same assumption that we did in Sections 1.5 and 2.9, namely, that all limits under consideration actually exist. Everything that we do—

algebraic manipulations, taking limits of products and quotients in the examples in this section—is predicated on this assumption.

Other limits of importance are

$$\lim_{x \to a} \sin x = \sin a, \tag{2}$$

$$\lim_{x \to a} \cos x = \cos a. \tag{3}$$

The results (2) and (3) are immediate consequences of the fact that $f(x) = \sin x$ and $g(x) = \cos x$ are continuous functions for all x. As we have seen in Section 4.3 the graphs of $\sin x$ and $\cos x$ are smooth and unbroken. For example, from (2),

$$\lim_{x \to \pi/6} \sin x = \sin\frac{\pi}{6} = \frac{1}{2} \tag{4}$$

and

$$\lim_{x \to 0} \sin x = \sin 0 = 0.$$

Also, from (3),

$$\lim_{x \to 0} \cos x = \cos 0 = 1. \tag{5}$$

The results in (1), (2), and (3) are used often to compute other limits. As in Section 1.5 many of the limits considered in this section are limits of fractional expressions where *both* the numerator and the denominator are approaching 0. Recall, these kinds of limits are said to have the **indeterminate form 0/0**. Note that the limit (1) is of this indeterminate form.

■ EXAMPLE 1   Using (1)

Find $\lim\limits_{x \to 0} \dfrac{10x - 3\sin x}{x}$.

Solution We rewrite the fractional expression as two fractions with the same denominator x:

$$\lim_{x \to 0} \frac{10x - 3\sin x}{x} = \lim_{x \to 0}\left[\frac{10x}{x} - \frac{3\sin x}{x} \right]$$

$$= \lim_{x \to 0} \frac{10x}{x} - 3\lim_{x \to 0}\frac{\sin x}{x} \qquad \leftarrow \text{cancel the } x \text{ in the first expression}$$

$$= \lim_{x \to 0} 10 - 3\lim_{x \to 0}\frac{\sin x}{x} \qquad \leftarrow \text{now use (1)}$$

$$= 10 - 3 \cdot 1$$

$$= 7. \qquad\qquad\qquad\qquad\qquad \equiv$$

■ EXAMPLE 2   Using the Double-Angle Formula

Find $\lim\limits_{x \to 0} \dfrac{\sin 2x}{x}$.

Solution To evaluate the given limit, we make use of the double-angle formula $\sin 2x = 2\sin x \cos x$ of Section 4.5 and the results in (1) and (5):

$$\overset{\text{from (5)} \quad \text{from (1)}}{\underset{\downarrow\qquad\downarrow}{}}$$

$$\lim_{x \to 0}\frac{\sin 2x}{x} = \lim_{x \to 0}\frac{2\cos x \sin x}{x} = 2\lim_{x \to 0}\cos x \cdot \frac{\sin x}{x} = 2 \cdot 1 \cdot 1 = 2.$$

Thus,

$$\lim_{x \to 0}\frac{\sin 2x}{x} = 2. \tag{6} \quad \equiv$$

☐ **Using a Substitution** We are often interested in limits similar to that considered in Example 2. But if we wish to find, say, $\lim\limits_{x \to 0} \dfrac{\sin 5x}{x}$, the procedure employed in Example 2 breaks down at a practical level since we have not developed a trigonometric identity for $\sin 5x$. There is an alternative procedure that allows us to quickly find $\lim\limits_{x \to 0} \dfrac{\sin kx}{x}$, where $k \neq 0$ is any real constant, by simply changing the variable by means of a **substitution**. If we let $t = kx$, then $x = t/k$. Notice that as $x \to 0$ then necessarily $t \to 0$. Thus we can write

$$\lim_{x \to 0} \frac{\sin kx}{x} = \lim_{t \to 0} \frac{\sin t}{t/k} = \lim_{t \to 0} \frac{\sin t}{1} \cdot \frac{k}{t} = k \overset{\underset{\text{this limit is 1 from (1)}}{\downarrow}}{\lim_{t \to 0} \frac{\sin t}{t}} = k.$$

Thus we have proved the general result

$$\lim_{x \to 0} \frac{\sin kx}{x} = k. \tag{7}$$

Hence $\lim\limits_{x \to 0} \dfrac{\sin 5x}{x} = 5$. See Problem 25 in Exercises 4.11.

▮ EXAMPLE 3 Trigonometric Limit

Find $\lim\limits_{x \to 0} \dfrac{\tan x}{x}$.

Solution Using the definition $\tan x = \sin x / \cos x$ we can write

$$\lim_{x \to 0} \frac{\tan x}{x} = \lim_{x \to 0} \frac{\dfrac{\sin x}{\cos x}}{x} = \lim_{x \to 0} \frac{1}{\cos x} \cdot \frac{\sin x}{x} = \lim_{x \to 0} \frac{1}{\cos x} \cdot \frac{\sin x}{x}.$$

From (5) and (1) we know that $\cos x \to 1$ and $(\sin x)/x \to 1$ as $x \to 0$, and so the preceding line becomes

$$\lim_{x \to 0} \frac{\tan x}{x} = \frac{1}{1} \cdot 1 = 1. \qquad \equiv$$

▮ EXAMPLE 4 Using a Pythagorean Identity

Find $\lim\limits_{x \to 0} \dfrac{1 - \cos x}{x}$.

Solution To compute this limit we start with a bit of algebraic cleverness by multiplying the numerator and denominator by the conjugate factor of the numerator. Next we use the fundamental Pythagorean identity $\sin^2 x + \cos^2 x = 1$ in the form $1 - \cos^2 x = \sin^2 x$:

$$\lim_{x \to 0} \frac{1 - \cos x}{x} = \lim_{x \to 0} \frac{1 - \cos x}{x} \cdot \frac{1 + \cos x}{1 + \cos x}$$

$$= \lim_{x \to 0} \frac{1 - \cos^2 x}{x(1 + \cos x)}$$

$$= \lim_{x \to 0} \frac{\sin^2 x}{x(1 + \cos x)}.$$

For the next step we resort back to algebra to rewrite the fractional expression as a product, then use the results in (1), (4), and (5):

$$\lim_{x \to 0} \frac{1 - \cos x}{x} = \lim_{x \to 0} \frac{\sin^2 x}{x(1 + \cos x)}$$

$$= \lim_{x \to 0} \frac{\sin x}{x} \cdot \frac{\sin x}{1 + \cos x}$$

$$= 1 \cdot \frac{0}{2} \qquad (8)$$

$$= 0.$$

That is,
$$\lim_{x \to 0} \frac{1 - \cos x}{x} = 0. \qquad \equiv$$

From (8) we obtain a limit result that is used in calculus to find the derivatives of the sine and cosine functions. Since the limit in (8) is equal to 0, we can write

$$\lim_{x \to 0} \frac{1 - \cos x}{x} = \lim_{x \to 0} \frac{-(\cos x - 1)}{x} = (-1)\lim_{x \to 0} \frac{\cos x - 1}{x} = 0.$$

Dividing by -1 then gives

$$\lim_{x \to 0} \frac{\cos x - 1}{x} = 0. \qquad (9)$$

☐ **The Calculus Connection** In Section 2.9 we saw that the derivative of a function $y = f(x)$ is the function $f'(x)$ defined by a limit of a difference quotient:

$$f'(x) = \lim_{h \to 0} \frac{f(x + h) - f(x)}{h}. \qquad (10)$$

In computing this limit we shrink h to zero but x is held fixed. Recall too, if a number $x = a$ is in the domains of f and f', then $f(a)$ is the y-coordinate of the point of tangency $(a, f(a))$ and $f'(a)$ is the slope of the tangent line at that point.

☐ **Derivatives of $f(x) = \sin x$ and $f(x) = \cos x$** To find the derivative of $f(x) = \sin x$ we use the four-step process illustrated in Example 3 of Section 2.9. In the first step we use from Section 4.6 the sum formula for the sine function:

$$\sin(x_1 + x_2) = \sin x_1 \cos x_2 + \cos x_1 \sin x_2. \qquad (11)$$

(*i*) With x and h playing the parts of x_1 and x_2, we have from (11):

$$f(x + h) = \sin(x + h) = \sin x \cos h + \cos x \sin h.$$

(*ii*) $f(x + h) - f(x) = \sin x \cos h + \cos x \sin h - \sin x$
$$= \sin x(\cos h - 1) + \cos x \sin h$$

As we see in the next line, we cannot cancel the h's in the difference quotient but we can rewrite the expression to make use of the limit results in (1) and (9).

(*iii*) $\dfrac{f(x + h) - f(x)}{h} = \dfrac{\sin x(\cos h - 1) + \cos x \sin h}{h}$

$$= \sin x \frac{\cos h - 1}{h} + \cos x \frac{\sin h}{h}$$

(*iv*) In this line, the symbol h plays the part of the symbol x in (1) and (9):

$$f'(x) = \lim_{h \to 0} \frac{f(x+h) - f(x)}{h} = \sin x \lim_{h \to 0} \frac{\cos h - 1}{h} + \cos x \lim_{h \to 0} \frac{\sin h}{h}.$$

From the limit results in (1) and (9), the last line is the same as

$$f'(x) = \lim_{h \to 0} \frac{f(x+h) - f(x)}{h} = \sin x \cdot 0 + \cos x \cdot 1 = \cos x.$$

In summary:

- the derivative of $f(x) = \sin x$ is $f'(x) = \cos x$.   (12)

It is left to you, the student, to show that

- the derivative of $f(x) = \cos x$ is $f'(x) = -\sin x$.   (13)

See Problems 23 and 24 in Exercises 4.11.

▮ EXAMPLE 5   Equation of a Tangent Line

Find an equation of the tangent line to the graph of $f(x) = \sin x$ at $x = 4\pi/3$.

Solution We start by finding the point of tangency. From

$$f\left(\frac{4\pi}{3}\right) = \sin\frac{4\pi}{3} = -\frac{\sqrt{3}}{2}$$

we see that the point of tangency is $\left(4\pi/3, -\sqrt{3}/2\right)$. The slope of the tangent line at that point is the derivative of $f(x) = \sin x$ evaluated at the x-coordinate. From (12) we know that $f'(x) = \cos x$ and so the slope at $\left(4\pi/3, -\sqrt{3}/2\right)$ is

$$f'\left(\frac{4\pi}{3}\right) = \cos\frac{4\pi}{3} = -\frac{1}{2}.$$

From the point-slope form of a line, an equation of the tangent line is

$$y + \frac{\sqrt{3}}{2} = -\frac{1}{2}\left(x - \frac{4\pi}{3}\right) \quad \text{or} \quad y = -\frac{1}{2}x + \frac{2\pi}{3} - \frac{\sqrt{3}}{2}.$$

See **FIGURE 4.11.1**.

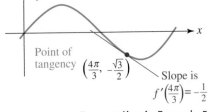

FIGURE 4.11.1 Tangent line in Example 5

In Problems 1–18, use the results in (1), (2), (3), (7), and (9) to find the indicated limit.

1. $\displaystyle\lim_{x \to 0} \frac{\sin\frac{1}{2}x}{x}$

2. $\displaystyle\lim_{x \to 0} \frac{\sin \pi x}{x}$

3. $\displaystyle\lim_{\theta \to 0} \frac{\sin(-\theta)}{\theta}$

4. $\displaystyle\lim_{t \to 0} \frac{\sin 3t}{4t}$

5. $\displaystyle\lim_{x \to 5\pi/6} \cos x$

6. $\displaystyle\lim_{x \to \pi/4} \sin x$

7. $\displaystyle\lim_{x \to \pi/2} (\cos x + 5\sin x)$

8. $\displaystyle\lim_{x \to \pi/6} \cos x \sin x$

9. $\displaystyle\lim_{x \to 0} \frac{\cos x - 1}{10x}$

10. $\displaystyle\lim_{\theta \to 0} \frac{8(1 - \cos\theta)}{\theta}$

11. $\displaystyle\lim_{x\to 0}\frac{4x^2-2\sin x}{x}$

12. $\displaystyle\lim_{x\to 0}\frac{2\sin 4x+1-\cos x}{x}$

13. $\displaystyle\lim_{x\to 0}\frac{\sin^2 x}{x}$

14. $\displaystyle\lim_{x\to 0}\frac{\sin^2 x}{x^2}$

15. $\displaystyle\lim_{x\to \pi/2}\frac{\cos x}{\cot x}$

16. $\displaystyle\lim_{x\to 0}\frac{\cos x\tan x}{x}$

17. $\displaystyle\lim_{x\to 0} x\cot x$

18. $\displaystyle\lim_{x\to \pi/4}\frac{\cos 2x}{\cos x-\sin x}$

In Problems 19–22, proceed as in Example 5 to find an equation of the tangent line to the graph of $f(x)=\sin x$ at the indicated value of x.

19. $x=0$

20. $x=\pi/2$

21. $x=\pi/6$

22. $x=2\pi/3$

23. Proceed as on page 271–272 and find the derivative of $f(x)=\cos x$.

24. Use the result of Problem 23 to find an equation of the tangent line to the graph of $f(x)=\cos x$ at $x=\pi/3$.

25. Use the facts that

$$\lim_{x\to 0}\frac{\cos 5x-1}{x}=0 \quad \text{and} \quad \lim_{x\to 0}\frac{\sin 5x}{x}=5$$

to find the derivative of $f(x)=\sin 5x$.

26. Use the result of Problem 25 to find an equation of the tangent line to the graph of $f(x)=\sin 5x$ at $x=\pi$.

Calculator/Computer Problems

In Problems 27 and 28, use a calculator or computer to estimate the given limit by completing each table. Round the entries in each table to eight decimal places.

27. $\displaystyle\lim_{x\to 0}\frac{1-\cos x}{x^2}$

$x\to 0^+$	0.1	0.01	0.001	0.0001	0.00001
$\dfrac{1-\cos x}{x^2}$					

Explain why we do not have to consider $x\to 0^-$.

28. $\displaystyle\lim_{x\to 2}\frac{x^2-4}{\sin(x-2)}$

$x\to 2^+$	2.1	2.01	2.001	2.0001	2.00001
$\dfrac{x^2-4}{\sin(x-2)}$					

$x\to 2^-$	1.9	1.99	1.999	1.9999	1.99999
$\dfrac{x^2-4}{\sin(x-2)}$					

For Discussion

In Problems 29–36, discuss how to use the result in (1) along with some clever algebra, trigonometry, or a substitution to find the given limit.

29. $\lim\limits_{x \to 0} \dfrac{x}{\sin 3x}$

30. $\lim\limits_{x \to 0} \dfrac{\sin 4x}{\sin 5x}$

31. $\lim\limits_{x \to 0} \dfrac{\sin x^2}{x^2}$

32. $\lim\limits_{x \to \pi} \dfrac{\sin x}{\pi - x}$

33. $\lim\limits_{x \to 0} \dfrac{x^2}{1 - \cos x}$

34. $\lim\limits_{x \to 0} \dfrac{\cos\left(x + \frac{1}{2}\pi\right)}{x}$

35. $\lim\limits_{x \to 0^+} \dfrac{\sin x}{\sqrt{x}}$

36. $\lim\limits_{x \to 1} \dfrac{\sin(x - 1)}{x^2 + 2x - 3}$

37. Using what you have learned in Problems 29 and 36, find the limit

$$\lim_{x \to 2} \frac{x^2 - 4}{\sin(x - 2)}$$

without the aid of the numerical table in Problem 28.

38. (a) Use a calculator to complete the following table.

$x \to 0^+$	0.1	0.01	0.001	0.0001	0.00001
$\dfrac{1 - \cos x^2}{x^4}$					

(b) Find the limit $\lim\limits_{x \to 0} \dfrac{1 - \cos x^2}{x^4}$ using the method given in Example 4.

(c) Discuss any differences that you observe between parts (a) and (b).

CHAPTER 4 **Review Exercises** Answers to selected odd-numbered problems begin on page ANS-16.

A. Fill in the Blanks

In Problems 1–25, fill in the blanks.

1. $\pi/5$ radians = _____ degrees.

2. 10 degrees = _____ radians.

3. The exact values of the coordinates of the point $P(t)$ on the unit circle corresponding to $t = 5\pi/6$ are _____.

4. The reference angle for $4\pi/3$ radians is _____ radians.

5. $\tan\dfrac{\pi}{3}$ = _____.

6. In standard position, the terminal side of the angle $8\pi/5$ radians lies in the _____ quadrant.

7. If $\sin\theta = -\frac{1}{3}$ and θ is in quadrant IV, then $\sec\theta$ = _____.

8. If $\tan t = 2$ and t is in quadrant III, then $\cos t$ = _____.

9. The y-intercept for the graph of the function $y = 2\sec(x + \pi)$ is _____.

10. The values of t in the interval $[0, 2\pi]$ that satisfy $\sin 2t = \frac{1}{2}$ are _____.

11. If $\sin u = \frac{3}{5}, 0 < u < \pi/2$, and $\cos v = 1/\sqrt{5}, 3\pi/2 < v < 2\pi$, then $\cos(u + v) = $ _____.

12. If $\cos t = -\frac{2}{3}, \pi < t < 3\pi/2$, then $\cos\frac{1}{2}t = $ _____.

13. A sine function with period 1 is _____.

14. The first vertical asymptote for the graph of $y = \tan\left(x - \dfrac{\pi}{4}\right)$ to the right of the y-axis is _____.

15. $\sin t + \cos t = $ _____ $\sin\left(t + \dfrac{\pi}{4}\right)$.

16. If $\sin t = \frac{1}{6}$, then $\cos\left(t - \dfrac{\pi}{2}\right) = $ _____.

17. The amplitude of $y = -10\cos\left(\dfrac{\pi}{3}x\right)$ is _____.

18. $\cos\left(\dfrac{\pi}{6} - \dfrac{5\pi}{4}\right) = $ _____

19. The exact value of $\arccos\left(\cos\dfrac{9\pi}{5}\right) = $ _____.

20. The period of the function $y = 2\sin\left(-\dfrac{\pi}{3}t\right)$ is _____.

21. The amplitude of the function $y = \sin x + 2\sqrt{2}\cos x$ is _____.

22. If $P(t) = \left(-\frac{1}{3}, \frac{2\sqrt{2}}{3}\right)$ is a point on the unit circle, then $\sin 2t = $ _____.

23. The fifth x-intercept of the graph of $y = \sin \pi x$ on the positive x-axis is _____.

24. The exact value of $\cos 70° \cos 40° + \sin 70° \sin 40°$ is _____.

25. $\lim\limits_{x \to 0} \dfrac{\sin 10x}{x} = $ _____

B. True/False

In Problems 1–25, answer true or false.

1. $\cos^2\theta - \sin^2\theta = 1 - 2\sin^2\theta$ _____

2. $\sin^2 55° + \sin^2 35° = 1$ _____

3. $\sec(-\pi) = \csc\left(\dfrac{3\pi}{2}\right)$ _____

4. There is no angle t such that $\sec t = \frac{1}{2}$. _____

5. $\sin(2\pi - t) = -\sin t$ _____

6. $1 + \sec^2\theta = \tan^2\theta$ _____

7. $\left(\frac{3}{2}, 0\right)$ is an x-intercept of the graph of $y = 3\cos \pi x$. _____

8. $\left(2\pi/3, -1/\sqrt{3}\right)$ is a point on the graph of $y = \cot x$. _____

9. The range of the function $y = \csc x$ is $(-\infty, -1] \cup [1, \infty)$. _____

10. The graph of $y = \csc x$ does not intersect the y-axis. _____

11. The line $x = \pi/2$ is a vertical asymptote for the graph of $y = \tan x$. _____

12. If $\tan(x + \pi) = 0.3$, then $\tan x = 0.3$. _____

13. For the sine function $y = -2\sin x$ we have $-2 \le y \le 2$. _____

14. $\sin 6x = 2\sin 3x\cos 3x$ _____

15. The graph of $y = \sin(2x - \pi/3)$ is the graph of $y = \sin 2x$ shifted $\pi/3$ units to the right. _____

16. Since $\tan(5\pi/4) = 1$, then $\arctan(1) = 5\pi/4$. _____

17. $\arccos\left(-\frac{1}{2}\right) = 2\pi/3$ _____

18. $f(x) = \arcsin x$ is not periodic. _____

19. $f(x) = x\sin x$ is 2π periodic. _____
20. $f(x) = \sin(\cos x)$ is an even function. _____
21. $\tan 8\pi = \tan 5\pi$ _____
22. The graph of $f(x) = 4\cos\dfrac{3x}{2}\sin 6x$ passes through the origin. _____
23. $\cos^2 15° - \sin^2 15° = \frac{1}{2}$ _____
24. If $\cos 210° = -\frac{1}{2}\sqrt{3}$, then $\cos 105° = -\frac{1}{4}\sqrt{3}$. _____
25. If $0 < x < \pi$, then necessarily $\cos\dfrac{x}{2} > 0$. _____

C. Review Exercises

In Problems 1–4, give two examples of the indicated trigonometric function such that each has the given properties.

1. sine function with period 4 and amplitude 6
2. cosine function with period π, amplitude 4, and phase shift $\frac{1}{2}$
3. sine function with period $\pi/2$, amplitude 3, and phase shift $\pi/4$
4. tangent function whose graph completes one cycle on the interval $(-\pi/8, \pi/8)$

In Problems 5–14, find all t in the interval $[0, 2\pi]$ that satisfy the given equation.

5. $\cos t\sin t - \cos t + \sin t - 1 = 0$ **6.** $\cos t - \sin t = 0$
7. $4\sin^2 t - 1 = 0$ **8.** $\sin t = 2\tan t$
9. $\cos 4x = -1$ **10.** $\tan t - 3\cot t = 2$
11. $-\sin 2t + \sin 4t = 0$ **12.** $\cos 9t - \cos 3t = 0$
13. $\sin t\cos t = \frac{1}{2}$ **14.** $\tan t = 4$

In Problems 15–22, find the indicated value without using a calculator.

15. $\cos^{-1}\left(-\frac{1}{2}\right)$ **16.** $\arcsin(-1)$
17. $\cot\left(\cos^{-1}\left(\frac{3}{4}\right)\right)$ **18.** $\cos\left(\arcsin \frac{2}{5}\right)$
19. $\sin^{-1}(\sin\pi)$ **20.** $\cos(\arccos 0.42)$
21. $\sin\left(\arccos \left(\frac{5}{13}\right)\right)$ **22.** $\arctan(\cos\pi)$

In Problems 23 and 24, write the given expression as an algebraic expression in x.

23. $\sin(\arccos x)$ **24.** $\sec(\tan^{-1}x)$

In Problems 25–28, the given graph can be interpreted as a rigid/nonrigid transformation of the graph of $y = \sin x$ and of the graph of $y = \cos x$. Find an equation of the graph using the sine function. Then find an equation of the same graph using the cosine function.

25.

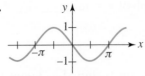

FIGURE 4.R.1 Graph for Problem 25

26.

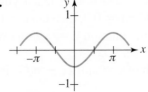

FIGURE 4.R.2 Graph for Problem 26

27.

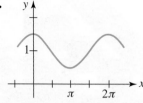

FIGURE 4.R.3 Graph for Problem 27

28.

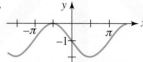

FIGURE 4.R.4 Graph for Problem 28

In Problems 29 and 30, verify the given trigonometric identity.

29. $(\tan x - \sec x)^2 = \dfrac{1 - \sin x}{1 + \sin x}$

30. $\cos^4 t + 1 - \sin^4 t = 2\cos^2 t$

5 Triangle Trigonometry

Chapter Outline

Right Triangle Trigonometry

≡ **Introduction** The word *trigonometry* (from the Greek *trigonon* meaning "triangle" and *metria* meaning "measurement") refers to the measurement of triangles. In Section 4.2 we defined the trigonometric functions using coordinates of points on the unit circle and by using radian measure we were able to define the trigonometric functions of any angle. In this section we will show that the trigonometric functions of an acute angle in a right triangle have an equivalent definition in terms of the lengths of the sides of the triangle.

☐ **Terminology** In FIGURE 5.1.1(a) we have drawn a right triangle with sides labeled a, b, and c (indicating their respective lengths) and one of the acute angles denoted by θ. From the Pythagorean theorem we know that $a^2 + b^2 = c^2$. The side opposite the right angle is called the **hypotenuse**; the remaining sides are referred to as the **legs** of the triangle. The legs labeled a and b are, in turn, said to be the side **adjacent** to the angle θ and the side **opposite** the angle θ. We will also use the abbreviations **hyp**, **adj**, and **opp** to denote the lengths of these sides.

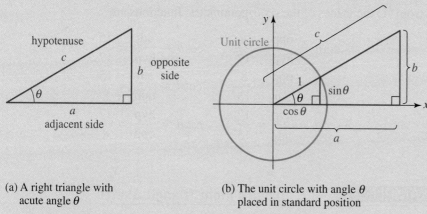

(a) A right triangle with acute angle θ

(b) The unit circle with angle θ placed in standard position

FIGURE 5.1.1 In (a) and (b) the right triangles are the same

If we place θ in standard position and draw a unit circle centered at the origin, we see from Figure 5.1.1(b) that there are two similar right triangles containing the same angle θ. Since corresponding sides of similar triangles are proportional, it follows that

$$\frac{\sin\theta}{1} = \frac{b}{c} = \frac{\text{opp}}{\text{hyp}} \quad \text{and} \quad \frac{\cos\theta}{1} = \frac{a}{c} = \frac{\text{adj}}{\text{hyp}}.$$

Also, we have

$$\frac{\tan\theta}{1} = \frac{\sin\theta}{\cos\theta} = \frac{b/c}{a/c} = \frac{b}{a} = \frac{\text{opp}}{\text{adj}}.$$

Then, applying the reciprocal identities (12) in Section 4.4, each trigonometric function of θ can be written as the ratio of the lengths of the sides of a right triangle as follows. See FIGURE 5.1.2.

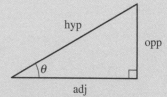

FIGURE 5.1.2 Defining the trigonometric functions of θ

DEFINITION 5.1.1 Trigonometric Functions of θ in a Right Triangle

For an acute angle θ in a right triangle as shown in Figure 5.1.2,

$$\begin{aligned} \sin\theta &= \frac{\text{opp}}{\text{hyp}} & \cos\theta &= \frac{\text{adj}}{\text{hyp}} \\ \tan\theta &= \frac{\text{opp}}{\text{adj}} & \cot\theta &= \frac{\text{adj}}{\text{opp}} \\ \sec\theta &= \frac{\text{hyp}}{\text{adj}} & \csc\theta &= \frac{\text{hyp}}{\text{opp}} \end{aligned} \tag{1}$$

EXAMPLE 1 Values of the Six Trigonometric Functions

Find the exact values of the six trigonometric functions of the angle θ in the right triangle shown in FIGURE 5.1.3.

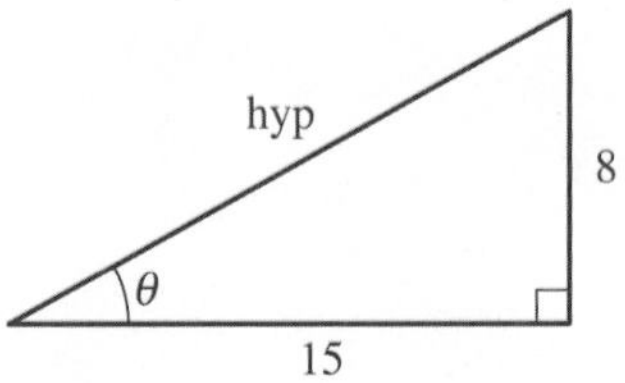

FIGURE 5.1.3 Right triangle in Example 1

Solution From Figure 5.1.3 we see that the side opposite θ has length 8 and the side adjacent has length 15. From the Pythagorean theorem the hypotenuse c is

$$c^2 = 8^2 + 15^2 = 289 \qquad \text{and so} \qquad c = \sqrt{289} = 17.$$

Thus from (1) the values of the six trigonometric functions are

$$\begin{aligned} \sin\theta &= \frac{\text{opp}}{\text{hyp}} = \frac{8}{17}, & \cos\theta &= \frac{\text{adj}}{\text{hyp}} = \frac{15}{17}, \\ \tan\theta &= \frac{\text{opp}}{\text{adj}} = \frac{8}{15}, & \cot\theta &= \frac{\text{adj}}{\text{opp}} = \frac{15}{8}, \\ \sec\theta &= \frac{\text{hyp}}{\text{adj}} = \frac{17}{15}, & \csc\theta &= \frac{\text{hyp}}{\text{opp}} = \frac{17}{8}. \end{aligned}$$

≡

EXAMPLE 2 Using a Right Triangle Sketch

If θ is an acute angle and $\sin\theta = \frac{2}{7}$, find the values of the other trigonometric functions of θ.

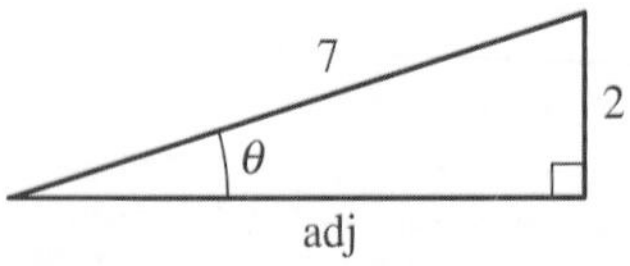

FIGURE 5.1.4 Right triangle in Example 2

Solution We sketch a right triangle with an acute angle θ satisfying $\sin\theta = \frac{2}{7}$, by making opp $= 2$ and hyp $= 7$ as shown in FIGURE 5.1.4. From the Pythagorean theorem we have

$$2^2 + (\text{adj})^2 = 7^2 \qquad \text{so that} \qquad (\text{adj})^2 = 7^2 - 2^2 = 45.$$

Thus,

$$\text{adj} = \sqrt{45} = 3\sqrt{5}.$$

The values of the remaining five trigonometric functions are obtained from the definitions in (1):

$$\begin{aligned} \cos\theta &= \frac{\text{adj}}{\text{hyp}} = \frac{3\sqrt{5}}{7}, & \sec\theta &= \frac{\text{hyp}}{\text{adj}} = \frac{7}{3\sqrt{5}} = \frac{7\sqrt{5}}{15}, \\ \tan\theta &= \frac{\text{opp}}{\text{adj}} = \frac{2}{3\sqrt{5}} = \frac{2\sqrt{5}}{15}, & \cot\theta &= \frac{\text{adj}}{\text{opp}} = \frac{3\sqrt{5}}{2}, \end{aligned}$$

$$\csc\theta = \frac{\text{hyp}}{\text{opp}} = \frac{7}{2}.$$

≡

□ **Solving Right Triangles** Applications of right triangle trigonometry in fields such as surveying and navigation involve **solving right triangles**. The expression "to solve a triangle" means that we wish to find the length of each side and the measure of each angle in the triangle. We can solve any right triangle if we know either two sides or one acute angle and one side. As the following examples will show, sketching and labeling the triangle is an essential part of the solution process. It will be our general practice to label a right triangle as shown in FIGURE 5.1.5. The three vertices will be denoted by A, B, and C, with C at the vertex of the right angle. We denote the angles at A and B by α and β and the lengths of the sides opposite these angles by a and b, respectively. The length of the side opposite the right angle at C is denoted by c.

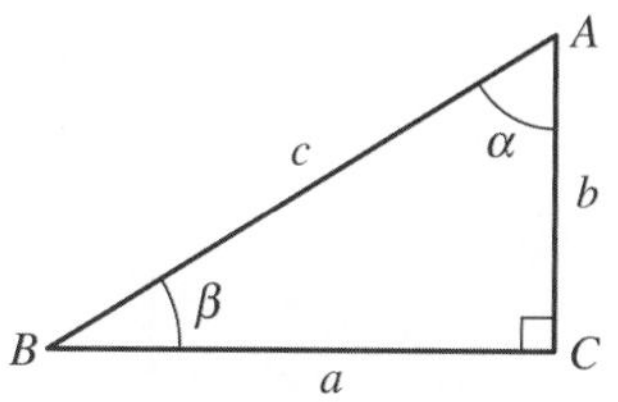

FIGURE 5.1.5 Standard labeling for a right triangle

EXAMPLE 3 Solving a Right Triangle

Solve the right triangle having a hypotenuse of length $4\sqrt{3}$ and one 60° angle.

Solution First we make a sketch of the triangle and label it as shown in FIGURE 5.1.6. We wish to find a, b, and β. Since α and β are complementary angles, $\alpha + \beta = 90°$ yields

$$\beta = 90° - \alpha = 90° - 60° = 30°.$$

We are given the length of the hypotenuse, namely, hyp $= 4\sqrt{3}$. To find a, the length of the side opposite the angle $\alpha = 60°$, we select the sine function. From $\sin\alpha = $ opp/hyp, we obtain

$$\sin 60° = \frac{a}{4\sqrt{3}} \qquad \text{or} \qquad a = 4\sqrt{3}\sin 60°.$$

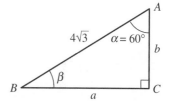

FIGURE 5.1.6 Right triangle in Example 3

Since $\sin 60° = \sqrt{3}/2$, we have

$$a = 4\sqrt{3}\sin 60° = 4\sqrt{3}\left(\frac{\sqrt{3}}{2}\right) = 6.$$

Finally, to find the length b of the side adjacent to the 60° angle, we select the cosine function. From $\cos\alpha = $ adj/hyp, we obtain

$$\cos 60° = \frac{b}{4\sqrt{3}} \qquad \text{or} \qquad b = 4\sqrt{3}\cos 60°.$$

Because $\cos 60° = \frac{1}{2}$, we find

$$b = 4\sqrt{3}\cos 60° = 4\sqrt{3}\left(\tfrac{1}{2}\right) = 2\sqrt{3}. \qquad \equiv$$

In Example 3 once we determined a, we could have found b by using either the Pythagorean theorem or the tangent function. In general, there are usually several ways to solve a triangle.

□ **Use of a Calculator** If angles other than 30°, 45°, or 60° are involved in a problem, we can obtain approximations of the desired trigonometric function values with a calculator. For the remainder of this chapter, whenever an approximation is used, we will round the final results to the nearest hundredth unless the problem specifies otherwise. To take full advantage of the calculator's accuracy, store the computed values of the trigonometric functions in the calculator for subsequent calculations. If, instead, a rounded version of a displayed value is written down and then later keyed back into the calculator, the accuracy of the final result may be diminished.

EXAMPLE 4   **Solving a Right Triangle**

Solve the right triangle with legs of length 4 and 5.

Solution After sketching and labeling the triangle as shown in FIGURE 5.1.7, we see that we need to find c, α, and β. From the Pythagorean theorem, the hypotenuse c is given by

$$c = \sqrt{5^2 + 4^2} = \sqrt{41} \approx 6.40.$$

To find β, we use $\tan\beta = $ opp/adj. (By choosing to work with the given quantities, we avoid error due to previous approximations.) Thus we have

$$\tan\beta = \tfrac{4}{5} = 0.8.$$

From a calculator set in degree mode, we find $\beta \approx 38.66°$. Since $\alpha = 90° - \beta$, we obtain $\alpha \approx 51.34°$. ≡

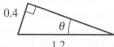

FIGURE 5.1.7 Right triangle in Example 4

| **5.1** | Exercises | Answers to selected odd-numbered problems begin on page ANS-16. |

In Problems 1–10, find the values of the six trigonometric functions of the angle θ in the given triangle.

1.

FIGURE 5.1.8 Triangle for Problem 1

2.
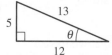
FIGURE 5.1.9 Triangle for Problem 2

3.

FIGURE 5.1.10 Triangle for Problem 3

4.
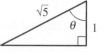
FIGURE 5.1.11 Triangle for Problem 4

5.
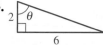
FIGURE 5.1.12 Triangle for Problem 5

6.
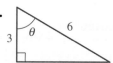
FIGURE 5.1.13 Triangle for Problem 6

7. 0.4
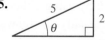
FIGURE 5.1.14 Triangle for Problem 7

8.

FIGURE 5.1.15 Triangle for Problem 8

9.

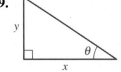

FIGURE 5.1.16 Triangle for Problem 9

10.

FIGURE 5.1.17 Triangle for Problem 10

In Problems 11–22, find the indicated unknowns. Each problem refers to the triangle shown in **FIGURE 5.1.18**.

11. $a = 4, \beta = 27°; b, c$
12. $c = 10, \beta = 49°; a, b$
13. $b = 8, \beta = 34.33°; a, c$
14. $c = 25, \alpha = 50°; a, b$
15. $b = 1.5, c = 3; \alpha, \beta, a$
16. $a = 5, b = 2; \alpha, \beta, c$
17. $a = 4, b = 10; \alpha, \beta, c$
18. $b = 4, \alpha = 58°; a, c$
19. $a = 9, c = 12; \alpha, \beta, b$
20. $b = 3, c = 6; \alpha, \beta, a$
21. $b = 20, \alpha = 23°; a, c$
22. $a = 11, \alpha = 33.5°; b, c$

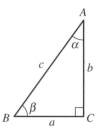

FIGURE 5.1.18 Triangle for Problems 11–22

In Problems 23 and 24, solve for x in the given triangle.

23.

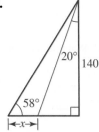

FIGURE 5.1.19 Triangle for Problem 23

24.

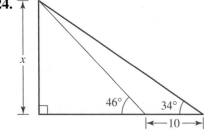

FIGURE 5.1.20 Triangle for Problem 24

For Discussion

For Problems 25 and 26, you should be familiar with the concepts and notation used in Section 4.11.

25. (a) A regular n-gon is an n-sided polygon inscribed in a circle; the polygon is formed by n equally spaced points on the circle. Suppose the polygon shown in **FIGURE 5.1.21** represents a regular n-gon inscribed in a circle of radius r. Use right triangle trigonometry to show that the area $A(n)$ of the n-gon is given by

$$A(n) = \frac{n}{2}r^2 \sin\left(\frac{2\pi}{n}\right).$$

(b) It stands to reason that the area $A(n)$ approaches the area of the circle as the number of sides of the n-gon increases. Compute A_{100} and A_{1000}.
(c) Let $x = 2\pi/n$ in $A(n)$ and note that as $n \to \infty$ then $x \to 0$. Use (1) of Section 4.11 to show that $\lim_{n\to\infty} A(n) = \pi r^2$.

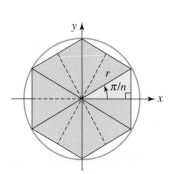

FIGURE 5.1.21 Inscribed n-gon in Problem 25

26. Consider a circle centered at the origin O with radius 1. As shown in **FIGURE 5.1.22(a)**, let the shaded region OPR be a sector of the circle with central angle t such that $0 < t < \pi/2$. We see from Figures 5.1.22(b)–(d) that

$$\text{area of } \triangle OPR < \text{area of sector } OPR < \text{area of } \triangle OQR. \quad (2)$$

(a) Use right triangle trigonometry to show that the area of $\triangle OPR$ is $\frac{1}{2}\sin t$ and that the area of $\triangle OQR$ is $\frac{1}{2}\tan t$.

(b) Since the area of a sector of a circle is $\frac{1}{2}r^2\theta$, where r is its radius and θ is measured in radians, it follows that the area of sector OPR is $\frac{1}{2}t$. Use this result, along with the areas in part (a), to show that the inequality in (2) yields

$$\cos t < \frac{\sin t}{t} < 1.$$

(c) Discuss how the preceding inequality proves (1) of Section 4.11 when we let $t \to 0^+$.

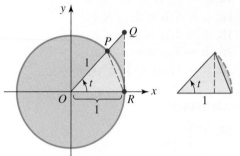

(a) Unit circle   (b) Triangle OPR (c) Sector OPR (d) Right triangle OQR

FIGURE 5.1.22 Unit circle in Problem 26

27. For the points $P(x, y)$ and $Q(x_1, 0)$, $x_1 < x$, in **FIGURE 5.1.23**, show that

$$d(O, Q) + d(Q, P) = y\tan\frac{\alpha}{2} + x.$$

[*Hint:* Use the identity in Problem 59 in Exercises 4.6.]

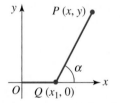

FIGURE 5.1.23 Points in Problem 27

5.2 Applications of Right Triangles

☰ **Introduction** Right triangle trigonometry can be used to solve many practical problems, particularly those involving lengths, heights, and distances.

EXAMPLE 1   Finding the Height of a Tree

A kite is caught in the top branches of a tree. If the 90-ft kite string makes an angle of 22° with the ground, estimate the height of the tree by finding the distance from the kite to the ground.

Solution Let h denote the height of the kite. From **FIGURE 5.2.1** we see that

$$\frac{h}{90} = \sin 22° \quad \text{or} \quad h = 90\sin 22°.$$

A calculator set in degree mode gives $h \approx 33.71$ ft.   ☰

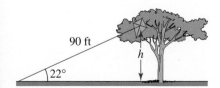

FIGURE 5.2.1 Tree in Example 1

EXAMPLE 2   Length of a Saw Cut

A carpenter cuts the end of a 4-in.-wide board on a 25° bevel from the vertical, starting at a point $1\frac{1}{2}$ in. from the end of the board. Find the lengths of the diagonal cut and the remaining side. See FIGURE 5.2.2.

Solution Let x, y, and z be the (unknown) dimensions, as labeled in Figure 5.2.2. It follows from the definition of the tangent function that

$$\frac{x}{4} = \tan 25° \qquad \text{so therefore} \qquad x = 4\tan 25° \approx 1.87 \text{ in.}$$

To find y we observe that

$$\frac{4}{y} = \cos 25° \qquad \text{so} \qquad y = \frac{4}{\cos 25°} \approx 4.41 \text{ in.}$$

Since $z = \frac{3}{2} + x$ and $x \approx 1.87$ in., we see that $z \approx 1.5 + 1.87 \approx 3.37$ in. $\equiv$

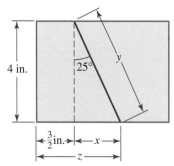

FIGURE 5.2.2 Saw cut in Example 2

☐ **Angles of Elevation and Depression** The angle between an observer's line of sight to an object and the horizontal is given a special name. As FIGURE 5.2.3 illustrates, if the line of sight is to an object above the horizontal, the angle is called an **angle of elevation**, whereas if the line of sight is to an object below the horizontal, the angle is called an **angle of depression**.

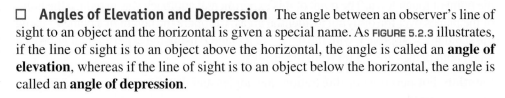

FIGURE 5.2.3 Angles of elevation and depression

EXAMPLE 3   Using Angles of Elevation

A surveyor uses an instrument called a theodolite to measure the angle of elevation between ground level and the top of a mountain. At one point the angle of elevation is measured to be 41°. A half kilometer farther from the base of the mountain, the angle of elevation is measured to be 37°. How high is the mountain?

Solution Let h represent the height of the mountain. FIGURE 5.2.4 shows that there are two right triangles sharing the common side h, so we obtain two equations in two unknowns z and h:

$$\frac{h}{z + 0.5} = \tan 37° \qquad \text{and} \qquad \frac{h}{z} = \tan 41°.$$

We can solve each of these for h, obtaining, respectively,

$$h = (z + 0.5)\tan 37° \qquad \text{and} \qquad h = z\tan 41°.$$

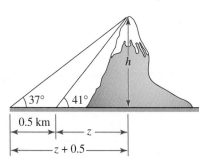

FIGURE 5.2.4 Mountain in Example 3

Equating the last two results gives an equation from which we can determine the distance z:

$$(z + 0.5)\tan 37° = z\tan 41°.$$

Solving for z gives us

$$z = \frac{-0.5\tan 37^\circ}{\tan 37^\circ - \tan 41^\circ}.$$

Using $h = z\tan 41^\circ$ we find the height h of the mountain to be

$$h = \frac{-0.5\tan 37^\circ \tan 41^\circ}{\tan 37^\circ - \tan 41^\circ} \approx 2.83 \text{ km.}$$

≡

EXAMPLE 4　　Glide Path

Most airplanes approach San Francisco International Airport (SFO) on a straight 3° glide path starting at a point 5.5 mi from the field. A few years ago, the FAA experimented with a computerized two-segment approach where a plane approaches the field on a 6° glide path starting at a point 5.5 mi out and then switches to a 3° glide path 1.5 mi from the point of touchdown. The point of this experimental approach was to reduce the noise of the planes over the outlying residential areas. Compare the height of a plane P' using the standard 3° approach with the height of a plane P using the experimental approach when both planes are 5.5 mi from the airport.

Solution For purposes of illustration, the angles and distances shown in FIGURE 5.2.5 are exaggerated.

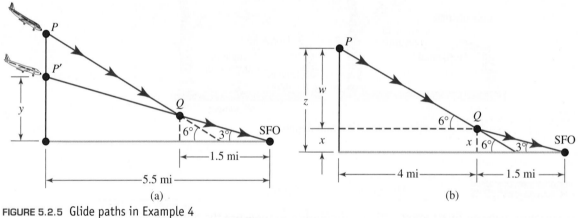

FIGURE 5.2.5 Glide paths in Example 4

First, suppose y is the height of plane P' on the standard approach when it is 5.5 mi out from the airport. As we see in Figure 5.2.5(a),

$$\frac{y}{5.5} = \tan 3^\circ \qquad \text{or} \qquad y = 5.5\tan 3^\circ.$$

Because distances from the airport are measured in miles, we convert y to feet

$$y = 5.5(5280)\tan 3^\circ \text{ ft} \approx 1522 \text{ ft.}$$

Now, suppose z is the height of plane P on the experimental approach when it is 5.5 mi out from the airport. As shown in Figure 5.2.5(b), $z = x + w$, so we use two right triangles to obtain

$$\frac{x}{1.5} = \tan 3^\circ \qquad \text{or} \qquad x = 1.5\tan 3^\circ.$$

and

$$\frac{w}{4} = \tan 6^\circ \qquad \text{or} \qquad w = 4\tan 6^\circ.$$

Hence the approximate height of plane P at a point 5.5 mi out from the airport is

$$
\begin{aligned}
z &= x + w \\
&= 1.5\tan 3^\circ + 4\tan 6^\circ \\
&= 1.5(5280)\tan 3^\circ + 4(5280)\tan 6^\circ \approx 2635 \text{ ft.}
\end{aligned}
$$

In other words, plane P is approximately 1113 ft higher than plane P'. ≡

☐ **Building a Function** Section 2.8 was devoted to setting up or constructing functions that were described or expressed in words. As emphasized in that section, this is a task that you will surely face in a course in calculus. Our final example illustrates a recommended procedure of sketching a figure and labeling quantities of interest with appropriate variables.

EXAMPLE 5 Functions That Involve Trigonometry

A plane flying horizontally at an altitude of 2 miles approaches a radar station as shown in FIGURE 5.2.6.

(a) Express the distance d between the plane and the radar station as a function of the angle of elevation θ.

(b) Express the angle of elevation θ of the plane as a function of the horizontal separation x between the plane and the radar station.

Solution As shown in Figure 5.2.6, θ is an acute angle in a right triangle.

(a) We can relate the distance d and the angle θ by $\sin \theta = 2/d$. Solving for d gives

$$
d(\theta) = \frac{2}{\sin \theta} \quad \text{or} \quad d(\theta) = 2\csc \theta,
$$

where $0 < \theta \le 90^\circ$.

(b) The horizontal separation x and θ are related by $\tan \theta = 2/x$. We make use of the inverse tangent function to solve for θ:

$$
\theta(x) = \tan^{-1}\frac{2}{x},
$$

where $0 < x < \infty$. ≡

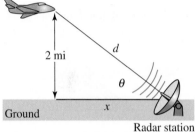
FIGURE 5.2.6 Plane in Example 5

5.2 Exercises Answers to selected odd-numbered problems begin on page ANS-16.

1. A building casts a shadow 20 m long. If the angle from the tip of the shadow to a point on top of the building is 69°, how high is the building?
2. Two trees are on opposite sides of a river, as shown in FIGURE 5.2.7. A baseline of 100 ft is measured from tree T_1, and from that position the angle β to T_2 is measured to be 29.7°. If the baseline is perpendicular to the line segment between T_1 and T_2, find the distance between the two trees.
3. A 50-ft tower is located on the edge of a river. The angle of elevation between the opposite bank and the top of the tower is 37°. How wide is the river?
4. A surveyor uses a geodometer to measure the straight-line distance from a point on the ground to a point on top of a mountain. Use the information given in FIGURE 5.2.8 to find the height of the mountain.
5. An observer on the roof of building A measures a 27° angle of depression between the horizontal and the base of building B. The angle of elevation from

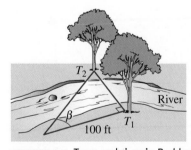
FIGURE 5.2.7 Trees and river in Problem 2

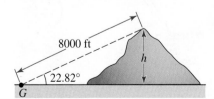
FIGURE 5.2.8 Mountain in Problem 4

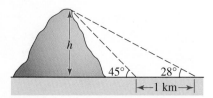

FIGURE 5.2.9 Mountain in Problem 6

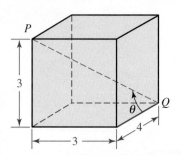

FIGURE 5.2.10 Box in Problem 10

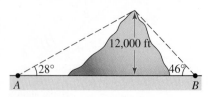

FIGURE 5.2.11 Mountain in Problem 11

the same point to the roof of the second building is 41.42°. What is the height of building B if the height of building A is 150 ft? Assume buildings A and B are on the same horizontal plane.

6. Find the height h of a mountain using the information given in FIGURE 5.2.9.

7. The top of a 20-ft ladder is leaning against the edge of the roof of a house. If the angle of inclination of the ladder from the horizontal is 51°, what is the approximate height of the house and how far is the bottom of the ladder from the base of the house?

8. An airplane flying horizontally at an altitude of 25,000 ft approaches a radar station located on a 2000-ft-high hill. At one instant in time, the angle between the radar dish pointed at the plane and the horizontal is 57°. What is the straight-line distance in miles between the airplane and the radar station at that particular instant?

9. A 5-mi straight segment of a road climbs a 4000-ft hill. Determine the angle that the road makes with the horizontal.

10. A box has dimensions as shown in FIGURE 5.2.10. Find the length of the diagonal between the corners P and Q. What is the angle θ formed between the diagonal and the bottom edge of the box?

11. Observers in two towns A and B on either side of a 12,000-ft mountain measure the angles of elevation between the ground and the top of the mountain. See FIGURE 5.2.11. Assuming that the towns and the mountaintop lie in the same vertical plane, find the horizontal distance between them.

12. A drawbridge* measures 7.5 m from shore to shore, and when completely open it makes an angle of 43° with the horizontal. See FIGURE 5.2.12(a). When the bridge is closed, the angle of depression from the shore to a point on the surface of the water below the opposite end is 27°. See Figure 5.2.12(b). When the bridge is fully open, what is the distance d between the highest point of the bridge and the water below?

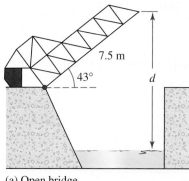

(a) Open bridge

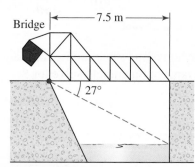

(b) Closed bridge

FIGURE 5.2.12 Drawbridge in Problem 12

13. A flagpole is located at the edge of a sheer 50-ft cliff at the bank of a river of width 40 ft. See FIGURE 5.2.13. An observer on the opposite side of the river measures an angle of 9° between her line of sight to the top of the flagpole and her line of sight to the top of the cliff. Find the height of the flagpole.

14. From an observation site 1000 ft from the base of Mt. Rushmore the angle of elevation to the top of the sculpted head of George Washington is measured to be 80.05°, whereas the angle of elevation to the bottom of his head is 79.946°. Determine the height of George Washington's head.

Bust of George Washington on Mt. Rushmore

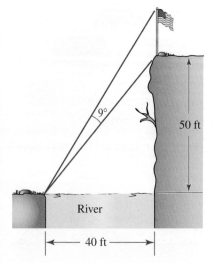

FIGURE 5.2.13 Flagpole in Problem 13

---

*The drawbridge shown in Figure 5.2.12, where the span is continuously balanced by a counterweight, is called a *bascule* bridge.

15. The length of a Boeing 747 airplane is 231 ft. What is the plane's altitude if it subtends an angle of $2°$ when it is directly above an observer on the ground? See FIGURE 5.2.14.

16. The height of a gnomon (pin) of a sundial is 4 in. If it casts a 6-in. shadow, what is the angle of elevation of the Sun?

17. Weather radar is capable of measuring both the angle of elevation to the top of a thunderstorm and its range (the horizontal distance to the storm). If the range of a storm is 90 km and the angle of elevation is $4°$, can a passenger plane that is able to climb to 10 km fly over the storm?

Sundial

18. Cloud ceiling is the lowest altitude at which solid cloud is present. The cloud ceiling at airports must be sufficiently high for safe takeoffs and landings. At night the cloud ceiling can be determined by illuminating the base of the clouds with a searchlight pointed vertically upward. If an observer is 1 km from the searchlight and the angle of elevation to the base of the illuminated cloud is $8°$, find the cloud ceiling. See FIGURE 5.2.15. (During the day cloud ceilings are generally estimated by sight. However, if an accurate reading is required, a balloon is inflated so that it will rise at a known constant rate. Then it is released and timed until it disappears into the cloud. The cloud ceiling is determined by multiplying the rate by the time of the ascent; trigonometry is not required for this calculation.)

19. Assuming that the Earth is a sphere, show that $C_\theta = C_e \cos \theta$, where C_θ is the circumference of the parallel of latitude at the latitude angle θ and C_e is the Earth's circumference at the equator. See FIGURE 5.2.16. [*Hint*: $R \cos \theta = r$.]

20. Use Problem 19 and the fact that the radius R of the Earth is 6400 km to find:
    (a) the circumference of the Arctic Circle, which lies at $66°33'$ N $(66.55°$ N) latitude, and
    (b) the distance "around the world" at the $58°40'$ N $(58.67°$ N) latitude.

21. The distance between the Earth and the Moon varies as the Moon revolves around the Earth. At a particular time the **geocentric parallax** angle shown in FIGURE 5.2.17 is measured to be $1°$. Calculate to the nearest hundred miles the distance between the center of the Earth and the center of the Moon at this instant. Assume that the radius of the Earth is 3963 miles.

22. The final length of a volcanic lava flow seems to decrease as the elevation of the lava vent from which it originates increases. An empirical study of Mt. Etna gives the final lava flow length L in terms of elevation h by the formula

$$L = 23 - 0.0053h,$$

where L is measured in kilometers and h is measured in meters. Suppose that a Sicilian village at elevation 750 m is on a $10°$ slope directly below a lava vent at 2500 m. See FIGURE 5.2.18. According to the formula, how close will the lava flow get to the village?

23. As shown in FIGURE 5.2.19, two tracking stations S_1 and S_2 sight a weather balloon between them at elevation angles α and β, respectively. Express the height h of the balloon in terms of α and β, and the distance c between the tracking stations.

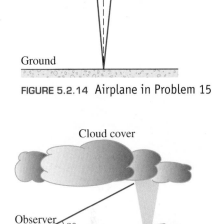
Ground

FIGURE 5.2.14 Airplane in Problem 15

Cloud cover

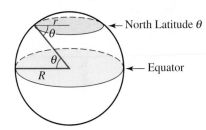
Observer
$8°$
Searchlight
$\vdash$—1 km—$\dashv$

FIGURE 5.2.15 Searchlight in Problem 18

r
θ
North Latitude θ
θ
R
Equator

FIGURE 5.2.16 Earth in Problem 19

3963 mi
Observer
Moon
$1°$
d
Earth

FIGURE 5.2.17 Angle in Problem 21

Mt. Etna

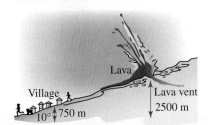
Lava
Village
Lava vent
$10°$ 750 m
2500 m

FIGURE 5.2.18 Lava flow in Problem 22

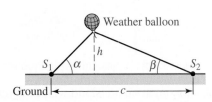
Weather balloon
h
S_1 α
β S_2
Ground $\vdash$—c—$\dashv$

FIGURE 5.2.19 Weather balloon in Problem 23

24. An entry in a soapbox derby rolls down a hill. Using the information given in FIGURE 5.2.20, find the total distance $d_1 + d_2$ that the soapbox travels.

25. Find the height and area of the isosceles trapezoid shown in FIGURE 5.2.21.

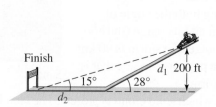

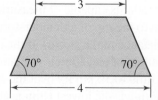

FIGURE 5.2.20 Soapbox in Problem 24   FIGURE 5.2.21 Trapezoid in Problem 25

26. An escalator between the first and second floors of a department store is 58 ft long and makes an angle of $20°$ with the first floor. See FIGURE 5.2.22. Find the vertical distance between the floors.

27. Recent History According to the online encyclopedia *Wikipedia*, a French helicopter flown by Jean Boulet attained the world's record height of 12,442 m in 1972. What would the angle of elevation to the helicopter have been from a point P on the ground 2000 m from the point directly beneath the helicopter?

28. Ancient History In an article from the online encyclopedia *Wikipedia*, the height h of the Lighthouse of Alexandria, one of the Seven Wonders of the Ancient World built between 280 and 247 B.C.E., is estimated to have been between 393 ft and 450 ft. The article goes on to say that there are ancient claims that the light could be seen on the ocean up to 29 miles away. Use the right triangle in FIGURE 5.2.23 along with the two given heights h to determine the accuracy of the 29 mile claim. Assume that the radius of the Earth is $r = 3963$ mi and s is distance measured in miles on the ocean. [*Hint*: Use 1 ft $= 1/5280$ mi and (7) of Section 4.1.]

Artist's rendering of the Lighthouse of Alexandria

In Problems 29–32, proceed as in Example 5 and translate the words into an appropriate function.

29. A tracking telescope, located 1.25 km from the point of a rocket launch, follows a vertically ascending rocket. Express the height h of the rocket as a function of the angle of elevation θ.

30. A searchlight one-half mile offshore illuminates a point P on the shore. Express the distance d from the searchlight to the point of illumination P as a function of the angle θ shown in FIGURE 5.2.24.

31. A statue is placed on a pedestal as shown in FIGURE 5.2.25. Express the viewing angle θ as a function of the distance x from the pedestal.

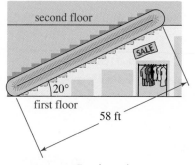

FIGURE 5.2.22 Escalator in Problem 26

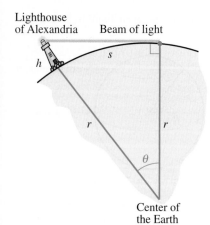

FIGURE 5.2.23 Lighthouse in Problem 28

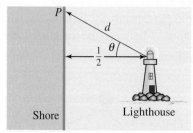

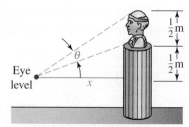

FIGURE 5.2.24 Searchlight in Problem 30   FIGURE 5.2.25 Viewing angle in Problem 31

32. A woman on an island wishes to reach a point R on a straight shore on the mainland from a point P on the island. The point P is 9 mi from the shore and 15 mi from point R. See FIGURE 5.2.26. If the woman rows a boat at a rate of 3 mi/h to a point Q on the mainland, then walks the rest of the way at a rate of 5 mi/h, express the total time it takes the woman to reach point R as a function of the indicated angle θ. [*Hint*: Distance = rate × time.]

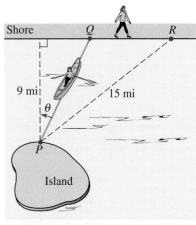

FIGURE 5.2.26 Woman rowing to shore in Problem 32

For Discussion

33. Consider the blue rectangle circumscribed around the red rectangle in FIGURE 5.2.27. With the aid of calculus it can be shown that the area of the blue rectangle is greatest when $\theta = \pi/4$. Find this area in terms of a and b.

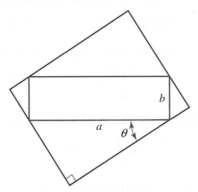

FIGURE 5.2.27 Rectangles in Problem 33

34. Home heating oil is often stored in a right circular cylindrical tank of diameter D that rests horizontally. As FIGURE 5.2.28 shows, the depth of the oil can be measured by inserting a dipstick down a vertical diameter. If the dipstick indicates that the depth of the oil is d inches, then show that the volume V of the oil is given by

$$V = \frac{V_0}{\pi}\left[\cos^{-1}\left(1 - \frac{2d}{D}\right) - 2\left(1 - \frac{2d}{D}\right)\sqrt{\left(1 - \frac{d}{D}\right)\frac{d}{D}}\right],$$

where V_0 is the volume of the tank. Although not necessary, assume for simplicity that the tank is less than half full as shown in the figure.

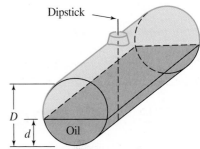

FIGURE 5.2.28 Oil tank in Problem 34

5.3 Law of Sines

≡ **Introduction** In Section 5.1 we saw how to solve *right* triangles. In this and the next section we consider two techniques for solving general triangles.

☐ **Law of Sines** Consider the triangle ABC, shown in FIGURE 5.3.1, with angles α, β, and γ, and corresponding opposite sides BC, AC, and AB. If we know the length of one side and two other parts of the triangle, we can then find the remaining three parts. One way of doing this is by the **Law of Sines**.

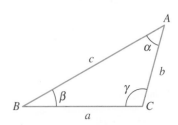

FIGURE 5.3.1 General triangle

THEOREM 5.3.1   The Law of Sines

Suppose angles α, β, and γ, and opposite sides of length a, b, and c are as shown in Figure 5.3.1. Then

$$\frac{\sin\alpha}{a} = \frac{\sin\beta}{b} = \frac{\sin\gamma}{c} \tag{1}$$

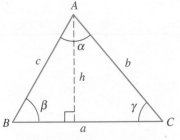

FIGURE 5.3.2 Acute triangle

PROOF: Although the Law of Sines is valid for any triangle, we will prove it only for acute triangles—that is, a triangle in which all three angles α, β, and γ, are less than 90°. As shown in FIGURE 5.3.2, let h be the length of the altitude from vertex A to side BC. Since the altitude is perpendicular to the base BC it determines two right triangles. Consequently, we can write

$$\frac{h}{c} = \sin\beta \qquad \text{and} \qquad \frac{h}{b} = \sin\gamma. \qquad (2)$$

Thus (2) gives

$$h = c\sin\beta \qquad \text{and} \qquad h = b\sin\gamma. \qquad (3)$$

Equating the two expressions in (3) gives $c\sin\beta = b\sin\gamma$ so that

$$\frac{\sin\beta}{b} = \frac{\sin\gamma}{c}. \qquad (4)$$

If we use the altitude from the vertex C to the side AB, it follows in the same manner that

$$\frac{\sin\alpha}{a} = \frac{\sin\beta}{b}. \qquad (5)$$

Combining (4) and (5) yields the result in (1). ≡

EXAMPLE 1　　　　Determining the Parts of a Triangle

Find the remaining parts of the triangle shown in FIGURE 5.3.3.

Solution Let $\beta = 20°$, $\alpha = 130°$, and $b = 6$. Because the sum of the angles in a triangle is 180° we have $\gamma + 20° + 130 = 180°$ and so $\gamma = 180° - 20° - 130° = 30°$. From (1) we then see that

$$\frac{\sin 130°}{a} = \frac{\sin 20°}{6} = \frac{\sin 30°}{c}. \qquad (6)$$

We use the first equality in (6) to solve for a:

$$a = 6\frac{\sin 130°}{\sin 20°} \approx 13.44.$$

The second equality in (6) gives c:

$$c = 6\frac{\sin 30°}{\sin 20°} \approx 8.77. \qquad ≡$$

FIGURE 5.3.3 Triangle in Example 1

EXAMPLE 2　　　　Height of a Building

A building is situated on the side of a hill that slopes downward at an angle of 15°. The Sun is uphill from the building at an angle of elevation of 42°. Find the building's height if it casts a shadow 36 ft long.

Solution Denote the height of the building on the downward slope by h and construct a right triangle QPS as shown in FIGURE 5.3.4. Now $\alpha + 15° = 42°$ so that $\alpha = 27°$. Since $\triangle QPS$ is a right triangle, $\gamma + 42° = 90°$ gives $\gamma = 90° - 42° = 48°$. From the Law of Sines (1),

$$\frac{\sin 27°}{h} = \frac{\sin 48°}{36} \qquad \text{so} \qquad h = 36\frac{\sin 27°}{\sin 48°} \approx 21.99 \text{ ft}. \qquad ≡$$

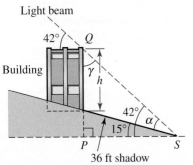

FIGURE 5.3.4 Triangle QPS in Example 2

In Examples 1 and 2, where we were given *two angles and a side opposite one of these angles*, each triangle had a unique solution. However, this may not always be true for triangles where we know *two sides and an angle opposite one of these sides*. The next example illustrates the latter situation.

EXAMPLE 3   Two Triangles

Find the remaining parts of the triangle with $\beta = 50°$, $b = 5$, and $c = 6$.

Solution From the Law of Sines, we have

$$\frac{\sin 50°}{5} = \frac{\sin \gamma}{6} \quad \text{or} \quad \sin \gamma = \frac{6}{5}\sin 50° \approx 0.9193.$$

From a calculator set in degree mode, we obtain $\gamma \approx 66.82°$. At this point it is essential to recall that the sine function is also positive for second quadrant angles. In other words, there is another angle satisfying $0° \le \gamma \le 180°$ for which $\sin \gamma \approx 0.9193$. Using $66.82°$ as a reference angle we find the second quadrant angle to be $180° - 66.82° = 113.18°$. Therefore, the two possibilities for γ are $\gamma_1 \approx 66.82°$ and $\gamma_2 \approx 113.18°$. Thus, as shown in FIGURE 5.3.5, there are two possible triangles ABC_1 and ABC_2 satisfying the given three conditions.

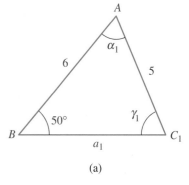

(a)

To complete the solution of triangle ABC_1 (Figure 5.3.5(a)), we first find $\alpha_1 = 180° - \gamma_1 - \beta$ or $\alpha_1 \approx 63.18°$. To find the side opposite this angle we use

$$\frac{\sin 63.18°}{a_1} = \frac{\sin 50°}{5} \quad \text{which gives} \quad a_1 = 5\left(\frac{\sin 63.18°}{\sin 50°}\right)$$

or $a_1 \approx 5.82$.

To complete the solution of triangle ABC_2 (Figure 5.3.5(b)), we find $\alpha_2 = 180° - \gamma_2 - \beta$ or $\alpha_2 \approx 16.82°$. Then from

$$\frac{\sin 16.82°}{a_2} = \frac{\sin 50°}{5} \quad \text{we find} \quad a_2 = 5\left(\frac{\sin 16.82°}{\sin 50°}\right)$$

or $a_2 \approx 1.89$.   ≡

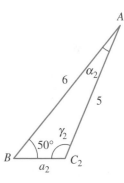

(b)

FIGURE 5.3.5 Triangles in Example 3

☐ **Ambiguous Case** When solving triangles, the situation where two sides and an angle opposite one of these sides are given is called the **ambiguous case**. We have just seen in Example 3 that the given information may determine two different triangles. In the ambiguous case other complications can arise. For instance, suppose that the length of sides AB and AC (that is, c and b, respectively) and the angle β in triangle ABC are specified. As shown in FIGURE 5.3.6, we draw the angle β and mark off side AB with length c to locate the vertices A and B. The third vertex C is located on the base by drawing an arc of a circle of radius b (the length of AC) with center A. As shown in FIGURE 5.3.7, there are four possible outcomes of this construction:

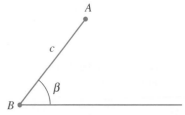

FIGURE 5.3.6 Horizontal base, the angle β, and side AB

- The arc does not intersect the base and no triangle is formed.
- The arc intersects the base in two distinct points C_1 and C_2 and two triangles are formed (as in Example 3).

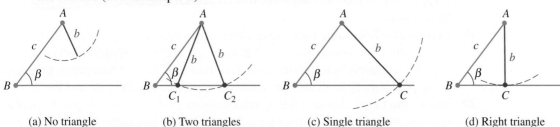

(a) No triangle   (b) Two triangles   (c) Single triangle   (d) Right triangle

FIGURE 5.3.7 Solution possibilities for the ambiguous case in the Law of Sines

- The arc intersects the base in one point and one triangle is formed.
- The arc is tangent to the base and a single right triangle is formed.

EXAMPLE 4   Determining the Parts of a Triangle

Find the remaining parts of the triangle with $\beta = 40°$, $b = 5$, and $c = 9$.

Solution From the Law of Sines (1), we have

$$\frac{\sin 40°}{5} = \frac{\sin \gamma}{9} \quad \text{and so} \quad \sin \gamma = \frac{9}{5}\sin 40° \approx 1.1570.$$

Since the sine of any angle must be between -1 and 1, $\sin \gamma \approx 1.1570$ is impossible. This means the triangle has no solution; the side with length b is not long enough to reach the base. This is the case illustrated in Figure 5.3.7(a).   ≡

5.3 Exercises
Answers to selected odd-numbered problems begin on page ANS-16.

In Problems 1–16, refer to Figure 5.3.1. ▶

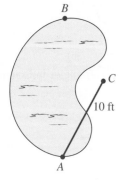

FIGURE 5.3.8 Pool in Problem 17

In Problems 1–16, use the Law of Sines to solve the triangle.

1. $\alpha = 80°$, $\beta = 20°$, $b = 7$
2. $\alpha = 60°$, $\beta = 15°$, $c = 30$
3. $\beta = 37°$, $\gamma = 51°$, $a = 5$
4. $\alpha = 30°$, $\gamma = 75°$, $a = 6$
5. $\beta = 72°$, $b = 12$, $c = 6$
6. $\alpha = 120°$, $a = 9$, $c = 4$
7. $\gamma = 62°$, $b = 7$, $c = 4$
8. $\beta = 110°$, $\gamma = 25°$, $a = 14$
9. $\gamma = 15°$, $a = 8$, $c = 5$
10. $\alpha = 55°$, $a = 20$, $c = 18$
11. $\gamma = 150°$, $b = 7$, $c = 5$
12. $\alpha = 35°$, $a = 9$, $b = 12$
13. $\beta = 30°$, $a = 10$, $b = 7$
14. $\alpha = 140°$, $\gamma = 20°$, $c = 12$
15. $\alpha = 20°$, $a = 8$, $c = 27$
16. $\alpha = 75°$, $\gamma = 45°$, $b = 8$

Miscellaneous Applications

17. **Length of a Pool** A 10-ft rope that is available to measure the length between two points A and B at opposite ends of a kidney-shaped swimming pool is not long enough. A third point C is found such that the distance from A to C is 10 ft. It is determined that angle ACB is $115°$ and angle ABC is $35°$. Find the distance from A to B. See FIGURE 5.3.8.

18. **Width of a River** Two points A and B lie on opposite sides of a river. Another point C is located on the same side of the river as B at a distance of 230 ft from B. If angle ABC is $105°$ and angle ACB is $20°$, find the distance across the river from A to B.

19. **Length of a Telephone Pole** A telephone pole makes an angle of $82°$ with the level ground. As shown in FIGURE 5.3.9, the angle of elevation of the Sun is $76°$. Find the length of the telephone pole if its shadow is 3.5 m. (Assume that the tilt of the pole is away from the Sun and in the same plane as the pole and the Sun.)

20. **Not on the Level** A man 5 ft 9 in. tall stands on a sidewalk that slopes down at a constant angle. A vertical street lamp directly behind him causes his shadow to be 25 ft long. The angle of depression from the top of the man to the tip of his

FIGURE 5.3.9 Telephone pole in Problem 19

CHAPTER 5 TRIANGLE TRIGONOMETRY

shadow is 31°. Find the angle α, as shown in FIGURE 5.3.10, that the sidewalk makes with the horizontal.

21. **How High?** If the man in Problem 20 is 20 ft down the sidewalk from the street lamp, find the height of the light above the sidewalk.

22. **Plane with an Altitude** Angles of elevation to an airplane are measured from the top and the base of a building that is 20 m tall. The angle from the top of the building is 38°, and the angle from the base of the building is 40°. Find the altitude of the airplane.

23. **Angle of Drive** The distance from the tee to the green on a particular golf hole is 370 yd. A golfer hits his drive and paces its distance off at 210 yd. From the point where the ball lies, he measures an angle of 160° between the tee and the green. Find the angle of his drive off the tee measured from the dashed line from the tee to the green shown in FIGURE 5.3.11.

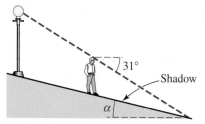

FIGURE 5.3.10 Sloping sidewalk in Problem 20

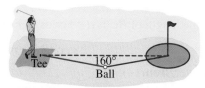

FIGURE 5.3.11 Angle of drive in Problem 23

24. In Problem 23, what is the distance from the ball to the green?

25. **Help!** One Coast Guard vessel is located 4 nautical miles due south of a second Coast Guard vessel when they receive a distress signal from a sailboat. To offer assistance, the first vessel sails on a bearing of S50°E at 5 knots and the second vessel sails S10°E at 10 knots. Which one of the Coast Guard vessels reaches the sailboat first? [*Hint*: The concept of bearing is reviewed on page 297.]

5.4 **Law of Cosines**

≣ **Introduction** Triangles for which we know either *three sides* or *two sides and the included angle* (that is, the angle formed by the given sides) cannot be solved directly using the Law of Sines. The method we consider next can be used to solve triangles in these two cases.

☐ **Pythagorean Theorem** In a right triangle, such as the one shown in FIGURE 5.4.1, the length c of the hypotenuse is related to the lengths a and b of the other two sides by the Pythagorean theorem

$$c^2 = a^2 + b^2. \tag{1}$$

This last equation is a special case of a general formula that relates the lengths of the sides of *any* triangle.

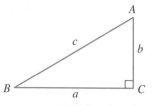

FIGURE 5.4.1 Right triangle

☐ **Law of Cosines** A generalization of (1) is called the **Law of Cosines**. Like the Law of Sines, (1) of Section 5.3, the Law of Cosines is valid for any triangle. But for convenience, we will prove the last two equations in (2) of Theorem 5.4.1 using again an acute triangle.

THEOREM 5.4.1   The Law of Cosines

Suppose angles α, β, and γ, and opposite sides of lengths a, b, and c are as shown in Figure 5.3.1. Then

$$
\begin{aligned}
a^2 &= b^2 + c^2 - 2bc\cos\alpha \\
b^2 &= a^2 + c^2 - 2ac\cos\beta \\
c^2 &= a^2 + b^2 - 2ab\cos\gamma
\end{aligned}
\tag{2}
$$

PROOF: Let P denote the point where the altitude from the vertex A intersects side BC. Then, since both $\triangle BPA$ and $\triangle CPA$ in FIGURE 5.4.2 are right triangles we have from (1),

$$c^2 = h^2 + (c\cos\beta)^2 \tag{3}$$

and

$$b^2 = h^2 + (b\cos\gamma)^2. \tag{4}$$

Now the length of BC is $a = c\cos\beta + b\cos\gamma$ so that

$$c\cos\beta = a - b\cos\gamma. \tag{5}$$

Moreover, from (4),

$$h^2 = b^2 - (b\cos\gamma)^2. \tag{6}$$

Substituting (5) and (6) into (3) and simplifying yields the third equation in (2):

$$
\begin{aligned}
c^2 &= b^2 - (b\cos\gamma)^2 + (a - b\cos\gamma)^2 \\
&= b^2 - b^2\cos^2\gamma + a^2 - 2ab\cos\gamma + b^2\cos^2\gamma
\end{aligned}
$$

or

$$c^2 = a^2 + b^2 - 2ab\cos\gamma. \tag{7}$$

Note that equation (7) reduces to the Pythagorean theorem (1) when $\gamma = 90°$.

Similarly, if we use $b\cos\gamma = a - c\cos\beta$ and $h^2 = c^2 - (c\cos\beta)^2$ to eliminate $b\cos\gamma$ and h^2 in (4), we obtain the second equation in (2). ≡

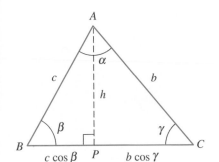

FIGURE 5.4.2 Acute triangle

▮ EXAMPLE 1   Determining the Parts of a Triangle

Find the remaining parts of the triangle shown in FIGURE 5.4.3.

Solution First, if we call the unknown side b and identify $a = 12$, $c = 10$, and $\beta = 26°$, then from the second equation in (2) we can write

$$b^2 = (12)^2 + (10)^2 - 2(12)(10)\cos 26°.$$

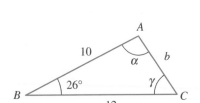

FIGURE 5.4.3 Triangle in Example 1

Therefore, $b^2 \approx 28.2894$ and so $b \approx 5.32$.

Next, we use the Law of Cosines to determine the remaining angles in the triangle in Figure 5.4.3. If γ is the angle at the vertex C, then the third equation in (2) gives

$$10^2 = 12^2 + (5.32)^2 - 2(12)(5.32)\cos\gamma \qquad \text{or} \qquad \cos\gamma \approx 0.5663.$$

With the aid of a calculator and the inverse cosine we find $\gamma \approx 55.51°$. Note that since the cosine of an angle between $90°$ and $180°$ is negative, there is no need to consider two possibilities as we did in Example 3 in Section 5.3. Finally, the angle at the vertex A is $\alpha = 180° - \beta - \gamma$ or $\alpha \approx 98.49°$. ≡

In Example 1, observe that after b is found, we know two sides and an angle opposite one of these sides. Hence we could have used the Law of Sines to find the angle γ.

In the next example we consider the case in which the lengths of the three sides of a triangle are given.

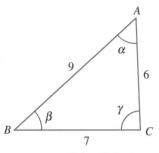

FIGURE 5.4.4 Triangle in Example 2

EXAMPLE 2   Determining the Angles in a Triangle

Find the angles α, β, and γ in the triangle shown in FIGURE 5.4.4.

Solution We use the Law of Cosines to find the angle opposite the longest side:

$$9^2 = 6^2 + 7^2 - 2(6)(7)\cos\gamma \qquad \text{or} \qquad \cos\gamma = \tfrac{1}{21}.$$

A calculator then gives $\gamma \approx 87.27°$. Although we could use the Law of Cosines, we choose to find β by the Law of Sines:

$$\frac{\sin\beta}{6} = \frac{\sin 87.27°}{9} \qquad \text{or} \qquad \sin\beta = \frac{6}{9}\sin 87.27° \approx 0.6659.$$

Since γ is the angle opposite the longest side it is the largest angle in the triangle, so β must be an acute angle. Thus, $\sin\beta \approx 0.6659$ yields $\beta \approx 41.75°$. Finally, from $\alpha = 180° - \beta - \gamma$ we find $\alpha \approx 50.98°$. ≡

☐ **Bearing** In navigation directions are given using bearings. A **bearing** designates the acute angle that a line makes with the north–south line. For example, FIGURE 5.4.5(a) illustrates a bearing of S40°W, meaning 40 degrees west of south. The bearings in Figures 5.4.5(b) and 5.4.5(c) are N65°E and S80°E, respectively.

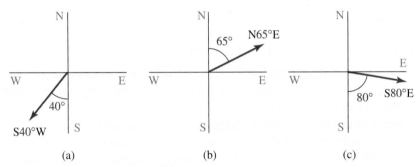

(a)   (b)   (c)

FIGURE 5.4.5 Three examples of bearings

EXAMPLE 3   Bearings of Two Ships

Two ships leave a port at 7:00 AM, one traveling at 12 knots (nautical miles per hour) and the other at 10 knots. If the faster ship maintains a bearing of N47°W and the other ship maintains a bearing of S20°W, what is their separation (to the nearest nautical mile) at 11:00 AM that day?

Solution Since the elapsed time is 4 hours, the faster ship has traveled $4 \cdot 12 = 48$ nautical miles from port and the slower ship $4 \cdot 10 = 40$ nautical miles. Using these distances and the given bearings, we can sketch the triangle (valid at 11:00 AM) shown in FIGURE 5.4.6. In the triangle, c denotes the distance separating the ships and γ is the angle opposite that side. Since $47° + \gamma + 20° = 180°$ we find $\gamma = 113°$. Finally, the Law of Cosines

$$c^2 = 48^2 + 40^2 - 2(48)(40)\cos 113°,$$

gives $c^2 \approx 5404.41$ or $c \approx 73.51$. Thus the distance between the ships (to the nearest nautical mile) is 74 nautical miles. ≡

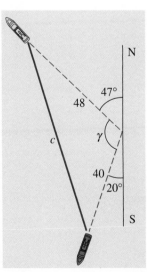

FIGURE 5.4.6 Ships in Example 3

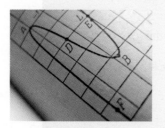

(*i*) An important first step in solving a triangle is determining which of the three approaches we have discussed to use: right triangle trigonometry, the Law of Sines, or the Law of Cosines. The following table describes the various types of problems and gives the most appropriate approach for each. The term *oblique* refers to any triangle that is not a right triangle.

Type of Triangle	Information Given	Technique
Right	Two sides or an angle and a side	Basic definitions of sine, cosine, and tangent; the Pythagorean theorem
Oblique	Three sides	Law of Cosines
Oblique	Two sides and the included angle	Law of Cosines
Oblique	Two angles and a side	Law of Sines
Oblique	Two sides and an angle opposite one of the sides	Law of Sines (if the given angle is acute; it is an ambiguous case)

(*ii*) Here are some additional bits of advice for solving triangles.

- Students will frequently use the Law of Sines when a right triangle trigonometric function could have been used. A right triangle approach is the simplest and most efficient.
- When three sides are given, check first to see whether the length of the longest side is greater than or equal to the sum of the lengths of the other two sides. If it is, there can be no solution (even though the given information indicates a Law of Cosines approach). This is because the shortest distance between two points is the length of the line segment joining them.
- In applying the Law of Sines, if you obtain a value greater than 1 for the sine of an angle, there is no solution.
- In the ambiguous case of the Law of Sines, when solving for the first unknown angle, you must consider *both the acute angle found from your calculator and its supplement as possible solutions*. The supplement will be a solution if the sum of the supplement and the angle given in the triangle is less than $180°$.

5.4 Exercises
Answers to selected odd-numbered problems begin on page ANS-16.

In Problems 1–16, refer to Figure 5.3,1. ▶

In Problems 1–16, use the Law of Cosines to solve the triangle.

1. $\gamma = 65°, a = 5, b = 8$
2. $\beta = 48°, a = 7, c = 6$
3. $a = 8, b = 10, c = 7$
4. $\gamma = 31.5°, a = 4, b = 8$
5. $\gamma = 97.33°, a = 3, b = 6$
6. $a = 7, b = 9, c = 4$
7. $a = 11, b = 9.5, c = 8.2$
8. $\alpha = 162°, b = 11, c = 8$

9. $a = 5, b = 7, c = 10$ **10.** $a = 6, b = 5, c = 7$
11. $a = 3, b = 4, c = 5$ **12.** $a = 5, b = 12, c = 13$
13. $a = 6, b = 8, c = 12$ **14.** $\beta = 130°, a = 4, c = 7$
15. $\alpha = 22°, b = 3, c = 9$ **16.** $\beta = 100°, a = 22.3, b = 16.1$

Miscellaneous Applications

17. How Far? A ship sails due west from a harbor for 22 nautical miles. It then sails S62°W for another 15 nautical miles. How far is the ship from the harbor?

18. How Far Apart? Two hikers leave their camp simultaneously, taking bearings of N42°W and S20°E, respectively. If they each average a rate of 5 km/h, how far apart are they after 1 h?

19. Bearings On a hiker's map point A is 2.5 in. due west of point B and point C is 3.5 in. from B and 4.2 in. from A, respectively. See **FIGURE 5.4.7**. Find **(a)** the bearing of A from C, and **(b)** the bearing of B from C.

20. How Long Will It Take? Two ships leave port simultaneously, one traveling at 15 knots and the other at 12 knots. They maintain bearings of S42°W and S10°E, respectively. After 3 h the first ship runs aground and the second ship immediately goes to its aid.
   (a) How long will it take the second ship to reach the first ship if it travels at 14 knots?
   (b) What bearing should it take?

21. A Robotic Arm A two-dimensional robot arm "knows" where it is by keeping track of a "shoulder" angle α and an "elbow" angle β. As shown in **FIGURE 5.4.8**, this arm has a fixed point of rotation at the origin. The shoulder angle is measured counterclockwise from the x-axis, and the elbow angle is measured counterclockwise from the upper to the lower arm. Suppose that the upper and lower arms are both of length 2 and that the elbow angle β is prevented from "hyperextending" beyond 180°. Find the angles α and β that will position the robot's hand at the point $(1, 2)$.

22. Which Way? Two lookout towers are situated on mountain tops A and B, 4 mi from each other. A helicopter firefighting team is located in a valley at point C, 3 mi from A and 2 mi from B. Using the line between A and B as a reference, a lookout spots a fire at an angle of 40° from tower A and 82° from tower B. See **FIGURE 5.4.9**. At what angle, measured from CB, should the helicopter fly in order to head directly for the fire?

23. Making a Kite For the kite shown in **FIGURE 5.4.10**, use the Law of Cosines to find the lengths of the two dowels required for the diagonal supports.

24. Distance Across a Canyon From the floor of a canyon it takes 62 ft of rope to reach the top of one canyon wall and 86 ft to reach the top of the opposite wall. See **FIGURE 5.4.11**. If the two ropes make an angle of 123°, what is the distance d from the top of one canyon wall to the other?

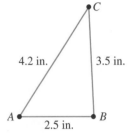

FIGURE 5.4.7 Triangle in Problem 19

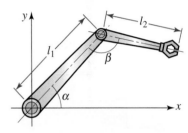

FIGURE 5.4.8 Robotic arm in Problem 21

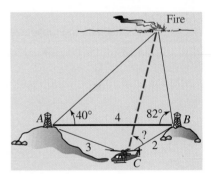

FIGURE 5.4.9 Fire in Problem 22

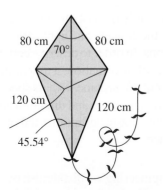

FIGURE 5.4.10 Kite in Problem 23

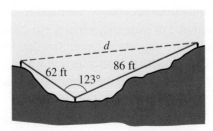

FIGURE 5.4.11 Canyon in Problem 24

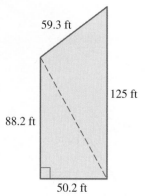

59.3 ft

88.2 ft

125 ft

50.2 ft

FIGURE 5.4.12 Corner lot in Problem 27

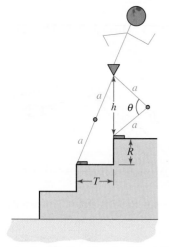

a

h θ

a

a

a

R

T

FIGURE 5.4.13 Blue man in Problem 29

For Discussion

25. Use the Law of Cosines to derive the formula
$$A = \sqrt{s(s-a)(s-b)(s-c)},$$
for the area of a triangle with sides a, b, c where $s = \frac{1}{2}(a+b+c)$. This formula is named after the Greek mathematician an inventor **Heron of Alexandria** (c. 20–62 C.E.) but should actually be credited to Archimedes.

26. Garden Plot Use Heron's formula in Problem 25 to find the area of a triangular garden plot if the lengths of the three sides are 25, 32, and 41 m, respectively.

27. Corner Lot Find the area of the irregular corner lot shown in FIGURE 5.4.12. [*Hint*: Divide the lot into two triangular lots as shown and then find the area of each triangle. Use Heron's formula in Problem 25 for the area of the acute triangle.]

28. Use Heron's formula in Problem 25 to find the area of a triangle with vertices located at $(3, 2)$, $(-3, -6)$, and $(0, 6)$ in a rectangular coordinate system.

29. Blue Man The effort in climbing a flight of stairs depends largely on the flexing angle of the leading knee. A simplified blue stick-figure model of a person walking up a staircase indicates that the maximum flexing of the knee occurs when the back leg is straight and the hips are directly over the heel of the front foot. See FIGURE 5.4.13. Show that
$$\cos\theta = \left(\frac{R}{a}\right)\sqrt{4 - \left(\frac{T}{a}\right)^2} + \frac{(T/a)^2 - (R/a)^2}{2} - 1,$$
where θ is the knee joint angle, $2a$ is the length of the leg, R is the rise of a single stair step, and T is the width of a step. [*Hint*: Let h be the vertical distance from hip to heel of the leading leg, as shown in the figure. Set up two equations involving h: one by applying the Pythagorean theorem to the right triangle outlined in color and the other by using the Law of Cosines on the angle θ. Then eliminate h and solve for $\cos\theta$.]

30. For a triangle with sides of lengths $a, b,$ and c and γ is the angle opposite c we have seen on page 296 that when γ is a right angle the Law of Cosines reduces to the Pythagorean theorem $c^2 = a^2 + b^2$. How is c^2 related to $a^2 + b^2$ when
(a) γ is an acute angle **(b)** γ is an obtuse angle?

5.5 Vectors and Dot Product

∫**Calculus PREVIEW**

≡ **Introduction** Approximately the last one-third of a typical course in calculus deals with the concept and applications of vectors in two and three dimensions. In order to describe certain physical quantities accurately, we must have two pieces of information: a magnitude and a direction. For example, when we discuss the flight of an airplane both its speed and its heading are important. Quantities that involve both magnitude and direction are represented by **vectors**. In this section we will survey some basic definitions and operations on vectors that lie in the coordinate plane or 2-space.

☐ **Terminology** In science, mathematics, and engineering we distinguish two important quantities: *scalars* and *vectors*. A **scalar** is simply a real number and is generally represented by a lowercase italicized letter, such as *a*, *k*, or *x*. Scalars may be used to represent magnitudes and may have specific units attached to them; for example, 80 feet, 10 lb, or 20° Celsius. On the other hand, a **vector**, or **displacement vector**, may be thought of as an arrow or directed line segment (a line with a direction specified by an arrowhead) connecting points *A* and *B* in 2-space. The *tail* of the arrow is called the **initial point** and the *tip* of the arrow is called the **terminal point**. As shown in FIGURE 5.5.1, a vector is usually denoted by a boldfaced letter such as **u** or **v**, or if we wish to emphasize the initial and terminal points *A* and *B*, by the symbol $\overrightarrow{AB}$. Thus, in contrast to a scalar which has only magnitude, a vector has both magnitude and direction. The **magnitude** of a vector is its length, that is, the distance between its initial and terminal points. The magnitude of vector is denoted by $|\mathbf{u}|$ or $|\overrightarrow{AB}|$. Two vectors $\overrightarrow{AB}$ and $\overrightarrow{CD}$ are said to be **equal**, written $\overrightarrow{AB} = \overrightarrow{CD}$, if they have both the same magnitude and the same direction, as shown in FIGURE 5.5.2. Thus vectors can be translated from one position to another so long as neither the magnitude nor the direction is changed.

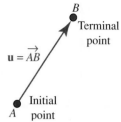

FIGURE 5.5.1 Directed line segment in 2-space

Because we can move a vector provided its magnitude and its direction are unchanged, we can place the initial point at the origin. Then, as shown in FIGURE 5.5.3, the terminal point *P* will have rectangular coordinates (x, y). Conversely, every ordered pair of real numbers (x, y) determines a vector $\overrightarrow{OP}$, where *P* has rectangular coordinates (x, y). Thus we have a one-to-one correspondence between vectors and ordered pairs of real numbers. We say that $\mathbf{u} = \overrightarrow{OP}$ is the **position vector** of the point $P(x, y)$ and is written

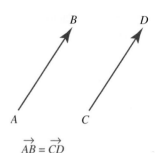

FIGURE 5.5.2 Equal vectors

$$\overrightarrow{OP} = \langle x, y \rangle.$$

In general, any vector in the plane can be identified with a unique position vector $\mathbf{u} = \langle a_1, a_2 \rangle$. The numbers a_1 and a_2 are said to be the **components** of the position vector **u** and the notation $\langle a_1, a_2 \rangle$ is called the **component form of a vector**.

Since the magnitude of $\langle a_1, a_2 \rangle$ is the distance from the point (a_1, a_2) to the origin, we define the **magnitude** $|\mathbf{u}|$ of the vector $\mathbf{u} = \langle a_1, a_2 \rangle$ to be

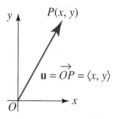

FIGURE 5.5.3 Position vector

$$|\mathbf{u}| = \sqrt{a_1^2 + a_2^2}. \tag{1}$$

The **zero vector**, denoted by **0**, is defined by the component form $\mathbf{0} = \langle 0, 0 \rangle$. The magnitude of the zero vector is zero. The zero vector is not assigned any direction.

Let $\mathbf{u} = \langle x, y \rangle$ be a nonzero vector. If θ is an angle in standard position formed by **u** and the positive *x*-axis, as shown in FIGURE 5.5.4, then θ is called a **direction angle** for **u**. Also any angle coterminal with θ is also a direction angle for **u**. But for the sake of definiteness we will choose θ such that in degrees $0° \leq \theta < 360°$ or in radians $0 \leq \theta < 2\pi$. Thus a vector **u** can be specified by giving either its components $\mathbf{u} = \langle x, y \rangle$ or by its magnitude $|\mathbf{u}|$ and a direction angle. From trigonometry, we have the following relationships between the components, magnitude, and the direction angle of a vector **u**.

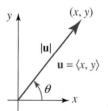

FIGURE 5.5.4 Direction angle of a vector

DEFINITION 5.5.1   Direction Angle

For any nonzero vector $\mathbf{u} = \langle x, y \rangle$ with direction angle θ:

$$\cos\theta = \frac{x}{|\mathbf{u}|}, \quad \sin\theta = \frac{y}{|\mathbf{u}|}, \quad \tan\theta = \frac{y}{x}, \quad x \neq 0 \tag{2}$$

where $|\mathbf{u}| = \sqrt{x^2 + y^2}$.

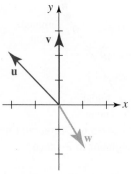

FIGURE 5.5.5 Vectors in Example 1

Sketch each of the following vectors. Find the magnitude and the direction angle θ of each vector.

(a) $\mathbf{u} = \langle -2, 2 \rangle$ (b) $\mathbf{v} = \langle 0, 3 \rangle$ (c) $\mathbf{w} = \langle 1, -\sqrt{3} \rangle$

Solution The three vectors are sketched using different colors in FIGURE 5.5.5.

(a) From (1), the magnitude of $\mathbf{u}$ is

$$|\mathbf{u}| = \sqrt{(-2)^2 + 2^2} = \sqrt{8} = 2\sqrt{2}$$

and from (2) its direction angle satisfies $\tan \theta = 2/(-2) = -1$. As we see in Figure 5.5.5, θ is a second quadrant angle and so we chose $\theta = \arctan(-1) + \pi = -\pi/4 + \pi$ or $\theta = 3\pi/4$.

(b) The magnitude of the vector $\mathbf{v}$ is $|\mathbf{v}| = \sqrt{0^2 + 3^2} = 3$ and from Figure 5.5.5 we see immediately that its direction angle is $\theta = \pi/2$.

(c) For the vector $\mathbf{w}$ we have $|\mathbf{w}| = \sqrt{1^2 + (-\sqrt{3})^2} = \sqrt{4} = 2$. From $\tan \theta = -\sqrt{3}$ we get $\arctan(-\sqrt{3}) = -\pi/3$ radians $= -60°$. Because we want $0 \le \theta < 360°$ we choose the direction angle to be $\theta = -60° + 360°$ or $\theta = 300°$. ≡

☐ **Vector Arithmetic** Vectors can be combined with other vectors by the arithmetic operation of addition. In addition, vectors can be combined with scalars through multiplication. Using the component form of a vector we give next the algebraic definitions of the **sum** of two vectors, the **scalar multiple** of a vector, and **equality** of two vectors.

DEFINITION 5.5.2   Operations on Vectors

Let $\mathbf{u} = \langle a_1, a_2 \rangle$ and $\mathbf{v} = \langle b_1, b_2 \rangle$ be vectors, and let k be a real number. Then we define the

Sum: $\mathbf{u} + \mathbf{v} = \langle a_1 + b_1, a_2 + b_2 \rangle$ (3)

Scalar Multiple: $k\mathbf{u} = \langle ka_1, ka_2 \rangle$ (4)

Equality: $\mathbf{u} = \mathbf{v}$ if and only if $a_1 = b_1, a_2 = b_2$ (5)

☐ **Subtraction** Using (4), we define the **negative** of a vector $\mathbf{u} = \langle a_1, a_2 \rangle$ by

$$-\mathbf{u} = (-1)\mathbf{u} = \langle -a_1, -a_2 \rangle.$$

We can then define **subtraction**, or the **difference**, of two vectors $\mathbf{u} = \langle a_1, a_2 \rangle$ and $\mathbf{v} = \langle b_1, b_2 \rangle$ as

$$\mathbf{u} - \mathbf{v} = \mathbf{u} + (-\mathbf{v}) = \langle a_1 - b_1, a_2 - b_2 \rangle. \qquad (6)$$

If $\mathbf{u} = \langle 2, 1 \rangle$ and $\mathbf{v} = \langle -1, 5 \rangle$ find $4\mathbf{u}$, $\mathbf{u} + \mathbf{v}$, and $3\mathbf{u} - 2\mathbf{v}$.

Solution From the definitions of addition, subtraction, and scalar multiples of vectors, we find

$$4\mathbf{u} = 4\langle 2, 1 \rangle = \langle 8, 4 \rangle \qquad \leftarrow \text{from (4)}$$
$$\mathbf{u} + \mathbf{v} = \langle 2, 1 \rangle + \langle -1, 5 \rangle = \langle 1, 6 \rangle \qquad \leftarrow \text{from (3)}$$
$$3\mathbf{u} - 2\mathbf{v} = 3\langle 2, 1 \rangle - 2\langle -1, 5 \rangle = \langle 6, 3 \rangle - \langle -2, 10 \rangle = \langle 8, -7 \rangle. \qquad \leftarrow \text{from (4) and (6)} \equiv$$

Operations (3), (4), and (6) possess the following properties.

THEOREM 5.5.1   Properties of Vector Operations

(i) $\mathbf{u} + \mathbf{v} = \mathbf{v} + \mathbf{u}$	(ii) $\mathbf{u} + (\mathbf{v} + \mathbf{w}) = (\mathbf{u} + \mathbf{v}) + \mathbf{w}$
(iii) $k(\mathbf{u} + \mathbf{v}) = k\mathbf{u} + k\mathbf{v}$	(iv) $(k_1 + k_2)\mathbf{u} = k_1\mathbf{u} + k_2\mathbf{u}$
(v) $k_1(k_2\mathbf{u}) = (k_1 k_2)\mathbf{u}$	(vi) $\mathbf{u} + \mathbf{0} = \mathbf{u}$
(vii) $\mathbf{u} + (-\mathbf{u}) = \mathbf{0}$	$(viii)$ $0\mathbf{u} = \mathbf{0}$
(ix) $1\mathbf{u} = \mathbf{u}$	(x) $|k\mathbf{u}| = |k||\mathbf{u}|$

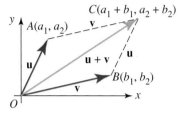

FIGURE 5.5.6 Sum of two vectors **u** and **v**

You should recognize properties (i) and (ii) of Theorem 5.5.1 as the commutative and associative laws of addition, respectively.

☐ **Geometric Interpretations**   The sum $\mathbf{u} + \mathbf{v}$ of two vectors can readily be interpreted geometrically in the plane using the concept of a position vector. If $\mathbf{u} = \langle a_1, a_2 \rangle$ and $\mathbf{v} = \langle b_1, b_2 \rangle$, then the three vectors $\mathbf{u}$, $\mathbf{v}$, and $\mathbf{u} + \mathbf{v}$ can be represented by directed line segments from the origin to the points $A(a_1, a_2)$, $B(b_1, b_2)$, and $C(a_1 + b_1, a_2 + b_2)$, respectively. As shown in FIGURE 5.5.6, if the vector $\mathbf{v}$ is translated so that its initial point is A, then its terminal point will be C. Thus a geometric representation of the sum $\mathbf{u} + \mathbf{v}$ can be obtained by placing the initial point of $\mathbf{v}$ on the terminal point of $\mathbf{u}$ and drawing the vector from the initial point of $\mathbf{u}$ to the terminal point of $\mathbf{v}$. By examining the coordinates of the quadrilateral $OACB$ in Figure 5.5.6, we see that it is a parallelogram formed by the vectors $\mathbf{u}$ and $\mathbf{v}$, with $\mathbf{u} + \mathbf{v}$ as one of its diagonals.

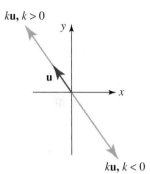

FIGURE 5.5.7 Scalar multiple of a vector **v**

We now consider a scalar multiple of the vector $\mathbf{u} = \langle x, y \rangle$. If the symbol k represents any real number, then

$$|k\mathbf{u}| = \sqrt{(kx)^2 + (ky)^2} = \sqrt{k^2(x^2 + y^2)}$$
$$= \sqrt{k^2}\sqrt{x^2 + y^2} = |k|\sqrt{x^2 + y^2} = |k||\mathbf{u}|.$$

We have derived the property of scalar multiplication given in part (x) of Theorem 5.5.1, that is,

$$|k\mathbf{u}| = |k||\mathbf{u}|. \tag{7}$$

This property states that in the scalar multiplication of a vector $\mathbf{u}$ by a real number k, the magnitude of $\mathbf{u}$ is multiplied by $|k|$. As shown in FIGURE 5.5.7, if $k > 0$, the direction of $\mathbf{u}$ does not change; but if $k > 0$, the direction of $\mathbf{u}$ is reversed. Figure 5.5.7 illustrates the case where $|k| > 1$. A vector $\mathbf{u}$ and its negative $-\mathbf{u}$ have the same length but opposite direction.

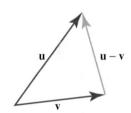

FIGURE 5.5.8 Difference of two vectors **u** and **v**

The geometric interpretation of the difference $\mathbf{u} - \mathbf{v}$ of two vectors is obtained by observing that $\mathbf{u} = \mathbf{v} + (\mathbf{u} - \mathbf{v})$. Thus, $\mathbf{u} - \mathbf{v}$ is the vector that when added to $\mathbf{v}$ yields $\mathbf{u}$. As we see in FIGURE 5.5.8, the initial point of $\mathbf{u} - \mathbf{v}$ will be at the terminal point of $\mathbf{v}$, and the terminal point of $\mathbf{u} - \mathbf{v}$ coincides with the terminal point of $\mathbf{u}$. Hence the vector $\mathbf{u} - \mathbf{v}$ is one diagonal of the parallelogram determined by $\mathbf{u}$ and $\mathbf{v}$, with $\mathbf{u} + \mathbf{v}$ being the other diagonal. See FIGURE 5.5.9.

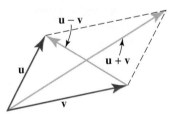

FIGURE 5.5.9 Sum and difference of vectors **u** and **v** as diagonals of a parallelogram

---

EXAMPLE 3   **Sum and Difference**

Let $\mathbf{u} = \langle -1, 1 \rangle$ and $\mathbf{v} = \langle 3, 2 \rangle$.

(a) Sketch the geometric interpretations of $\mathbf{u} + \mathbf{v}$ and $\mathbf{u} - \mathbf{v}$.

(b) Sketch $\mathbf{u} + \mathbf{v}$ and $\mathbf{u} - \mathbf{v}$ as position vectors.

Solution (a) To interpret these vectors geometrically, we form the parallelogram with two sides determined by the vectors **u** and **v** and identify **u** + **v** and **u** − **v** as the diagonals shown in gold in FIGURE 5.5.10(a).

(b) From (3) and (6) we have, in turn,

$$\mathbf{u} + \mathbf{v} = \langle -1, 1 \rangle + \langle 3, 2 \rangle = \langle -1 + 3, 1 + 2 \rangle = \langle 2, 3 \rangle$$
$$\mathbf{u} - \mathbf{v} = \langle -1, 1 \rangle - \langle 3, 2 \rangle = \langle -1 - 3, 1 - 2 \rangle = \langle -4, -1 \rangle.$$

As position vectors, we plot the points $(2, 3)$ and $(-4, -1)$ and then draw a vector (in red) stemming from the origin to each point. See Figure 5.5.10(b).

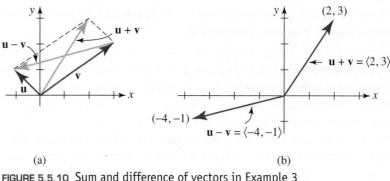

(a)   (b)

FIGURE 5.5.10 Sum and difference of vectors in Example 3   ≡

▌EXAMPLE 4   Scalar Multiples

Let $\mathbf{u} = \langle 1, 2 \rangle$. Find $2\mathbf{u}$ and $-\mathbf{u}$ and give geometric interpretations of the vectors.

Solution The scalar multiple is $2\mathbf{u} = 2\langle 1, 2 \rangle = \langle 2, 4 \rangle$. The negative of the vector **u** is $-\mathbf{u} = (-1)\langle 1, 2 \rangle = \langle -1, -2 \rangle$. Geometrically, the vector $2\mathbf{u}$ has the same direction as **u** but is twice as long. The negative $-\mathbf{u}$ has the same length as **u** but has the opposite direction. See FIGURE 5.5.11.   ≡

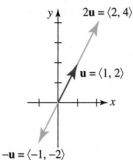

FIGURE 5.5.11 Scalar multiples and negative of the vector **u** in Example 4

☐ **Unit Vectors** Any vector with magnitude 1 is called a **unit vector**. We can obtain a unit vector **u** in the same direction as a nonzero vector **v** by multiplying **v** by the positive scalar $k = 1/|\mathbf{v}|$ (reciprocal of its magnitude). In this case we say that

$$\mathbf{u} = \left(\frac{1}{|\mathbf{v}|}\right)\mathbf{v} = \frac{\mathbf{v}}{|\mathbf{v}|} \tag{8}$$

is the **normalization** of the vector **v**. It follows from (7) that the normalization of a vector **v** is a unit vector because

$$|\mathbf{u}| = \left|\frac{1}{|\mathbf{v}|}\mathbf{v}\right| = \frac{1}{|\mathbf{v}|}|\mathbf{v}| = 1.$$

▌EXAMPLE 5   Unit Vector

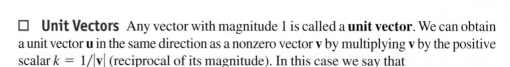

Given $\mathbf{v} = \langle 2, -1 \rangle$, find a unit vector **(a)** in the same direction as **v**, and **(b)** in the opposite direction of **v**.

Solution First, we find the magnitude of the vector **v**:

$$|\mathbf{v}| = \sqrt{4 + (-1)^2} = \sqrt{5}.$$

(a) From (8), a unit vector in the same direction as **v** is then

$$\mathbf{u} = \frac{1}{\sqrt{5}}\mathbf{v} = \frac{1}{\sqrt{5}}\langle 2, -1 \rangle = \left\langle \frac{2}{\sqrt{5}}, \frac{-1}{\sqrt{5}} \right\rangle.$$

(b) A unit vector in the opposite direction of **v** is the negative of **u**:

$$-\mathbf{u} = \left\langle -\frac{2}{\sqrt{5}}, \frac{1}{\sqrt{5}} \right\rangle. \qquad \equiv$$

☐ **i, j Vectors** The unit vectors in the direction of the positive x- and y-axes, denoted by

$$\mathbf{i} = \langle 1, 0 \rangle \qquad \text{and} \qquad \mathbf{j} = \langle 0, 1 \rangle, \tag{9}$$

are of special importance. See FIGURE 5.5.12. The two unit vectors in (9) are called the **standard basis vectors** for the vectors in 2-space because every vector can be expressed in terms of **i** and **j**. To see why this is so we use the definitions of vector addition and scalar multiplication to rewrite $\mathbf{u} = \langle a_1, a_2 \rangle$ as

$$\mathbf{u} = \langle a_1, 0 \rangle + \langle 0, a_2 \rangle = a_1\langle 1, 0 \rangle + a_2\langle 0, 1 \rangle$$

or

$$\mathbf{u} = \langle a_1, a_2 \rangle = a_1\mathbf{i} + a_2\mathbf{j}.$$

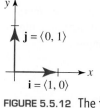

FIGURE 5.5.12 The **i** and **j** vectors

As shown in FIGURE 5.5.13, since **i** and **j** are unit vectors, the vectors $a_1\mathbf{i}$ and $a_2\mathbf{j}$ are horizontal and vertical vectors of length and $|a_1|$ and $|a_2|$, respectively. For this reason, a_1 is called the **horizontal component** of **u**, and a_2 is called the **vertical component**. The vector $a_1\mathbf{i} + a_2\mathbf{j}$ is often referred to as a **linear combination** of **i** and **j**. Using this notation for the vectors $\mathbf{u} = a_1\mathbf{i} + a_2\mathbf{j}$ and $\mathbf{v} = b_1\mathbf{i} + b_2\mathbf{j}$, we can write the definition of the sum, difference, and scalar multiples of **u** and **v** in the following manner:

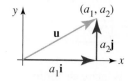

FIGURE 5.5.13 A vector **u** is a linear combination of **i** and **j**

Sum: $(a_1\mathbf{i} + a_2\mathbf{j}) + (b_1\mathbf{i} + b_2\mathbf{j}) = (a_1 + b_1)\mathbf{i} + (a_2 + b_2)\mathbf{j}$ (10)

Difference: $(a_1\mathbf{i} + a_2\mathbf{j}) - (b_1\mathbf{i} + b_2\mathbf{j}) = (a_1 - b_1)\mathbf{i} + (a_2 - b_2)\mathbf{j}$ (11)

Scalar multiple: $k(a_1\mathbf{i} + a_2\mathbf{j}) = (ka_1)\mathbf{i} + (ka_2)\mathbf{j}$ (12)

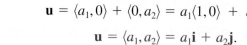

EXAMPLE 6 Difference of Vectors

If $\mathbf{u} = 3\mathbf{i} + \mathbf{j}$ and $\mathbf{v} = 5\mathbf{i} - 2\mathbf{j}$, find $4\mathbf{u} - 2\mathbf{v}$.

Solution We use (12) followed by (11) to obtain

$$\begin{aligned}
4\mathbf{u} - 2\mathbf{v} &= 4(3\mathbf{i} + \mathbf{j}) - 2(5\mathbf{i} - 2\mathbf{j}) \\
&= (12\mathbf{i} + 4\mathbf{j}) - (10\mathbf{i} - 4\mathbf{j}) \\
&= (12 - 10)\mathbf{i} + (4 - (-4))\mathbf{j} \\
&= 2\mathbf{i} + 8\mathbf{j}.
\end{aligned} \qquad \equiv$$

☐ **Trigonometric Form of a Vector** There is yet another way of representing vectors. For a nonzero vector $\mathbf{u} = \langle x, y \rangle$ with direction angle θ, we see from (2) that $x = |\mathbf{u}|\cos\theta$ and $y = |\mathbf{u}|\sin\theta$. Thus,

$$\mathbf{u} = x\mathbf{i} + y\mathbf{j} = |\mathbf{u}|\cos\theta\,\mathbf{i} + |\mathbf{u}|\sin\theta\,\mathbf{j},$$

or

$$\mathbf{u} = |\mathbf{u}|(\cos\theta\,\mathbf{i} + \sin\theta\,\mathbf{j}). \tag{13}$$

This latter representation is called the **trigonometric form** of the vector **u**.

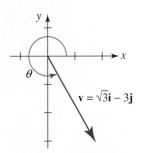

FIGURE 5.5.14 Vector and direction angle in Example 7

EXAMPLE 7   Trigonometric Form

Express the vector $\mathbf{u} = \sqrt{3}\mathbf{i} - 3\mathbf{j}$ in trigonometric form.

Solution To write $\mathbf{u}$ in trigonometric form, we must find the magnitude $|\mathbf{u}|$ and its direction angle θ. From (1) and (2) we find

$$|\mathbf{u}| = \sqrt{(\sqrt{3})^2 + (-3)^2} = \sqrt{12} = 2\sqrt{3}$$

$$\tan \theta = -\frac{3}{\sqrt{3}} = -\sqrt{3}.$$

To determine θ, we sketch $\mathbf{u}$ and observe that the terminal side of the angle θ lies in the fourth quadrant. See FIGURE 5.5.14. Thus, with $|\mathbf{u}| = 2\sqrt{3}$ and $\theta = 5\pi/3$, (13) gives the trigonometric form of $\mathbf{u}$:

$$\mathbf{u} = 2\sqrt{3}\left(\cos\frac{5\pi}{3}\mathbf{i} + \sin\frac{5\pi}{3}\mathbf{j}\right). \qquad \equiv$$

FIGURE 5.5.15 Velocity vector in Example 8

EXAMPLE 8   Velocity as a Vector

Given that an airplane is flying at 200 mi/h on a bearing of N20°E, express its velocity as a vector.

Solution The desired velocity vector $\mathbf{v}$ is shown in FIGURE 5.5.15. Measured from the positive x-axis we see that the direction angle θ of $\mathbf{v}$ is $\theta = 90° - 20° = 70°$. Then using $|\mathbf{v}| = 200$, we have the vector

$$\mathbf{v} = 200(\cos 70°\mathbf{i} + \sin 70°\mathbf{j}) \approx 68.4\mathbf{i} + 187.9\mathbf{j}. \qquad \equiv$$

In Example 8 we see that velocity is a vector quantity. The magnitude $|\mathbf{v}|$ of the velocity $\mathbf{v}$ is a scalar quantity called **speed**.

In physics it is shown that when two forces act simultaneously at the same point P on an object, the object reacts as though a single force equal to the vector sum of the two forces is acting on the object at P. This single force is called the **resultant force**.

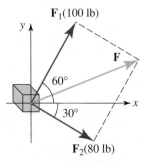

FIGURE 5.5.16 Resultant force (gold) in Example 9

EXAMPLE 9   Resultant Force

Two people push on a crate with forces $\mathbf{F}_1$ and $\mathbf{F}_2$, whose magnitudes and directions are shown in FIGURE 5.5.16. Find the magnitude and the direction of the resultant force.

Solution From the figure, we see that the direction angles for the two forces $\mathbf{F}_1$ and $\mathbf{F}_2$ are $\theta_1 = 60°$ and $\theta_2 = 330°$, respectively. Thus,

$$\mathbf{F}_1 = 100(\cos 60°\mathbf{i} + \sin 60°\mathbf{j}) = 50\mathbf{i} + 50\sqrt{3}\mathbf{j}$$
$$\mathbf{F}_2 = 80(\cos 330°\mathbf{i} + \sin 330°\mathbf{j}) = 40\sqrt{3}\mathbf{i} - 40\mathbf{j}.$$

The resultant force $\mathbf{F}$ can then be found by vector addition:

$$\mathbf{F} = \mathbf{F}_1 + \mathbf{F}_2 = (50\mathbf{i} + 50\sqrt{3}\mathbf{j}) + (40\sqrt{3}\mathbf{i} - 40\mathbf{j})$$
$$= (50 + 40\sqrt{3})\mathbf{i} + (50\sqrt{3} - 40)\mathbf{j}.$$

Thus the magnitude $|\mathbf{F}|$ of the resultant force is

$$|\mathbf{F}| = \sqrt{(50 + 40\sqrt{3})^2 + (50\sqrt{3} - 40)^2} \approx 128.06.$$

If θ is a direction angle for $\mathbf{F}$, then we know from (2) that

$$\tan \theta = \frac{50\sqrt{3} - 40}{50 + 40\sqrt{3}}.$$

Since θ is a first quadrant angle, we find with the help of a calculator that $\theta \approx 21.34°$.

☐ **Dot Product** Up to now we have considered two kinds of vector operations on vectors, addition and scalar multiplication, which produced another vector. We now consider a special kind of product between vectors that originated in the study of mechanics. This product, known as the **dot product**, or **inner product**, of vectors $\mathbf{u}$ and $\mathbf{v}$ is denoted by $\mathbf{u} \cdot \mathbf{v}$ and is a real number, or scalar, defined in terms of the components of the vectors.

DEFINITION 5.5.3   Dot Product

In 2-space the **dot product** of two vectors $\mathbf{u} = \langle a_1, a_2 \rangle$ and $\mathbf{v} = \langle b_1, b_2 \rangle$ is the number

$$\mathbf{u} \cdot \mathbf{v} = a_1 b_1 + a_2 b_2 \qquad (14)$$

EXAMPLE 10   **Dot Product Using (14)**

Suppose $\mathbf{u} = \langle -2, 5 \rangle$, $\mathbf{v} = \langle \frac{1}{2}, 4 \rangle$, and $\mathbf{w} = \langle 8, -1 \rangle$. Find:
(a) $\mathbf{u} \cdot \mathbf{v}$ **(b)** $\mathbf{w} \cdot \mathbf{u}$ **(c)** $\mathbf{v} \cdot \mathbf{w}$

Solution It follows from (14) that

(a) $\mathbf{u} \cdot \mathbf{v} = \langle -2, 5 \rangle \cdot \langle \frac{1}{2}, 4 \rangle = (-2)(\frac{1}{2}) + (5)(4) = -1 + 20 = 19$
(b) $\mathbf{w} \cdot \mathbf{u} = \langle 8, -1 \rangle \cdot \langle -2, 5 \rangle = (8)(-2) + (-1)(5) = -16 - 5 = -21$
(c) $\mathbf{v} \cdot \mathbf{w} = \langle \frac{1}{2}, 4 \rangle \cdot \langle 8, -1 \rangle = (\frac{1}{2})(8) + (4)(-1) = 4 - 4 = 0.$ ≡

☐ **Properties** The dot product possesses the following properties.

THEOREM 5.5.2   Properties of the Dot Product

 (i) $\mathbf{u} \cdot \mathbf{v} = 0$ if $\mathbf{u} = \mathbf{0}$ or $\mathbf{v} = \mathbf{0}$
 (ii) $\mathbf{u} \cdot \mathbf{v} = \mathbf{v} \cdot \mathbf{u}$ ← commutative law
 (iii) $\mathbf{u} \cdot (\mathbf{v} + \mathbf{w}) = \mathbf{u} \cdot \mathbf{v} + \mathbf{u} \cdot \mathbf{w}$ ← distributive law
 (iv) $\mathbf{u} \cdot (k\mathbf{v}) = (k\mathbf{u}) \cdot \mathbf{v} = k(\mathbf{u} \cdot \mathbf{v})$, k a scalar
 (v) $\mathbf{u} \cdot \mathbf{u} \geq 0$
 (vi) $\mathbf{u} \cdot \mathbf{u} = |\mathbf{u}|^2$

PROOF: We prove parts (ii) and (vi). The remaining proofs are straightforward and left for the reader. To prove part (ii) we let $\mathbf{u} = \langle a_1, a_2 \rangle$ and $\mathbf{v} = \langle b_1, b_2 \rangle$. Then

$$\begin{aligned}
\mathbf{u} \cdot \mathbf{v} &= \langle a_1, a_2 \rangle \cdot \langle b_1, b_2 \rangle \\
&= a_1 b_1 + a_2 b_2 \\
&= b_1 a_1 + b_2 a_2 \\
&= \langle b_1, b_2 \rangle \cdot \langle a_1, a_2 \rangle = \mathbf{v} \cdot \mathbf{u}.
\end{aligned}$$

← $\begin{cases} \text{since multiplication} \\ \text{of real numbers is} \\ \text{commutative} \end{cases}$

To prove part (vi) we note that

$$\mathbf{u} \cdot \mathbf{u} = \langle a_1, a_2 \rangle \cdot \langle a_1, a_2 \rangle = a_1^2 + a_2^2 = |\mathbf{u}|^2.$$ ≡

EXAMPLE 11   **Dot Products**

Let $\mathbf{u} = \langle 3, 2 \rangle$ and $\mathbf{v} = \langle -4, -5 \rangle$. Find **(a)** $(\mathbf{u} \cdot \mathbf{v})\mathbf{u}$ **(b)** $\mathbf{u} \cdot \left(\frac{1}{2}\mathbf{v}\right)$ **(c)** $|\mathbf{v}|$.

Solution **(a)** From (14),

$$\mathbf{u} \cdot \mathbf{v} = \langle 3, 2 \rangle \cdot \langle -4, -5 \rangle$$
$$= 3(-4) + 2(-5)$$
$$= -22.$$

Because $\mathbf{u} \cdot \mathbf{v}$ is a scalar we have from (4) of Definition 5.5.2,

$$(\mathbf{u} \cdot \mathbf{v})\mathbf{u} = (-22)\langle 3, 2 \rangle = \langle -66, -44 \rangle.$$

(b) From (*iv*) of Theorem 5.5.2 and part (a),

$$\mathbf{u} \cdot \left(\tfrac{1}{2}\mathbf{v}\right) = \tfrac{1}{2}(\mathbf{u} \cdot \mathbf{v}) = \tfrac{1}{2}(-22) = -11.$$

(c) Part (*vi*) of Theorem 5.5.2 relates the magnitude of a vector with the dot product. From (14) we have

$$\mathbf{v} \cdot \mathbf{v} = \langle -4, -5 \rangle \cdot \langle -4, -5 \rangle$$
$$= (-4)(-4) + (-5)(-5)$$
$$= 41.$$

Therefore, $\mathbf{v} \cdot \mathbf{v} = |\mathbf{v}|^2$ implies $|\mathbf{v}| = \sqrt{\mathbf{v} \cdot \mathbf{v}} = \sqrt{41}$.   ≡

☐ **Alternative Form** The dot product of two vectors can also be expressed in terms of the lengths of the vectors and the angle between them.

THEOREM 5.5.3   Alternative Form of the Dot Product

The dot product of two vectors $\mathbf{u}$ and $\mathbf{v}$ is

$$\mathbf{u} \cdot \mathbf{v} = |\mathbf{u}||\mathbf{v}|\cos\theta \qquad (15)$$

where θ is the angle between the vectors such that $0 \le \theta \le \pi$.

▶ This more geometric form is what is generally used as the definition of the dot product in a physics course.

PROOF: Suppose θ is the angle between the vectors $\mathbf{u} = a_1\mathbf{i} + a_2\mathbf{j}$ and $\mathbf{v} = b_1\mathbf{i} + b_2\mathbf{j}$. Then the vector

$$\mathbf{w} = \mathbf{v} - \mathbf{u} = (b_1 - a_1)\mathbf{i} + (b_2 - a_2)\mathbf{j}$$

is the third side of the triangle indicated in FIGURE 5.5.17. By the Law of Cosines, (2) of Section 5.4, we can write

$$|\mathbf{w}|^2 = |\mathbf{v}|^2 + |\mathbf{u}|^2 - 2|\mathbf{v}||\mathbf{u}|\cos\theta \quad \text{or} \quad |\mathbf{v}||\mathbf{u}|\cos\theta = \tfrac{1}{2}\left(|\mathbf{v}|^2 + |\mathbf{u}|^2 - |\mathbf{w}|^2\right). \quad (16)$$

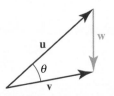

FIGURE 5.5.17 The vector $\mathbf{w}$ in the proof of Theorem 5.5.3

Using
$$|\mathbf{u}|^2 = a_1^2 + a_2^2, \quad |\mathbf{v}|^2 = b_1^2 + b_2^2,$$
and
$$|\mathbf{w}|^2 = |\mathbf{v} - \mathbf{u}|^2 = (b_1 - a_1)^2 + (b_2 - a_2)^2,$$

the right-hand side of the second equation in (16) simplifies to $a_1b_1 + a_2b_2$. Since this is the definition of the dot product given in (14), we see that $|\mathbf{u}||\mathbf{v}|\cos\theta = \mathbf{u} \cdot \mathbf{v}$.   ≡

☐ **Angle Between Vectors** FIGURE 5.5.18 illustrates three cases of the angle θ in (15). If the vectors **u** and **v** are not parallel, then θ is the *smaller* of the two possible angles between them. Solving for $\cos\theta$ in (15) and then using the definition of the dot product in (14) we have a formula for the cosine of the angle between two vectors:

$$\cos\theta = \frac{\mathbf{u} \cdot \mathbf{v}}{|\mathbf{u}||\mathbf{v}|} = \frac{a_1 b_1 + a_2 b_2}{|\mathbf{u}||\mathbf{v}|}. \qquad (17)$$

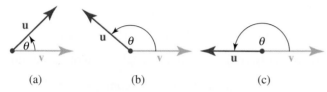

(a) (b) (c)

FIGURE 5.5.18 The angle θ in the dot product

EXAMPLE 12   Angle Between Two Vectors

Find the angle between $\mathbf{u} = 2\mathbf{i} + 5\mathbf{j}$ and $\mathbf{v} = 5\mathbf{i} - 4\mathbf{j}$.

Solution We have $|\mathbf{u}| = \sqrt{\mathbf{u} \cdot \mathbf{u}} = \sqrt{29}$, $|\mathbf{v}| = \sqrt{\mathbf{v} \cdot \mathbf{v}} = \sqrt{41}$, and $\mathbf{u} \cdot \mathbf{v} = -10$. Hence, (17) gives

$$\cos\theta = \frac{-10}{\sqrt{29}\sqrt{41}},$$

and so $\theta = \cos^{-1}\left(\dfrac{-10}{\sqrt{29}\sqrt{41}}\right) \approx 1.8650$ radians or $\theta \approx 106.86°$. See FIGURE 5.5.19. ≡

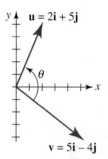

FIGURE 5.5.19 Angle between the vectors in Example 12

☐ **Orthogonal Vectors** If **u** and **v** are nonzero vectors, then Theorem 5.5.3 implies that

(*i*) $\mathbf{u} \cdot \mathbf{v} > 0$ if θ is acute,
(*ii*) $\mathbf{u} \cdot \mathbf{v} < 0$ if θ is obtuse, and
(*iii*) $\mathbf{u} \cdot \mathbf{v} = 0$ if $\cos\theta = 0$.

But in the last case, the only number in the interval $[0, \pi]$ for which $\cos\theta = 0$ is $\theta = \pi/2$. When $\theta = \pi/2$, we say that the vectors are **orthogonal** or **perpendicular**. Thus, we are led to the following result.

THEOREM 5.5.4   Criterion for Orthogonal Vectors

Two nonzero vectors **u** and **v** are orthogonal if and only if $\mathbf{u} \cdot \mathbf{v} = 0$.

As seen in Figure 5.5.12 the standard basis vectors **i** and **j** are orthogonal. Moreover, because $\mathbf{i} = \langle 1, 0\rangle$, $\mathbf{j} = \langle 0, 1\rangle$ we have

$$\mathbf{i} \cdot \mathbf{j} = \langle 1, 0\rangle \cdot \langle 0, 1\rangle = (1)(0) + (0)(1) = 0$$

and so from Theorem 5.5.4 the vectors **i** and **j** are orthogonal. Inspection of the result in part (c) of Example 10 shows that the two vectors $\mathbf{v} = \langle \frac{1}{2}, 4\rangle$, and $\mathbf{w} = \langle 8, -1\rangle$ are orthogonal.

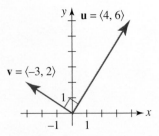

FIGURE 5.5.20 Orthogonal vectors in Example 13

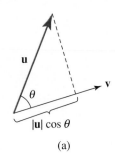

(a)

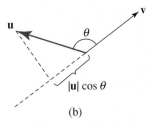

(b)

FIGURE 5.5.21 Component of vector **u** on vector **v**

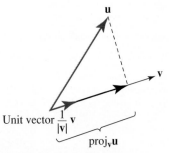

Unit vector $\dfrac{1}{|\mathbf{v}|}\mathbf{v}$

$\text{proj}_\mathbf{v}\mathbf{u}$

FIGURE 5.5.22 Projection of vector **u** onto vector **v**

If $\mathbf{u} = \langle 4, 6\rangle$ and $\mathbf{v} = \langle -3, 2\rangle$, then

$$\mathbf{u} \cdot \mathbf{v} = (4)(-3) + (6)(2) = -12 + 12 = 0.$$

From Theorem 5.5.4, we conclude that **u** and **v** are orthogonal. See FIGURE 5.5.20.   ≡

☐ **Component of u on v** Parts (*ii*), (*iii*), and (*vi*) of Theorem 5.5.2 enable us to express the components of a vector $\mathbf{u} = a_1\mathbf{i} + a_2\mathbf{j}$ in terms of a dot product:

$$\mathbf{i} \cdot \mathbf{u} = \mathbf{u} \cdot \mathbf{i} = (a_1\mathbf{i} + a_2\mathbf{j}) \cdot \mathbf{i} = a_1\overbrace{(\mathbf{i} \cdot \mathbf{i})}^{1} + a_2\overbrace{(\mathbf{j} \cdot \mathbf{i})}^{0} = a_1.$$

That is, $\mathbf{u} \cdot \mathbf{i} = a_1$. Similarly, $\mathbf{u} \cdot \mathbf{j} = a_2$. Symbolically, we write these components of **u** as

$$\text{comp}_\mathbf{i}\mathbf{u} = \mathbf{u} \cdot \mathbf{i} \qquad \text{and} \qquad \text{comp}_\mathbf{j}\mathbf{u} = \mathbf{u} \cdot \mathbf{j}. \tag{18}$$

We shall now see that the procedure indicated in (18) carries over to finding the **component of u on a vector v**. Note that in either of the two cases shown in FIGURE 5.5.21,

$$\text{comp}_\mathbf{v}\mathbf{u} = |\mathbf{u}|\cos\theta. \tag{19}$$

In Figure 5.5.21(a), $\text{comp}_\mathbf{v}\mathbf{u} \geq 0$ since $0 < \theta \leq \pi/2$, whereas in Figure 5.5.21(b), $\text{comp}_\mathbf{v}\mathbf{u} < 0$ since $\pi/2 < \theta \leq \pi$. Now, by writing (19) as

$$\text{comp}_\mathbf{v}\mathbf{u} = \frac{|\mathbf{u}||\mathbf{v}|\cos\theta}{|\mathbf{v}|} = \frac{\mathbf{u} \cdot \mathbf{v}}{|\mathbf{v}|},$$

we see that

$$\text{comp}_\mathbf{v}\mathbf{u} = \mathbf{u} \cdot \left(\frac{\mathbf{v}}{|\mathbf{v}|}\right). \tag{20}$$

In other words:

*To find the component of vector **u** on vector **v**, we dot **u** with a unit vector in the direction of* **v**.

■■■ EXAMPLE 14   Component of a Vector on Another

Let $\mathbf{u} = 2\mathbf{i} + 3\mathbf{j}$ and $\mathbf{v} = \mathbf{i} + \mathbf{j}$. Find $\text{comp}_\mathbf{v}\mathbf{u}$.

Solution We first form a unit vector in the direction of **v**:

$$|\mathbf{v}| = \sqrt{2} \quad \text{so} \quad \frac{\mathbf{v}}{|\mathbf{v}|} = \frac{1}{\sqrt{2}}(\mathbf{i} + \mathbf{j}).$$

Then from (20) we have

$$\text{comp}_\mathbf{v}\mathbf{u} = (2\mathbf{i} + 3\mathbf{j}) \cdot \frac{1}{\sqrt{2}}(\mathbf{i} + \mathbf{j}) = \frac{5}{\sqrt{2}}. \qquad ≡$$

☐ **Projection of u onto v** The *projection* of a vector **u** in any of the directions determined by **i** and **j**, is the *vector* formed by multiplying the component of $\mathbf{u} = a_1\mathbf{i} + a_2\mathbf{j}$ in the specified direction with a unit vector in that direction; for example,

$$\text{proj}_\mathbf{i}\mathbf{u} = (\text{comp}_\mathbf{i}\mathbf{u})\mathbf{i} = (\mathbf{u} \cdot \mathbf{i})\mathbf{i} = a_1\mathbf{i},$$

and so on. FIGURE 5.5.22 shows the general case of the **projection of u onto v**:

$$\text{proj}_\mathbf{v}\mathbf{u} = (\text{comp}_\mathbf{v}\mathbf{u})\overset{\substack{\text{scalar} \quad \text{unit vector} \\ \downarrow \qquad \downarrow}}{\frac{\mathbf{v}}{|\mathbf{v}|}} \tag{21}$$

That is,

> *To find the projection of vector* **u** *onto a vector* **v**, *we multiply a unit vector in the direction of* **v** *by the component of* **u** *on* **v**.

If desired, the result in (21) can be expressed in terms of two dot products. Using (20)

$$\text{proj}_v\mathbf{u} = \left(\mathbf{u}\cdot\underset{\uparrow}{\left(\frac{\mathbf{v}}{|\mathbf{v}|}\right)}\right)\frac{\mathbf{v}}{\underset{\uparrow}{|\mathbf{v}|}} = \left(\frac{\mathbf{u}\cdot\mathbf{v}}{|\mathbf{v}|^2}\right)\mathbf{v}$$

scalar   unit vector

or

$$\text{proj}_v\mathbf{u} = \left(\frac{\mathbf{u}\cdot\mathbf{v}}{\mathbf{v}\cdot\mathbf{v}}\right)\mathbf{v}. \quad \leftarrow |\mathbf{v}|^2 = \mathbf{v}\cdot\mathbf{v} \text{ by } (vi) \text{ of Theorem 5.5.2}$$

▮ EXAMPLE 15   Projection of u onto v

Find the projection of $\mathbf{u} = 4\mathbf{i} + \mathbf{j}$ onto the vector $\mathbf{v} = 2\mathbf{i} + 3\mathbf{j}$. Graph.

Solution First, we find the component of **u** on **v**. A unit vector in the direction of **v** is

$$\frac{1}{\sqrt{13}}(2\mathbf{i} + 3\mathbf{j}).$$

and so the component of **u** on **v** is the number

$$\text{comp}_v\mathbf{u} = (4\mathbf{i} + \mathbf{j})\cdot\frac{1}{\sqrt{13}}(2\mathbf{i} + 3\mathbf{j}) = \frac{11}{\sqrt{13}}.$$

Thus, from (21)

$$\text{proj}_v\mathbf{u} = \underbrace{\left(\frac{11}{\sqrt{13}}\right)}_{}\underbrace{\frac{1}{\sqrt{13}}(2\mathbf{i}+3\mathbf{j})}_{} = \tfrac{22}{13}\mathbf{i} + \tfrac{33}{13}\mathbf{j}.$$

component of u   unit vector in
in the direction   the direction
of v   of v

The graph of this vector is shown in gold color in FIGURE 5.5.23.

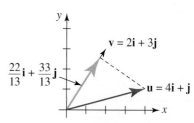

FIGURE 5.5.23 Projection of **u** onto **v** in Example 15

☐ Physical Interpretation of the Dot Product

When a constant force of magnitude F moves an object a distance d in the same direction of the force, the work done is defined to be

$$W = Fd. \tag{22}$$

However, if a constant force **F** applied to a body acts at an angle θ to the direction of motion, then the work done by **F** is defined to be the product of the component of **F** in the direction of the displacement and the distance $|\mathbf{d}|$ that the body moves:

$$W = (|\mathbf{F}|\cos\theta)|\mathbf{d}| = |\mathbf{F}||\mathbf{d}|\cos\theta.$$

See FIGURE 5.5.24. It follows from Theorem 5.5.3 that if **F** causes a displacement **d** of a body, then the work done is

$$W = \mathbf{F}\cdot\mathbf{d}. \tag{23}$$

Note that (23) reduces to (22) when $\theta = 0$.

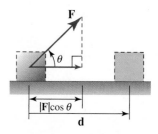

FIGURE 5.5.24 Work done by a force acting at an angle θ to the direction of motion

▮ EXAMPLE 16   Work Done by a Force at an Angle

Find the work done by a constant force $\mathbf{F} = 2\mathbf{i} + 4\mathbf{j}$ on a block that moves from $P_1(1, 1)$ to $P_2(4, 6)$. Assume that $|\mathbf{F}|$ is measured in pounds and $|\mathbf{d}|$ is measured in feet.

Solution The displacement vector of the block is given by

$$\mathbf{d} = \overrightarrow{P_1 P_2} = \overrightarrow{OP_2} - \overrightarrow{OP_1} = 3\mathbf{i} + 5\mathbf{j}.$$

It follows from (23) that the work done is

$$W = \mathbf{F} \cdot \mathbf{d} = (2\mathbf{i} + 4\mathbf{j}) \cdot (3\mathbf{i} + 5\mathbf{j}) = 26 \text{ ft-lb}.$$

≡

NOTES FROM THE CLASSROOM

You should not draw the conclusion from the preceding discussion that all vector quantities can be pictured as arrows. Many applications of vectors in advanced mathematics do not lend themselves to this interpretation. However, for purposes in this text, we find this interpretation both convenient and useful.

5.5 | **Exercises** Answers to selected odd-numbered problems begin on page ANS-16.

In Problems 1–8, sketch the given vector. Find the magnitude and the smallest positive direction angle of each vector.

1. $\langle \sqrt{3}, -1 \rangle$ **2.** $\langle 4, -4 \rangle$

3. $\langle 5, 0 \rangle$ **4.** $\langle -2, 2\sqrt{3} \rangle$

5. $-4\mathbf{i} + 4\sqrt{3}\mathbf{j}$ **6.** $\mathbf{i} - \mathbf{j}$

7. $-10\mathbf{i} + 10\mathbf{j}$ **8.** $-3\mathbf{j}$

In Problems 9–14, find $\mathbf{u} + \mathbf{v}$, $\mathbf{u} - \mathbf{v}$, $-3\mathbf{u}$, and $3\mathbf{u} - 4\mathbf{v}$.

9. $\mathbf{u} = \langle 2, 3 \rangle$, $\mathbf{v} = \langle 1, -1 \rangle$ **10.** $\mathbf{u} = \langle 4, -2 \rangle$, $\mathbf{v} = \langle 10, 2 \rangle$

11. $\mathbf{u} = \langle -4, 2 \rangle$, $\mathbf{v} = \langle 4, 1 \rangle$ **12.** $\mathbf{u} = \langle -1, -5 \rangle$, $\mathbf{v} = \langle 8, 7 \rangle$

13. $\mathbf{u} = \langle -5, -7 \rangle$, $\mathbf{v} = \langle \frac{1}{2}, -\frac{1}{4} \rangle$ **14.** $\mathbf{u} = \langle 0.1, 0.2 \rangle$, $\mathbf{v} = \langle -0.3, 0.4 \rangle$

In Problems 15–20, find $\mathbf{u} - 4\mathbf{v}$ and $2\mathbf{u} + 5\mathbf{v}$.

15. $\mathbf{u} = \mathbf{i} - 2\mathbf{j}$, $\mathbf{v} = 8\mathbf{i} + 3\mathbf{j}$ **16.** $\mathbf{u} = \mathbf{j}$, $\mathbf{v} = 4\mathbf{i} - \mathbf{j}$

17. $\mathbf{u} = \frac{1}{2}\mathbf{i} - \frac{3}{2}\mathbf{j}$, $\mathbf{v} = 2\mathbf{i}$ **18.** $\mathbf{u} = 2\mathbf{i} - 3\mathbf{j}$, $\mathbf{v} = 3\mathbf{i} - 2\mathbf{j}$

19. $\mathbf{u} = 0.2\mathbf{i} + 0.1\mathbf{j}$, $\mathbf{v} = -1.4\mathbf{i} - 2.1\mathbf{j}$ **20.** $\mathbf{u} = 5\mathbf{i} - 10\mathbf{j}$, $\mathbf{v} = -10\mathbf{i}$

In Problems 21–24, sketch the vectors $\mathbf{u} + \mathbf{v}$ and $\mathbf{u} - \mathbf{v}$.

21. $\mathbf{u} = 2\mathbf{i} + 3\mathbf{j}$, $\mathbf{v} = -\mathbf{i} + 2\mathbf{j}$ **22.** $\mathbf{u} = -4\mathbf{i} + \mathbf{j}$, $\mathbf{v} = 2\mathbf{i} + 2\mathbf{j}$

23. $\mathbf{u} = 5\mathbf{i} - \mathbf{j}$, $\mathbf{v} = 4\mathbf{i} - 3\mathbf{j}$ **24.** $\mathbf{u} = 2\mathbf{i} - 7\mathbf{j}$, $\mathbf{v} = -7\mathbf{i} - 3\mathbf{j}$

In Problems 25–28, sketch the vectors $2\mathbf{v}$ and $-2\mathbf{v}$.

25. $\mathbf{v} = \langle -2, 1 \rangle$ **26.** $\mathbf{v} = \langle 4, 7 \rangle$

27. $\mathbf{v} = 3\mathbf{i} - 5\mathbf{j}$ **28.** $\mathbf{v} = -\frac{1}{2}\mathbf{i} + \frac{3}{2}\mathbf{j}$

In Problems 29–32, if $\mathbf{u} = 3\mathbf{i} - \mathbf{j}$ and $\mathbf{v} = 2\mathbf{i} + 4\mathbf{j}$, find the horizontal and the vertical components of the indicated vector.

29. $2\mathbf{u} - \mathbf{v}$ **30.** $3(\mathbf{u} + \mathbf{v})$

31. $\mathbf{v} - 4\mathbf{u}$ **32.** $4(\mathbf{u} + 3\mathbf{v})$

In Problems 33–36, express the given vector **(a)** in trigonometric form and **(b)** as a linear combination of the unit vectors **i** and **j**.

33. $\langle -\sqrt{2}, \sqrt{2} \rangle$ **34.** $\langle 7, 7\sqrt{3} \rangle$
35. $\langle -3\sqrt{3}, 3 \rangle$ **36.** $\langle -4, -4 \rangle$

In Problems 37–40, find a unit vector **(a)** in the same direction as **v**, and **(b)** in the opposite direction of **v**.

37. $\mathbf{v} = \langle 2, 2 \rangle$ **38.** $\mathbf{v} = \langle -3, 4 \rangle$
39. $\mathbf{v} = \langle 0, -5 \rangle$ **40.** $\mathbf{v} = \langle 1, -\sqrt{3} \rangle$

In Problems 41 and 42, normalize the given vector when $\mathbf{v} = \langle 2, 8 \rangle$ and $\mathbf{w} = \langle 3, 4 \rangle$.

41. $\mathbf{v} + \mathbf{w}$ **42.** $2\mathbf{v} - 3\mathbf{w}$

43. Two forces $\mathbf{F}_1$ and $\mathbf{F}_2$ of magnitudes 4 N and 7 N, respectively, act on a point. If the angle between the forces is 47°, find the magnitude of the resultant force **F** and the angle between $\mathbf{F}_1$ and **F**.

44. The resultant **F** of two forces $\mathbf{F}_1$ and $\mathbf{F}_2$ has a magnitude of 100 lb and direction as shown in FIGURE 5.5.25. If $\mathbf{F}_1 = -200\mathbf{i}$, find the horizontal and the vertical components of $\mathbf{F}_2$.

FIGURE 5.5.25 Resultant in Problem 44

In Problems 45–48, find the dot product **u · v**.

45. $\mathbf{u} = \langle 4, 2 \rangle$, $\mathbf{v} = \langle 3, -1 \rangle$ **46.** $\mathbf{u} = \langle 1, -2 \rangle$, $\mathbf{v} = \langle 4, 0 \rangle$
47. $\mathbf{u} = 3\mathbf{i} - 2\mathbf{j}$, $\mathbf{v} = \mathbf{i} + \mathbf{j}$ **48.** $\mathbf{u} = 4\mathbf{i}$, $\mathbf{v} = -3\mathbf{j}$

In Problems 49–62, $\mathbf{u} = \langle 2, -3 \rangle$, $\mathbf{v} = \langle -1, 5 \rangle$, and $\mathbf{w} = \langle 3, -2 \rangle$. Find the indicated scalar or vector.

49. $\mathbf{u} \cdot \mathbf{v}$ **50.** $\mathbf{v} \cdot \mathbf{w}$
51. $\mathbf{u} \cdot \mathbf{w}$ **52.** $\mathbf{v} \cdot \mathbf{v}$
53. $\mathbf{w} \cdot \mathbf{w}$ **54.** $\mathbf{u} \cdot (\mathbf{v} + \mathbf{w})$
55. $\mathbf{u} \cdot (4\mathbf{v})$ **56.** $\mathbf{v} \cdot (\mathbf{u} - \mathbf{w})$
57. $(-\mathbf{v}) \cdot \left(\frac{1}{2}\mathbf{w}\right)$ **58.** $(2\mathbf{v}) \cdot (3\mathbf{w})$
59. $\mathbf{u} \cdot (\mathbf{u} + \mathbf{v} + \mathbf{w})$ **60.** $(2\mathbf{u}) \cdot (\mathbf{u} - 2\mathbf{v})$
61. $\left(\dfrac{\mathbf{u} \cdot \mathbf{v}}{\mathbf{v} \cdot \mathbf{v}}\right)\mathbf{v}$ **62.** $(\mathbf{w} \cdot \mathbf{v})\mathbf{u}$

In Problems 63 and 64, find the dot product **u · v** if the smaller angle between **u** and **v** is as given.

63. $|\mathbf{u}| = 10$, $|\mathbf{v}| = 5$, $\theta = \pi/4$ **64.** $|\mathbf{u}| = 6$, $|\mathbf{v}| = 12$, $\theta = \pi/6$

In Problems 65–68, find the angle between the given pair of vectors. Round your answer to two decimal places.

65. $\langle 1, 4 \rangle$, $\langle 2, -1 \rangle$ **66.** $\langle 3, 5 \rangle$, $\langle -4, -2 \rangle$
67. $\mathbf{i} - \mathbf{j}$, $3\mathbf{i} + \mathbf{j}$ **68.** $2\mathbf{i} - \mathbf{j}$, $4\mathbf{i} + \mathbf{j}$

In Problems 69–72, determine whether the given vectors are orthogonal.

69. $\mathbf{u} = \langle -5, -4 \rangle$, $\mathbf{v} = \langle -6, 8 \rangle$ **70.** $\mathbf{u} = \langle 3, -2 \rangle$, $\mathbf{v} = \langle -6, -9 \rangle$
71. $4\mathbf{i} - 5\mathbf{j}$, $\mathbf{i} + \frac{4}{5}\mathbf{j}$ **72.** $\frac{1}{2}\mathbf{i} + \frac{3}{4}\mathbf{j}$, $-\frac{2}{5}\mathbf{i} + \frac{4}{5}\mathbf{j}$

In Problems 73 and 74, find a scalar c so that the given vectors are orthogonal.

73. $\mathbf{u} = 2\mathbf{i} - c\mathbf{j}$, $\mathbf{v} = 3\mathbf{i} + 2\mathbf{j}$ **74.** $\mathbf{u} = 4c\mathbf{i} - 8\mathbf{j}$, $\mathbf{v} = c\mathbf{i} + 2\mathbf{j}$

75. Verify that the vector

$$\mathbf{w} = \mathbf{v} - \frac{\mathbf{v} \cdot \mathbf{u}}{|\mathbf{u}|^2}\mathbf{u}$$

is orthogonal to the vector $\mathbf{u}$.

76. Find a scalar c so that the angle between the vectors $\mathbf{u} = \mathbf{i} + c\mathbf{j}$ and $\mathbf{v} = \mathbf{i} + \mathbf{j}$ is $45°$.

In Problems 77–80, $\mathbf{u} = \langle 1, -1 \rangle$ and $\mathbf{v} = \langle 2, 6 \rangle$. Find the indicated number.

77. $\text{comp}_\mathbf{v}\,\mathbf{u}$ **78.** $\text{comp}_\mathbf{u}\,\mathbf{v}$

79. $\text{comp}_\mathbf{u}(\mathbf{v} - \mathbf{u})$ **80.** $\text{comp}_{2\mathbf{v}}(\mathbf{u} + \mathbf{v})$

In Problems 81 and 82, find **(a)** $\text{proj}_\mathbf{v}\mathbf{u}$, and **(b)** $\text{proj}_\mathbf{u}\mathbf{v}$.

81. $\mathbf{u} = -5\mathbf{i} + 5\mathbf{j}, \mathbf{v} = -3\mathbf{i} + 4\mathbf{j}$ **82.** $\mathbf{u} = 4\mathbf{i} + 2\mathbf{j}, \mathbf{v} = -3\mathbf{i} + \mathbf{j}$

In Problems 83 and 84, $\mathbf{u} = 4\mathbf{i} + 3\mathbf{j}$ and $\mathbf{v} = -\mathbf{i} + \mathbf{j}$. Find the indicated vector.

83. $\text{proj}_{\mathbf{u}+\mathbf{v}}\mathbf{u}$ **84.** $\text{proj}_{\mathbf{u}-\mathbf{v}}\mathbf{v}$

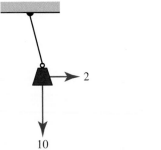

FIGURE 5.5.26 Block in Problem 86

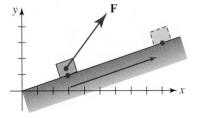

FIGURE 5.5.27 Block in Problem 87

85. A sled is pulled horizontally over ice by a rope attached to its front. A 20-lb force acting at an angle of $60°$ with the horizontal moves the sled 100 ft. Find the work done.

86. A block with weight $\mathbf{w}$ is pulled along a frictionless horizontal surface by a constant force $\mathbf{F}$ of magnitude 30 lb in the direction by the vector $\mathbf{d}$. See FIGURE 5.5.26.
   (a) What is the work done by the weight $\mathbf{w}$?
   (b) What is the work done by the force $\mathbf{F}$ if $\mathbf{d} = 4\mathbf{i} + 3\mathbf{j}$?

87. A constant force $\mathbf{F}$ of magnitude 3 lb is applied to the block shown in FIGURE 5.5.27. The force $\mathbf{F}$ has the same direction as the vector $\mathbf{u} = 3\mathbf{i} + 4\mathbf{j}$. Find the work done in the direction of motion if the block moves from $P_1(3, 1)$ to $P_2(9, 3)$. Assume distance is measured in feet.

Miscellaneous Applications

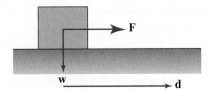

FIGURE 5.5.28 Hanging mass in Problem 89

88. Resultant Force A small boat is pulled along a canal by two tow ropes on opposite sides of the canal. The angle between the ropes is $50°$. If one rope is pulled with a force of 250 lb and the other with a force of 400 lb, find the magnitude of the resultant force and the angle it makes with the 250-lb force.

89. Resultant Force A mass weighing 10 lb is hanging from a rope. A 2-lb force is applied horizontally to the weight, moving the rope from its horizontal position. See FIGURE 5.5.28. Find the resultant of this force and the force due to gravity.

90. In What Direction? As a freight train, traveling at 10 mi/h, passes a landing, a mail sack is tossed out perpendicular to the train with a velocity of 15 feet per second. In what direction does the mail sack slide on the landing?

91. Actual Direction The current in a river that is 0.5 mi across is 6 mi/h. A swimmer heads out from shore perpendicular to the current at 2 mi/h. In what direction is the swimmer actually going?

92. Getting One's Bearings A hiker walks 1.0 mi to the northeast, then 1.5 mi to the east, and then 2.0 mi to the southeast. What are the hiker's distance and bearing from the starting point? [*Hint*: Each part of the journey can be represented by a vector. Find the vector sum.]

93. What Is the Speed? In order for an airplane to fly due north at 300 mi/h, it must set a course $10°$ west of north (N10°W) because of a strong wind blowing due east. What is the speed of the wind?

For Discussion

94. A 200-lb traffic light supported by two cables hangs in static equilibrium. A condition of static equilibrium is that the object is at rest and that the sum of forces acting on the object is the zero vector **0**. As shown in **FIGURE 5.5.29(b)**, let the weight of the light be represented by **w** and forces in the two cables by $\mathbf{F}_1$ and $\mathbf{F}_2$. From the Figure 5.5.29(b) we then have

$$\mathbf{w} + \mathbf{F}_1 + \mathbf{F}_2 = \mathbf{0}, \tag{24}$$

where the vectors on the left-hand side of the equality in trigonometric form are

$$\mathbf{w} = 200(\cos 270° \mathbf{i} + \sin 270° \mathbf{j})$$
$$\mathbf{F}_1 = |\mathbf{F}_1|(\cos 20° \mathbf{i} + \sin 20° \mathbf{j})$$
$$\mathbf{F}_2 = |\mathbf{F}_2|(\cos 165° \mathbf{i} + \sin 165° \mathbf{j})$$

Use (24) to determine the magnitude of $\mathbf{F}_1$ and $\mathbf{F}_2$. [*Hint*: Use (5) of Definition 5.5.2 and let $\mathbf{v} = \mathbf{0} = \langle 0, 0 \rangle$.]

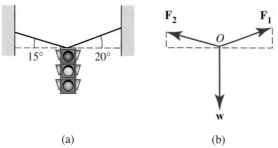

(a)   (b)

FIGURE 5.5.29 Hanging mass in Problem 94

| CHAPTER 5 | Review Exercises | Answers to selected odd-numbered problems begin on page ANS–17. |

A. Fill in the Blanks

In Problems 1–12, fill in the blanks.

1. To solve a triangle in which you know two angles and a side of opposite one of these angles, you would use the Law of _____ first.
2. The ambiguous case refers to solving a triangle when _____ are given.
3. To solve a triangle in which you know two sides and the included angle, you would use the Law of _____ first.
4. In an isosceles triangle, if a is the length of one of the two equal sides and θ is one of the two equal angles, then the area of the triangle in terms of a and θ is _____.
5. The largest angle in the triangle whose sides are 5.3, 4.4, and 4.1 is _____.
6. If θ is one of the acute angles in a right triangle and $\sin \theta = \frac{1}{3}$, then $\tan \theta =$ _____.
7. A 6-ft-tall man walking along a level beach climbs onto a 4-ft-tall tree stump and looks over the water to the horizon. Assuming that the Earth is a perfect sphere of radius $r = 3963$ mi, then the distance measured along the surface of the Earth from the man to the horizon is _____ miles.
8. The difference between a *scalar* and *vector* is _____.
9. A unit vector in the opposite direction of $\mathbf{v} = \langle 12, -5 \rangle$ is _____.

10. If $\mathbf{u} = 4\mathbf{i} - 6\mathbf{j}$ and $\mathbf{v} = -3\mathbf{i} + 10\mathbf{j}$, then $5\mathbf{u} - 6\mathbf{v} = $ _____.
11. The angle between the vectors $\mathbf{u} = 5\mathbf{i}$ and $\mathbf{v} = -2\mathbf{j}$ is _____.
12. If $|\mathbf{u}| = 4$, $|\mathbf{v}| = 3$ and the angle between $\mathbf{u}$ and $\mathbf{v}$ is $\theta = 2\pi/3$, then $\mathbf{u} \cdot \mathbf{v} = $ _____.

B. True/False

In Problems 1–12, answer true or false.

1. In a right triangle, if $\tan\theta = \frac{3}{4}$, then $\sin\theta = 3$ and $\cos\theta = 4$. _____
2. In a right triangle, if $\sin\theta = \frac{11}{61}$, then $\cot\theta = \frac{60}{11}$. _____
3. For an acute angle θ in a right triangle, $\csc\theta = \frac{\text{opp}}{\text{hyp}}$. _____
4. The Pythagorean theorem is a special case of the Law of Cosines. _____
5. In a right triangle, the hypotenuse is always the longest side. _____
6. If α and β are the acute angles in a right triangle, then $\sin\alpha = \cos\beta$. _____
7. If $\tan\theta = \sqrt{15}$ for an acute angle θ in a right triangle, then $\cos\theta = \frac{1}{4}$. _____
8. A rowboat departs from a point on a straight beach that coincides with a north-south line. If the rowboat travels at a rate of 2.5 mi/h with a bearing of N35°W, then after 4 h the rowboat is 10 mi from the beach. _____
9. A 20 ft extension ladder rests against the side of a vertical wall. If the base of the ladder is on flat ground 5.2 ft from the wall, then the angle the ladder makes with the ground is 66.3°. _____
10. The vector $\mathbf{v} = \langle \sqrt{3}, \sqrt{5} \rangle$ is twice as long as the vector $\mathbf{u} = \langle -1, 1 \rangle$. _____
11. If $\mathbf{u}$ is a unit vector, then $\mathbf{u} \cdot \mathbf{u} = 1$. _____
12. If $\mathbf{u}$ and $\mathbf{v}$ are unit vectors, then $\mathbf{u} + \mathbf{v}$ and $\mathbf{u} - \mathbf{v}$ are orthogonal. _____

C. Exercises

In Problems 1–4, solve the triangle satisfying the given conditions.

1. $\alpha = 30°$, $\beta = 70°$, $b = 10$
2. $\gamma = 145°$, $a = 25$, $c = 20$
3. $\alpha = 51°$, $b = 20$, $c = 10$
4. $a = 4$, $b = 6$, $c = 3$

5. A surveyor 100 m from the base of an overhanging cliff measures a 28° angle of elevation from that point to the top of the cliff. See FIGURE 5.R.1. If the cliff makes an angle of 65° with the horizontal ground, determine its height h.

6. A rocket is launched from ground level at an angle of elevation of 43°. If the rocket hits a drone target plane flying at 20,000 ft, find the horizontal distance between the rocket launch site and the point directly beneath the plane. What is the straight-line distance between the rocket launch site and the target plane?

7. A competition water skier leaves a ramp at point R and lands at point S. See FIGURE 5.R.2. A judge at point J measures an $\angle RJS$ as 47°. If the distance from the ramp to the judge is 110 ft, find the length of the jump. Assume that $\angle SRJ$ is 90°.

8. The angle between two sides of a parallelogram is 40°. If the lengths of the sides are 5 and 10 cm, find the lengths of the two diagonals.

9. A weather satellite orbiting the equator of the Earth at a height of $H = 36{,}000$ km spots a thunderstorm to the north at P at an angle of $\theta = 6.5°$ from its vertical. See FIGURE 5.R.3.
   (a) Given that the Earth's radius is approximately $R = 6370$ km, find the latitude ϕ of the thunderstorm.
   (b) Show that angles θ and ϕ are related by

$$\tan\theta = \frac{R\sin\phi}{H + R(1 - \cos\phi)}.$$

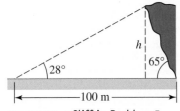

FIGURE 5.R.1 Cliff in Problem 5

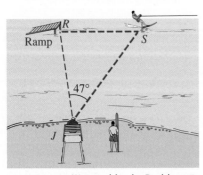

FIGURE 5.R.2 Water skier in Problem 7

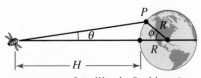

FIGURE 5.R.3 Satellite in Problem 9

CHAPTER 5 TRIANGLE TRIGONOMETRY

10. It can be shown that a basketball of diameter d approaching the basket from an angle θ to the horizontal will pass through a hoop of diameter D if $D \sin \theta > d$, where $0° \leq \theta \leq 90°$. See FIGURE 5.R.4. If the basketball has diameter 24.6 cm and the hoop has diameter 45 cm, what range of approach angles θ will result in a basket?

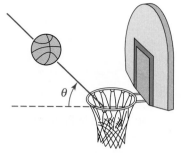

FIGURE 5.R.4 Basketball in Problem 10

11. Each of the 24 NAVSTAR Global Positioning System (GPS) satellites orbits the Earth at an altitude of $h = 20{,}200$ km. Using this network of satellites, an inexpensive handheld GPS receiver can determine its position on the surface of the Earth to within 10 m. Find the greatest distance s (in km) on the surface of the Earth that can be observed from a single GPS satellite. See FIGURE 5.R.5. Take the radius of the Earth to be 6370 km. [*Hint*: Find the central angle θ subtended by s.]

12. An airplane flying horizontally at a speed of 400 miles per hour is climbing at an angle of 6° from the horizontal. When the airplane passes directly over a car traveling 60 miles per hour, it is 2 miles above the car. Assuming that the airplane and the car remain in the same vertical plane, find the angle of elevation from the car to the airplane after 30 minutes.

13. A house measures 45 ft from front to back. The roof measures 32 ft from the front of the house to the peak and 18 ft from the peak to the back of the house. See FIGURE 5.R.6. Find the angles of elevation of the front and back parts of the roof.

14. The angle between two sides of a parallelogram is 40°. If the lengths of the sides are 5 and 10 cm, find the lengths of the two diagonals.

15. **Help is Coming** From two lifeguard towers A and B, a swimmer in distress is sighted on bearings of N46°E and N27°W, respectively. If tower B is 250 ft due east of tower A, what is the distance from each tower to the swimmer?

16. **Navigator's Error** An airplane is supposed to fly 500 mi due west to a refueling rendezvous point. If a 5° error is made in the heading, how far is the plane from the rendezvous point after flying 400 mi? Through what angle must the airplane turn in order to correct its course at that point?

In Problems 17–26, translate the words into an appropriate function.

17. A 20-ft-long water trough has ends in the form of isosceles triangles with sides that are 4 ft long. See Figure 2.8.21 in Exercises 2.8. As shown in FIGURE 5.R.7, let θ denote the angle between the vertical and one of the sides of a triangular end. Express the volume of the trough as a function of 2θ.

18. A person driving a car approaches a freeway sign as shown in FIGURE 5.R.8. Let θ be her viewing angle of the sign and let x represent her horizontal distance (measured in feet) to that sign. Express θ as a function of x.

19. As shown in FIGURE 5.R.9, a plank is supported by a sawhorse so that one end rests on the ground and the other end rests against a building. Express the length of the plank as a function of the indicated angle θ.

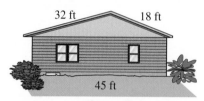

FIGURE 5.R.5 GPS satellite in Problem 11

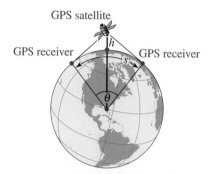

FIGURE 5.R.6 House in Problem 13

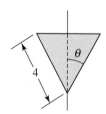

FIGURE 5.R.7 End of water trough in Problem 17

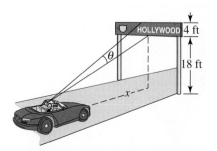

FIGURE 5.R.8 Freeway sign in Problem 18

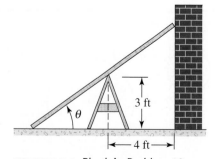

FIGURE 5.R.9 Plank in Problem 19

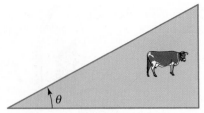

FIGURE 5.R.10 Pasture in Problem 20

20. A farmer wishes to enclose a pasture in the form of a right triangle using 2000 ft of fencing on hand. See FIGURE 5.R.10. Show that the area of the pasture as a function of the indicated angle θ is

$$A(\theta) = \frac{1}{2}\cot\theta \cdot \left(\frac{2000}{1 + \cot\theta + \csc\theta}\right)^2.$$

21. Express the volume of the box shown in FIGURE 5.R.11 as a function of the indicated angle θ.

22. A corner of an 8.5-in. × 11-in. piece of paper is folded over to the other edge of the paper as shown in FIGURE 5.R.12. Express the length L of the crease as a function of the angle θ shown in the figure.

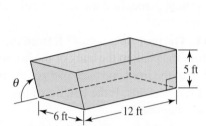

FIGURE 5.R.11 Box in Problem 21

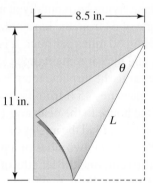

FIGURE 5.R.12 Folded paper in Problem 22

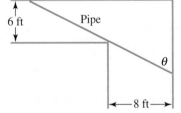

FIGURE 5.R.13 Gutter in Problem 23

23. A gutter is to be made from a sheet of metal 30 cm wide by turning up the edges of width 10 cm along each side so that the sides make equal angles ϕ with the vertical. See FIGURE 5.R.13. Express the cross-sectional area of the gutter as a function of the angle ϕ.

24. A metal pipe is to be carried horizontally around a right-angled corner from a hallway 8 feet wide into a hallway that is 6 feet wide. See FIGURE 5.R.14. Express the length L of the pipe as a function of the angle θ shown in the figure.

25. In FIGURE 5.R.15 the blue, green, and red circles are of radii 3, 4, and 6, respectively. The dots represent the centers of the circles.

    (a) Express the distance d between the centers of the blue and red circles as a function of the angle θ shown in the figure.

    (b) Use the function is part (a) to determine the value of θ corresponding to $d = 14$.

26. The container shown in FIGURE 5.R.16 consists of an inverted cone (open at its top) attached to the bottom of a right circular cylinder (open at its top and bottom) of fixed radius R. The container has a fixed volume V. Express the total surface area S of the container as a function of the indicated angle θ. [*Hint*: See Appendix B for the lateral surface area of a cone.]

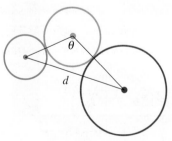

FIGURE 5.R.14 Pipe in Problem 24

FIGURE 5.R.15 Circle in Problem 25

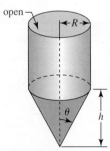

FIGURE 5.R.16 Container in Problem 26

In Problems 27–42, $\mathbf{u} = -2\mathbf{i} + 3\mathbf{j}$, $\mathbf{v} = \mathbf{i} + \mathbf{j}$, and $\mathbf{w} = \mathbf{i} - 4\mathbf{j}$. Find the indicated vector or scalar.

27. $-5\mathbf{u} + 3\mathbf{v}$ **28.** $\mathbf{u} - 10\mathbf{v}$

29. $\mathbf{u} + (2\mathbf{v} + 3\mathbf{w})$ **30.** $4\mathbf{u} - (3\mathbf{v} + \mathbf{w})$

31. $(\mathbf{u} \cdot \mathbf{v})\mathbf{w} + (\mathbf{w} \cdot \mathbf{v})\mathbf{u}$ **32.** $(\mathbf{u} - \mathbf{v}) \cdot (\mathbf{v} + \mathbf{w})$

33. $\text{comp}_{\mathbf{w}}\mathbf{v}$ **34.** $\text{comp}_{\mathbf{u}}(-\mathbf{v})$

35. $\text{proj}_{\mathbf{v}}(2\mathbf{u})$ **36.** $\text{proj}_{\mathbf{w}}(\mathbf{u} + \mathbf{v})$

37. $|\mathbf{u}| + |2\mathbf{v}|$ **38.** $|\mathbf{u} + \mathbf{v}|$

39. trigonometric form of $2\mathbf{v}$

40. horizontal component of $-2(\mathbf{u} + \mathbf{w})$

41. a unit vector in the opposite direction of $\mathbf{w}$

42. the angle between $\mathbf{v}$ and $\mathbf{w}$

43. Two forces $\mathbf{F}_1$ and $\mathbf{F}_2$ act at a point such that the resultant force $\mathbf{F}$ has a magnitude of 5 lb and is orthogonal to $\mathbf{F}_1$. If $|\mathbf{F}_1| = 5$ lb then find the magnitude of the vector $\mathbf{F}_2$ and the angle between $\mathbf{F}_1$ and $\mathbf{F}_2$ in degrees.

44. A baby elephant weighing 315 lb is standing still on a loading ramp shown in FIGURE 5.R.17. Assume that the origin of the rectangular coordinate system is at O and that the ramp makes an angle of $20°$ with the horizontal.

(a) Express the vectors $\mathbf{w} = \overrightarrow{OB}$, $\mathbf{u} = \overrightarrow{OA}$ and $\mathbf{v} = \overrightarrow{OC}$ in trigonometric form. In each case use a direction angle that is positive and measured from the positive x-axis. The unknown quantities $|\mathbf{u}|$ and $|\mathbf{v}|$ are the magnitudes of the components of the weight vector $\mathbf{w} = \overrightarrow{OB}$ in the direction parallel to the ramp and perpendicular to the ramp, respectively.

(b) Find $|\mathbf{u}|$ and $|\mathbf{v}|$ by using (20) of Section 5.5 to find $\text{comp}_{\mathbf{u}}\, \mathbf{w}$ and $\text{comp}_{\mathbf{v}}\, \mathbf{w}$.

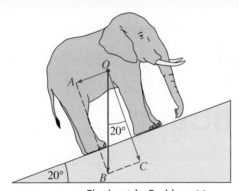

FIGURE 5.R.17 Elephant in Problem 44

6 Exponential and Logarithmic Functions

Chapter Outline

6.1 Exponential Functions

≡ **Introduction** In the preceding chapters we considered functions such as $f(x) = x^2$, that is, a function with a variable base x and constant power or exponent 2. We now examine functions having a constant base b and a variable exponent x.

DEFINITION 6.1.1 Exponential Function

If $b > 0$ and $b \neq 1$, then an **exponential function** $y = f(x)$ is a function of the form

$$f(x) = b^x. \tag{1}$$

The number b is called the **base** and x is called the **exponent**.

The **domain** of an exponential function f defined in (1) of Definition 6.1.1 is the set of all real numbers $(-\infty, \infty)$.

In (1) the base b is restricted to positive numbers in order to guarantee that b^x is always a real number. For example, with this restriction we avoid complex numbers such as $(-4)^{1/2}$. Also, the base $b = 1$ is of little interest to us since it can be shown that f is the constant function $f(x) = 1^x = 1$. Moreover, for $b > 0$, we have $f(0) = b^0 = 1$.

☐ **Exponents** As just mentioned, the domain of an exponential function (1) is the set of all real numbers. This means that the exponent x can be either a rational or an irrational number. For example, if the base $b = 3$ and the exponent x is a *rational number*, say, $x = \frac{1}{5}$ and $x = 1.4$, then

$$3^{1/5} = \sqrt[5]{3} \qquad \text{and} \qquad 3^{1.4} = 3^{14/10} = 3^{7/5} = \sqrt[5]{3^7}.$$

For an exponent x that is an *irrational number*, b^x is defined, but its precise definition is beyond the scope of this text. We can, however, suggest a procedure for defining a number such as $3^{\sqrt{2}}$. From the decimal representation $\sqrt{2} = 1.414213562\ldots$ we see that the rational numbers

$$1, 1.4, 1.41, 1.414, 1.4142, 1.41421, \ldots$$

are successively better approximations to $\sqrt{2}$. By using these rational numbers as exponents, we would expect that the numbers

$$3^1, 3^{1.4}, 3^{1.41}, 3^{1.414}, 3^{1.4142}, 3^{1.41421}, \ldots$$

are then successively better approximations to $3^{\sqrt{2}}$. In fact, this can be shown to be true with a precise definition of b^x for an irrational value of x. But on a practical level, we can use the $\boxed{y^x}$ key on a calculator to obtain the approximation 4.728804388 to $3^{\sqrt{2}}$.

☐ **Laws of Exponents** In most algebra texts the laws of exponents are stated first for integer exponents and then for rational exponents. Since b^x can be defined for all real numbers x when $b > 0$, it can be proved that these same **laws of exponents** hold for all real number exponents.

THEOREM 6.1.1 Laws of Exponents

If $a > 0$, $b > 0$ and x, x_1, and x_2 denote real numbers, then

(i) $b^{x_1} \cdot b^{x_2} = b^{x_1 + x_2}$

(ii) $\dfrac{b^{x_1}}{b^{x_2}} = b^{x_1 - x_2}$

(iii) $\dfrac{1}{b^x} = b^{-x}$

(iv) $(b^{x_1})^{x_2} = b^{x_1 x_2}$

(v) $(ab)^x = a^x b^x$

(vi) $\left(\dfrac{a}{b}\right)^x = \dfrac{a^x}{b^x}$

EXAMPLE 1   Rewriting a Function

At times, we will use the laws of exponents to rewrite a function in a different form. For example, neither $f(x) = 2^{3x}$ nor $g(x) = 4^{-2x}$ has the precise form of the exponential function defined in (1). However, by the laws of exponents given in Theorem 6.1.1, f can be rewritten as $f(x) = 8^x$ ($b = 8$ in (1)), and g can be recast as $g(x) = \left(\frac{1}{16}\right)^x$ ($b = \frac{1}{16}$ in (1)). The details are shown below:

$$f(x) = 2^{3x} \overset{\text{by } (iv)}{=} (2^3)^x \overset{\text{form is now } b^x}{=} 8^x$$

$$g(x) = 4^{-2x} \overset{\text{by } (iv)}{=} (4^{-2})^x \overset{\text{by } (iii)}{=} \left(\frac{1}{4^2}\right)^x \overset{\text{form is now } b^x}{=} \left(\frac{1}{16}\right)^x.$$

$\equiv$

☐ **Graphs** We distinguish two types of graphs for (1) depending on whether the base b satisfies $b > 1$ or $0 < b < 1$. The next two examples illustrate, in turn, the graphs of $f(x) = 3^x$ and $f(x) = \left(\frac{1}{3}\right)^x$. Before graphing, we can make some intuitive observations about both functions. Since the bases $b = 3$ and $b = \frac{1}{3}$ are positive, the values of 3^x and $\left(\frac{1}{3}\right)^x$ are *positive* for every real number x. As a consequence, there are no real numbers x_1 and x_2 for which 3^{x_1} and $\left(\frac{1}{3}\right)^{x_2}$ are zero. Graphically, this means that the graphs of $f(x) = 3^x$ and $f(x) = \left(\frac{1}{3}\right)^x$ have no x-intercepts. Also, $3^0 = 1$ and $\left(\frac{1}{3}\right)^0 = 1$, and so $f(0) = 1$ in each case. This means that the graphs of $f(x) = 3^x$ and $f(x) = \left(\frac{1}{3}\right)^x$ have the same y-intercept $(0, 1)$.

EXAMPLE 2   Graph for $b > 1$

Graph the function $f(x) = 3^x$.

Solution We first construct a table of some function values corresponding to preselected values of x. As shown in **FIGURE 6.1.1**, we plot the corresponding points obtained from the table and connect them with a continuous curve. The graph shows that f is an increasing function on the interval $(-\infty, \infty)$.

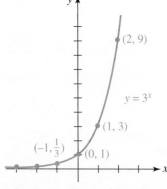

FIGURE 6.1.1 Graph of function in Example 2

x	-3	-2	-1	0	1	2
$f(x)$	$\frac{1}{27}$	$\frac{1}{9}$	$\frac{1}{3}$	1	3	9

$\equiv$

EXAMPLE 3   **Graph for $0 < b < 1$**

Graph the function $f(x) = \left(\frac{1}{3}\right)^x$.

Solution Proceeding as in Example 2, we construct a table of some function values corresponding to preselected values of x. Note, for example, by the laws of exponents

$$f(-2) = \left(\tfrac{1}{3}\right)^{-2} = (3^{-1})^{-2} = 3^2 = 9.$$

As shown in FIGURE 6.1.2, we plot the corresponding points obtained from the table and connect them with a continuous curve. In this case the graph shows that f is a decreasing function on the interval $(-\infty, \infty)$.

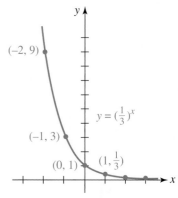

FIGURE 6.1.2 Graph of function in Example 3

x	-3	-2	-1	0	1	2
$f(x)$	27	9	3	1	$\frac{1}{3}$	$\frac{1}{9}$

$\equiv$

☐ **Reflections** Exponential functions with bases satisfying $0 < b < 1$, such as $b = \frac{1}{3}$, are frequently written in an alternative manner. We note that $y = \left(\frac{1}{3}\right)^x$ is the same as $y = 3^{-x}$. From this last result we see that the graph of $y = 3^{-x}$ is simply the graph of $y = 3^x$ reflected in the y-axis.

◀ Review Theorem 2.2.3 in Section 2.2 for reflections in the x- and y-axes.

☐ **Horizontal Asymptote** FIGURE 6.1.3 illustrates the two general shapes that the graph of an exponential function $f(x) = b^x$ can have; but there is one more important aspect of all such graphs. Observe in Figure 6.1.3 that for $b > 1$,

$$f(x) = b^x \to 0 \quad \text{as} \quad x \to -\infty, \quad \leftarrow \text{blue graph}$$

whereas for $0 < b < 1$,

$$f(x) = b^x \to 0 \quad \text{as} \quad x \to \infty. \quad \leftarrow \text{red graph}$$

In other words, the line $y = 0$ (the x-axis) is a **horizontal asymptote** for both types of exponential graphs.

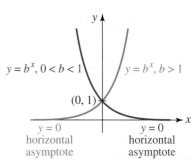

FIGURE 6.1.3 f increasing for $b > 1$; f decreasing for $0 < b < 1$

☐ **Properties** The following list summarizes some of the important properties of the exponential function $f(x) = b^x$. Reexamine the graphs in Figures 6.1.1–6.1.3 as you read this list.

Properties of the Exponential Function

- The domain of f is the set of real numbers, that is, $(-\infty, \infty)$.
- The range of f is the set of positive real numbers, that is, $(0, \infty)$.
- The y-intercept of f is $(0, 1)$. The graph of f has no x-intercepts.
- The function f is increasing for $b > 1$ and decreasing for $0 < b < 1$.
- The x-axis, that is, $y = 0$, is a horizontal asymptote for the graph of f.
- The function f is continuous on $(-\infty, \infty)$.
- The function f is one-to-one.

Although the graphs $y = b^x$ in the case, say, when $b > 1$, all share the same basic shape and all pass through the same point $(0, 1)$, there are subtle differences. The larger the base b the more steeply the graph rises as x increases. In FIGURE 6.1.4 we compare the graphs of $y = 5^x$, $y = 3^x$, $y = 2^x$, and $y = (1.2)^x$ in green, blue, gold, and red, respectively, on the same coordinate axes. We see from its graph that the values of $y = (1.2)^x$ increase slowly as x increases. For example, for $y = (1.2)^x$, $f(3) = (1.2)^3 = 1.728$, whereas, for $y = 5^x$, $f(3) = 5^3 = 125$.

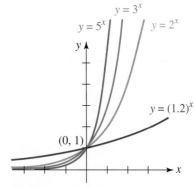

FIGURE 6.1.4 Graphs of $y = b^x$ for $b = 1.2, 2, 3, 5$

6.1 Exponential Functions   **323**

The fact that (1) is a one-to-one function, follows from the horizontal line test discussed in Section 2.7. Note in Figures 6.1.1–6.1.4 that a horizontal line can cross or intersect an exponential graph in at most one point.

Of course, we can obtain other kinds of graphs by rigid and nonrigid transformations, or when an exponential function is combined with other functions by either an arithmetic operation or by function composition. In the next several examples we examine variations of the exponential graph.

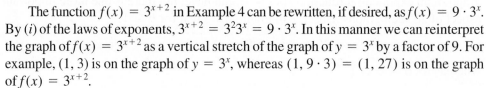

EXAMPLE 4   Horizontally Shifted Graph

Graph the function $f(x) = 3^{x+2}$.

Solution From the discussion in Section 2.2 you should recognize that the graph of $f(x) = 3^{x+2}$ is the graph of $y = 3^x$ shifted 2 units to the left. Recall, since the shift is a rigid transformation to the left, the points on the graph of $f(x) = 3^{x+2}$ are the points on the graph of $y = 3^x$ moved horizontally 2 units to the left. This means that the y-coordinates of points (x, y) on the graph of $y = 3^x$ remain unchanged but 2 is subtracted from all the x-coordinates of the points. Thus we see from FIGURE 6.1.5 that the points $(0, 1)$ and $(2, 9)$ on the graph of $y = 3^x$ are moved, in turn, to the points $(-2, 1)$ and $(0, 9)$ on the graph of $f(x) = 3^{x+2}$.

≡

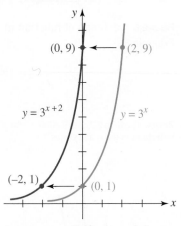

FIGURE 6.1.5 Shifted graph in Example 4

The function $f(x) = 3^{x+2}$ in Example 4 can be rewritten, if desired, as $f(x) = 9 \cdot 3^x$. By (*i*) of the laws of exponents, $3^{x+2} = 3^2 3^x = 9 \cdot 3^x$. In this manner we can reinterpret the graph of $f(x) = 3^{x+2}$ as a vertical stretch of the graph of $y = 3^x$ by a factor of 9. For example, $(1, 3)$ is on the graph of $y = 3^x$, whereas $(1, 9 \cdot 3) = (1, 27)$ is on the graph of $f(x) = 3^{x+2}$.

☐ **The Number *e*** Most every student of mathematics has heard of, and has likely worked with, the famous irrational number $\pi = 3.141592654. \ldots$ Recall, that an irrational number is a nonrepeating and nonterminating decimal. In calculus and applied mathematics the irrational number

$$e = 2.718281828459\ldots$$

arguably plays a role more important than the number π. The usual definition of the number *e* is the number that the function $f(x) = (1 + 1/x)^x$ approaches as we let x become large without bound in the positive direction, that is, $f(x) \rightarrow e$ as $x \rightarrow \infty$. Using the limit notation introduced in Sections 1.5 and 2.9, we write

$$e = \lim_{x \to \infty} \left(1 + \frac{1}{x} \right)^x. \tag{3}$$

See Problems 47 and 49 in Exercises 6.1. You will often see an alternative definition of the number *e*. If we let $h = 1/x$ in (3), then as $x \rightarrow \infty$ we have simultaneously $h \rightarrow 0$. Hence an equivalent form of (3) is

$$e = \lim_{h \to 0}(1 + h)^{1/h}. \tag{4}$$

See Problems 48 and 50 in Exercises 6.1. Of course, advancing (3) and (4) as *definitions* of the number *e* raises the obvious question: Where do these strange limits come from? An unsatisfying partial answer is: Definitions (3) and (4) come from calculus. While we cannot prove in this course that the limits in (3) and (4) exist, we will, however, discuss the origins of *e* in Section 6.5.

     CHAPTER 6 EXPONENTIAL AND LOGARITHMIC FUNCTIONS

☐ **The Natural Exponential Function** When the base in (1) is chosen to be $b = e$, the function

$$f(x) = e^x \tag{5}$$

is called the **natural exponential function**. Since $b = e > 1$ and $b = 1/e < 1$, the graphs of $y = e^x$ and $y = e^{-x}$ (or $y = (1/e)^x = 1/e^x$) are given in FIGURE 6.1.6.

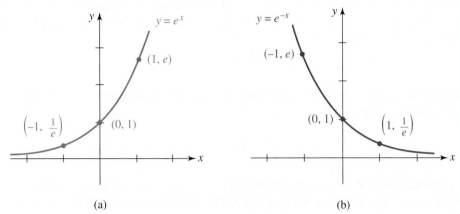

(a)   (b)

FIGURE 6.1.6 Graphs of the natural exponential function (in (a)) and its reciprocal (in (b))

On the face of it, the natural exponential function (5) possesses no noticeable graphical characteristic that distinguishes it from, say, the function $f(x) = 3^x$, and has no special properties other than the ones given in (2), above. Questions as to why (5) is a "natural" and frankly, the most important exponential function, can only be answered fully in courses in calculus and beyond. We will explore some of the importance of the number e in Sections 6.4 and 6.5.

EXAMPLE 5   Reflection and Vertical Shift

Graph the function $f(x) = 2 - e^{-x}$. State the range.

Solution We first draw the graph of $y = e^{-x}$ as shown in FIGURE 6.1.7(a). Then we reflect the first graph in the x-axis to obtain the graph of $y = -e^{-x}$ in Figure 6.1.7(b). Finally, the graph in Figure 6.1.7(c) is obtained by shifting the graph in part (b) upward 2 units.

The y-intercept $(0, -1)$ of $y = -e^{-x}$ when shifted upward 2 units returns us to the original y-intercept in Figure 6.1.7(a). Finally, because of the vertical shift the horizontal asymptote, which was $y = 0$ in parts (a) and (b) of the figure, becomes $y = 2$ in Figure 6.1.7(c). From the last graph we can conclude that the range of the function $f(x) = 2 - e^{-x}$ is the set of real numbers defined by $y < 2$, that is, the interval $(-\infty, 2)$ on the y-axis.

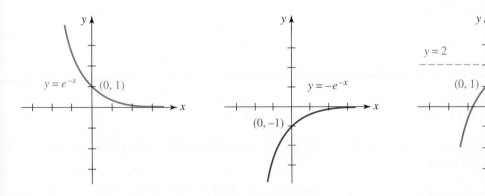

(a) Start with graph of $y = e^{-x}$ (b) Graph in (a) reflected in x-axis   (c) Graph in (b) shifted upward 2 units

FIGURE 6.1.7 Graph of function in Example 5

≡

In the next example we graph the function composition of the natural exponential function $y = e^x$ with the simple quadratic polynomial function $y = -x^2$.

■ EXAMPLE 6   A Function Composition

Graph the function $f(x) = e^{-x^2}$.

Solution Because $f(0) = e^{-0^2} = e^0 = 1$, the y-intercept of the graph is $(0, 1)$. Also, $f(x) \neq 0$ since $e^{-x^2} \neq 0$ for every real number x. This means that the graph of f has no x-intercepts. Then from

$$f(-x) = e^{-(-x)^2} = e^{-x^2} = f(x)$$

we conclude that f is an even function and so its graph is symmetric with respect to the y-axis. Lastly, observe that

$$f(x) = \frac{1}{e^{x^2}} \to 0 \quad \text{as} \quad x \to \infty.$$

By symmetry we can also conclude that $f(x) \to 0$ as $x \to -\infty$. This shows that $y = 0$ is a horizontal asymptote for the graph of f. The graph of f is given in FIGURE 6.1.8. ≡

Bell-shaped graphs such as that given in Figure 6.1.8 are very important in the study of probability and statistics.

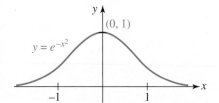

FIGURE 6.1.8 Graph of function in Example 6

6.1   Exercises   Answers to selected odd-numbered problems begin on page ANS–17.

In Problems 1–12, sketch the graph of the given function f. Find the y-intercept and the horizontal asymptote of the graph. State whether the function is increasing or decreasing.

1. $f(x) = \left(\frac{3}{4}\right)^x$

2. $f(x) = \left(\frac{4}{3}\right)^x$

3. $f(x) = -2^x$

4. $f(x) = -2^{-x}$

5. $f(x) = 2^{x+1}$

6. $f(x) = 2^{2-x}$

7. $f(x) = -5 + 3^x$

8. $f(x) = 2 + 3^{-x}$

9. $f(x) = 3 - \left(\frac{1}{5}\right)^x$

10. $f(x) = 9 - e^x$

11. $f(x) = -1 + e^{x-3}$

12. $f(x) = -3 - e^{x+5}$

In Problems 13–18, find an exponential function $f(x) = b^x$ such that the graph of f passes through the given point.

13. $(3, 216)$

14. $(-1, 5)$

15. $(-1, e^2)$

16. $(2, e)$

17. $(-2, 9)$

18. $\left(\frac{1}{2}, 6\right)$

In Problems 19–22, determine the range of the given function.

19. $f(x) = 5 + e^{-x}$

20. $f(x) = 4 - 2^{-x}$

21. $f(x) = 3^x - 2$

22. $f(x) = -e^x - 3$

In Problems 23–28, find the x- and y-intercepts of the graph of the given function. Do not graph.

23. $f(x) = 2^x - 4$

24. $f(x) = -3^{2x} + 9$

25. $f(x) = xe^x + 10e^x$

26. $f(x) = x^2 2^x - 2^x$

27. $f(x) = x^3 8^x + 5x^2 8^x + 6x 8^x$

28. $f(x) = 4^x x^4 - 4^{x+1}$

In Problems 29–32, use a graph to solve the given inequality.

29. $2^x > 16$
30. $e^x \leq 1$
31. $e^{x-2} < 1$
32. $\left(\frac{1}{2}\right)^x \geq 8$

In Problems 33 and 34, use the graph in Figure 6.1.8 to sketch the graph of the given function f.

33. $f(x) = e^{-(x-3)^2}$
34. $f(x) = 3 - e^{-(x+1)^2}$

In Problems 35 and 36, use $f(-x) = f(x)$ to demonstrate that the given function is even. Sketch the graph of f.

35. $f(x) = e^{x^2}$
36. $f(x) = e^{-|x|}$

In Problems 37–40, use the graphs obtained in Problems 35 and 36 as an aid in sketching the graph of the given function f.

37. $f(x) = 1 - e^{x^2}$
38. $f(x) = 2 + 3e^{|x|}$
39. $f(x) = -e^{|x-3|}$
40. $f(x) = e^{(x+2)^2}$

41. Show that $f(x) = 2^x + 2^{-x}$ is an even function. Sketch the graph of f.
42. Show that $f(x) = 2^x - 2^{-x}$ is an odd function. Sketch the graph of f.

In Problems 43 and 44, sketch the graph of the given piecewise-defined function f.

43. $f(x) = \begin{cases} -e^x, & x < 0 \\ -e^{-x}, & x \geq 0 \end{cases}$
44. $f(x) = \begin{cases} e^{-x}, & x \leq 0 \\ -e^x, & x > 0 \end{cases}$

45. Find an equation of the red line in FIGURE 6.1.9.
46. Find the total area of the shaded region in FIGURE 6.1.10.

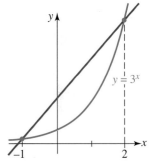

FIGURE 6.1.9 Graph for Problem 45

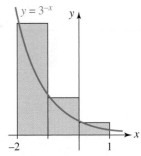

FIGURE 6.1.10 Graph for Problem 46

Calculator Problems

In Problems 47 and 48, use a calculator to fill out the given table.

47.

x	10	100	1000	10,000	100,000	1,000,000
$(1 + 1/x)^x$						

48.

h	0.1	0.01	0.001	0.0001	0.00001	0.000001
$(1 + h)^{1/h}$						

49. (a) Use a graphing utility to graph the functions $f(x) = (1 + 1/x)^x$ and $g(x) = e$ on the same set of coordinate axes. Use the intervals $(0, 10]$, $(0, 100]$, $(0, 1000]$. Describe the behavior of f for large values of x. In graphical terms, what is $g(x) = e$?
  (b) Graph the function f in part (a) on the interval $[-10, 0)$. Superimpose that graph with the graph of f on $(0, 10]$ obtained in part (a). Is f a continuous function?
50. Use a graphing utility to graph the function $f(x) = (1 + x)^{1/x}$ on the intervals $[0.1, 1]$, $[0.01, 1]$, and $[0.001, 1]$. Describe the behavior of f near $x = 0$.

In Problems 51 and 52, use a graphing utility as an aid in approximating the x-coordinates of the points of intersection of the graphs of the functions f and g.

51. $f(x) = x^2$, $g(x) = 2^x$
52. $f(x) = x^3$, $g(x) = 3^x$

For Discussion

53. Suppose $2^t = a$ and $6^t = b$. Using the laws of exponents given in this section, find the value of the given expression in terms of a and b.

   (a) 12^t **(b)** 3^t

   (c) 6^{-t} **(d)** 6^{3t}

   (e) $2^{-3t}2^{7t}$ **(f)** 18^t

54. Discuss: What does the graph of $y = e^{e^x}$ look like? Do not use a graphing utility.

55. If $f(x) = b^x$, $b > 0$, then show that $f(x_1 + x_2) = f(x_1)f(x_2)$.

6.2   Logarithmic Functions

≡ **Introduction** Since an exponential function $y = b^x$ is one-to-one, we know that it has an inverse function. To find this inverse, we interchange the variables x and y to obtain $x = b^y$. This last formula defines y as a function of x:

        y is that exponent of the base b that produces x.

By replacing the word *exponent* with the word *logarithm*, we can rephrase the preceding line as

        y is that logarithm of the base b that produces x.

This last line is abbreviated by the notation $y = \log_b x$ and is called the logarithmic function.

DEFINITION 6.2.1   Logarithmic Function

The **logarithmic function** with base $b > 0$, $b \neq 1$, is defined by

$$y = \log_b x \qquad \text{if and only if} \qquad x = b^y. \qquad (1)$$

For $b > 0$ there is no real number y for which b^y can be either 0 or negative. It then follows from $x = b^y$ that $x > 0$. In other words, the **domain** of a logarithmic function $y = \log_b x$ is the set of positive real numbers $(0, \infty)$.

For emphasis, all that is being said in the preceding sentences is:

The logarithmic expression $y = \log_b x$ and the exponential expression $x = b^y$ are equivalent.

That is, both symbols mean the same thing. As a consequence, within a specific context such as solving a problem, we can use whichever form happens to be more convenient. The following table lists several examples of equivalent logarithmic and exponential statements.

Logarithmic Form	Exponential Form
$\log_3 9 = 2$	$9 = 3^2$
$\log_8 2 = \frac{1}{3}$	$2 = 8^{1/3}$
$\log_{10} 0.001 = -3$	$0.001 = 10^{-3}$
$\log_b 5 = -1$	$5 = b^{-1}$

☐ **Graphs** Recall from Section 2.7 that the graph of an inverse function can be obtained by reflecting the graph of the original function in the line $y = x$. This technique was used to obtain the red graphs from the blue graphs in FIGURE 6.2.1. As you inspect the two graphs in Figure 6.2.1(a) and in Figure 6.2.1(b), remember that the domain $(-\infty, \infty)$ and range $(0, \infty)$ of $y = b^x$ become, in turn, the range $(-\infty, \infty)$ and domain $(0, \infty)$ of $y = \log_b x$. Also note that the y-intercept $(0, 1)$ for the exponential function (blue graphs) becomes the x-intercept $(1, 0)$ for the logarithmic function (red graphs).

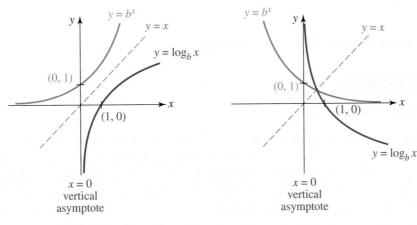

(a) Base $b > 1$ (b) Base $0 < b < 1$

FIGURE 6.2.1 Graphs of logarithmic functions

☐ **Vertical Asymptote** When the exponential function is reflected in the line $y = x$, the horizontal asymptote $y = 0$ for the graph of $y = b^x$ becomes a vertical asymptote for the graph of $y = \log_b x$. In Figure 6.2.1 we see that for $b > 1$,

$$\log_b x \to -\infty \quad \text{as} \quad x \to 0^+, \quad \leftarrow \text{red graph in (a)}$$

whereas for $0 < b < 1$,

$$\log_b x \to \infty \quad \text{as} \quad x \to 0^+. \quad \leftarrow \text{red graph in (b)}$$

From (7) of Section 3.6 we conclude that $x = 0$, which is the equation of the y-axis, is a **vertical asymptote** for the graph of $y = \log_b x$.

☐ **Properties** The following list summarizes some of the important properties of the logarithmic function $f(x) = \log_b x$.

Properties of the Logarithmic Function

- The domain of f is the set of positive real numbers, that is, $(0, \infty)$.
- The range of f is the set of real numbers, that is, $(-\infty, \infty)$.
- The x-intercept of f is $(1, 0)$. The graph of f has no y-intercept.
- The function f is increasing for $b > 1$ and decreasing for $0 < b < 1$.   (2)
- The y-axis, that is, $x = 0$, is a vertical asymptote for the graph of f.
- The function f is continuous on $(0, \infty)$.
- The function f is one-to-one.

We would like to call attention to the third entry in the foregoing list (2) for special emphasis:

$$\log_b 1 = 0 \qquad \text{since} \qquad b^0 = 1. \tag{3}$$

Also,
$$\log_b b = 1 \qquad \text{since} \qquad b^1 = b. \tag{4}$$

Thus, in addition to (1, 0) the graph of any logarithmic function (1) with base b also contains the point $(b, 1)$. The equivalence of $y = \log_b x$ and $x = b^y$ also yields two sometimes-useful identities. By substituting $y = \log_b x$ into $x = b^y$, and then $x = b^y$ into $y = \log_b x$ gives

$$x = b^{\log_b x} \qquad \text{and} \qquad y = \log_b b^y. \tag{5}$$

For example, from (5), $8^{\log_8 10} = 10$ and $\log_{10} 10^5 = 5$.

▌EXAMPLE 1   Logarithmic Graph for $b > 1$

Graph $f(x) = \log_{10}(x + 10)$.

Solution This is the graph of $y = \log_{10} x$, which has the shape shown in Figure 6.2.1(a), shifted 10 units to the left. To reinforce the fact that the domain of a logarithmic function $y = \log_{10} x$ is the set of positive real numbers, that is, $x > 0$, we can obtain the domain of $f(x) = \log_{10}(x + 10)$ by replacing x by $x + 10$ and requiring that $x + 10 > 0$ or $x > -10$. In interval notation, the domain of f is $(-10, \infty)$. In the short accompanying table, we have chosen convenient values of x in order to plot a few points.

x	-9	0	90
$f(x)$	0	1	2

Notice,

$$f(-9) = \log_{10} 1 = 0 \qquad \leftarrow \text{by (3)}$$
$$f(0) = \log_{10} 10 = 1. \qquad \leftarrow \text{by (4)}$$

The vertical asymptote $x = 0$ for the graph of $y = \log_{10} x$ becomes $x = -10$ for the shifted graph. This asymptote is the red dashed vertical line in FIGURE 6.2.2. ≡

□ **Natural Logarithm** Logarithms with base $b = 10$ are called **common logarithms** and logarithms with base $b = e$ are called **natural logarithms**. Furthermore, it is customary to write the natural logarithm

$$\log_e x \qquad \text{as} \qquad \ln x.$$

The symbol "$\ln x$" is usually read phonetically as "ell-en of x." Since $b = e > 1$, the graph of $y = \ln x$ has the characteristic logarithmic shape shown in Figure 6.2.1(a). See FIGURE 6.2.3. For base $b = e$, (1) of Definition 6.2.1 becomes

$$y = \ln x \qquad \text{if and only if} \qquad x = e^y. \tag{6}$$

The analogs of properties (3) and (4) for the natural logarithm are

$$\ln 1 = 0 \qquad \text{since} \qquad e^0 = 1, \tag{7}$$
$$\ln e = 1 \qquad \text{since} \qquad e^1 = e. \tag{8}$$

The identities in (5) become

$$x = e^{\ln x} \qquad \text{and} \qquad y = \ln e^y. \tag{9}$$

For example, from (9), $e^{\ln 13} = 13$.

Common and natural logarithms can be found on all calculators. Often the symbol for the common logarithm is written without a subscript, that is, $\log_{10} x$ is simply

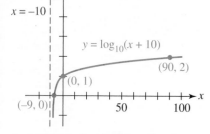

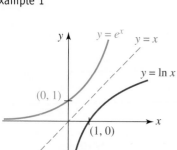

FIGURE 6.2.2 Graph of function in Example 1

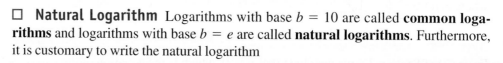

FIGURE 6.2.3 Graph of the natural logarithm is shown in red

written $\log x$. But in this text we will continue to use $\log_{10} x$. See (*iv*) in *Notes from the Classroom*.

☐ **Laws of Logarithms** The laws of exponents given in Theorem 6.1.1 can be restated in an equivalent manner as the laws of logarithms. To see this, suppose we write $M = b^{x_1}$ and $N = b^{x_2}$. Then by (1), $x_1 = \log_b M$ and $x_2 = \log_b N$.

Product: By (*i*) of Theorem 6.1.1, $MN = b^{x_1 + x_2}$. Expressed as a logarithm this is $x_1 + x_2 = \log_b MN$. Substituting for x_1 and x_2 gives

$$\log_b M + \log_b N = \log_b MN.$$

Quotient: By (*ii*) of Theorem 6.1.1, $M/N = b^{x_1 - x_2}$. Expressed as a logarithm this is $x_1 - x_2 = \log_b(M/N)$. Substituting for x_1 and x_2 gives

$$\log_b M - \log_b N = \log_b(M/N).$$

Power: By (*iv*) of Theorem 6.1.1, $M^c = b^{cx_1}$. Expressed as a logarithm this is $cx_1 = \log_b M^c$. Substituting for x_1 gives

$$c \log_b M = \log_b M^c.$$

For convenience and future reference, we summarize these product, quotient, and power laws of logarithms next.

THEOREM 6.2.1 Laws of Logarithms

For any base $b > 0$, $b \neq 1$, and positive numbers M and N:

   (*i*) $\log_b MN = \log_b M + \log_b N$

   (*ii*) $\log_b\left(\dfrac{M}{N}\right) = \log_b M - \log_b N$

   (*iii*) $\log_b M^c = c \log_b M$, for c any real number

■ EXAMPLE 2   **Using the Laws of Logarithms**

Simplify and write as a single logarithm

$$\tfrac{1}{2}\ln 36 + 2\ln 4 - \ln 4.$$

Solution There are several ways to approach this problem. Note, for example, that the second and third terms can be combined arithmetically as

$$2\ln 4 - \ln 4 = \ln 4. \quad \leftarrow \text{analogous to } 2x - x = x$$

Alternatively, we can use (*iii*) followed by (*ii*) of Theorem 6.2.1 to combine these terms:

$$\begin{aligned}
2\ln 4 - \ln 4 &= \ln 4^2 - \ln 4 \\
&= \ln 16 - \ln 4 \\
&= \ln \tfrac{16}{4} \\
&= \ln 4.
\end{aligned}$$

Hence,
$$\begin{aligned}
\tfrac{1}{2}\ln 36 + 2\ln 4 - \ln 4 &= \ln(36)^{1/2} + \ln 4 \quad \leftarrow \text{by (\textit{iii}) of Theorem 6.2.1} \\
&= \ln 6 + \ln 4 \\
&= \ln 24. \quad \leftarrow \text{by (\textit{i}) of Theorem 6.2.1} \ \ \equiv
\end{aligned}$$

EXAMPLE 3　　　　**Rewriting Logarithmic Expressions**

Use the laws of logarithms to rewrite each expression and evaluate.

(a) $\ln\sqrt{e}$　　　**(b)** $\ln 5e$　　　**(c)** $\ln\dfrac{1}{e}$

Solution **(a)** Since $\sqrt{e} = e^{1/2}$ we have from (*iii*) of Theorem 6.2.1:

$$\ln\sqrt{e} = \ln e^{1/2} = \tfrac{1}{2}\ln e = \tfrac{1}{2}.\quad \leftarrow\text{from (8), } \ln e = 1$$

(b) From (*i*) of Theorem 6.2.1 and a calculator:

$$\ln 5e = \ln 5 + \ln e = \ln 5 + 1 \approx 2.6094.$$

(c) From (*ii*) of the Theorem 6.2.1:

$$\ln\frac{1}{e} = \ln 1 - \ln e = 0 - 1 = -1.\quad \leftarrow\text{from (7) and (8)}$$

Note that (*iii*) of the Theorem 6.2.1 can also be used here:

$$\ln\frac{1}{e} = \ln e^{-1} = (-1)\ln e = -1.\quad \leftarrow \ln e = 1 \qquad \equiv$$

EXAMPLE 4　　　　**Value of a Logarithm**

If $\log_b 2 = 0.4307$ and $\log_b 3 = 0.6826$, then find $\log_b\sqrt[3]{18}$.

Solution We begin by rewriting $\sqrt[3]{18}$ as $(18)^{1/3}$. Then by the laws of logarithms

$$
\begin{aligned}
\log_b(18)^{1/3} &= \tfrac{1}{3}\log_b 18 &&\leftarrow\text{by (}iii\text{) of Theorem 6.2.1}\\
&= \tfrac{1}{3}\log_b(2\cdot 3^2)\\
&= \tfrac{1}{3}[\log_b 2 + \log_b 3^2] &&\leftarrow\text{by (}i\text{) of Theorem 6.2.1}\\
&= \tfrac{1}{3}[\log_b 2 + 2\log_b 3] &&\leftarrow\text{by (}iii\text{) of Theorem 6.2.1}\\
&= \tfrac{1}{3}[0.4307 + 2(0.6826)]\\
&= 0.5986.
\end{aligned}
$$

$\equiv$

NOTES FROM THE CLASSROOM

(*i*) Students often struggle with the concept of a *logarithm*. It may help if you repeat to yourself a few dozen times, "A logarithm is an exponent." It may also help if you begin reading a statement such as $3 = \log_{10}1000$ as "3 is the exponent of 10 that. . . ."

(*ii*) Be *very* careful applying the laws of logarithms. The logarithm does *not* distribute over addition,

$$\log_b(M + N) \ne \log_b M + \log_b N.$$

In other words, the exponent of a sum is not the sum of the exponents.

Also,　　　　$\dfrac{\log_b M}{\log_b N} \ne \log_b M - \log_b N.$

In general, there is no way that we can rewrite either

$$\log_b(M + N) \qquad \text{or} \qquad \frac{\log_b M}{\log_b N}.$$

(*iii*) In calculus, the first step in a procedure known as *logarithmic differentiation* requires the student to take the natural logarithm of both sides of a complicated function such as $y = \dfrac{x^{10}\sqrt{x^2 + 5}}{\sqrt[3]{8x^3 + 2}}$. The idea is to use the laws of logarithms to transform powers into constant multiples, products into sums, and quotients into differences. See Problems 63–66 in Exercises 6.2.

(*iv*) You may see different notations for the natural exponential function and for the natural logarithm. For example, on some calculators you may see $y = \exp x$ instead of $y = e^x$. In the computer algebra system *Mathematica* the natural exponential function is written $\text{Exp}[x]$ and the natural logarithm is written $\text{Log}[x]$.

6.2 Exercises — Answers to selected odd-numbered problems begin on page ANS-18.

In Problems 1–6, rewrite the given exponential expression as an equivalent logarithmic expression.

1. $4^{-1/2} = \frac{1}{2}$
2. $9^0 = 1$
3. $10^4 = 10{,}000$
4. $10^{0.3010} = 2$
5. $t^{-s} = v$
6. $(a + b)^2 = a^2 + 2ab + b^2$

In Problems 7–12, rewrite the given logarithmic expression as an equivalent exponential expression.

7. $\log_2 128 = 7$
8. $\log_5 \frac{1}{25} = -2$
9. $\log_{\sqrt{3}} 81 = 8$
10. $\log_{16} 2 = \frac{1}{4}$
11. $\log_b u = v$
12. $\log_b b^2 = 2$

In Problems 13–18, find the exact value of the given logarithm.

13. $\log_{10}(0.0000001)$
14. $\log_4 64$
15. $\log_2(2^2 + 2^2)$
16. $\log_9 \frac{1}{3}$
17. $\ln e^e$
18. $\ln(e^4 e^9)$

In Problems 19–22, find the exact value of the given expression.

19. $10^{\log_{10} 6^2}$
20. $25^{\log_5 8}$
21. $e^{-\ln 7}$
22. $e^{\frac{1}{2}\ln \pi}$

In Problems 23 and 24, find a logarithmic function $f(x) = \log_b x$ such that the graph of f passes through the given point.

23. $(49, 2)$
24. $\left(4, \frac{1}{3}\right)$

In Problems 25–32, find the domain of the given function f. Find the x-intercept and the vertical asymptote of the graph. Use transformations to graph the given function f.

25. $f(x) = -\log_2 x$
26. $f(x) = -\log_2(x + 1)$
27. $f(x) = \log_2(-x)$
28. $f(x) = \log_2(3 - x)$
29. $f(x) = 3 - \log_2(x + 3)$
30. $f(x) = 1 - 2\log_4(x - 4)$
31. $f(x) = -1 + \ln x$
32. $f(x) = 1 + \ln(x - 2)$

In Problems 33 and 34, use a graph to solve the given inequality.

33. $\ln(x + 1) < 0$ **34.** $\log_{10}(x + 3) > 1$

35. Show that $f(x) = \ln|x|$ is an even function. Rewrite f as a piecewise-defined function and sketch its graph. Find the x-intercepts and the vertical asymptote of the graph.

36. Use the graph obtained in Problem 35 to sketch the graph of $y = \ln|x - 2|$. Find the x-intercept and the vertical asymptote of the graph.

In Problems 37 and 38, sketch the graph of the given function f.

37. $f(x) = |\ln x|$ **38.** $f(x) = |\ln(x + 1)|$

In Problems 39–44, find the domain of the given function f.

39. $f(x) = \ln(2x - 3)$ **40.** $f(x) = \ln(3 - x)$
41. $f(x) = \ln(9 - x^2)$ **42.** $f(x) = \ln(x^2 - 2x)$
43. $f(x) = \sqrt{\ln x}$ **44.** $f(x) = \dfrac{1}{\ln x}$

In Problems 45–50, use the laws of logarithms in Theorem 6.2.1 to rewrite the given expression as one logarithm.

45. $\log_{10} 2 + 2\log_{10} 5$ **46.** $\frac{1}{2}\log_5 49 - \frac{1}{3}\log_5 8 + 13\log_5 1$

47. $\ln(x^4 - 4) - \ln(x^2 + 2)$ **48.** $\ln\left(\dfrac{x}{y}\right) - 2\ln x^3 - 4\ln y$

49. $\ln 5 + \ln 5^2 + \ln 5^3 - \ln 5^6$ **50.** $5\ln 2 + 2\ln 3 - 3\ln 4$

In Problems 51–62, use $\log_b 4 = 0.6021$ and $\log_b 5 = 0.6990$ to evaluate the given logarithm. Round your answer to four decimal places.

51. $\log_b 2$ **52.** $\log_b 20$
53. $\log_b 64$ **54.** $\log_b 625$
55. $\log_b \sqrt{5}$ **56.** $\log_b \frac{5}{4}$
57. $\log_b \sqrt[3]{4}$ **58.** $\log_b 80$
59. $\log_b 0.8$ **60.** $\log_b 3.2$
61. $\log_4 b$ **62.** $\log_5 5b$

In Problems 63–66, use the laws of logarithms in Theorem 6.2.1 so that $\ln y$ contains no products, quotients, or powers.

63. $y = \dfrac{x^{10}\sqrt{x^2 + 5}}{\sqrt[3]{8x^3 + 2}}$ **64.** $y = \sqrt{\dfrac{(2x + 1)(3x + 2)}{4x + 3}}$

65. $y = \dfrac{(x^3 - 3)^5(x^4 + 3x^2 + 1)^8}{\sqrt{x}(7x + 5)^9}$ **66.** $y = 64x^6\sqrt{x + 1}\sqrt[3]{x^2 + 2}$

In Problems 67–70, verify the given identity.

67. $\ln|\sec x| = -\ln|\cos x|$
68. $\ln|\cot x| = -\ln|\tan x|$
69. $\ln|\sec x - \tan x| = -\ln|\sec x + \tan x|$
70. $\ln|1 + \cos x| + \ln|1 - \cos x| = 2\ln|\sin x|$

For Discussion

71. In science it is sometimes useful to display data using logarithmic coordinates. Which of the following equations determines the graph shown in FIGURE 6.2.4?

(*i*) $y = 2x + 1$ (*ii*) $y = e + x^2$

(*iii*) $y = ex^2$ (*iv*) $x^2 y = e$

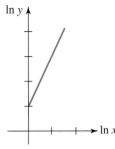

FIGURE 6.2.4 Graph for Problem 71

72. (**a**) Use a graphing utility to obtain the graph of the function $f(x) = \ln(x + \sqrt{x^2 + 1})$.

(**b**) Show that f is an odd function, that is, $f(-x) = -f(x)$.

73. If $a > 0$ and $b > 0$, $a \neq b$, then $\log_a x$ is a constant multiple of $\log_b x$. That is, $\log_a x = k \log_b x$. Find k.

74. Show that $(\log_{10} e)(\log_e 10) = 1$. Can you generalize this result?

75. Discuss: How can the graphs of the given function be obtained from the graph of $f(x) = \ln x$ by means of a rigid transformation (a shift or a reflection)?

(**a**) $y = \ln 5x$ (**b**) $y = \ln \dfrac{x}{4}$

(**c**) $y = \ln x^{-1}$ (**d**) $y = \ln(-x)$

76. Using correct mathematical notation, rewrite the statement:

c is the exponent of 5 that gives the number N,

in two different, but equivalent, ways.

77. Find the zeros of the function $f(x) = 5 - \log_2 |-x + 4|$. Check your answers.

78. The following question appeared on an examination:

Find the domain of the function $f(x) = \ln\left(\dfrac{x-3}{x}\right)$.

One student reasoned that using the laws of logarithms the function f could be rewritten as

$$f(x) = \ln(x - 3) - \ln x.$$

Because the domain of $\ln(x - 3)$ is the interval $(3, \infty)$ and the domain of $\ln x$ is the interval $(0, \infty)$, the domain of f is the intersection $(0, \infty) \cap (3, \infty) = (3, \infty)$. Discuss: Is the student's reasoning valid?

79. Find the vertical asymptotes for the graph of $f(x) = \ln\left(\dfrac{x-3}{x}\right)$. Sketch the graph of f. Do not use a calculator or a computer.

80. If $f(x) = \log_b x$, $b > 0$, $x > 0$, then show that $f(x_1 x_2) = f(x_1) + f(x_2)$.

6.3 Exponential and Logarithmic Equations

≡ **Introduction** Since exponential and logarithmic functions appear in the context of many different applications, we are often called upon to solve equations that involve these functions. While we postpone applications until Section 6.4, we examine in the present section some of the ways that can be used to solve a variety of exponential and logarithmic equations.

☐ **Solving Equations** Here is a brief list of equation-solving strategies.

Solving Exponential and Logarithmic Equations

     (*i*) Rewrite an exponential expression as a logarithmic expression.
     (*ii*) Rewrite a logarithmic expression as an exponential expression.
     (*iii*) Use the one-to-one properties of b^x and $\log_b x$.
     (*iv*) For equations for the form $a^{x_1} = b^{x_2}$, where $a \neq b$, take the natural logarithm of both sides of the equality and simplify using (*iii*) of the laws of logarithms given in Theorem 6.2.1 of Section 6.2.

    Of course, this list is not comprehensive and does not reflect the fact that in solving equations involving exponential and logarithmic functions we may also have to employ standard algebraic procedures such as *factoring* and using the *quadratic formula*.
    In the first two examples we use the equivalence

$$y = \log_b x \qquad \text{if and only if} \qquad x = b^y \tag{1}$$

to toggle between logarithmic and exponential expressions.

EXAMPLE 1   Rewriting an Exponential Expression

Solve $e^{10k} = 7$ for k.

Solution We use (1), with $b = e$, to rewrite the given exponential expression as a logarithmic expression:

$$e^{10k} = 7 \quad \text{means} \quad 10k = \ln 7.$$

Therefore, with the aid of a calculator $k = \frac{1}{10}\ln 7 \approx 0.1946$.   ≡

EXAMPLE 2   Rewriting a Logarithmic Expression

Solve $\log_2 x = 5$ for x.

Solution We use (1), with $b = 2$, to rewrite the logarithmic expression in its equivalent exponential form:

$$x = 2^5 = 32.$$ ≡

☐ **One-to-One Properties** Recall from (1) of Section 2.7 that a one-to-one function f possesses the property that if $f(x_1) = f(x_2)$, then necessarily $x_1 = x_2$. We have seen in Sections 6.1 and 6.2 that both the exponential function $y = b^x$, $b > 0$, $b \neq 1$, and the logarithmic function $y = \log_b x$ are one-to-one. As a consequence we have:

$$\text{If } b^{x_1} = b^{x_2}, \text{ then } x_1 = x_2. \tag{2}$$
$$\text{If } \log_b x_1 = \log_b x_2, \text{ then } x_1 = x_2. \tag{3}$$

EXAMPLE 3   Using the One-to-One Property (2)

Solve $2^{x-3} = 8^{x+1}$ for x.

Solution Observe on the right-hand side of the given equality that 8 can be written as a power of 2, that is, $8 = 2^3$. Furthermore, by (*iv*) of the laws of exponents given in Theorem 6.1.1,

multiply exponents
↓ ↓
$$8^{x+1} = (2^3)^{x+1} = 2^{3x+3}.$$

Thus, the equation is the same as

$$2^{x-3} = 2^{3x+3}.$$

From the one-to-one property (2) it follows that the exponents are equal, that is, $x - 3 = 3x + 3$. Solving for x then gives $2x = -6$ or $x = -3$. You are encouraged to check this answer by substituting -3 for x in the original equation. ≡

◼ EXAMPLE 4　　Using the One-to-One Property (2)

Solve $7^{2(x+1)} = 343$ for x.

Solution By noting that $343 = 7^3$, we have the same base on both sides of the equality:

$$7^{2(x+1)} = 7^3.$$

Thus by (2) we can equate exponents and solve for x:

$$
\begin{aligned}
2(x + 1) &= 3 \\
2x + 2 &= 3 \\
2x &= 1 \\
x &= \frac{1}{2}.
\end{aligned}
$$
　　≡

◼ EXAMPLE 5　　Using the One-to-One Property (3)

Solve $\ln 2 + \ln(4x - 1) = \ln(2x + 5)$ for x.

Solution By (i) of the laws of logarithms in Theorem 6.2.1, the left-hand side of the equation can be written

$$\ln 2 + \ln(4x - 1) = \ln 2(4x - 1) = \ln(8x - 2).$$

The original equation is then

$$\ln(8x - 2) = \ln(2x + 5).$$

Since two logarithms with the same base are equal, it follows immediately from the one-to-one property (3) that $8x - 2 = 2x + 5$ or $6x = 7$ or $x = \frac{7}{6}$. ≡

☐ **Extraneous Solutions** For logarithmic equations, especially of the kind in Example 5, you should get accustomed to checking your answer by substituting it back into the original equation. It is possible for a logarithmic equation to have an **extraneous solution**.

◼ EXAMPLE 6　　An Extraneous Solution

Solve $\log_2 x + \log_2(x - 2) = 3$.

Solution We start using again that the sum of logarithms on the left-hand side of the equation is the logarithm of a product:

$$\log_2 x(x - 2) = 3.$$

With $b = 2$ we use (1) to rewrite the last equation in the equivalent exponential form

$$x(x - 2) = 2^3.$$

By ordinary algebra we then have

$$
\begin{aligned}
x^2 - 2x &= 8 \\
x^2 - 2x - 8 &= 0 \\
(x - 4)(x + 2) &= 0.
\end{aligned}
$$

From the last equation we conclude that either $x = 4$ or $x = -2$. However, we must rule out $x = -2$ as a solution. In other words, the number $x = -2$ is an extraneous solution because, when substituted into the original equation, the very first term, $\log_2(-2)$, is not defined. Thus the only solution of the given equation is $x = 4$.

Check:
$$\log_2 4 + \log_2 2 = \log_2 2^2 + \log_2 2$$
$$= \log_2 2^3 = 3\log_2 2 = 3 \cdot 1 = 3. \qquad \equiv$$

When we use the phrase "take the logarithm of both sides of an equality" we are actually using the property that if M and N are two positive numbers such that $M = N$, then $\log_b M = \log_b N$.

■ EXAMPLE 7   Taking the Natural Logarithm of Both Sides

Solve $e^{2x} = 3^{x-4}$.

Solution Since the bases of the exponential expression on each side of the equality are different, one way to proceed is to take the natural logarithm (the common logarithm could also be used) of both sides. From the equality

$$\ln e^{2x} = \ln 3^{x-4}$$

and (*iii*) of the laws of logarithms in Theorem 6.2.1, we get

$$2x \ln e = (x - 4)\ln 3.$$

Now using $\ln e = 1$ and the distributive law, the last equation becomes

$$2x = x \ln 3 - 4 \ln 3.$$

Gathering the terms involving the symbol x to one side of the equality then gives

$$\overbrace{2x - x \ln 3}^{\substack{\text{factor } x \text{ out} \\ \text{of these terms}}} = -4\ln 3 \quad \text{or} \quad (2 - \ln 3)x = -4\ln 3 \quad \text{or} \quad x = \frac{-4\ln 3}{2 - \ln 3}.$$

You are encouraged to verify the calculation that $x \approx -4.8752$. $\equiv$

■ EXAMPLE 8   Using the Quadratic Formula

Solve $5^x - 5^{-x} = 2$.

Solution Because $5^{-x} = 1/5^x$, the equation is

$$5^x - \frac{1}{5^x} = 2.$$

Multiplying both sides of the foregoing equation by 5^x then gives

$$(5^x)^2 - 1 = 2(5^x) \quad \text{or} \quad (5^x)^2 - 2(5^x) - 1 = 0.$$

If we let $X = 5^x$, then the last equation can be interpreted as a quadratic equation $X^2 - 2X - 1 = 0$. Using the quadratic formula to solve for X yields

$$X = \frac{2 \pm \sqrt{4 + 4}}{2} = 1 \pm \sqrt{2} \quad \text{or} \quad 5^x = 1 \pm \sqrt{2}.$$

Because $1 - \sqrt{2}$ is a negative number and 5^x is positive for every real number x there are no real solutions of $5^x = 1 - \sqrt{2}$ and so

$$5^x = 1 + \sqrt{2}. \qquad\qquad (4)$$

Now by taking the natural logarithm of both sides of the equality we obtain

$$\ln 5^x = \ln(1 + \sqrt{2})$$
$$x \ln 5 = \ln(1 + \sqrt{2})$$
$$x = \frac{\ln(1 + \sqrt{2})}{\ln 5}. \qquad (5)$$

Using the $\boxed{\ln}$ key of a calculator, the division yields $x \approx 0.5476$. ≡

☐ **Change of Base** In (4) of Example 8 it follows from (1) that a perfectly valid solution of the equation $5^x - 5^{-x} = 2$ is $x = \log_5(1 + \sqrt{2})$. But from a computational viewpoint (that is, expressing x as a number), the last answer is not desirable since no calculator has a logarithmic function with base 5. But by equating $x = \log_5(1 + \sqrt{2})$ with the result in (5) we have discovered that logarithm with base 5 can be expressed in terms of the natural logarithm:

$$\log_5(1 + \sqrt{2}) = \frac{\ln(1 + \sqrt{2})}{\ln 5}. \qquad (6)$$

The result given in (6) is just a special case of a more general result known as the **change-of-base formula**.

THEOREM 6.3.1 Change-of-Base Formula

If $a \neq 1$, $b \neq 1$, and M are positive numbers, then

$$\log_a M = \frac{\log_b M}{\log_b a} \qquad (7)$$

PROOF: If we let $y = \log_a M$, then from (1), $a^y = M$. Then

$$\log_b a^y = \log_b M$$
$$y \log_b a = \log_b M \qquad \leftarrow \text{by (iii) of Theorem 6.2.1}$$
$$y = \frac{\log_b M}{\log_b a} \qquad \leftarrow \text{by assumption } y = \log_a M$$
$$\log_a M = \frac{\log_b M}{\log_b a}. \qquad ≡$$

In order to obtain the numerical value of a logarithm using a calculator, we usually choose $b = 10$ or $b = e$ in (7):

$$\log_a M = \frac{\log_{10} M}{\log_{10} a} \quad \text{or} \quad \log_a M = \frac{\ln M}{\ln a}. \qquad (8)$$

EXAMPLE 9 **Changing the Base**

Find the numerical value of $\log_2 50$.

Solution We can use either formula in (8). If we choose the first formula in (8) with $M = 50$ and $a = 2$, we have

$$\log_2 50 = \frac{\log_{10} 50}{\log_{10} 2}.$$

Using the [log] key to calculate the two common logarithms and then dividing yields the approximation

$$\log_2 50 \approx 5.6439.$$

Alternatively, the second formula in (8) gives the same result:

$$\log_2 50 = \frac{\ln 50}{\ln 2} \approx 5.6439.$$ ≡

   We can check the answer in Example 9 on a calculator by using the [y^x] key. You are urged to verify that $2^{5.6439} \approx 50$.

▮ EXAMPLE 10   Changing the Base

Find the x in the domain of $f(x) = 6^x$ for which $f(x) = 73$.

Solution We must find a solution of the equation $6^x = 73$. One way of proceeding is to rewrite the exponential expression as an equivalent logarithmic expression:

$$x = \log_6 73.$$

Then with the identification $a = 6$ it follows from the second equation in (8) and the aid of a calculator that

$$x = \log_6 73 = \frac{\ln 73}{\ln 6} \approx 2.3946.$$

You should verify that $f(2.3946) = 6^{2.3946} \approx 73$. ≡

6.3   Exercises   Answers to selected odd-numbered problems begin on page ANS-18.

In Problems 1–20, solve the given exponential equation.

1. $5^{x-2} = 1$ **2.** $3^x = 27^{x^2}$

3. $10^{-2x} = \dfrac{1}{10{,}000}$ **4.** $27^x = \dfrac{9^{2x-1}}{3^x}$

5. $e^{5x-2} = 30$ **6.** $\left(\dfrac{1}{e}\right)^x = e^3$

7. $2^x \cdot 3^x = 36$ **8.** $\dfrac{4^x}{3^x} = \dfrac{9}{16}$

9. $2^{x^2} = 8^{2x-3}$ **10.** $\frac{1}{4}(10^{-2x}) = 25(10^x)$

11. $5 - 10^{2x} = 0$ **12.** $7^{-x} = 9$

13. $3^{2(x-1)} = 7^2$ **14.** $\left(\frac{1}{2}\right)^{-x+2} = 8(2^{x-1})^3$

15. $\dfrac{1}{3} = (2^{|x|-2} - 1)^{-1}$ **16.** $\left(\frac{1}{3}\right)^x = 9^{1-2x}$

17. $5^{|x|-1} = 25$ **18.** $(e^2)^{x^2} - \dfrac{1}{e^{5x+3}} = 0$

19. $4^x = 5^{2x+1}$ **20.** $3^{x+4} = 2^{x-16}$

In Problems 21–40, solve the given logarithmic equation.

21. $\log_3 5x = \log_3 160$ **22.** $\ln(10 + x) = \ln(3 + 4x)$

23. $\ln x = \ln 5 + \ln 9$

24. $3 \log_8 x = \log_8 36 + \log_8 12 - \log_8 2$

25. $\log_{10} \dfrac{1}{x^2} = 2$

26. $\log_3 \sqrt{x^2 + 17} = 2$

27. $\log_2(\log_3 x) = 2$

28. $\log_5 |1 - x| = 1$

29. $\log_3 81^x - \log_3 3^{2x} = 3$

30. $\dfrac{\log_2 8^x}{\log_2 \frac{1}{4}} = \dfrac{1}{2}$

31. $\log_{10} x = 1 + \log_{10} \sqrt{x}$

32. $\log_2(x - 3) - \log_2(2x + 1) = -\log_2 4$

33. $\log_2 x + \log_2(10 - x) = 4$

34. $\log_8 x + \log_8 x^2 = 1$

35. $\log_6 2x - \log_6(x + 1) = 0$

36. $\log_{10} 54 - \log_{10} 2 = 2 \log_{10} x - \log_{10} \sqrt{x}$

37. $\log_9 \sqrt{10x + 5} - \dfrac{1}{2} = \log_9 \sqrt{x + 1}$

38. $\log_{10} x^2 + \log_{10} x^3 + \log_{10} x^4 - \log_{10} x^5 = \log_{10} 16$

39. $\ln 3 + \ln(2x - 1) = \ln 4 + \ln(x + 1)$

40. $\ln(x + 3) + \ln(x - 4) - \ln x = \ln 3$

In Problems 41–50, either use factoring or the quadratic formula to solve the given equation.

41. $(5^x)^2 - 26(5^x) + 25 = 0$

42. $64^x - 10(8^x) + 16 = 0$

43. $\log_4 x^2 = (\log_4 x)^2$

44. $(\log_{10} x)^2 + \log_{10} x = 2$

45. $(5^x)^2 - 2(5^x) - 1 = 0$

46. $2^{2x} - 12(2^x) + 35 = 0$

47. $(\ln x)^2 + \ln x = 2$

48. $(\log_{10} 2x)^2 = \log_{10}(2x)^2$

49. $2^x + 2^{-x} = 2$

50. $10^{2x} - 103(10^x) + 300 = 0$

In Problems 51–56, find the x-intercepts of the graph of the given function.

51. $f(x) = e^{x+4} - e$

52. $f(x) = 1 - \frac{1}{5}(0.1)^x$

53. $f(x) = 4^{x-1} - 3$

54. $f(x) = -3^{2x} + 5$

55. $f(x) = \dfrac{10}{2 + e^{-2x}} - 1$

56. $f(x) = \dfrac{2^x - 6 + 2^{3-x}}{x + 2}$

In Problems 57 and 58, find the x- and y-intercepts of the given graphs.

57.

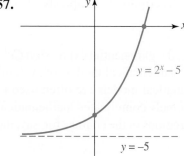

FIGURE 6.3.1 Graph for Problem 57

58.

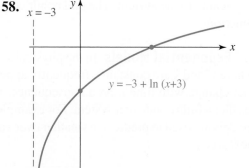

FIGURE 6.3.2 Graph for Problem 58

In Problems 59–64, graph the given functions. Determine the approximate x-coordinates of the points of intersection of their graphs.

59. $f(x) = 4e^x$, $\quad g(x) = 3^{-x}$

60. $f(x) = 2^x$, $\quad g(x) = 3 - 2^x$

61. $f(x) = 3^{x^2}$, $\quad g(x) = 2(3^x)$

62. $f(x) = \dfrac{1}{3} \cdot 2^{x^2}$, $\quad g(x) = 2^{x^2} - 1$

63. $f(x) = \log_{10} \dfrac{10}{x}$, $\quad g(x) = \log_{10} x$

64. $f(x) = \log_{10} \dfrac{x}{2}$, $\quad g(x) = \log_2 x$

In Problems 65–68, solve the given equation.

65. $x^{\ln x} = e^9$

66. $x^{\log_{10} x} = \dfrac{1000}{x^2}$

67. $\log_x 81 = 2$

68. $\log_5 125^x = -2$

In Problems 69 and 70, use the natural logarithm to find x in the domain of the given function for which f takes on the indicated value.

69. $f(x) = 6^x;$ $f(x) = 51$

70. $f(x) = \left(\tfrac{1}{2}\right)^x;$ $f(x) = 7$

For Discussion

71. Discuss: How would you find the x-intercepts of the graph of the function $f(x) = \log_2 x + \log_4 x - 6$?

72. Use a graphing utility to obtain the graph of the function $f(x) = \log_{x+2}(3 - x)$. Give the domain of the function f.

73. Discuss: Are the given two equations equivalent, that is, do they have the same solution set?
   (a) $\log_5(x - 2)^2 = 2;$ $2\log_5(x - 2) = 2$
   (b) $\log_5(x - 2)^3 = 3;$ $3\log_5(x - 2) = 3$

6.4 Exponential and Logarithmic Models

≡ **Introduction** In this section we consider some **mathematical models** utilizing exponential or logarithmic functions. Roughly speaking, a mathematical model is a mathematical description of something that we will call a *system*. To construct a mathematical model we start with a set of reasonable assumptions about the system that we are trying to describe. These assumptions include any empirical laws that are applicable to the system. The end result could be a description as simple as a single function.

☐ **Exponential Models** In the physical sciences, the exponential expression Ce^{kt}, where C and k are constants, frequently appears in mathematical models of systems that change with time t. As a consequence, mathematical models are often used to predict a future state of a system. For example, extremely complicated mathematical models are used to predict the weather over various regions of the country for, say, the next week.

☐ **Population Growth** In one model of a growing population, it is assumed that the *rate* of growth of the population is proportional to the *number present* at time t. If $P(t)$ denotes the population or number present at time t, then with the aid of calculus it can be shown that this assumption gives rise to

$$P(t) = P_0 e^{kt}, \quad k > 0, \tag{1}$$

where t is time, and P_0 and k are constants. The function (1) is used to describe the growth of populations of bacteria, small animals, and, in some rare circumstances, humans. Setting $t = 0$ gives $P(0) = P_0$, and so P_0 is called the **initial population**. The constant $k > 0$ is called the **growth constant** or **growth rate**. Since e^{kt}, $k > 0$, is an increasing function on the interval $[0, \infty)$, the model in (1) describes uninhibited growth.

EXAMPLE 1   **Bacterial Growth**

It is known that the doubling time* of *E. Coli* bacteria, which reside in the large intestine (colon) of healthy people, is just 20 minutes. Use the exponential growth model (1) to find the number of *E. Coli* bacteria in a culture after 6 hours.

Solution Let us use hours as our unit of time, so that 20 min $= \frac{1}{3}$ h. Because the initial number of *E. Coli* in the culture is not specified, we will simply denote the initial size of the culture as P_0. Now using (1), a function interpretation of the first sentence in this example is $P\left(\frac{1}{3}\right) = 2P_0$. This means $P_0 e^{k/3} = 2P_0$ or $e^{k/3} = 2$. Solving this last equation for k gives the growth constant

$$\frac{k}{3} = \ln 2 \qquad \text{or} \qquad k = 3\ln 2 \approx 2.0794.$$

A model for the size of the culture after t hours is then $P(t) = P_0 e^{2.0794t}$. Setting $t = 6$ gives $P(6) = P_0 e^{2.0794(6)} \approx 262,144 P_0$. Put another way, if the culture consists of only *one* bacterium at $t = 0$, then (with $P_0 = 1$) the model predicts that there will be 262,144 cells 6 hours later. ≡

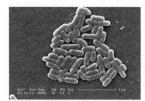

E. Coli bacteria

◀ When working problems such as this, be sure to store the value of k in the memory of your calculator.

In the early nineteenth century the English clergyman and economist Thomas R. Malthus used the growth model (1) to predict the world population. For specific values of P_0 and k, the function values $P(t)$ were actually reasonable approximations to the world population for a period of time during the nineteenth century. Since $P(t)$ is an increasing function, Malthus predicted that the future population growth would surpass the world's ability to produce food. As a consequence he also predicted wars and worldwide famine. More a doomsayer than a seer, Malthus failed to foresee that the food supply would keep pace with the increased population through simultaneous advances in science and technology.

In 1840, a more realistic model for predicting human populations in small countries was advanced by the Belgian mathematician/biologist **P. F. Verhulst** (1804–1849). The so-called **logistic function**

$$P(t) = \frac{K}{1 + ce^{rt}}, \quad r < 0, \tag{2}$$

Thomas R. Malthus
(1776–1834)

where K, c, and r are constants, has over the years proved to be an accurate growth model for populations of protozoa, bacteria, fruit flies, water fleas, fish, and animals confined to limited spaces. In contrast to uninhibited growth of the Malthusian model (1), (2) exhibits bounded growth. More specifically, the population predicted by (2) will not increase beyond the number K, called the **carrying capacity** of the ecosystem. For $r < 0$, $e^{rt} \to 0$ and $P(t) \to K$ as $t \to \infty$. You are asked to graph a special case of (2) in Problem 7 in Exercises 5.4.

☐ **Radioactive Decay** Element 88, better known as **radium**, was discovered by Pierre and Marie Curie in 1898. Radium is a radioactive element, which means that a radium atom spontaneously **decays**, or disintegrates, by emitting radiation in the form of alpha particles, beta particles, and gamma rays. When an atom disintegrates in this manner, its nucleus is transmuted into a nucleus of another element. For example, the nucleus of an atom of the most stable isotope of radium, Ra-226, is transmuted into the nucleus of a radon atom Rn-222. Radon is a heavy, odorless, colorless, and highly dangerous radioactive gas that usually originates in the ground. Because it can penetrate a

Pierre and Marie Curie

*In biology the doubling time is sometimes referred to as the **generation time**.

sealed concrete floor, radon frequently accumulates in the basements of some new and highly insulated homes. Some medical organizations have claimed that after cigarette smoking, exposure to radon gas is the second leading cause of lung cancer.

If it is assumed that the rate of decay of a radioactive substance is proportional to the amount remaining or present at time t, then we arrive at basically the same model as in (1). The important difference is that $k < 0$. If $A(t)$ represents the amount of the decaying substance that remains at time t, then

$$A(t) = A_0 e^{kt}, \quad k < 0, \tag{3}$$

where A_0 is the initial amount of the substance present, that is, $A(0) = A_0$. The constant $k < 0$ in (3) is called the **decay constant** or **decay rate**.

▮ EXAMPLE 2　　　Decay of Radium

Suppose there are 20 grams of radium on hand initially. After t years the amount remaining is modeled by the function $A(t) = 20e^{-0.000418t}$. Find the amount of radium remaining after 100 years. What percent of the original 20 grams has decayed after 100 years?

Solution Using a calculator, we find that after 100 years there remains

$$A(100) = 20e^{-0.000418(100)} \approx 19.18 \text{ g}.$$

Thus, only

$$\frac{20 - 19.18}{20} \times 100\% \approx 4.1\%$$

of the initial 20 grams has decayed. ≡

□ **Half-Life** The **half-life** of a radioactive substance is the time T it takes for one-half of a given amount of that element to disintegrate and change into a new element. See FIGURE 6.4.1. Half-life is a measure of the stability of an element; that is, the shorter the half-life, the more unstable the element. For example, the half-life of the highly radioactive strontium-90, Sr-90, produced in nuclear explosions, is 29 days, whereas the half-life of the uranium isotope U-238 is 4,560,000 years. The half-life of californium, Cf-244, first discovered in 1950, is only 45 minutes. Polonium, Po-213, has a half-life of 0.000001 second.

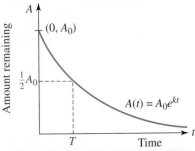

FIGURE 6.4.1 Time T is the half-life

▮ EXAMPLE 3　　　Half-Life of Radium

Use the exponential model in Example 2 to determine the half-life of radium.

Solution If $A(t) = 20e^{-0.000418t}$, then we must find the time T for which

one-half the initial amount
↓
$$A(T) = \tfrac{1}{2}(20) = 10.$$

From $20e^{-0.000418T} = 10$ we get $e^{-0.000418T} = \tfrac{1}{2}$. By rewriting the last expression in the logarithmic form $-0.000418\,T = \ln\tfrac{1}{2}$ we can solve for T:

$$T = \frac{\ln\tfrac{1}{2}}{-0.000418} \approx 1660 \text{ years.}$$ ≡

A careful reading of Example 3 reveals that the initial amount present plays no part in the actual calculation of the half-life. Since the solution of $A(T) = A_0 e^{-0.000418T} = \tfrac{1}{2}A_0$

leads to $e^{-0.000418T} = \frac{1}{2}$, we see that T is independent of A_0. In other words, the half-life of 1 gram, 20 grams, or 10,000 grams of radium is the same. It takes about 1660 years for one-half of *any* given quantity of radium to transmute into radon.

Medications also have half-lives. In this case, the half-life of a drug is the time T that it takes for the body to eliminate, by metabolism or excretion, one-half of the amount of the drug taken. For example, the most popular NSAIDs (nonsteroidal anti-inflammatory drugs such as aspirin and ibuprofen) taken for the relief of continuing pain, have relatively short half-lives of a few hours and as a consequence must be taken several times a day. The NSAID naproxen has a longer half-life and is usually taken once every 12 hours. See Problem 31 in Exercises 6.4.

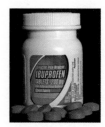
Ibuprofen is an NSAID

☐ **Carbon Dating** The approximate age of fossils of once-living matter can sometimes be determined by a method known as **carbon dating**. The radioactive isotope of carbon, carbon-14 or C-14, is formed presumably at a constant rate in the atmosphere by the interaction of cosmic rays on nitrogen-14. The carbon-dating method, invented by the chemist Willard Libby around 1950, is based on the fact that a plant or an animal absorbs C-14 through the process of breathing and eating, and ceases to absorb C-14 when it dies. As the next example shows, the carbon-dating procedure is based on the knowledge that the half-life of C-14 is about 5730 years. Carbon-14 decays back to the original nitrogen-14.

Libby won the 1960 Nobel Prize in chemistry for his work. Libby's method has been used to date wooden furniture found in Egyptian tombs, the Dead Sea scrolls written on papyrus and animal skin, the famous linen Shroud of Turin, and a recently discovered copy of the Gnostic Gospel of Judas written on papyrus.

Willard Libby (1908–1980)

The Psalms scroll

EXAMPLE 4   Carbon Dating a Fossil

A fossilized bone is found to contain $\frac{1}{1000}$ of the initial amount of C-14 that the organism contained while it was alive. Determine the approximate age of the fossil.

Solution If A_0 denotes an initial amount A_0, measured in grams, of C-14 in the organism, then t years after its death there are $A(t) = A_0 e^{kt}$ grams remaining. When $t = 5730$, $A(5730) = \frac{1}{2}A_0$, and so $\frac{1}{2}A_0 = A_0 e^{5730k}$. Solving this last equation for the decay constant k gives

$$e^{5730k} = \frac{1}{2} \quad \text{and so} \quad k = \frac{\ln\frac{1}{2}}{5730} \approx -0.00012097.$$

Hence a model for the amount of C-14 remaining is $A(t) = A_0 e^{-0.00012097t}$. Using this model, we now solve $A(t) = \frac{1}{1000}A_0$ for t:

$$A_0 e^{-0.00012097t} = \frac{1}{1000}A_0 \quad \text{implies} \quad t = \frac{\ln\frac{1}{1000}}{-0.00012097} \approx 57{,}100 \text{ years.} \equiv$$

The age determined in the last example is actually beyond the border of accuracy for the carbon-14-dating method. After 9 half-lives of the isotope, or about 52,000 years, about 99.7% of carbon-14 has decayed making its measurement in a fossil nearly impossible.

☐ **Newton's Law of Cooling/Warming** Suppose an object or body is placed in a medium (air, water, etc.) that is held at constant temperature T_m, called the **ambient temperature**. If the initial temperature T_0 of the body or object at the moment it is

We take this moment to correspond to the ▶
time $t = 0$.

placed into the medium is greater than the ambient temperature T_m, then the body will cool. On the other hand, if T_0 is less than T_m, then it will warm up. For example, in an office kept at, say, 70°F, a steaming cup of coffee will cool off, whereas a glass of ice water will warm up. The usual cooling/warming assumption is that the rate at which an object cools/warms is proportional to the difference $T(t) - T_m$, where $T(t)$ represents the temperature of the object at time t. In either case, cooling or warming, this assumption leads to $T(t) - T_m = (T_0 - T_m)e^{kt}$, where k is a negative constant. Observe that since $e^{kt} \to 0$ for $k < 0$, the last expression is consistent with one's intuitive expectation that $T(t) - T_m \to 0$, or equivalently $T(t) \to T_m$, as $t \to \infty$ (the coffee cools to room temperature; the ice water warms to room temperature). Solving for $T(t)$ we obtain a function for the temperature of the object,

$$T(t) = T_m + (T_0 - T_m)e^{kt}, \quad k < 0. \tag{4}$$

The mathematical model in (4), named after its discoverer, is called **Newton's law of cooling/warming**. Note that $T(0) = T_0$.

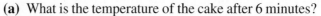

EXAMPLE 5   Cooling of a Cake

A cake is removed from an oven where the temperature was 350°F into a kitchen where the temperature is 75°F. One minute later the temperature of the cake is measured to be 300°F. Assume that the temperature of the cake in the kitchen is given by (4).

(a) What is the temperature of the cake after 6 minutes?
(b) At what time is the temperature of the cake 80°F?
(c) Graph $T(t)$.

Solution (a) When the cake is removed from the oven its temperature is also 350°F, that is, $T_0 = 350$. The ambient temperature is the temperature of the kitchen $T_m = 75$. Thus (4) becomes $T(t) = 75 + 275e^{kt}$. The measurement that $T(1) = 300$ is the condition that determines the value of k. From $T(1) = 75 + 275e^k = 300$ we find

$$e^k = \frac{225}{275} = \frac{9}{11} \qquad \text{or} \qquad k = \ln\frac{9}{11} \approx -0.2007.$$

The mathematical model $T(t) = 75 + 275e^{-0.2007t}$ then predicts that the temperature of the cake in 6 minutes after it is removed from the oven will be

$$T(6) = 75 + 275e^{-0.2007(6)} \approx 157.5°. \tag{5}$$

(b) To determine when the temperature of the cake will be 80°F, we solve the equation $T(t) = 80$ for t. Rewriting $T(t) = 75 + 275e^{-0.2007t} = 80$ as

$$e^{-0.2007t} = \frac{5}{275} = \frac{1}{55} \qquad \text{we find} \qquad t = \frac{\ln\frac{1}{55}}{-0.2007} \approx 20 \text{ min.}$$

(c) With the aid of a graphing utility we obtain the graph of $T(t)$ shown in blue in **FIGURE 6.4.2**. Since $T(t) = 75 + 275e^{-0.2007t} \to 75$ as $t \to \infty$, $T = 75$, shown in red in Figure 6.4.2, is a horizontal asymptote for the graph of $T(t) = 75 + 275e^{-0.2007t}$. ≡

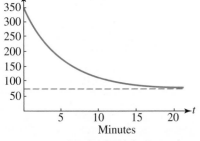

Cake will cool off to room temperature

FIGURE 6.4.2 Graph of $T(t)$ in Example 5

☐ **Compound Interest** Investments such as savings accounts pay an annual rate of interest that can be compounded annually, quarterly, monthly, weekly, daily, and so on. In general, if a principal of P dollars is invested at an annual rate r of interest that is compounded n times a year, then the amount S accrued at the end of t years is given by

$$S = P\left(1 + \frac{r}{n}\right)^{nt}. \tag{6}$$

S is called the **future value** of the principal P. If the number n is increased without bound, then interest is said to be **compounded continuously**. To find the future value of P in this case, we let $m = n/r$. Then $n = mr$ and

$$\left(1 + \frac{r}{n}\right)^{nt} = \left(1 + \frac{1}{m}\right)^{mrt} = \left[\left(1 + \frac{1}{m}\right)^{m}\right]^{rt}.$$

Since $n \to \infty$ implies that $m \to \infty$, we see from page 324 of Section 6.1 that $(1 + 1/m)^m \to e$. The right-hand side of (6) becomes

$$P\left[\left(1 + \frac{1}{m}\right)^{m}\right]^{rt} \to P[e]^{rt} \quad \text{as} \quad m \to \infty.$$

Thus, if an annual rate r of interest is compounded continuously, the future value S of a principal P in t years is

$$S = Pe^{rt}. \tag{7}$$

<hr/>

EXAMPLE 6　　　　**Comparison of Future Values**

Suppose that $1000 is deposited in a savings account whose annual rate of interest is 3%. Compare the future value of this principal in 10 years **(a)** if interest is compounded monthly and **(b)** if interest is compounded continuously.

Solution (a) Since there are 12 months in a year, we identify $n = 12$. Furthermore, with $P = 1000$, $r = 0.03$, and $t = 10$, (6) becomes

$$S = 1000\left(1 + \frac{0.03}{12}\right)^{12(10)} = 1000(1.0025)^{120} \approx \$1{,}349.35.$$

(b) From (7),

$$S = 1000e^{(0.03)(10)} = 1000e^{0.3} \approx \$1{,}349.86.$$

Thus over 10 years we have gained only $0.51 by compounding continuously rather than monthly. ≡

☐ **Logarithmic Models** Probably the most famous application of the base 10 logarithm, or common logarithm, is the **Richter scale**. In 1935, the American seismologist Charles F. Richter devised a logarithmic scale for comparing the energies of different earthquakes. The magnitude M of an earthquake is defined by

$$M = \log_{10}\frac{A}{A_0}, \tag{8}$$

where A is the amplitude of the largest seismic wave of the earthquake and A_0 is a reference amplitude that corresponds to the magnitude $M = 0$. The number M is calculated to one decimal place. Earthquakes of magnitude 6 or greater are considered potentially destructive.

Since 1979 the USGS has used the **moment magnitude scale** to assign a magnitude to earthquakes that range from strong to massive. This scale devised by the seismologists Thomas C. Hanks and Hiroo Kanamori corrects certain deficiencies in the Richter scale at that level. The moment magnitude M_w is defined by

$$M_w = \tfrac{2}{3}\log_{10}M_0 - 10.7,$$

where M_0 denotes the magnitude of the seismic moment. For moderate earthquakes the magnitudes given by moment magnitude scale and Richter scale are about the same. Using the moment magnitude scale the USGS assigned a magnitude of $M_w = 9.0$ to the Japan earthquake on March 11, 2011.

Charles F. Richter
(1900–1985)

EXAMPLE 7   Comparing Intensities

The earthquake on December 26, 2004, off the west coast of Northern Sumatra, which spawned a tsunami causing over 200,000 deaths, was initially classified as a 9.3 on the Richter scale. On March 28, 2005, an aftershock in the same area was classified as an 8.7 on the Richter scale. How many times more intense was the 2004 earthquake?

Solution From (8) we have

$$9.3 = \log_{10}\left(\frac{A}{A_0}\right)_{2004} \quad \text{and} \quad 8.7 = \log_{10}\left(\frac{A}{A_0}\right)_{2005}.$$

This means, in turn, that

$$\left(\frac{A}{A_0}\right)_{2004} = 10^{9.3} \quad \text{and} \quad \left(\frac{A}{A_0}\right)_{2005} = 10^{8.7}.$$

Now, since $9.3 = 0.6 + 8.7$, it follows from the laws of exponents that

$$\left(\frac{A}{A_0}\right)_{2004} = 10^{9.3} = 10^{0.6}10^{8.7} = 10^{0.6}\left(\frac{A}{A_0}\right)_{2005} \approx 3.98\left(\frac{A}{A_0}\right)_{2005}.$$

Thus the original earthquake in 2004 was approximately 4 times as intense as the aftershock in 2005.   ≡

You can see from Example 7 that if, say, one earthquake is a 6.0 and another is a 4.0 on the Richter scale, then the 6.0 earthquake is $10^2 = 100$ times more intense than the 4.0 earthquake.

☐ **pH of a Solution** In chemistry, the hydrogen potential, or **pH**, of a solution is defined as

$$pH = -\log_{10}[H^+], \tag{9}$$

where the symbol $[H^+]$ denotes the concentration of hydrogen ions in a solution measured in moles per liter. The pH scale was invented in 1909 by the Danish biochemist Søren Sørensen. Solutions are classified according to their pH value as *acidic*, *base*, or *neutral*. A solution with a pH in the range $0 < pH < 7$ is said to be acidic; when $pH > 7$, the solution is base (or alkaline). In the case when $pH = 7$, the solution is neutral. Water, if uncontaminated by other solutions or by acid rain, is an example of a neutral solution, whereas undiluted lemon juice is highly acidic and has a pH in the range $pH \leq 3$. A solution with $pH = 6$ is ten times more acidic than a neutral solution. See Problems 47–50 in Exercises 6.4.

As the next example illustrates, pH values are usually calculated to one decimal place.

Søren Sørensen (1868–1939)

EXAMPLE 8   pH of Human Blood

The concentration of hydrogen ions in the blood of a healthy person is found to be $[H^+] = 3.98 \times 10^{-8}$ moles/liter. Find the pH of blood.

Solution From (9) and the laws of logarithms (Theorem 6.2.1),

$$\begin{aligned} pH &= -\log_{10}[3.98 \times 10^{-8}] \\ &= -[\log_{10}3.98 + \log_{10}10^{-8}] \\ &= -[\log_{10}3.98 - 8\log_{10}10] \qquad \leftarrow \log_{10}10 = 1 \\ &= -[\log_{10}3.98 - 8]. \end{aligned}$$

With the help of the base 10 log key on a calculator, we find that

$$pH \approx -[0.5999 - 8] \approx 7.4.$$ ≡

Human blood in usually a base solution. The pH values of blood usually fall within the rather narrow range $7.2 < \text{pH} < 7.6$. A person with a blood pH outside these limits can suffer illness and even death.

<hr>

6.4 Exercises Answers to selected odd-numbered problems begin on page ANS-18.

Population Growth

1. After 2 hours the number of bacteria in a culture is observed to have doubled.
   (a) Find an exponential model (1) for the number of bacteria in the culture at time t.
   (b) Find the number of bacteria present in the culture after 5 hours.
   (c) Find the time that it takes the culture to grow to 20 times its initial size.

2. A model for the number of bacteria in a culture after t hours is given by (1).
   (a) Find the growth constant k if it is known that after 1 hour the colony has expanded to 1.5 times its initial population.
   (b) Find the time that it takes for the culture to quadruple in size.

3. A model for the population in a small community is given by $P(t) = 1500e^{kt}$. If the initial population increases by 25% in 10 years, what will the population be in 20 years?

4. A model for the population in a small community after t years is given by (1).
   (a) If the initial population has doubled in 5 years, how long will it take to triple? To quadruple?
   (b) If the population of the community in part (a) is 10,000 after 3 years, what was the initial population?

5. A model for the number of bacteria in a culture after t hours is given by $P(t) = P_0 e^{kt}$. After 3 hours it is observed that 400 bacteria are present. After 10 hours 2000 bacteria are present. What was the initial number of bacteria?

6. In genetic research a small colony of *drosophila* (small two-winged fruit flies) is grown in a laboratory environment. After 2 days it is observed that the population of flies in the colony has increased to 200. After 5 days the colony has 400 flies.
   (a) Find a model $P(t) = P_0 e^{kt}$ for the population of the fruit-fly colony after t days.
   (b) What will be the population of the colony in 10 days?
   (c) When will the population of the colony be 5000 fruit flies?

7. A student sick with a flu virus returns to an isolated college campus of 2000 students. A model for the number of students infected with the flu t days after the student's return is given by the logistic function

$$P(t) = \frac{2000}{1 + 1999e^{-0.8905t}}.$$

   (a) According to this model, how many students will be infected with the flu after 5 days?
   (b) How long will it take for one-half of the student population to become infected?
   (c) How many students does the model predict will become infected after a very long period of time?
   (d) Sketch a graph of $P(t)$.

8. In 1920, Pearl and Reed proposed a logistic model for the population of the United States based on the years 1790, 1850, and 1910. The logistic function they proposed was

$$P(t) = \frac{2930.3009}{0.014854 + e^{-0.0313395t}},$$

   where P is measured in thousands and t represents the number of years past 1780.

(a) The model agrees quite well with the census figures between 1790 and 1910. Determine the population figures for 1790, 1850, and 1910.

(b) What does this model predict for the population of the United States after a very long time? How does this prediction compare with the 2000 census population of 281 million?

Radioactive Decay and Half-Life

9. Initially 200 milligrams of a radioactive substance was present. After 6 hours the mass had decreased by 3%. Construct an exponential model $A(t) = A_0 e^{kt}$ for the amount remaining of the decaying substance after t hours. Find the amount remaining after 24 hours.

10. Determine the half-life of the substance in Problem 9.

11. Do this problem without using the exponential model (3). Initially there are 400 grams of a radioactive substance on hand. If the half-life of the substance is 8 hours, give an educated guess of how much remains (approximately) after 17 hours. After 23 hours. After 33 hours.

12. Construct an exponential model $A(t) = A_0 e^{kt}$ for the amount remaining of the decaying substance in Problem 11. Compare the predicted values $A(17)$, $A(23)$, and $A(33)$ with your guesses.

13. Iodine-131, used in nuclear medicine procedures, is radioactive and has a half-life of 8 days. Find the decay constant k for iodine-131. If the amount remaining of an initial sample after t days is given by the exponential model $A(t) = A_0 e^{kt}$, how long will it take for 95% of the sample to decay?

14. The amount remaining of a radioactive substance after t hours is given by $A(t) = 100 e^{kt}$. After 12 hours, the initial amount has decreased by 7%. How much remains after 48 hours? What is the half-life of the substance?

15. The half-life of polonium-210, Po-210, is 140 days. If $A(t) = A_0 e^{kt}$ represents the amount of Po-210 remaining after t days, what is the amount remaining after 80 days? After 300 days?

16. Strontium-90 is a dangerous radioactive substance found in acid rain. As such it can make its way into the food chain by polluting the grass in a pasture on which milk cows graze. The half-life of strontium-90 is 29 years.

(a) Find an exponential model (3) for the amount remaining after t years.

(b) Suppose a pasture is found to contain Str-90 that is 3 times a safe level A_0. How long will it be before the pasture can be used again for grazing cows?

Carbon Dating

Charcoal drawing in Problem 17

17. Charcoal drawings were discovered on walls and ceilings in a cave in Lascaux, France. Determine the approximate age of the drawings, if it was found that 86% of C-14 in a piece of charcoal found in the cave had decayed through radioactivity.

18. Analysis on an animal bone fossil at an archeological site reveals that the bone has lost between 90% and 95% of C-14. Give an interval for the possible ages of the bone.

19. The Shroud of Turin shows the negative image of the body of a man who appears to have been crucified. It is believed by many to be the burial shroud of Jesus of Nazareth. In 1988 the Vatican granted permission to have the shroud carbon dated. Several independent scientific laboratories analyzed the cloth and the consensus opinion was that the shroud is approximately 660 years old, an age consistent with its historical appearance. This age has been disputed by many scholars. Using this age, determine what percentage of the original amount of C-14 remained in the cloth as of 1988.

Shroud image in Problem 19

20. In 1991 hikers found a preserved body of a man partially frozen in a glacier in the Austrian Alps. Through carbon-dating techniques it was found that the body of Ötzi—the iceman, as he came to be called—contained 53% as much C-14 as found in a living person. What is the approximate date of his death?

The iceman in Problem 20

Newton's Law of Cooling/Warming

21. Suppose a pizza is removed from an oven at 400°F into a kitchen whose temperature is a constant 80°F. Three minutes later the temperature of the pizza is found to be 275°F.
    (a) What is the temperature $T(t)$ of the pizza after 5 minutes?
    (b) Determine the time when the temperature of the pizza is 150°F.
    (c) After a very long period of time, what is the approximate temperature of the pizza?

22. A glass of cold water is removed from a refrigerator whose interior temperature is 39°F into a room maintained at 72°F. One minute later the temperature of the water is 43°F. What is the temperature of the water after 10 minutes? After 25 minutes?

23. A thermometer is brought from the outside, where the air temperature is −20°F, into a room where the air temperature is a constant 70°F. After 1 minute inside the room the thermometer reads 0°F. How long will it take for the thermometer to read 60°F?

24. A thermometer is taken from inside a house to the outside, where the air temperature is 5°F. After 1 minute outside the thermometer reads 59°F, and after 5 minutes it reads 32°F. What is the temperature inside the house?

Thermometer in Problem 24

25. A dead body was found within a closed room of a house where the temperature was a constant 70°F. At the time of discovery, the core temperature of the body was determined to be 85°F. One hour later a second measurement showed that the core temperature of the body was 80°F. Assume that the time of death corresponds to $t = 0$ and that the core temperature at that time was 98.6°F. Determine how many hours elapsed before the body was found.

26. Repeat Problem 25 if evidence indicated that the dead person was running a fever of 102°F at the time of death.

Compound Interest

27. Suppose that 1¢ is deposited in a savings account paying 1% annual interest compounded continuously. How much money will have accrued in the account after 2000 years? What is the future value of 1¢ in 2000 years if the account pays 2% annual interest compounded continuously?

28. Suppose that $100,000 is invested at an annual interest rate of 5%. Use (6) and (7) to compare the future values of that amount in 1 year by completing the following table.

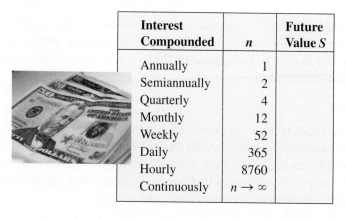

Interest Compounded	n	Future Value S
Annually	1	
Semiannually	2	
Quarterly	4	
Monthly	12	
Weekly	52	
Daily	365	
Hourly	8760	
Continuously	$n \to \infty$	

6.4 Exponential and Logarithmic Models

29. Suppose that $5000 is deposited in a savings account paying 6% annual interest compounded continuously. How much interest will be earned in 8 years?

30. **Present Value** If (7) is solved for P, that is, $P = Se^{-rt}$, we obtain the amount that should be invested now at an annual rate r of interest in order to be worth S dollars after t years. We say that P is the **present value** of the amount S. What is the present value of $100,000 at an annual rate of 3% compounded continuously for 30 years?

Miscellaneous Exponential Models

31. **Effective Half-life** Radioactive substances are removed from living organisms by two processes: natural physical decay and biological metabolism. Each process contributes to an effective half-life E that is defined by

$$1/E = 1/P + 1/B,$$

where P is the physical half-life of the radioactive substance and B is the biological half-life.
   (a) Radioactive iodine, I-131, is used to treat hyperthyroidism (overactive thyroid). It is known that for human thyroids, $P = 8$ days and $B = 24$ days. Find the effective half-life of I-131 in the thyroid.
   (b) Suppose the amount of I-131 in the human thyroid after t days is modeled by $A(t) = A_0 e^{kt}$, $k < 0$. Use the effective half-life found in part (a) to determine the percentage of radioactive iodine remaining in the human thyroid gland two weeks after its ingestion.

32. **Newton's Law of Cooling Revisited** The rate at which a body cools also depends on its exposed surface area S. If S is a constant, then a modification of (4) is

$$T(t) = T_m + (T_0 - T_m)e^{kSt}, \quad k < 0.$$

Suppose two cups A and B are filled with coffee at the same time. Initially the temperature of the coffee is 150°F. The exposed surface area of the coffee in cup B is twice the surface area of the coffee in cup A. After 30 min, the temperature of the coffee in cup A is 100°F. If $T_m = 70°F$, what is the temperature of the coffee in cup B after 30 min?

33. **Series Circuit** In a simple series circuit consisting of a constant voltage E, an inductance of L henries, and a resistance of R ohms, it can be shown that the current $I(t)$ is given by

$$I(t) = \frac{E}{R}(1 - e^{-(R/L)t}).$$

Solve for t in terms of the other symbols.

34. **Drug Concentration** Under some conditions the concentration of a drug at time t after injection is given by

$$C(t) = \frac{a}{b} + \left(C_0 - \frac{a}{b}\right)e^{-bt}.$$

Here a and b are positive constants and C_0 is the concentration of the drug at $t = 0$. Determine the steady-state concentration of a drug, that is, the limiting value of $C(t)$ as $t \to \infty$. Determine the time t at which $C(t)$ is one-half the steady-state concentration.

Richter Scale

Marina district in San Francisco, 1989

35. Two of the most devastating earthquakes in the San Francisco Bay area occurred in 1906 along the San Andreas fault and in 1989 in the Santa Cruz Mountains near Loma Prieta peak. The 1906 and 1989 earthquakes measured 8.5 and 7.1 on the Richter scale, respectively. How much greater was the intensity of the 1906 earthquake compared to the 1989 earthquake?

36. How much greater was the intensity of the 2004 Northern Sumatra earthquake (Example 7) compared to the 1964 Alaskan earthquake of magnitude 8.9?

37. If an earthquake has a magnitude 4.2 on the Richter scale, what is the magnitude on the Richter scale of an earthquake that has an intensity 20 times greater? [*Hint*: First solve the equation $10^x = 20$.]

38. Show that the Richter scale defined in (8) of this section can be written

$$M = \frac{\ln A - \ln A_0}{\ln 10}.$$

pH of a Solution

In Problems 39–42, determine the pH of a solution with the given hydrogen-ion concentration $[H^+]$.

39. 10^{-6} **40.** 4×10^{-7} **41.** 2.8×10^{-8} **42.** 5.1×10^{-5}

In Problems 43–46, determine the hydrogen-ion concentration $[H^+]$ of a solution with the given pH.

43. 3.3   **44.** 7.3   **45.** 6.6   **46.** 8.1

In Problems 47–50, determine how many more times acidic the first substance is compared to the second substance.

47. lemon juice: pH = 2.3;   vinegar, pH = 3.3
48. battery acid: pH = 1;   lye, pH = 13
49. clean rain: pH = 5.6;   acidic rain, pH = 3.8
50. NaOH: $[H^+] = 10^{-14}$;   HCl, $[H^+] = 1$

Miscellaneous Logarithmic Models

51. Richter Scale and Energy **(a)** Charles Richter working with Beno Gutenberg developed the model

$$M = \tfrac{2}{3}[\log_{10} E - 11.8]$$

that relates the Richter magnitude M of an earthquake and its seismic energy E (measured in ergs). Calculate the seismic energy E of the 2004 Northern Sumatra earthquake where $M = 9.3$.

(b) Show that the proportional energy difference $f_{\Delta E} = E_1/E_2$ between two different earthquakes of Richter magnitudes M_1 and M_2 is given by

$$f_{\Delta E} = E_1/E_2 = 10^{\frac{3}{2}[M_1 - M_2]}.$$

(c) Use part (b) to show that if M_1 is one unit more than M_2 then the seismic energy E_1 of an earthquake of magnitude M_1 is approximately 32 times the seismic energy E_2 of the earthquake of magnitude M_2. Repeat the calculation if M_1 is two units more than M_2.

52. Intensity Level The **intensity level** b of a sound measured in decibels (dB) is defined by

$$b = 10 \log_{10} \frac{I}{I_0}, \tag{10}$$

where I is the **intensity of the sound** measured in watts/cm^2 and $I_0 = 10^{-16}$ watts/cm^2 is the intensity of the faintest sound that can be heard (0 dB). Use (10) and complete the following table.

Sound	Intensity I (watts/cm^2)	Intensity Level b (dB)
Whisper	10^{-14}	
Conversation	10^{-11}	
TV commercials	10^{-10}	
Smoke alarm	10^{-9}	
Jet plane taking off	10^{-7}	
Rock band	10^{-4}	

53. **Threshold of Pain** The threshold of pain is generally taken to be around 140 dB. Find the intensity of sound I corresponding to 140 dB.

54. **Intensity Levels** The intensity of sound I is inversely proportional to the square of the distance d from its source, that is,

$$I = \frac{k}{d^2},\qquad(11)$$

where k is the constant of proportionality. Suppose d_1 and d_2 are distances from a source of sound, and that the corresponding intensity levels of the sounds are b_1 and b_2. Use (11) in (10) to show that b_1 and b_2 are related by

$$b_2 = b_1 + 20\log_{10}\frac{d_1}{d_2}.\qquad(12)$$

55. **Intensity Level** When a plane P_1, flying at an altitude of 1500 ft, passed over a point on the ground its intensity level b_1 was measured as 70 dB. Use (12) to find the intensity level b_2 of a second plane P_2, flying at an altitude of 2600 ft, when it passed over the same point.

56. **Talking Politics** At a distance of 4 ft, the intensity level of an animated political conversation is 60 dB. Use (12) to find the intensity level 14 ft from the conversation.

57. **Pupil of the Eye** An empirical model devised by DeGroot and Gebhard relates the diameter d of the pupil of the eye (measured in millimeters, mm) to the luminance B of light source (measured in millilambert's, mL):

$$\log_{10}d = 0.8558 - 0.000401(8.1 + \log_{10}B)^3.$$

See FIGURE 6.4.3.

(a) The average luminance of clear sky is approximately $B = 255$ mL. Find the corresponding pupil diameter.

(b) The luminance of the Sun varies from approximately $B = 190{,}000$ mL at sunrise to $B = 51{,}000{,}000$ mL at noon. Find the corresponding pupil diameters.

(c) Find the luminance B corresponding to a pupil diameter of 7 mm.

58. **Body Surface Area** A mathematical model for estimating body surface area S (in square meters) is given by

$$\log_{10}S = -0.69364 + (0.425)\log_{10}w + (0.725)\log_{10}h,$$

where w and h are a person's weight (in kilograms) and height (in meters), respectively. This empirical formula, due to D. Dubois and E. F. Dubois, first published in the *Archives of Internal Medicine* in 1916, is still used today by medical researchers.

(a) Estimate the body surface area of a person whose weight is $w = 70$ kg and who is $h = 1.75$ m tall.

(b) Determine your weight and height and estimate your own body surface area.

(c) Eliminate the logarithms in the surface area formula and express S in terms of w and h.

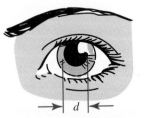

FIGURE 6.4.3 Pupil diameter in Problem 57

The Hyperbolic Functions

∫**Calculus** **PREVIEW**

≡ **Introduction** Logarithms were invented in the late sixteenth century by the Scottish lord—and nonmathematician—**John Napier** (1550–1617). It was he who coined the word "logarithm" from the two Greek words *logos*, meaning ratio, and *arithmos*, meaning number or power. But it took almost two centuries and the genius of the Swiss mathematician **Leonhard Euler** (1707–1783) before the mathematical community became fully aware of the irrational number *e* and its importance. It is his work that we emulate below in showing why the number *e* is the natural choice of base for the exponential and logarithmic functions.

Leonhard Euler

☐ **Difference Quotient Revisited** We return to the difference quotient concept first introduced in Section 2.9. Recall that we compute

$$\frac{f(x + h) - f(x)}{h} \qquad (1)$$

in three steps. For the exponential function $f(x) = b^x$, we have

    (*i*) $f(x + h) = b^{x+h} = b^x b^h$ ← laws of exponents

    (*ii*) $f(x + h) - f(x) = b^{x+h} - b^x$ ← $\begin{cases} \text{law of exponents} \\ \text{and factoring} \end{cases}$

                     $= b^x b^h - b^x = b^x(b^h - 1)$

    (*iii*) $\dfrac{f(x + h) - f(x)}{h} = \dfrac{b^x(b^h - 1)}{h} = b^x \dfrac{b^h - 1}{h}$

In the fourth step, the calculus step, we let $h \to 0$ but, unlike all the problems given in Exercises 2.9, there is no apparent way of canceling the *h* in (*iii*). Nonetheless, the derivative of $f(x) = b^x$ is

$$f'(x) = \lim_{h \to 0} b^x \cdot \frac{b^h - 1}{h}. \qquad (2)$$

Since b^x does not depend on the variable *h*, we can rewrite (2) as

$$f'(x) = b^x \cdot \lim_{h \to 0} \frac{b^h - 1}{h}. \qquad (3)$$

Now here are the amazing results. The limit in (3),

$$\lim_{h \to 0} \frac{b^h - 1}{h}, \qquad (4)$$

can be shown to exist for every positive base *b*. However, as one might expect, we will get a different answer for each base *b*. So let's denote the expression in (4) by the symbol $m(b)$. The derivative of $f(x) = b^x$ in (3) is then

$$f'(x) = b^x m(b). \qquad (5)$$

You are asked to approximate the value of $m(b)$ in the four cases $b = 1.5, 2, 3,$ and 5 in Problems 25–28 of Exercises 6.5. For example, it can be shown that $m(10) \approx 2.302585 \ldots$, and as a consequence the derivative of $f(x) = 10^x$ is

$$f'(x) = (2.302585 \ldots) 10^x. \qquad (6)$$

We can get a better understanding of what $m(b)$ is by evaluating (5) at $x = 0$. Since $b^0 = 1$, we have $f'(0) = m(b)$. In other words, $m(b)$ is the slope of the tangent line to

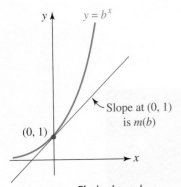

FIGURE 6.5.1 Find a base b so that the slope $m(b)$ of tangent line at $(0, 1)$ is 1

the graph of $f(x) = b^x$ at $x = 0$, that is, at the y-intercept $(0, 1)$. See FIGURE 6.5.1. Given that we have to calculate a different $m(b)$ for each base b, and that $m(b)$ is likely to be an "ugly" number as in (6), over time the following question arose naturally:

$$\text{Is there a base } b \text{ for which } m(b) = 1? \tag{7}$$

☐ **The Answer** To answer the question posed in (7), we must return to the definitions of e given in Section 6.1. Specifically, (4) of Section 6.1,

$$e = \lim_{h \to 0} (1 + h)^{1/h}, \tag{8}$$

provides the means for answering the question posed in (7). If you have studied Sections 1.5, 2.9, and 4.11 you should have an intuitive understanding that the equality in (8) means that as h gets closer and closer to 0 then $(1 + h)^{1/h}$ can be made arbitrarily close to the number e. Thus for values of h near 0, we have the approximation $(1 + h)^{1/h} \approx e$, and so it follows that $1 + h \approx e^h$. By rewriting the last expression in the form

$$\frac{e^h - 1}{h} \approx 1 \tag{9}$$

we can conclude that

$$1 = \lim_{h \to 0} \frac{e^h - 1}{h}. \tag{10}$$

Since the right-hand side of (10) is $m(e)$, we have the answer to the question in (7):

$$\text{The base } b \text{ for which } m(b) = 1 \text{ is } b = e. \tag{11}$$

In addition, from (3) we have discovered a wonderfully simple result: The derivative of $f(x) = e^x$ is

$$f'(x) = e^x. \tag{12}$$

The result in (12) is the same as

$$f'(x) = f(x).$$

Moreover, the only other nonzero function f in calculus whose derivative is equal to itself is $f(x) = ce^x$, where $c \neq 0$ is a constant.

☐ **What's Next?** Since $y = \log_b x$ and $y = b^x$ are inverse functions, one would expect that since the simplest derivative of $y = b^x$ is obtained when $b = e$ that the simplest derivative of $y = \log_b x$ also occurs for that base. That is indeed the case. You are encouraged to reexamine (3) of Section 6.1 and then work Problems 1–4 in Exercises 6.5.

☐ **Hyperbolic Functions** We have already seen in Section 6.4 the usefulness of the exponential function e^x in various mathematical models. As a further application, consider a long rope or a flexible wire, such as a telephone wire hanging only under its own weight between two fixed supports. It can be shown that under certain conditions the hanging wire assumes the shape of the graph of the function

$$f(x) = c\frac{e^{x/c} + e^{-x/c}}{2}. \tag{13}$$

The symbol c stands for a positive constant that depends on the physical characteristics of the wire. Functions such as (13), consisting of certain combinations of e^x and e^{-x}, appear in so many applications that mathematicians have given them names.

Telephone wires

CHAPTER 6 EXPONENTIAL AND LOGARITHMIC FUNCTIONS

☐ **Hyperbolic Functions** In particular, when $c = 1$ in (13), the resulting function $f(x) = \dfrac{e^x + e^{-x}}{2}$ is called the **hyperbolic cosine**.

DEFINITION 6.5.1 Hyperbolic Functions

For any real number x, the **hyperbolic sine** of x, denoted $\sinh x$ is

$$\sinh x = \frac{e^x - e^{-x}}{2}, \qquad (14)$$

and the **hyperbolic cosine** of x, denoted $\cosh x$, is

$$\cosh x = \frac{e^x + e^{-x}}{2}. \qquad (15)$$

Analogous to the trigonometric functions $\tan x$, $\cot x$, $\sec x$, and $\csc x$ that are defined in terms of $\sin x$ and $\cos x$, there are four additional hyperbolic functions $\tanh x$, $\coth x$, $\operatorname{sech} x$, and $\operatorname{csch} x$ that are defined in terms of $\sinh x$ and $\cosh x$:

$$\tanh x = \frac{\sinh x}{\cosh x} = \frac{e^x - e^{-x}}{e^x + e^{-x}} \qquad \coth x = \frac{1}{\tanh x} = \frac{e^x + e^{-x}}{e^x - e^{-x}} \qquad (16)$$

$$\operatorname{sech} x = \frac{1}{\cosh x} = \frac{2}{e^x + e^{-x}} \qquad \operatorname{csch} x = \frac{1}{\sinh x} = \frac{2}{e^x - e^{-x}}. \qquad (17)$$

☐ **Graphs** The graph of the hyperbolic cosine, shown in FIGURE 6.5.2, is called a **catenary**. The word *catenary* derives from the Latin word for a chain, *catena*. The shape of the famous 630-ft tall Gateway Arch in St. Louis, Missouri, is an inverted catenary. Compare the shape in Figure 6.5.2 with that in the accompanying photo. The graph of $y = \sinh x$ is given in FIGURE 6.5.3.

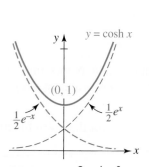

FIGURE 6.5.2 Graph of $y = \cosh x$

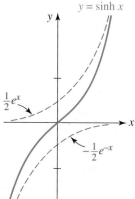

FIGURE 6.5.3 Graph of $y = \sinh x$

Gateway Arch in St. Louis, MO

The graphs of the hyperbolic tangent, cotangent, secant, and cosecant are given in FIGURE 6.5.4. Observe that $y = 1$ and $y = -1$ are horizontal asymptotes for the graphs of $y = \tanh x$ and $y = \coth x$ and that $x = 0$ is a vertical asymptote for the graphs of $y = \coth x$ and $y = \operatorname{csch} x$.

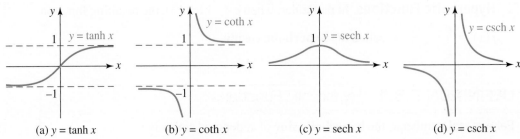

(a) $y = \tanh x$ (b) $y = \coth x$ (c) $y = \operatorname{sech} x$ (d) $y = \operatorname{csch} x$

FIGURE 6.5.4 Graphs of the hyperbolic tangent (a), cotangent (b), secant (c), and cosecant (d)

☐ **Identities** Although the hyperbolic functions are not periodic, they possess identities that are very similar to trigonometric identities. Analogous to the basic Pythagorean identity of trigonometry $\cos^2 x + \sin^2 x = 1$, for the hyperbolic sine and cosine we have

$$\cosh^2 x - \sinh^2 x = 1. \tag{18}$$

See Problems 9–14 in Exercises 6.5.

6.5 Exercises Answers to selected odd-numbered problems begin on page ANS–19.

---

1. Use the laws of logarithms to show that for $f(x) = \log_b x$,

$$\frac{f(x+h) - f(x)}{h} = \frac{1}{h}\log_b\left(1 + \frac{h}{x}\right) = \frac{1}{x}\log_b\left(1 + \frac{h}{x}\right)^{x/h}.$$

2. From Problem 1, the derivative of $f(x) = \log_b x$ is

$$f'(x) = \lim_{h \to 0} \frac{f(x+h) - f(x)}{h} = \frac{1}{x}\lim_{h \to 0}\log_b\left(1 + \frac{h}{x}\right)^{x/h}.$$

Let us assume that the limiting process can be taken inside the logarithm:

$$f'(x) = \frac{1}{x}\log_b\left[\lim_{h \to 0}\left(1 + \frac{h}{x}\right)^{x/h}\right].$$

Rewrite the foregoing result using the substitution $n = x/h$. Notice that since x is held fixed, as $h \to 0$ we must have $n \to \infty$. Give the precise value of $f'(x)$.

In Problems 3 and 4, use the result of Problem 2 to find $f'(x)$ for the given function.

3. $f(x) = \log_{10} x$

4. $f(x) = \ln x$

In Problems 5 and 6, use the result of Problems 1 and 2 to find $f'(x)$ for the given function. Before using the difference quotient, use the laws of logarithms to rewrite the function.

5. $f(x) = \ln\dfrac{x}{9}$

6. $f(x) = \log_{10} 6x$

In Problems 7 and 8, compute $\dfrac{f(x + h) - f(x)}{h}$ for the given function.

7. $f(x) = e^{5x}$ **8.** $f(x) = e^{-x+4}$

In Problems 9–14, use the definitions of $\sinh x$ and $\cosh x$ in (14) and (15) to verify the given identity.

9. $\cosh^2 x - \sinh^2 x = 1$ **10.** $1 - \tanh^2 x = \operatorname{sech}^2 x.$
11. $\cosh(-x) = \cosh x$ **12.** $\sinh(-x) = -\sinh x$
13. $\sinh 2x = 2 \sinh x \cosh x$ **14.** $\cosh 2x = \cosh^2 x + \sinh^2 x$

15. (a) If $\sinh x = -\frac{3}{2}$, use the identity given in Problem 9 to find the value of $\cosh x$.
   (b) Use the result of part (a) to find the numerical values of $\tanh x$, $\coth x$, $\operatorname{sech} x$, and $\operatorname{csch} x$.
16. (a) If $\tanh x = \frac{1}{2}$, use the identity given in Problem 10 to find the value of $\operatorname{sech} x$.
   (b) Use the result of part (a) to find the numerical values of $\cosh x$, $\sinh x$, $\coth x$, and $\operatorname{csch} x$.
17. As can be seen in Figure 6.5.3, the hyperbolic sine function $y = \sinh x$ is one-to-one. Use the definition of the hyperbolic sine in (14) in the form $e^x - 2y - e^{-x} = 0$ to show that the **inverse hyperbolic sine** $\sinh^{-1} x$ can be expressed in terms of the natural logarithm:

$$\sinh^{-1} x = \ln\left(x + \sqrt{x^2 + 1}\right).$$

[*Hint*: Interchange x and y in $e^x - 2y - e^{-x} = 0$ and solve for y.]

18. (a) Use the graph of $y = \sinh x$ in Figure 6.5.3 to sketch the graph of the inverse hyperbolic sine $y = \sinh^{-1} x$ defined in Problem 17.
   (b) Give the domain and range of $y = \sinh^{-1} x$.
19. The function $y = \cosh x$ is one-to-one on the restricted domain $[0, \infty)$. Proceed as in Problem 17 to show that the **inverse hyperbolic cosine** $\cosh^{-1} x$ can be expressed in terms of the natural logarithm:

$$\cosh^{-1} x = \ln\left(x + \sqrt{x^2 - 1}\right).$$

20. (a) Use the graph of $y = \cosh x$ in Figure 6.5.2 to sketch the graph of the inverse hyperbolic cosine $y = \cosh^{-1} x$ defined in Problem 19.
   (b) Give the domain and range of $y = \cosh^{-1} x$.
21. The function $y = \tanh x$ is one-to-one. Use the graph in Figure 6.5.4(a) to sketch the graph of the **inverse hyperbolic tangent** $y = \tanh^{-1} x$. Give the domain and range of $y = \tanh^{-1} x$.
22. The function $y = \operatorname{sech} x$ is one-to-one on the restricted domain $[0, \infty)$. Use the graph in Figure 6.5.4(c) to sketch the graph of the **inverse hyperbolic secant** $y = \operatorname{sech}^{-1} x$. Give the domain and range of $y = \operatorname{sech}^{-1} x$.

Calculator Problems

23. Use a calculator to investigate $\lim\limits_{h \to 0} \dfrac{f(x + h) - f(x)}{h}$ for the function in
   Problem 7. Determine $f'(x)$.

24. Use a calculator to investigate $\lim\limits_{h \to 0} \dfrac{f(x + h) - f(x)}{h}$ for the function in
   Problem 8. Determine $f'(x)$.

In Problems 25–28, use a calculator to estimate the value $m(b) = \lim\limits_{h \to 0} \dfrac{b^h - 1}{h}$ for $b = 1.5, b = 2, b = 3$, and $b = 5$ by filling out the given table.

25.

$h \to 0$	0.1	0.01	0.001	0.0001	0.00001	0.000001
$\dfrac{(1.5)^h - 1}{h}$						

26.

$h \to 0$	0.1	0.01	0.001	0.0001	0.00001	0.000001
$\dfrac{2^h - 1}{h}$						

27.

$h \to 0$	0.1	0.01	0.001	0.0001	0.00001	0.000001
$\dfrac{3^h - 1}{h}$						

28.

$h \to 0$	0.1	0.01	0.001	0.0001	0.00001	0.000001
$\dfrac{5^h - 1}{h}$						

29. Fill out a table of the kind in Problems 25–28, but this time use $\dfrac{e^h - 1}{h}$.

30. A Curiosity The logarithm developed by John Napier (see page 355) was actually

$$10^7 \log_{1/e}\left(\frac{x}{10^7}\right).$$

Use (8) of Section 6.3 to express this logarithm in terms of the natural logarithm.

CHAPTER 6 | **Review Exercises** | Answers to selected odd-numbered problems begin on page ANS-19.

A. Fill in the Blanks

In Problems 1–22, fill in the blanks.

1. The graph of $y = 6 - e^{-x}$ has the y-intercept _____ and horizontal asymptote $y =$ _____.

2. The x-intercept of the graph of $y = -10 + 10^{5x}$ is _____.

3. The graph of $y = \ln(x + 4)$ has the x-intercept _____ and vertical asymptote $x =$ _____.

4. The y-intercept of the graph of $y = \log_8(x + 2)$ is _____.

5. $\log_5 2 - \log_5 10 =$ _____

6. $6\ln e + 3\ln\dfrac{1}{e} =$ _____

7. $e^{3\ln 10} = $ _____

8. $10^{\log_{10} 4.89} = $ _____

9. $\log_4(4 \cdot 4^2 \cdot 4^3) = $ _____

10. $\dfrac{\log_5 625}{\log_5 125} = $ _____

11. If $\log_3 N = -2$, then $N = $ _____.

12. If $\log_b 6 = \frac{1}{2}$, then $b = $ _____.

13. If $\ln e^3 = y$, then $y = $ _____.

14. If $\ln 3 + \ln(x - 1) = \ln 2 + \ln x$, then $x = $ _____.

15. If $-1 + \ln(x - 3) = 0$, then $x = $ _____.

16. If $\ln(\ln x) = 1$, then $x = $ _____.

17. If $100 - 20e^{-0.15t} = 35$, then to four rounded decimals $t = $ _____.

18. If $3^x = 5$, then $3^{-2x} = $ _____.

19. $f(x) = 4^{3x} = ($ _____ $)^x$

20. $f(x) = (e^2)^{x/6} = ($ _____ $)^x$

21. If the graph of $y = e^{x-2} + C$ passes through $(2, 9)$, then $C = $ _____.

22. By rigid transformations, the point $(0, 1)$ on the graph of $y = e^x$ is moved to the point _____ on the graph of $y = 4 + e^{x-3}$.

B. True/False

In Problems 1–22, answer true or false.

1. $y = \ln x$ and $y = e^x$ are inverse functions. _____

2. The point $(b, 1)$ is on the graph of $f(x) = \log_b x$. _____

3. $y = 10^{-x}$ and $y = (0.1)^x$ are the same function. _____

4. If $f(x) = e^{x^2} - 1$, then $f(x) = 1$ when $x = \pm \ln \sqrt{2}$. _____

5. $4^{x/2} = 2^x$ _____

6. $\dfrac{2^{x^2}}{2^x} = 2^x$ _____

7. $2^x + 2^{-x} = (2 + 2^{-1})^x$

8. $2^{3+3x} = 8^{1+x}$ _____

9. $-\ln 2 = \ln(\frac{1}{2})$ _____

10. $\ln\dfrac{e^a}{e^b} = a - b$ _____

11. $\ln(\ln e) = 1$ _____

12. $\ln\sqrt{43} = \dfrac{\ln 43}{2}$ _____

13. $\ln(e + e) = 1 + \ln 2$ _____

14. $\log_6(36)^{-1} = -2$ _____

15. $\ln\dfrac{1}{e} = -1$ _____

16. $\dfrac{\ln 10}{\ln 2} = \ln 5$ _____

17. The range of the function $f(x) = 4^{-x} - 5$ is the interval $(-5, \infty)$ on the y-axis. _____

18. The point $(-5, 40)$ is on the graph of $f(x) = (\frac{1}{2})^x + 8$. _____

19. If $f(x) = b^x$, then $f(nx) = f(x)^n$. _____

20. $f(x) = \log_b x$, then $f(x^n) = nf(x)$. _____

21. $\ln y = 2\ln x + \ln 5$, then $y = 2x + 5$. _____

22. $\ln e \cdot \ln e^2 \cdots \ln e^n = n!$ _____

C. Review Exercises

In Problems 1 and 2, rewrite the given exponential expression as an equivalent logarithmic expression.

1. $5^{-1} = 0.2$

2. $\sqrt[3]{512} = 8$

In Problems 3 and 4, rewrite the given logarithmic expression as an equivalent exponential expression.

3. $\log_9 27 = 1.5$ **4.** $\log_6 (36)^{-2} = -4$

In Problems 5–12, solve for x.

5. $2^{1-x} = 8$ **6.** $3^{2x} = 81$

7. $e^{1-2x} = e^2$ **8.** $e^{x^2} - e^5 e^{x-1} = 0$

9. $2^{1-x} = 7$ **10.** $3^x = 7^{x-1}$

11. $e^{x+2} = 6$ **12.** $3e^x = 4e^{-3x}$

In Problems 13 and 14, solve for the indicated variable.

13. $P = Se^{-rm}$; for m **14.** $P = \dfrac{K}{1 + ce^{rt}}$; for t

In Problems 15 and 16, graph the given functions on the same coordinate axes.

15. $y = 4^x$, $y = \log_4 x$ **16.** $y = \left(\frac{1}{2}\right)^x$, $y = \log_{1/2} x$

17. Match the letter of the graph in FIGURE 6.R.1 with the appropriate function.

     (i) $f(x) = b^x, b > 2$ (ii) $f(x) = b^x, 1 < b < 2$

     (iii) $f(x) = b^x, \frac{1}{2} < b < 1$ (iv) $f(x) = b^x, 0 < b < \frac{1}{2}$

18. In FIGURE 6.R.2, fill in the blanks for the coordinates of the points on each graph.

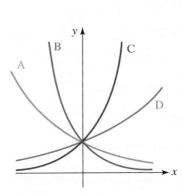

FIGURE 6.R.1 Graph for
Problem 17

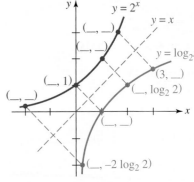

FIGURE 6.R.2 Graph for
Problem 18

In Problems 19 and 20, find the slope of the line L given in each figure.

19. $f(x) = 3^{-(x+1)}$

FIGURE 6.R.3 Graph for
Problem 19

20. $f(x) = \ln x$

FIGURE 6.R.4 Graph for
Problem 20

CHAPTER 6 EXPONENTIAL AND LOGARITHMIC FUNCTIONS

In Problems 21–26, match each of the following functions with one of the given graphs.

(i) $y = \ln(x - 2)$ (ii) $y = 2 - \ln x$

(iii) $y = 2 + \ln(x + 2)$ (iv) $y = -2 - \ln(x + 2)$

(v) $y = -\ln(2x)$ (vi) $y = 2 + \ln(-x + 2)$

21.

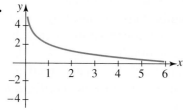

FIGURE 6.R.5 Graph for Problem 21

22.

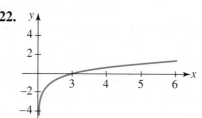

FIGURE 6.R.6 Graph for Problem 22

23.

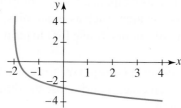

FIGURE 6.R.7 Graph for Problem 23

24.

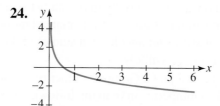

FIGURE 6.R.8 Graph for Problem 24

25.

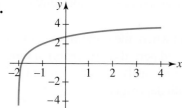

FIGURE 6.R.9 Graph for Problem 25

26.

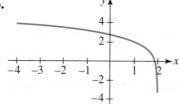

FIGURE 6.R.10 Graph for Problem 26

In Problems 27–28, in words describe the graph of the function f in terms of a transformation of the graph of $y = \ln x$.

27. $f(x) = \ln ex$ **28.** $f(x) = \ln x^3$

29. Find a function $f(x) = Ae^{kx}$ if $(0, 5)$ and $(6, 1)$ are points on the graph of f.

30. Find a function $f(x) = A10^{kx}$ if $f(3) = 8$ and $f(0) = \frac{1}{2}$.

31. Find a function $f(x) = a + b^x, 0 < b < 1$, if $f(1) = 5.5$ and the graph of f has a horizontal asymptote $y = 5$.

32. Find a function $f(x) = a + \log_3(x - c)$ if $f(11) = 10$ and the graph of f has a vertical asymptote $x = 2$.

33. Doubling Time If the initial number of bacteria present in a culture doubles after 9 hours, how long will it take for the number of bacteria in the culture to double again?

34. Got Bait? A commercial fishing lake is stocked with 10,000 fingerlings. Find a model $P(t) = P_0 e^{kt}$ for the fish population of the lake at time t if the owner of the lake estimates that there will be 5,000 fish left after six months. After how many months does the model predict that there will be 1000 fish left?

35. Radioactive Decay Tritium, an isotope of hydrogen, has a half-life of 12.5 years. How much of an initial quantity of this element remains after 50 years?

36. Old Bones It is found that 97% of C-14 has been lost in a human skeleton found at an archeological site. What is the approximate age of the skeleton?

37. Wishful Thinking A person facing retirement invests $650,000 in a savings account. She wants the account to be worth $1,000,000 in 10 years. What annual rate r of interest compounded continuously will achieve this dream?

38. Light Intensity According to the **Bouguer-Lambert law**, the intensity I (measured in lumens) of a vertical beam of light passing through a transparent substance decreases according to the exponential function $I(x) = I_0 e^{kx}$, $k < 0$, where I_0 is the intensity of the incident beam and x is the depth measured in meters. If the intensity of light 1 meter below the surface of water is 30% of I_0, what is the intensity 3 meters below the surface?

39. The **Gompertz function** $y = ae^{-be^{-ct}}$, where a, b, and c are positive constants, is named after the self-educated mathematician **Benjamin Gompertz** (1779–1865) and was used initially in the study of population demographics. Today Gompertz's function is used as a mathematical model in diverse areas such as economics, statistics, and oncological studies of the growth of tumors. Solve for t in terms of the other symbols.

40. The graph of a Gompertz function is naturally called a **Gompertz curve**. Sketch the Gompertz curve in the following cases. In each case, superimpose the three graphs on the same rectangular coordinate system.
  (a) $y = ae^{-e^{-t}}$, $a = \frac{1}{2}$, $a = 1$, $a = 2$
  (b) $y = e^{-be^{-t}}$, $b = \frac{1}{2}$, $b = 1$, $b = 2$
  (c) $y = e^{-e^{-ct}}$, $c = \frac{1}{2}$, $c = 1$, $c = 2$ [*Hint*: The graph has two horizontal asymptotes.]

41. Seriously Saving An annuity is a savings plan where the same amount of money P is deposited into an account at n equally spaced periods (say, years) of time. If the annual rate r of interest is compounded continuously, then the amount S accrued in the account immediately after the nth deposit is

$$S = P + Pe^r + Pe^{2r} + \cdots + Pe^{(n-1)r}.$$

What is the value of such an annuity in 15 years if $P = \$3000$ and the annual rate of interest is 2%.

42. If $a > 0$ and $b > 0$, then show that $\log_{a^2} b^2 = \log_a b$.

In Problems 43–52, eliminate the logarithms in the given expression. Express your answer in the form $y = f(x)$.

43. $\ln y = \ln(x - 3)$

44. $\ln y = 2\ln(x + 5)$

45. $\ln y = \ln x + 3\ln 2$

46. $\ln y = 4\ln x - \ln 5$

47. $\ln(y - 6) = \ln(x - 1) + \ln x + 2\ln 2$

48. $\ln y - \ln(x - 1) = \ln(x + 4) - \ln 2$

49. $\ln(y - 3) = \ln y - \ln(x^2 + 1) + 2\ln x$

50. $\ln(y/4) = \frac{1}{2}\ln(x^2 + 1) + \ln x$

51. $\ln y = -3x + \ln 5$

52. $\ln(y - x) = x^2 + 2\ln x - \ln\frac{1}{2}$

7 Conic Sections

Chapter Outline

The Parabola

≡ **Introduction** **Hypatia** is the first woman in the history of mathematics about whom we have considerable knowledge. Born in 370 C.E. in Alexandria, she was renowned as a mathematician and philosopher. Among her writings is *On the Conics of Apollonius*, which popularized **Apollonius'** (200 B.C.E.) work on curves that can be obtained by intersecting a double-napped cone with a plane: the circle, parabola, ellipse, and hyperbola. Note in FIGURE 7.1.1 that the plane does not pass through the vertex of the cone. When the plane passes through the vertex, the resulting figures: a single point, a single line, or two intersecting lines are commonly called **degenerate conics**. With the close of the Greek period, interest in conic sections waned; after Hypatia the study of these curves was neglected for over 1000 years.

Hypatia

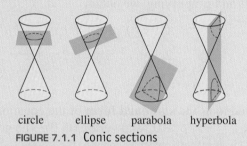

circle   ellipse   parabola   hyperbola

FIGURE 7.1.1 Conic sections

In the seventeenth century, the Italian physicist and mathematician **Galileo Galilei** (1564–1642) showed that in the absence of air resistance the path of a projectile follows a parabolic arc. At about the same time, the German mathematician, astronomer, and astrologist **Johannes Kepler** (1571–1630) hypothesized that the orbits of planets around the Sun are ellipses with the Sun at one focus. This was later verified by the English mathematician **Sir Isaac Newton** (1642–1726), using the methods of the newly developed calculus. Kepler also experimented with the reflecting properties of parabolic mirrors; these investigations sped the development of the reflecting telescope. The Greeks had known little of these practical applications. They had studied the conics for their beauty and fascinating properties. In the first three sections of this chapter, we will examine both the ancient properties and the modern applications of these curves. Rather than using a cone, we shall see how the parabola, ellipse, and hyperbola are defined by means of distance. Using a rectangular coordinate system and the distance formula, we obtain equations for the conics. Each of these equations will be in the form of a quadratic equation in variables x and y:

$$Ax^2 + Bxy + Cy^2 + Dx + Ey + F = 0,$$

Solar System

where A, B, C, D, E, and F are constants. We have already studied the special case $y = ax^2 + bx + c$ of the foregoing equation in Section 2.4.

DEFINITION 7.1.1 Parabola

A **parabola** is the set of points $P(x, y)$ in the plane that are equidistant from a fixed line L, called the **directrix**, and a fixed point F, called the **focus**.

A parabola is shown in FIGURE 7.1.2. The line through the focus perpendicular to the directrix is called the **axis** of the parabola. The point of intersection of the parabola and the axis is called the **vertex**, denoted by V in Figure 7.1.2.

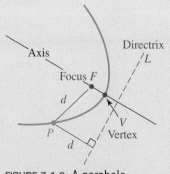

FIGURE 7.1.2 A parabola

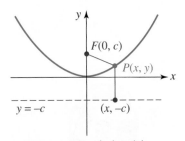

FIGURE 7.1.3 Parabola with vertex (0, 0) and focus on the y-axis

□ **Parabola with Vertex (0, 0)** To describe a parabola analytically, we use a rectangular coordinate system where the directrix is a horizontal line $y = -c$, where $c > 0$, and the focus is the point $F(0, c)$. Then we see that the axis of the parabola is along the y-axis, as FIGURE 7.1.3 shows. The origin is necessarily the vertex, since it lies on the axis c units from both the focus and the directrix. The distance from a point $P(x, y)$ to the directrix is

$$y - (-c) = y + c.$$

Using the distance formula, the distance from P to the focus F is

$$d(P, F) = \sqrt{(x - 0)^2 + (y - c)^2}.$$

From the definition of the parabola it follows that $d(P, F) = y + c$, or

$$\sqrt{(x - 0)^2 + (y - c)^2} = y + c.$$

By squaring both sides and simplifying, we obtain

$$x^2 + (y - c)^2 = (y + c)^2$$

$$x^2 + y^2 - 2cy + c^2 = y^2 + 2cy + c^2$$

or $$x^2 = 4cy. \tag{1}$$

Equation (1) is referred to as the **standard form** of the equation of a parabola with focus $(0, c)$, directrix $y = -c, c > 0$, and vertex $(0, 0)$. The graph of any parabola with standard form (1) is symmetric with respect to the y-axis.

Equation (1) does not depend on the assumption that $c > 0$. However, the direction in which the parabola opens does depend on the sign of c. Specifically, if $c > 0$ the parabola opens *upward* as in Figure 7.1.3; if $c < 0$, the parabola opens *downward*.

If the focus of a parabola is assumed to lie on the x-axis at $F(c, 0)$ and the directrix is $x = -c$, then the x-axis is the axis of the parabola and the vertex is $(0, 0)$. If $c > 0$ the parabola opens to the right; if $c < 0$, it opens to the left. In either case, the **standard form** of the equation is

$$y^2 = 4cx. \tag{2}$$

The graph of any parabola with standard form (2) is symmetric with respect to the x-axis.

A summary of all this information for equations (1) and (2) is given in FIGURES 7.1.4 and 7.1.5, respectively. You may be surprised to see in Figure 7.1.4(b) that the directrix above the x-axis is labeled $y = -c$ and the focus on the negative y-axis has coordinates $F(0, c)$. Bear in mind that in this case the assumption is that $c < 0$ and so $-c > 0$. A similar remark holds for Figure 7.1.5(b).

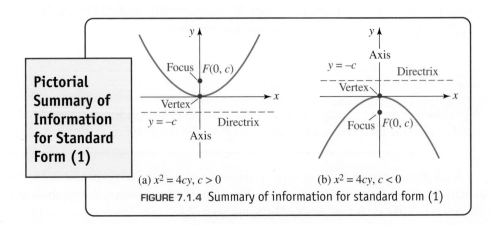

Pictorial Summary of Information for Standard Form (1)

(a) $x^2 = 4cy, c > 0$

(b) $x^2 = 4cy, c < 0$

FIGURE 7.1.4 Summary of information for standard form (1)

CHAPTER 7 CONIC SECTIONS

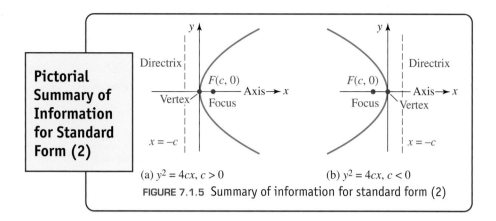

Pictorial Summary of Information for Standard Form (2)

(a) $y^2 = 4cx, c > 0$ (b) $y^2 = 4cx, c < 0$

FIGURE 7.1.5 Summary of information for standard form (2)

■ **EXAMPLE 1**   **The Simplest Parabola**

We first encountered the graph of $y = x^2$ in Section 2.2. By comparing this equation with (1) we see

$$x^2 = \overset{\overset{4c}{\downarrow}}{1} \cdot y$$

and so $4c = 1$ or $c = \frac{1}{4}$. Therefore the graph of $y = x^2$ is a parabola with vertex at the origin, focus at $\left(0, \frac{1}{4}\right)$, and directrix $y = -\frac{1}{4}$. These details are indicated in the graph in **FIGURE 7.1.6**. ≡

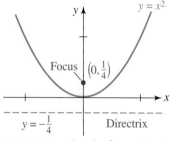

FIGURE 7.1.6 Graph of equation in Example 1

◀ Graphing tip for equations (1) and (2).

Knowing the basic parabolic shape, all we need to know to sketch a *rough* graph of either equation (1) or (2) is the fact that the graph passes through its vertex $(0, 0)$ and the direction in which the parabola opens. To add more accuracy to the graph it is convenient to use the number c determined by the standard form equation to plot two additional points. Note that if we choose $y = c$ in (1), then $x^2 = 4c^2$ implies $x = \pm 2c$. Thus $(2c, c)$ and $(-2c, c)$ lie on the graph of $x^2 = 4cy$. Similarly, the choice $x = c$ in (2) implies $y = \pm 2c$, and so $(c, 2c)$ and $(c, -2c)$ are points on the graph of $y^2 = 4cx$. The *line segment* through the focus with endpoints $(2c, c)$, $(-2c, c)$ for equations with standard form (1), and $(c, 2c)$, $(c, -2c)$ for equations with standard form (2) is called the **focal chord**. For example, in Figure 7.1.6, if we choose $y = \frac{1}{4}$, then $x^2 = \frac{1}{4}$ implies $x = \pm\frac{1}{2}$. Endpoints of the horizontal focal chord for $y = x^2$ are $\left(-\frac{1}{2}, \frac{1}{4}\right)$ and $\left(\frac{1}{2}, \frac{1}{4}\right)$.

■ **EXAMPLE 2**   **Finding an Equation of a Parabola**

Find the equation in standard form of the parabola with directrix $x = 2$ and focus $(-2, 0)$. Graph.

Solution In **FIGURE 7.1.7** we have graphed the directrix and the focus. We see from their placement that the equation we seek is of the form $y^2 = 4cx$. Since $c = -2$, the parabola opens to the left and so

$$y^2 = 4(-2)x \qquad \text{or} \qquad y^2 = -8x.$$

As mentioned in the discussion preceding this example, if we substitute $x = c$, or in this case $x = -2$, into the equation $y^2 = -8x$ we can find two points on its graph. From

$y^2 = -8(-2) = 16$ we get $y = \pm 4$. As shown in FIGURE 7.1.8, the graph passes through $(0, 0)$ as well as through the endpoints $(-2, -4)$ and $(-2, 4)$ of the focal chord.

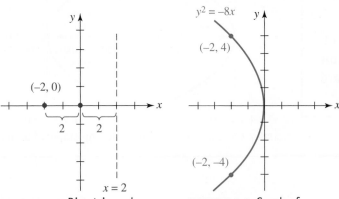

FIGURE 7.1.7 Directrix and focus in Example 2

FIGURE 7.1.8 Graph of parabola in Example 2

$\equiv$

☐ **Parabola with Vertex (h, k)** Suppose that a parabola is shifted both horizontally and vertically so that its vertex is at the point (h, k) and its axis is the vertical line $x = h$. The **standard form** of the equation of the parabola is then

$$(x - h)^2 = 4c(y - k). \tag{3}$$

Similarly, if its axis is the horizontal line $y = k$, the **standard form** of the equation of the parabola with vertex (h, k) is

$$(y - k)^2 = 4c(x - h). \tag{4}$$

☐ **Shifts** The parabolas defined by these equations are identical in shape to the parabolas defined by equations (1) and (2) because equations (3) and (4) represent rigid transformations (shifts up, down, left, and right) of the graphs of (1) and (2). For example, the parabola

$$(x + 1)^2 = 8(y - 5)$$

has vertex $(-1, 5)$. Its graph is the graph of $x^2 = 8y$ shifted horizontally one unit to the left followed by an upward vertical shift of five units.

For each of the equations, (1) and (2) or (3) and (4), the *distance* from the vertex to the focus, as well as the distance from the vertex to the directrix, is $|c|$.

<hr>

EXAMPLE 3   Finding an Equation of a Parabola

Find the equation in standard form of the parabola with vertex $(-3, -1)$ and directrix $y = 3$.

Solution We begin by graphing the vertex at $(-3, -1)$ and the directrix $y = 3$. From FIGURE 7.1.9 we can see that the parabola must open downward, and so its standard form is (3). This fact, plus the observation that the vertex lies 4 units below the directrix, indicates that the appropriate solution of $|c| = 4$ is $c = -4$. Substituting $h = -3, k = -1$, and $c = -4$ into (3) gives

$$[x - (-3)]^2 = 4(-4)[y - (-1)] \quad \text{or} \quad (x + 3)^2 = -16(y + 1). \quad \equiv$$

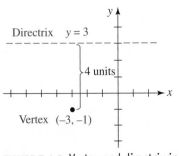

FIGURE 7.1.9 Vertex and directrix in Example 3

CHAPTER 7 CONIC SECTIONS

EXAMPLE 4   **Find Everything**

Find the vertex, focus, directrix, intercepts, and graph of the parabola

$$y^2 - 4y - 8x - 28 = 0. \qquad (5)$$

Solution In order to write the equation in one of the standard forms we complete the square in y:

$$y^2 - 4y + 4 = 8x + 28 + 4 \quad \leftarrow \text{add 4 to both sides}$$

$$(y - 2)^2 = 8x + 32.$$

Thus the standard form of equation (5) is $(y - 2)^2 = 8(x + 4)$. Comparing this equation with (4) we conclude that the vertex is $(-4, 2)$ and that $4c = 8$ or $c = 2$. Thus the parabola opens to the right. From $c = 2 > 0$, the focus is 2 units to the right of the vertex at $(-4 + 2, 2)$ or $(-2, 2)$. The directrix is the vertical line 2 units to the left of the vertex, $x = -4 - 2$ or $x = -6$. Knowing the parabola opens to the right from the point $(-4, 2)$ also tells us that the graph has intercepts. To find the x-intercept we set $y = 0$ in (5) and find immediately that $x = -\frac{28}{8} = -\frac{7}{2}$. The x-intercept is $\left(-\frac{7}{2}, 0\right)$. To find the y-intercepts we set $x = 0$ in (5) and find from the quadratic formula that $y = 2 \pm 4\sqrt{2}$ or $y \approx 7.66$ and $y \approx -3.66$. The y-intercepts are $\left(0, 2 - 4\sqrt{2}\right)$ and $\left(0, 2 + 4\sqrt{2}\right)$. Putting all this information together we get the graph in FIGURE 7.1.10. ≡

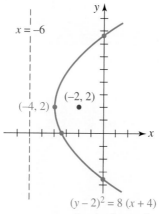

FIGURE 7.1.10 Graph of equation in Example 4

☐ **Applications of the Parabola** The parabola has many interesting properties that make it suitable for certain applications. Reflecting surfaces are often designed to take advantage of a reflection property of parabolas. Such surfaces, called **paraboloids**, are three-dimensional and are formed by rotating a parabola about its axis. As illustrated in FIGURE 7.1.11(a), rays of light (or electronic signals) from a point source located at the focus of a parabolic reflecting surface will be reflected along lines parallel to the axis. This is the idea behind the design of searchlights, some flashlights, and on-location satellite dishes. Conversely, if the incoming rays of light are parallel to the axis of a parabola, they will be reflected off the surface along lines passing through the focus. See Figure 7.1.11(b). Beams of light from a distant object such as a galaxy are essentially parallel, and so when these beams enter a reflecting telescope they are reflected by the parabolic mirror to the focus, where a camera is usually placed to capture the image over time. A parabolic home satellite dish operates on the same principle as the reflecting telescope; the digital signal from a TV satellite is captured at the focus of the dish antenna by a receiver.

Searchlight

TV satellite dish

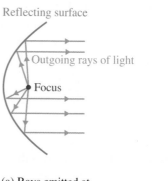

(a) Rays emitted at focus are reflected as parallel rays

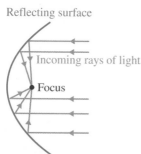

(b) Incoming rays reflected to focus

FIGURE 7.1.11 Parabolic reflecting surface

The Brooklyn bridge is a suspension bridge

Parabolas are also important in the design of suspension bridges. It can be shown that if the weight of the bridge is distributed uniformly along its length, then a support cable in the shape of a parabola will bear the load evenly.

The trajectory of an obliquely launched projectile—say, a basketball thrown from the free throw line—will travel in a parabolic arc.

Tuna, which prey on smaller fish, have been observed swimming in schools of 10 to 20 fish arrayed approximately in a parabolic shape. One possible explanation for this is that the smaller fish caught in the school of tuna will try to escape by "reflecting" off the parabola. As a result, they are concentrated at the focus and become easy prey for the tuna. See FIGURE 7.1.12.

The ball travels in a parabolic arc

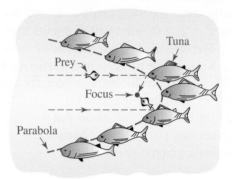

FIGURE 7.1.12 Tuna hunting in a Parabolic arc

| **7.1** | Exercises | Answers to selected odd-numbered problems begin on page ANS–19. |

In Problems 1–24, find the vertex, focus, directrix, and axis of the given parabola. Graph the parabola.

1. $y^2 = 4x$ **2.** $y^2 = \frac{7}{2}x$

3. $y^2 = -\frac{4}{3}x$ **4.** $y^2 = -10x$

5. $x^2 = -16y$ **6.** $x^2 = \frac{1}{10}y$

7. $x^2 = 28y$ **8.** $x^2 = -64y$

9. $(y - 1)^2 = 16x$ **10.** $(y + 3)^2 = -8(x + 2)$

11. $(x + 5)^2 = -4(y + 1)$ **12.** $(x - 2)^2 + y = 0$

13. $y^2 + 12y - 4x + 16 = 0$ **14.** $x^2 + 6x + y + 11 = 0$

15. $x^2 + 5x - \frac{1}{4}y + 6 = 0$ **16.** $x^2 - 2x - 4y + 17 = 0$

17. $y^2 - 8y + 2x + 10 = 0$ **18.** $y^2 - 4y - 4x + 3 = 0$

19. $4x^2 = 2y$ **20.** $3(y - 1)^2 = 9x$

21. $-2x^2 + 12x - 8y - 18 = 0$ **22.** $4y^2 + 16y - 6x - 2 = 0$

23. $6y^2 - 12y - 24x - 42 = 0$ **24.** $3x^2 + 30x - 8y + 75 = 0$

In Problems 25–44, find an equation of the parabola that satisfies the given conditions.

25. Focus $(0, 7)$, directrix $y = -7$ **26.** Focus $(0, -5)$, directrix $y = 5$

27. Focus $(-4, 0)$, directrix $x = 4$ **28.** Focus $\left(\frac{3}{2}, 0\right)$, directrix $x = -\frac{3}{2}$

29. Focus $\left(\frac{5}{2}, 0\right)$, vertex $(0, 0)$ **30.** Focus $(0, -10)$, vertex $(0, 0)$

31. Focus $(2, 3)$, directrix $y = -3$ **32.** Focus $(1, -7)$, directrix $x = -5$
33. Focus $(-1, 4)$, directrix $x = 5$ **34.** Focus $(-2, 0)$, directrix $y = \frac{3}{2}$
35. Focus $(1, 5)$, vertex $(1, -3)$ **36.** Focus $(-2, 3)$, vertex $(-2, 5)$
37. Focus $(8, -3)$, vertex $(0, -3)$ **38.** Focus $(1, 2)$, vertex $(7, 2)$
39. Vertex $(0, 0)$, directrix $y = -\frac{7}{4}$ **40.** Vertex $(0, 0)$, directrix $x = 6$
41. Vertex $(5, 1)$, directrix $y = 7$ **42.** Vertex $(-1, 4)$, directrix $x = 0$
43. Vertex $(0, 0)$, through $(-2, 8)$, axis along the y-axis
44. Vertex $(0, 0)$, through $\left(1, \frac{1}{4}\right)$, axis along the x-axis

In Problems 45–48, find the x- and y-intercepts of the given parabola.

45. $(y + 4)^2 = 4(x + 1)$ **46.** $(x - 1)^2 = -2(y - 1)$
47. $x^2 + 2y - 18 = 0$ **48.** $y^2 - 8y - x + 15 = 0$

Miscellaneous Applications

49. Spotlight A large spotlight is designed so that a cross section through its axis is a parabola and the light source is at the focus. Find the position of the light source if the spotlight is 4 ft across at the opening and 2 ft deep.

50. Reflecting Telescope A reflecting telescope has a parabolic mirror that is 20 ft across at the top and 4 ft deep at the center. Where should the eyepiece be located?

51. Light Ray Suppose that a light ray emanating from the focus of the parabola $y^2 = 4x$ strikes the parabola at $(1, -2)$. What is the equation of the reflected ray?

52. Suspension Bridge Suppose that two towers of a suspension bridge are 350 ft apart and the vertex of the parabolic cable is tangent to the road midway between the towers. If the cable is 1 ft above the road at a point 20 ft from the vertex, find the height of the towers above the road.

53. Another Suspension Bridge Two 75-ft towers of a suspension bridge with a parabolic cable are 250 ft apart. The vertex of the parabola is tangent to the road midway between the towers. Find the height of the cable above the roadway at a point 50 ft from one of the towers.

54. Drainpipe Assume that the water gushing from the end of a horizontal pipe follows a parabolic arc with vertex at the end of the pipe. The pipe is 20 m above the ground. At a point 2 m below the end of the pipe, the horizontal distance from the water to a vertical line through the end of the pipe is 4 m. See FIGURE 7.1.13. Where does the water strike the ground?

55. A Bull's Eye A dart thrower releases a dart 5 ft above the ground. The dart is thrown horizontally and follows a parabolic path. It hits the ground $10\sqrt{10}$ ft from the dart thrower. At a distance of 10 ft from the dart thrower, how high should a bull's eye be placed in order for the dart to hit it?

56. Path of a Projectile The vertical position of a projectile is given by the equation $y = -16t^2$ and the horizontal position by $x = 40t$ for $t \geq 0$. By eliminating t between the two equations, show that the path of the projectile is a parabolic arc. Graph the path of the projectile.

57. Focal Width The focal width of a parabola is the length of the focal chord, that is, the line segment through the focus perpendicular to the axis, with endpoints on the parabola. See FIGURE 7.1.14.
(a) Find the focal width of the parabola $x^2 = 8y$.
(b) Show that the focal width of the parabola $x^2 = 4cy$ and $y^2 = 4cx$ is $4|c|$.

58. Parabolic Orbit The orbit of a comet is a parabola with the Sun at the focus. When the comet is 50,000,000 km from the Sun, the line from the comet to the

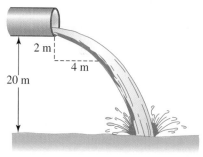

FIGURE 7.1.13 Pipe in Problem 54

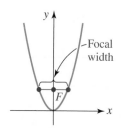

FIGURE 7.1.14 Focal width in Problem 57

Sun is perpendicular to the axis of the parabola. Use the result of Problem 57(b) to write an equation of the comet's path. (A comet with a parabolic path will not return to the Solar System.)

For Discussion

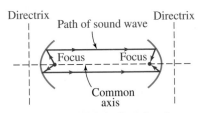

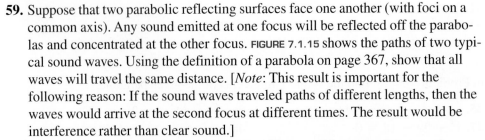

Directrix   Directrix

Path of sound wave

Focus   Focus

Common
axis

FIGURE 7.1.15 Parabolic reflecting surfaces in Problem 59

59. Suppose that two parabolic reflecting surfaces face one another (with foci on a common axis). Any sound emitted at one focus will be reflected off the parabolas and concentrated at the other focus. **FIGURE 7.1.15** shows the paths of two typical sound waves. Using the definition of a parabola on page 367, show that all waves will travel the same distance. [*Note*: This result is important for the following reason: If the sound waves traveled paths of different lengths, then the waves would arrive at the second focus at different times. The result would be interference rather than clear sound.]

60. The point closest to the focus is the vertex. How would you go about proving this? Carry out your ideas.

61. For the comet in Problem 58, use the result of Problem 60 to determine the shortest distance between the Sun and the comet.

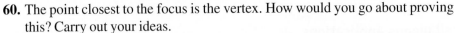

7.2 The Ellipse

≡ **Introduction** The ellipse occurs frequently in astronomy. For example, the paths of the planets around the Sun are elliptical with the Sun located at one focus. Similarly, communication satellites, the Hubble Space Telescope, and the International Space Station revolve around the Earth in elliptical orbits with the Earth at one focus. In this section we define the ellipse and study some of its properties and applications.

$d_1 + d_2$ = constant

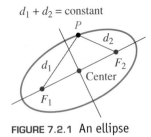

FIGURE 7.2.1 An ellipse

DEFINITION 7.2.1 Ellipse

An **ellipse** is the set of points $P(x, y)$ in the plane such that the sum of the distances between P and two fixed points F_1 and F_2 is constant. The fixed points F_1 and F_2 are called **foci** (plural for **focus**). The midpoint of the line segment joining points F_1 and F_2 is called the **center** of the ellipse.

As shown in **FIGURE 7.2.1**, if P is a point on the ellipse and if $d_1 = d(F_1, P)$ and $d_2 = d(F_2, P)$ are the distances from the foci to P, then the preceding definition asserts that

$$d_1 + d_2 = k, \tag{1}$$

where $k > 0$ is some constant.

On a practical level, equation (1) suggests a way of generating an ellipse. **FIGURE 7.2.2** shows that if a string of length k is attached to a piece of paper by two tacks, then an ellipse can be traced out by inserting a pencil against the string and moving it in such a manner that the string remains taut.

FIGURE 7.2.2 A way to draw an ellipse

Focus   Focus

□ **Ellipse with Center (0, 0)** We now derive an equation of the ellipse. For algebraic convenience, let us choose $k = 2a > 0$ and put the foci on the x-axis with coordinates $F_1(-c, 0)$ and $F_2(c, 0)$ as shown in FIGURE 7.2.3. It follows from (1) that

$$\sqrt{(x + c)^2 + y^2} + \sqrt{(x - c)^2 + y^2} = 2a$$

or

$$\sqrt{(x + c)^2 + y^2} = 2a - \sqrt{(x - c)^2 + y^2}. \tag{2}$$

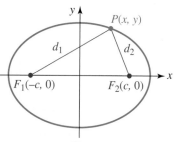

FIGURE 7.2.3 Ellipse with center (0, 0) and foci on the x-axis

We square both sides of the second equation in (2) and simplify,

$$(x + c)^2 + y^2 = 4a^2 - 4a\sqrt{(x - c)^2 + y^2} + (x - c)^2 + y^2$$

$$a\sqrt{(x - c)^2 + y^2} = a^2 - cx.$$

Squaring a second time gives,

$$a^2[(x - c)^2 + y^2] = a^4 - 2a^2cx + c^2x^2$$

or

$$(a^2 - c^2)x^2 + a^2y^2 = a^2(a^2 - c^2). \tag{3}$$

Referring to Figure 7.2.3, we see that the points F_1, F_2, and P form a triangle. Because the sum of the lengths of any two sides of a triangle is greater than the remaining side, we must have $2a > 2c$ or $a > c$. Hence, $a^2 - c^2 > 0$. When we let $b^2 = a^2 - c^2$, then (3) becomes $b^2x^2 + a^2y^2 = a^2b^2$. Dividing this last equation by a^2b^2 gives

$$\frac{x^2}{a^2} + \frac{y^2}{b^2} = 1. \tag{4}$$

Equation (4) is called the **standard form** of the equation of an ellipse centered at $(0, 0)$ with foci $(-c, 0)$ and $(c, 0)$, where c is defined by $b^2 = a^2 - c^2$, and $a > b > 0$.

If the foci are placed on the y-axis, then a repetition of the above analysis leads to

$$\frac{x^2}{b^2} + \frac{y^2}{a^2} = 1. \tag{5}$$

Equation (5) is called the **standard form** of the equation of an ellipse centered at $(0, 0)$ with foci $(0, -c)$ and $(0, c)$, where c is defined by $b^2 = a^2 - c^2$ and $a > b > 0$.

□ **Major and Minor Axes** The **major axis** of an ellipse is the line segment through its center, containing the foci, and with endpoints on the ellipse. For an ellipse with standard form equation (4) the major axis is horizontal, whereas for (5) the major axis is vertical. The line segment through the center, perpendicular to the major axis, and with endpoints on the ellipse, is called the **minor axis**. The two endpoints of the major axis are called the **vertices** of the ellipse. For (4) the vertices are the x-intercepts. Setting $y = 0$ in (4) gives $x = \pm a$. The vertices are then $(-a, 0)$ and $(a, 0)$. For (5) the vertices are the y-intercepts $(0, -a)$ and $(0, a)$. For equation (4), the endpoints of the minor axis are $(0, -b)$ and $(0, b)$; for (5) the endpoints are $(-b, 0)$ and $(b, 0)$. For either (4) or (5), the **length of the major axis** is $a - (-a) = 2a$; the **length of the minor axis** is $2b$. Since $a > b$, the major axis of an ellipse is always longer than its minor axis.

A summary of all this information for equations (4) and (5) is given in FIGURE 7.2.4.

Pictorial Summary of Information for Standard Forms (4) and (5)

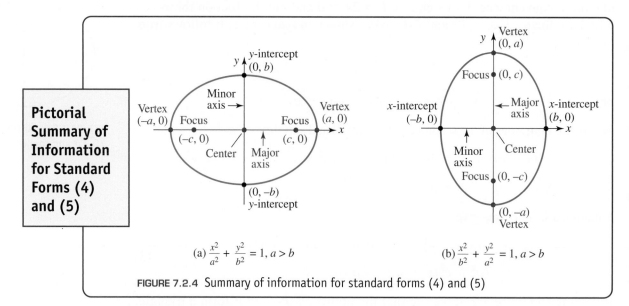

(a) $\dfrac{x^2}{a^2} + \dfrac{y^2}{b^2} = 1, a > b$ (b) $\dfrac{x^2}{b^2} + \dfrac{y^2}{a^2} = 1, a > b$

FIGURE 7.2.4 Summary of information for standard forms (4) and (5)

■ EXAMPLE 1   Vertices and Foci

Find the vertices and foci of the ellipse whose equation is $3x^2 + y^2 = 9$. Graph.

Solution By dividing both sides of the equality by 9 the standard form of the equation is

$$\frac{x^2}{3} + \frac{y^2}{9} = 1.$$

We see that $9 > 3$ and so we identify the equation with (5). From $a^2 = 9$ and $b^2 = 3$, we see that $a = 3$ and $b = \sqrt{3}$. The major axis is vertical with vertices $(0, -3)$ and $(0, 3)$. The minor axis is horizontal with endpoints $\left(-\sqrt{3}, 0\right)$ and $\left(\sqrt{3}, 0\right)$. Of course, the vertices are also the y-intercepts and the endpoints of the minor axis are the x-intercepts. Now, to find the foci we use $b^2 = a^2 - c^2$ or $c^2 = a^2 - b^2$ to write $c = \sqrt{a^2 - b^2}$. With $a = 3, b = \sqrt{3}$, we get $c = \sqrt{9 - 3} = \sqrt{6}$. Hence, the foci are on the y-axis at $\left(0, -\sqrt{6}\right)$ and $\left(0, \sqrt{6}\right)$. The graph is given in FIGURE 7.2.5. ≡

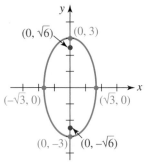

FIGURE 7.2.5 Ellipse in Example 1

■ EXAMPLE 2   Finding an Equation of an Ellipse

Find an equation of the ellipse with a focus $(2, 0)$ and an x-intercept $(5, 0)$.

Solution Since the given focus is on the x-axis, we can find an equation in standard form (4). Consequently, $c = 2, a = 5, a^2 = 25$, and $b^2 = a^2 - c^2$ or $b^2 = 5^2 - 2^2 = 21$. The desired equation is

$$\frac{x^2}{25} + \frac{y^2}{21} = 1.$$
≡

□ **Ellipse with Center (h, k)** When the center is at (h, k), the **standard form** for the equation of an ellipse is either

$$\frac{(x - h)^2}{a^2} + \frac{(y - k)^2}{b^2} = 1 \qquad (6)$$

or

$$\frac{(x - h)^2}{b^2} + \frac{(y - k)^2}{a^2} = 1. \qquad (7)$$

The ellipses defined by these equations are identical in shape to the ellipses defined by equations (4) and (5) since equations (6) and (7) represent rigid transformations of the graphs of (4) and (5). For example, the ellipse

$$\frac{(x-1)^2}{9} + \frac{(y+3)^2}{16} = 1$$

has center $(1, -3)$. Its graph is the graph of $x^2/9 + y^2/16 = 1$ shifted horizontally one unit to the right followed by a downward vertical shift of three units.

It is not a good idea to memorize formulas for the vertices and foci of an ellipse with center (h, k). Everything is the same as before: a, b, and c are positive and $a > b$, $a > c$. You can locate vertices, foci, and endpoints of the minor axis using the fact that a is the distance from the center to a vertex, b is the distance from the center to an endpoint on the minor axis, and c is the distance from the center to a focus. Also, the number c is still defined by the equation $b^2 = a^2 - c^2$.

EXAMPLE 3   Ellipse Centered at (h, k)

Find the center, vertices, and foci of the ellipse $4x^2 + 16y^2 - 8x - 96y + 84 = 0$. Graph.

Solution To write the given equation in one of the standard forms (6) or (7) we must complete the square in x and in y. Recall that in order to complete the square we want the coefficients of the quadratic terms x^2 and y^2 to be 1. To do this we factor 4 from both x^2 and x and factor 16 from both y^2 and y:

$$4(x^2 - 2x\ \ ) + 16(y^2 - 6y\ \ ) = -84.$$

Then from

4 · 1 and 16 · 9 are added to both sides

$$4(x^2 - 2x + 1) + 16(y^2 - 6y + 9) = -84 + 4 \cdot 1 + 16 \cdot 9$$

we obtain

$$4(x - 1)^2 + 16(y - 3)^2 = 64$$

or

$$\frac{(x-1)^2}{16} + \frac{(y-3)^2}{4} = 1. \qquad (8)$$

From (8) we see that the center of the ellipse is $(1, 3)$. Since the last equation has the standard form (6), we identify $a^2 = 16$ or $a = 4$ and $b^2 = 4$ or $b = 2$. The major axis is horizontal and lies on the horizontal line $y = 3$ passing through $(1, 3)$. This is the red horizontal dashed line segment in FIGURE 7.2.6. By measuring $a = 4$ units to the left and then to the right of the center along the line $y = 3$, we arrive at the vertices $(-3, 3)$ and $(5, 3)$. By measuring $b = 2$ units both down and up the vertical line $x = 1$ through the center, we arrive at the endpoints of the minor axis $(1, 1)$ and $(1, 5)$. The minor axis is the black dashed vertical line segment in Figure 7.2.6. Because $c^2 = a^2 - b^2 = 16 - 4 = 12$, $c = 2\sqrt{3}$. Finally, by measuring $c = 2\sqrt{3}$ units to the left and right of the center along $y = 3$, we obtain the foci $\left(1 - 2\sqrt{3}, 3\right)$ and $\left(1 + 2\sqrt{3}, 3\right)$. ☰

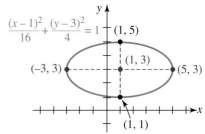

FIGURE 7.2.6 Ellipse in Example 3

EXAMPLE 4   Two Functions

In Example 1, if we solve the equation of the ellipse $3x^2 + y^2 = 9$ for the variable y in terms of x we obtain two functions,

◀ Review the dicussion following Example 6 in Section 2.1.

$$y = \sqrt{9 - 3x^2} \qquad \text{and} \qquad y = -\sqrt{9 - 3x^2}.$$

The domain of each of these functions is the interval $[-\sqrt{3}, \sqrt{3}]$. The graphs of $y = \sqrt{9 - 3x^2}$ and $y = -\sqrt{9 - 3x^2}$ are, in turn, the upper half-ellipse given in **FIGURE 7.2.7(a)** and the lower half-ellipse in Figure 7.2.7(b).

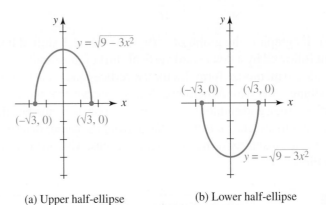

(a) Upper half-ellipse   (b) Lower half-ellipse

FIGURE 7.2.7 Two half-ellipses defined by the functions in Example 4

To graph an ellipse on a calculator you may have to resort to superimposing the graphs of two half-ellipses defined by two functions in order to obtain the graph of the complete ellipse.

EXAMPLE 5   **Finding an Equation of an Ellipse**

Find an equation of the ellipse with center $(2, -1)$, vertical major axis of length 6, and minor axis of length 3.

Solution The length of the major axis is $2a = 6$; hence $a = 3$. Similarly, the length of the minor axis is $2b = 3$, so $b = \frac{3}{2}$. By sketching the center and the axes, we see from **FIGURE 7.2.8** that the vertices are $(2, 2)$ and $(2, -4)$ and the endpoints of the minor axis are $\left(\frac{1}{2}, -1\right)$ and $\left(\frac{7}{2}, -1\right)$. Because the major axis is vertical, the standard equation of this ellipse is

$$\frac{(x - 2)^2}{\left(\frac{3}{2}\right)^2} + \frac{(y - (-1))^2}{3^2} = 1 \qquad \text{or} \qquad \frac{(x - 2)^2}{\frac{9}{4}} + \frac{(y + 1)^2}{9} = 1.$$

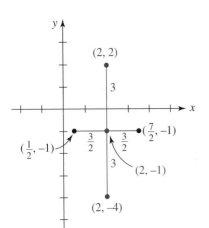

FIGURE 7.2.8 Graphical interpretation of data in Example 5

☐ **Eccentricity** Associated with each conic section is a number e called its **eccentricity**. The eccentricity of an ellipse is defined to be

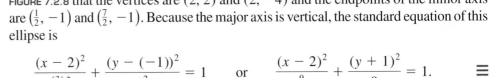
$$e = \frac{c}{a},$$

where $c = \sqrt{a^2 - b^2}$. Since $0 < \sqrt{a^2 - b^2} < a$ implies $0 < \dfrac{\sqrt{a^2 - b^2}}{a} < 1$, the eccentricity of an ellipse satisfies $0 < e < 1$.

EXAMPLE 6   **Example 3 Revisited**

Determine the eccentricity of the ellipse in Example 3.

Solution In the solution of Example 3 we found that $a = 4$ and $c = 2\sqrt{3}$. Hence, the eccentricity of the ellipse is $e = (2\sqrt{3})/4 = \sqrt{3}/2 \approx 0.87$.

Eccentricity is an indicator of the shape of an ellipse. When $e \approx 0$, that is, e is close to zero, the ellipse is nearly circular, and when $e \approx 1$ the ellipse is flattened or elongated. To see this, observe that if e is close to 0, it follows from $e = \sqrt{a^2 - b^2}/a$ that $c = \sqrt{a^2 - b^2} \approx 0$ and consequently $a \approx b$. As you can see from the standard equations in (4) and (5), this means that the shape of the ellipse is close to circular. Also, because c is the distance from the center of the ellipse to a focus, the two foci are close together near the center. See FIGURE 7.2.9(a). On the other hand, if $e \approx 1$ or $\sqrt{a^2 - b^2}/a \approx 1$, then $c = \sqrt{a^2 - b^2} \approx a$ and so $b \approx 0$. Also, $c \approx a$ means that the foci are far apart; each focus is close to a vertex. Thus, the ellipse is elongated as shown in Figure 7.2.9(b).

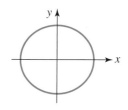

(a) e close to zero

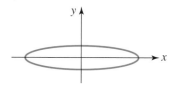

(b) e close to 1

FIGURE 7.2.9 Effect of eccentricity on the shape of an ellipse

□ **Applications of the Ellipse** Ellipses have a reflection property analogous to the one discussed in Section 7.1 for the parabola. It can be shown that if a light or sound source is placed at one focus of an ellipse, then all rays or waves will be reflected off the ellipse to the other focus. See FIGURE 7.2.10. For example, if a pool table is constructed in the form of an ellipse with a pocket at one focus, then any shot originating at the other focus will never miss the pocket.

Statuary Hall in Washington, DC

Similarly, if a ceiling is elliptical with two foci on (or near) the floor but considerably distant from each other, then anyone whispering at one focus will be heard at the other. Some famous "whispering galleries" are the Statuary Hall at the Capitol in Washington, DC, the Mormon Tabernacle in Salt Lake City, and St. Paul's Cathedral in London.

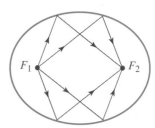

FIGURE 7.2.10 Reflection property of an ellipse

Using his Law of Universal Gravitation and the newly developed calculus, Isaac Newton was the first to prove Kepler's first law of planetary motion:

The orbit of each planet about the Sun is an ellipse with the Sun at one focus.

EXAMPLE 7	**Eccentricity of Earth's Orbit**

The perihelion distance of the Earth (the least distance between the Earth and the Sun) is approximately 9.16×10^7 miles, and its aphelion distance (the greatest distance between the Earth and the Sun) is approximately 9.46×10^7 miles. What is the eccentricity of Earth's orbit?

Solution Let us assume that the orbit of the Earth is as shown in FIGURE 7.2.11. From the figure we see that

$$a - c = 9.16 \times 10^7$$
$$a + c = 9.46 \times 10^7.$$

Solving this system of equations gives $a = 9.31 \times 10^7$ and $c = 0.15 \times 10^7$. Thus the eccentricity $e = c/a$ is

$$e = \frac{0.15 \times 10^7}{9.31 \times 10^7} \approx 0.016.$$ ≡

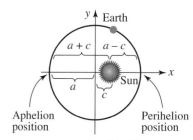

FIGURE 7.2.11 Graphical interpretation of data in Example 7

The orbits of seven of the planets have eccentricities less than 0.1 and, hence, the orbits are not far from circular. The planet Mercury is the exception. The orbit of the dwarf planet Pluto has the eccentricity 0.25. Many of the asteroids and comets have highly eccentric orbits. The orbit of the asteroid Hildago is one of the most eccentric, with $e = 0.66$. Another notable case is the orbit of Comet Halley. See Problem 47 in Exercises 7.2.

In Problems 1–20, find the center, foci, vertices, endpoints of the minor axis, and eccentricity of the given ellipse. Graph the ellipse.

1. $\dfrac{x^2}{25} + \dfrac{y^2}{9} = 1$

2. $\dfrac{x^2}{16} + \dfrac{y^2}{4} = 1$

3. $x^2 + \dfrac{y^2}{16} = 1$

4. $\dfrac{x^2}{4} + \dfrac{y^2}{10} = 1$

5. $9x^2 + 16y^2 = 144$

6. $2x^2 + y^2 = 4$

7. $9x^2 + 4y^2 = 36$

8. $x^2 + 4y^2 = 4$

9. $\dfrac{(x-1)^2}{49} + \dfrac{(y-3)^2}{36} = 1$

10. $\dfrac{(x+1)^2}{25} + \dfrac{(y-2)^2}{36} = 1$

11. $(x+5)^2 + \dfrac{(y+2)^2}{16} = 1$

12. $\dfrac{(x-3)^2}{64} + \dfrac{(y+4)^2}{81} = 1$

13. $4x^2 + \left(y + \tfrac{1}{2}\right)^2 = 4$

14. $36(x+2)^2 + (y-4)^2 = 72$

15. $5(x-1)^2 + 3(y+2)^2 = 45$

16. $6(x-2)^2 + 8y^2 = 48$

17. $25x^2 + 9y^2 - 100x + 18y - 116 = 0$

18. $9x^2 + 5y^2 + 18x - 10y - 31 = 0$

19. $x^2 + 3y^2 + 18y + 18 = 0$

20. $12x^2 + 4y^2 - 24x - 4y + 1 = 0$

In Problems 21–40, find an equation of the ellipse that satisfies the given conditions.

21. Vertices $(\pm 5, 0)$, foci $(\pm 3, 0)$

22. Vertices $(\pm 9, 0)$, foci $(\pm 2, 0)$

23. Vertices $(0, \pm 3)$, foci $(0, \pm 1)$

24. Vertices $(0, \pm 7)$, foci $(0, \pm 3)$

25. Vertices $(0, \pm 3)$, endpoints of minor axis $(\pm 1, 0)$

26. Vertices $(\pm 4, 0)$, endpoints of minor axis $(0, \pm 2)$

27. Vertices $(-3, -3)$, $(5, -3)$, endpoints of minor axis $(1, -1)$, $(1, -5)$

28. Vertices $(1, -6)$, $(1, 2)$, endpoints of minor axis $(-2, -2)$, $(4, -2)$

29. One focus $(0, -2)$, center at origin, $b = 3$

30. One focus $(1, 0)$, center at origin, $a = 3$

31. Foci $\left(\pm \sqrt{2}, 0\right)$, length of minor axis 6

32. Foci $\left(0, \pm \sqrt{5}\right)$, length of major axis 16

33. Foci $(0, \pm 3)$, passing through $\left(-1, 2\sqrt{2}\right)$

34. Vertices $(\pm 5, 0)$, passing through $\left(\sqrt{5}, 4\right)$

35. Vertices $(\pm 4, 1)$, passing through $\left(2\sqrt{3}, 2\right)$

36. Center $(1, -1)$, one focus $(1, 1)$, $a = 5$

37. Center $(1, 3)$, one focus $(1, 0)$, one vertex $(1, -1)$

38. Center $(5, -7)$, length of vertical major axis 8, length of minor axis 6

39. Endpoints of minor axis $(0, 5)$, $(0, -1)$, one focus $(6, 2)$

40. Endpoints of major axis $(2, 4)$, $(13, 4)$, one focus $(4, 4)$

In Problems 41–44, find a function $y = f(x)$ that defines the indicated half-ellipse. Give the domain of each function. The equations are from Problems 1, 3, 9, and 12.

41. $\dfrac{x^2}{25} + \dfrac{y^2}{9} = 1$; lower half-ellipse

42. $x^2 + \dfrac{y^2}{16} = 1$; upper half-ellipse

43. $\dfrac{(x-1)^2}{49} + \dfrac{(y-3)^2}{36} = 1$; upper half-ellipse

44. $\dfrac{(x-3)^2}{64} + \dfrac{(y+4)^2}{81} = 1$; lower half-ellipse

45. The orbit of the planet Mercury is an ellipse with the Sun at one focus. The length of the major axis of this orbit is 72 million miles and the length of the minor axis is 70.4 million miles. What is the least distance (perihelion) between Mercury and the Sun? What is the greatest distance (aphelion)?

46. What is the eccentricity of the orbit of Mercury in Problem 45?

47. The orbit of Comet Halley is an ellipse whose major axis is 3.34×10^9 miles long, and whose minor axis is 8.5×10^8 miles long. What is the eccentricity of the comet's orbit?

48. A satellite orbits the Earth in an elliptical path with the center of the Earth at one focus. It has a minimum altitude of 200 mi and a maximum altitude of 1000 mi above the surface of the Earth. If the radius of the Earth is 4000 mi, what is an equation of the satellite's orbit?

Miscellaneous Applications

49. Archway A semielliptical archway has a vertical major axis. The base of the arch is 10 ft across and the highest part of the arch is 15 ft. Find the height of the arch above the point on the base of the arch 3 ft from the center.

50. Gear Design An elliptical gear rotates about its center and is always kept in mesh with a circular gear that is free to move horizontally. See FIGURE 7.2.12. If the origin of the xy-coordinate system is placed at the center of the ellipse, then the equation of the ellipse in its present position is $x^2 + 3y^2 = 8$. The diameter of the circular gear equals the length of the minor axis of the elliptical gear. Given that the units are centimeters, how far does the center of the circular gear move horizontally during the rotation from one vertex of the elliptical gear to the next?

FIGURE 7.2.12 Elliptical and circular gears in Problem 50

51. Carpentry A carpenter wishes to cut an elliptical top for a coffee table from a rectangular piece of wood that is 4 ft by 3 ft utilizing the entire length and width available. If the ellipse is to be drawn using the string and tack method illustrated in Figure 7.2.2, how long should the piece of string be and where should the tacks be placed?

52. Park Design The Ellipse is a park in Washington, DC. It is bounded by an elliptical path with a major axis of length 458 m and a minor axis of length 390 m. Find the distance between the foci of this ellipse.

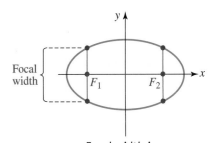

The Ellipse Park in Washington, DC

53. Whispering Gallery Suppose that a room is constructed on a flat elliptical base by rotating a semiellipse 180° about its major axis. Then, by the reflection property of the ellipse, anything whispered at one focus will be distinctly heard at the other focus. If the height of the room is 16 ft and the length is 40 ft, find the location of the whispering and listening posts.

54. Focal Width The focal width of the ellipse is the length of a focal chord, that is, a line segment, perpendicular to the major axis, through a focus with endpoints on the ellipse. See FIGURE 7.2.13.

(a) Find the focal width of the ellipse $x^2/9 + y^2/4 = 1$.

(b) Show that, in general, the focal width of the ellipse $x^2/a^2 + y^2/b^2 = 1$ is $2b^2/a$.

55. Find an equation of the ellipse with foci $(0, 2)$ and $(8, 6)$ and fixed distance sum $2a = 12$. [*Hint*: Here the major axis is neither horizontal nor vertical; thus none of the standard forms from this section apply. Use the definition of the ellipse.]

56. Proceed as in Problem 55, and find an equation of the ellipse with foci $(-1, -3)$ and $(-5, 7)$ and fixed distance sum $2a = 20$.

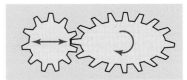

FIGURE 7.2.13 Focal width in Problem 54

For Discussion

57. The graph of the ellipse $x^2/4 + (y-1)^2/9 = 1$ is shifted 4 units to the right. What are the center, foci, vertices, and endpoints of the minor axis for the shifted graph?

58. The graph of the ellipse $(x-1)^2/9 + (y-4)^2 = 1$ is shifted 5 units to the left and 3 units up. What are the center, foci, vertices, and endpoints of the minor axis for the shifted graph?

59. In engineering the eccentricity of an ellipse is often expressed only in terms of a and b. Show that $e = \sqrt{1 - b^2/a^2}$.

60. Look up the definition of **semi-ellipse**. Are the half-ellipses in Figure 7.2.7 semi-ellipses?

7.3 The Hyperbola

☰ **Introduction** The definition of a hyperbola is basically the same as the definition of the ellipse with only one exception: the word *sum* is replaced by the word *difference*.

DEFINITION 7.3.1 Hyperbola

A **hyperbola** is the set of points $P(x, y)$ in the plane such that the difference of the distances between P and two fixed points F_1 and F_2 is constant. The fixed points F_1 and F_2 are called **foci** (plural for **focus**). The midpoint of the line segment joining points F_1 and F_2 is called the **center** of the hyperbola.

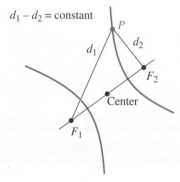

$d_1 - d_2 = \text{constant}$

FIGURE 7.3.1 A hyperbola

As shown in **FIGURE 7.3.1**, a hyperbola consists of two **branches**. If P is a point on the hyperbola, then

$$|d_1 - d_2| = k, \tag{1}$$

where $d_1 = d(F_1, P)$ and $d_2 = d(F_2, P)$.

☐ **Hyperbola with Center (0, 0)** Proceeding as for the ellipse, we place the foci on the x-axis at $F_1(-c, 0)$ and $F_2(c, 0)$ as shown in **FIGURE 7.3.2** and choose the constant k to be $2a$ for algebraic convenience. It follows from (1) that

$$d_1 - d_2 = \pm 2a. \tag{2}$$

As drawn in Figure 7.3.2, P is on the right branch of the hyperbola and so $d_1 - d_2 = 2a > 0$. If P is on the left branch then the difference is $-2a$. Writing (2) as

$$\sqrt{(x+c)^2 + y^2} - \sqrt{(x-c)^2 + y^2} = \pm 2a$$

or

$$\sqrt{(x+c)^2 + y^2} = \pm 2a + \sqrt{(x-c)^2 + y^2}$$

we square, simplify, and square again:

$$(x+c)^2 + y^2 = 4a^2 \pm 4a\sqrt{(x-c)^2 + y^2} + (x-c)^2 + y^2$$

$$\pm a\sqrt{(x-c)^2 + y^2} = cx - a^2$$

$$a^2[(x-c)^2 + y^2] = c^2 x^2 - 2a^2 cx + a^4$$

$$(c^2 - a^2)x^2 - a^2 y^2 = a^2(c^2 - a^2). \tag{3}$$

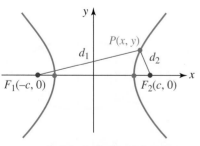

FIGURE 7.3.2 Hyperbola with center (0, 0) and foci on the x-axis

From Figure 7.3.2, we see that the triangle inequality gives

$$d_1 < d_2 + 2c \quad \text{and} \quad d_2 < d_1 + 2c,$$

or

$$d_1 - d_2 < 2c \quad \text{and} \quad d_2 - d_1 < 2c.$$

Using $d_1 - d_2 = \pm 2a$, the last two inequalities imply that $2a < 2c$ or $a < c$. Since $c > a > 0$, $c^2 - a^2$ is a positive constant. If we let $b^2 = c^2 - a^2$, (3) becomes $b^2x^2 - a^2y^2 = a^2b^2$ or, after dividing by a^2b^2,

$$\frac{x^2}{a^2} - \frac{y^2}{b^2} = 1. \tag{4}$$

Equation (4) is called the **standard form** of the equation of a hyperbola centered at $(0, 0)$ with foci $(-c, 0)$ and $(c, 0)$, where c is defined by $b^2 = c^2 - a^2$.

When the foci lie on the y-axis, a repetition of the foregoing algebra leads to

$$\frac{y^2}{a^2} - \frac{x^2}{b^2} = 1. \tag{5}$$

Equation (5) is the **standard form** of the equation of a hyperbola centered at $(0, 0)$ with foci $(0, -c)$ and $(0, c)$. Here again, $c > a$ and $b^2 = c^2 - a^2$.

For the hyperbola (unlike the ellipse), bear in mind that in (4) and (5) there is no relationship between the relative sizes of a and b; rather, a^2 is always the denominator of the *positive term* and the intercepts *always* have $\pm a$ as a coordinate.

◀ Note of Caution

☐ **Transverse and Conjugate Axes** The line segment with endpoints on the hyperbola and lying on the line through the foci is called the **transverse axis**; its endpoints are called the **vertices** of the hyperbola. For the hyperbola described by equation (4), the transverse axis lies on the x-axis. Therefore, the coordinates of the vertices are the x-intercepts. Setting $y = 0$ gives $x^2/a^2 = 1$, or $x = \pm a$. Thus, as shown in FIGURE 7.3.3 the vertices are $(-a, 0)$ and $(a, 0)$; the **length of the transverse axis** is $2a$. Notice that by setting $y = 0$ in (4), we get $-y^2/b^2 = 1$ or $y^2 = -b^2$, which has no real solutions. Hence the graph of any equation in that form has no y-intercepts. Nonetheless, the numbers $\pm b$ are important. The line segment through the center of the hyperbola perpendicular to the transverse axis and with endpoints $(0, -b)$ and $(0, b)$ is called the **conjugate axis**. Similarly, the graph of an equation in standard form (5) has no x-intercepts. The conjugate axis for (5) is the line segment with endpoints $(-b, 0)$ and $(b, 0)$.

This information for equations (4) and (5) is summarized in Figure 7.3.3.

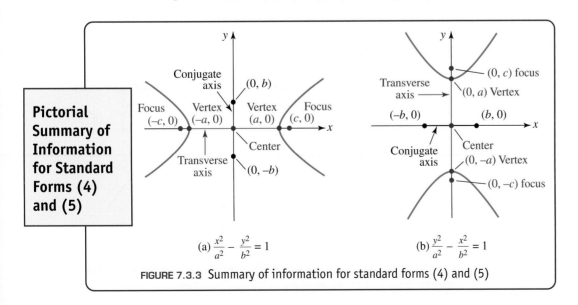

Pictorial Summary of Information for Standard Forms (4) and (5)

(a) $\dfrac{x^2}{a^2} - \dfrac{y^2}{b^2} = 1$

(b) $\dfrac{y^2}{a^2} - \dfrac{x^2}{b^2} = 1$

FIGURE 7.3.3 Summary of information for standard forms (4) and (5)

☐ **Asymptotes** Every hyperbola possesses a pair of slant asymptotes that pass through its center. These asymptotes are indicative of end behavior, and as such are an invaluable aid in sketching the graph of a hyperbola. Solving (4) for y in terms of x gives

$$y = \pm \frac{b}{a} x \sqrt{1 - \frac{a^2}{x^2}}.$$

As $x \to -\infty$ or as $x \to \infty$, $a^2/x^2 \to 0$, and thus $\sqrt{1 - a^2/x^2} \to 1$. Therefore, for large values of $|x|$, points on the graph of the hyperbola are close to the points on the lines

$$y = \frac{b}{a} x \qquad \text{and} \qquad y = -\frac{b}{a} x. \tag{6}$$

By a similar analysis we find that the slant asymptotes for (5) are

$$y = \frac{a}{b} x \qquad \text{and} \qquad y = -\frac{a}{b} x. \tag{7}$$

Each pair of asymptotes intersect at the origin, which is the center of the hyperbola. Note, too, in FIGURE 7.3.4(a) that the asymptotes are simply the *extended diagonals* of a rectangle of width $2a$ (the length of the transverse axis) and height $2b$ (the length of the conjugate axis); in Figure 7.3.4(b) the asymptotes are the extended diagonals of a rectangle of width $2b$ and height $2a$. This rectangle is referred to as the **auxiliary rectangle**.

We recommend that you *do not* memorize the equations in (6) and (7). There is an easy method for obtaining the asymptotes of a hyperbola. For example, since $y = \pm \frac{b}{a} x$ is equivalent to

$$\frac{x^2}{a^2} = \frac{y^2}{b^2}$$

the asymptotes of the hyperbola given in (4) are obtained from a single equation

$$\frac{x^2}{a^2} - \frac{y^2}{b^2} = 0. \tag{8}$$

Note that (8) factors as the difference of two squares:

This is a mnemonic, or memory device. It ▶ has no geometric significance.

$$\left(\frac{x}{a} - \frac{y}{b} \right) \left(\frac{x}{a} + \frac{y}{b} \right) = 0.$$

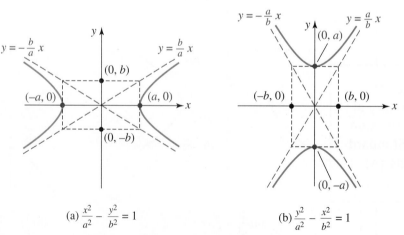

(a) $\dfrac{x^2}{a^2} - \dfrac{y^2}{b^2} = 1$ (b) $\dfrac{y^2}{a^2} - \dfrac{x^2}{b^2} = 1$

FIGURE 7.3.4 Hyperbolas (4) and (5) with slant asymptotes (in red) as the extended diagonals of the auxiliary rectangles (in black)

Setting each factor equal to zero and solving for y gives an equation of an asymptote. You do not even have to memorize (8) because it is simply the left-hand side of the standard form of the equation of a hyperbola given in (4). In like manner, to obtain the asymptotes for (5) just replace 1 by 0 in the standard form, factor $y^2/a^2 - x^2/b^2 = 0$, and solve for y.

▉ EXAMPLE 1   Hyperbola Centered at (0, 0)

Find the vertices, foci, and asymptotes of the hyperbola $9x^2 - 25y^2 = 225$. Graph.

Solution We first put the equation into standard form by dividing the left-hand side by 225:

$$\frac{x^2}{25} - \frac{y^2}{9} = 1. \tag{9}$$

From this equation we see that $a^2 = 25$ and $b^2 = 9$, and so $a = 5$ and $b = 3$. Therefore the vertices are $(-5, 0)$ and $(5, 0)$. Since $b^2 = c^2 - a^2$ implies $c^2 = a^2 + b^2$, we have $c^2 = 34$, and so the foci are $(-\sqrt{34}, 0)$ and $(\sqrt{34}, 0)$. To find the slant asymptotes we use the standard form (9) with 1 replaced by 0:

$$\frac{x^2}{25} - \frac{y^2}{9} = 0 \qquad \text{factors as} \qquad \left(\frac{x}{5} - \frac{y}{3}\right)\left(\frac{x}{5} + \frac{y}{3}\right) = 0.$$

Setting each factor equal to zero and solving for y gives the asymptotes $y = \pm 3x/5$. We plot the vertices and graph the two lines through the origin. Both branches of the hyperbola must become arbitrarily close to the asymptotes as $x \to \pm\infty$. See **FIGURE 7.3.5**. ▣

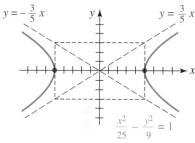

FIGURE 7.3.5 Hyperbola in Example 1

▉ EXAMPLE 2 ▉   Finding an Equation of a Hyperbola

Find an equation of the hyperbola with vertices $(0, -4)$, $(0, 4)$ and asymptotes $y = -\frac{1}{2}x$, $y = \frac{1}{2}x$.

Solution The center of the hyperbola is $(0, 0)$. This is revealed by the fact that the asymptotes intersect at the origin. Moreover, the vertices are on the y-axis and are 4 units on either side of the origin. Thus the equation we seek is of form (5). From (7) or Figure 7.3.4(b), the asymptotes must be of the form $y = \pm\frac{a}{b}x$ so that $a/b = 1/2$. From the given vertices we identify $a = 4$, and so

$$\frac{4}{b} = \frac{1}{2} \qquad \text{implies} \qquad b = 8.$$

The equation of the hyperbola is then

$$\frac{y^2}{4^2} - \frac{x^2}{8^2} = 1 \qquad \text{or} \qquad \frac{y^2}{16} - \frac{x^2}{64} = 1. \qquad\qquad ▣$$

☐ **Hyperbola with Center (h, k)** When the center of the hyperbola is (h, k) the **standard form** analogues of equations (4) and (5) are, in turn,

$$\frac{(x - h)^2}{a^2} - \frac{(y - k)^2}{b^2} = 1 \tag{10}$$

and

$$\frac{(y - k)^2}{a^2} - \frac{(x - h)^2}{b^2} = 1. \tag{11}$$

As in (4) and (5), the numbers a^2, b^2, and c^2 are related by $b^2 = c^2 - a^2$.

You can locate vertices and foci using the fact that a is the distance from the center to a vertex and c is the distance from the center to a focus. The slant asymptotes for (10) can be obtained by factoring

$$\frac{(x-h)^2}{a^2} - \frac{(y-k)^2}{b^2} = 0$$

as

$$\left(\frac{x-h}{a} - \frac{y-k}{b}\right)\left(\frac{x-h}{a} + \frac{y-k}{b}\right) = 0.$$

Similarly, the asymptotes for (11) can be obtained from factoring

$$\frac{(y-k)^2}{a^2} - \frac{(x-h)^2}{b^2} = 0,$$

setting each factor equal to zero and solving for y in terms of x. As a check on your work, remember that (h, k) must be a point that lies on each asymptote.

▌ EXAMPLE 3　　　Hyperbola Centered at (h, k)

Find the center, vertices, foci, and asymptotes of the hyperbola
$$4x^2 - y^2 - 8x - 4y - 4 = 0.$$

Graph.

Solution Before completing the square in x and y, we factor 4 from the two x-terms and factor -1 from the two y-terms so that the leading coefficient in each expression is 1. Then we have

$$4(x^2 - 2x \quad) + (-1)(y^2 + 4y \quad) = 4$$
$$4(x^2 - 2x + 1) - (y^2 + 4y + 4) = 4 + 4 \cdot 1 + (-1) \cdot 4$$
$$4(x - 1)^2 - (y + 2)^2 = 4$$
$$\frac{(x - 1)^2}{1} - \frac{(y + 2)^2}{4} = 1.$$

We see now that the center is $(1, -2)$. Since the term in the standard form involving x has the positive coefficient, the transverse axis is horizontal along the line $y = -2$, and we identify $a = 1$ and $b = 2$. The vertices are 1 unit to the left and to the right of the center at $(0, -2)$ and $(2, -2)$, respectively. From $b^2 = c^2 - a^2$, we have

$$c^2 = a^2 + b^2 = 1 + 4 = 5,$$

and so $c = \sqrt{5}$. Hence the foci are $\sqrt{5}$ units to the left and the right of the center $(1, -2)$ at $\left(1 - \sqrt{5}, -2\right)$ and $\left(1 + \sqrt{5}, -2\right)$.

To find the asymptotes, we solve

$$\frac{(x - 1)^2}{1} - \frac{(y + 2)^2}{4} = 0 \quad \text{or} \quad \left(x - 1 - \frac{y + 2}{2}\right)\left(x - 1 + \frac{y + 2}{2}\right) = 0$$

for y. From $y + 2 = \pm 2(x - 1)$ we find that the asymptotes are $y = -2x$ and $y = 2x - 4$. Observe that by substituting $x = 1$, both equations give $y = -2$, which means that both lines pass through the center. We then locate the center, plot the vertices, and graph the asymptotes. As shown in FIGURE 7.3.6, the graph of the hyperbola passes through the vertices and becomes closer and closer to the asymptotes as $x \to \pm \infty$.　　　　　　≡

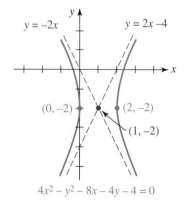

FIGURE 7.3.6 Hyperbola in Example 3

EXAMPLE 4   **Finding an Equation of a Hyperbola**

Find an equation of the hyperbola with center $(2, -3)$, passing through the point $(4, 1)$, and having one vertex $(2, 0)$.

Solution Since the distance from the center to one vertex is a, we have $a = 3$. From the location of the center and the vertex, it follows that the transverse axis is vertical and lies along the line $x = 2$. Therefore, the equation of the hyperbola must be of form (11):

$$\frac{(y + 3)^2}{3^2} - \frac{(x - 2)^2}{b^2} = 1, \tag{12}$$

where b^2 is yet to be determined. Since the point $(4, 1)$ is on the graph on the hyperbola, its coordinates must satisfy equation (12). From

$$\frac{(1 + 3)^2}{3^2} - \frac{(4 - 2)^2}{b^2} = 1$$
$$\frac{16}{9} - \frac{4}{b^2} = 1$$
$$\frac{7}{9} = \frac{4}{b^2}$$

we find $b^2 = \frac{36}{7}$. We conclude that the desired equation is

$$\frac{(y + 3)^2}{3^2} - \frac{(x - 2)^2}{\frac{36}{7}} = 1.$$

≡

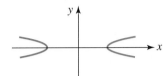

(a) e close to 1

□ **Eccentricity** Like the ellipse, the equation that defines the **eccentricity** of a hyperbola is $e = c/a$. Except in this case the number c is given by $c = \sqrt{a^2 + b^2}$. Since $0 < a < \sqrt{a^2 + b^2}$, the eccentricity of a hyperbola satisfies $e > 1$. As with the ellipse, the magnitude of the eccentricity of a hyperbola is an indicator of its shape. **FIGURE 7.3.7** shows examples of two extreme cases: $e \approx 1$ and e much greater than 1.

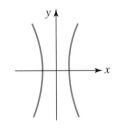

(b) e much greater than 1

FIGURE 7.3.7 Effect of eccentricity on the shape of a hyperbola

EXAMPLE 5   **Eccentricity of a Hyperbola**

Find the eccentricity of the hyperbola $\dfrac{y^2}{2} - \dfrac{(x - 1)^2}{36} = 1$.

Solution Identifying $a^2 = 2$ and $b^2 = 36$, we get $c^2 = 2 + 36 = 38$. Thus the eccentricity of the given hyperbola is

$$e = \frac{c}{a} = \frac{\sqrt{38}}{\sqrt{2}} = \sqrt{19} \approx 4.36.$$

We conclude that the hyperbola is one whose branches open widely as in Figure 7.3.7(b).

≡

□ **Applications of the Hyperbola** The hyperbola has several important applications involving sounding techniques. In particular, several navigational systems utilize hyperbolas as follows. Two fixed radio transmitters at a known distance from each other transmit synchronized signals. The difference in reception times by a navigator determines the difference $2a$ of the distances from the navigator to the two transmitters. This information locates the navigator somewhere on the hyperbola with foci at the transmitters and fixed difference in distances from the foci equal to $2a$. By using two sets of signals obtained from a single master station paired with each of two secondary stations, the long-range navigation system LORAN locates a ship or plane at the intersection of two hyperbolas. See **FIGURE 7.3.8**.

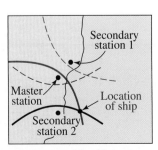

FIGURE 7.3.8 The idea behind LORAN

The next example illustrates the use of a hyperbola in another situation involving sounding techniques.

▮ **EXAMPLE 6**   **Locating a Big Blast**

The sound of a dynamite blast is heard at different times by two observers at points A and B. Knowing that the speed of sound is approximately 1100 ft/s or 335 m/s, it is determined that the blast occurred 1000 meters closer to point A than to point B. If A and B are 2600 meters apart, show that the location of the blast lies on a branch of a hyperbola. Find an equation of the hyperbola.

Solution In FIGURE 7.3.9, we have placed the points A and B on the x-axis at $(1300, 0)$ and $(-1300, 0)$, respectively. If $P(x, y)$ denotes the location of the blast, then

$$d(P, B) - d(P, A) = 1000.$$

From the definition of the hyperbola on page 382 and the derivation following it, we see that this is the equation for the right branch of a hyperbola with fixed distance difference $2a = 1000$ and $c = 1300$. Thus the equation has the form

$$\frac{x^2}{a^2} - \frac{y^2}{b^2} = 1, \text{ where } x \geq 0,$$

or after solving for x,

$$x = a\sqrt{1 + \frac{y^2}{b^2}}.$$

With $a = 500$ and $c = 1300$, $b^2 = (1300)^2 - (500)^2 = (1200)^2$. Substituting in the foregoing equation gives

$$x = 500\sqrt{1 + \frac{y^2}{(1200)^2}} \qquad \text{or} \qquad x = \frac{5}{12}\sqrt{(1200)^2 + y^2}. \qquad \equiv$$

To find the exact location of the blast in Example 6 we would need another observer hearing the blast at a third point C. Knowing the time between when this observer hears the blast and when the observer at A hears the blast, we find a second hyperbola. The actual point of detonation is a point of intersection of the two hyperbolas.

There are many other applications of the hyperbola. As shown in FIGURE 7.3.10(a), a plane flying at a supersonic speed parallel to level ground leaves a hyperbolic sonic "footprint" on the ground. Like the parabola and ellipse, a hyperbola also possesses a reflecting property. The Cassegrain reflecting telescope shown in Figure 7.3.10(b)

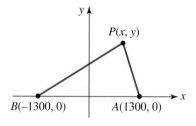

FIGURE 7.3.9 Graph for Example 6

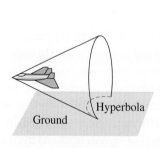

(a) Sonic footprint

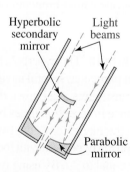

(b) Cassegrain telescope

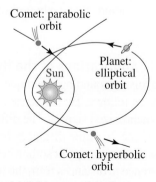

(c) Orbits around the Sun

FIGURE 7.3.10 Applications of hyperbolas

utilizes a convex hyperbolic secondary mirror to reflect a ray of light back through a hole to an eyepiece (or camera) behind the parabolic primary mirror. This telescope construction makes use of the fact that a beam of light directed along a line through one focus of a hyperbolic mirror will be reflected on a line through the other focus. The most famous telescope in the world (or out of the world), the Hubble Space Telescope, is an example of a Cassegrain telescope. See the website:

http://hubblesite.org/the_telescope/hubble_essentials/

Hubble Space Telescope

Orbits of objects in the universe can be parabolic, elliptic, or hyperbolic. When an object passes close to the Sun (or a planet), it is not necessarily captured by the gravitational field of the larger body. Under certain conditions, the object picks up a fractional amount of orbital energy of this much larger body and the resulting "slingshot-effect" orbit of the object as it passes the Sun is hyperbolic. See Figure 7.3.10(c).

| 7.3 | Exercises | Answers to selected odd-numbered problems begin on page ANS-21. |

In Problems 1–20, find the center, foci, vertices, asymptotes, and eccentricity of the given hyperbola. Graph the hyperbola.

1. $\dfrac{x^2}{16} - \dfrac{y^2}{25} = 1$

2. $\dfrac{x^2}{4} - \dfrac{y^2}{4} = 1$

3. $\dfrac{y^2}{64} - \dfrac{x^2}{9} = 1$

4. $\dfrac{y^2}{6} - 4x^2 = 1$

5. $4x^2 - 16y^2 = 64$

6. $5x^2 - 5y^2 = 25$

7. $y^2 - 5x^2 = 20$

8. $9x^2 - 16y^2 + 144 = 0$

9. $\dfrac{(x-5)^2}{4} - \dfrac{(y+1)^2}{49} = 1$

10. $\dfrac{(x+2)^2}{10} - \dfrac{(y+4)^2}{25} = 1$

11. $\dfrac{(y-4)^2}{36} - x^2 = 1$

12. $\dfrac{(y-\frac{1}{4})^2}{4} - \dfrac{(x+3)^2}{9} = 1$

13. $25(x-3)^2 - 5(y-1)^2 = 125$

14. $10(x+1)^2 - 2(y-\frac{1}{2})^2 = 100$

15. $8(x+4)^2 - 5(y-7)^2 + 40 = 0$

16. $9(x-1)^2 - 81(y-2)^2 = 9$

17. $5x^2 - 6y^2 - 20x + 12y - 16 = 0$

18. $16x^2 - 25y^2 - 256x - 150y + 399 = 0$

19. $4x^2 - y^2 - 8x + 6y - 4 = 0$

20. $2y^2 - 9x^2 - 18x + 20y + 5 = 0$

In Problems 21–44, find an equation of the hyperbola that satisfies the given conditions.

21. Foci $(\pm 5, 0)$, $a = 3$

22. Foci $(\pm 10, 0)$, $b = 2$

23. Foci $(0, \pm 4)$, one vertex $(0, -2)$

24. Foci $(0, \pm 3)$, one vertex $\left(0, -\frac{3}{2}\right)$

25. Foci $(\pm 4, 0)$, length of transverse axis 6

26. Foci $(0, \pm 7)$, length of transverse axis 10

27. Center $(0, 0)$, one vertex $\left(0, \frac{5}{2}\right)$, one focus $(0, -3)$

28. Center $(0, 0)$, one vertex $(7, 0)$, one focus $(9, 0)$

29. Center $(0, 0)$, one vertex $(-2, 0)$, one focus $(-3, 0)$

30. Center $(0, 0)$, one vertex $(1, 0)$, one focus $(5, 0)$

31. Vertices $(0, \pm 8)$, asymptotes $y = \pm 2x$

32. Foci $(0, \pm 3)$, asymptotes $y = \pm \frac{3}{2}x$

33. Vertices $(\pm 2, 0)$, asymptotes $y = \pm \frac{4}{3}x$

34. Foci $(\pm 5, 0)$, asymptotes $y = \pm \frac{3}{5}x$

35. Center $(1, -3)$, one focus $(1, -6)$, one vertex $(1, -5)$

36. Center $(2, 3)$, one focus $(0, 3)$, one vertex $(3, 3)$

37. Foci $(-4, 2)$, $(2, 2)$, one vertex $(-3, 2)$

38. Vertices $(2, 5)$, $(2, -1)$, one focus $(2, 7)$

39. Vertices $(\pm 2, 0)$, passing through $(2\sqrt{3}, 4)$

40. Vertices $(0, \pm 3)$, passing through $(\frac{16}{5}, 5)$

41. Center $(-1, 3)$, one vertex $(-1, 4)$, passing through $(-5, 3 + \sqrt{5})$

42. Center $(3, -5)$, one vertex $(3, -2)$, passing through $(1, -1)$

43. Center $(2, 4)$, one vertex $(2, 5)$, one asymptote $2y - x - 6 = 0$

44. Eccentricity $\sqrt{10}$, endpoints of conjugate axis $(-5, 4)$, $(-5, 10)$

In Problems 45–48, find a function $y = f(x)$ by solving the given equation for the variable y and taking the nonnegative root. Give the domain of the function. Describe the graph of $y = f(x)$ in words. The equations are from Problems 1, 3, 11, and 16.

45. $\dfrac{x^2}{16} - \dfrac{y^2}{25} = 1$

46. $\dfrac{y^2}{64} - \dfrac{x^2}{9} = 1$

47. $\dfrac{(y-4)^2}{36} - x^2 = 1$

48. $9(x-1)^2 - 81(y-2)^2 = 9$

49. Three points are located at $A(-10, 16)$, $B(-2, 0)$, and $C(2, 0)$, where the units are kilometers. An artillery gun is known to lie on the line segment between A and C, and using sounding techniques it is determined that the gun is 2 km closer to B than to C. Find the point where the gun is located.

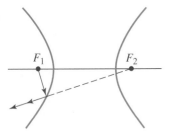

FIGURE 7.3.11 Reflecting property in Problem 50

50. It can be shown that a ray of light emanating from one focus of a hyperbola will be reflected back along the line from the opposite focus. See **FIGURE 7.3.11**. A light ray from the left focus of the hyperbola $x^2/16 - y^2/20 = 1$ strikes the hyperbola at $(-6, -5)$. Find an equation of the reflected ray.

51. Find an equation of the hyperbola with foci $(0, -2)$ and $(8, 4)$ and fixed distance difference $2a = 8$. [*Hint*: See Problem 55 in Exercises 7.2.]

52. Focal Width The **focal width** of a hyperbola is the length of a focal chord, that is, a line segment, perpendicular to the line containing the transverse axis and through a focus, with endpoints on the hyperbola. See **FIGURE 7.3.12**.

(a) Find the focal width of the hyperbola $x^2/4 - y^2/9 = 1$.

(b) Show that, in general, the focal width of the hyperbola $x^2/a^2 - y^2/b^2 = 1$ is $2b^2/a$.

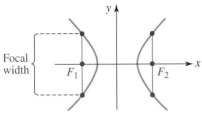

FIGURE 7.3.12 Focal width in Problem 52

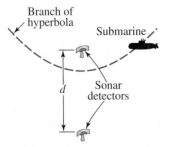

FIGURE 7.3.13 Sonic detectors in Problem 53

For Discussion

53. Sub Hunting Two sonar detectors are located at a distance d from one another. Suppose that a sound (such as a sneeze aboard a submarine) is heard at the two detectors with a time delay h between them. See **FIGURE 7.3.13**. Assume that sound travels in straight lines to the two detectors with speed v.

(a) Explain why h cannot be larger than d/v.

(b) Explain why, for given values of d, v, and h, the source of the sound can be determined to lie on one branch of a hyperbola. [*Hint*: Where do you suppose that the foci might be?]

(c) Find an equation for the hyperbola in part (b), assuming that the detectors are at the points $(0, d/2)$ and $(0, -d/2)$. Express the answer in the standard form $y^2/a^2 - x^2/b^2 = 1$.

54. The hyperbolas

$$\frac{x^2}{a^2} - \frac{y^2}{b^2} = 1 \quad \text{and} \quad \frac{y^2}{b^2} - \frac{x^2}{a^2} = 1$$

are said to be **conjugates** of each other.

(a) Find the equation of the hyperbola that is conjugate to

$$\frac{x^2}{25} - \frac{y^2}{144} = 1.$$

(b) Discuss how the graphs of conjugate hyperbolas are related.

55. A **rectangular hyperbola** is one for which the asymptotes are perpendicular.

(a) Show that $y^2 - x^2 + 5y + 3x = 1$ is a rectangular hyperbola.

(b) Which of the hyperbolas given in Problems 1–20 are rectangular?

56. Suppose a hyperbola is rectangular. See Problem 55.

(a) How are the constants a and b related?

(b) Show that all rectangular hyperbolas have the same eccentricity.

7.4 Rotation of Axes

≡ **Introduction** In the introduction to Section 7.1 we pointed out that equations of conic sections are special cases of the general second-degree equation

$$Ax^2 + Bxy + Cy^2 + Dx + Ey + F = 0. \tag{1}$$

When $B = 0$, we obtain the standard forms of equations of circles, parabolas, ellipses, and hyperbolas studied in preceding sections from equations of the form

$$Ax^2 + Cy^2 + Dx + Ey + F = 0 \tag{2}$$

by completion of the square. Since each standard form is a second-degree equation we must have $A \neq 0$ or $C \neq 0$ in (2).

In addition to the familiar conics, equation (1) could also represent two intersecting lines, one line, a single point, two parallel lines, or no graph at all. These are referred to as the **degenerate cases** of equation (1). See Problems 29–31 in Exercises 7.4.

When $B \neq 0$, we will see in the discussion that follows that it is possible to remove the xy-term in equation (1) by a **rotation of axes**. In other words, it is always possible to select the angle of rotation θ so that any equation of the form (1) can be transformed into an equation in x' and y' with no $x'y'$-term:

$$A'(x')^2 + C'(y')^2 + D'x' + E'y' + F' = 0. \tag{3}$$

Proceeding as we would for equation (2), we can recast (3) into a standard form and thereby enabling us to identify the conic and graph it in the $x'y'$-coordinate plane.

☐ **Rotation of Axes** We begin with an xy-coordinate system with origin O and rotate the x- and y-axes about O through an angle θ, as shown in FIGURE 7.4.1. In their rotated

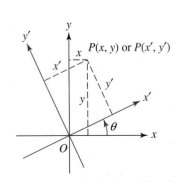

FIGURE 7.4.1 Rotated axes in red

position we will denote the x- and y-axes by the symbols x' and y', respectively. In this manner, any point P in the plane has two sets of coordinates: (x, y) in terms of the original xy-coordinate system and (x', y') in terms of the $x'y'$-coordinate system. It is a straightforward exercise in trigonometry to show that the xy-coordinates of P can be converted to the new $x'y'$-coordinates by

$$\begin{aligned} x' &= x\cos\theta + y\sin\theta \\ y' &= -x\sin\theta + y\cos\theta. \end{aligned} \tag{4}$$

Conversely, by solving (4) for x and y we obtain a set of equations that allow us to convert the $x'y'$-coordinates of P to xy-coordinates:

$$\begin{aligned} x &= x'\cos\theta - y'\sin\theta \\ y &= x'\sin\theta + y'\cos\theta. \end{aligned} \tag{5}$$

See Problem 32 in Exercises 7.4. Of the two sets of **rotation equations**, (4) and (5), the set given in (5) is the more important for our purposes.

███ EXAMPLE 1   **Coordinates**

Suppose that the x-axis is rotated by an angle of $60°$. Find

(a) the $x'y'$-coordinates of the point whose xy-coordinates are $(4, 4)$,
(b) the xy-coordinates of the point whose $x'y'$-coordinates are $(3, -5)$.

Solutions **(a)** The point $(4, 4)$ is indicated by the black dot in FIGURE 7.4.2. With $\theta = 60°$, $x = 4$, and $y = 4$ the equations in (4) give

$$x' = 4\cos 60° + 4\sin 60° = 4\left(\frac{1}{2}\right) + 4\left(\frac{\sqrt{3}}{2}\right) = 2 + 2\sqrt{3}$$

$$y' = -4\sin 60° + 4\cos 60° = -4\left(\frac{\sqrt{3}}{2}\right) + 4\left(\frac{1}{2}\right) = 2 - 2\sqrt{3}.$$

The $x'y'$-coordinates of $(4, 4)$ are $\left(2 + 2\sqrt{3}, 2 - 2\sqrt{3}\right)$ or approximately $(5.46, -1.46)$.

(b) The point $(3, -5)$ is indicated by the red dot in Figure 7.4.2. With $\theta = 60°$, $x' = 3$, and $y' = -5$ the equations in (5) give

$$x = 3\cos 60° - (-5)\sin 60° = 3\left(\frac{1}{2}\right) + 5\left(\frac{\sqrt{3}}{2}\right) = \frac{3 + 5\sqrt{3}}{2}$$

$$y = 3\sin 60° + (-5)\cos 60° = 3\left(\frac{\sqrt{3}}{2}\right) - 5\left(\frac{1}{2}\right) = \frac{3\sqrt{3} - 5}{2}.$$

Thus the xy-coordinates of $(3, -5)$ are $\left(\frac{1}{2}(3 + 5\sqrt{3}), \frac{1}{2}(3\sqrt{3} - 5)\right)$ or approximately $(5.83, 0.10)$.   ≡

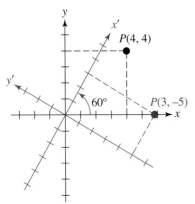

FIGURE 7.4.2 Rotated axes in Example 1

Using the rotation equations in (5) it is possible to determine an angle of rotation θ so that any equation of form (1) where $B \neq 0$, can be transformed into an equation in x' and y' with no $x'y'$-term. Substituting the equations in (5) for x and y in (1),

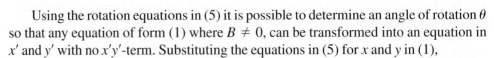

and simplifying, we discover that the resulting equation can be written

$$A'(x')^2 + B'x'y' + C'(y')^2 + D'x' + E'y' + F' = 0. \tag{6}$$

The coefficients A', B', C', D', E', F' depend on A, B, C, D, E, F and on $\sin\theta$, $\cos\theta$. In particular, the coefficient of the $x'y'$-term in (6) is

$$B' = 2(C - A)\sin\theta\cos\theta + B(\cos^2\theta - \sin^2\theta).$$

Thus, in order to eliminate the $x'y'$-term in (6), we can select any angle θ so that $B' = 0$, that is,

$$2(C - A)\sin\theta\cos\theta + B(\cos^2\theta - \sin^2\theta) = 0.$$

By the double-angle formulas for sine and cosine, the last equation is equivalent to

◀ See (11) and (12) in Section 4.6.

$$(C - A)\sin 2\theta + B\cos 2\theta = 0 \quad \text{or} \quad \cot 2\theta = \frac{A - C}{B}.$$

We have proved the following result.

THEOREM 7.4.1 Elimination of the xy-Term

The xy-term can be eliminated from the general second-degree equation

$$Ax^2 + Bxy + Cy^2 + Dx + Ey + F = 0,$$

where $B \neq 0$, by a rotation of axes through an angle θ that satisfies

$$\cot 2\theta = \frac{A - C}{B} \qquad (7)$$

Although equation (7) possesses an infinite number of solutions, it suffices to take a solution such that $0° < \theta < 90°$. This inequality comes from the three cases:

- If $\cot 2\theta = 0$, then $2\theta = 90°$ and therefore $\theta = 45°$.
- If $\cot 2\theta > 0$, then we can take $0° < 2\theta < 90°$ or $0° < \theta < 45°$.
- If $\cot 2\theta < 0$, then $90° < 2\theta < 180°$ or $45° < \theta < 90°$.

EXAMPLE 2 An $x'y'$-Equation

The simple equation $xy = 1$ can be written in terms of x' and y' without the product xy. By comparing $xy = 1$ with equation (1) we see that $A = 0$, $C = 0$, and $B = 1$. Thus (7) shows $\cot 2\theta = 0$. Using $\theta = 45°$, $\cos 45° = \sin 45° = \sqrt{2}/2$, the rotation equations in (5) become

$$x = x'\cos 45° - y'\sin 45° = \frac{\sqrt{2}}{2}(x' - y')$$

$$y = x'\sin 45° + y'\cos 45° = \frac{\sqrt{2}}{2}(x' + y').$$

Substituting these expressions for x and y in $xy = 1$, we obtain

$$\frac{\sqrt{2}}{2}(x' - y') \cdot \frac{\sqrt{2}}{2}(x' + y') = 1$$

or

$$\frac{x'^2}{2} - \frac{y'^2}{2} = 1.$$

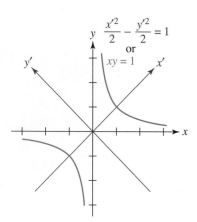

$$\frac{x'^2}{2} - \frac{y'^2}{2} = 1$$
or
$$xy = 1$$

FIGURE 7.4.3 Rotated axes in Example 2

We recognize this as the standard equation of a hyperbola with vertices on the x'-axis at the $x'y'$-points $(\pm\sqrt{2}, 0)$. The asymptotes of the hyperbola are $y' = -x'$ and $y' = x'$ (which are simply the original x- and y-axes). See **FIGURE 7.4.3**. ≡

We do not actually have to determine the value of θ if we simply want to obtain an $x'y'$-equation to identify the conic. When $\cot 2\theta \neq 0$, it is clear that in order to use (5) we only need to know both $\sin\theta$ and $\cos\theta$. To do this, we use the value of $\cot 2\theta$ to find the value of $\cos 2\theta$ and then use the half-angle formulas

$$\sin\theta = \sqrt{\frac{1 - \cos 2\theta}{2}} \qquad \text{and} \qquad \cos\theta = \sqrt{\frac{1 + \cos 2\theta}{2}}. \qquad (8)$$

However, if we wish to sketch the conic, then we need to find θ to determine the position of the x' and y' axes. The next example illustrates these ideas.

EXAMPLE 3   Eliminating the xy-Term

After a suitable rotation of axes, identify and sketch the graph of

$$5x^2 + 3xy + y^2 = 44.$$

Solutions With $A = 5$, $B = 3$, and $C = 1$, (7) shows that the desired rotation angle satisfies

$$\cot 2\theta = \frac{5 - 1}{3} = \frac{4}{3}. \qquad (9)$$

From the discussion following (7), since $\cot 2\theta$ is positive, we can choose 2θ such that $0 < \theta < 45°$. From the identity $1 + \cot^2 2\theta = \csc^2 2\theta$ we find $\csc 2\theta = \frac{5}{3}$ and so $\sin 2\theta = \frac{3}{5}$. Then $\cot 2\theta = \cos 2\theta / \sin 2\theta = \frac{4}{3}$ yields $\cos 2\theta = \frac{4}{5}$. Now from the half-angle formulas in (8), we find

$$\begin{aligned} \sin\theta &= \sqrt{\frac{1 - \cos 2\theta}{2}} = \sqrt{\frac{1 - \frac{4}{5}}{2}} = \frac{1}{\sqrt{10}} \\ \cos\theta &= \sqrt{\frac{1 + \cos 2\theta}{2}} = \sqrt{\frac{1 + \frac{4}{5}}{2}} = \frac{3}{\sqrt{10}}. \end{aligned} \qquad (10)$$

Thus, the equations in (5) become

$$x = \frac{3}{\sqrt{10}}x' - \frac{1}{\sqrt{10}}y' = \frac{1}{\sqrt{10}}(3x' - y')$$

$$y = \frac{1}{\sqrt{10}}x' + \frac{3}{\sqrt{10}}y' = \frac{1}{\sqrt{10}}(x' + 3y').$$

Substituting these into the given equation, we have

$$5\left(\frac{1}{\sqrt{10}}\right)^2 (3x' - y')^2 + 3\frac{1}{\sqrt{10}}(3x' - y') \cdot \frac{1}{\sqrt{10}}(x' + 3y') + \left(\frac{1}{\sqrt{10}}\right)^2 (x' + 3y')^2 = 44$$

$$\frac{5}{10}(9x'^2 - 6x'y' + y'^2) + \frac{3}{10}(3x'^2 + 8x'y' - 3y'^2) + \frac{1}{10}(x'^2 + 6x'y' + 9y'^2) = 44$$

$$45x'^2 - 30x'y' + 5y'^2 + 9x'^2 + 24x'y' - 9y'^2 + x'^2 + 6x'y' + 9y'^2 = 440.$$

The last equation simplifies to

$$\frac{x'^2}{8} + \frac{y'^2}{88} = 1. \tag{11}$$

We recognize this as the standard equation of an ellipse. Now from (10) we have $\sin\theta = 1/\sqrt{10}$ and so with the aid of a calculator we find $\theta \approx 18.4°$. This rotation angle is shown in FIGURE 7.4.4 and we use the new axes to sketch the ellipse. ≡

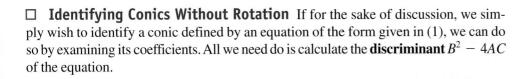

FIGURE 7.4.4 Rotated axes in Example 3

Don't be misled by the last two examples. After using the rotation equations in (5) the conic section may not be immediately identifiable without some extra work. For example, after an appropriate rotation of axes, the equation

$$11x^2 + 16\sqrt{2}xy + 19y^2 - 24\sqrt{3}x - 24\sqrt{6}y + 45 = 0 \tag{12}$$

is transformed into

$$9x'^2 - 24x' + y'^2 + 15 = 0.$$

After completing the square in x', $(3x' - 4)^2 + y'^2 = 1$, we recognize that the equation (12) defines an ellipse.

☐ **Identifying Conics Without Rotation** If for the sake of discussion, we simply wish to identify a conic defined by an equation of the form given in (1), we can do so by examining its coefficients. All we need do is calculate the **discriminant** $B^2 - 4AC$ of the equation.

THEOREM 7.4.2   Identifying Conics

Excluding the degenerate cases, the graph of the second-degree equation (1) is

   (*i*) a **parabola** when $B^2 - 4AC = 0$,
  (*ii*) an **ellipse** when $B^2 - 4AC < 0$, or
 (*iii*) a **hyperbola** when $B^2 - 4AC > 0$.

EXAMPLE 4   Identification

Identify the conic defined by the given equation.

(a) $9x^2 + 12xy + 4y^2 + 2x - 3y = 0$ **(b)** $3x^2 - 5y^2 + 8x - y + 2 = 0$

Solutions (a) With $A = 9$, $B = 12$, $C = 4$ the discriminant

$$B^2 - 4AC = (12)^2 - 4(9)(4) = 144 - 144 = 0$$

indicates that the equation defines a parabola.

(b) With $A = 3$, $B = 0$, $C = -5$ the discriminant is

$$B^2 - 4AC = (0)^2 - 4(3)(-5) = 60 > 0.$$

The equation defines a hyperbola. ≡

Answers to selected odd-numbered problems begin on page ANS-22.

In Problems 1–4, use (4) to find the $x'y'$-coordinates of the given xy-point. Use the specified angle of rotation θ.

1. $(6, 2), \theta = 45°$

2. $(-2, 8), \theta = 30°$

3. $(-1, -1), \theta = 60°$

4. $(5, 3), \theta = 15°$

In Problems 5–10, use (5) to find the xy-coordinates of the given $x'y'$-point. Use the specified angle of rotation θ.

5. $(2, -8), \theta = 30°$

6. $(-5, 7), \theta = 45°$

7. $(0, 4), \theta = \dfrac{\pi}{2}$

8. $(3, 0), \theta = \dfrac{\pi}{3}$

9. $(4, 6), \theta = 15°$

10. $(1, 1), \theta = 75°$

In Problems 11–16, use rotation of axes to eliminate the xy-term in the given equation. Identify the conic and graph.

11. $x^2 + xy + y^2 = 4$

12. $2x^2 - 3xy - 2y^2 = 5$

13. $x^2 - 2xy + y^2 = 8x + 8y$

14. $3x^2 + 4xy = 16$

15. $x^2 + 4xy - 2y^2 - 6 = 0$

16. $x^2 + 4xy + 4y^2 = 16\sqrt{5}x - 8\sqrt{5}y$

In Problems 17–20, use rotation of axes to eliminate the xy-term in the given equation. Identify the conic.

17. $4x^2 - 4xy + 7y^2 + 12x + 6y - 9 = 0$

18. $-x^2 + 6\sqrt{3}xy + 5y^2 - 8\sqrt{3}x + 8y = 12$

19. $8x^2 - 8xy + 2y^2 + 10\sqrt{5}x = 5$

20. $x^2 - xy + y^2 - 4x - 4y = 20$

21. Given $3x^2 + 2\sqrt{3}xy + y^2 + 2x - 2\sqrt{3}y = 0$.
  (a) By rotation of axes show that the graph of the equation is a parabola.
  (b) Find the $x'y'$-coordinates of the focus. Use this information to find the xy-coordinates of the focus.
  (c) Find an equation of the directrix in terms of the $x'y'$-coordinates. Use this information to find an equation of the directrix in terms of the xy-coordinates.

22. Given $13x^2 - 8xy + 7y^2 = 30$.
  (a) By rotation of axes show that the graph of the equation is an ellipse.
  (b) Find the $x'y'$-coordinates of the foci. Use this information to find the xy-coordinates of the foci.
  (c) Find the xy-coordinates of the vertices.

In Problems 23–28, use the discriminant to determine the type of graph without actually graphing.

23. $x^2 - 3xy + y^2 = 5$

24. $2x^2 - 2xy + 2y^2 = 1$

25. $4x^2 - 4xy + y^2 - 6 = 0$

26. $x^2 + \sqrt{3}xy - \frac{1}{2}y^2 = 0$

27. $x^2 + xy + y^2 - x + 2y + 1 = 0$

28. $3x^2 + 2\sqrt{3}xy + y^2 - 2x + 2\sqrt{3}y - 4 = 0$

For Discussion

29. In (2), show that if A and C have the same signs, then the graph of the equation is either an ellipse, a circle, or a point, or does not exist. Give an example of each type of equation.

30. In (2), show that if A and C have opposite signs, then the graph of the equation is either a hyperbola or a pair of intersecting lines. Give an example of each type of equation.

31. In (2), show that if either $A = 0$ or $C = 0$, then the graph of the equation is either a parabola, two parallel lines, or one line, or does not exist. Give an example of each type of equation.

32. (a) Use FIGURE 7.4.5 to show that

$$x = r\cos\phi, \quad y = r\sin\phi$$

and

$$x' = r\cos(\phi - \theta), \quad y' = r\sin(\phi - \theta).$$

(b) Use the results from part (a) to derive the rotation equations in (4).
(c) Use (4) to find the rotation equations in (5).

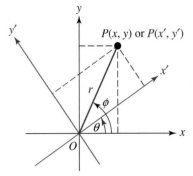

FIGURE 7.4.5 Rotated axes in Problem 32

7.5 3-Space

∫**Calculus PREVIEW**

≡ **Introduction** In the plane, or **2-space**, one way of describing the position of a point P is to assign to it coordinates relative to two perpendicular coordinate axes called the x- and y-axes. The intersection of the two axes is called the origin and denoted by O. Recall,

- a vertical line $x = a$ consists of all points of the form (a, y), and
- a horizontal line $y = b$ consists of all points of the form (x, b).   (1)

If P is the point of intersection of the vertical line $x = a$ (perpendicular to the x-axis) and the horizontal line $y = b$ (perpendicular to the y-axis), then the ordered pair (a, b) is said to be the rectangular or Cartesian coordinates of the point. See FIGURE 7.5.1. In this section we extend this method of representation of a point to three dimensions.

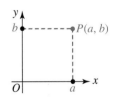

FIGURE 7.5.1 Point in 2-space

☐ **Rectangular Coordinate System in 3-Space** In three dimensions, or **3-space**, a rectangular coordinate system is constructed using three mutually perpendicular **coordinate axes**. The point at which these axes intersect is called the **origin** O. The axes drawn as solid lines in FIGURE 7.5.2(a) represent the positive axes, and are labeled in accordance with the so-called **right-hand rule** illustrated in Figure 7.5.2(b):

- *If the fingers of the right hand, pointing in the direction of the positive x-axis, are curled toward the positive y-axis, then the thumb will point in the direction of a new axis perpendicular to the plane of the x- and y-axes. This new axis is labeled the z-axis.*

The dashed lines in Figure 7.5.2(a), represent the negative axes. If the x- and y-axes are interchanged in Figure 7.5.2, the coordinate system is said to be **left-handed**.

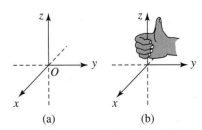

FIGURE 7.5.2 Three-dimensional coordinate axes

In 3-space, the graph of the equations $x = a$, $y = b$, and $z = c$ consist of all **ordered triples** or points of the form (a, y, z), (x, b, z), and (x, y, c), respectively. The graphs of the equations $x = a$, $y = b$, and $z = c$ are, in turn, planes perpendicular to the x-, y-, and z-axes. The point P at which these planes intersect can be represented by an **ordered triple** of numbers (a, b, c) said to be the **rectangular**, or **Cartesian**, **coordinates** of the point. The numbers a, b, and c are called the x-, y-, and z-coordinates of $P(a, b, c)$, respectively. See **FIGURE 7.5.3**.

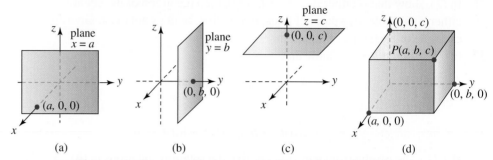

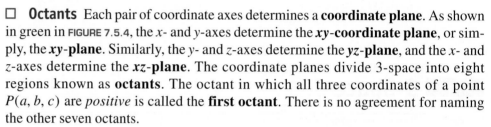

(a)   (b)   (c)   (d)

FIGURE 7.5.3 Three mutually perpendicular planes intersect in a point

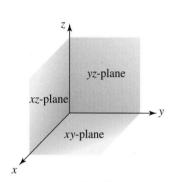

FIGURE 7.5.4 Coordinate planes

□ **Octants** Each pair of coordinate axes determines a **coordinate plane**. As shown in green in **FIGURE 7.5.4**, the x- and y-axes determine the **xy-coordinate plane**, or simply, the **xy-plane**. Similarly, the y- and z-axes determine the **yz-plane**, and the x- and z-axes determine the **xz-plane**. The coordinate planes divide 3-space into eight regions known as **octants**. The octant in which all three coordinates of a point $P(a, b, c)$ are *positive* is called the **first octant**. There is no agreement for naming the other seven octants.

The following table summarizes the coordinates of a point either on a coordinate axis or in a coordinate plane. As seen in the table, we can also describe, say, the xy-plane by the simple equation $z = 0$. Similarly, the xz-plane is $y = 0$ and the yz-plane is $x = 0$. A point on a coordinate axes is not considered to be in any octant.

Axes	Coordinates	Plane	Coordinates
x	$(x, 0, 0)$	xy	$(x, y, 0)$
y	$(0, y, 0)$	xz	$(x, 0, z)$
z	$(0, 0, z)$	yz	$(0, y, z)$

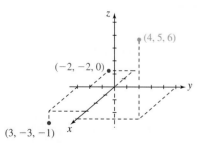

FIGURE 7.5.5 Points in Example 1

EXAMPLE 1   **Graphing Points in 3-Space**

Graph the points $(4, 5, 6)$, $(3, -3, -1)$, and $(-2, -2, 0)$.

Solution Of the three points shown in **FIGURE 7.5.5** only $(4, 5, 6)$ is in the first octant. The point $(-2, -2, 0)$ lies in the xy-plane.   ≡

□ **Distance Formula** To find the **distance** between two points $P_1(x_1, y_1, z_1)$ and $P_2(x_2, y_2, z_2)$ in 3-space, let us first consider their projections onto the xy-plane. As seen in **FIGURE 7.5.6**, the distance between $(x_1, y_1, 0)$ and $(x_2, y_2, 0)$ follows from the usual distance formula in the plane and is $\sqrt{(x_2 - x_1)^2 + (y_2 - y_1)^2}$. Hence, from the Pythagorean theorem applied to the right triangle $P_1 P_3 P_2$, we have

$$[d(P_1, P_2)]^2 = [\sqrt{(x_2 - x_1)^2 + (y_2 - y_1)^2}]^2 + |z_2 - z_1|^2$$

or $\quad\quad d(P_1, P_2) = \sqrt{(x_2 - x_1)^2 + (y_2 - y_1)^2 + (z_2 - z_1)^2}.$ (1)

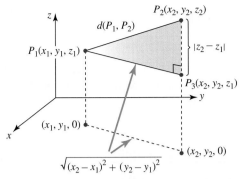

FIGURE 7.5.6 Distance between two points in 3-space

EXAMPLE 2　　　　**Distance Between Points in 3-Space**

Find the distance between $(2, -3, 6)$ and $(-1, -7, 4)$.

Solution　From (1), the distance is

$$d = \sqrt{(2 - (-1))^2 + (-3 - (-7))^2 + (6 - 4)^2} = \sqrt{29}.$$　　≡

☐ **Midpoint Formula**　The distance formula can be used to show that the coordinates of the **midpoint M of the line segment** in 3-space connecting the distinct points $P_1(x_1, y_1, z_1)$ and $P_2(x_2, y_2, z_2)$ are

$$M = \left(\frac{x_1 + x_2}{2}, \frac{y_1 + y_2}{2}, \frac{z_1 + z_2}{2} \right). \tag{2}$$

See Problem 90 in Exercises 7.5.

EXAMPLE 3　　　　**Midpoint in 3-Space**

Find the coordinates of the midpoint M of the line segment between the two points in Example 2.

Solution　From (2) we find that the coordinates of M are

$$\left(\frac{2 + (-1)}{2}, \frac{-3 + (-7)}{2}, \frac{6 + 4}{2} \right) \quad \text{or} \quad \left(\tfrac{1}{2}, -5, 5 \right).$$　　≡

☐ **Sphere**　Like a circle, a sphere can be defined in terms of the distance formula.　　◀ Review Section 1.4.

DEFINITION 7.5.1　Sphere

A **sphere** is the set of all points $P(x, y, z)$ in 3-space that are a given fixed distance r, called the **radius**, from a given fixed point C called the **center**.

If the center is $P_1(h, k, l)$, then a point $P(x, y, z)$ is on the sphere if and only if P_1 and P satisfy $[d(P_1, P)]^2 = r^2$, or

$$(x - h)^2 + (y - k)^2 + (z - l)^2 = r^2. \tag{3}$$

Equation (3) is called the **standard form** of the equation of a sphere.

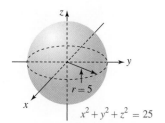

FIGURE 7.5.7 Sphere in Example 4

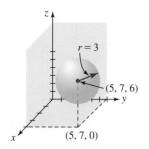

FIGURE 7.5.8 Sphere in Example 5

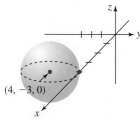

FIGURE 7.5.9 Sphere tangent to plane $y = 0$ in Example 6

■ EXAMPLE 4   Graph of a Sphere

Graph $x^2 + y^2 + z^2 = 25$.

Solution We identify $h = 0, k = 0, l = 0$, and $r^2 = 25 = 5^2$ in (3), and so the graph of $x^2 + y^2 + z^2 = 25$ is a sphere of radius 5 whose center is at the origin. The graph of the equation is given in FIGURE 7.5.7. ≡

■ EXAMPLE 5   Graph of a Sphere

Graph $(x - 5)^2 + (y - 7)^2 + (z - 6)^2 = 9$.

Solution In this case we identify $h = 5, k = 7, l = 6$, and $r^2 = 9$. From (3) we see that the graph of $(x - 5)^2 + (y - 7)^2 + (z - 6)^2 = 3^2$ is a sphere with center $(5, 7, 6)$ and radius 3. Its graph lies entirely in the first octant and is shown in FIGURE 7.5.8. ≡

■ EXAMPLE 6   Equation of a Sphere

Find an equation of the sphere whose center is $(4, -3, 0)$ that is tangent to the xz-plane.

Solution The perpendicular distance from the point $(4, -3, 0)$ to the xz-plane ($y = 0$), and hence the radius of the sphere, is the absolute value of the y-coordinate, $|-3| = 3$. Thus, the standard form of the equation of the sphere is

$$(x - 4)^2 + (y + 3)^2 + z^2 = 3^2.$$

See FIGURE 7.5.9. ≡

To put an equation of a sphere in the standard form (3) it is necessary to complete the square in the three variables x, y, and z.

■ EXAMPLE 7   Center and Radius

Find the center and radius of the sphere whose equation is

$$2x^2 + 2y^2 + 2z^2 - 2x + y - 4z + 2 = 0.$$

Solution We first divide by 2, group like terms together, and then complete the square in x, y, and z:

$$(x^2 - x \quad) + \left(y^2 + \tfrac{1}{2}y \quad\right) + (z^2 - 2z \quad) = -1$$
$$\left[x^2 - x + \left(-\tfrac{1}{2}\right)^2\right] + \left[y^2 + \tfrac{1}{2}y + \left(\tfrac{1}{4}\right)^2\right] + \left[z^2 - 2z + (-1)^2\right]$$
$$= -1 + \left(-\tfrac{1}{2}\right)^2 + \left(\tfrac{1}{4}\right)^2 + (-1)^2$$
$$\left(x - \tfrac{1}{2}\right)^2 + \left(y + \tfrac{1}{4}\right)^2 + (z - 1)^2 = \tfrac{5}{16}.$$

From the last equation we see that the center and radius of the sphere are $\left(\tfrac{1}{2}, -\tfrac{1}{4}, 1\right)$ and $\tfrac{1}{4}\sqrt{5}$, respectively. ≡

□ **Linear Equation in Three Variables** In the introduction to Section 2.3 we defined a linear equation in two variables to be $Ax + By + C = 0$. For various choices of the coefficients A, B, and C, the graph of a linear equation is a line in 2-space. The graph of a **linear equation in three variables**

$$Ax + By + Cz + D = 0, \tag{4}$$

A, B, C not all zero, is a **plane** in 3-space. We note that the simple equations $x = x_0$, $y = y_0$, and $z = z_0$, where x_0, y_0, and z_0 are constants, are special cases of (4). Here are two guidelines for graphing planes:

- The graphs of $x = x_0$, $y = y_0$, and $z = z_0$ are planes perpendicular to the x-, y-, and z-axes, respectively. See Figure 7.5.3.
- To graph a linear equation (4), find the x-, y-, and z-intercepts, or if necessary, find the trace of a the plane in the coordinate planes.

A **trace** of a plane in a coordinate plane is the line of intersection of the plane with the coordinate plane. For example, by setting $z = 0$ we see that the trace of the plane $2x + 3y + 6z = 18$ in the xy-plane is the line $2x + 3y = 18$. To find the **intercepts** of a plane we use the fact that points on the x-, y-, and z-axes are of the form $(x, 0, 0)$, $(0, y, 0)$, and $(0, 0, z)$, respectively.

EXAMPLE 8   Graph

Graph the equation $2x + 3y + 6z = 18$.

Solution Setting:

$$y = 0, z = 0 \quad \text{gives} \quad x = 9$$
$$x = 0, z = 0 \quad \text{gives} \quad y = 6$$
$$x = 0, y = 0 \quad \text{gives} \quad z = 3.$$

As shown in FIGURE 7.5.10, we use the x-, y-, and z-intercepts $(9, 0, 0)$, $(0, 6, 0)$, and $(0, 0, 3)$ to draw the graph of the portion of the plane in the first octant. ≡

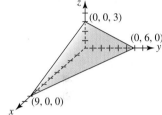

FIGURE 7.5.10 Plane in Example 8

EXAMPLE 9   Graph

Graph the equation $x + y - z = 0$.

Solution First observe that the plane passes through the origin $(0, 0, 0)$. Now, the trace of the plane in the xz-plane ($y = 0$) is $z = x$, whereas its trace in the yz-plane ($x = 0$) is $z = y$. The traces are the two black lines in FIGURE 7.5.11. ≡

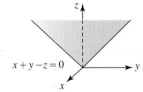

FIGURE 7.5.11 Plane in Example 9

□ **Missing Variables** If *two* of the variables are missing in equation (4), then as we have already seen the plane is perpendicular to the coordinate axis corresponding to the variable present. For example, $z = 1$ is the equation of the plane perpendicular to the z-axis at the point $(0, 0, 1)$. Alternatively, we can interpret $z = 1$ as the equation of the plane through the point $(0, 0, 1)$ parallel to the coordinate plane corresponding to the missing two variables, in this case, parallel to the xy-plane. If *one* of the variables is missing in equation (4), the equation is then the same as the equation of the trace of the plane in the appropriate coordinate plane. The plane then is parallel to the coordinate axis corresponding to the missing variable. The following example illustrates this last idea.

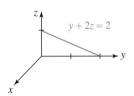

(a) Trace in yz-plane

EXAMPLE 10   Graph

Graph the equation $y + 2z = 2$.

Solution In 3-space the graph of the equation $y + 2z = 2$ is the graph of the set of ordered triples:

$$\{(x, y, z) \mid y + 2z = 2, x \text{ arbitrary}\}.$$

All we need do is draw the line $y + 2z = 2$ in the yz-plane. Because the x variable is missing in the given equation the plane is drawn parallel to the x-axis. Necessarily the plane is perpendicular to the yz-plane. See FIGURE 7.5.12. ≡

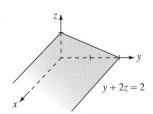

(b) Plane parallel to x-axis

FIGURE 7.5.12 Trace and plane in Example 10

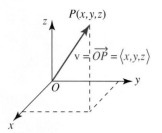

FIGURE 7.5.13 Position vector in 3-space

☐ **Vectors in 3-Space** In Section 5.5 we saw that a **position vector** of a point $P(x, y)$ in 2-space is a vector whose initial point is the origin O and whose terminal point is P. **FIGURE 7.5.13** shows a position vector

$$\overrightarrow{OP} = \langle x, y, z \rangle$$

of a point $P(x, y, z)$ in 3-space. Any vector in 3-space can be identified with a unique position vector $\mathbf{u} = \langle a_1, a_2, a_3 \rangle$, where the real numbers $a_1, a_2,$ and a_3 are called the **components** of the vector $\mathbf{u}$. Written in component form the **zero vector** is one in which all components are 0, that is, $\mathbf{0} = \langle 0, 0, 0 \rangle$.

The component definitions of addition, subtraction, scalar multiplication, and the dot product, and so on, are natural generalizations of those given in 2-space. For convenience we summarize some of these important vector concepts in 3-space.

Vector Operations and Concepts in 3-Space

Let $\mathbf{u} = \langle a_1, a_2, a_3 \rangle$ and $\mathbf{v} = \langle b_1, b_2, b_3 \rangle$ be two vectors in 3-space.

(*i*) The **sum** of $\mathbf{u}$ and $\mathbf{v}$ is

$$\mathbf{u} + \mathbf{v} = \langle a_1, a_2, a_3 \rangle + \langle b_1, b_2, b_3 \rangle = \langle a_1 + b_1, a_2 + b_2, a_3 + b_3 \rangle.$$

(*ii*) The **difference** of $\mathbf{u}$ and $\mathbf{v}$ is

$$\mathbf{u} - \mathbf{v} = \langle a_1, a_2, a_3 \rangle - \langle b_1, b_2, b_3 \rangle = \langle a_1 - b_1, a_2 - b_2, a_3 - b_3 \rangle.$$

(*iii*) For a real number k, the **scalar multiple** of a vector $\mathbf{u}$ is

$$k\mathbf{u} = \langle ka_1, ka_2, ka_3 \rangle.$$

(*iv*) The **magnitude** of a vector $\mathbf{u}$ is

$$|\mathbf{u}| = \sqrt{a_1^2 + a_2^2 + a_3^2}.$$

(*v*) A **unit vector** $\mathbf{u}$ in the same direction as the vector $\mathbf{v}$ is

$$\mathbf{u} = \frac{1}{|\mathbf{v}|}\mathbf{v}, \quad \mathbf{v} \neq \mathbf{0}.$$

(*vi*) The **dot product** of $\mathbf{u}$ and $\mathbf{v}$ is

$$\mathbf{u} \cdot \mathbf{v} = a_1 b_1 + a_2 b_2 + a_3 b_3.$$

(*vii*) Nonzero vectors $\mathbf{u}$ and $\mathbf{v}$ are **orthogonal** if and only if $\mathbf{u} \cdot \mathbf{v} = 0$.

☐ **i, j, k Vectors** In Section 5.5 we saw that the set of two unit vectors $\mathbf{i} = \langle 1, 0 \rangle$ and $\mathbf{j} = \langle 0, 1 \rangle$ constitute a basis for the system of two-dimensional vectors. That is, any vector $\mathbf{u}$ in 2-space can be written as a linear combination of $\mathbf{i}$ and $\mathbf{j}$: $\mathbf{u} = a_1\mathbf{i} + a_2\mathbf{j}$. Likewise any vector $\mathbf{u} = \langle a_1, a_2, a_3 \rangle$ in 3-space can be expressed as a linear combination of the unit vectors

$$\mathbf{i} = \langle 1, 0, 0 \rangle, \quad \mathbf{j} = \langle 0, 1, 0 \rangle, \quad \mathbf{k} = \langle 0, 0, 1 \rangle.$$

To see this we use (*iii*) of the above summary:

$$\mathbf{u} = \langle a_1, a_2, a_3 \rangle = a_1\langle 1, 0, 0 \rangle + a_2\langle 0, 1, 0 \rangle + a_3\langle 0, 0, 1 \rangle$$

that is, $\qquad\qquad \mathbf{u} = a_1\mathbf{i} + a_2\mathbf{j} + a_3\mathbf{k}.$

The vectors **i**, **j**, and **k** illustrated in FIGURE 7.5.14(a) are called the **standard basis** for the system of three-dimensional vectors. In Figure 7.5.14(b) we see that a position vector $\mathbf{u} = a_1\mathbf{i} + a_2\mathbf{j} + a_3\mathbf{k}$ is the sum of three unit vectors $a_1\mathbf{i}$, $a_2\mathbf{j}$, and $a_3\mathbf{k}$, which lie along the x-, y-, and z-coordinate axes, respectively, and have the origin as a common initial point.

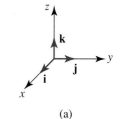

(a)

EXAMPLE 11   Using the i, j, k Vectors

Consider the vectors $\mathbf{u} = \langle -3, -1, 4 \rangle$ and $\mathbf{v} = \langle 2, 14, 5 \rangle$.

(a) Expressed in terms of the **i**, **j**, **k** basis vectors **u** and **v** are the same as $\mathbf{u} = -3\mathbf{i} - \mathbf{j} + 4\mathbf{k}$ and $\mathbf{v} = 2\mathbf{i} + 14\mathbf{j} + 5\mathbf{k}$.

(b) The linear combination $5\mathbf{u} - 2\mathbf{v}$ is

$$
\begin{aligned}
5\mathbf{u} - 2\mathbf{v} &= 5(-3\mathbf{i} - \mathbf{j} + 4\mathbf{k}) - 2(2\mathbf{i} + 14\mathbf{j} + 5\mathbf{k}) \\
&= (-15\mathbf{i} - 5\mathbf{j} + 20\mathbf{k}) - (4\mathbf{i} + 28\mathbf{j} + 10\mathbf{k}) \\
&= (-15 - 4)\mathbf{i} + (-5 - 28)\mathbf{j} + (20 - 10)\mathbf{k} \\
&= -19\mathbf{i} - 33\mathbf{j} + 10\mathbf{k}.
\end{aligned}
$$

(c) The dot product of **u** and **v** is the constant

$$
\mathbf{u} \cdot \mathbf{v} = (-3)(2) + (-1)(14) + (4)(5) = -20 + 20 = 0.
$$

≡

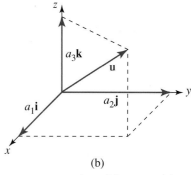

FIGURE 7.5.14 A position vector in terms of **i**, **j**, **k**

Because the dot product of the vector **u** and **v** in part (c) of Example 11 is 0 we can conclude that **u** and **v** are orthogonal.

NOTES FROM THE CLASSROOM

(*i*) We have included this brief section on 3-space because in a typical three-semester course in calculus, the third semester deals primarily with calculus and vectors in three dimensions.

(*ii*) As we know in 2-space, two distinct points determine a line. Analogously, in 3-space, three noncollinear points (x_1, y_1, z_1), (x_2, y_2, z_2), (x_3, y_3, z_3) determine a plane. One way of finding an equation of the plane is by substituting the coordinates of the three points in (4) and solving the three simultaneous linear equations

$$
\begin{aligned}
Ax_1 + By_1 + Cz_1 + D &= 0 \\
Ax_2 + By_2 + Cz_2 + D &= 0 \\
Ax_3 + By_3 + Cz_3 + D &= 0
\end{aligned}
\tag{5}
$$

for A, B, and C in terms of D. We can then choose D to be any nonzero real number. See Problems 59–64 in Exercises 7.5.

(*iii*) The natural follow-up to this introduction to 3-space is to consider **multivariable functions**. The graph of a function f of one independent variable $y = f(x)$ is a *curve* in 2-space, whereas the graph of a function $z = f(x, y)$ of two independent variables is a *surface* in 3-space. For example, if we solve for z in Example 8 the resulting equation $z = -\frac{1}{3}x - \frac{1}{2}y + 3$ is a **linear function** whose graph, as we have seen, is a plane. If we solve the equation in Example 4 for z, then one result is a function $z = \sqrt{25 - x^2 - y^2}$ that describes the upper hemisphere of the sphere.

Graphing a function $z = f(x, y)$ can be difficult and may, at times, require a computer. If you have access to a computer lab, check to see whether the computers have 3D graphing software. The graphs of the polynomial function $z = 2x^2 - 2y^2 + 2$ and the trigonometric function $z = \sin(xy)$ shown in FIGURE 7.5.15 and FIGURE 7.5.16 were obtained using the computer algebra system (CAS) *Mathematica*.

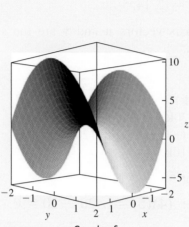

FIGURE 7.5.15 Graph of
$z = 2x^2 - 2y^2 + 2$ for
$-2 \le x \le 2, -2 \le y \le 2$

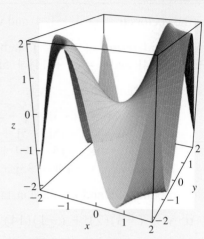

FIGURE 7.5.16 Graph of
$z = \sin(xy)$ for
$-2 \le x \le 2, -2 \le y \le 2$

7.5 Exercises Answers to selected odd-numbered problems begin on page ANS–22.

In Problems 1–6, graph the given point.

1. $(1, 1, 5)$ **2.** $(0, 0, 4)$ **3.** $(3, 4, 0)$
4. $(6, 0, 0)$ **5.** $(6, -2, 0)$ **6.** $(5, -4, 3)$

In Problems 7–10, describe geometrically all points $P(x, y, z)$ whose coordinates satisfy the given conditions.

7. $z = 5$ **8.** $x = 1$
9. $x = 2, y = 3$ **10.** $x = 4, y = -1, z = 7$

11. Give the coordinates of the vertices of the rectangular parallelepiped whose sides are on the coordinate planes and the planes $x = 2, y = 5, z = 8$.

12. In FIGURE 7.5.17, two vertices are shown of a rectangular parallelepiped having sides parallel to the coordinate planes. Find the coordinates of the remaining six vertices.

13. Consider the point $P(-2, 5, 4)$.
   (a) If lines are drawn from P perpendicular to the coordinate planes, what are the coordinates of the point at the base of each perpendicular?
   (b) If a line is drawn from P to the plane $z = -2$, what are the coordinates of the point at the base of the perpendicular?
   (c) Find the point in the plane $x = 3$ that is closest to P.

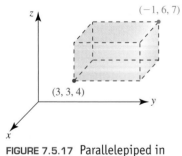

FIGURE 7.5.17 Parallelepiped in Problem 12

14. Determine an equation of a plane parallel to a coordinate plane that contains the given pair of points.
  (a) $(3, 4, -5), (-2, 8, -5)$
  (b) $(1, -1, 1), (1, -1, -1)$
  (c) $(-2, 1, 2), (2, 4, 2)$

In Problems 15–20, describe the set of points $P(x, y, z)$ in 3-space whose coordinates satisfy the given equation.

15. $xyz = 0$
16. $x^2 + y^2 + z^2 = 0$
17. $(x + 1)^2 + (y - 2)^2 + (z + 3)^2 = 0$
18. $(x - 2)(z - 8) = 0$
19. $z^2 - 25 = 0$
20. $x = y = z$

In Problems 21 and 22, find the distance between the given points.
21. $(3, -1, 2), (6, 4, 8)$
22. $(-1, -3, 5), (0, 4, 3)$

23. Find the distance from the point $(7, -3, -4)$ to:
  (a) the yz-plane   (b) the x-axis.
24. Find the distance from the point $(-6, 2, -3)$ to:
  (a) the xz-plane   (b) the origin.

In Problems 25–28, the given three points form a triangle. Determine which triangles are isosceles and which are right triangles.

25. $(0, 0, 0), (3, 6, -6), (2, 1, 2)$
26. $(0, 0, 0), (1, 2, 4), \left(3, 2, 2\sqrt{2}\right)$
27. $(1, 2, 3), (4, 1, 3), (4, 6, 4)$
28. $(1, 1, -1), (1, 1, 1), (0, -1, 1)$

In Problems 29–32, use the distance formula to determine whether the given points are collinear.

29. $P_1(1, 2, 0), P_2(-2, -2, -3), P_3(7, 10, 6)$
30. $P_1(1, 2, -1), P_2(0, 3, 2), P_3(1, 1, -3)$
31. $P_1(1, 0, 4), P_2(-4, -3, 5), P_3(-7, -4, 8)$
32. $P_1(2, 3, 2), P_2(1, 4, 4), P_3(5, 0, -4)$

In Problems 33 and 34, solve for the unknown.

33. $P_1(x, 2, 3), P_2(2, 1, 1); d(P_1, P_2) = \sqrt{21}$
34. $P_1(x, x, 1), P_2(0, 3, 5); d(P_1, P_2) = 5$

In Problems 35 and 36, find the coordinates of the midpoint of the line segment between the given points.

35. $\left(1, 3, \frac{1}{2}\right), \left(7, -2, \frac{5}{2}\right)$
36. $(0, 5, -8), (4, 1, -6)$

37. The coordinates of the midpoint of the line segment between $P_1(x_1, y_1, z_1)$ and $P_2(2, 3, 6)$ are $(-1, -4, 8)$. Find the coordinates of P_1.
38. Let P_3 be the midpoint of the line segment between $P_1(-3, 4, 1)$ and $P_2(-5, 8, 3)$. Find the coordinates of the midpoint of the line segment:
  (a) between P_1 and P_3 (b) between P_3 and P_2.

In Problems 39–42, sketch the graph of the given equation.

39. $x^2 + y^2 + z^2 = 9$
40. $x^2 + y^2 + (z - 3)^2 = 16$
41. $(x - 1)^2 + (y - 1)^2 + (z - 1)^2 = 1$
42. $(x + 3)^2 + (y + 4)^2 + (z - 5)^2 = 4$

In Problems 43–46, complete the square in x, y, and z to find the center and radius of the given sphere.

43. $x^2 + y^2 + z^2 + 8x - 6y - 4z - 7 = 0$
44. $4x^2 + 4y^2 + 4z^2 + 4x - 12z + 9 = 0$
45. $x^2 + y^2 + z^2 - 16z = 0$
46. $x^2 + y^2 + z^2 - x + y = 0$

In Problems 47–52, find an equation of a sphere that satisfies the given conditions.

47. Center $(-1, 4, 6)$; radius $\sqrt{3}$
48. Center $(0, -3, 0)$; diameter $\frac{5}{2}$
49. Center $(1, 1, 4)$; tangent to the xy-plane
50. Center $(5, 2, -2)$; tangent to the yz-plane
51. Center on the positive y-axis; radius 2; tangent to $x^2 + y^2 + z^2 = 36$
52. Center $(-3, 1, 2)$; passing through the origin

In Problems 53–58, graph the plane whose equation is given.

53. $5x + 2y + z = 10$
54. $3x + 2z = 9$
55. $3x + z - 6 = 0$
56. $3x + 4y - 2z - 12 = 0$
57. $-x + 2y + z = 4$
58. $3x - y - 6 = 0$

In Problems 59–64, use (4) and (5) to find an equation of a plane that contains the given points.

59. $(3, 5, 2)$, $(2, 3, 1)$, $(-1, -1, 4)$
60. $(0, 1, 0)$, $(0, 1, 1)$, $(1, 3, -1)$
61. $(-1, -1, 2)$, $(1, 1, 1)$, $(3, 2, -1)$
62. $(0, 0, 3)$, $(0, -1, 0)$, $(6, 0, 0)$
63. $(1, 2, -1)$, $(4, 3, 1)$, $(7, 4, 1)$
64. $(2, 1, 2)$, $(4, 1, 0)$, $(5, 2, -5)$

In Problems 65–76, $\mathbf{u} = \langle 1, -3, 2 \rangle$, $\mathbf{v} = \langle -1, 1, 1 \rangle$, and $\mathbf{w} = \langle 2, 6, 9 \rangle$. Find the indicated vector or scalar.

65. $\mathbf{u} + (\mathbf{v} + \mathbf{w})$
66. $2\mathbf{u} - (\mathbf{v} - \mathbf{w})$
67. $\mathbf{v} + 2(\mathbf{u} - 3\mathbf{w})$
68. $4(\mathbf{u} + 2\mathbf{w}) - 6\mathbf{v}$
69. $|\mathbf{u} + \mathbf{w}|$
70. $|\mathbf{w}||2\mathbf{v}|$
71. $\left|\dfrac{\mathbf{u}}{|\mathbf{u}|}\right| + 5\left|\dfrac{\mathbf{v}}{|\mathbf{v}|}\right|$
72. $|\mathbf{v}|\mathbf{u} + |\mathbf{u}|\mathbf{v}$
73. $\frac{1}{2}\mathbf{u} \cdot \mathbf{v}$
74. $(\mathbf{v} \cdot \mathbf{w})\mathbf{u}$
75. $(\mathbf{u} + \mathbf{v}) \cdot \mathbf{w}$
76. $(\mathbf{u} - \mathbf{v}) \cdot (\mathbf{v} + \mathbf{w})$

77. Find a unit vector in the opposite direction of $\mathbf{v} = \langle 10, -5, 10 \rangle$.
78. Find a unit vector in the same direction as $\mathbf{v} = \mathbf{i} - 3\mathbf{j} + 2\mathbf{k}$.
79. Find a vector $\mathbf{u}$ that is four times as long as $\mathbf{v} = \mathbf{i} - \mathbf{j} + \mathbf{k}$ in the same direction as $\mathbf{v}$.
80. Find a vector $\mathbf{v}$ for which $|\mathbf{v}| = \frac{1}{2}$ that is in the opposite direction of $\mathbf{w} = \langle -6, 3, -2 \rangle$.

The **cross product** of two three-dimensional vectors $\mathbf{u} = \langle a_1, a_2, a_3 \rangle$ and $\mathbf{v} = \langle b_1, b_2, b_3 \rangle$ is a vector that can be written as the 3×3 determinant

If you are unfamiliar on how to expand a ▶
3×3 determinant, review Section 9.2.

$$\mathbf{u} \times \mathbf{v} = \begin{vmatrix} \mathbf{i} & \mathbf{j} & \mathbf{k} \\ a_1 & a_2 & a_3 \\ b_1 & b_2 & b_3 \end{vmatrix}. \tag{6}$$

In Problems 81–84, use (6) to find the cross product of the given vectors.

81. $\mathbf{u} = \langle 4, -2, 5 \rangle, \mathbf{v} = \langle 3, 1, -1 \rangle$
82. $\mathbf{u} = \langle 1, -3, 1 \rangle, \mathbf{v} = \langle 2, 0, 4 \rangle$
83. $\mathbf{u} = 2\mathbf{i} + \mathbf{j} - \mathbf{k}, \mathbf{v} = -6\mathbf{i} - 3\mathbf{j} + 3\mathbf{k}$
84. $\mathbf{u} = 8\mathbf{i} + \mathbf{j} - 6\mathbf{k}, \mathbf{v} = \mathbf{i} - 2\mathbf{j} + 10\mathbf{k}$

In Problems 85 and 86, verify that the cross product (6) of the given vectors is orthogonal to each vector. It can be shown that $\mathbf{u} \times \mathbf{v}$ is perpendicular to the plane determined by the vectors $\mathbf{u}$ and $\mathbf{v}$, and as a consequence $\mathbf{u} \times \mathbf{v}$ is orthogonal to $\mathbf{u}$ and orthogonal to $\mathbf{v}$.

85. $\mathbf{u} = \langle 2, 7, -4 \rangle, \mathbf{v} = \langle 1, 1, -1 \rangle$ **86.** $\mathbf{u} = \langle -1, -2, 4 \rangle, \mathbf{v} = \langle 4, -1, 0 \rangle$

For Discussion

In Problems 87 and 88, discuss how the procedure used in Problems 59–64 can be used to find an equation of a plane that contains the given points. Carry out your ideas.

87. $(0, 0, 0), (1, 1, -1), (3, 2, 1)$ **88.** $(0, 0, 1), (0, 0, 5), (0, 2, 1)$

89. If you have ever sat at a four-legged table that rocks, you might consider replacing it with a three-legged table. Why?
90. Use the distance formula to prove that (2) is the midpoint of the line segment between $P_1(x_1, y_1, z_1)$ and $P_2(x_2, y_2, z_2)$. [*Hint*: Show that $d(P_1, M) = d(M, P_2)$ and $d(P_1, P_2) = d(P_1, M) + d(M, P_2)$.]

In Problems 91–96, describe geometrically all points in 3-space whose coordinates satisfy the given condition(s).

91. $x^2 + y^2 + (z - 1)^2 = 4, 1 \leq z \leq 3$
92. $x^2 + y^2 + (z - 1)^2 = 4, z = 2$
93. $x^2 + y^2 + z^2 \geq 1$
94. $0 < (x - 1)^2 + (y - 2)^2 + (z - 3)^2 < 1$
95. $1 \leq x^2 + y^2 + z^2 \leq 9$
96. $1 \leq x^2 + y^2 + z^2 \leq 9, z \leq 0$

In Problems 97 and 98, describe the surface in 3-space defined by the given set of points.

97. $\{(x, y, z) \mid x^2 + y^2 = 1\}$ **98.** $\{(x, y, z) \mid z = 1 - y^2\}$

99. Determine whether the cross product (6) of two vectors is commutative. That is, does $\mathbf{u} \times \mathbf{v} = \mathbf{v} \times \mathbf{u}$?
100. If $\mathbf{u} = \langle 1, 2, 3 \rangle, \mathbf{v} = \langle 4, 5, 6 \rangle, \mathbf{w} = \langle 7, 8, 3 \rangle$, use (6) to find $(\mathbf{u} \times \mathbf{v}) \times \mathbf{w}$.

| CHAPTER 7 | Review Exercises | Answers to selected odd-numbered problems begin on page ANS–23. |

A. Fill in the Blanks

In Problems 1–20, fill in the blanks.

1. An equation in the standard form $y^2 = 4cx$ of a parabola with focus $(5, 0)$ is _____.
2. An equation in the standard form $x^2 = 4cy$ of a parabola through $(2, 6)$ is _____.

3. A rectangular equation of a parabola with focus $(1, -3)$ and directrix $y = -7$ is _____.

4. The directrix and vertex of a parabola are $x = -3$ and $(-1, -2)$, respectively. The focus of the parabola is _____.

5. The focus and directrix of a parabola are $\left(0, \frac{1}{4}\right)$ and $y = -\frac{1}{4}$, respectively. The vertex of the parabola is _____.

6. The vertex and focus of the parabola $8(x + 4)^2 = y - 2$ are _____.

7. After the graph of $8(x + 4)^2 = y - 2$ is moved rigidly 4 units to the right its equation is _____.

8. The center and vertices of the ellipse $\dfrac{(x - 2)^2}{16} + \dfrac{(y + 5)^2}{4} = 1$ are _____.

9. The center and vertices of the hyperbola $y^2 - \dfrac{(x + 3)^2}{4} = 1$ are _____.

10. The asymptotes of the hyperbola $y^2 - (x - 1)^2 = 1$ are _____.

11. The y-intercepts of the hyperbola $y^2 - (x - 1)^2 = 1$ are _____.

12. The eccentricity of the ellipse $9x^2 + y^2 = 1$ is _____.

13. If the graph of an ellipse is very elongated, then its eccentricity e is close to _____. (Fill in with 0 or 1.)

14. The line segment with endpoints on a hyperbola and lying on the line through the foci is called the _____.

15. The length of the minor axes of the ellipse $4x^2 + 9y^2 = 25$ is _____.

16. The function $y = 2 - 3\sqrt{x + 5}$ defines a portion of the graph of a conic section. The conic section is a(n)_____. Describe the portion of the graph defined by this function: _____.

17. An equation of the directrix for the parabola $y^2 = -2x$ is _____.

18. Because $B^2 - 4AC$ _____ (Fill in with <0, $=0$, or >0), the conic $3x^2 - xy - y^2 + 1 = 0$ is a _____.

19. A horizontal line through the focus intersects the graph of the parabola $(x - 1)^2 = 16y$ at the points _____.

20. The center of a hyperbola with asymptotes $y = -\frac{5}{4}x + \frac{3}{2}$ and $y = \frac{5}{4}x + \frac{13}{2}$ is _____.

B. True/False

In Problems 1–20, answer true or false.

1. The axis of the parabola $x^2 = -4y$ is vertical. _____

2. The foci of an ellipse lie on its graph. _____

3. After simplifying the squares of $x = \sqrt{y^2 - 5}$ and $y = \sqrt{5 + x^2}$ we obtain the hyperbola $y^2 - x^2 = 5$. Therefore the graphs of all three equations are the same. _____

4. The minor axis of an ellipse bisects the major axis. _____

5. The point $(-2, 5)$ is on the ellipse $x^2/8 + y^2/50 = 1$. _____

6. The graphs of $y = x^2$ and $y^2 - x^2 = 1$ have at most two points in common. _____

7. The eccentricity of the hyperbola $x^2 - y^2 = 1$ is $\sqrt{2}$. _____

8. For an ellipse, the length of the major axis is always greater than the length of the minor axis. _____

9. The vertex and focus are both on the axis of symmetry of a parabola. _____

10. The asymptotes for $(x - h)^2/a^2 - (y - k)^2/b^2 = 1$ must pass through (h, k). _____

11. An ellipse with eccentricity $e = 0.01$ is nearly circular. _____

12. The transverse axis of the hyperbola $x^2/9 - y^2/49 = 1$ is vertical. _____

13. The two hyperbolas $x^2 - y^2/25 = 1$ and $y^2/25 - x^2 = 1$ have the same pair of slant asymptotes. _____

14. The major axis of the ellipse $4(x + 1)^2 + 25(y - 3)^2 = 100$ lies on the line $y = 3$. _____

15. If P is a point on a parabola, then the perpendicular distance between P and the directrix equals the distance between P and the vertex. _____

16. If $y = \pm 5x$ are the asymptotes of a hyperbola, its center is necessarily $(0, 0)$. _____

17. The graph of $x^2/a^2 - y^2/b^2 = 1$ cannot cross its asymptotes $y = \pm bx/a$. _____

18. If $y = 3x + 8$ is an asymptote of a hyperbola, then the slope of the other asymptote is $m = -3$. _____

19. The graph of $x^2 - 5y^2 = 5$ is symmetric about the x-axis, the y-axis, and the origin. _____

20. The xy-term can be eliminated from the equation $4x^2 + \sqrt{3}xy + 3y^2 = 1$ by a rotation of axes through the angle $\theta = 30°$. _____

C. Review Exercises

In Problems 1–4, find the vertex, focus, directrix, and axis of the given parabola. Graph the parabola.

1. $(y - 3)^2 = -8x$

2. $8(x + 4)^2 = y - 2$

3. $x^2 - 2x + 4y + 1 = 0$

4. $y^2 + 10y + 8x + 41 = 0$

In Problems 5–8, find an equation of the parabola that satisfies the given conditions.

5. Focus $(1, -3)$, directrix $y = -7$

6. Focus $(3, -1)$, vertex $(0, -1)$

7. Vertex $(1, 2)$, vertical axis, passing through $(4, 5)$

8. Vertex $(-1, -4)$, directrix $x = 2$

In Problems 9–12, find the center, vertices, and foci of the given ellipse. Graph the ellipse.

9. $\dfrac{x^2}{3} + \dfrac{(y + 5)^2}{25} = 1$

10. $\dfrac{(x - 2)^2}{16} + \dfrac{(y + 5)^2}{4} = 1$

11. $4x^2 + y^2 + 8x - 6y + 9 = 0$

12. $5x^2 + 9y^2 - 20x + 54y + 56 = 0$

In Problems 13–16, find an equation of the ellipse that satisfies the given conditions.

13. Endpoints of minor axis $(0, \pm 4)$, foci $(\pm 5, 0)$

14. Foci $\left(2, -1 \pm \sqrt{2}\right)$, one vertex $\left(2, -1 + \sqrt{6}\right)$

15. Vertices $(\pm 2, -2)$, passing through $\left(1, -2 + \frac{1}{2}\sqrt{3}\right)$

16. Center $(2, 4)$, one focus $(2, 1)$, one vertex $(2, 0)$

In Problems 17–20, find the center, vertices, foci, and asymptotes of the given hyperbola. Graph the hyperbola.

17. $(x - 1)(x + 1) = y^2$

18. $y^2 - \dfrac{(x + 3)^2}{4} = 1$

19. $9x^2 - y^2 - 54x - 2y + 71 = 0$

20. $16y^2 - 9x^2 - 64y - 80 = 0$

In Problems 21–24, find an equation of the hyperbola that satisfies the given conditions.

21. Center $(0, 0)$, one vertex $(6, 0)$, and one focus $(8, 0)$

22. Foci $(2, \pm 3)$, one vertex $\left(2, -\frac{3}{2}\right)$

23. Foci $(\pm 2\sqrt{5}, 0)$, asymptotes $y = \pm 2x$

24. Vertices $(-3, 2)$ and $(-3, 4)$, one focus $\left(-3, 3 + \sqrt{2}\right)$

In Problems 25 and 26, perform a suitable rotation of axes so that the resulting $x'y'$-equation has no $x'y'$-term. Sketch the graph.

25. $xy = -8$

26. $8x^2 - 4xy + 5y^2 = 36$

27. Find an equation of the ellipse when the center of $4x^2 + y^2 = 4$ is rigidly translated to the point $(-5, 2)$.

28. Carefully describe the graphs of the given functions.
  (a) $f(x) = \sqrt{36 - 9x^2}$ **(b)** $f(x) = -\sqrt{36 + 9x^2}$

29. Distance from a Satellite A satellite orbits the planet Neptune in an elliptical orbit with the center of the planet at one focus. If the length of the major axis of the orbit is 2×10^9 m and the length of the minor axis is 6×10^8 m, find the maximum distance between the satellite and the center of the planet.

30. Find an equation of the ellipse (in red) inscribed in the auxiliary rectangle (dashed black) of the hyperbola shown in FIGURE 7.R.1.

31. Mirror, Mirror. . . A parabolic mirror has a depth of 7 cm at its center and the distance across the top of the mirror is 20 cm. Find the distance from the vertex to the focus.

32. Identify the conic section that appears in the drawing of a wooden pencil in FIGURE 7.R.2.

33. Find an equation of a sphere that has a diameter with endpoints $(0, -4, 7)$ and $(2, 12, -3)$.

Parabolic mirror in Problem 31

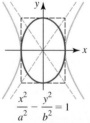

FIGURE 7.R.1 Graphs for Problem 30

FIGURE 7.R.2 Pencil in Problem 32

34. Find an equation of the plane such that the points (x, y, z) on the plane are equidistant from $(1, -2, 3)$ and $(2, 5, -1)$.

35. Find an equation of the parabola (blue) and of the ellipse (red) that have a common vertex $(6, 0)$ and common focus $(4, 0)$ as shown in FIGURE 7.R.3. Explain why this information does not determine a unique hyperbola (green). Find equations of two different hyperbolas that have a vertex $(6, 0)$ and a focus $(4, 0)$.

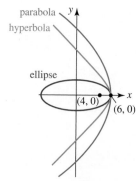

FIGURE 7.R.3 Graphs for Problem 35

8 Polar Coordinates

Chapter Outline

8.1 The Polar Coordinate System

≡ **Introduction** So far we have used the rectangular coordinate system to specify a point P in the plane. We can regard this system as a grid of horizontal and vertical lines. The coordinates (a, b) of a point P are determined by the intersection of two lines: one line $x = a$ is perpendicular to the horizontal reference line called the x-axis, and the other $y = b$ is perpendicular to the vertical reference line called the y-axis. See FIGURE 8.1.1(a). Another system for locating points in the plane is the **polar coordinate system**.

□ **Terminology** To set up a **polar coordinate system**, we use a system of circles centered at a point O, called the **pole**, and straight lines or rays emanating from O. We take as a reference axis a horizontal half-line directed to the right of the pole and call it the **polar axis**. By specifying a directed (signed) distance r from O and an angle θ whose initial side is the polar axis and whose terminal side is the ray OP, we label the point P by (r, θ). We say that the ordered pair (r, θ) are the **polar coordinates** of P. See Figures 8.1.1(b) and 8.1.1(c) where we have assumed $r > 0$.

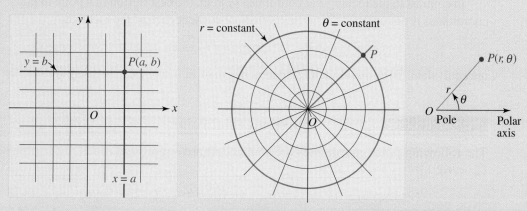

(a) Rectangular coordinate system   (b) Polar coordinate system   (c) Polar coordinates of P

FIGURE 8.1.1 Comparison of rectangular and polar coordinates of a point P

Although the measure of the angle θ can be either in degrees or radians, in calculus radian measure is used almost exclusively. Consequently, we shall use only radian measure in this discussion.

In the polar coordinate system we adopt the following conventions.

DEFINITION 8.1.1 Conventions in Polar Coordinates

(*i*) Angles $\theta > 0$ are measured counterclockwise from the polar axis, whereas angles $\theta < 0$ are measured clockwise.

(*ii*) To graph a point (r, θ), where $r < 0$, measure $|r|$ units along the ray $\theta + \pi$.

(*iii*) The coordinates of the pole O are $(0, \theta)$, where θ is any angle.

EXAMPLE 1   Plotting Polar Points

Plot the points whose polar coordinates are given.
(a) $(4, \pi/6)$ **(b)** $(2, -\pi/4)$ **(c)** $(-3, 3\pi/4)$

Solution **(a)** Measure 4 units along the ray $\pi/6$ as shown in FIGURE 8.1.2(a).

(b) Measure 2 units along the ray $-\pi/4$. See Figure 8.1.2(b).

(c) Measure 3 units along the ray $3\pi/4 + \pi = 7\pi/4$. Equivalently, we can measure 3 units along the ray $3\pi/4$ extended *backward* through the pole. Note carefully in Figure 8.1.2(c) that the point $(-3, 3\pi/4)$ is not in the same quadrant as the terminal side of the given angle.

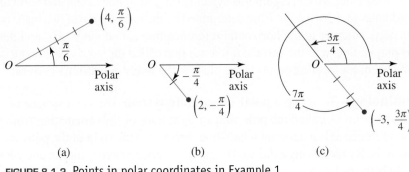

(a)   (b)   (c)

FIGURE 8.1.2 Points in polar coordinates in Example 1

In contrast to the rectangular coordinate system, the description of a point in polar coordinates is not unique. This is an immediate consequence of the fact that

$$(r, \theta) \quad \text{and} \quad (r, \theta + 2n\pi), \quad n \text{ an integer,}$$

are equivalent. To compound the problem, negative values of r can be used.

■ EXAMPLE 2   Equivalent Polar Points

The following polar coordinates are some alternative representations of the point $(2, \pi/6)$:

$$(2, 13\pi/6), \quad (2, -11\pi/6), \quad (-2, 7\pi/6), \quad (-2, -5\pi/6).$$

□ **Conversion of Polar Coordinates to Rectangular** By superimposing a rectangular coordinate system on a polar coordinate system, as shown in **FIGURE 8.1.3**, we can convert a polar description of a point to rectangular coordinates by using

$$x = r\cos\theta, \quad y = r\sin\theta. \tag{1}$$

These conversion formulas hold true for any values of r and θ in an equivalent polar representation of (r, θ).

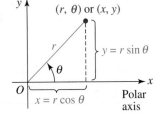

FIGURE 8.1.3 Relating polar and rectangular coordinates

■ EXAMPLE 3   Polar to Rectangular

Convert $(2, \pi/6)$ in polar coordinates to rectangular coordinates.

Solution With $r = 2, \theta = \pi/6$, we have from (1),

$$x = 2\cos\frac{\pi}{6} = 2\left(\frac{\sqrt{3}}{2}\right) = \sqrt{3}$$

$$y = 2\sin\frac{\pi}{6} = 2\left(\frac{1}{2}\right) = 1.$$

Thus, $(2, \pi/6)$ is equivalent to $\left(\sqrt{3}, 1\right)$ in rectangular coordinates.

☐ **Conversion of Rectangular Coordinates to Polar** It should be evident from Figure 8.1.3 that x, y, r, and θ are also related by

$$r^2 = x^2 + y^2, \quad \tan\theta = \frac{y}{x}. \tag{2}$$

The equations in (2) are used to convert the rectangular coordinates (x, y) to the polar coordinates (r, θ).

▮ EXAMPLE 4   Rectangular to Polar

Convert $(-1, 1)$ in rectangular coordinates to polar coordinates.

Solution With $x = -1$, $y = 1$, we have from (2)

$$r^2 = 2 \quad \text{and} \quad \tan\theta = -1.$$

Now, $r^2 = 2$ or $r = \pm\sqrt{2}$, and two of many angles that satisfy $\tan\theta = -1$ are $3\pi/4$ and $7\pi/4$. From FIGURE 8.1.4 we see that two polar representations for $(-1, 1)$ are $(\sqrt{2}, 3\pi/4)$ and $(-\sqrt{2}, 7\pi/4)$. ≡

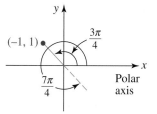

FIGURE 8.1.4 Point in Example 4

In Example 4, observe that we cannot pair just *any* angle θ and *any* value r that satisfy (2); these solutions must also be consistent with (1). Because the points $(-\sqrt{2}, 3\pi/4)$ and $(\sqrt{2}, 7\pi/4)$ lie in the fourth quadrant, they are not polar representations of the second-quadrant point $(-1, 1)$.

There are instances in calculus when a rectangular equation must be expressed as a polar equation $r = f(\theta)$. The next example shows how to do this using the conversion formulas in (1).

▮ EXAMPLE 5   Rectangular Equation to Polar Equation

Find a polar equation that has the same graph as the circle $x^2 + y^2 = 8x$.

Solution Substituting $x = r\cos\theta$, $y = r\sin\theta$, into the given equation we find

$$\begin{aligned}
r^2\cos^2\theta + r^2\sin^2\theta &= 8r\cos\theta \\
r^2(\cos^2\theta + \sin^2\theta) &= 8r\cos\theta \qquad \leftarrow \cos^2\theta + \sin^2\theta = 1 \\
r(r - 8\cos\theta) &= 0.
\end{aligned}$$

The last equation implies that

$$r = 0 \quad \text{or} \quad r = 8\cos\theta.$$

Since $r = 0$ determines only the pole O, we conclude that a polar equation of the circle is $r = 8\cos\theta$. Note that the circle $x^2 + y^2 = 8x$ passes through the origin since $x = 0$ and $y = 0$ satisfy the equation. Relative to the polar equation $r = 8\cos\theta$ of the circle, the origin or pole corresponds to the polar coordinates $(0, \pi/2)$. ≡

▮ EXAMPLE 6   Rectangular Equation to Polar Equation

Find a polar equation that has the same graph as the parabola $x^2 = 8(2 - y)$.

Solution We replace x and y in the given equation by $x = r\cos\theta$, $y = r\sin\theta$ and solve for r in terms of θ:

$$\begin{aligned}
r^2\cos^2\theta &= 8(2 - r\sin\theta) \\
r^2(1 - \sin^2\theta) &= 16 - 8r\sin\theta \\
r^2 &= r^2\sin^2\theta - 8r\sin\theta + 16 \qquad \leftarrow \left\{\begin{array}{l}\text{right side is a}\\\text{perfect square}\end{array}\right. \\
r^2 &= (r\sin\theta - 4)^2 \\
r &= \pm(r\sin\theta - 4).
\end{aligned}$$

Solving for r gives two equations,

$$r = \frac{4}{1 + \sin\theta} \qquad \text{or} \qquad r = \frac{-4}{1 - \sin\theta}.$$

Now recall that, by convention (ii) in Definition 8.1.1, (r, θ) and $(-r, \theta + \pi)$ represent the same point. You should verify that if (r, θ) is replaced by $(-r, \theta + \pi)$ in the second of these two equations, we obtain the first equation. In other words, the equations are equivalent and so we may simply take the polar equation of the parabola to be $r = 4/(1 + \sin\theta)$.

≡

EXAMPLE 7   Polar Equation to Rectangular Equation

Find a rectangular equation that has the same graph as the polar equation $r^2 = 9\cos 2\theta$.

Solution First, we use the trigonometric identity for the cosine of a double angle:

$$r^2 = 9(\cos^2\theta - \sin^2\theta). \qquad \leftarrow \cos 2\theta = \cos^2\theta - \sin^2\theta$$

Then, from $r^2 = x^2 + y^2$, $\cos\theta = x/r$, $\sin\theta = y/r$, we have

$$x^2 + y^2 = 9\left(\frac{x^2}{x^2 + y^2} - \frac{y^2}{x^2 + y^2}\right) \quad \text{or} \quad (x^2 + y^2)^2 = 9(x^2 - y^2).$$

≡

The next section will be devoted to graphing polar equations.

8.1   Exercises   Answers to selected odd-numbered problems begin on page ANS-23.

In Problems 1–6, plot the point with the given polar coordinates.

1. $(3, \pi)$
2. $(2, -\pi/2)$
3. $\left(-\frac{1}{2}, \pi/2\right)$
4. $(-1, \pi/6)$
5. $(-4, -\pi/6)$
6. $\left(\frac{2}{3}, 7\pi/4\right)$

In Problems 7–14, find alternative polar coordinates that satisfy

(a) $r > 0, \theta < 0$
(b) $r > 0, \theta > 2\pi$
(c) $r < 0, \theta > 0$
(d) $r < 0, \theta < 0$

for each point with the given polar coordinates.

7. $(2, 3\pi/4)$
8. $(5, \pi/2)$
9. $(4, \pi/3)$
10. $(3, \pi/4)$
11. $(1, \pi/6)$
12. $(3, 7\pi/6)$
13. $(9, 3\pi/2)$
14. $(5, \pi)$

In Problems 15–24, find the rectangular coordinates for each point with the given polar coordinates.

15. $\left(\frac{1}{2}, 2\pi/3\right)$
16. $(-1, 7\pi/4)$
17. $(-6, -\pi/3)$
18. $\left(\sqrt{2}, 11\pi/6\right)$
19. $(4, 5\pi/4)$
20. $(-5, \pi/2)$
21. $(-1, -5\pi/6)$
22. $(10, -4\pi/3)$
23. $(4, \pi/8)$
24. $(-8, 5\pi/12)$

In Problems 25–32, find polar coordinates that satisfy

(a) $r > 0, -\pi < \theta \le \pi$ (b) $r < 0, -\pi < \theta \le \pi$

for each point with the given rectangular coordinates.

25. $(-2, -2)$ **26.** $(0, -4)$
27. $(1, -\sqrt{3})$ **28.** $(\sqrt{6}, \sqrt{2})$
29. $(7, 0)$ **30.** $(1, 2)$
31. $(-3, 4)$ **32.** $(1, -1)$

In Problems 33–38, sketch the region on the plane that consists of points (r, θ) whose polar coordinates satisfy the given conditions.

33. $2 \le r < 4, 0 \le \theta \le \pi$ **34.** $2 < r \le 4$
35. $0 \le r \le 2, -\pi/2 \le \theta \le \pi/2$ **36.** $r \ge 0, \pi/4 < \theta < 3\pi/4$
37. $-1 \le r \le 1, 0 \le \theta \le \pi/2$ **38.** $-2 \le r < 4, \pi/3 \le \theta \le \pi$

In Problems 39–50, find a polar equation that has the same graph as the given rectangular equation.

39. $y = 5$ **40.** $x + 1 = 0$
41. $y = 7x$ **42.** $3x + 8y + 6 = 0$
43. $y^2 = -4x + 4$ **44.** $x^2 - 12y - 36 = 0$
45. $x^2 + y^2 = 36$ **46.** $x^2 - y^2 = 1$
47. $x^2 + y^2 + x = \sqrt{x^2 + y^2}$ **48.** $x^3 + y^3 - xy = 0$
49. $x^2 + y^2 = 8x$ **50.** $2xy = 5$

In Problems 51–62, find a rectangular equation that has the same graph as the given polar equation.

51. $r = 2\sec\theta$ **52.** $r\cos\theta = -4$
53. $r = 6\sin 2\theta$ **54.** $2r = \tan\theta$
55. $r^2 = 4\sin 2\theta$ **56.** $r^2\cos 2\theta = 16$
57. $r + 5\sin\theta = 0$ **58.** $r = 2 + \cos\theta$

59. $r = \dfrac{2}{1 + 3\cos\theta}$ **60.** $r(4 - \sin\theta) = 10$

61. $r = \dfrac{5}{3\cos\theta + 8\sin\theta}$ **62.** $r = 3 + 3\sec\theta$

For Discussion

63. How would you express the distance d between two points (r_1, θ_1) and (r_2, θ_2) in terms of their polar coordinates?

64. You know how to find a rectangular equation of a line through two points with rectangular coordinates. How would you find a polar equation of a line through two points with polar coordinates (r_1, θ_1) and (r_2, θ_2)? Carry out your ideas by finding a polar equation of the line through $(3, 3\pi/4)$ and $(1, \pi/4)$. Find the polar coordinates of the x- and y-intercepts of the line.

65. In rectangular coordinates the x-intercepts of the graph of a function $y = f(x)$ are determined from the solutions of the equation $f(x) = 0$. In the next section we will graph polar equations $r = f(\theta)$. What is the significance of the solutions of the equation $f(\theta) = 0$?

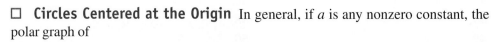

8.2 Graphs of Polar Equations

☰ **Introduction** The graph of a polar equation $r = f(\theta)$ is the set of points P with *at least* one set of polar coordinates that satisfies the equation. Since it is most likely that your classroom does not have a polar coordinate grid, to facilitate graphing and discussion of graphs of a polar equation $r = f(\theta)$, we will, as in the preceding section, superimpose a rectangular coordinate system over the polar coordinate system.

We begin with some simple polar graphs.

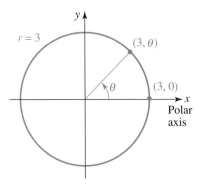

FIGURE 8.2.1 Circle in Example 1

■ EXAMPLE 1   A Circle Centered at the Origin

Graph $r = 3$.

Solution Since θ is not specified, the point $(3, \theta)$ lies on the graph of $r = 3$ for any value of θ and is 3 units from the origin. We see in **FIGURE 8.2.1** that the graph is the circle of radius 3 centered at the origin.

Alternatively, we know from (2) of Section 8.1 that $r = \pm\sqrt{x^2 + y^2}$ so that $r = 3$ yields the familiar rectangular equation $x^2 + y^2 = 3^2$ of a circle of radius 3 centered at the origin. ☰

□ **Circles Centered at the Origin** In general, if a is any nonzero constant, the polar graph of

$$r = a \qquad (1)$$

is a circle of radius $|a|$ with center at the origin.

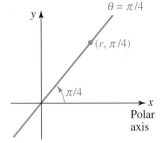

FIGURE 8.2.2 Line in Example 2

■ EXAMPLE 2   A Ray Through the Origin

Graph $\theta = \pi/4$.

Solution Since r is not specified, the point $(r, \pi/4)$ lies on the graph for any value of r. If $r > 0$, then this point lies on the half-line in the first quadrant; if $r < 0$, then the point lies on the half-line in the third quadrant. For $r = 0$, the point $(0, \pi/4)$ is the pole or origin. Therefore, the polar graph of $\theta = \pi/4$ is the line through the origin that makes an angle of $\pi/4$ with the polar axis or positive x-axis. See **FIGURE 8.2.2**. ☰

□ **Lines Through the Origin** In general, if α is any real constant, the polar graph of

$$\theta = \alpha \qquad (2)$$

is a line through the origin that makes an angle of α radians with the polar axis. Lines described by (2) are called **radial lines**.

■ EXAMPLE 3   A Spiral

Graph $r = \theta$.

Solution As $\theta \geq 0$ increases, r increases and the points (r, θ) wind around the pole in a counterclockwise manner. This is illustrated by the blue portion of the graph in **FIGURE 8.2.3**. The red portion of the graph is obtained by plotting points for $\theta < 0$.

CHAPTER 8 POLAR COORDINATES

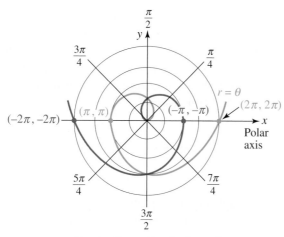

FIGURE 8.2.3 Graph of equation in Example 3

□ **Spirals** Many graphs in polar coordinates are given special names. The graph in Example 3 is a special case of

$$r = a\theta, \tag{3}$$

where a is a constant. A graph of this equation is called a **spiral of Archimedes**. You are asked to graph other types of spiral curves in Problems 31 and 32 in Exercises 8.2.

In addition to basic point plotting, symmetry can often be utilized to graph a polar equation.

□ **Symmetry** As shown in FIGURE 8.2.4, a polar graph can have three types of symmetry. A polar graph is **symmetric with respect to the y-axis** if whenever (r, θ) is a point on the graph, $(r, \pi - \theta)$ is also a point on the graph. A polar graph is **symmetric with respect to the x-axis** if whenever (r, θ) is a point on the graph, $(r, -\theta)$ is also a point on the graph. Finally, a polar graph is **symmetric with respect to the origin** if whenever (r, θ) is on the graph, $(-r, \theta)$ is also a point on the graph.

Symmetries of a snowflake

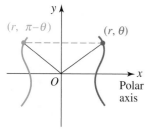

(a) Symmetry with respect to y-axis

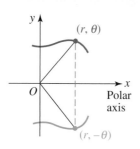

(b) Symmetry with respect to x-axis

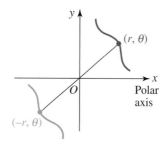

(c) Symmetry with respect to origin

FIGURE 8.2.4 Symmetries of a polar graph

We have the following tests for symmetries in polar coordinates.

THEOREM 8.2.1   Tests for Symmetry

The graph of a polar equation is:

   (*i*) **symmetric with respect to the y-axis** if replacing (r, θ) by $(r, \pi - \theta)$ results in the same equation;
   (*ii*) **symmetric with respect to the x-axis** if replacing (r, θ) by $(r, -\theta)$ results in the same equation;
   (*iii*) **symmetric with respect to the origin** if replacing (r, θ) by $(-r, \theta)$ results in the same equation.

Because the polar description of a point is not unique, the graph of a polar equation may still have a particular type of symmetry even though the test for it may fail. For example, if replacing (r, θ) by $(r, -\theta)$ fails to give the original polar equation, the graph of that equation may still possess symmetry with respect to the x-axis. Therefore, if one of the replacement tests in (i)–(iii) of Theorem 8.2.1 fails to give the same polar equation, the best we can say is "no conclusion."

◼ EXAMPLE 4　　　　　Graphing a Polar Equation

Graph $r = 1 - \cos\theta$.

Solution One way of graphing this equation is to plot a few well-chosen points corresponding to $0 \le \theta \le 2\pi$. As the following table shows

θ	0	$\pi/4$	$\pi/2$	$3\pi/4$	π	$5\pi/4$	$3\pi/2$	$7\pi/4$	2π
r	0	0.29	1	1.71	2	1.71	1	0.29	0

as θ advances from $\theta = 0$ to $\theta = \pi/2$, r increases from $r = 0$ (the origin) to $r = 1$. See FIGURE 8.2.5(a). As θ advances from $\theta = \pi/2$ to $\theta = \pi$, r continues to increase from $r = 1$ to its maximum value of $r = 2$. See Figure 8.2.5(b). Then, for $\theta = \pi$ to $\theta = 3\pi/2$, r begins to decrease from $r = 2$ to $r = 1$. For $\theta = 3\pi/2$ to $\theta = 2\pi$, r continues to decrease and we end up again at the origin $r = 0$. See Figures 8.2.5(c) and 8.2.5(d).

By taking advantage of symmetry we could have simply plotted points for $0 \le \theta \le \pi$. From the trigonometric identity for the cosine function $\cos(-\theta) = \cos\theta$ it follows from (ii) of Theorem 8.2.1 that the graph of $r = 1 - \cos\theta$ is symmetric with respect to the x-axis. We can obtain the complete graph of $r = 1 - \cos\theta$ by reflecting in the x-axis that portion of the graph given in Figure 8.2.5(b).

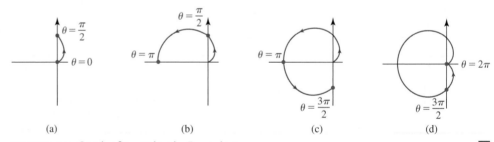

(a)　　　　　(b)　　　　　(c)　　　　　(d)

FIGURE 8.2.5 Graph of equation in Example 4　　　　　≡

☐ **Cardioids** The polar equation in Example 4 is a member of a family of equations that all have a "heart-shaped" graph that passes through the origin. A graph of any polar equation of the form

$$r = a \pm a\sin\theta \qquad \text{or} \qquad r = a \pm a\cos\theta \qquad (4)$$

is called a **cardioid**. The only difference in the graph of these four equations is their symmetry with respect to the y-axis ($r = a \pm a\sin\theta$) or symmetry with respect to the x-axis ($r = a \pm a\cos\theta$). See FIGURE 8.2.6 on page 421.

By knowing the basic shape and orientation of a cardioid, you can obtain a quick and accurate graph by plotting the four points corresponding to $\theta = 0$, $\theta = \pi/2$, $\theta = \pi$, and $\theta = 3\pi/2$.

$r = a(1 + \cos \theta)$

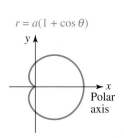

$r = a(1 - \cos \theta)$

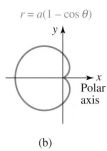

$r = a(1 + \sin \theta)$

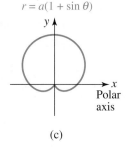

$r = a(1 - \sin \theta)$

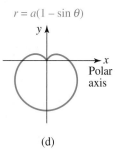

(a)   (b)   (c)   (d)

FIGURE 8.2.6 Cardioids

☐ **Limaçons** Cardioids are special cases of polar curves known as **limaçons**:

$$r = a \pm b\sin\theta \qquad \text{or} \qquad r = a \pm b\cos\theta. \qquad (5)$$

The shape of a limaçon depends on the relative magnitudes of a and b. Let us assume that $a > 0$ and $b > 0$. For $a/b < 1$, we get a **limaçon with an interior loop** as shown in FIGURE 8.2.7(a). When $a = b$ or equivalently $a/b = 1$ we get a **cardioid**. For $1 < a/b < 2$, we get a **dimpled limaçon** as shown in Figure 8.2.7(b). For $a/b \geq 2$, the curve is called a **convex limaçon**. See Figure 8.2.7(c).

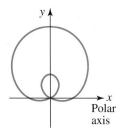

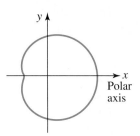

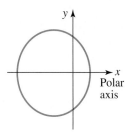

(a) Limaçon with interior loop   (b) Dimpled limaçon   (c) Convex limaçon

FIGURE 8.2.7 Three kinds of limaçons

EXAMPLE 5   A Limaçon

The graph of $r = 3 - \sin\theta$ is a convex limaçon, since $a = 3, b = 1$, and $a/b = 3 > 2$.   ☰

EXAMPLE 6   A Limaçon

The graph of $r = 1 + 2\cos\theta$ is a limaçon with an interior loop, since $a = 1, b = 2$, and $a/b = \frac{1}{2} < 1$. For $\theta \geq 0$, notice in FIGURE 8.2.8 the limaçon starts at $\theta = 0$ or $(3, 0)$. The graph passes through the y-axis at $(1, \pi/2)$ and then enters the origin $(r = 0)$ for the first angle for which $r = 0$ or $1 + 2\cos\theta = 0$ or $\cos\theta = -\frac{1}{2}$. This implies that $\theta = 2\pi/3$. At $\theta = \pi$, the curve passes through $(-1, \pi)$. The remainder of the graph can then be completed using the fact that it is symmetric with respect to the x-axis.   ☰

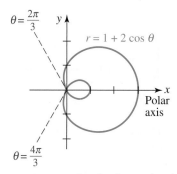

FIGURE 8.2.8 Graph of equation in Example 6

EXAMPLE 7   A Rose Curve

Graph $r = 2\cos 2\theta$.

Solution Since

$$\cos(-2\theta) = \cos 2\theta \qquad \text{and} \qquad \cos 2(\pi - \theta) = \cos 2\theta$$

we conclude by (*i*) and (*ii*) of the tests for symmetry in Theorem 8.2.1 that the graph is symmetric with respect to both the x- and the y-axes. A moment of reflection should

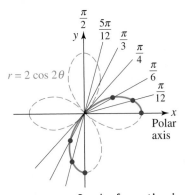

$r = 2 \cos 2\theta$

FIGURE 8.2.9 Graph of equation in Example 7

convince you that we need only consider $0 \le \theta \le \pi/2$. Using the data in the following table, we see that the dashed portion of the graph given in FIGURE 8.2.9 is that completed by symmetry. The graph is called a **rose curve with four petals**.

θ	0	$\pi/12$	$\pi/6$	$\pi/4$	$\pi/3$	$5\pi/12$	$\pi/2$
r	2	1.7	1	0	-1	-1.7	-2

☐ **Rose Curves** In general, if n is a positive integer, the graphs of

$$r = a \sin n\theta \qquad \text{or} \qquad r = a \cos n\theta, \quad n \ge 2 \tag{6}$$

are called **rose curves**, although as you can see in FIGURE 8.2.10 the curve looks more like a daisy. When n is odd, the number of **loops** or **petals** of the curve is n; if n is even, the curve has $2n$ petals. To graph a rose curve we can start by graphing one petal. To begin, we find an angle θ for which r is a maximum. This gives the center line of the petal. We then find corresponding values of θ for which the rose curve enters the origin ($r = 0$). To complete the graph we use the fact that the center lines of the petals are spaced $2\pi/n$ radians ($360/n$ degrees) apart if n is odd, and $2\pi/2n = \pi/n$ radians ($180/n$ degrees) apart if n is even. In Figure 8.2.10 we have drawn the graph of $r = a \sin 5\theta$, $a > 0$. The spacing between the center lines of the five petals is $2\pi/5$ radians ($72°$).

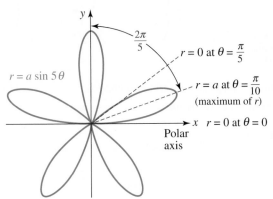

$r = a \sin 5\theta$

$r = 0$ at $\theta = \dfrac{\pi}{5}$

$r = a$ at $\theta = \dfrac{\pi}{10}$ (maximum of r)

$r = 0$ at $\theta = 0$

FIGURE 8.2.10 Rose curve with five petals

In Example 5 in Section 8.1 we saw that the polar equation $r = 8\cos\theta$ is equivalent to the rectangular equation $x^2 + y^2 = 8x$. By completing the square in x in the rectangular equation, we recognize

$$(x - 4)^2 + y^2 = 16$$

as a circle of radius 4 centered at $(4, 0)$ on the x-axis. Polar equations such as $r = 8\cos\theta$ or $r = 8\sin\theta$ are circles and are also special cases of rose curves. See FIGURE 8.2.11.

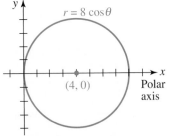

$r = 8\cos\theta$

$(4, 0)$

FIGURE 8.2.11 Graph of polar equation $r = 8\cos\theta$

☐ **Circles with Centers on an Axis** When $n = 1$ in (6) we get

$$r = a \sin\theta \qquad \text{or} \qquad r = a \cos\theta, \tag{7}$$

which are polar equations of circles passing through the origin with diameter $|a|$ with centers $(a/2, 0)$ on the x-axis ($r = a\cos\theta$), or with centers $(0, a/2)$ on the y-axis ($r = a\sin\theta$). FIGURE 8.2.12 illustrates the graphs of the equations in (7) in the cases when $a > 0$ and $a < 0$.

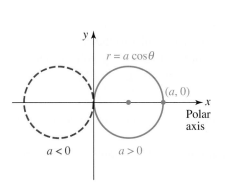

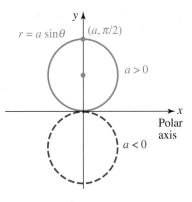

(a) Centers on the x-axis

(b) Centers on the y-axis

FIGURE 8.2.12 Circles through the origin with centers on an axis

☐ **Lemniscates** If n is a positive integer, the graphs of

$$r^2 = a\cos 2\theta \qquad \text{or} \qquad r^2 = a\sin 2\theta \qquad (8)$$

where $a > 0$, are called **lemniscates**. By (*iii*) of the tests for symmetry you can see the graphs of both of the equations in (8) are symmetric with respect to the origin. Moreover, by (*ii*) of the tests for symmetry the graph of $r^2 = a\cos 2\theta$ is symmetric with respect to the x-axis. FIGURES 8.2.13(a) and 8.2.13(b) show typical graphs of the equations $r^2 = a\cos 2\theta$ and $r^2 = a\sin 2\theta$, respectively.

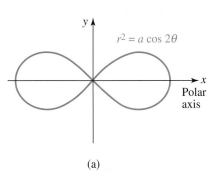

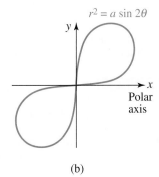

(a)

(b)

FIGURE 8.2.13 Lemniscates

☐ **Points of Intersection** In rectangular coordinates we can find the points (x, y) where the graphs of two functions $y = f(x)$ and $y = g(x)$ intersect by equating the y-values. The real solutions of the equation $f(x) = g(x)$ correspond to *all* the x-coordinates of the points where the graphs intersect. In contrast, problems may arise in polar coordinates when we try the same method to determine where the graphs of two polar equations $r = f(\theta)$ and $r = g(\theta)$ intersect.

EXAMPLE 8   Intersecting Circles

FIGURE 8.2.14 shows that the circles $r = \sin\theta$ and $r = \cos\theta$ have two points of intersection. By equating the r values, the equation $\sin\theta = \cos\theta$ leads to $\theta = \pi/4$. Substituting this value into either equation yields $r = \sqrt{2}/2$. Thus we have found only a single polar point $\left(\sqrt{2}/2, \pi/4\right)$ where the graphs intersect. From the figure, it is apparent that the graphs also intersect at the origin. But the problem here is that the origin or pole is $(0, \pi/2)$ on the graph of $r = \cos\theta$ but is $(0, 0)$ on the graph of $r = \sin\theta$. This situation is analogous to the curves reaching the same point at different times. ≡

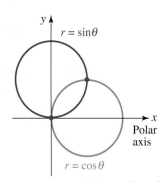

FIGURE 8.2.14 Intersecting circles in Example 8

□ **Rotation of Polar Graphs** In Section 2.2 we saw that if $y = f(x)$ is the rectangular equation of a function, then the graphs of $y = f(x - c)$ and $y = f(x + c)$, $c > 0$, are obtained by *shifting* the graph of f horizontally c units to the right and to the left, respectively. In contrast, if $r = f(\theta)$ is a polar equation, then the graphs of $f(\theta - \gamma)$ and $f(\theta + \gamma)$, where $\gamma > 0$, can be obtained by *rotating* the graph of f by an amount γ. Specifically:

- The graph of $r = f(\theta - \gamma)$ is the graph of $r = f(\theta)$ rotated *counterclockwise* about the origin by an amount γ.
- The graph of $r = f(\theta + \gamma)$ is the graph of $r = f(\theta)$ rotated *clockwise* about the origin by an amount γ.

For example, the graph of the cardioid $r = a(1 + \cos\theta)$ is shown in Figure 8.2.6(a). The graph of $r = a(1 + \cos(\theta - \pi/2))$ is the graph of $r = a(1 + \cos\theta)$ rotated counterclockwise about the origin by an amount $\pi/2$. Its graph then must be that given in Figure 8.2.6(c). This makes sense, because the difference formula of the cosine gives

See the identity in (5) of Section 4.6. ▶ $$r = a[1 + \cos(\theta - \pi/2)] = a[1 + \cos\theta\cos(\pi/2) + \sin\theta\sin(\pi/2)] = a(1 + \sin\theta).$$

Similarly, rotating $r = a(1 + \cos\theta)$ counterclockwise about the origin by an amount π gives the equation

$$r = a[1 + \cos(\theta - \pi)] = a[1 + \cos\theta\cos\pi + \sin\theta\sin\pi] = a(1 - \cos\theta)$$

whose graph is given in Figure 8.2.6(b). As another example, take a look again at Figure 8.2.13. From

$$r^2 = a\cos 2\left(\theta - \frac{\pi}{4}\right) = a\cos\left(2\theta - \frac{\pi}{2}\right) = a\sin 2\theta$$

we see that the graph of the lemniscate in Figure 8.2.13(b) is the graph in Figure 8.2.13(a) rotated counterclockwise about the origin by an amount $\pi/4$.

■ **EXAMPLE 9**   **Rotated Polar Graphs**

Graph $r = 1 + 2\sin(\theta + \pi/4)$.

Solution The graph of the given equation is the graph of the limaçon $r = 1 + 2\sin\theta$ rotated clockwise about the origin by an amount $\pi/4$. In **FIGURE 8.2.15** the blue graph is that of $r = 1 + 2\sin\theta$ and the red graph is the rotated graph. ≡

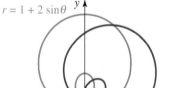

$r = 1 + 2\sin\theta$

FIGURE 8.2.15 Graphs of polar equations in Example 9

NOTES FROM THE CLASSROOM

(*i*) Example 8 illustrates one of several frustrating difficulties of working in polar coordinates:

A point can be on the graph of a polar equation even though its coordinates do not satisfy the equation.

You should verify that $(2, \pi/2)$ is an alternative polar description of the point $(-2, 3\pi/2)$. Moreover, verify that $(-2, 3\pi/2)$ is a point on the graph of $r = 1 + 3\sin\theta$ by showing that the coordinates satisfy the equation. However, note that the alternative coordinates $(2, \pi/2)$ do not satisfy the equation.

(*ii*) The four-petal rose curve in Example 7 is obtained by plotting r for θ-values satisfying $0 \le \theta \le 2\pi$. See **FIGURE 8.2.16**. Do not assume this is true for every rose curve. Indeed, the five-petal rose curve discussed in Figure 8.2.10 is obtained using θ-values satisfying $0 \leq \theta \leq \pi$. In general, a rose curve $r = a \sin n\theta$ or $r = a \cos n\theta$ is traced out exactly once for $0 \leq \theta \leq 2\pi$ if n is even and once for $0 \leq \theta \leq \pi$ if n is odd.

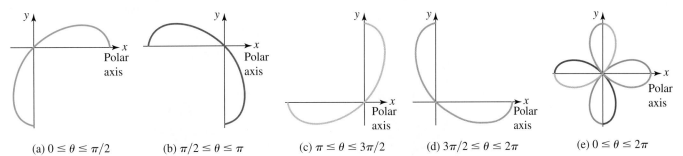

(a) $0 \leq \theta \leq \pi/2$ (b) $\pi/2 \leq \theta \leq \pi$ (c) $\pi \leq \theta \leq 3\pi/2$ (d) $3\pi/2 \leq \theta \leq 2\pi$ (e) $0 \leq \theta \leq 2\pi$

FIGURE 8.2.16 Plotting $r = 2\cos 2\theta$

8.2 Exercises Answers to selected odd-numbered problems begin on page ANS-24.

In Problems 1–30, identify by name the graph of the given polar equation. Then sketch the graph of the equation.

1. $r = 6$
2. $r = -1$
3. $\theta = \pi/3$
4. $\theta = 5\pi/6$
5. $r = 2\theta, \theta \leq 0$
6. $r = 3\theta, \theta \geq 0$
7. $r = 1 + \cos\theta$
8. $r = 5 - 5\sin\theta$
9. $r = 2(1 + \sin\theta)$
10. $2r = 1 - \cos\theta$
11. $r = 1 - 2\cos\theta$
12. $r = 2 + 4\sin\theta$
13. $r = 4 - 3\sin\theta$
14. $r = 3 + 2\cos\theta$
15. $r = 4 + \cos\theta$
16. $r = 4 - 2\sin\theta$
17. $r = \sin 2\theta$
18. $r = 3\sin 4\theta$
19. $r = 3\cos 3\theta$
20. $r = 2\sin 3\theta$
21. $r = \cos 5\theta$
22. $r = 2\sin 9\theta$
23. $r = 6\cos\theta$
24. $r = -2\cos\theta$
25. $r = -3\sin\theta$
26. $r = 5\sin\theta$
27. $r^2 = 4\sin 2\theta$
28. $r^2 = 4\cos 2\theta$
29. $r^2 = -25\cos 2\theta$
30. $r^2 = -9\sin 2\theta$

In Problems 31 and 32, the graph of the given equation is a spiral. Sketch its graph.

31. $r = 2^\theta, \theta \geq 0$ (logarithmic)   **32.** $r\theta = \pi, \theta > 0$ (hyperbolic)

In Problems 33–38, find an equation of the given polar graph.

33.

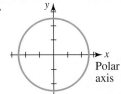

FIGURE 8.2.17 Graph for Problem 33

34.

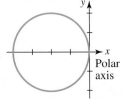

FIGURE 8.2.18 Graph for Problem 34

35.

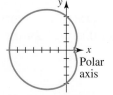

FIGURE 8.2.19 Graph for
Problem 35

36.

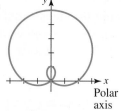

FIGURE 8.2.20 Graph for
Problem 36

37.

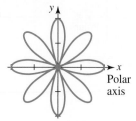

FIGURE 8.2.21 Graph for
Problem 37

38.

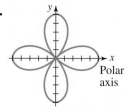

FIGURE 8.2.22 Graph for
Problem 38

In Problems 39–42, find the points of intersection of the graphs of the given pair of polar equations.

39. $r = 2, r = 4\sin\theta$

40. $r = \sin\theta, r = \sin 2\theta$

41. $r = 1 - \cos\theta, r = 1 + \cos\theta$

42. $r = 3 - 3\cos\theta, r = 3\cos\theta$

43. Suppose the red circle in Figure 8.2.14 is rotated clockwise about the origin by an amount $\pi/4$. Show that a polar equation of the rotated graph is given by

$$r = \frac{\sqrt{2}}{2}(\sin\theta + \cos\theta).$$

44. (a) Sketch the circle whose equation is given in Problem 43.
  (b) Find the polar coordinates of all intercepts.
  (c) Find the rectangular coordinates of the center of the circle.
  (d) Find the rectangular coordinates of the points at the end of the diameter that passes through the origin and the center.

Calculator/Computer Problems

45. Use a graphing utility to obtain the graph of the **bifolium** $r = 4\sin\theta\cos^2\theta$ and the circle $r = \sin\theta$ on the same axes. Find all points of intersection of the graphs.

46. Use a graphing utility to verify that the cardioid $r = 1 + \cos\theta$ and the lemniscate $r^2 = 4\cos\theta$ intersect at four points. Find these points of intersection of the graphs.

In Problems 47 and 48, the graphs of the equations (a)–(d) represent a rotation of the graph of the given equation. Try sketching these graphs by hand. If you have difficulties, then use a calculator or CAS.

47. $r = 1 + \sin\theta$
  (a) $r = 1 + \sin(\theta - \pi/2)$ **(b)** $r = 1 + \sin(\theta + \pi/2)$
  (c) $r = 1 + \sin(\theta - \pi/6)$ **(d)** $r = 1 + \sin(\theta + \pi/4)$

48. $r = 2 + 4\cos\theta$
  (a) $r = 2 + 4\cos(\theta + \pi/6)$ **(b)** $r = 2 + 4\cos(\theta - 3\pi/2)$
  (c) $r = 2 + 4\cos(\theta + \pi)$ **(d)** $r = 2 + 4\cos(\theta - \pi/8)$

49. Use a CAS to obtain graphs of the polar equation $r = a + \cos\theta$ for $a = 0, \frac{1}{4}, \frac{1}{2}, \frac{3}{4}, 1, \frac{5}{4}, \ldots . 3$.

50. Identify all the curves in Problem 49. What happens to the graphs as $a \to \infty$?

For Discussion

In Problems 51 and 52, suppose $r = f(\theta)$ is a polar equation. Graphically interpret the given property.

51. $f(-\theta) = f(\theta)$ (even function)   **52.** $f(-\theta) = -f(\theta)$ (odd function)

8.3 Conic Sections in Polar Coordinates

☰ **Introduction** In Chapter 7 we derived equations for the parabola, ellipse, and hyperbola using the distance formula in rectangular coordinates. By using polar coordinates and the concept of eccentricity, we can now give one general definition of a conic section that encompasses all three curves.

> **DEFINITION 8.3.1** Conic Section
>
> Let L be a fixed line in the plane, and let F be a point not on the line. A **conic section** is the set of points P in the plane for which the distance from P to F divided by the distance from P to L is a constant.

The fixed line L is called a **directrix**, and the point F is a **focus**. The fixed constant is the **eccentricity** e of the conic. As FIGURE 8.3.1 shows, the point P lies on the conic if and only if

$$\frac{d(P, F)}{d(P, Q)} = e, \tag{1}$$

where Q denotes the foot of the perpendicular from P to L. In (1), if

- $e = 1$, the conic is a **parabola**,
- $0 < e < 1$, the conic is an **ellipse**, and
- $e > 1$, the conic is a **hyperbola**.

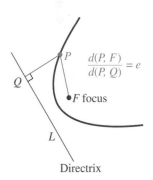

FIGURE 8.3.1 Geometric interpretation of (1)

☐ **Polar Equations of Conics** Equation (1) is readily interpreted using polar coordinates. Suppose F is placed at the pole and L is p units ($p > 0$) to the left of F perpendicular to the extended polar axis. We see from FIGURE 8.3.2 that (1) written as $d(P, F) = ed(P, Q)$ is the same as

$$r = e(p + r\cos\theta) \qquad \text{or} \qquad r - er\cos\theta = ep. \tag{2}$$

Solving for r yields

$$r = \frac{ep}{1 - e\cos\theta}. \tag{3}$$

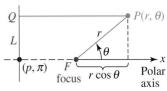

FIGURE 8.3.2 Polar coordinate interpretation of (1)

To see that (3) yields the familiar equations of the conics, let us superimpose a rectangular coordinate system on the polar coordinate system with origin at the pole and the positive x-axis coinciding with the polar axis. We then express the first equation in (2) in rectangular coordinates and simplify:

$$\pm\sqrt{x^2 + y^2} = ex + ep$$
$$x^2 + y^2 = e^2 x^2 + 2e^2 px + e^2 p^2$$
$$(1 - e^2)x^2 - 2e^2 px + y^2 = e^2 p^2. \tag{4}$$

Choosing $e = 1$, (4) becomes

$$-2px + y^2 = p^2 \qquad \text{or} \qquad y^2 = 2p\left(x + \tfrac{1}{2}p\right).$$

The last equation is the standard form of a parabola whose axis is the x-axis, vertex is at $\left(-\tfrac{1}{2}p, 0\right)$ and, consistent with the placement of F, whose focus is at the origin.

It is a good exercise in algebra to show that (3) yields standard form equations of an ellipse in the case $0 < e < 1$ and a hyperbola in the case $e > 1$. See Problem 45 in Exercises 8.3. Thus, depending on the value of e, the polar equation (3) can have three possible graphs as shown in FIGURE 8.3.3.

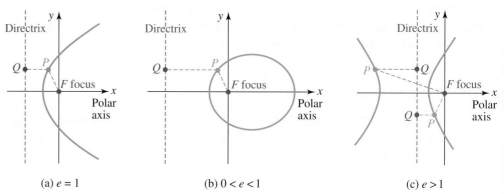

(a) $e = 1$ (b) $0 < e < 1$ (c) $e > 1$

FIGURE 8.3.3 Graphs of equation (3)

If we had placed the focus F to the *left* of the directrix in our derivation of the polar equation (3), then the equation $r = ep/(1 + e\cos\theta)$ would be obtained. When the directrix L is chosen parallel to the polar axis (that is, horizontal), then the equation of the conic is found to be either $r = ep/(1 - e\sin\theta)$ or $r = ep/(1 + e\sin\theta)$. A summary of the preceding discussion is given next.

THEOREM 8.3.1   Polar Equations of Conics

Any polar equation of the form

$$r = \frac{ep}{1 \pm e\cos\theta} \tag{5}$$

or

$$r = \frac{ep}{1 \pm e\sin\theta} \tag{6}$$

is a conic section with focus at the origin and axis along a coordinate axis. The axis of the conic section is along the x-axis for equations of the form (5) and along the y-axis for equations of the form (6). The conic is a **parabola** if $e = 1$, an **ellipse** if $0 < e < 1$, and a **hyperbola** if $e > 1$.

EXAMPLE 1   Identifying Conics

Identify each of the following conics (a) $r = \dfrac{2}{1 - 2\sin\theta}$ (b) $r = \dfrac{3}{4 + \cos\theta}$.

Solution (a) A term-by-term comparison of the given equation with the polar form $r = ep/(1 - e\sin\theta)$ enables us to make the identification $e = 2$. Hence the conic is a hyperbola.

(b) In order to identify the conic section, we divide the numerator and the denominator of the given equation by 4. This puts the equation into the form

$$r = \frac{\frac{3}{4}}{1 + \frac{1}{4}\cos\theta}.$$

Then by comparison with $r = ep/(1 + e\cos\theta)$ we see that $e = \frac{1}{4}$. Hence the conic is an ellipse. ≡

☐ **Graphs** A rough graph of a conic defined by (5) or (6) can be obtained by knowing the orientation of its axis, finding the x- and y-intercepts, and finding the vertices. In the case of (5),

- the two vertices of the **ellipse** or a **hyperbola** occur at $\theta = 0$ and $\theta = \pi$; the vertex of a **parabola** can occur at only one of the values: $\theta = 0$ or $\theta = \pi$.

For (6),

- the two vertices of an **ellipse** or a **hyperbola** occur at $\theta = \pi/2$ and $\theta = 3\pi/2$; the vertex of a **parabola** can occur at only one of the values: $\theta = \pi/2$ or $\theta = 3\pi/2$.

EXAMPLE 2   Graphing a Conic

Graph $r = \dfrac{4}{3 - 2\sin\theta}$.

Solution By writing the equation as $r = \dfrac{\frac{4}{3}}{1 - \frac{2}{3}\sin\theta}$ we see that the eccentricity is

$e = \frac{2}{3}$ and so the conic is an ellipse. Moreover, because the equation is of the form given in (6), we know that the axis of the ellipse is vertical along the y-axis. Now in view of the discussion preceding this example, we obtain:

$$vertices: \ (4, \pi/2), \ \left(\tfrac{4}{5}, 3\pi/2\right)$$
$$x\text{-}intercepts: \ \left(\tfrac{4}{3}, 0\right), \ \left(\tfrac{4}{3}, \pi\right).$$

The graph of the equation is given in **FIGURE 8.3.4**. ≡

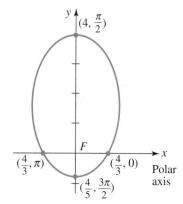

FIGURE 8.3.4 Graph of polar equation in Example 2

EXAMPLE 3   Graphing a Conic

Graph $r = \dfrac{1}{1 - \cos\theta}$.

Solution Inspection of the equation shows that it is of the form given in (5) with $e = 1$. Hence the conic section is a parabola whose axis is horizontal along the x-axis. Since r is undefined at $\theta = 0$, the vertex of the parabola occurs at $\theta = \pi$:

$$vertex: \ \left(\tfrac{1}{2}, \pi\right)$$
$$y\text{-}intercepts: \ (1, \pi/2), \ (1, 3\pi/2).$$

The graph of the equation is given in **FIGURE 8.3.5**. ≡

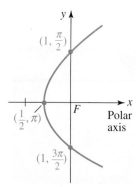

FIGURE 8.3.5 Graph of polar equation in Example 3

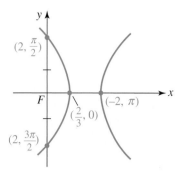

FIGURE 8.3.6 Graph of polar equation in Example 4

Graph $r = \dfrac{2}{1 + 2\cos\theta}$.

Solution From (5) we see that $e = 2$ and so the conic section is a hyperbola whose axis is horizontal along the x-axis. The vertices, the endpoints of the transverse axis of the hyperbola, occur at $\theta = 0$ and at $\theta = \pi$:

$$\textit{vertices:}\quad \left(\tfrac{2}{3}, 0\right), (-2, \pi)$$
$$\textit{y-intercepts:}\quad (2, \pi/2), (2, 3\pi/2).$$

The graph of the equation is given in **FIGURE 8.3.6**.   ≡

☐ **Rotated Conics** We saw in Section 8.2 that graphs of $r = f(\theta - \gamma)$ and $r = f(\theta + \gamma), \gamma > 0$, are rotations of the graph of the polar equation $r = f(\theta)$ about the origin by an amount γ. Thus

$$\left. \begin{aligned} r &= \frac{ep}{1 \pm e\cos(\theta - \gamma)} \\[2mm] r &= \frac{ep}{1 \pm e\sin(\theta - \gamma)} \end{aligned} \right\} \begin{array}{l} \text{conics rotated} \\ \text{counterclockwise} \\ \text{about the origin} \end{array} \qquad \left. \begin{aligned} r &= \frac{ep}{1 \pm e\cos(\theta + \gamma)} \\[2mm] r &= \frac{ep}{1 \pm e\sin(\theta + \gamma)} \end{aligned} \right\} \begin{array}{l} \text{conics rotated} \\ \text{clockwise} \\ \text{about the origin} \end{array}$$

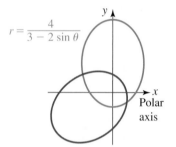

$r = \dfrac{4}{3 - 2\sin\theta}$

FIGURE 8.3.7 Graphs of polar equations in Example 5

In Example 2 we saw that the graph of the polar equation $r = \dfrac{4}{3 - 2\sin\theta}$ is an ellipse with major axis along the y-axis. This is the blue graph in **FIGURE 8.3.7**. The graph of $r = \dfrac{4}{3 - 2\sin(\theta - 2\pi/3)}$ is the red graph in Figure 8.3.7 and is a counterclockwise rotation of the blue graph by the amount $2\pi/3$ radians (or $120°$) about the origin. The major axis of the red graph lies along the line radial $\theta = 7\pi/6$.   ≡

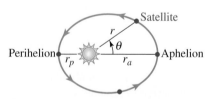

FIGURE 8.3.8 Orbit of satellite around the Sun

☐ **Applications** Equations of the type in (5) and (6) are well suited to describe a closed orbit of satellite around the Sun (Earth or Moon) since such an orbit is an ellipse with the Sun (Earth or Moon) at one focus. Suppose that an equation of the orbit is given by $r = ep/(1 - e\cos\theta), 0 < e < 1$, and r_p is the value of r at perihelion (perigee or perilune) and r_a is the value of r at aphelion (apogee or apolune). These are the points in the orbit, occurring on the x-axis, at which the satellite is closest and farthest, respectively, from the Sun (Earth or Moon). See **FIGURE 8.3.8**. It is left as an exercise to show that the eccentricity e of the orbit is related to r_p and r_a by

$$e = \frac{r_a - r_p}{r_a + r_p}. \tag{7}$$

Mercury is the closest planet to the Sun

Find a polar equation of the orbit of the planet Mercury around the Sun if $r_p = 2.85 \times 10^7$ miles and $r_a = 4.36 \times 10^7$ miles.

Solution From (7), the eccentricity of Mercury's orbit is

$$e = \frac{4.36 \times 10^7 - 2.85 \times 10^7}{4.36 \times 10^7 + 2.85 \times 10^7} = 0.21.$$

Hence
$$r = \frac{0.21p}{1 - 0.21\cos\theta}.$$ (8)

All we need to do now is to solve for the quantity $0.21p$. To do this we use the fact that aphelion occurs at $\theta = 0$:

$$4.36 \times 10^7 = \frac{0.21p}{1 - 0.21}.$$

The last equation yields $0.21p = 3.44 \times 10^7$. Hence a polar equation of Mercury's orbit is

$$r = \frac{3.44 \times 10^7}{1 - 0.21\cos\theta}. \qquad \equiv$$

8.3 **Exercises** Answers to selected odd-numbered problems begin on page ANS-25.

In Problems 1–10, determine the eccentricity, identify the conic, and sketch its graph.

1. $r = \dfrac{2}{1 - \sin\theta}$
2. $r = \dfrac{2}{2 - \cos\theta}$

3. $r = \dfrac{16}{4 + \cos\theta}$
4. $r = \dfrac{5}{2 + 2\sin\theta}$

5. $r = \dfrac{4}{1 + 2\sin\theta}$
6. $r = \dfrac{-4}{\cos\theta - 1}$

7. $r = \dfrac{18}{3 - 6\cos\theta}$
8. $r = \dfrac{4\csc\theta}{3\csc\theta + 2}$

9. $r = \dfrac{6}{1 - \cos\theta}$
10. $r = \dfrac{2}{2 + 5\cos\theta}$

In Problems 11–14, determine the eccentricity e of the given conic. Then convert the polar equation to a rectangular equation and verify that $e = c/a$.

11. $r = \dfrac{6}{1 + 2\sin\theta}$
12. $r = \dfrac{10}{2 - 3\cos\theta}$

13. $r = \dfrac{12}{3 - 2\cos\theta}$
14. $r = \dfrac{2\sqrt{3}}{\sqrt{3} + \sin\theta}$

In Problems 15–20, find a polar equation of the conic with focus at the origin that satisfies the given conditions.

15. $e = 1$, directrix $x = 3$
16. $e = \frac{3}{2}$, directrix $y = 2$
17. $e = \frac{2}{3}$, directrix $y = -2$
18. $e = \frac{1}{2}$, directrix $x = 4$
19. $e = 2$, directrix $x = 6$
20. $e = 1$, directrix $y = -2$

21. Find a polar equation of the conic in Problem 15 if the graph is rotated clockwise about the origin by an amount $2\pi/3$.
22. Find a polar equation of the conic in Problem 16 if the graph is rotated counterclockwise about the origin by an amount $\pi/6$.

In Problems 23–28, find a polar equation of the parabola with focus at the origin and the given vertex.

23. $\left(\frac{3}{2}, 3\pi/2\right)$ **24.** $(2, \pi)$

25. $\left(\frac{1}{2}, \pi\right)$ **26.** $(2, 0)$

27. $\left(\frac{1}{4}, 3\pi/2\right)$ **28.** $\left(\frac{3}{2}, \pi/2\right)$

In Problems 29–32, identify the given rotated conic. Find the polar coordinates of its vertex or vertices.

29. $r = \dfrac{4}{1 + \cos(\theta - \pi/4)}$ **30.** $r = \dfrac{5}{3 + 2\cos(\theta - \pi/3)}$

31. $r = \dfrac{10}{2 - \sin(\theta + \pi/6)}$ **32.** $r = \dfrac{6}{1 + 2\sin(\theta + \pi/3)}$

33. Perigee Distance A communications satellite is 12,000 km above the Earth at its apogee. The eccentricity of its elliptical orbit is 0.2. Use (7) to find its perigee distance.

34. Orbit Find a polar equation $r = ep/(1 - e\cos\theta)$ of the orbit of the satellite in Problem 33.

35. Earth's Orbit Find a polar equation of the orbit of the Earth around the Sun if $r_p = 1.47 \times 10^8$ km and $r_a = 1.52 \times 10^8$ km.

36. Comet Halley (a) The eccentricity of the elliptical orbit of Comet Halley is 0.97 and the length of the major axis of its orbit is 3.34×10^9 mi. Find a polar equation of its orbit of the form $r = ep/(1 - e\cos\theta)$.
(b) Use the equation in part (a) to obtain r_p and r_a for the orbit of Comet Halley.

Calculator/Computer Problems

In Problems 37–40, use a calculator or a computer to superimpose the graphs of the given two polar equations on the same coordinate axes.

37. $r = \dfrac{4}{4 + 3\cos\theta}$; $r = \dfrac{4}{4 + 3\cos(\theta - \pi/2)}$

38. $r = \dfrac{4}{6 - 3\sin\theta}$; $r = \dfrac{4}{6 - 3\sin(\theta - \pi)}$

39. $r = \dfrac{2}{1 - \sin\theta}$; $r = \dfrac{2}{1 - \sin(\theta + 3\pi/4)}$

40. $r = \dfrac{8}{3 + 5\cos\theta}$; $r = \dfrac{8}{3 + 5\cos(\theta - 2\pi/3)}$

The orbital characteristics (eccentricity, perigee, and major axis) of a satellite near the Earth gradually degrade over time due to many small forces acting on the satellite other than the gravitational force of the Earth. These forces include atmospheric drag, the gravitational attractions of the Sun and the Moon, and magnetic forces. Approximately once a month tiny rockets are activated for a few seconds in order to "boost" the orbital characteristics back into the desired range. Rockets are turned on longer to make a major change in the orbit of a satellite. The most fuel-efficient way to move from an inner orbit to an outer orbit, called a **Hohmann transfer**, is to add velocity in the direction of flight at the time the satellite reaches perigee on the inner orbit, follow the Hohmann transfer ellipse halfway around to its apogee, and add velocity again to achieve the outer orbit. A similar process (subtracting velocity at apogee on the outer orbit and subtracting velocity at perigee on the Hohmann transfer orbit) moves a satellite from an outer orbit to an inner orbit.

(continues)

Next visit of Comet Halley to the Solar System will be in 2061

In Problems 41–44, use a graphing calculator or computer to superimpose the graphs of the given three polar equations on the same coordinate axes. Print out your result and use a colored pencil to trace out the Hohmann transfer.

41. Inner orbit $r = \dfrac{24}{1 + 0.2\cos\theta}$, Hohmann transfer $r = \dfrac{32}{1 + 0.6\cos\theta}$,

outer orbit $r = \dfrac{56}{1 + 0.3\cos\theta}$

42. Inner orbit $r = \dfrac{5.5}{1 + 0.1\cos\theta}$, Hohmann transfer $r = \dfrac{7.5}{1 + 0.5\cos\theta}$,

outer orbit $r = \dfrac{13.5}{1 + 0.1\cos\theta}$

43. Inner orbit $r = 9$, Hohmann transfer $r = \dfrac{15.3}{1 + 0.7\cos\theta}$,

outer orbit $r = 51$

44. Inner orbit $r = \dfrac{73.5}{1 + 0.05\cos\theta}$, Hohmann transfer $r = \dfrac{77}{1 + 0.1\cos\theta}$,

outer orbit $r = \dfrac{84.7}{1 + 0.01\cos\theta}$

For Discussion

45. Show that (2) yields standard form equations of an ellipse in the case $0 < e < 1$ and a hyperbola in the case $e > 1$.
46. Use the equation $r = ep/(1 - e\cos\theta)$ to derive the result in (7).

8.4 Parametric Equations

≡ **Introduction** Rectangular equations and polar equations are not the only, and often not the most convenient, ways of describing curves in the coordinate plane. In this section we will consider a different way of representing a curve that is important in many applications. Let's consider one example. The motion of a particle along a curve, in contrast to a straight line, is called *curvilinear motion*. If it is assumed that a golf ball is hit off the ground, perfectly straight (no hook or slice), and that its path stays in a coordinate plane as shown in FIGURE 8.4.1, then it can be shown that its *x*- and *y*-coordinates at time *t* are given by

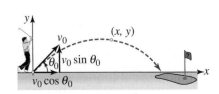

FIGURE 8.4.1 Fore!

$$x = (v_0\cos\theta_0)t, \quad y = -\tfrac{1}{2}gt^2 + (v_0\sin\theta_0)t, \qquad (1)$$

where θ_0 is the launch angle, v_0 is its initial velocity, and $g = 32$ ft/s^2 is the acceleration due to gravity. These equations, which give the golf ball's position in the coordinate plane at time t, are said to be **parametric equations**. The third variable t in (1) is called a **parameter** and is restricted to some interval defined by $0 \le t \le T$, where $t = 0$ gives the origin $(0, 0)$, and $t = T$ is the time the ball hits the ground.

In general, a curve in a coordinate plane can be *defined* in terms of parametric equations.

DEFINITION 8.4.1 Plane Curve

A **plane curve** is a set C of ordered pairs $(f(t), g(t))$, where f and g are functions defined on a common interval I. The equations

$$x = f(t), \quad y = g(t), \quad \text{for } t \text{ in } I,$$

are called **parametric equations** for C. The variable t is called a **parameter** and I is called the **parameter interval**.

It is also common practice to refer to $x = f(t), y = g(t)$, for t in I, as a **parameterization** for C.

The **graph** of a plane curve C is the set of all points (x, y) in the coordinate plane corresponding to the ordered pairs $(f(t), g(t))$. Hereafter, we will refer to a plane curve as a **curve** or as a **parameterized curve**.

EXAMPLE 1 Graph of a Parametric Curve

Graph the curve C that has the parametric equations

$$x = t^2, \quad y = t^3, \quad -1 \le t \le 2.$$

Solution As shown in the accompanying table, for any choice of t in the interval $[-1, 2]$, we obtain a single ordered pair (x, y). By connecting the points with a curve, we obtain the graph in **FIGURE 8.4.2**.

t	-1	$-\frac{1}{2}$	0	$\frac{1}{2}$	1	$\frac{3}{2}$	2
x	1	$\frac{1}{4}$	0	$\frac{1}{4}$	1	$\frac{9}{4}$	4
y	-1	$-\frac{1}{8}$	0	$\frac{1}{8}$	1	$\frac{27}{8}$	8

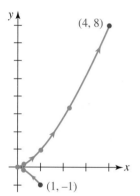

FIGURE 8.4.2 Curve in Example 1

In Example 1, if we think in terms of motion and t as time, then as t increases from -1 to 2, a point P defined as (t^2, t^3) starts from $(1, -1)$, advances up the lower branch to the origin $(0, 0)$, passes to the upper branch, and finally stops at $(4, 8)$. In general, as we plot points corresponding to *increasing values* of the parameter, the curve C is traced out by $(f(t), g(t))$ in a certain *direction* indicated by the arrowheads on the curve in Figure 8.4.2. This direction is called the **orientation** of the curve C.

A parameter need have no relation to time. When the interval I over which f and g in (1) are defined is a closed interval $[a, b]$, we say that $(f(a), g(a))$ is the **initial point** of the curve C and that $(f(b), g(b))$ is its **terminal point**. In Example 1 the initial point is $(1, -1)$ and the terminal point is $(4, 8)$. If the terminal point is the same as the initial point, that is,

$$(f(a), g(a)) = (f(b), g(b)),$$

then C is a **closed curve**. If C is closed and does not cross itself, then it is called a **simple closed curve**. In FIGURE 8.4.3, A and B represent the initial and terminal points, respectively.

(a) Plane curve   (b) Simple closed curve   (c) Closed but not simple

FIGURE 8.4.3 Some plane curves

The next example illustrates a simple closed curve.

EXAMPLE 2   A Parameterization of a Circle

Find a parameterization for the circle $x^2 + y^2 = a^2$.

Solution The circle has center at the origin and radius a. If t represents the central angle, that is, an angle with vertex at the origin and initial side coinciding with the positive x-axis, then as shown in FIGURE 8.4.4 the equations

$$x = a\cos t, \quad y = a\sin t, \quad 0 \le t \le 2\pi \tag{2}$$

give every point P on the circle. For example, at $t = 0$ we get $x = a$ and $y = 0$; in other words, the initial point is $(a, 0)$. The terminal point corresponds to $t = 2\pi$ and is also $(a, 0)$. Since the initial and terminal points are the same, this proves the obvious: the curve C defined by the parametric equations in (2) is a closed curve. Note the orientation of C in Figure 8.4.4; as t increases from 0 to 2π, the point P traces out C in a counterclockwise direction. ≡

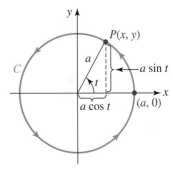

FIGURE 8.4.4 Circle in Example 2

☐ **Changing the Parameter Interval** In Example 2, if we wish to describe *two* complete counterclockwise revolutions around the circle, we modify the parameter interval by writing

$$x = a\cos t, \quad y = a\sin t, \quad 0 \le t \le 4\pi.$$

On the other hand, we can describe *portions* of the circle by again modifying the parameter interval. For example, the upper semicircle $x^2 + y^2 = a^2$, $0 \le y \le a$, is defined parametrically by restricting the parameter interval to $[0, \pi]$,

$$x = a\cos t, \quad y = a\sin t, \quad 0 \le t \le \pi.$$

Observe that when $t = \pi$, the terminal point is now $(-a, 0)$. As seen in FIGURE 8.4.5(a) the orientation of the curve is counterclockwise. The same equations with a different parameter interval

$$x = a\cos t, \quad y = a\sin t, \quad -\pi/2 \le t \le \pi/2,$$

define the semicircle shown in Figure 8.4.5(b). The value $t = -\pi/2$ gives the initial point $(0, -a)$ and $t = \pi/2$ gives the terminal point $(0, a)$. Unlike the semicircle in Figure 8.4.5(a) the orientation of the curve in Figure 8.4.5(b) is seen to be clockwise.

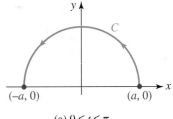

(a) $0 \le t \le \pi$

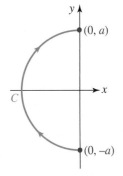

(b) $-\pi/2 \le t \le \pi/2$

FIGURE 8.4.5 Semicircles obtained from $x = a\cos t$, $y = a\sin t$ by restricting the parameter interval

☐ **Ellipse** It can also be seen from Figure 8.4.4 that a parameterization of an ellipse $x^2/a^2 + y^2/b^2 = 1$ is

$$x = a\cos t, \quad y = b\sin t, \quad 0 \le t \le 2\pi.$$

Portions of the ellipse or changing the orientation of the parameterized curve can be done as illustrated in the preceding discussion.

☐ **Eliminating the Parameter** Given a set of parametric equations, we sometimes desire to eliminate or clear the parameter to obtain a rectangular equation for the curve. There is no well-defined method of eliminating the parameter; the method is dictated by the parametric equations. For example, to eliminate the parameter in (2), we simply square x and y and add the two equations:

$$x^2 + y^2 = a^2\cos^2 t + a^2\sin^2 t \qquad \text{implies} \qquad x^2 + y^2 = a^2$$

since $\sin^2 t + \cos^2 t = 1$. In the next example, we illustrate a substitution method.

EXAMPLE 3   Eliminating the Parameter

(a) In the first equation in (1) we can solve for t in terms of x and then substitute $t = x/(v_0\cos\theta_0)$ into the second equation. This gives an equation involving only the variables x and y:

$$y = -\frac{g}{2(v_0\cos\theta_0)^2}x^2 + (\tan\theta_0)x.$$

Since v_0, θ_0, and g are constants, the last equation has the form $y = ax^2 + bx$ and so the trajectory of any projectile launched at the angle $0 < \theta_0 < \pi/2$ is a parabolic arc.

(b) In Example 1, we can eliminate the parameter by solving the second equation for t in terms of y and then substituting in the first equation. We find

$$t = y^{1/3} \quad \text{and so} \quad x = (y^{1/3})^2 = y^{2/3}.$$

The curve shown in Figure 8.4.2 is only a portion of the graph of $x = y^{2/3}$. For $-1 \le t \le 2$, we have correspondingly $-1 \le y \le 8$. Thus, a rectangular equation for the curve in Example 1 is given by $x = y^{2/3}$, $-1 \le y \le 8$.   ≡

▶ A curve C can have many different parameterizations.

A curve C can have more than one parameterization. For example, an alternative parameterization for the circle in Example 2 is

$$x = a\cos 2t, \quad y = a\sin 2t, \quad 0 \le t \le \pi.$$

Note that the parameter interval is now $[0, \pi]$. We see that as t increases from 0 to π, the new angle $2t$ increases from 0 to 2π.

EXAMPLE 4   Alternative Parameterizations

(a) Consider the curve C that has the parametric equations $x = t, y = 2t^2$, $-\infty < t < \infty$. We can eliminate the parameter by using $t = x$ and substituting in $y = 2t^2$. This gives the rectangular equation $y = 2x^2$ that we recognize as a parabola. Moreover, since $-\infty < t < \infty$ is equivalent to $-\infty < x < \infty$, the point $(t, 2t^2)$ traces out the complete parabola $y = 2x^2$, $-\infty < x < \infty$.

(b) An alternative parameterization of the curve C in part (a) is given by $x = t^3/4$, $y = t^6/8$, $-\infty < t < \infty$. Using $t^3 = 4x$ and substituting in $y = t^6/8$ or $y = (t^3 \cdot t^3)/8$ gives $y = (4x)^2/8 = 2x^2$. Moreover, $-\infty < t < \infty$ implies $-\infty < t^3 < \infty$ and so $-\infty < x < \infty$.   ≡

We note in Example 4 that a point on the curve C need not correspond to the same value of the parameter in each set of parametric equations for C. For example, $(1, 2)$ is obtained for $t = 1$ in $x = t, y = 2t^2$, but $t = \sqrt[3]{4}$ yields $(1, 2)$ in $x = t^3/4, y = t^6/8$.

EXAMPLE 5   **Example 4 Revisited**

One has to be careful when working with parametric equations. Eliminating the parameter in $x = t^2$, $y = 2t^4$, $-\infty < t < \infty$, would seem to yield the same parabola $y = 2x^2$
as in Example 4. However, this is *not* the case because for any value of t, satisfying
$-\infty < t < \infty$, we have $t^2 \geq 0$ and so $x \geq 0$. In other words, the last set of equations
$x = t^2$, $y = 2t^4$, $-\infty < t < \infty$ is a parametric representation of only the right-hand
branch of the parabola, that is, $y = 2x^2$, $0 \leq x < \infty$.   ≡

◀ Proceed with caution when eliminating
the parameter.

EXAMPLE 6   **Eliminating the Parameter**

Consider the curve C defined parametrically by
$$x = \sin t, \quad y = \cos 2t, \quad 0 \leq t \leq \pi/2.$$
Eliminate the parameter and obtain a rectangular equation for C.

Solution Using the double-angle formula $\cos 2t = \cos^2 t - \sin^2 t$, we can write
$$\begin{aligned} y &= \cos^2 t - \sin^2 t \\ &= (1 - \sin^2 t) - \sin^2 t \\ &= 1 - 2\sin^2 t \quad \leftarrow \text{substitute } \sin t = x \\ &= 1 - 2x^2. \end{aligned}$$

Now the curve C described by the parametric equations does not consist of the complete
parabola, that is, $y = 1 - 2x^2$, $-\infty < x < \infty$. See FIGURE 8.4.6(a). For $0 \leq t \leq \pi/2$ we
have $0 \leq \sin t \leq 1$ and $-1 \leq \cos 2t \leq 1$. This means that C is only that portion of the
parabola for which the coordinates of a point $P(x, y)$ satisfy $0 \leq x \leq 1$ and
$-1 \leq y \leq 1$. The curve C, along with its orientation, is shown in Figure 8.4.5(b). A
rectangular equation for C is $y = 1 - 2x^2$ with the restricted domain $0 \leq x \leq 1$.   ≡

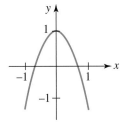

(a) $y = 1 - 2x^2$

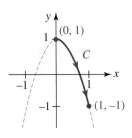

(b) $x = \sin t$, $y = \cos 2t$,
    $0 \leq t \leq \pi/2$

FIGURE 8.4.6 Curve C in
Example 6

☐ **Intercepts** We can get intercepts of a curve C without finding its rectangular
equation. For instance, in Example 6 we can find the x-intercept by finding the value of t
in the parameter interval for which $y = 0$. The equation $\cos 2t = 0$ yields $2t = \pi/2$ so
that $t = \pi/4$. The corresponding point at which C crosses the x-axis is $(\sqrt{2}/2, 0)$.
Similarly, the y-intercept of C is found by solving $x = 0$. From $\sin t = 0$ we immediately conclude $t = 0$ and so the y-intercept is $(0, 1)$.

☐ **Applications of Parametric Equations** Cycloidal curves were a popular topic of
study by mathematicians in the seventeenth century. Suppose a point $P(x, y)$, marked on
a circle of radius a, is at the origin when its diameter lies along the y-axis. As the circle rolls
along the x-axis, the point P traces out a curve C that is called a **cycloid**. See FIGURE 8.4.7.*

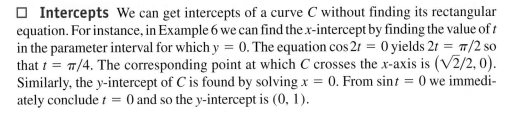

(a) Circle rolling on x-axis

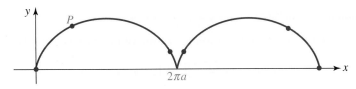

(b) Point P on the circle traces out this curve
FIGURE 8.4.7 Curve in red is a cycloid

*For an animation of the rolling circle go to http://mathworld.wolfram.com/Cycloid.html.

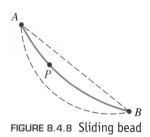

Two problems were extensively studied in the seventeenth century. Consider a flexible (frictionless) wire fixed at points A and B and a bead free to slide down the wire starting at P. See FIGURE 8.4.8. Is there a particular shape of the wire so that, regardless of where the bead starts, the time to slide down the wire to B will be the same? Also, what would the shape of the wire be so that the bead slides from P to B in the shortest time? The so-called **tautochrone** (same time) and **brachistochrone** (least time) were shown to be an inverted half-arch of a cycloid.

EXAMPLE 7   Parameterization of a Cycloid

Find a parameterization for the cycloid shown in Figure 8.4.7(b).

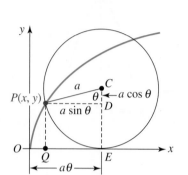

FIGURE 8.4.9 Cycloid in Example 7

Solution A circle of radius a whose diameter initially lies along the y-axis rolls along the x-axis without slipping. We take as a parameter the angle θ (in radians) through which the circle has rotated. The point $P(x, y)$ starts at the origin, which corresponds to $\theta = 0$. As the circle rolls through an angle θ, its distance from the origin is the arc $PE = \overline{OE} = a\theta$. From FIGURE 8.4.9 we then see that the x-coordinate of P is

$$x = \overline{OE} - \overline{QE} = a\theta - a\sin\theta.$$

Now the y-coordinate of P is seen to be

$$y = \overline{CE} - \overline{CD} = a - a\cos\theta.$$

Hence parametric equations for the cycloid are

$$x = a\theta - a\sin\theta, \quad y = a - a\cos\theta.$$

As shown in Figure 8.4.7(a), one arch of a cycloid is generated by one rotation of the circle and corresponds to the parameter interval $0 \le \theta \le 2\pi$. ≡

☐ **Parameterizations of Rectangular and Polar Curves** A curve C described by a continuous function $y = f(x)$ can always be parameterized by letting $x = t$. Parametric equations for C are then

$$x = t, \quad y = f(t). \tag{3}$$

Also, it is sometimes convenient to use parametric equations to plot the graphs of polar equations. This can be done using the conversion formulas $x = r\cos\theta, y = r\sin\theta$. Similarly, if $r = f(\theta), \alpha \le \theta \le \beta$ describes a polar curve C, then a parameterization for C is given by

$$x = f(\theta)\cos\theta, \quad y = f(\theta)\sin\theta, \quad \alpha \le \theta \le \beta. \tag{4}$$

EXAMPLE 8   Parameterization of a Rectangular Curve

Find a parameterization of $f(x) = \sqrt{x - 1}$.

Solution If we let $x = t$, then a parameterization of the curve defined by f is

$$x = t, \quad y = \sqrt{t - 1}, \quad 1 \le t < \infty.$$

Alternative Solution A curve can have many different parameterizations. If we let $x - 1 = t$, then another parameterization of the curve defined by f is

Note that the parameter ▶ interval is also changed.

$$x = t + 1, \quad y = \sqrt{t}, \quad 0 \le t < \infty.$$ ≡

EXAMPLE 9   **Parameterization of a Polar Curve**

Find a parameterization of the cardioid $r = 1 + \sin\theta$.

Solution   With the identification $f(\theta) = 1 + \sin\theta$ and the knowledge that a complete cardioid is generated by allowing the values of θ range over the interval $[0, 2\pi]$, it follows that a parameterization of the polar curve defined by f is

◀ See the comment preceding Figure 8.2.6 and (*ii*) of the *Notes from the Classroom* in Section 8.2.

$$x = (1 + \sin\theta)\cos\theta, \quad y = (1 + \sin\theta)\sin\theta, \quad 0 \le \theta \le 2\pi. \qquad \equiv$$

NOTES FROM THE CLASSROOM

In this section we have focused on **plane curves**, curves C defined parametrically in two dimensions. In the study of multivariable calculus you will see curves and surfaces in three dimensions that are defined by means of parametric equations. For example, a **space curve** C consists of a set of ordered triples $(f(t), g(t), h(t))$, where f, g, and h are defined on a common interval. Parametric equations for C are $x = f(t), y = g(t), z = h(t)$. For example, the **circular helix** such as shown in FIGURE 8.4.10 is a space curve whose parametric equations are

$$x = a\cos t, \quad y = a\cos t, \quad z = bt, \quad t \ge 0. \qquad (4)$$

Surfaces in three dimensions can be represented by a set of parametric equations involving *two* parameters, $x = f(u, v), y = g(u, v), z = h(u, v)$.

For example, the **circular helicoid** shown in FIGURE 8.4.11 arises from the study of minimal surfaces and is defined by the set of parametric equations similar to those in (4):

$$x = u\cos v, \quad y = u\sin v, \quad z = bv,$$

where b is a constant. The circular helicoid has a circular helix as its boundary. You might recognize the helicoid as the model for the rotating curved blade in machinery such as post-hole diggers, ice augers, and snow blowers.

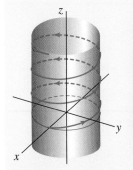
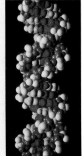
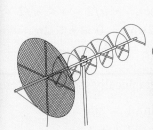
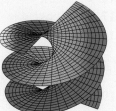

FIGURE 8.4.10 Circular helix

DNA is a double helix

Helical antenna

FIGURE 8.4.11 Circular helicoid

Exercises Answers to selected odd-numbered problems
begin on page ANS-25.

In Problems 1 and 2, fill in the table for the given set of parametric equations. Find the
x- and y-intercepts . Sketch the curve and indicate its orientation.

1. $x = t + 2, y = 3 + \frac{1}{2}t, \quad -\infty < t < \infty$

t	-3	-2	-1	0	1	2	3
x							
y							

2. $x = 2t + 1, y = t^2 + t, \quad -\infty < t < \infty$

t	-3	-2	-1	0	1	2	3
x							
y							

In Problems 3–10, sketch the curve that has the given set of parametric equations.

3. $x = t - 1, y = 2t - 1, \quad -1 \leq t \leq 5$
4. $x = t^2 - 1, y = 3t, \quad -2 \leq t \leq 3$
5. $x = \sqrt{t}, y = 5 - t, \quad t \geq 0$
6. $x = t^3 + 1, y = t^2 - 1, \quad -2 \leq t \leq 2$
7. $x = 3\cos t, y = 5\sin t, \quad 0 \leq t \leq 2\pi$
8. $x = 3 + 2\sin t, y = 4 + \sin t, \quad -\pi/2 \leq t \leq \pi/2$
9. $x = e^t, y = e^{3t}, \quad 0 \leq t \leq \ln 2$
10. $x = -e^t, y = e^{-t}, \quad t \geq 0$

In Problems 11–18, eliminate the parameter from the given set of parametric
equations and obtain a rectangular equation that has the same graph.

11. $x = t^2, y = t^4 + 3t^2 - 1$
12. $x = t^3 + t + 4, y = -2(t^3 + t)$
13. $x = \cos 2t, y = \sin t, \quad -\pi/2 \leq t \leq \pi/2$
14. $x = e^t, y = \ln t, \quad t > 0$
15. $x = t^3, y = 3\ln t, \quad t > 0$
16. $x = \tan t, y = \sec t, \quad -\pi/2 < t < \pi/2$
17. $x = 4\cos t, y = 2\sin t, \quad 0 \leq t \leq 2\pi$
18. $x = -1 + \cos t, y = 2 + \sin t, \quad 0 \leq t \leq 2\pi$

In Problems 19–24, graphically show the difference between the given curves.

19. $y = x$ and $x = \sin t, y = \sin t$
20. $y = x^2$ and $x = -\sqrt{t}, y = t$
21. $y = \frac{1}{4}x^2 - 1$ and $x = 2t, y = t^2 - 1, \quad -1 \leq t \leq 2$
22. $y = -x^2$ and $x = e^t, y = -e^{2t}, \quad t \geq 0$
23. $x^2 - y^2 = 1$ and $x = \cosh t, y = \sinh t$ ← See (14) and (15) in Section 6.5.
24. $y = 2x - 2$ and $x = t^2 - 1, y = 2t^2 - 4$

In Problems 25–28 graphically show the difference between the given curves. Assume that $a > 0$ and $b > 0$,

25. $x = a\cos t, y = a\sin t, \quad 0 \le t \le \pi$
$\quad x = a\sin t, y = a\cos t, \quad 0 \le t \le \pi$
26. $x = a\cos t, y = b\sin t, a > b, \quad \pi \le t \le 2\pi$
$\quad x = a\sin t, y = b\cos t, a > b, \quad \pi \le t \le 2\pi$
27. $x = a\cos t, y = a\sin t, \quad -\pi/2 \le t \le \pi/2$
$\quad x = a\cos 2t, y = a\sin 2t, \quad -\pi/2 \le t \le \pi/2$
28. $x = a\cos\dfrac{t}{2}, y = a\sin\dfrac{t}{2}, \quad 0 \le t \le \pi$

$\quad x = a\cos\left(-\dfrac{t}{2}\right), y = a\sin\left(-\dfrac{t}{2}\right), \quad -\pi \le t \le 0$

In Problems 29 and 30, find the x- and y-intercepts of the given curves.

29. $x = t^2 - 2t, y = t + 1, \quad -2 \le t < 4$
30. $x = t^2 + t, y = t^2 + t - 6, \quad -5 \le t < 5$

31. Show that parametric equations for a line through (x_1, y_1) and (x_2, y_2) are

$$x = x_1 + (x_2 - x_1)t, \quad y = y_1 + (y_2 - y_1)t, \quad -\infty < t < \infty.$$

What do these equations represent when $0 \le t \le 1$?

32. (a) Use the result of Problem 31 to find parametric equations of the line through $(-2, 5)$ to $(4, 8)$.
(b) Eliminate the parameter in part (a) to obtain a rectangular equation for the line.
(c) Find parametric equations for the line segment with $(-2, 5)$ as the initial point and $(4, 8)$ as the terminal point.

33. A famous golfer can generate a club head speed of approximately 130 mi/h or $v_0 = 190$ ft/s. If the golf ball leaves the ground at an angle $\theta_0 = 45°$, use (1) to find parametric equations for the path of the ball. What are the coordinates of the ball at $t = 2$ s?

34. Use the parametric equations obtained in Problem 33 to determine
(a) how long the golf ball is in the air,
(b) its maximum height, and
(c) the horizontal distance that the golf ball travels.

35. As shown in FIGURE 8.4.12, a piston is attached by means of a rod of length L to a circular crank mechanism of radius r. Parameterize the coordinates of the point P in terms of the angle ϕ shown in the figure.

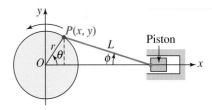

FIGURE 8.4.12 Crank mechanism in Problem 35

36. Consider a circle of radius a, which is tangent to the x-axis at the origin. Let B be a point on the horizontal line $y = 2a$ and let the line segment OB cut the circle at point A. As shown in FIGURE 8.4.13, the projection of AB on the vertical gives the line segment BP. Using the angle θ in the figure as a parameter, find parametric equations of the curve traced by the point P as A varies around the circle. The curve, more historically famous than useful, is called the **witch of Agnesi.*

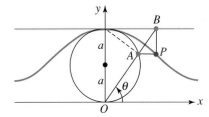

FIGURE 8.4.13 Witch of Agnesi in Problem 36

---

*No, the curve has nothing to do with witches and goblins. This curve, called *versoria*, which is Latin for a kind of rope, was included in a text on analytic geometry written in 1748 by the Italian mathematician **Maria Agnesi** (1718–1799). A translator of the text confused *versoria* with the Italian word *versiera*, which means *female goblin*. In English, *female goblin* became a *witch*.

Calculator/Computer Problems

In Problems 37–42, use a graphing utility to obtain the graph of the given set of parametric equations.

37. $x = 4\sin 2t, y = 2\sin t, \quad 0 \le t \le 2\pi$
38. $x = 6\cos 3t, y = 4\sin 2t, \quad 0 \le t \le 2\pi$
39. $x = 6\sin 4t, y = 4\sin t, \quad 0 \le t \le 2\pi$
40. $x = \cos t + t\sin t, y = \sin t - t\cos t, \quad 0 \le t \le 3\pi$
41. $x = 4\cos t - \cos 4t, y = 4\sin t - \sin 4t, \quad 0 \le t \le 2\pi$
42. $x = \cos^3 t, y = \sin^3 t, \quad 0 \le t \le 2\pi$

In Problems 43–46, use (4) to parameterize the curve whose polar equation is given. Use a graphing utility to obtain the graph of the resulting set of parametric equations.

43. $r = 2\sin\dfrac{\theta}{2}, \quad 0 \le \theta \le 4\pi$ **44.** $r = 2\sin\dfrac{\theta}{4}, \quad 0 \le \theta \le 8\pi$

45. $r = 2\cos\dfrac{\theta}{5}, \quad 0 \le \theta \le 5\pi$ **46.** $r = 2\cos\dfrac{3\theta}{2}, \quad 0 \le \theta \le 4\pi$

| CHAPTER 8 | Review Exercises | Answers to selected odd-numbered problems begin on page ANS–26. |

A. Fill in the Blanks

In Problems 1–16, fill in the blanks.

1. The rectangular coordinates of the point with polar coordinates $\left(-\sqrt{2}, 5\pi/4\right)$ are _____.
2. Approximate polar coordinates of the point with rectangular coordinates $(-1, 3)$ are _____.
3. Polar coordinates of the point with rectangular coordinates $(0, -10)$ are _____.
4. On the graph of the polar equation $r = 4\cos\theta$, two pairs of coordinates of the pole or origin are _____.
5. The radius of the circle $r = \cos\theta$ is _____.
6. If $a > 0$, the center of the circle $r = -2a\sin\theta$ is _____.
7. The conic section $r = \dfrac{1}{2 + 5\cos\theta}$ is a _____.
8. In polar coordinates, the polar graph of $\theta = \pi/3$ is a _____.
9. The name of the polar graph of $r = 2 + \cos\theta$ is _____.
10. $r = \dfrac{12}{2 + \cos\theta}$, center _____, foci _____, vertices _____.
11. If the points (r, θ) and $(-r, \pi - \theta)$ are on the graph of the polar equation $r = f(\theta)$, then the graph is symmetric with respect to _____.
12. The graph of the polar equation $r = 2 + 2\sin(\theta - \pi/4)$ is a _____.
13. A polar equation of the circle passing through $(0, 0)$ and center $(10, \pi/2)$ is _____.
14. The point P with polar coordinates $(0, 1)$ is the _____.

15. The equations $x = t + 2$, $y = 3 + \frac{1}{2}t$, $-\infty < t < \infty$, are a parameterization of a _____.

16. The point on the curve C defined by the parametric equations $x = -4 + 2\cos t$, $y = 2 + \sin t$, $-\infty < t < \infty$, corresponding to $t = 5\pi/2$ is _____.

B. True/False

In Problems 1–18, answer true or false.

1. Rectangular coordinates of a point in the plane are unique. _____

2. The graph of the polar equation $r = 5\sec\theta$ is a line. _____

3. $(3, \pi/6)$ and $(-3, -5\pi/6)$ are polar coordinates of the same point. _____

4. The graph of the ellipse $r = \dfrac{90}{15 - \sin\theta}$ is nearly circular. _____

5. The graph of the rose curve $r = 5\sin 6\theta$ has 6 petals. _____

6. The graph of $r = 2 + 4\sin\theta$ is a limaçon with an interior loop. _____

7. The graph of the polar $r^2 = 4\sin 2\theta$ is symmetric with respect to the origin. _____

8. The graphs of the cardioids $r = 3 + 3\cos\theta$ and $r = -3 + 3\cos\theta$ are the same. _____

9. The point $(4, 3\pi/2)$ is not on the graph of $r = 4\cos 2\theta$, because its coordinates do not satisfy the polar equation. _____

10. The polar equation $r^2\sin 2\theta = 1$ has the same graph as the rectangular equation $xy = \frac{1}{2}$. _____

11. The terminal side of the angle θ is always in the same quadrant as the point (r, θ). _____

12. The eccentricity e of a parabola satisfies $0 < e < 1$. _____

13. The graph of the polar equation $\tan\theta = 7$ is a line through the origin. _____

14. The polar coordinates of the point of intersection of the lines $r\cos\theta = 1$ and $r\sin\theta = 1$ are $(1, 1)$. _____

15. The graphs of the polar equations $r = -5$ and $r = 5$ are the same. _____

16. The transverse axis of the hyperbola $r = \dfrac{5}{2 + 3\cos\theta}$ lies along the x-axis. _____

17. The curve C with parametric equations $x = e^t$, $y = e^t - 5$, $-\infty < t < \infty$ is the same as the graph of the rectangular equation $y = x - 5$. _____

18. The curve C with parametric equations $x = 1 + \cos t$, $y = 1 + \sin t$, $0 \le t \le 2\pi$, is a circle of radius 1 centered at $(1, 1)$. _____

C. Review Exercises

In Problems 1 and 2, find a rectangular equation that has the same graph as the given polar equation.

1. $r = \cos\theta + \sin\theta$

2. $r(\cos\theta + \sin\theta) = 1$

In Problems 3 and 4, find a polar equation that has the same graph as the given rectangular equation.

3. $x^2 + y^2 - 4y = 0$

4. $(x^2 + y^2 - 2x)^2 = 9(x^2 + y^2)$.

5. Determine the rectangular coordinates of the vertices of the ellipse whose polar equation is $r = 2/(2 - \sin\theta)$.

6. Find a polar equation of the hyperbola with focus at the origin, vertices (in rectangular coordinates) $\left(0, -\frac{4}{3}\right)$ and $(0, -4)$, and eccentricity 2.

In Problems 7 and 8, find polar coordinates satisfying **(a)** $r > 0$, $-\pi < \theta \leq \pi$, and **(b)** $r < 0$, $-\pi < \theta \leq \pi$, for each point given in rectangular coordinates.

7. $\left(\sqrt{3}, -\sqrt{3}\right)$ **8.** $\left(-\frac{1}{4}, \frac{1}{4}\right)$

In Problems 9–20, identify and sketch the graph of the given polar equation.

9. $r = 5$ **10.** $\theta = -\pi/3$
11. $r = 5\sin\theta$ **12.** $r = -4\cos\theta$
13. $r = 4 - 4\cos\theta$ **14.** $r = 1 + \sin\theta$
15. $r = 2 + \sin\theta$ **16.** $r = 1 - 2\cos\theta$
17. $r = \sin 3\theta$ **18.** $r = 3\sin 4\theta$
19. $r = \dfrac{8}{3 - 2\cos\theta}$ **20.** $r = \dfrac{1}{1 + \cos\theta}$

In Problems 21 and 22, find an equation of the given polar graph.

21.

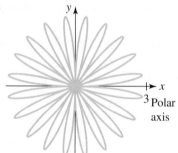

FIGURE 8.R.1 Graph for Problem 21

22.

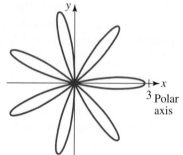

FIGURE 8.R.2 Graph for Problem 22

In Problems 23 and 24, the graph of the given polar equation is rotated about the origin by the indicated amount. **(a)** Find a polar equation of the new graph. **(b)** Find a rectangular equation for the new graph.

23. $r = 2\cos\theta$; counterclockwise, $\pi/4$
24. $r = 1/(1 + \cos\theta)$; clockwise, $\pi/6$

25. (a) Show that the graph of the polar equation

$$r = a\sin\theta + b\cos\theta$$

for $a \neq 0$ and $b \neq 0$, is a circle.
(b) Determine the center and radius of the circle in part (a).
(c) Does the case $a = b$ have any graphical significance?

26. (a) Find a rectangular equation that has the same graph as the given polar equation: $r\cos\theta = 1$, $r\cos(\theta - \pi/3) = 1$, $r = 1$. Sketch the graph of each equation.

(b) How are the graphs of $r\cos\theta = 1$ and $r\cos(\theta - \pi/3) = 1$ related?

(c) Show that rectangular point $\left(\frac{1}{2}, \frac{\sqrt{3}}{2}\right)$ is on the graphs of $r\cos(\theta - \pi/3) = 1$ and $r = 1$.

(d) Use the information in parts (a) and (c) to explain how the graphs of $r\cos(\theta - \pi/3) = 1$ and $r = 1$ are related.

9 Systems of Equations and Inequalities

Chapter Outline

9.1 Systems of Linear Equations

≡ Introduction

Recall from Section 2.3 that a **linear equation in two variables** x and y is any equation that can be put in the form $ax + by = c$, where a and b are real numbers and not both zero. In general, a **linear equation in n variables** $x_1, x_2, \ldots, x_n$ is an equation of the form

$$a_1x_1 + a_2x_2 + \cdots + a_nx_n = b, \tag{1}$$

where the real numbers $a_1, a_2, \ldots, a_n$ are not all zero. The number b is called the **constant term** of the equation. The equation in (1) is also called a **first-degree equation** in that the exponent of each of the n variables is 1. In this and the next section we examine solution methods for systems of equations.

☐ Terminology

A **system of equations** consists of two or more equations with each equation containing at least one variable. If each equation in a system is linear, we say that it is a **system of linear equations** or simply a **linear system**. Whenever possible, we will use the familiar symbols x, y, and z to represent variables in a system. For example,

$$\begin{cases} 2x + y - z = 0 \\ x + 3y + z = 2 \\ -x - y + 5z = 14 \end{cases} \tag{2}$$

is a linear system of three equations in three variables. The brace in (2) is just a way of reminding us that we are trying to solve a system of equations and that the equations must be dealt with simultaneously. A **solution** of a system of n equations in n variables consists of values of the variables that satisfy each equation in the system. A solution of such a system is also written as an **ordered n-tuple**. For example, as we see $x = 2$, $y = -1$, and $z = 3$ satisfy each equation in the linear system (2):

$$\begin{cases} 2x + y - z = 0 \\ x + 3y + z = 2 \\ -x - y + 5z = 14 \end{cases} \xrightarrow[\text{and } z = 3]{\substack{\text{substituting} \\ x = 2, y = -1,}} \begin{cases} 2 \cdot 2 + (-1) - 3 = 4 - 4 = 0 \\ 2 + 3(-1) + 3 = 5 - 3 = 2 \\ -2 - (-1) + 5 \cdot 3 = 16 - 2 = 14 \end{cases}$$

and so these values constitute a solution. Alternatively, this solution can be written as the **ordered triple** $(2, -1, 3)$. To **solve** a system of equations we find all solutions of the system. Often to solve a system of equations we perform operations on the system to transform it into an equivalent set of equations. Two systems of equations are said to be **equivalent** if they have precisely the same **solution sets**.

☐ Linear Systems in Two Variables

The simplest linear system consists of two equations in two variables:

$$\begin{cases} a_1x + b_1y = c_1 \\ a_2x + b_2y = c_2. \end{cases} \tag{3}$$

Because the graph of a linear equation $ax + by = c$ is a straight line, the system determines two straight lines in the xy-plane.

☐ Consistent and Inconsistent Systems

As shown in FIGURE 9.1.1 there are three possible cases for the graphs of the equations in system (3):

- The lines intersect in a single point.   ← Figure 9.1.1(a)
- The equations describe coincident lines.   ← Figure 9.1.1(b)
- The two lines are parallel.   ← Figure 9.1.1(c)

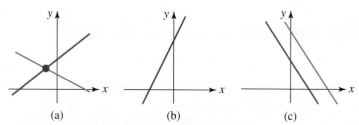

FIGURE 9.1.1 Two lines in the plane

In these three cases we say, respectively:

- The system is **consistent** and the equations are **independent**. The system has exactly one solution, that is, the ordered pair of real numbers corresponding to the point of intersection of the lines.
- The system is **consistent**, but the equations are **dependent**. The system has infinitely many solutions, that is, all the ordered pairs of real numbers corresponding to the points on the one line.
- The system is **inconsistent**. The lines are parallel and so there are no solutions.

For example, the equations in the linear system

$$\begin{cases} x - y = 0 \\ x - y = 3 \end{cases}$$

are parallel lines as in Figure 9.1.1(c). Hence the system is inconsistent.

To solve a system of linear equations, we can use either the method of substitution or the method of elimination.

☐ **Method of Substitution** The first solution technique considered is called the **method of substitution**.

Method of Substitution

(*i*) Use one of the equations in the system to solve for one variable in terms of the other variables.

(*ii*) Substitute this expression into the other equations.

(*iii*) If one of the equations obtained in step (*ii*) contains one variable, then solve it. Otherwise repeat (*i*) until one equation in one variable is obtained.

(*iv*) Finally, use back-substitution to find the values of the remaining variables.

EXAMPLE 1   Method of Substitution

Solve the linear system

$$\begin{cases} 3x + 4y = -5 \\ 2x - y = 4. \end{cases}$$

Solution Solving the second equation for y yields

$$y = 2x - 4.$$

We substitute this expression into the first equation and solve for x:

$$3x + 4(2x - 4) = -5 \quad \text{or} \quad 11x = 11 \quad \text{or} \quad x = 1.$$

We then substitute this value *back* into the first equation:

◀ Back-substitution

$$3(1) + 4y = -5 \quad \text{or} \quad 4y = -8 \quad \text{or} \quad y = -2.$$

Thus the only solution of the system is $(1, -2)$. The system is consistent and the equations are independent.

◀ Solution written as an ordered pair.

$\equiv$

□ **Linear Systems in Three Variables** In Section 7.5 we saw that the graph of a **linear equation in three variables**,

$$ax + by + cz = d,$$

where a, b, and c are not all zero, is a *plane* in three-dimensional space. As we have seen in (2), a solution of a system of three equations in three variables

$$\begin{cases} a_1x + b_1y + c_1z = d_1 \\ a_2x + b_2y + c_2z = d_2 \\ a_3x + b_3y + c_3z = d_3 \end{cases} \qquad (4)$$

is an ordered triple of the form (x, y, z); an ordered triple of numbers represents a point in three-dimensional space. The intersection of the three planes described by the system (4) may be

- a single point,
- infinitely many points, or
- no points.

As before, to each of these cases we apply the terms *consistent and independent*, *consistent and dependent*, and *inconsistent*, respectively. Each is illustrated in FIGURE 9.1.2.

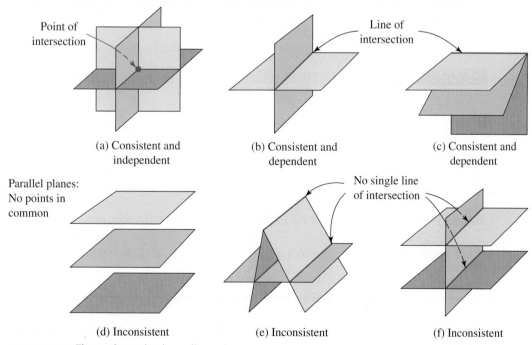

FIGURE 9.1.2 Three planes in three dimensions

□ **Method of Elimination** The next method that we illustrate uses **elimination operations**. When applied to a system of equations, these operations yield an equivalent system of equations.

Method of Elimination

(*i*) Interchange any two equations in a system.

(*ii*) Multiply an equation by a nonzero constant.

(*iii*) Add a nonzero constant multiple of an equation in a system to another equation in the same system.

We often add a nonzero constant multiple of one equation to other equations in a system with the intention of eliminating a variable from those equations.

For convenience, we represent these operations by the following symbols, where the letter E stands for the word *equation*:

$$E_i \leftrightarrow E_j: \quad \text{Interchange the } i\text{th equation with the } j\text{th equation.}$$
$$kE_i: \quad \text{Multiply the } i\text{th equation by a constant } k.$$
$$kE_i + E_j: \quad \text{Multiply the } i\text{th equation by } k \text{ and add to the } j\text{th equation.}$$

Reading a linear system from the top, E_1 represents the first equation, E_2 represents the second equation, and so on.

Using the method of elimination it is possible to reduce the system (4) of three linear equations in three variables to an equivalent system in triangular form,

$$\begin{cases} a_1'x + b_1'y + c_1'z = d_1' \\ \qquad\quad b_2'y + c_2'z = d_2' \\ \qquad\qquad\qquad c_3'z = d_3'. \end{cases}$$

A solution of the system (if one exists) can be readily obtained by **back-substitution**. The next example illustrates the procedure.

EXAMPLE 2   Elimination and Back-Substitution

Solve the linear system

$$\begin{cases} x + 2y + z = -6 \\ 4x - 2y - z = -4 \\ 2x - y + 3z = 19. \end{cases}$$

Solution We begin by eliminating x from the second and third equations:

$$\left. \begin{array}{l} x + 2y + z = -6 \\ 4x - 2y - z = -4 \\ 2x - y + 3z = 19 \end{array} \right\} \xrightarrow[-2E_1 + E_3]{-4E_1 + E_2} \begin{cases} x + 2y + z = -6 \\ \quad -10y - 5z = 20 \\ \quad -5y + z = 31. \end{cases} \tag{5}$$

We then eliminate y from the third equation and obtain an equivalent system in triangular form:

$$\left. \begin{array}{l} x + 2y + z = -6 \\ \quad -10y - 5z = 20 \\ \quad -5y + z = 31 \end{array} \right\} \xrightarrow{-\frac{1}{2}E_2 + E_3} \begin{cases} x + 2y + z = -6 \\ \quad -10y - 5z = 20 \\ \qquad\qquad \frac{7}{2}z = 21. \end{cases} \tag{6}$$

We arrive at another triangular form that is equivalent to the original system by multiplying the third equation by $\frac{2}{7}$:

$$\left. \begin{array}{rcl} x + 2y + z &=& -6 \\ -10y - 5z &=& 20 \\ \frac{7}{2}z &=& 21 \end{array} \right\} \xrightarrow{\frac{2}{7}E_3} \left\{ \begin{array}{rcl} x + 2y + z &=& -6 \\ y + \frac{1}{2}z &=& -2 \\ z &=& 6. \end{array} \right.$$

From this last system it is evident that $z = 6$. Using this value and substituting back into the second equation gives

$$y = -\tfrac{1}{2}z - 2 = -\tfrac{1}{2}(6) - 2 = -5.$$

Finally, by substituting $y = -5$ and $z = 6$ back into the first equation, we obtain

$$x = -2y - z - 6 = -2(-5) - 6 - 6 = -2.$$

Therefore the solution of the system is $(-2, -5, 6)$.   ≡

◀ The answer indicates that the three planes intersect at a point as in Figure 9.1.2 (a).

EXAMPLE 3   Elimination and Back-Substitution

Solve the linear system

$$\left\{ \begin{array}{rcl} x + y + z &=& 2 \\ 5x - 2y + 2z &=& 0 \\ 8x + y + 5z &=& 6. \end{array} \right. \tag{7}$$

Solution Using the first equation to eliminate the variable x from the second and third equations, we get the equivalent system

$$\left. \begin{array}{rcl} x + y + z &=& 2 \\ 5x - 2y + 2z &=& 0 \\ 8x + y + 5z &=& 6 \end{array} \right\} \xrightarrow[-8E_1 + E_3]{-5E_1 + E_2} \left\{ \begin{array}{rcl} x + y + z &=& 2 \\ -7y - 3z &=& -10 \\ -7y - 3z &=& -10. \end{array} \right.$$

This system, in turn, is equivalent to the system in triangular form:

$$\left. \begin{array}{rcl} x + y + z &=& 2 \\ -7y - 3z &=& -10 \\ -7y - 3z &=& -10 \end{array} \right\} \xrightarrow[-E_2 + E_3]{-E_2} \left\{ \begin{array}{rcl} x + y + z &=& 2 \\ 7y + 3z &=& 10 \\ 0z &=& 0. \end{array} \right. \tag{8}$$

In this system we cannot determine unique values for x, y, and z. At best we can solve for two variables in terms of the remaining variable. For example, from the second equation in (8), we obtain y in terms of z:

$$y = -\tfrac{3}{7}z + \tfrac{10}{7}.$$

Substituting this equation for y in the first equation for x gives

$$x + \left(-\tfrac{3}{7}z + \tfrac{10}{7} \right) + z = 2 \quad \text{or} \quad x = -\tfrac{4}{7}z + \tfrac{4}{7}.$$

Thus in the solutions for y and x, we can choose z *arbitrarily*. If we denote z by the symbol α, where α represents a real number, then the solutions of the system are all ordered triples of the form $\left(-\tfrac{4}{7}\alpha + \tfrac{4}{7}, -\tfrac{3}{7}\alpha + \tfrac{10}{7}, \alpha \right)$. We emphasize that for any real number α, we obtain a solution of (7). For example, by choosing α to be, say, 0, 1, and 2, we obtain the solutions $\left(\tfrac{4}{7}, \tfrac{10}{7}, 0 \right)$, $(0, 1, 1)$, and $\left(-\tfrac{4}{7}, \tfrac{4}{7}, 2 \right)$, respectively. In other words, the system is consistent and has infinitely many solutions.   ≡

◀ The answer indicates that the two planes intersect in a line as in Figure 9.1.2 (b).

In Example 3 there is nothing special about solving (8) for x and y in terms of z. For instance, by solving (8) for x and z in terms of y, we obtain the solution $\left(\frac{4}{3}\beta - \frac{4}{3}, \beta, -\frac{7}{3}\beta + \frac{10}{3}\right)$, where β is any real number. Note that by setting β equal to $\frac{10}{7}$, 1, and $\frac{4}{7}$, we get the same solutions in Example 3 corresponding, in turn, to $\alpha = 0$, $\alpha = 1$, and $\alpha = 2$.

EXAMPLE 4   No Solution

Solve the linear system

$$\begin{cases} 2x - y - z = 0 \\ 2x + 3y = 1 \\ 8x - 3z = 4. \end{cases}$$

Solution The elimination method,

$$\left.\begin{array}{l} 2x - y - z = 0 \\ 2x + 3y = 1 \\ 8x - 3z = 4 \end{array}\right\} \xrightarrow[\substack{-E_1 + E_2 \\ -4E_1 + E_3}]{} \left\{\begin{array}{l} 2x - y - z = 0 \\ 4y + z = 1 \\ 4y + z = 4 \end{array}\right.$$

$$\left.\begin{array}{l} 2x - y - z = 0 \\ 4y + z = 1 \\ 4y + z = 4 \end{array}\right\} \xrightarrow[]{-E_2 + E_3} \left\{\begin{array}{l} 2x - y - z = 0 \\ 4y + z = 1 \\ 0z = 3, \end{array}\right.$$

shows that the last equation $0z = 3$ is *never* satisfied for any number z since $0 \neq 3$. Thus, the system is inconsistent and so has no solutions. ≡

EXAMPLE 5   Elimination and Back-Substitution

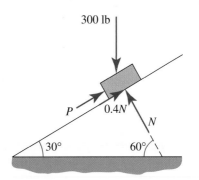

300 lb

P 0.4N

N

30°   60°

FIGURE 9.1.3 Inclined plane in Example 5

A force of smallest magnitude P is applied to a 300-lb block on an inclined plane in order to keep it from sliding down the plane. See FIGURE 9.1.3. If the coefficient of friction between the block and the surface is 0.4, then the magnitude of the frictional force is 0.4N, where N is the magnitude of the normal force exerted on the block by the plane. Since the system is in equilibrium, the horizontal and the vertical components of the forces must be zero:

$$\begin{cases} P\cos 30° + 0.4N\cos 30° - N\cos 60° = 0 \\ P\sin 30° + 0.4N\sin 30° + N\sin 60° - 300 = 0. \end{cases}$$

Solve this system for P and N.

Solution Using $\sin 30° = \cos 60° = \frac{1}{2}$ and $\sin 60° = \cos 30° = \sqrt{3}/2$, we simplify the system above to

$$\begin{cases} \sqrt{3}P + \left(0.4\sqrt{3} - 1\right)N = 0 \\ P + \left(0.4 + \sqrt{3}\right)N = 600. \end{cases}$$

By elimination,

$$\left.\begin{array}{l} \sqrt{3}P + \left(0.4\sqrt{3} - 1\right)N = 0 \\ P + \left(0.4 + \sqrt{3}\right)N = 600 \end{array}\right\} \xrightarrow{E_1 - \sqrt{3}E_2} \left\{\begin{array}{l} \sqrt{3}P + \left(0.4\sqrt{3} - 1\right)N = 0 \\ -4N = -600\sqrt{3} \end{array}\right.$$

$$\left.\begin{array}{l} \sqrt{3}P + \left(0.4\sqrt{3} - 1\right)N = 0 \\ -4N = -600\sqrt{3} \end{array}\right\} \xrightarrow{-\frac{1}{4}E_2} \left\{\begin{array}{l} \sqrt{3}P + \left(0.4\sqrt{3} - 1\right)N = 0 \\ N = 150\sqrt{3}. \end{array}\right.$$

The second equation of the last system gives $N = 150\sqrt{3} \approx 259.81$ lb. By substituting this value into the first equation we obtain $P = 150\left(1 - 0.4\sqrt{3}\right) \approx 46.08$ lb. ≡

□ **Homogeneous Systems** A linear system in which all the constant terms are zero, such as

$$\begin{cases} a_1x + b_1y = 0 \\ a_2x + b_2y = 0 \end{cases} \tag{9}$$

or

$$\begin{cases} a_1x + b_1y + c_1z = 0 \\ a_2x + b_2y + c_2z = 0 \\ a_3x + b_3y + c_3z = 0, \end{cases} \tag{10}$$

is said to be **homogeneous**. Note that systems (9) and (10) have the solutions $(0, 0)$ and $(0, 0, 0)$, respectively. A solution of a system of equations in which each of its variables is zero is called the **zero solution** or the **trivial solution**. Because a homogeneous linear system always possesses at least the zero solution, such a system is *always consistent*. In addition to the zero solution, however, there *may* exist infinitely many nonzero solutions. These solutions can be found by proceeding exactly as in Example 3.

◀ A homogeneous system is consistent even in the case where the linear system consists of m equations in n variables, where $m \neq n$.

EXAMPLE 6 A Homogeneous System

The same steps used to solve the system in Example 3 can be used to solve the related homogeneous system

$$\begin{cases} x + 2y + z = 0 \\ 5x - 2y + 2z = 0 \\ 8x + 2y + 5z = 0. \end{cases}$$

In this case the elimination steps yield

$$\begin{cases} x + y + z = 0 \\ 7y + 3z = 0 \\ 0z = 0. \end{cases}$$

Choosing $z = \alpha$, where α is a real number, we find from the second equation of the last system that $y = -\frac{3}{7}\alpha$. Then using the first equation, we obtain $x = -\frac{4}{7}\alpha$. Thus, the solutions of the system consist of all ordered triples of the form $\left(-\frac{4}{7}\alpha, -\frac{3}{7}\alpha, \alpha\right)$. Note that for $\alpha = 0$, we obtain the trivial solution $(0, 0, 0)$ but for, say, $\alpha = -7$, we obtain the non-trivial solution $(4, 3, -7)$. ≡

The two techniques in this section are also applicable to systems of n linear equations in n variables for $n > 3$. See Problems 25 and 26 in Exercises 9.1. In addition, these techniques are applicable to linear systems where the number of equations is not the same as the number of variables. See Problems 27–30 in Exercises 9.1. In the following example, we consider a system of 2 equations in 3 variables.

◀ Note

EXAMPLE 7 Two Equations in Three Variables

Use method of elimination to solve the linear system

$$\begin{cases} x + 2y - 4z = 6 \\ 5x - y + 2z = -3. \end{cases}$$

Solution We have

$$
\begin{aligned}
\left.\begin{array}{l} x + 2y - 4z = 6 \\ 5x - y + 2z = -3 \end{array}\right\} & \xrightarrow{-5E_1 + E_2} \left\{\begin{array}{l} x + 2y - 4z = 6 \\ -11y + 22z = -33 \end{array}\right. \\[2mm]
\left.\begin{array}{l} x + 2y - 4z = 6 \\ -11y + 22z = -33 \end{array}\right\} & \xrightarrow{-\frac{1}{11}E_2} \left\{\begin{array}{l} x + 2y - 4z = 6 \\ y - 2z = 3 \end{array}\right. \\[2mm]
\left.\begin{array}{l} x + 2y - 4z = 6 \\ y - 2z = 3 \end{array}\right\} & \xrightarrow{-2E_2 + E_1} \left\{\begin{array}{l} x \qquad\qquad = 0 \\ y - 2z = 3. \end{array}\right.
\end{aligned}
$$

The last system indicates that $x = 0$ and $y = 2z + 3$. As in Examples 3 and 6, we may now assign any value to z. Hence the solutions of the system are all ordered triples of the form $(0, 2\alpha + 3, \alpha)$ where α is any real number. $\equiv$

A homogeneous linear system, regardless of the number equations and variables is consistent. In our last example we consider an example from chemistry where we must solve a homogeneous system of 3 equations in 4 variables.

EXAMPLE 8   Balancing a Chemical Equation

Balance the chemical equation $C_2H_6 + O_2 \rightarrow CO_2 + H_2O$.

Solution We seek positive integers x, y, z, and w so that the balanced equation is

$$x C_2H_6 + y O_2 \rightarrow z CO_2 + w H_2O.$$

Because the number of atoms of each element must be the same on each side of the last equation, we obtain the homogeneous system of 3 equations in 4 variables:

$$
\begin{array}{lll}
\text{carbon (C):} & 2x = z \\
\text{hydrogen (H):} & 6x = 2w \qquad \text{or} \\
\text{oxygen (O):} & 2y = 2z + w
\end{array}
\left\{\begin{array}{l}
2x + 0y - z + 0w = 0 \\
6x + 0y + 0z - 2w = 0 \\
0x + 2y - 2z - w = 0.
\end{array}\right.
$$

Since the last system is homogeneous it must be consistent. Using elementary row operations, we find

$$
\left.\begin{array}{l}
2x + 0y - z + 0w = 0 \\
6x + 0y + 0z - 2w = 0 \\
0x + 2y - 2z - w = 0
\end{array}\right\}
\xrightarrow[\text{operations}]{\text{elimination}}
\left\{\begin{array}{l}
x \qquad\qquad - \frac{1}{3}w = 0 \\
y \qquad - \frac{7}{6}w = 0 \\
z - \frac{2}{3}w = 0
\end{array}\right.
$$

and so a solution of the system is $x = \frac{1}{3}\alpha$, $y = \frac{7}{6}\alpha$, $z = \frac{2}{3}\alpha$, $w = \alpha$. In this case α must be a positive integer chosen in such a manner so that x, y, z, and w are positive integers. To accomplish this we pick $\alpha = 6$. This gives $x = 2$, $y = 7$, $z = 4$, and $w = 6$. The balanced equation is then

$$2C_2H_6 + 7O_2 \rightarrow 4CO_2 + 6H_2O. \qquad \equiv$$

9.1   Exercises   Answers to selected odd-numbered problems begin on page ANS-26.

In Problems 1–26, solve the given linear system. State whether the system is consistent, with independent or dependent equations, or whether it is inconsistent.

1. $\begin{cases} 2x + y = 2 \\ 3x - 2y = -4 \end{cases}$ **2.** $\begin{cases} 2x - 2y = 1 \\ 3x + 5y = 11 \end{cases}$

3. $\begin{cases} 4x - y + 1 = 0 \\ x + 3y + 9 = 0 \end{cases}$

4. $\begin{cases} x - 4y + 1 = 0 \\ 3x + 2y - 1 = 0 \end{cases}$

5. $\begin{cases} x - 2y = 6 \\ -0.5x + y = 1 \end{cases}$

6. $\begin{cases} 6x - 4y = 9 \\ -3x + 2y = -4.5 \end{cases}$

7. $\begin{cases} x - y = 2 \\ x + y = 1 \end{cases}$

8. $\begin{cases} 2x + y = 4 \\ 2x + y = 0 \end{cases}$

9. $\begin{cases} -x - 2y + 4 = 0 \\ 5x + 10y - 20 = 0 \end{cases}$

10. $\begin{cases} 7x - 3y - 14 = 0 \\ x + y - 1 = 0 \end{cases}$

11. $\begin{cases} x + y - z = 0 \\ x - y + z = 2 \\ 2x + y - 4z = -8 \end{cases}$

12. $\begin{cases} x + y + z = 8 \\ x - 2y + z = 4 \\ x + y - z = -4 \end{cases}$

13. $\begin{cases} 2x + 6y + z = -2 \\ 3x + 4y - z = 2 \\ 5x - 2y - 2z = 0 \end{cases}$

14. $\begin{cases} x + 7y - 4z = 1 \\ 2x + 3y + z = -3 \\ -x - 18y + 13z = 2 \end{cases}$

15. $\begin{cases} 2x + y + z = 1 \\ x - y + 2z = 5 \\ 3x + 4y - z = -2 \end{cases}$

16. $\begin{cases} x + y - 5z = -1 \\ 4x - y + 3z = 1 \\ 5x - 5y + 21z = 5 \end{cases}$

17. $\begin{cases} x - 5y + z = 0 \\ 10x + y + 3z = 0 \\ 4x + 2y - 5z = 0 \end{cases}$

18. $\begin{cases} -5x + y + z = 0 \\ 4x - y = 0 \\ 2x - y + 2z = 0 \end{cases}$

19. $\begin{cases} x - 3y = 22 \\ y + 6z = -3 \\ \frac{1}{3}x + 2z = 3 \end{cases}$

20. $\begin{cases} 2x - z = 12 \\ x + y = 7 \\ 5x + 4z = -9 \end{cases}$

21. $\begin{cases} -x + 3y + 2z = 2 \\ \frac{1}{2}x - \frac{3}{2}y - z = -1 \\ -\frac{1}{3}x + y + \frac{2}{3}z = \frac{2}{3} \end{cases}$

22. $\begin{cases} x + 6y + z = 9 \\ 3x + y - 2z = 7 \\ -6x + 3y + 7z = -2 \end{cases}$

23. $\begin{cases} x + y - z = 0 \\ 2x + 2y - 2z = 1 \\ 5x + 5y - 5z = 2 \end{cases}$

24. $\begin{cases} x + y + z = 4 \\ 2x - y + 2z = 11 \\ 4x + 3y - 6z = -18 \end{cases}$

25. $\begin{cases} 2x - y + 3z - w = 8 \\ x + y - z + w = 3 \\ x - y + 5z - 3w = -1 \\ 6x + 2y + z - w = -2 \end{cases}$

26. $\begin{cases} x - 2y + z - 3w = 0 \\ 8x - 8y - z - 5w = 16 \\ -x - y + 3w = -6 \\ 4x - 7y + 3z - 10w = 2 \end{cases}$

In Problems 27–30, solve the given linear system.

27. $\begin{cases} x - y + 4z = 1 \\ 6x + y - z = 2 \end{cases}$

28. $\begin{cases} 4x - 2y + z = 9 \\ y - z = 2 \end{cases}$

29. $\begin{cases} 2x - 3y = 2 \\ x + 2y = 1 \\ 3x + 2y = -1 \end{cases}$

30. $\begin{cases} x - y + z = 0 \\ x + y - z = 0 \end{cases}$

In Problems 31–36, use the procedure illustrated in Example 8 to balance the given chemical equation.

31. $Na + H_2O \rightarrow NaOH + H_2$

32. $KClO_3 \rightarrow KCl + O_2$

33. $Fe_3O_4 + C \rightarrow Fe + CO$

34. $C_5H_8 + O_2 \rightarrow CO_2 + H_2O$

35. $Cu + HNO_3 \rightarrow Cu(NO_3)_2 + H_2O + NO$

36. $Ca_3(PO_4)_2 + H_3PO_4 \rightarrow Ca(H_2PO_4)_2$

In Problems 37–40, each system is nonlinear in the given variables. Use substitutions to convert the system into one that is linear in the new variables. Solve, and then give the solution of the original system.

37. $\begin{cases} \dfrac{1}{x} - \dfrac{1}{y} = \dfrac{1}{6} \\ \dfrac{4}{x} + \dfrac{3}{y} = 3 \end{cases}$

38. $\begin{cases} \dfrac{1}{x} - \dfrac{1}{y} + \dfrac{2}{z} = 3 \\ \dfrac{2}{x} + \dfrac{1}{y} - \dfrac{4}{z} = -1 \\ \dfrac{3}{x} + \dfrac{1}{y} + \dfrac{1}{z} = \dfrac{5}{2} \end{cases}$

39. $\begin{cases} 3\log_{10} x + \log_{10} y = 2 \\ 5\log_{10} x + 2\log_{10} y = 1 \end{cases}$

40. $\begin{cases} \cos x - \sin y = 1 \\ 2\cos x + \sin y = -1 \end{cases}$

41. The magnitudes T_1 and T_2 of the tensions in the two cables shown in **FIGURE 9.1.4** satisfy the system of equations

$$\begin{cases} T_1\cos 25° - T_2\cos 15° = 0 \\ T_1\sin 25° + T_2\sin 15° - 200 = 0. \end{cases}$$

Find T_1 and T_2.

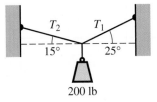

FIGURE 9.1.4 Cables in Problem 41

42. If we change the direction of the frictional force in Figure 9.1.3 of Example 5, then the system of equations becomes

$$\begin{cases} P\cos 30° - 0.4N\cos 30° - N\cos 60° = 0 \\ P\sin 30° - 0.4N\sin 30° + N\sin 60° - 300 = 0. \end{cases}$$

In this case, P represents the magnitude of the force that is just enough to start the block up the plane. Find P and N.

Miscellaneous Applications

43. **Speed** An airplane flies 3300 mi from Hawaii to California in 5.5 h with a tailwind. From California to Hawaii, flying against a wind of the same velocity, the trip takes 6 h. Determine the speed of the plane and the speed of the wind.

44. **How Many Coins?** A person has 20 coins, consisting of dimes and quarters, which total $4.25. Determine how many of each coin the person has.

45. **Number of Gallons** A 100-gal tank is full of water in which 50 lb of salt is dissolved. A second tank contains 200 gal of water with 75 lb of salt. How much should be removed from both tanks and mixed together in order to make a solution of 90 gal with $\frac{4}{9}$ lb of salt per gallon?

46. **Playing with Numbers** The sum of three numbers is 20. The difference of the first two numbers is 5, and the third number is 4 times the sum of the first two. Find the numbers.

47. **How Long?** Three pumps P_1, P_2, and P_3 working together can fill a tank in 2 hours. Pumps P_1 and P_2 can fill the same tank in 3 hours, whereas pumps P_2 and P_3 can fill it in 4 hours. Determine how long it would take each pump working alone to fill the tank.

48. **Parabola Through Three Points** The parabola $y = ax^2 + bx + c$ passes through the points $(1, 10)$, $(-1, 12)$, and $(2, 18)$. Find a, b, and c.

49. **Area** Find the area of the right triangle shown in **FIGURE 9.1.5**.

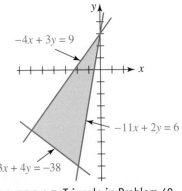

FIGURE 9.1.5 Triangle in Problem 49

50. Current According to Kirchhoff's law of voltages, the currents i_1, i_2, and i_3 in the parallel circuit shown in FIGURE 9.1.6 satisfy the equations

$$\begin{cases} i_1 + 2(i_1 - i_2) + 0i_3 & = 6 \\ 3i_2 + 4(i_2 - i_3) + 2(i_2 - i_1) & = 0 \\ 2i_3 + 4(i_3 - i_2) + 0i_1 & = 12. \end{cases}$$

Solve for i_1, i_2, and i_3.

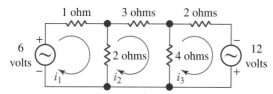

FIGURE 9.1.6 Circuit in Problem 50

51. The A, B, C's When Beth graduated from college, she had completed 40 courses, in which she received grades of A, B, and C. Her final GPA (grade point average) was 3.125. Her GPA in only those courses in which she received grades of A and B was 3.8. Assume that A, B, and C grades are worth four points, three points, and two points, respectively. Determine the number of A's, B's, and C's that Beth received.

52. Conductivity Cosmic rays are deflected toward the poles by the Earth's magnetic field, so that only the most energetic rays can penetrate the equatorial regions. See FIGURE 9.1.7. As a result, the ionization rate, and hence the conductivity σ of the stratosphere, is greater near the poles than it is near the equator. Conductivity can be approximated by the formula

$$\sigma = (A + B \sin^4\phi)^{1/2},$$

where ϕ is latitude and A and B are constants that must be chosen to fit the physical data. Balloon measurements made in the southern hemisphere indicated a conductivity of approximately 3.8×10^{-12} siemens/meter at $35.5°$ south latitude and 5.6×10^{-12} siemens/meter at $51°$ south latitude. (A *siemen* is the reciprocal of an *ohm*, which is a unit of electrical resistance.) Determine the constants A and B. What is the conductivity at $42°$ south latitude?

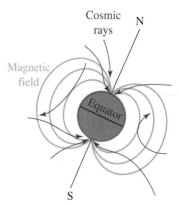

FIGURE 9.1.7 Earth's magnetic field in Problem 52

For Discussion

53. Determine conditions on a_1, a_2, b_1, and b_2 so that the linear system (9) has only the trivial solution.

54. Determine a value of k such that the linear system

$$\begin{cases} 2x - 3y = 10 \\ 6x - 9y = k \end{cases}$$

is **(a)** inconsistent, and **(b)** dependent.

55. Devise a system of two linear equations whose solution is $(2, -5)$.

Determinants and Cramer's Rule

≡ **Introduction** If a linear system of equations has the same number of equations as it does variables, say two equations and two variables, then it may be possible to solve the system using determinants. If $a_{11}, a_{12}, \ldots$ represent real numbers, then the symbols

$$\begin{vmatrix} a_{11} & a_{12} \\ a_{21} & a_{22} \end{vmatrix} \quad \text{and} \quad \begin{vmatrix} a_{11} & a_{12} & a_{13} \\ a_{21} & a_{22} & a_{23} \\ a_{31} & a_{32} & a_{33} \end{vmatrix} \tag{1}$$

are called **determinants of order 2** and **order 3**, respectively. Although a determinant can be any order n, n a positive integer, most of the elementary applications of determinants in calculus utilize second- and third-order determinants.

☐ **Determinant of Order 2** A determinant is a number. In the case of a determinant of order 2 we have the following definition.

DEFINITION 9.2.1 Determinant of Order 2

A determinant of order 2 is the number

$$\begin{vmatrix} a_{11} & a_{12} \\ a_{21} & a_{22} \end{vmatrix} = a_{11}a_{22} - a_{12}a_{21}. \tag{2}$$

EXAMPLE 1 Determinant of Order 2

Evaluate the determinant $\begin{vmatrix} 2 & 3 \\ 4 & 5 \end{vmatrix}$.

Solution From (2) of Definition 9.2.1,

$$\begin{vmatrix} 2 & 3 \\ 4 & 5 \end{vmatrix} = 2(5) - 3(4) = 10 - 12 = -2. \qquad ≡$$

As a mnemonic for the formula in (2), remember that the determinant is the difference of the products of the diagonal entries:

$$\begin{matrix} \text{multiply} & \text{multiply} & \overset{\text{subtract}}{\underset{\downarrow}{\text{products}}} \\ \begin{vmatrix} a_{11} & a_{12} \\ a_{21} & a_{22} \end{vmatrix} & = & a_{11}a_{22} - a_{12}a_{21}. \end{matrix}$$

Even though a determinant is a *number* it is convenient to think of it as a square array. Thus determinants of orders 2 and 3 are also referred to as 2×2 (read "two by two") and 3×3 ("three by three") determinants, respectively.

Determinants of order 2 play a fundamental role in the evaluating of determinants of order n, $n > 2$. In general, a determinant of order n can be expressed in terms of determinants of order $n - 1$. Thus, for example, a 3×3 determinant can be expressed in terms of determinants of order 2. In preparation for a method for finding the value of a 3×3 determinant we need to introduce the notion of a cofactor determinant.

☐ **Minor and Cofactor** If a_{ij} denotes the entry in the ith row and jth column of a determinant of order n, then the **minor** M_{ij} of a_{ij} is defined to be the determinant of order $n - 1$ obtained by deleting the ith row and the jth column of the determinant of order n. Thus for the 3×3 determinant

$$\begin{vmatrix} 1 & 5 & 3 \\ 2 & 4 & 5 \\ 1 & 2 & 3 \end{vmatrix} \tag{3}$$

the minors of $a_{11} = 1$, $a_{12} = 5$, $a_{22} = 4$, and $a_{32} = 2$ are, in turn, the determinants

delete first column ↓

delete first row →
$$M_{11} = \begin{vmatrix} 1 & 5 & 3 \\ 2 & 4 & 5 \\ 1 & 2 & 3 \end{vmatrix} = \begin{vmatrix} 4 & 5 \\ 2 & 3 \end{vmatrix} = 4(3) - 5(2) = 2,$$

delete second column ↓

delete first row →
$$M_{12} = \begin{vmatrix} 1 & 5 & 3 \\ 2 & 4 & 5 \\ 1 & 2 & 3 \end{vmatrix} = \begin{vmatrix} 2 & 5 \\ 1 & 3 \end{vmatrix} = 2(3) - 5(1) = 1,$$

delete second row →
$$M_{22} = \begin{vmatrix} 1 & 5 & 3 \\ 2 & 4 & 5 \\ 1 & 2 & 3 \end{vmatrix} = \begin{vmatrix} 1 & 3 \\ 1 & 3 \end{vmatrix} = 1(3) - 3(1) = 0,$$

$$M_{32} = \begin{vmatrix} 1 & 5 & 3 \\ 2 & 4 & 5 \\ 1 & 2 & 3 \end{vmatrix} = \begin{vmatrix} 1 & 3 \\ 2 & 5 \end{vmatrix} = 1(5) - 3(2) = -1,$$

delete third row →

The **cofactor** A_{ij} of the entry a_{ij} is defined to be the minor M_{ij} multiplied by $(-1)^{i+j}$, that is,

$$A_{ij} = (-1)^{i+j} M_{ij}. \tag{4}$$

◄ $(-1)^{i+j}$ is 1 if $i + j$ is even, $(-1)^{i+j}$ is -1 if $i + j$ is odd.

Thus for the determinant in (3) the cofactors associated with the foregoing minor determinants are

$$A_{11} = (-1)^{1+1} M_{11} = 2,$$
$$A_{12} = (-1)^{1+2} M_{12} = -1,$$
$$A_{22} = (-1)^{2+2} M_{22} = 0,$$
$$A_{32} = (-1)^{3+2} M_{32} = -(-1) = 1,$$

and so on. For a 3×3 determinant, the coefficient $(-1)^{i+j}$ of the minor M_{ij} follows the pattern

$$\begin{vmatrix} + & - & + \\ - & + & - \\ + & - & + \end{vmatrix}.$$

This "checkerboard" pattern of signs extends to determinants of greater order as well.

EXAMPLE 2   Cofactors

Find the cofactor of the given entry: **(a)** 0   **(b)** 7   **(c)** -1 for the determinant

$$\begin{vmatrix} -2 & 1 & 0 \\ 5 & -3 & 7 \\ -1 & 6 & -5 \end{vmatrix}.$$

Solution (a) The number 0 is the entry of the first row ($i = 1$) and third column ($j = 3$). From (4) the cofactor of 0 is the determinant

$$A_{13} = (-1)^{1+3}M_{13} = (1)\begin{vmatrix} 5 & -3 \\ -1 & 6 \end{vmatrix} = 30 - 3 = 27.$$

(b) The number 7 is the entry in the second row ($i = 2$) and third column ($j = 3$). Thus the cofactor is

$$A_{23} = (-1)^{2+3}M_{23} = (-1)\begin{vmatrix} -2 & 1 \\ -1 & 6 \end{vmatrix} = (-1)\cdot[-12 - (-1)] = 11.$$

(c) Finally, because -1 is the entry in the third row ($i = 3$) and first column ($j = 1$), its cofactor is

$$A_{31} = (-1)^{3+1}M_{31} = (1)\begin{vmatrix} 1 & 0 \\ -3 & 7 \end{vmatrix} = 7 - 0 = 7. \qquad \equiv$$

We are now in a position to find the value of a determinant of order 3. Although the next theorem is stated for 3×3 determinants, it is equally valid for any determinant of order n.

THEOREM 9.2.1 Expansion Theorem

The value of a 3×3 determinant can be found by multiplying each entry in any row (or column) by its cofactor and adding the results.

When we use Theorem 9.2.1 to find the value of a determinant we say that we have **expanded the determinant of A by a given row or by a given column.** For example, the expansion of the determinant of the 3×3 matrix in (1) by the first row is

$$\begin{vmatrix} a_{11} & a_{12} & a_{13} \\ a_{21} & a_{22} & a_{23} \\ a_{31} & a_{32} & a_{33} \end{vmatrix} = a_{11}A_{11} + a_{12}A_{12} + a_{13}A_{13}$$

$$= a_{11}(-1)^{1+1}\begin{vmatrix} a_{22} & a_{23} \\ a_{32} & a_{33} \end{vmatrix} + a_{12}(-1)^{1+2}\begin{vmatrix} a_{21} & a_{23} \\ a_{31} & a_{33} \end{vmatrix} + a_{13}(-1)^{1+3}\begin{vmatrix} a_{21} & a_{22} \\ a_{31} & a_{32} \end{vmatrix}$$

or
$$\begin{vmatrix} a_{11} & a_{12} & a_{13} \\ a_{21} & a_{22} & a_{23} \\ a_{31} & a_{32} & a_{33} \end{vmatrix} = a_{11}\begin{vmatrix} a_{22} & a_{23} \\ a_{32} & a_{33} \end{vmatrix} - a_{12}\begin{vmatrix} a_{21} & a_{23} \\ a_{31} & a_{33} \end{vmatrix} + a_{13}\begin{vmatrix} a_{21} & a_{22} \\ a_{31} & a_{32} \end{vmatrix}. \qquad (5)$$

◼ **EXAMPLE 3** **Expansion by the First Row**

Evaluate the 3×3 determinant

$$\begin{vmatrix} 6 & 5 & 3 \\ 2 & 4 & 5 \\ 1 & 2 & -3 \end{vmatrix}.$$

Solution Using the expansion by the first row given in (5) we have

$$\begin{vmatrix} 6 & 5 & 3 \\ 2 & 4 & 5 \\ 1 & 2 & -3 \end{vmatrix} = 6 \begin{vmatrix} 4 & 5 \\ 2 & -3 \end{vmatrix} - 5 \begin{vmatrix} 2 & 5 \\ 1 & -3 \end{vmatrix} + 3 \begin{vmatrix} 2 & 4 \\ 1 & 2 \end{vmatrix}$$

$$= 6 \cdot (-22) - 5 \cdot (-11) + 3 \cdot 0 = -77. \qquad \equiv$$

Theorem 9.2.1 states that a determinant can be expanded by *any* row or *any* column. For example, the expansion of the 3×3 determinant in (1) by, say, the second row is

$$\begin{vmatrix} a_{11} & a_{12} & a_{13} \\ a_{21} & a_{22} & a_{23} \\ a_{31} & a_{32} & a_{33} \end{vmatrix} = a_{21}A_{21} + a_{22}A_{22} + a_{23}A_{23}$$

$$= (-1)^{2+1}a_{21} \begin{vmatrix} a_{12} & a_{13} \\ a_{32} & a_{33} \end{vmatrix} + (-1)^{2+2}a_{22} \begin{vmatrix} a_{11} & a_{13} \\ a_{31} & a_{33} \end{vmatrix} + (-1)^{2+3}a_{23} \begin{vmatrix} a_{11} & a_{12} \\ a_{31} & a_{32} \end{vmatrix}$$

$$= -a_{21} \begin{vmatrix} a_{12} & a_{13} \\ a_{32} & a_{33} \end{vmatrix} + a_{22} \begin{vmatrix} a_{11} & a_{13} \\ a_{31} & a_{33} \end{vmatrix} + a_{23} \begin{vmatrix} a_{11} & a_{12} \\ a_{31} & a_{32} \end{vmatrix}$$

EXAMPLE 4   Example 3 Revisited

The expansion of the determinant in Example 3 by the third column is

$$\begin{vmatrix} 6 & 5 & 3 \\ 2 & 4 & 5 \\ 1 & 2 & -3 \end{vmatrix} = 3(-1)^{1+3} \begin{vmatrix} 2 & 4 \\ 1 & 2 \end{vmatrix} + 5(-1)^{2+3} \begin{vmatrix} 6 & 5 \\ 1 & 2 \end{vmatrix} + (-3)(-1)^{3+3} \begin{vmatrix} 6 & 5 \\ 2 & 4 \end{vmatrix}$$

$$= 3 \begin{vmatrix} 2 & 4 \\ 1 & 2 \end{vmatrix} + 5(-1) \begin{vmatrix} 6 & 5 \\ 1 & 2 \end{vmatrix} + (-3) \begin{vmatrix} 6 & 5 \\ 2 & 4 \end{vmatrix}$$

$$= 3 \cdot 0 + 5 \cdot (-1) \cdot 7 + (-3) \cdot 14 = -77. \qquad \equiv$$

In the expansion of a determinant, since the entries in a row (or column) multiply the cofactors of that row (or column), it makes sense that if a determinant has a row (or column) with several 0 entries that we expand the determinant by that row (or column). It also follows, that if a determinant has an entire row (or column) of 0 entries, then the value of the determinant is 0.   ◀ Note

□ **Cramer's Rule** Suppose the linear equations

$$\begin{cases} a_1x + b_1y = c_1 \\ a_2x + b_2y = c_2 \end{cases} \tag{6}$$

are independent. If we multiply the first equation by b_2 and the second by $-b_1$, we obtain the equivalent system

$$\begin{cases} a_1b_2x + b_1b_2y = c_1b_2 \\ -a_2b_1x - b_2b_1y = -c_2b_1. \end{cases}$$

Then we can eliminate the y-variable and solve for x by adding the two equations:

$$x = \frac{c_1b_2 - b_1c_2}{a_1b_2 - b_1a_2}. \tag{7}$$

Similarly, by eliminating the x-variable, we find

$$y = \frac{a_1 c_2 - c_1 a_2}{a_1 b_2 - b_1 a_2}. \tag{8}$$

The numerators and the common denominator in (7) and (8) can be written as 2×2 determinants. If we denote these determinants by

$$D = \begin{vmatrix} a_1 & b_1 \\ a_2 & b_2 \end{vmatrix}, \quad D_x = \begin{vmatrix} c_1 & b_1 \\ c_2 & b_2 \end{vmatrix}, \quad D_y = \begin{vmatrix} a_1 & c_1 \\ a_2 & c_2 \end{vmatrix}, \tag{9}$$

then we can summarize the discussion in a compact fashion.

THEOREM 9.2.2 Two Equations in Two Variables

If $D \neq 0$, then the system (6) has the unique solution

$$x = \frac{D_x}{D}, \; y = \frac{D_y}{D} \tag{10}$$

By comparing the determinant D in (9) with the system (6) we see that D is the determinant of the coefficients of x and y. Moreover, a careful inspection of D_x and D_y reveals that these determinants are, in turn, D with the x-coefficients and the y-coefficients replaced by the numbers c_1 and c_2.

EXAMPLE 5　　　　**Using (10)**

Solve the linear system

$$\begin{cases} 3x - y = -3 \\ -2x + 4y = 6. \end{cases}$$

Solution Since

$$D = \begin{vmatrix} 3 & -1 \\ -2 & 4 \end{vmatrix} = 10 \neq 0,$$

Theorem 9.2.2 guarantees that the system has a unique solution. Continuing, we find

$$D_x = \begin{vmatrix} -3 & -1 \\ 6 & 4 \end{vmatrix} = -6, \; D_y = \begin{vmatrix} 3 & -3 \\ -2 & 6 \end{vmatrix} = 12.$$

From (10) the solution is given by

$$x = \frac{-6}{10} = -\frac{3}{5} \quad \text{and} \quad y = \frac{12}{10} = \frac{6}{5}. \qquad \equiv$$

In like manner, the solution (10) can be extended to larger systems of linear equations. In particular, for a system of three equations in three variables,

$$\begin{cases} a_1 x + b_1 y + c_1 z = d_1 \\ a_2 x + b_2 y + c_2 z = d_2 \\ a_3 x + b_3 y + c_3 z = d_3 \end{cases} \tag{11}$$

the determinants that correspond to those in (9) are

$$D = \begin{vmatrix} a_1 & b_1 & c_1 \\ a_2 & b_2 & c_2 \\ a_3 & b_3 & c_3 \end{vmatrix}, D_x = \begin{vmatrix} d_1 & b_1 & c_1 \\ d_2 & b_2 & c_2 \\ d_3 & b_3 & c_3 \end{vmatrix}, D_y = \begin{vmatrix} a_1 & d_1 & c_1 \\ a_2 & d_2 & c_2 \\ a_3 & d_3 & c_3 \end{vmatrix}, D_z = \begin{vmatrix} a_1 & b_1 & d_1 \\ a_2 & b_2 & d_2 \\ a_3 & b_3 & d_3 \end{vmatrix}. \quad (12)$$

As in (10), the determinants D_x, D_y, and D_z are obtained from the determinant D of the coefficients of the system, by replacing the x-, y-, and z-coefficients, respectively, by the numbers d_1, d_2, and d_3. The solution of (11) that is analogous to (10) is given next.

THEOREM 9.2.3 Three Equations in Three Variables

If $D \neq 0$, then the system (11) has the unique solution

$$x = \frac{D_x}{D}, \ y = \frac{D_y}{D}, \ z = \frac{D_z}{D} \quad (13)$$

The solutions in (10) and (13) are special cases of a more general method known as **Cramer's Rule**, named after the Swiss mathematician **Gabriel Cramer** (1704–1752) who was the first to publish these results.

EXAMPLE 6   Using Cramer's Rule

Solve the linear system

$$\begin{cases} -x + 2y + 4z = 9 \\ x - y + 6z = -2 \\ 4x + 6y - 2z = -1. \end{cases}$$

Solution We must evaluate four determinants using the cofactor expansion (4). We begin by finding the value of the determinant of the coefficients of the variables in the system:

$$D = \begin{vmatrix} -1 & 2 & 4 \\ 1 & -1 & 6 \\ 4 & 6 & -2 \end{vmatrix} = 126 \neq 0.$$

The fact that this determinant is nonzero is sufficient to indicate that the system is consistent and has a unique solution. Continuing, we find

$$D_x = \begin{vmatrix} 9 & 2 & 4 \\ -2 & -1 & 6 \\ -1 & 6 & -2 \end{vmatrix} = -378, \ D_y = \begin{vmatrix} -1 & 9 & 4 \\ 1 & -2 & 6 \\ 4 & -1 & -2 \end{vmatrix} = 252, \ D_z = \begin{vmatrix} -1 & 2 & 9 \\ 1 & -1 & -2 \\ 4 & 6 & -1 \end{vmatrix} = 63.$$

From (13), the solution of the system is then

$$x = \frac{-378}{126} = -3, \quad y = \frac{252}{126} = 2, \quad z = \frac{63}{126} = \frac{1}{2}. \qquad \equiv$$

EXAMPLE 7   **Using Cramer's Rule**

Solve the linear system

$$\begin{cases} 2x + 3y + z = 3 \\ \qquad\quad y - 2z = -8 \\ \qquad -3y + 2z = 4. \end{cases}$$

Solution As in Example 6 we begin by finding the value of the determinant of the coefficients of the variables:

$$D = \begin{vmatrix} 2 & 3 & 1 \\ 0 & 1 & -2 \\ 0 & -3 & 2 \end{vmatrix}.$$

Because the first column of this determinant has two zero entries we expand D by the first column:

$$D = 2 \begin{vmatrix} 1 & -2 \\ -3 & 2 \end{vmatrix} = 2(2 - (-2)(-3)) = -8.$$

Because $D \neq 0$ we continue and find the values

$$D_x = \begin{vmatrix} 3 & 3 & 1 \\ -8 & 1 & -2 \\ 4 & -3 & 2 \end{vmatrix} = 32, \; D_y = \begin{vmatrix} 2 & 3 & 1 \\ 0 & -8 & -2 \\ 0 & 4 & 2 \end{vmatrix} = -16, \; D_z = \begin{vmatrix} 2 & 3 & 3 \\ 0 & 1 & -8 \\ 0 & -3 & 4 \end{vmatrix} = -40.$$

From (13), the solution of the system is then

$$x = \frac{32}{-8} = -4, \quad y = \frac{-16}{-8} = 2, \quad z = \frac{-40}{-8} = 5. \qquad \equiv$$

   When the determinant D of the coefficients of the variables in a linear system is 0, Cramer's Rule cannot be used. As we see in the next example, this does *not* mean the system has no solution.

EXAMPLE 8   **Consistent System**

For the linear system

$$\begin{cases} 4x - 16y = 3 \\ -x + 4y = -0.75 \end{cases}$$

we see that

$$\begin{vmatrix} 4 & -16 \\ -1 & 4 \end{vmatrix} = 16 - 16 = 0.$$

Although we cannot apply (10), the method of elimination would show us that the system is consistent but that the equations in the system are dependent. $\equiv$

   As mentioned previously, Cramer's Rule can be extended to systems of n linear equations in n variables for $n > 3$. But as a practical matter, Cramer's Rule is seldom used on systems with a large number of equations simply because evaluating the determinants by hand becomes a Herculean task.

In Problems 1–4, find the minor and cofactor determinants for each entry in the given determinant.

1. $\begin{vmatrix} 4 & 0 \\ 3 & -2 \end{vmatrix}$

2. $\begin{vmatrix} 6 & -2 \\ 5 & 1 \end{vmatrix}$

3. $\begin{vmatrix} 1 & -7 & 8 \\ 2 & 1 & 0 \\ -3 & 0 & 5 \end{vmatrix}$

4. $\begin{vmatrix} 4 & -3 & 0 \\ 2 & -1 & 6 \\ -5 & 4 & 1 \end{vmatrix}$

In Problems 5–18, evaluate the given determinant. In Problem 10, assume that $a \neq 0, b \neq 0$.

5. $\begin{vmatrix} \frac{5}{3} & \frac{1}{2} \\ 6 & 18 \end{vmatrix}$

6. $\begin{vmatrix} 0 & -1 \\ 8 & 0 \end{vmatrix}$

7. $\begin{vmatrix} 4 & 2 \\ 0 & 3 \end{vmatrix}$

8. $\begin{vmatrix} 3 & -4 \\ 5 & 6 \end{vmatrix}$

9. $\begin{vmatrix} a & -b \\ b & a \end{vmatrix}$

10. $\begin{vmatrix} a & b \\ \frac{1}{b} & \frac{1}{a} \end{vmatrix}$

11. $\begin{vmatrix} -3 & 4 & 1 \\ 2 & -6 & 1 \\ 6 & 8 & -4 \end{vmatrix}$

12. $\begin{vmatrix} 6 & 2 & 1 \\ 0 & 3 & -4 \\ 1 & 0 & 2 \end{vmatrix}$

13. $\begin{vmatrix} 4 & 6 & 1 \\ 3 & 2 & 3 \\ 0 & -1 & 7 \end{vmatrix}$

14. $\begin{vmatrix} 5 & 4 & 0 \\ 3 & -6 & 1 \\ 2 & 0 & 3 \end{vmatrix}$

15. $\begin{vmatrix} 5 & 9 & 1 \\ 1 & 2 & -3 \\ 0 & 0 & 0 \end{vmatrix}$

16. $\begin{vmatrix} 1 & 0 & 6 \\ 2 & 4 & 3 \\ -2 & 5 & 2 \end{vmatrix}$

17. $\begin{vmatrix} a & b & c \\ 0 & d & e \\ 0 & 0 & f \end{vmatrix}$

18. $\begin{vmatrix} 1 & 2 & 4 \\ 5 & 1 & -1 \\ 1 & 2 & 4 \end{vmatrix}$

In Problems 19–34, use Cramer's Rule, if applicable, to solve the given linear system.

19. $\begin{cases} x - y = 7 \\ 3x + 2y = 6 \end{cases}$

20. $\begin{cases} -x + 2y = 0 \\ 4x - 2y = 3 \end{cases}$

21. $\begin{cases} 2x - y = -3 \\ -x + 3y = 19 \end{cases}$

22. $\begin{cases} 4x + y = 1 \\ 8x - 2y = 2 \end{cases}$

23. $\begin{cases} 2x - 5y = 5 \\ -x + 10y = -15 \end{cases}$

24. $\begin{cases} -x - 3y = -7 \\ -2x + 6y = -9 \end{cases}$

25. $\begin{cases} 2x - y = -1 \\ 12x + 3y = 0 \end{cases}$

26. $\begin{cases} -x + 2y = 3 \\ 4x - 8y = 1 \end{cases}$

27. $\begin{cases} x + y - z = 5 \\ 2x - y + 3z = -3 \\ 2x + 3y = -4 \end{cases}$

28. $\begin{cases} 2x + y - z = -1 \\ 3x + 3y + z = 9 \\ x - 2y + 4z = 8 \end{cases}$

29. $\begin{cases} 2x - y + 3z = 13 \\ 3y + z = 5 \\ x - 7y + z = -1 \end{cases}$

30. $\begin{cases} 2x + y - 2z = 4 \\ 4x - y + 2z = -1 \\ 2x + 3y + 8z = 3 \end{cases}$

31. $\begin{cases} x + y + z = 2 \\ 4x - 8y + 3z = -2 \\ 2x - 2y + 2z = 1 \end{cases}$

32. $\begin{cases} 2x - y + z = 0 \\ x + 2y + z = 10 \\ 3x + y = 0 \end{cases}$

33. $\begin{cases} -3x - 6y + 9z = 2 \\ x - y + 5z = 0 \\ x + 2y - 3z = 1 \end{cases}$

34. $\begin{cases} 2x + 3y + 4z = 0 \\ 2x - y + 3z = 0 \\ x + y - z = 0 \end{cases}$

In Problems 35 and 36, solve for x.

35. $\begin{vmatrix} x & 1 & -2 \\ 1 & -1 & 1 \\ -1 & 0 & 2 \end{vmatrix} = 7$

36. $\begin{vmatrix} x & 1 & -1 \\ 1 & x & 1 \\ -1 & x & 2 \end{vmatrix} = 0$

In Problems 37–40, verify the given identity by evaluating each determinant.

37. $\begin{vmatrix} c & d \\ a & b \end{vmatrix} = -\begin{vmatrix} a & b \\ c & d \end{vmatrix}$

38. $\begin{vmatrix} ka & kb \\ c & d \end{vmatrix} = k\begin{vmatrix} a & b \\ c & d \end{vmatrix}$

39. $\begin{vmatrix} a & b \\ a & b \end{vmatrix} = 0$

40. $\begin{vmatrix} a & b \\ ka + c & kb + d \end{vmatrix} = \begin{vmatrix} a & b \\ c & d \end{vmatrix}$

41. Show that

$$\begin{vmatrix} 1 & a & a^2 \\ 1 & b & b^2 \\ 1 & c & c^2 \end{vmatrix} = (b - a)(c - a)(c - b).$$

42. Verify that an equation of the line through the points (x_1, y_1), (x_2, y_2) is given by

$$\begin{vmatrix} x & y & 1 \\ x_1 & y_1 & 1 \\ x_2 & y_2 & 1 \end{vmatrix} = 0.$$

In Problems 43–46, find the values of λ for which given determinant is 0.

43. $\begin{vmatrix} 1 - \lambda & 2 \\ 3 & -4 - \lambda \end{vmatrix}$

44. $\begin{vmatrix} 6 - \lambda & 3 \\ -11 & -6 - \lambda \end{vmatrix}$

45. $\begin{vmatrix} -1 - \lambda & 1 & 0 \\ 1 & 2 - \lambda & 1 \\ 0 & 3 & -1 - \lambda \end{vmatrix}$

46. $\begin{vmatrix} 13 - \lambda & 0 & 0 \\ 0 & \frac{1}{2} - \lambda & 0 \\ 0 & 0 & -7 - \lambda \end{vmatrix}$

Miscellaneous Applications

47. Echo Sounding This problem shows how the depth of an ocean and the speed of sound in water can be measured by a procedure known as **echo sounding**. Suppose that an oceanographic vessel emits sonar signals and that the arrival times of the signals reflected from the flat ocean floor are recorded at two trailing sonobuoys. See FIGURE 9.2.1. Using the relation distance = rate × time, we see from the figure that $2l_1 = vt_1$ and $2l_2 = vt_2$, where v is the speed of sound in water, t_1 and t_2 are the arrival times of the signals at the two sonobuoys, and l_1 and l_2 are the indicated distances.

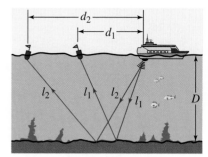

FIGURE 9.2.1 Echo sounding procedure in Problem 47

(a) Show that the speed of sound in water v and ocean depth D satisfy the system of equations

$$\begin{cases} t_1^2 v^2 - 4D^2 = d_1^2 \\ t_2^2 v^2 - 4D^2 = d_2^2. \end{cases}$$

[*Hint*: Use the Pythagorean theorem to relate l_1, d_1, and D, and l_2, d_2, and D.]

(b) Use Cramer's Rule to solve the system of equations in part (a) to obtain formulas for v^2 and D^2. Then express v and D in terms of the measurable quantities d_1, d_2, t_1, and t_2.

(c) The sonobuoys, trailing at 1000 m and 2000 m, record the arrival times of the reflected signals at 1.4 s and 1.8 s, respectively. Find the depth of the ocean and the speed of sound in water.

48. Take Your Vitamins The United States recommended daily allowance (U.S. RDA), in percent of vitamin content per ounce of food groups X, Y, and Z, is given in the following table.

	X	Y	Z
Vitamin A	9	5	4
Vitamin B$_1$	3	5	0
Vitamin C	24	10	5

Use Cramer's Rule to determine how many ounces of each food group one must consume each day in order to get 100% of the daily recommended allowance of vitamin A. 30% of the daily recommended allowance of vitamin B$_1$, and 200% of the daily recommended allowance of vitamin C.

For Discussion

In Problems 49 and 50, evaluate each determinant given that

$$\begin{vmatrix} a_{11} & a_{12} & a_{13} \\ a_{21} & a_{22} & a_{23} \\ a_{31} & a_{32} & a_{33} \end{vmatrix} = 3.$$

49. $\begin{vmatrix} a_{31} & a_{32} & a_{33} \\ a_{21} & a_{22} & a_{23} \\ a_{11} & a_{12} & a_{13} \end{vmatrix}$

50. $\begin{vmatrix} a_{11} & a_{12} & a_{13} \\ -2a_{31} & -2a_{32} & -2a_{33} \\ -a_{21} & -a_{22} & -a_{23} \end{vmatrix}$

In Problems 51 and 52, extend Theorem 9.2.1 to 4 × 4 determinants and then evaluate the given determinant.

51. $\begin{vmatrix} 6 & -1 & 0 & 4 \\ 3 & 3 & -2 & 0 \\ 0 & 1 & 8 & 6 \\ 2 & 3 & 0 & 4 \end{vmatrix}$

52. $\begin{vmatrix} -5 & 0 & 4 & 2 \\ -9 & 6 & -2 & 18 \\ -2 & 1 & 0 & 3 \\ 0 & 3 & 6 & 8 \end{vmatrix}$

9.3 Systems of Nonlinear Equations

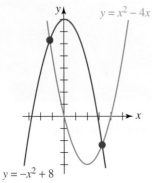

FIGURE 9.3.1 Intersection of two parabolas

≡ **Introduction** As FIGURE 9.3.1 illustrates, the graphs of the parabolas $y = x^2 - 4x$ and $y = -x^2 + 8$ intersect at two points. Thus the coordinates of the points of intersection must satisfy *both* equations,

$$\begin{cases} y = x^2 - 4x \\ y = -x^2 + 8. \end{cases} \tag{1}$$

Recall from Sections 2.3 and 9.1 that any equation that can be put in the form $ax + by + c = 0$ is called a **linear equation** in two variables. A **nonlinear equation** is simply one that is not linear. For example, in system (1) both equations $y = x^2 - 4x$ and $y = -x^2 + 8$ are nonlinear. A system of equations in which at least one of the equations is nonlinear will be referred to as a **system of nonlinear equations** or simply a **nonlinear system**.

In the examples that follow, will use the *methods of substitution* and *elimination* introduced in Section 9.1 to solve nonlinear systems.

EXAMPLE 1   Solution of (1)

Find solutions of system (1).

Solution Since the first equation already expresses y in terms of x, we substitute this expression for y into the second equation to get a single equation in one variable:

$$x^2 - 4x = -x^2 + 8.$$

Simplifying the last equation we get a quadratic equation $x^2 - 2x - 4 = 0$ that we solve using the quadratic formula: $x = 1 - \sqrt{5}$ and $x = 1 + \sqrt{5}$. We then substitute each of these numbers *back* into the first equation in (1) to solve for the corresponding values of y. This gives

$$y = (1 - \sqrt{5})^2 - 4(1 - \sqrt{5}) = 2 + 2\sqrt{5}$$

and

$$y = (1 + \sqrt{5})^2 - 4(1 + \sqrt{5}) = 2 - 2\sqrt{5}.$$

Thus, $(1 - \sqrt{5}, 2 + 2\sqrt{5})$ and $(1 + \sqrt{5}, 2 - 2\sqrt{5})$ are solutions of the system. ≡

EXAMPLE 2   Solving a Nonlinear System

Find solutions of the nonlinear system

$$\begin{cases} x^4 - 2(10^{2y}) - 3 = 0 \\ x - 10^y = 0. \end{cases}$$

Solution From the second equation, we have $x = 10^y$, and therefore $x^2 = 10^{2y}$. Substituting this last result into the first equation gives

$$x^4 - 2x^2 - 3 = 0,$$

or

$$(x^2 - 3)(x^2 + 1) = 0. \tag{2}$$

Since $x^2 + 1 > 0$ for all real numbers x, it follows that $x^2 = 3$ or $x = \pm\sqrt{3}$. But $x = 10^y > 0$ for all y; therefore, we must take $x = \sqrt{3}$. Solving $\sqrt{3} = 10^y$ for y gives

$$y = \log_{10}\sqrt{3} \qquad \text{or} \qquad y = \tfrac{1}{2}\log_{10}3.$$

Hence, $x = \sqrt{3}$, $y = \tfrac{1}{2}\log_{10}3$ is the only real solution of the system. $\equiv$

◀ Solution written by specifying values of the variables.

Nonlinear systems of equations can have complex solutions. Observe in Example 2 that equation (2) is also satisfied when $x^2 + 1 = 0$ or for $x = \pm i$, where $i = \sqrt{-1}$. The corresponding values of y are also complex but will not be given since that entails working with the logarithm of a complex number. For the rest of this section we are concerned only with finding the real solutions of nonlinear systems.

EXAMPLE 3 Dimensions of a Rectangle

Consider a rectangle in the first quadrant bounded by the x- and y-axes and the graph of $y = 20 - x^2$. See FIGURE 9.3.2. Find the dimensions of such a rectangle if its area is 16 square units.

Solution Let (x, y) be the coordinates of the point P on the graph of $y = 20 - x^2$ shown in the figure. Then the

$$\text{area of the rectangle} = xy \qquad \text{or} \qquad 16 = xy.$$

Thus we obtain the system of equations

$$\begin{cases} xy = 16 \\ y = 20 - x^2. \end{cases}$$

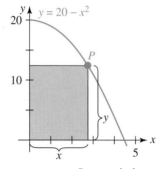

FIGURE 9.3.2 Rectangle in Example 3

The first equation of the system yields $y = 16/x$. After substituting this expression for y in the second equation, we get

$$\frac{16}{x} = 20 - x^2 \qquad \leftarrow \text{multiply this equation by } x$$

or $\qquad 16 = 20x - x^3 \qquad$ or $\qquad x^3 - 20x + 16 = 0.$

Now from the Rational Zeros Theorem in Section 3.4 the only possible rational roots of the last equation are ± 1, ± 2, ± 4, ± 8, and ± 16. Testing these numbers by synthetic division eventually shows that

$$\begin{array}{r|rrrr} 4 & 1 & 0 & -20 & 16 \\ & & 4 & 16 & -16 \\ \hline & 1 & 4 & -4 & \boxed{0} = r \end{array}$$

and so 4 is a solution. But the division above gives the factorization

$$x^3 - 20x + 16 = (x - 4)(x^2 + 4x - 4).$$

Applying the quadratic formula to $x^2 + 4x - 4 = 0$ reveals two more real roots:

$$x = \frac{-4 \pm \sqrt{32}}{2} = -2 \pm 2\sqrt{2}.$$

The positive number $-2 + 2\sqrt{2}$ is another solution. Since dimensions are positive, we reject the negative number $-2 - 2\sqrt{2}$. In other words, there are two rectangles with area 16 square units.

To find y, we use $y = 16/x$. If $x = 4$, then $y = 4$, and if $x = -2 + 2\sqrt{2} \approx 0.83$, then $y = 16/(-2 + 2\sqrt{2}) \approx 19.31$. Thus the dimensions of the two rectangles are

$$4 \times 4 \quad \text{and} \quad 0.83 \times 19.31 \text{ (approximately)}. \qquad \equiv$$

Note: In Example 3 observe that the equation $16 = 20x - x^3$ was obtained by multiplying the equation preceding it by x. Remember, when equations are multiplied by a variable, there is the possibility of introducing an extraneous solution. To make sure that this is not the case, you should check each solution.

■ EXAMPLE 4   Solving a Nonlinear System

Solve the nonlinear system

$$\begin{cases} x^2 + y^2 = 4 \\ -2x^2 + 7y^2 = 7. \end{cases}$$

Solution In preparation for eliminating an x^2-term, we begin by multiplying the first equation by 2. The system

$$\begin{cases} 2x^2 + 2y^2 = 8 \\ -2x^2 + 7y^2 = 7 \end{cases} \tag{3}$$

is equivalent to the given system. Now, by adding the first equation of this last system to the second, we obtain yet another system equivalent to the original system. In this case, we have eliminated x^2 from the second equation:

$$\begin{cases} 2x^2 + 2y^2 = 8 \\ \qquad\quad 9y^2 = 15. \end{cases}$$

From the last equation, we see that $y = \pm\frac{1}{3}\sqrt{15}$. Substituting these two values of y into $x^2 + y^2 = 4$ then gives

$$x^2 + \frac{15}{9} = 4 \quad \text{or} \quad x^2 = \frac{21}{9}$$

so that $x = \pm\frac{1}{3}\sqrt{21}$. Thus, $\left(\frac{1}{3}\sqrt{21}, \frac{1}{3}\sqrt{15}\right)$, $\left(-\frac{1}{3}\sqrt{21}, \frac{1}{3}\sqrt{15}\right)$, $\left(\frac{1}{3}\sqrt{21}, -\frac{1}{3}\sqrt{15}\right)$, and $\left(-\frac{1}{3}\sqrt{21}, -\frac{1}{3}\sqrt{15}\right)$ are all solutions. The graphs of the given equations and the four points corresponding to the ordered pairs are indicated by the red dots in **FIGURE 9.3.3**. $\qquad \equiv$

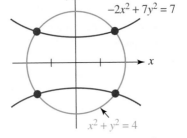

FIGURE 9.3.3 Intersection of a circle and a hyperbola in Example 4

In Example 4 we note that the system can also be solved by the substitution method by substituting, say, $y^2 = 4 - x^2$ into the second equation.

In the next example, we use the third elimination operation to simplify the system *before* applying the substitution method.

■ EXAMPLE 5   Solving a Nonlinear System

Solve the nonlinear system

$$\begin{cases} x^2 - 2x + y^2 = 0 \\ x^2 - 2y + y^2 = 0. \end{cases}$$

Solution By multiplying the first equation by -1 and adding the result to the second, we eliminate x^2 and y^2 from that equation:

$$\begin{cases} x^2 - 2x + y^2 = 0 \\ \qquad 2x - 2y = 0. \end{cases}$$

The second equation of the latter system implies that $y = x$. Substituting this expression into the first equation then yields

$$x^2 - 2x + x^2 = 0 \qquad \text{or} \qquad 2x(x - 1) = 0.$$

It follows that $x = 0$, $x = 1$ and, correspondingly, $y = 0$, $y = 1$. Thus solutions of the system are $(0, 0)$ and $(1, 1)$.   ≡

By completing the square in x and y, we can write the system in Example 5 as

$$\begin{cases} (x - 1)^2 + y^2 = 1 \\ x^2 + (y - 1)^2 = 1. \end{cases}$$

From this system we see that both equations describe circles of radius $r = 1$. The circles and their points of intersection are illustrated in **FIGURE 9.3.4**.

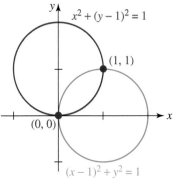

FIGURE 9.3.4 Intersecting circles in Example 5

9.3 **Exercises** Answers to selected odd-numbered problems begin on page ANS-26.

In Problems 1–6, determine graphically whether the given nonlinear system has any solutions.

1. $\begin{cases} x = 5 \\ x = y^2 \end{cases}$

2. $\begin{cases} y = 3 \\ (x + 1)^2 + y^2 = 10 \end{cases}$

3. $\begin{cases} -x^2 + y = -1 \\ x^2 + y = 4 \end{cases}$

4. $\begin{cases} x + y = 5 \\ x^2 + y^2 = 1 \end{cases}$

5. $\begin{cases} x^2 + y^2 = 1 \\ x^2 - 4x + y^2 = -3 \end{cases}$

6. $\begin{cases} y = 2^x - 1 \\ y = \log_2(x + 2) \end{cases}$

In Problems 7–42, solve the given nonlinear system.

7. $\begin{cases} y = x \\ y^2 = x + 2 \end{cases}$

8. $\begin{cases} y = 3x \\ x^2 + y^2 = 4 \end{cases}$

9. $\begin{cases} y = 2x - 1 \\ y = x^2 \end{cases}$

10. $\begin{cases} x + y = 1 \\ x^2 - 2y = 0 \end{cases}$

11. $\begin{cases} 64x + y = 1 \\ x^3 - y = -1 \end{cases}$

12. $\begin{cases} y - x = 3 \\ x^2 + y^2 = 9 \end{cases}$

13. $\begin{cases} x = \sqrt{y} \\ x^2 = \dfrac{6}{y} + 1 \end{cases}$

14. $\begin{cases} y = 2\sqrt{2}x^2 \\ y = \sqrt{x} \end{cases}$

15. $\begin{cases} xy = 1 \\ x + y = 1 \end{cases}$

16. $\begin{cases} xy = 3 \\ x + y = 4 \end{cases}$

17. $\begin{cases} xy = 5 \\ x^2 + y^2 = 10 \end{cases}$

18. $\begin{cases} xy = 1 \\ x^2 = y^2 + 2 \end{cases}$

19. $\begin{cases} 16x^2 - y^4 = 16y \\ y^2 + y = x^2 \end{cases}$

20. $\begin{cases} x^3 + 3y = 26 \\ y = x(x + 1) \end{cases}$

21. $\begin{cases} x^2 - y^2 = 4 \\ 2x^2 + y^2 = 1 \end{cases}$

22. $\begin{cases} 3x^2 + 2y^2 = 4 \\ x^2 + 4y^2 = 1 \end{cases}$

23. $\begin{cases} x^2 + y^2 = 4 \\ x^2 - 4x + y^2 - 2y = 4 \end{cases}$

24. $\begin{cases} x^2 + y^2 - 6y = -9 \\ x^2 + 4x + y^2 = -1 \end{cases}$

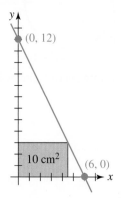

FIGURE 9.3.5 Rectangle in Problem 44

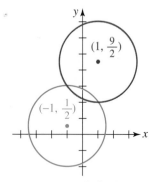

FIGURE 9.3.6 Circles in Problem 46

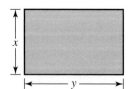

FIGURE 9.3.7 Rectangle in Problem 47

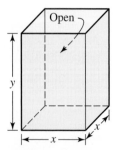

FIGURE 9.3.8 Open box in Problem 49

25. $\begin{cases} x^2 + y^2 = 5 \\ y = x^2 - 5 \end{cases}$

26. $\begin{cases} y = x(x^2 - 6x + 8) \\ y + 4 = (x - 2)^2 \end{cases}$

27. $\begin{cases} (x - y)^2 = 4 \\ (x + y)^2 = 12 \end{cases}$

28. $\begin{cases} (x - y)^2 = 0 \\ (x + y)^2 = 1 \end{cases}$

29. $\begin{cases} y = \sin x \\ y = \cos x \end{cases}$

30. $\begin{cases} y = \cos x \\ 2y \tan x = \sqrt{3} \end{cases}$

31. $\begin{cases} 2y \sin x = 1 \\ y = 2 \sin x \end{cases}$

32. $\begin{cases} y = \sin 2x \\ y = \sin x \end{cases}$

33. $\begin{cases} y = \log_{10} x \\ y^2 = 5 + 4 \log_{10} x \end{cases}$

34. $\begin{cases} x + \log_{10} y = 2 \\ y + 15 = 10^x \end{cases}$

35. $\begin{cases} \log_{10}(x^2 + y)^2 = 8 \\ y = 2x + 1 \end{cases}$

36. $\begin{cases} \log_{10} x = y - 5 \\ 7 = y - \log_{10}(x + 6) \end{cases}$

37. $\begin{cases} x = 3^y \\ x = 9^y - 20 \end{cases}$

38. $\begin{cases} y = 2^{x^2} \\ \sqrt{5}x = \log_2 y \end{cases}$

39. $\begin{cases} 2x + \lambda = 0 \\ 2y + \lambda = 0 \\ xy - 3 = 0 \end{cases}$

40. $\begin{cases} -2x + \lambda = 0 \\ y - y\lambda = 0 \\ y^2 - x = 0 \end{cases}$

41. $\begin{cases} y^2 = 2x\lambda \\ 2xy = 2y\lambda \\ x^2 + y^2 - 1 = 0 \end{cases}$

42. $\begin{cases} 8x + 5y = 2xy\lambda \\ 5x = x^2\lambda \\ x^2y - 1000 = 0 \end{cases}$

Miscellaneous Applications

43. **Dimensions of a Corral** The perimeter of a rectangular corral is 260 ft and its area is 4000 ft². What are its dimensions?

44. **Inscribed Rectangle** Find the dimensions of the rectangle(s) with area 10 cm² inscribed in the triangle consisting of the blue line and the two coordinate axes shown in **FIGURE 9.3.5**.

45. **Sum of Areas** The sum of the radii of two circles is 8 cm. Find the radii if the sum of the areas of the circles is 32π cm².

46. **Intersecting Circles** Find the two points of intersection of the circles shown in **FIGURE 9.3.6** if the radius of each circle is $\frac{5}{2}$.

47. **Golden Ratio** The **golden ratio** for the rectangle shown in **FIGURE 9.3.7** is defined by

$$\frac{x}{y} = \frac{y}{x + y}.$$

This ratio is often used in architecture and in paintings. Find the dimensions of a rectangular sheet of paper containing 100 in.² that satisfy the golden ratio.

48. **Length** The hypotenuse of a right triangle is 20 cm. Find the lengths of the remaining two sides if the shorter side is one-half the length of the longer side.

49. **Topless Box** A box is to be made with a square base and no top. See **FIGURE 9.3.8**. The volume of the box is to be 32 ft³, and the combined areas of the sides and bottom are to be 68 ft². Find the dimensions of the box.

50. **Dimensions of a Cylinder** The volume of a right circular cylinder is 63π in.³, and its height h is 1 in. greater than twice its radius r. Find the dimensions of the cylinder.

For Discussion

51. A **tangent to an ellipse** is defined exactly as it was for the circle, namely, a straight line that touches the ellipse at only one point (x_1, y_1). See Problem 52 in

Exercises 2.3. It can be shown (see Problem 52 that follows) that an equation of the tangent line at a given point (x_1, y_1) on an ellipse $x^2/a^2 + y^2/b^2 = 1$ is

$$\frac{xx_1}{a^2} + \frac{yy_1}{b^2} = 1. \tag{4}$$

(a) Find the equation of the tangent line to the ellipse $x^2/50 + y^2/8 = 1$ at the point $(5, -2)$.

(b) Write your answer in the form of $y = mx + b$.

(c) Sketch the ellipse and the tangent line.

52. In this problem, you are guided through the steps to derive equation (4).

(a) An alternative form of the equation $x^2/a^2 + y^2/b^2 = 1$ is

$$b^2x^2 + a^2y^2 = a^2b^2.$$

Since the point (x_1, y_1) is on the ellipse, its coordinates must satisfy the foregoing equation:

$$b^2x_1^2 + a^2y_1^2 = a^2b^2.$$

Show that

$$b^2(x^2 - x_1^2) + a^2(y^2 - y_1^2) = 0.$$

(b) Using the point-slope form of a line, the tangent line at (x_1, y_1) is $y - y_1 = m(x - x_1)$. Use substitution in the system

$$\begin{cases} b^2(x^2 - x_1^2) + a^2(y^2 - y_1^2) = 0 \\ y - y_1 = m(x - x_1) \end{cases}$$

to show that

$$b^2(x^2 - x_1^2) + a^2m^2(x - x_1)^2 + 2a^2my_1(x - x_1) = 0. \tag{5}$$

The last equation is a quadratic equation in x. Explain why x_1 is a repeated root or a root of multiplicity 2.

(c) By factoring, (5) becomes

$$(x - x_1)[(b^2(x + x_1) + a^2m^2(x - x_1) + 2a^2my_1] = 0.$$

and so we must have

$$b^2(x + x_1) + a^2m^2(x - x_1) + 2a^2my_1 = 0.$$

Use the last equation to find the slope m of the tangent line at (x_1, y_1). Finish the problem by finding the equation of the tangent as given in (4).

9.4 Systems of Inequalities

≡ **Introduction** In Chapter 1 we solved linear and nonlinear inequalities involving a *single* variable x and then graphed the solution set of the inequality on the number line. In this section our focus will be on inequalities involving *two* variables x and y. For example,

$$x + 2y - 4 > 0, \qquad y \le x^2 + 1, \qquad x^2 + y^2 \ge 1$$

are inequalities in two variables. A **solution** of an inequality in two variables is any ordered pair of real numbers (x_0, y_0) that satisfies the inequality—that is, results in a true statement—when x_0 and y_0 are substituted for x and y, respectively. A **graph** of the solution set of an inequality in two variables is made up of all points in the plane whose coordinates satisfy the inequality.

Many results obtained in calculus are valid only in a specialized region either in the xy-plane or in three-dimensional space, and these regions are often defined by means of **systems of inequalities** in two or three variables. In this section we consider only systems of inequalities involving two variables x and y.

We begin with linear inequalities in two variables.

□ **Half-Planes** A **linear inequality in two variables** x and y is any inequality that has one of the forms

$$ax + by + c < 0, \quad ax + by + c > 0, \tag{1}$$
$$ax + by + c \leq 0, \quad ax + by + c \geq 0. \tag{2}$$

Since the inequalities in (1) and (2) have infinitely many solutions, the notation

$$\{(x, y)\,|\, ax + by + c < 0\}, \quad \{(x, y)\,|\, ax + by + c \geq 0\},$$

and so on, is used to denote a set of solutions. Geometrically, each of these sets describes a **half-plane**. As shown in FIGURE 9.4.1, the graph of the linear equation $ax + by + c = 0$ divides the xy-plane into two regions, or half-planes. One of these half-planes is the graph of the set of solutions of the linear inequality. If the inequality is strict, as in (1), then we draw the graph of $ax + by + c = 0$ as a dashed line, because the points on the line are not in the set of solutions of the inequality. See Figure 9.4.1(a). On the other hand, if the inequality is nonstrict, as in (2), the set of solutions includes the points satisfying $ax + by + c = 0$, and so we draw the graph of the equation as a solid line. See Figure 9.4.1(b).

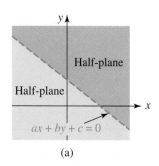

(a)

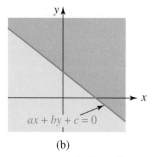

(b)

FIGURE 9.4.1 A single line determines two half-planes

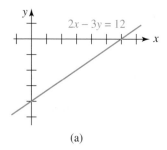

(a)

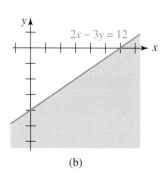

(b)

FIGURE 9.4.2 Half-plane in Example 1

■ **EXAMPLE 1**   **Graph of a Linear Inequality**

Graph the linear inequality $2x - 3y \geq 12$.

Solution First, we graph the line $2x - 3y = 12$, as shown in FIGURE 9.4.2(a). Solving the given inequality for y gives

$$y \leq \tfrac{2}{3}x - 4. \tag{3}$$

Since the y-coordinate of any point (x, y) on the graph of $2x - 3y \geq 12$ must satisfy (3), we conclude that the point (x, y) must lie on or below the graph of the line. This solution set is the region that is shaded blue in Figure 9.4.2(b).

Alternatively, we know that the set

$$\{(x, y)\,|\, 2x - 3y - 12 \geq 0\}$$

describes a half-plane. Thus we can determine whether the graph of the inequality includes the region above or below the line $2x - 3y = 12$ by determining whether a test point not on the line, such as $(0, 0)$, satisfies the original inequality. Substituting $x = 0$, $y = 0$ into $2x - 3y \geq 12$ gives $0 \geq 12$. This false statement implies that the graph of the inequality is the region on the other side of the line $2x - 3y = 12$, that is, the side that does *not* contain the origin. Note that the blue half-plane in Figure 9.4.2(b) does not contain the point $(0, 0)$.   ≡

In general, given a linear inequality of the forms in (1) or (2), we can graph the solutions by proceeding in the following manner.

- Graph the line $ax + by + c = 0$.
- Select a **test point** not on this line.
- Shade the half-plane containing the test point if its coordinates satisfy the original inequality. If they do not satisfy the inequality, shade the other half-plane.

■ EXAMPLE 2   Graph of a Linear Inequality

Graph the linear inequality $3x + y - 2 < 0$.

Solution In FIGURE 9.4.3 we draw the graph of $3x + y = 2$ as a dashed line, since it will not be part of the solution set of the inequality. Then we select $(0, 0)$ as a test point that is not on the line. Because substituting $x = 0$, $y = 0$ into $3x + y - 2 < 0$ gives the true statement $-2 < 0$ we shade that region of the plane containing the origin. ≡

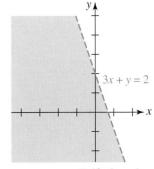

FIGURE 9.4.3 Half-plane in Example 2

☐ **Systems of Inequalities** We say (x_0, y_0) is a **solution of a system of inequalities** when it is a member of the set of solutions *common* to all inequalities. In other words, the **solution set** of a system of inequalities is the intersection of the solution sets of the individual inequalities in the system.

   In the next two examples we graph the solution set of a system of linear inequalities.

■ EXAMPLE 3   System of Linear Inequalities

Graph the system of linear inequalities

$$\begin{cases} x \geq 1 \\ y \leq 2. \end{cases}$$

Solution The sets

$$\{(x, y)\,|\, x \geq 1\} \qquad \text{and} \qquad \{(x, y)\,|\, y \leq 2\}$$

denote the sets of solutions for each inequality. These sets are illustrated in FIGURE 9.4.4 by the blue and the red shading, respectively. The solutions of the given system are the ordered pairs in the intersection

$$\{(x, y)\,|\, x \geq 1\} \cap \{(x, y)\,|\, y \leq 2\} = \{(x, y)\,|\, x \geq 1 \text{ and } y \leq 2\}.$$

This last set is the region of darker color (overlapping red and blue colors) shown in the figure. ≡

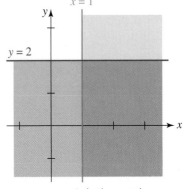

FIGURE 9.4.4 Solution set in Example 3

■ EXAMPLE 4   System of Linear Inequalities

Graph the system of linear inequalities

$$\begin{cases} x + y \leq 1 \\ -x + 2y \geq 4. \end{cases} \qquad (4)$$

Solution Substitution of $(0, 0)$ into the first inequality in (4) gives the true statement $0 \leq 1$, which implies that the graph of the solutions of $x + y \leq 1$ is the half-plane *below* (and including) the line $x + y = 1$. This is the shaded blue region in FIGURE 9.4.5(a). Similarly, substituting $(0, 0)$ into the second inequality gives the false statement $0 \geq 4$, and so the graph of the solutions of $-x + 2y \geq 4$ is the half-plane *above* (and including) the line $-x + 2y = 4$. This is the shaded red region in Figure 9.4.5(b). The graph of the solutions of the system of inequalities is then the intersection of the graphs of these two solution sets. This intersection is the darker region of overlapping colors shown in Figure 9.4.5(c).

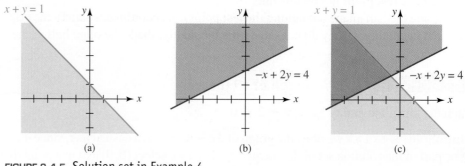

FIGURE 9.4.5 Solution set in Example 4

Often we are interested in the solutions of a system of linear inequalities subject to the restrictions that $x \geq 0$ and $y \geq 0$. This means that the graph of the solutions is a subset of the set consisting of the points in the first quadrant and on the nonnegative coordinate axes. For example, inspection of Figure 9.4.5(c) reveals that the system of inequalities (4) subject to the added requirements that $x \geq 0$, $y \geq 0$, has no solutions.

EXAMPLE 5   System of Linear Inequalities

The graph of the solutions of the system of linear inequalities

$$\begin{cases} -2x + y \leq 2 \\ x + 2y \leq 8 \end{cases}$$

is the region shown in FIGURE 9.4.6(a). The graph of the solutions of

$$\begin{cases} -2x + y \leq 2 \\ x + 2y \leq 8 \\ x \geq 0, \ y \geq 0 \end{cases}$$

is the region in the first quadrant along with portions of the two lines and portions of the coordinate axes illustrated in Figure 9.4.6(b).

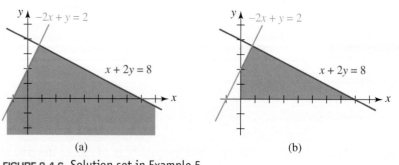

FIGURE 9.4.6 Solution set in Example 5

☐ **Nonlinear Inequalities** Graphing **nonlinear inequalities** in two variables x and y is basically the same as graphing linear inequalities. In the next example we again utilize the notion of a test point.

EXAMPLE 6　　　　Graph of a Nonlinear Inequality

To graph the nonlinear inequality

$$x^2 + y^2 - 4 \geq 0$$

we begin by drawing the circle $x^2 + y^2 = 4$ using a solid line. Since $(0, 0)$ lies in the interior of the circle we can use it for a test point. Substituting $x = 0$ and $y = 0$ in the inequality gives the false statement $-4 \geq 0$ and so the solution set of the given inequality consists of all the points either on the circle or in its exterior. See FIGURE 9.4.7. ≡

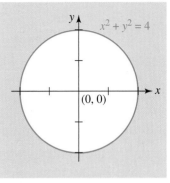

FIGURE 9.4.7 Solution set in Example 6

EXAMPLE 7　　　　System of Inequalities

Graph the system of inequalities

$$\begin{cases} y \leq 4 - x^2 \\ y > x. \end{cases}$$

Solution Substitution of the coordinates of $(0, 0)$ into the first inequality gives the true statement $0 \leq 4$ and so the graph of $y \leq 4 - x^2$ is the shaded blue region in FIGURE 9.4.8 below the parabola $y = 4 - x^2$. Note that we cannot use $(0, 0)$ as a test point for the second inequality since $(0, 0)$ is a point on the line $y = x$. However, if we use $(1, 2)$ as a test point, the second inequality gives the true statement $2 > 1$. Thus the graph of the solutions of $y > x$ is the shaded red half-plane above the line $y = x$ in Figure 9.4.8. The line itself is dashed because of the strict inequality. The intersection of these two colored regions is the darker region in the figure. ≡

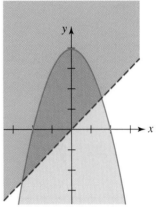

FIGURE 9.4.8 Solution set in Example 7

9.4　Exercises
Answers to selected odd-numbered problems begin on page ANS-27.

In Problems 1–12, graph the given inequality.

1. $x + 3y \geq 6$
2. $x - y \leq 4$
3. $x + 2y < -x + 3y$
4. $2x + 5y > x - y + 6$
5. $-y \geq 2(x + 3) - 5$
6. $x \geq 3(x + 1) + y$
7. $y \geq (x - 1)^2$
8. $x^2 + \frac{1}{4}y^2 < 1$
9. $y - 1 \leq \sqrt{x}$
10. $y \geq \sqrt{x + 1}$
11. $y \geq |x + 2|$
12. $xy \geq 3$

In Problems 13–36, graph the given system of inequalities.

13. $\begin{cases} y \leq x \\ x \geq 2 \end{cases}$
14. $\begin{cases} y \geq x \\ y \geq 0 \end{cases}$

15. $\begin{cases} x - y > 0 \\ x + y > 1 \end{cases}$
16. $\begin{cases} x + y < 1 \\ -x + y < 1 \end{cases}$

17. $\begin{cases} x + 2y \leq 4 \\ -x + 2y \geq 6 \\ x \geq 0 \end{cases}$
18. $\begin{cases} 4x + y \geq 12 \\ -2x + y \leq 0 \\ y \geq 0 \end{cases}$

19. $\begin{cases} x - 3y > -9 \\ x \geq 0, y \geq 0 \end{cases}$
20. $\begin{cases} x + y > 4 \\ x \geq 0, y \geq 0 \end{cases}$

21. $\begin{cases} y < x + 2 \\ 1 \leq x \leq 3 \\ y \geq 1 \end{cases}$
22. $\begin{cases} 4y > x \\ x \geq 2 \\ y \leq 5 \end{cases}$

23. $\begin{cases} x + y \leq 4 \\ \quad y \geq -x \\ \quad y \leq 2x \end{cases}$

24. $\begin{cases} 2x + 3y \geq 6 \\ \quad x - \quad y \geq -6 \\ 2x + \quad y \leq 6 \end{cases}$

25. $\begin{cases} -2x + \quad y \leq 2 \\ \quad x + 3y \leq 10 \\ \quad x - \quad y \leq 5 \\ x \geq 0, y \geq 0 \end{cases}$

26. $\begin{cases} -x + \quad y \leq 0 \\ -x + 3y \geq 0 \\ \quad x + \quad y - 8 \geq 0 \\ \quad\quad\quad y - 2 \leq 0 \end{cases}$

27. $\begin{cases} x^2 + \quad y^2 \geq 1 \\ \frac{1}{9}x^2 + \frac{1}{4}y^2 \leq 1 \end{cases}$

28. $\begin{cases} x^2 + y^2 \leq 25 \\ x + y \geq 5 \end{cases}$

29. $\begin{cases} y \leq x^2 + 1 \\ y \geq -x^2 \end{cases}$

30. $\begin{cases} x^2 + y^2 \leq 4 \\ y \leq x^2 - 1 \end{cases}$

31. $\begin{cases} y \geq |x| \\ x^2 + y^2 \leq 2 \end{cases}$

32. $\begin{cases} y \leq e^x \\ y \geq x - 1 \\ x \geq 0 \end{cases}$

33. $\begin{cases} \frac{1}{9}x^2 - \frac{1}{4}y^2 \geq 1 \\ y \geq 0 \end{cases}$

34. $\begin{cases} y < \ln x \\ y > 0 \end{cases}$

35. $\begin{cases} y \leq x^3 + 1 \\ x \geq 0 \\ x \leq 1 \\ y \geq 0 \end{cases}$

36. $\begin{cases} y \geq x^4 \\ y \leq 2 \\ x \geq -1 \\ x \leq 1 \end{cases}$

In Problems 37–40, find a system of linear inequalities whose graph is the region shown in the given figure.

37.

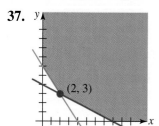

FIGURE 9.4.9 Region for Problem 37

38.

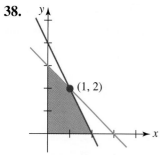

FIGURE 9.4.10 Region for Problem 38

39.

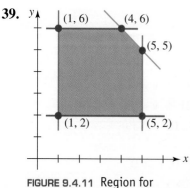

FIGURE 9.4.11 Region for Problem 39

40.

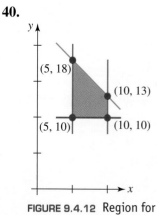

FIGURE 9.4.12 Region for Problem 40

For Discussion

In Problems 41 and 42, graph the given inequality.

41. $-1 \le x + y \le 1$ **42.** $-x \le y \le x$

Project

43. Ancient History and USPS Some years ago the restrictions on first-class envelope size were a bit more confusing than they are today. Consider the rectangular envelope of length x and height y shown in FIGURE 9.4.13 and the following postal regulation of November 1978:

> All first-class items weighing one ounce or less and all single-piece third-class items weighing two ounces or less are subject to an extra mailing fee when the height is greater than $6\frac{1}{8}$ in., or the length is greater than $11\frac{1}{2}$ in., or the length is less than 1.3 times the height, or the length is greater than 2.5 times the height.

In parts (a)–(c) assume that the weight specification is satisfied.
- **(a)** Using x and y, interpret the above regulation as a system of linear inequalities.
- **(b)** Graph the region that describes envelope sizes that are *not* subject to an extra mailing fee.
- **(c)** Under this regulation does an envelope of length 8 in. and height 4 in. require an extra fee?
- **(d)** Do some research and compare the 2010 first-class regulation against the one just given.

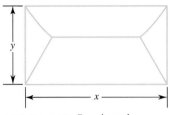

FIGURE 9.4.13 Envelope in Problem 43

9.5 Partial Fractions

∫ **Calculus PREVIEW** ≡ **Introduction** When two rational functions, say, $f(x) = \dfrac{2}{x + 5}$ and $g(x) = \dfrac{1}{x + 1}$ are added, the terms are combined by means of a common denominator:

$$\frac{2}{x+5} + \frac{1}{x+1} = \frac{2}{x+5}\left(\frac{x+1}{x+1}\right) + \frac{1}{x+1}\left(\frac{x+5}{x+5}\right). \tag{1}$$

Adding numerators on the right-hand side of (1) yields the single rational expression

$$\frac{2x+7}{(x+5)(x+1)}. \tag{2}$$

An important procedure in the study of integral calculus requires that we be able to reverse the process; in other words, starting with a rational expression such as (2) break it down, or *decompose* it, into simpler component fractions $2/(x + 5)$ and $1/(x + 1)$ called **partial fractions**.

☐ **Terminology** The algebraic process for breaking down a rational expression such as (2) into partial fractions is known as **partial fraction decomposition**. For convenience we will assume that the rational function $P(x)/Q(x)$, $Q(x) \ne 0$, is a **proper fraction** or **proper rational expression**; that is, the degree of $P(x)$ is less than the degree of $Q(x)$. We will also assume once again that the polynomials $P(x)$ and $Q(x)$ have no common factors.

In the discussion that follows we consider four cases of partial fraction decomposition of $P(x)/Q(x)$. The cases depend on the factors in the denominator $Q(x)$. When the polynomial $Q(x)$ is factored as a product of $(ax + b)^n$ and $(ax^2 + bx + c)^m$, $n = 1, 2, \ldots, m = 1, 2, \ldots$, where the coefficients a, b, c are real numbers and the quadratic polynomial $ax^2 + bx + c$ is **irreducible** over the real numbers (that is, does not factor using real numbers), the rational expression $P(x)/Q(x)$ can be decomposed into a sum of partial fractions of the form

$$\frac{C_k}{(ax + b)^k} \quad \text{and} \quad \frac{A_k x + B_k}{(ax^2 + bx + c)^k}.$$

CASE 1: $Q(x)$ Contains Only Nonrepeated Linear Factors

We state the following fact from algebra without proof. If the denominator can be factored completely into linear factors,

$$Q(x) = (a_1 x + b_1)(a_2 x + b_2) \cdots (a_n x + b_n),$$

where all the $a_i x + b_i$, $i = 1, 2, \ldots, n$ are distinct (that is, no two factors are the same), then unique real constants $C_1, C_2, \ldots, C_n$ can be found such that

$$\frac{P(x)}{Q(x)} = \frac{C_1}{a_1 x + b_1} + \frac{C_2}{a_2 x + b_2} + \cdots + \frac{C_n}{a_n x + b_n}. \tag{3}$$

In practice we will use the letters A, B, C, $\ldots$ in place of the subscripted coefficients $C_1, C_2, C_3, \ldots$. The next example illustrates this first case.

EXAMPLE 1   **Distinct Linear Factors**

To decompose $\dfrac{2x + 1}{(x - 1)(x + 3)}$ into individual partial fractions we make the assumption, based on the form given in (3), that the rational function can be written as

$$\frac{2x + 1}{(x - 1)(x + 3)} = \frac{A}{x - 1} + \frac{B}{x + 3}. \tag{4}$$

We now clear (4) of fractions; this can be done by either combining the terms on the right-hand side of the equality over a least common denominator and equating numerators or by simply multiplying both sides of the equality by the denominator $(x - 1)(x + 3)$ on the left-hand side. Either way, we arrive at

$$2x + 1 = A(x + 3) + B(x - 1). \tag{5}$$

Multiplying out the right-hand side of (5) and grouping by powers of x gives

$$2x + 1 = A(x + 3) + B(x - 1) = (A + B)x + (3A - B). \tag{6}$$

Each of the equations (5) and (6) is an identity, which means that the equality is true for *all* real values of x. As a consequence, the coefficients of x on the left-hand side of (6) must be the same as the coefficients of the corresponding powers of x on the right-hand side, that is,

$$2x + 1x^0 = (A + B)x + (3A - B)x^0.$$

equal

equal

The result is a system of two linear equations in two variables A and B:

$$\begin{cases} 2 = A + B \\ 1 = 3A - B. \end{cases} \tag{7}$$

By adding the two equations we get $3 = 4A$ and so we find that $A = \frac{3}{4}$. Substituting this value into either equation in (7) then yields $B = \frac{5}{4}$. Hence the desired decomposition is

$$\frac{2x + 1}{(x - 1)(x + 3)} = \frac{\frac{3}{4}}{x - 1} + \frac{\frac{5}{4}}{x + 3}.$$

You are encouraged to verify the foregoing result by combining the terms on the right-hand side of the last equation by means of a common denominator. ≡

□ **A Shortcut Worth Knowing** If the denominator contains, say, three linear factors such as in $\dfrac{4x^2 - x + 1}{(x - 1)(x + 3)(x - 6)}$, then the partial fraction decomposition looks like this:

$$\frac{4x^2 - x + 1}{(x - 1)(x + 3)(x - 6)} = \frac{A}{x - 1} + \frac{B}{x + 3} + \frac{C}{x - 6}.$$

By following the same steps as in Example 1, we would find that the analog of (7) is now three equations in the three unknowns A, B, and C. The point is this: The more linear factors in the denominator the larger the system of equations we must solve. There is a procedure worth learning that can cut down on some of the algebra. To illustrate, let's return to the identity (5). Since the equality is true for every value of x, it holds for $x = 1$ and $x = -3$, *the zeros of the denominator*. Setting $x = 1$ in (5) gives $3 = 4A$, from which it follows immediately that $A = \frac{3}{4}$. Similarly, by setting $x = -3$ in (5), we obtain $-5 = (-4)B$ or $B = \frac{5}{4}$.

CASE 2: $Q(x)$ **Contains Repeated Linear Factors**

If the denominator $Q(x)$ contains a repeated linear factor $(ax + b)^n$, $n > 1$, then unique real constants $C_1, C_2, \ldots, C_n$ can be found such that the partial fraction decomposition of $P(x)/Q(x)$ contains the terms

$$\frac{C_1}{ax + b} + \frac{C_2}{(ax + b)^2} + \cdots + \frac{C_n}{(ax + b)^n}. \tag{8}$$

EXAMPLE 2   **Repeated Linear Factors**

To decompose $\dfrac{6x - 1}{x^3(2x - 1)}$ into partial fractions we first observe that the denominator consists of the repeated linear factor x and the nonrepeated linear factor $2x - 1$. Based on the forms in (3) and (8) we assume that

according to Case 2   according to Case 1

$$\frac{6x - 1}{x^3(2x - 1)} = \overbrace{\frac{A}{x} + \frac{B}{x^2} + \frac{C}{x^3}} + \overbrace{\frac{D}{2x - 1}}. \tag{9}$$

Multiplying (9) by $x^3(2x - 1)$ clears it of fractions and yields

$$6x - 1 = Ax^2(2x - 1) + Bx(2x - 1) + C(2x - 1) + Dx^3 \tag{10}$$

or $\qquad 6x - 1 = (2A + D)x^3 + (-A + 2B)x^2 + (-B + 2C)x - C. \tag{11}$

Now the zeros of the denominator in the original expression are $x = 0$ and $x = \frac{1}{2}$. If we then set $x = 0$ and $x = \frac{1}{2}$ in (10), we find, in turn, that $C = 1$ and $D = 16$. Because the denominator of the original expression has only two distinct zeros, we can find A and B by equating the corresponding coefficients of x^3 and x^2 in (11):

◄ The coefficients of x^3 and x^2 on the left-hand side of (11) are both 0.

$$0 = 2A + D, \qquad 0 = -A + 2B.$$

Using the known value of D, the first equation yields $A = -D/2 = -8$. The second then gives $B = A/2 = -4$. The partial fraction decomposition is

$$\frac{6x - 1}{x^3(2x - 1)} = -\frac{8}{x} - \frac{4}{x^2} + \frac{1}{x^3} + \frac{16}{2x - 1}. \qquad \equiv$$

CASE 3: $Q(x)$ Contains Nonrepeated Irreducible Quadratic Factors

If the denominator $Q(x)$ contains nonrepeated irreducible quadratic factors $a_i x^2 + b_i x + c_i$, then unique real constants $A_1, A_2, \ldots, A_n, B_1, B_2, \ldots, B_n$ can be found such that the partial fraction decomposition of $P(x)/Q(x)$ contains the terms

$$\frac{A_1 x + B_1}{a_1 x^2 + b_1 x + c_1} + \frac{A_2 x + B_2}{a_2 x^2 + b_2 x + c_2} + \cdots + \frac{A_n x + B_n}{a_n x^2 + b_n x + c_n}. \qquad (12)$$

▌EXAMPLE 3   Irreducible Quadratic Factors

To decompose $\dfrac{4x}{(x^2 + 1)(x^2 + 2x + 3)}$ into partial fractions we first observe that the

◄ Use the quadratic formula. For either factor you will find that $b^2 - 4ac < 0$.

quadratic polynomials $x^2 + 1$ and $x^2 + 2x + 3$ are irreducible over the real numbers. Hence by (12) we assume that

$$\frac{4x}{(x^2 + 1)(x^2 + 2x + 3)} = \frac{Ax + B}{x^2 + 1} + \frac{Cx + D}{x^2 + 2x + 3}.$$

After clearing fractions in the preceding line, we find

$$\begin{aligned} 4x &= (Ax + B)(x^2 + 2x + 3) + (Cx + D)(x^2 + 1) \\ &= (A + B)x^3 + (2A + B + D)x^2 + (3A + 2B + C)x + (3B + D). \end{aligned}$$

Because the denominator of the original fraction has no real zeros, we have no recourse except to form a system of equations by comparing coefficients of all powers of x:

$$\begin{cases} 0 = A + C \\ 0 = 2A + B + D \\ 4 = 3A + 2B + C \\ 0 = 3B + D. \end{cases}$$

Using $C = -A$ and $D = -3B$ from the first and fourth equations we can eliminate C and D in the second and third equations:

$$\begin{cases} 0 = A - B \\ 2 = A + B. \end{cases}$$

Solving this simpler system of equations yields $A = 1$ and $B = 1$. Hence, $C = -1$ and $D = -3$. The partial fraction decomposition is

$$\frac{4x}{(x^2 + 1)(x^2 + 2x + 3)} = \frac{x + 1}{x^2 + 1} - \frac{x + 3}{x^2 + 2x + 3}. \qquad \equiv$$

CASE 4: $Q(x)$ **Contains Repeated Irreducible Quadratic Factors**

If the denominator $Q(x)$ contains a repeated irreducible quadratic factor $(ax^2 + bx + c)^n$, $n > 1$, then unique real constants $A_1, A_2, \ldots, A_n, B_1, B_2, \ldots, B_n$ can be found such that the partial fraction decomposition of $P(x)/Q(x)$ contains the terms

$$\frac{A_1 x + B}{ax^2 + bx + c} + \frac{A_2 x + B_2}{(ax^2 + bx + c)^2} + \cdots + \frac{A_n x + B_n}{(ax^2 + bx + c)^n}. \qquad (13)$$

EXAMPLE 4   Repeated Quadratic Factor

Decompose $\dfrac{x^2}{(x^2 + 4)^2}$ into partial fractions.

Solution The denominator contains only the repeated irreducible quadratic factor $x^2 + 4$. As indicated in (13) we assume a decomposition of the form

$$\frac{x^2}{(x^2 + 4)^2} = \frac{Ax + B}{x^2 + 4} + \frac{Cx + D}{(x^2 + 4)^2}.$$

Clearing fractions by multiplying both sides of the preceding equality by $(x^2 + 4)^2$ gives

$$x^2 = (Ax + B)(x^2 + 4) + Cx + D. \qquad (14)$$

As in Example 3, the denominator of the original has no real zeros and so we must solve a system of four linear equations for A, B, C, and D. To that end we rewrite (14) as

$$0x^3 + 1x^2 + 0x + 0x^0 = Ax^3 + Bx^2 + (4A + C)x + (4B + D)x^0$$

and compare coefficients of like powers (match the colors) to obtain

$$\begin{cases} 0 = A \\ 1 = B \\ 0 = 4A + C \\ 0 = 4B + D. \end{cases}$$

From this system we find that $A = 0$, $B = 1$, $C = 0$, and $D = -4$. The required partial fraction decomposition is then

$$\frac{x^2}{(x^2 + 4)^2} = \frac{1}{x^2 + 4} - \frac{4}{(x^2 + 4)^2}. \qquad \equiv$$

EXAMPLE 5   Combination of Cases

Determine the form of the decomposition of $\dfrac{x + 3}{(x - 5)(x + 2)^2(x^2 + 1)^2}$.

Solution The denominator contains a single linear factor $x - 5$, a repeated linear factor $x + 2$, and a repeated irreducible quadratic factor $x^2 + 1$. By Cases 1, 2, and 4 the assumed form of the partial fraction decomposition is

$$\frac{x + 3}{(x - 5)(x + 2)^2(x^2 + 1)^2} = \overbrace{\frac{A}{x - 5}}^{\text{Case 1}} + \overbrace{\frac{B}{x + 2} + \frac{C}{(x + 2)^2}}^{\text{Case 2}} + \overbrace{\frac{Dx + E}{x^2 + 1} + \frac{Fx + G}{(x^2 + 1)^2}}^{\text{Case 4}}.$$

$$\equiv$$

NOTES FROM THE CLASSROOM

We assumed throughout the foregoing discussion that the degree of the numerator $P(x)$ was less than the degree of the denominator $Q(x)$. If, however, the degree of $P(x)$ is greater than or equal to the degree of $Q(x)$, then $P(x)/Q(x)$ is an **improper fraction**. We can still do partial fraction decomposition but the process starts with long division until a polynomial quotient and a proper fraction is attained. For example, long division gives

$$\underset{\uparrow}{\overset{\text{improper fraction}}{}}\qquad\qquad\qquad\underset{}{\overset{\text{proper fraction}}{\downarrow}}$$

$$\frac{x^3 + x - 1}{x^2 - 3x} = x + 3 + \frac{10x - 1}{x(x - 3)}.$$

Then by using Case 1 we finish the problem with the decomposition of the proper fraction term in the last equality:

$$\frac{x^3 + x - 1}{x^2 - 3x} = x + 3 + \frac{10x - 1}{x(x - 3)} = x + 3 + \frac{\frac{1}{3}}{x} + \frac{\frac{29}{3}}{x - 3}.$$

See Problems 25–30 in Exercises 9.5.

9.5 **Exercises** Answers to selected odd-numbered problems begin on page ANS-28.

In Problems 1–24, find the partial fraction decomposition of the given rational expression.

1. $\dfrac{1}{x(x + 2)}$

2. $\dfrac{2}{x(4x - 1)}$

3. $\dfrac{-9x + 27}{x^2 - 4x - 5}$

4. $\dfrac{-5x + 18}{x^2 + 2x - 63}$

5. $\dfrac{2x^2 - x}{(x + 1)(x + 2)(x + 3)}$

6. $\dfrac{1}{x(x - 2)(2x - 1)}$

7. $\dfrac{3x}{x^2 - 16}$

8. $\dfrac{10x - 5}{25x^2 - 1}$

9. $\dfrac{5x - 6}{(x - 3)^2}$

10. $\dfrac{5x^2 - 25x + 28}{x^2(x - 7)}$

11. $\dfrac{1}{x^2(x + 2)^2}$

12. $\dfrac{-4x + 6}{(x - 2)^2(x - 1)^2}$

13. $\dfrac{3x - 1}{x^3(x - 1)(x + 3)}$

14. $\dfrac{x^2 - x}{x(x + 4)^3}$

15. $\dfrac{6x^2 - 7x + 11}{(x - 1)(x^2 + 9)}$

16. $\dfrac{2x + 10}{2x^3 + x}$

17. $\dfrac{4x^2 + 4x - 6}{(2x - 3)(x^2 - x + 1)}$

18. $\dfrac{2x^2 - x + 7}{(x - 6)(x^2 + x + 5)}$

19. $\dfrac{t + 8}{t^4 - 1}$

20. $\dfrac{y^2 + 1}{y^3 - 1}$

21. $\dfrac{x^3}{(x^2 + 2)(x^2 + 1)}$

22. $\dfrac{x - 15}{(x^2 + 2x + 5)(x^2 + 6x + 10)}$

23. $\dfrac{(x + 1)^2}{(x^2 + 1)^2}$

24. $\dfrac{2x^2}{(x - 2)(x^2 + 4)^2}$

In Problems 25–30, first use long division followed by partial fraction decomposition.

25. $\dfrac{x^5}{x^2 - 1}$

26. $\dfrac{(x + 2)^2}{x(x + 3)}$

27. $\dfrac{x^2 - 4x + 1}{2x^2 + 5x + 2}$

28. $\dfrac{x^4 + 3x}{x^2 + 2x + 1}$

29. $\dfrac{x^6}{x^3 - 2x^2 + x - 2}$

30. $\dfrac{x^3 + x^2 - x + 1}{x^3 + 3x^2 + 3x + 1}$

| CHAPTER 9 | Review Exercises | Answers to selected odd-numbered problems begin on page ANS-28. |

A. Fill in the Blanks

In Problems 1–10, fill in the blanks.

1. The linear system

$$\begin{cases} 1x - 2y = 3 \\ -\tfrac{1}{2}x + y = b \end{cases}$$

is consistent for $b =$ _____.

2. $\begin{vmatrix} a & a + 1 \\ a + 2 & a + 3 \end{vmatrix} =$ _____.

3. If $\begin{vmatrix} x & 1 & 1 \\ 1 & 1 & x \\ 1 & x & 1 \end{vmatrix} = 0$, then $x =$ _____.

4. By Cramer's Rule the solution of the linear system

$$\begin{cases} \alpha x - \beta y = 1 \\ \beta x + \alpha y = 1 \end{cases}$$

$\alpha \neq 0, \beta \neq 0$, is _____.

5. The graph of a single linear inequality in two variables represents a _____ in the plane.

6. A solution of the nonlinear system

$$\begin{cases} y = \ln x \\ y = 1 - x \end{cases}$$

is _____.

7. The solution of the linear system

$$\begin{cases} 3x + y + z = 2 \\ y + 2z = 1 \\ 4z = -8 \end{cases}$$

is _____.

8. If the system of two linear equations in two variables has an infinite number of solutions, then the equations are said to be _____.

9. If the graph of $y = ax^2 + bx$ passes through $(1, 1)$ and $(2, 1)$, then $a =$ _____ and $b =$ _____.

10. The graph of the nonlinear system of inequalities

$$\begin{cases} x^2 + y^2 \le 25 \\ y - 1 > 0 \\ x + 1 < 0 \end{cases}$$

lies in the _____ quadrant.

B. True/False

In Problems 1–10, answer true or false.

1. The graphs of $2x + 7y = 6$ and $x^4 + 8xy - 3y^6 = 0$ intersect at $(-4, 2)$. _____

2. The homogeneous linear system

$$\begin{cases} x + 2y - 3z = 0 \\ x + y + z = 0 \\ -2x - 4y + 6z = 0 \end{cases}$$

possesses only the zero solution $(0, 0, 0)$. _____

3. The nonlinear system

$$\begin{cases} y = mx \\ x^2 + y^2 = k \end{cases}$$

always has two solutions when $m \ne 0$ and $k > 0$. _____

4. If the determinant of the coefficients in a system of three linear equations and three variables is 0, then Cramer's Rule indicates that the system has no solution. _____

5. The nonlinear systems

$$\begin{cases} y = \sqrt{x} \\ y = \sqrt{4 - x} \end{cases} \quad \text{and} \quad \begin{cases} y^2 = x \\ y^2 = 4 - x \end{cases}$$

are equivalent. _____

6. $(1, -2)$ is a solution of the inequality $4x - 3y + 5 \le 0$. _____

7. The origin is in the half-plane determined by $4x - 3y < 6$. _____

8. The system of linear inequalities

$$\begin{cases} x + y > 4 \\ x + y < -1 \end{cases}$$

has no solutions. _____

9. The system of nonlinear equations

$$\begin{cases} x^2 + y^2 = 25 \\ x^2 - y = 5 \end{cases}$$

has exactly three solutions. _____

10. The form of the partial-fraction decomposition of $\dfrac{1}{x^2(x + 1)^2}$ is $\dfrac{A}{x^2} + \dfrac{B}{(x + 1)^2}$. _____

C. Review Exercises

In Problems 1–14, find all real solutions of the given system of equations.

1. $\begin{cases} x + y + z = 0 \\ x + 2y + 3z = 0 \\ x - y - z = 0 \end{cases}$

2. $\begin{cases} x + 5y + 6z = 1 \\ 4x - y + 2z = 4 \\ 2x - 11y + 14z = 2 \end{cases}$

3. $\begin{cases} 2x + y - z = 7 \\ x + y + z = -2 \\ 4x + 2y + 2z = -6 \end{cases}$

4. $\begin{cases} 5x - 2y = 10 \\ x + 4y = 4 \end{cases}$

5. $\begin{cases} x^2 - 4x + y = 5 \\ x + y = -1 \end{cases}$

6. $\begin{cases} 101y = 10^x + 10^{-x} \\ y - 10^x = 0 \end{cases}$

7. $\begin{cases} 4x^2 + y^2 = 16 \\ x^2 + 4y^2 = 16 \end{cases}$

8. $\begin{cases} xy = 12 \\ -\dfrac{1}{x} + \dfrac{1}{y} = \dfrac{1}{3} \end{cases}$

9. $\begin{cases} y - \log_{10} x = 0 \\ y^2 - 4\log_{10} x + 4 = 0 \end{cases}$

10. $\begin{cases} x^2 y = 63 \\ y = 16 - x^2 \end{cases}$

11. $\begin{cases} 2\ln x + \ln y = 3 \\ 5\ln x + 2\ln y = 8 \end{cases}$

12. $\begin{cases} x^2 + y^2 = 4 \\ xy = 1 \end{cases}$

13. $\begin{cases} e^{x+y} - e^x = 0 \\ e^{-y+x} - e^{-y+1} = 0 \end{cases}$

14. $\begin{cases} y = x^2 \\ x^2 - y^2 = 4 \end{cases}$

15. **Playing with Numbers** In a two-digit number, the units digit is 1 greater than 3 times the tens digit. When the digits are reversed, the new number is 45 more than the old number. Find the old number.

16. **Lengths** A right triangle has an area of 24 cm². If its hypotenuse has a length of 10 cm, find the lengths of the remaining two sides of the triangle.

17. **Got a Wire Cutter?** A wire 1 m long is cut into two pieces. One piece is bent into a circle and the other piece is bent into a square. The sum of the areas of the circle and the square is $\frac{1}{16}$ m². What are the lengths of the sides of the square and the radius of the circle?

18. **Coordinates** Find the coordinates of the point P of intersection of the line and the parabola shown in FIGURE 9.R.1.

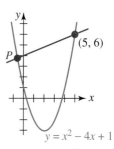

FIGURE 9.R.1 Graphs for Problem 18

In Problems 19–22, find the partial fraction decomposition of the given rational expression.

19. $\dfrac{2x - 1}{x(x^2 + 2x - 3)}$

20. $\dfrac{1}{x^4(x^2 + 5)}$

21. $\dfrac{x^2}{(x^2 + 4)^2}$

22. $\dfrac{x^5 - x^4 + 2x^3 + 5x - 1}{(x - 1)^2}$

In Problems 23–28, graph the given system of inequalities.

23. $\begin{cases} y - x \le 0 \\ y + x \le 0 \\ y \ge -1 \end{cases}$

24. $\begin{cases} x + y \le 4 \\ 2x - 3y \ge -6 \\ 3x - 2y \le 12 \end{cases}$

25. $\begin{cases} x + y \le 5 \\ x + y \ge 1 \\ -x + y \le 7 \end{cases}$

26. $\begin{cases} 1 \le x \le 4 \\ 2 \le y \le 6 \\ x + y \ge 5 \\ -x + y \le 9 \end{cases}$

27. $\begin{cases} x^2 + y^2 \leq 4 \\ x^2 + y^2 - 4y \leq 0 \end{cases}$ **28.** $\begin{cases} y \leq -x^2 - x + 6 \\ y \geq x^2 - 2x \end{cases}$

29. In words, describe the graph of the inequality $1 \leq x - y \leq 4$.

30. Using the equations $y = 9 - x^2$ and $y = 4 - x^2$, give:
    (a) a system of two inequalities that has no solution,
    (b) a system of two inequalities for which $(1, 9)$ is a member of the solution set.

In Problems 31–34, use the functions $y = x^2$ and $y = 2 - x$ to form a system of inequalities whose graph is given in the figure.

31.

FIGURE 9.R.2 Graphs for
Problem 31

32.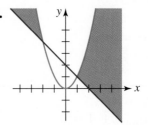

FIGURE 9.R.3 Graphs for
Problem 32

33.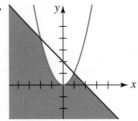

FIGURE 9.R.4 Graphs for
Problem 33

34.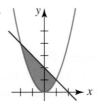

FIGURE 9.R.5 Graphs for
Problem 34

In Problems 35–40, match one of the systems of equations given in (a)–(f) with the graphs given in the figure.

(a) $\begin{cases} xy = 1 \\ x^3 - y = 0 \end{cases}$

(b) $\begin{cases} y - x = 1 \\ y - \sqrt{x} = 2 \end{cases}$

(c) $\begin{cases} x^2 + y^2 = 4 \\ x^2 - x^2y = 1 \end{cases}$

(d) $\begin{cases} x + 2y = 2 \\ 2x - y = 3 \end{cases}$

(e) $\begin{cases} 16x^2 + 9y^2 = 144 \\ x + y^2 = 4 \end{cases}$

(f) $\begin{cases} x - y^3 = 0 \\ y = x^3 - x \end{cases}$

35.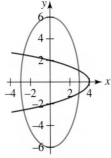

FIGURE 9.R.6 Graphs for
Problem 35

36.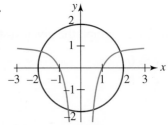

FIGURE 9.R.7 Graphs for
Problem 36

37.

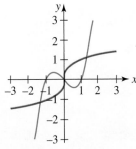

FIGURE 9.R.8 Graphs for Problem 37

38.

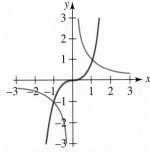

FIGURE 9.R.9 Graphs for Problem 38

39.

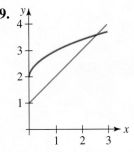

FIGURE 9.R.10 Graphs for Problem 39

40.

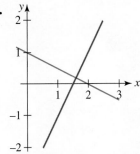

FIGURE 9.R.11 Graphs for Problem 40

41. Give a system of inequalities whose graph is given in **FIGURE 9.R.12**.

42. Give an example of a nonlinear system of two equations that has four solutions but the graphs of the equations intersect at only two points.

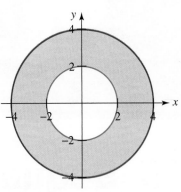

FIGURE 9.R.12 Graphs for Problem 41

10 Sequences and Series

Chapter Outline

10.1 Sequences

≡ **Introduction** Most people have heard the phrases "sequence of cards," "sequence of events," and "sequence of car payments." Intuitively, we can describe a **sequence** as a list of objects, events, or numbers that come one after the other, that is, a list of things given in some definite order. The months of the year listed in the order in which they occur,

$$\text{January, February, March, } \ldots, \text{ December} \tag{1}$$

and

$$3, 4, 5, \ldots, 12 \tag{2}$$

are two examples of sequences. Each object in the list is called a **term** of the sequence. The lists in (1) and (2) are **finite sequences**: The sequence in (1) has 12 terms and the sequence in (2) has 10 terms. A sequence such as

$$1, \tfrac{1}{2}, \tfrac{1}{3}, \tfrac{1}{4}, \ldots, \tag{3}$$

where no last term is indicated, is understood to be an **infinite sequence**. The three dots in (1), (2), and (3) is called an *ellipsis* and indicates that succeeding terms follow the same pattern as that set by the terms given.

In this chapter, unless otherwise specified, we will use the word *sequence* to mean *infinite sequence*. ◀ Note

The terms of a sequence can be put into a one-to-one correspondence with the set N of positive integers. For example, a natural correspondence for the sequence in (3) is

$$
\begin{array}{cccc}
1, & \tfrac{1}{2}, & \tfrac{1}{3}, & \tfrac{1}{4}, \ldots \\
\uparrow & \uparrow & \uparrow & \uparrow \cdots \\
1 & 2 & 3 & 4
\end{array}
$$

Because of this correspondence property, we can give a precise mathematical definition of a sequence.

DEFINITION 10.1.1 Sequence

A **sequence** is a function f with domain the set N of positive integers $\{1, 2, 3, \ldots\}$.

You should be aware that in some instances it is convenient to take the domain of a sequence to be the set of nonnegative integers $\{0, 1, 2, 3, \ldots\}$. A **finite sequence** is also a function and its domain is some subset $\{1, 2, 3, \ldots, n\}$ of N.

□ **Terminology** The elements in the **range** of a sequence are simply the terms of the sequence. We will assume hereafter that the range of a sequence is some set of real numbers. The number $f(1)$ is taken to be the first term of the sequence, the second term is $f(2)$, and, in general, the **nth term** is $f(n)$. Rather than using function notation, we commonly represent the terms of a sequence using subscripts: $f(1) = a_1, f(2) = a_2, \ldots,$ and so on. The nth term $f(n) = a_n$ is also called the **general term** of the sequence. We denote a sequence

$$a_1, a_2, a_3, \ldots, a_n, \ldots$$

by the notation $\{a_n\}$. If we identify the general term in (3) as $1/n$, the sequence $1, \tfrac{1}{2}, \tfrac{1}{3}, \ldots,$ can then be written compactly as $\{1/n\}$.

EXAMPLE 1   Three Sequences

List the first five terms of the given sequence.

(a) $\{2^n\}$ **(b)** $\left\{\dfrac{1}{n^2}\right\}$ **(c)** $\left\{\dfrac{(-1)^n n}{n+1}\right\}$

Solution Letting n take on the values 1, 2, 3, 4, and 5, the first five terms of the (infinite) sequences are

(a) $2^1, 2^2, 2^3, 2^4, 2^5, \ldots$ or $2, 4, 8, 16, 32, \ldots$.

(b) $\dfrac{1}{1^2}, \dfrac{1}{2^2}, \dfrac{1}{3^2}, \dfrac{1}{4^2}, \dfrac{1}{5^2}, \ldots$ or $1, \dfrac{1}{4}, \dfrac{1}{9}, \dfrac{1}{16}, \dfrac{1}{25}, \ldots$.

(c) $\dfrac{(-1)^1 \cdot 1}{1+1}, \dfrac{(-1)^2 \cdot 2}{2+1}, \dfrac{(-1)^3 \cdot 3}{3+1}, \dfrac{(-1)^4 \cdot 4}{4+1}, \dfrac{(-1)^5 \cdot 5}{5+1}, \ldots$

or $-\dfrac{1}{2}, \dfrac{2}{3}, -\dfrac{3}{4}, \dfrac{4}{5}, -\dfrac{5}{6}, \ldots$.   ≡

☐ **Sequences Defined Recursively** Instead of giving the general term of a sequence $a_1, a_2, a_3, \ldots, a_n, a_{n+1}, \ldots$, sequences are often defined using a rule or formula in which a_{n+1} is expressed using the preceding terms. For example, if we set $a_1 = 1$ and define successive terms by $a_{n+1} = a_n + 2$ for $n = 1, 2, \ldots$, then

given
$$\downarrow$$
$$a_2 = a_1 + 2 = 1 + 2 = 3,$$
$$a_3 = a_2 + 2 = 3 + 2 = 5,$$
$$a_4 = a_3 + 2 = 5 + 2 = 7,$$

and so on. Sequences such as this are said to be defined **recursively**. In this example, the rule that $a_{n+1} = a_n + 2$ is called a **recursion formula**.

EXAMPLE 2   Sequence Defined Recursively

List the first five terms of the sequence defined by $a_1 = 2$ and $a_{n+1} = (n+2)a_n$.

Solution We are given $a_1 = 2$. From the recursion formula we have, respectively, for $n = 1, 2, 3, 4, \ldots$

given
$$\downarrow$$
$$a_2 = (1+2)a_1 = 3(2) = 6,$$
$$a_3 = (2+2)a_2 = 4(6) = 24,$$
$$a_4 = (3+2)a_3 = 5(24) = 120,$$
$$a_5 = (4+2)a_4 = 6(120) = 720,$$

and so on. Including $a_1 = 2$ the first five terms of the sequence are

$$2, 6, 24, 120, 720, \ldots.$$ ≡

Of course, if we choose a different value for a_1 in Example 2 we would obtain an entirely different sequence.

For the remainder of this section we will examine two special types of recursively defined sequences.

☐ **Arithmetic Sequence** In the sequence $1, 3, 5, 7, \ldots$, note that each term after the first is obtained by adding the number 2 to the term preceding it. In other words, successive terms in the sequence differ by 2. A sequence of this type is known as an **arithmetic sequence**.

DEFINITION 10.1.2 Arithmetic Sequence

A sequence such that the successive terms a_{n+1} and a_n, for $n = 1, 2, 3, \ldots$, have a fixed difference $a_{n+1} - a_n = d$ is called an **arithmetic sequence**. The number d is called the **common difference** of the sequence.

From $a_{n+1} - a_n = d$, we obtain the recursion formula

$$a_{n+1} = a_n + d \tag{4}$$

for an arithmetic sequence with common difference d.

EXAMPLE 3   **An Arithmetic Sequence**

The first several terms of the recursive sequence defined by $a_1 = 3$ and $a_{n+1} = a_n + 4$ are

$$a_1 = 3$$
$$a_2 = a_1 + 4 = 3 + 4 = 7,$$
$$a_3 = a_2 + 4 = 7 + 4 = 11,$$
$$a_4 = a_3 + 4 = 11 + 4 = 15,$$
$$a_5 = a_4 + 4 = 15 + 4 = 19,$$
$$\vdots$$

or $3, 7, 11, 15, 19, \ldots$. This is an arithmetic sequence with common difference 4. $\equiv$

If we let a_1 be the first term of an arithmetic sequence having common difference d, we find from the recursion formula (4) that

$$a_2 = a_1 + d$$
$$a_3 = a_2 + d = a_1 + 2d$$
$$a_4 = a_3 + d = a_1 + 3d$$
$$\vdots$$
$$a_n = a_{n-1} + d = a_1 + (n - 1)d$$

and so on. In general, an arithmetic sequence with first term a_1 and common difference d is given by

$$\{a_1 + (n - 1)d\}. \tag{5}$$

EXAMPLE 4   **Arithmetic Sequence Using (5)**

A woman decides to jog a particular distance each week according to the following schedule. The first week she will jog 1000 m per day. Each succeeding week she will jog 250 m per day farther than she did the preceding week.
(a) How far will she jog per day in the 26th week?
(b) In which week will she jog 10,000 m per day?

Solution The example describes an arithmetic sequence with $a_1 = 1000$ and $d = 250$.
(a) To find the distance the woman jogs per day in the 26th week, we set $n = 26$ and compute a_{26} using (5):

$$a_{26} = 1000 + (26 - 1)(250) = 1000 + 6250 = 7250.$$

Thus she will jog 7250 m per day in the 26th week.

(b) Here we are given $a_n = 10{,}000$ and we need to find n. From (5) we have $10{,}000 = 1000 + (n-1)(250)$ or $9000 = (n-1)(250)$. Solving for n gives

$$n - 1 = \frac{9000}{250} = 36 \quad \text{or} \quad n = 37.$$

Therefore, she will jog 10,000 m per day in the 37th week. ≡

EXAMPLE 5   Find the First Term

The common difference in an arithmetic sequence is -2 and the sixth term is 3. Find the first term of the sequence.

Solution From (5) the sixth term of the sequence is

$$a_6 = a_1 + (6-1)d.$$

Setting $a_6 = 3$ and $d = -2$, we have $3 = a_1 + 5(-2)$, or $a_1 = 3 + 10$. Thus the first term is $a_1 = 13$.

Check: The sequence with $a_1 = 13$ and $d = -2$ is $13, 11, 9, 7, 5, 3, \ldots$. The sixth term of this sequence is 3. ≡

□ **Geometric Sequence** In the sequence $1, 2, 4, 8, \ldots$, each term after the first is obtained by multiplying the term preceding it by the number 2. In this case, we observe that the ratio of a term to the term preceding it is a constant, namely, 2. A sequence of this type is said to be a **geometric sequence**.

DEFINITION 10.1.3   Geometric Sequence

A sequence such that the successive terms a_{n+1} and a_n, for $n = 1, 2, 3, \ldots$, have a fixed ratio $a_{n+1}/a_n = r$, is called a **geometric sequence**. The number r is called the **common ratio** of the sequence.

From $a_{n+1}/a_n = r$ in Definition 10.1.3, we see that a geometric sequence with a common ratio r is defined recursively by the formula

$$a_{n+1} = a_n r. \tag{6}$$

EXAMPLE 6   Geometric Sequence Using (6)

The sequence defined recursively by $a_1 = 2$ and $a_{n+1} = -3a_n$ is

$$2, -6, 18, -54, \ldots.$$

This is a geometric sequence with common ratio $r = -3$. ≡

If we let $a_1 = a$ be the first term of a geometric sequence with common ratio r, we find from the recursion formula (6) that

$$a_2 = a_1 r = ar$$
$$a_3 = a_2 r = ar^2$$
$$a_4 = a_3 r = ar^3$$
$$\vdots$$
$$a_n = a_{n-1} r = ar^{n-1}$$

and so on. In general, a geometric sequence with first term a and common ratio r is

$$\{ar^{n-1}\}. \tag{7}$$

EXAMPLE 7 **Find the Third Term**

Find the third term of a geometric sequence with common ratio $\frac{2}{3}$ and sixth term $\frac{128}{81}$.

Solution We first find a. Since $a_6 = \frac{128}{81}$ and $r = \frac{2}{3}$, we have from (7) that

$$\frac{128}{81} = a\left(\frac{2}{3}\right)^{6-1}.$$

Solving for a, we find

$$a = \frac{\frac{128}{81}}{\left(\frac{2}{3}\right)^5} = \frac{2^7}{3^4}\left(\frac{3^5}{2^5}\right) = 12.$$

Applying (7) again with $n = 3$, we have

$$a_3 = 12\left(\frac{2}{3}\right)^{3-1} = 12\left(\frac{4}{9}\right) = \frac{16}{3}.$$

The third term of the sequence is $a_3 = \frac{16}{3}$. ≡

☐ **Compound Interest** An initial amount of money deposited in a savings account is called the **principal** and is denoted by P. Suppose that the annual **rate of interest** for the account is r. If interest is *compounded annually,* then at the end of the first year the interest on P is Pr and the amount A_1 accumulated in the account at the end of the first year is principal plus interest:

$$A_1 = P + Pr = P(1 + r).$$

The interest earned on this amount at the end of the second year is $P(1 + r)r$. If this amount is deposited, then at the end of the second year the account contains

$$\begin{aligned} A_2 &= P(1 + r) + P(1 + r)r \\ &= P(1 + 2r + r^2) = P(1 + r)^2. \end{aligned}$$

Continuing in this fashion, we can construct the following table.

Year	Amount at the End of the Year
1	$P(1 + r)$
2	$P(1 + r)^2$
3	$P(1 + r)^3$
4	$P(1 + r)^4$
$\vdots$	$\vdots$

The amounts in the second column of the table form a geometric sequence with first term $P(1 + r)$ and common ratio $1 + r$. Thus from (7) we conclude that the amount in the savings account at the end of the nth year is $A_n = [P(1 + r)](1 + r)^{n-1}$ or

$$A_n = P(1 + r)^n. \tag{8}$$

EXAMPLE 8 **Compound Interest**

On January 1, 2010, a principal of \$500 was deposited in an account drawing 4% interest compounded annually. Find the amount in the account on January 1, 2024.

Solution We make the identification $P = 500$ and $r = 0.04$. As of January 1, 2024, the principal will have drawn interest for 14 years. Using (8) and a calculator, we find

$$A_{14} = 500(1 + 0.04)^{14}$$
$$= 500(1.04)^{14}$$
$$\approx 865.84.$$

To the nearest dollar amount, the account will contain $866 at the end of 14 years. ≡

| 10.1 | Exercises | Answers to selected odd-numbered problems begin on page ANS-28. |

In Problems 1–10, list the first five terms of the given sequence.

1. $\{(-1)^n\}$

2. $\left\{\dfrac{n}{n+3}\right\}$

3. $\{\frac{1}{2}n(n+1)\}$

4. $\left\{\dfrac{(-2)^n}{n^2}\right\}$

5. $\left\{\dfrac{1}{n^2+1}\right\}$

6. $\left\{\dfrac{n+1}{n+2}\right\}$

7. $\{n\cos n\pi\}$

8. $\left\{\dfrac{1}{n^3}\sin\dfrac{n\pi}{2}\right\}$

9. $\left\{\dfrac{n+(-1)^n}{1+4n}\right\}$

10. $\{(-1)^{n-1}(1+n)^2\}$

In Problems 11 and 12, list the first six terms of a sequence whose general term is given.

11. $a_n = \begin{cases} -2^n, & n \text{ odd,} \\ n^2, & n \text{ even} \end{cases}$

12. $a_n = \begin{cases} \sqrt{n}, & n \text{ odd,} \\ 1/n, & n \text{ even} \end{cases}$

In Problems 13 and 14, discern a pattern for the given sequence and determine the next three terms.

13. $1, 2, \frac{1}{9}, 4, \frac{1}{25}, 6, \ldots$

14. $2, 3, 5, 8, 12, 17, \ldots$

In Problems 15–22, list the first five terms of the sequence defined recursively.

15. $a_1 = 3$, $a_n = \dfrac{(-1)^n}{a_{n-1}}$

16. $a_1 = \frac{1}{2}$, $a_n = (-1)^n(a_{n-1})^2$

17. $a_1 = 0$, $a_n = 2 + 3a_{n-1}$

18. $a_1 = 2$, $a_n = \frac{1}{3}na_{n-1}$

19. $a_1 = 1$, $a_n = \dfrac{1}{n}a_{n-1}$

20. $a_1 = 0$, $a_2 = 1$, $a_n = a_{n-1} - a_{n-2}$

21. $a_1 = 7$, $a_{n+1} = a_n + 2$

22. $a_1 = -6$, $a_{n+1} = \frac{2}{3}a_n$

In Problems 23–32, the given sequence is either an arithmetic or a geometric sequence. Find either the common difference or the common ratio. Write the general term and the recursion formula of the sequence.

23. $4, -1, -6, -11, \ldots$

24. $\frac{1}{16}, \frac{1}{8}, \frac{1}{4}, \frac{1}{2}, \ldots$

25. $4, -3, \frac{9}{4}, -\frac{27}{16}, \ldots$

26. $\frac{1}{2}, 1, \frac{3}{2}, 2, \ldots$

27. $2, -9, -20, -31, \ldots$

28. $-\frac{1}{3}, 1, -3, 9, \ldots$

29. $0.1, 0.01y, 0.001y^2, 0.0001y^3, \ldots$

30. $4x, 7x, 10x, 13x, \ldots$

31. $\frac{3}{8}, -\frac{1}{4}, \frac{1}{6}, -\frac{1}{9}, \ldots$

32. $\log_3 2, \log_3 4, \log_3 8, \log_3 16, \ldots$

33. Find the twentieth term of the sequence $-1, 5, 11, 17, \ldots$.
34. Find the fifteenth term of the sequence $2, 6, 10, 14, \ldots$.
35. Find the fifth term of a geometric sequence with first term 8 and common ratio $r = -\frac{1}{2}$.
36. Find the eighth term of the sequence $\frac{1}{1024}, \frac{1}{128}, \frac{1}{16}, \frac{1}{2}, \ldots$
37. Find the first term of a geometric sequence with third and fourth terms 2 and 8, respectively.
38. Find the first term of an arithmetic sequence with fourth and fifth terms 5 and -3, respectively.
39. Find the seventh term of an arithmetic sequence with first and third terms 357 and 323, respectively.
40. Find the tenth term of a geometric sequence with fifth and sixth terms 2 and 3, respectively.
41. Find an arithmetic sequence whose first term is 4 such that the sum of the second and third terms is 17.
42. Find a geometric sequence whose second term is 1 such that $a_5/a_3 = 64$.
43. If $1000 is invested at 7% interest compounded annually, find the amount in the account after 20 years.
44. Find the amount that must be deposited in an account drawing 5% interest compounded annually in order to have $10,000 in the account 30 years later.
45. At what rate of interest compounded annually should $450 be deposited in order to have $750 in 8 years?
46. At 6% interest compounded annually, how long will it take an initial investment to double?

Miscellaneous Applications

47. **Cookie-Jar Savings** A couple decides to set aside $5 each month the first year of their marriage, $15 each month the second year, $25 each month the third year, and so on, increasing the monthly amount by $10 each year. Find the amount they will set aside each month of the fifteenth year.
48. **Cookie-Jar Savings—Continued** In Problem 47, find a formula for the amount the couple will set aside each month of the nth year.
49. **Population Growth** The population of a certain community is observed to grow geometrically by a factor of $\frac{3}{2}$ each year. If the population at the beginning of the first year is 1000, find the population at the beginning of the eleventh year.
50. **Profit** A small company expects its profits to increase at a rate of $10,000 per year. If its profit after the first year is $6000, how much profit can the company expect after 15 years of operation?
51. **Family Tree** Everyone has two parents. Determine how many great-great-great-grandparents a person will have.
52. **How Many Rabbits?** Besides its famous leaning bell tower, the city of Pisa, Italy, is also noted as the birthplace of **Leonardo Pisano**, aka **Leonardo Fibonacci** (1170–1250). Fibonacci was the first in Europe to introduce the Hindu–Arabic place-valued decimal system and the use of Arabic numerals. His book *Liber Abacci*, published in 1202, is basically a text on how to do arithmetic in this decimal system. But in Chapter 12 of *Liber Abacci*, Fibonacci poses and solves the following problem on the reproduction of rabbits:

> *How many pairs of rabbits will be produced in a year beginning with a single pair, if in every month each pair bears a new pair that become productive from the second month on?*

Statue of
Fibonacci in
Pisa, Italy

Discern the pattern of the solution of this problem and complete the following table.

	Start	After each month											
		1	2	3	4	5	6	7	8	9	10	11	12
Adult pairs	1	1	2	3	5	8	13	21					
Baby pairs	0	1	1	2	3	5	8	13					
Total pairs	1	2	3	5	8	13	21	34					

53. Write out five terms, after the initial two, of the sequence defined recursively by $F_{n+1} = F_n + F_{n-1}$, $F_1 = 1$, $F_2 = 1$. This sequence is called the **Fibonacci sequence** and the terms of the sequence are called **Fibonacci numbers**. Reexamine Problem 52.

For Discussion

54. Verify that the general term of the sequence defined in Problem 53 is

$$F_n = \frac{1}{\sqrt{5}}\left(\frac{1 + \sqrt{5}}{2}\right)^n - \frac{1}{\sqrt{5}}\left(\frac{1 - \sqrt{5}}{2}\right)^n$$

by showing that this result satisfies the recursion formula.

55. Find two different values of x such that $-\frac{3}{2}, x, -\frac{8}{27}, \ldots$ is a geometric sequence.

56. If $\{a_n\}$ and $\{b_n\}$ are geometric sequences, then show that $\{a_n b_n\}$ is a geometric sequence.

10.2 Series

≡ **Introduction** In the following discussion, we will be concerned with the sum of the terms of a sequence. Of special interest are the sums of the first n terms of arithmetic and geometric sequences. We begin by reviewing a special notation that is used as a convenient shorthand for an indicated sum of terms.

☐ **Summation Notation** Suppose we are interested in the sum of the first n terms of a sequence $\{a_n\}$. Rather than writing

$$a_1 + a_2 + \cdots + a_n,$$

mathematicians have invented a notation for representing such sums in a compact manner:

$$\sum_{k=1}^{n} a_k = a_1 + a_2 + \cdots + a_n.$$

Because $\sum$ is the capital Greek letter *sigma*, the notation $\sum_{k=1}^{n} a_k$ is referred to as

See Section 3.7. ▶
summation or **sigma notation** and is read "the sum from $k = 1$ to $k = n$ of a sub k." The subscript k is called the **index of summation** and takes on the successive values $1, 2, \ldots, n$:

sum ends with this number
↓
$$\sum_{k=1}^{n} a_k.$$
↑
sum starts with this number

EXAMPLE 1   **Summation Notation**

Write out each sum.

(a) $\displaystyle\sum_{k=1}^{4} k^2$ **(b)** $\displaystyle\sum_{k=1}^{20} (3k+1)$ **(c)** $\displaystyle\sum_{k=1}^{n} (-1)^{k+1} a_k$

Solution **(a)** $\displaystyle\sum_{k=1}^{4} k^2 = 1^2 + 2^2 + 3^2 + 4^2 = 1 + 4 + 9 + 16$

(b) $\displaystyle\sum_{k=1}^{20} (3k+1) = (3(1)+1) + (3(2)+1) + (3(3)+1) + \cdots + (3(20)+1)$

$$= 4 + 7 + 10 + \cdots + 61$$

(c) $\displaystyle\sum_{k=1}^{n} (-1)^{k+1} a_k = (-1)^{1+1} a_1 + (-1)^{2+1} a_2 + (-1)^{3+1} a_3 + \cdots + (-1)^{n+1} a_n$

$$= a_1 - a_2 + a_3 - \cdots + (-1)^{n+1} a_n \qquad\qquad \equiv$$

The choice of the letter used as the index of summation is arbitrary. Although we will consistently use the letter k, we note that

$$\sum_{k=1}^{n} a_k = \sum_{j=1}^{n} a_j = \sum_{m=1}^{n} a_m,$$

and so on. Also, as we see in the next example, we may sometimes allow the index of summation to start at a value other than $k = 1$.

EXAMPLE 2   **Using Summation Notation**

Write $1 - \frac{1}{2} + \frac{1}{4} - \frac{1}{8} + \cdots + \frac{1}{256}$ using summation notation.

Solution We observe that the kth term of the sequence $1, -\frac{1}{2}, \frac{1}{4}, -\frac{1}{8}, \ldots$ can be written as $(-1)^k \frac{1}{2^k}$, where $k = 0, 1, 2, \ldots$. We note too that $\frac{1}{256} = \frac{1}{2^8}$. Therefore,

$$\sum_{k=0}^{8} (-1)^k \frac{1}{2^k} = 1 - \frac{1}{2} + \frac{1}{4} - \frac{1}{8} + \cdots + \frac{1}{256}. \qquad\qquad \equiv$$

☐ **Properties** Some properties of summation notation are listed in the theorem that follows next.

THEOREM 10.2.1   Properties of Summation Notation

Suppose c is a constant (that is, does not depend on k), then

(i) $\displaystyle\sum_{k=1}^{n} ca_k = c\sum_{k=1}^{n} a_k$ (ii) $\displaystyle\sum_{k=1}^{n} c = nc$

(iii) $\displaystyle\sum_{k=1}^{n} (a_k + b_k) = \sum_{k=1}^{n} a_k + \sum_{k=1}^{n} b_k$ (iv) $\displaystyle\sum_{k=1}^{n} (a_k - b_k) = \sum_{k=1}^{n} a_k - \sum_{k=1}^{n} b_k$

Property (i) of Theorem 10.2.1 is simply factoring a common term from a sum:

$$\sum_{k=1}^{n} ca = ca_1 + ca_2 + \cdots + ca_n = c(a_1 + a_2 + \cdots + a_n) = c\sum_{k=1}^{n} a_k.$$

To understand property (*ii*) of Theorem 10.2.1, consider the following simple examples:

$$\overbrace{2 + 2 + 2}^{\text{three 2's}} = 3 \cdot 2 = 6 \quad \text{and} \quad \overbrace{7 + 7 + 7 + 7}^{\text{four 7's}} = 4 \cdot 7 = 28.$$

Thus, if $a_k = c$ is a real constant for $k = 1, 2, \ldots, n$, then

$$a_1 = c, \quad a_2 = c, \ldots, \quad a_n = c.$$

Consequently,

$$\sum_{k=1}^{n} c = \overbrace{c + c + c + \cdots + c}^{n \text{ terms}} = nc.$$

For example, $\sum_{k=1}^{10} 6 = 10 \cdot 6 = 60$.

☐ **Arithmetic Series** Recall, we saw in (5) of Section 10.1 that an arithmetic sequence could be written as $\{a_1 + (n - 1)d\}$. The addition of the first n terms of an arithmetic sequence,

$$S_n = \sum_{k=1}^{n} (a_1 + (k - 1)d) = a_1 + (a_1 + d) + (a_1 + 2d) + \cdots + (a_1 + (n - 1)d) \quad (1)$$

is called an **arithmetic series**. If we denote the last term of the series in (1) by a_n, then S_n can be written as

$$S_n = (a_n - (n - 1)d) + \cdots + (a_n - 2d) + (a_n - d) + a_n. \quad (2)$$

Reversing the terms in (1), we have

$$S_n = (a_1 + (n - 1)d) + \cdots + (a_1 + 2d) + (a_1 + d) + a_1. \quad (3)$$

Adding (2) and (3) gives

$$2S_n = (a_1 + a_n) + (a_1 + a_n) + \cdots + (a_1 + a_n) + (a_1 + a_n) = n(a_1 + a_n).$$

Thus,
$$S_n = n\left(\frac{a_1 + a_n}{2}\right). \quad (4)$$

In other words, the sum of the first n terms of an arithmetic sequence is the number of terms n times the average of the first term a_1 and the nth term a_n of the sequence.

EXAMPLE 3 Arithmetic Series

Find the sum of the first seven terms of the arithmetic sequence $\{5 - 4(n - 1)\}$.

Solution The first term of the sequence is 5 and the seventh term is -19. By identifying $a_1 = 5$, $a_7 = -19$, and $n = 7$ it follows from (4) that the sum of the seven terms in the arithmetic series

$$5 + 1 + (-3) + (-7) + (-11) + (-15) + (-19)$$

is
$$S_7 = 7\left(\frac{5 + (-19)}{2}\right) = 7(-7) = -49. \qquad \equiv$$

EXAMPLE 4 **Sum of the First 100 Positive Integers**

Find the sum of the first 100 positive integers.

Solution The sequence of positive integers $\{n\}$,

$$1, 2, 3, \ldots,$$

is an arithmetic sequence with common difference 1. Thus, from (4) the value of $S_{100} = 1 + 2 + 3 + \cdots + 100$ is given by

$$S_{100} = 100\left(\frac{1 + 100}{2}\right) = 50(101) = 5050. \qquad \equiv$$

An alternative form for the sum of an arithmetic series can be obtained by substituting $a_1 + (n - 1)d$ for a_n in (4). We then have

$$S_n = n\left(\frac{2a_1 + (n - 1)d}{2}\right), \tag{5}$$

which expresses the sum of an arithmetic series in terms of the first term, the number of terms, and the common difference.

EXAMPLE 5 **Paying Off a Loan**

A woman wishes to pay off an interest-free loan of $1300 by paying $10 the first month and increasing her payments by $15 each succeeding month. How many months will it take to pay off the entire loan? Find the amount of the final payment.

Solution The monthly payments form an arithmetic sequence with first term $a_1 = 10$ and common difference $d = 15$. Since the sum of the arithmetic series formed by the sequence of payments is $1300, we let $S_n = 1300$ in (5) and solve for n:

$$1300 = n\left(\frac{2(10) + (n - 1)15}{2}\right)$$
$$= n\left(\frac{5 + 15n}{2}\right)$$
$$2600 = 5n + 15n^2.$$

By dividing by 5 the last equation simplifies to $3n^2 + n - 520 = 0$ or $(3n + 40)(n - 13) = 0$. Thus, $n = -\frac{40}{3}$ or $n = 13$. Since n must be a positive integer, we conclude that it will take 13 months to pay off the loan. The final payment will be

$$a_{13} = 10 + (13 - 1)15 = 10 + 180 = 190 \text{ dollars.} \qquad \equiv$$

☐ **Geometric Series** The addition of the first n terms of a geometric sequence $\{ar^{n-1}\}$ is

$$S_n = \sum_{k=1}^{n} ar^{k-1} = a + ar + ar^2 + \cdots + ar^{n-1} \tag{6}$$

and is called a **finite geometric series**. Multiplying (6) by the common ratio r gives

$$rS_n = ar + ar^2 + ar^3 + \cdots + ar^n. \tag{7}$$

Subtracting (7) from (6) and simplifying gives

$$S_n - rS_n = (a + ar + ar^2 + \cdots + ar^{n-1}) - (ar + ar^2 + \cdots + ar^n) = a - ar^n,$$

or
$$(1 - r)S_n = a(1 - r^n).$$

Solving this equation for S_n, we obtain a formula for the **sum** of a geometric series containing n terms:

$$S_n = \frac{a(1 - r^n)}{1 - r}. \tag{8}$$

EXAMPLE 6   Sum of a Geometric Series

Compute the sum $3 + \frac{3}{2} + \frac{3}{4} + \frac{3}{8} + \frac{3}{16} + \frac{3}{32}$.

Solution This geometric series is the sum of the first six terms of the geometric sequence $\{3(\frac{1}{2})^{n-1}\}$. Identifying the first term $a = 3$, the common ratio $r = \frac{1}{2}$, and $n = 6$ in (8), we have

$$S_6 = \frac{3\left(1 - \left(\frac{1}{2}\right)^6\right)}{1 - \frac{1}{2}} = \frac{3\left(1 - \frac{1}{64}\right)}{\frac{1}{2}} = 6\left(\frac{63}{64}\right) = \frac{189}{32}. \qquad \equiv$$

EXAMPLE 7   Sum of a Geometric Series

A developer constructed one house in 2002. With his profits, he was able to build two houses in 2003. With the profits from these, he constructed four houses in 2004. Assuming that he is able to continue doubling the number of houses he builds each year, find the total number of houses he will have constructed by the end of 2012.

Solution The total number of houses he constructs in the 11 years from 2002 through 2012 is the sum of the geometric series with first term $a = 1$ and common ratio $r = 2$. From (8) the total number of houses constructed is

$$S_{11} = \frac{1 \cdot (1 - 2^{11})}{1 - 2} = \frac{1 - 2048}{-1} = 2047. \qquad \equiv$$

We will return to the subject of geometric series in Section 10.7.

10.2   Exercises Answers to selected odd-numbered problems begin on page ANS-28.

In Problems 1–6, compute the given sum.

1. $\displaystyle\sum_{k=1}^{4} (k - 1)^2$ **2.** $\displaystyle\sum_{k=1}^{3} (-1)^k 2^k$

3. $\displaystyle\sum_{k=0}^{5} (k - k^2)$ **4.** $\displaystyle\sum_{k=1}^{15} 3$

5. $\displaystyle\sum_{k=2}^{6} (-1)^k \frac{30}{k}$ **6.** $\displaystyle\sum_{k=0}^{3} (1 - k)^3$

In Problems 7–10, write out the terms of the given sum.

7. $\displaystyle\sum_{k=1}^{5} \sqrt{k}$ **8.** $\displaystyle\sum_{k=1}^{5} k a_k$

9. $\displaystyle\sum_{k=0}^{3} (-1)^n$ **10.** $\displaystyle\sum_{k=0}^{4} k^2 f(k)$

In Problems 11–16, write the given series in summation notation.

11. $3 + 5 + 7 + 9 + 11$

12. $\frac{1}{2} + \frac{4}{3} + \frac{9}{4} + \frac{16}{5} + \frac{25}{6} + \frac{36}{7}$

13. $\frac{1}{3} - \frac{1}{6} + \frac{1}{12} - \frac{1}{24} + \frac{1}{48} - \frac{1}{96}$

14. $\frac{3}{5} + \frac{5}{6} + \frac{7}{7} + \frac{9}{8} + \frac{11}{9}$

15. $\frac{3}{2} + \frac{5}{4} + \frac{9}{8} + \frac{17}{16} + \frac{33}{32}$

16. $a_0 + \frac{1}{3}a_2 + \frac{1}{5}a_4 + \frac{1}{7}a_6 + \cdots + \dfrac{1}{2n + 1}a_{2n}$

In Problems 17–22, find the sum of the given arithmetic series.

17. $1 + 4 + 7 + 10 + 13$

18. $131 + 111 + 91 + 71 + 51 + 31$

19. $\displaystyle\sum_{k=1}^{12}[3 + (k - 1)8]$

20. $\displaystyle\sum_{k=1}^{20}[-6 + (k - 1)3]$

21. $12 + 5 - 2 - \cdots - 100$

22. $-5 - 3 - 1 + \cdots + 25$

In Problems 23–28, find the sum of the given geometric series.

23. $\frac{1}{3} + \frac{1}{9} + \frac{1}{27} + \frac{1}{81}$

24. $7 + 14 + 28 + 56 + 112 + 224$

25. $60 + 6 + 0.6 + 0.06 + 0.006$

26. $1 - \frac{2}{3} + \frac{4}{9} - \frac{8}{27} + \frac{16}{81} - \frac{32}{243}$

27. $\displaystyle\sum_{k=1}^{8}\left(-\frac{1}{2}\right)^{k-1}$

28. $\displaystyle\sum_{k=1}^{5}4\left(\frac{1}{5}\right)^{k-1}$

29. If $\{a_n\}$ is an arithmetic sequence with $d = 2$ such that $S_{10} = 135$, find a_1 and a_{10}.

30. If $\{a_n\}$ is an arithmetic sequence with $a_1 = 4$ such that $S_8 = 86$, find a_8 and d.

31. Suppose that $a_1 = 5$ and $a_n = 45$ are the first and nth terms, respectively, of an arithmetic series for which $S_n = 2000$. Find n.

32. If $\{a_n\}$ is a geometric sequence with $r = \frac{1}{2}$ such that $S_6 = \frac{65}{8}$, find the first term a.

33. The sum of the first n terms of the geometric sequence $\{2^n\}$ is $S_n = 8190$. Find n.

34. Find the sum of the first 10 terms of the arithmetic sequence

$$y, \frac{x + 3y}{2}, x + 2y, \ldots.$$

35. Find the sum of the first 15 terms of the geometric sequence

$$\frac{x}{y}, -1, \frac{y}{x}, \ldots.$$

36. Find a formula for the sum of the first n positive integers:

$$1 + 2 + 3 + \cdots + n.$$

37. Find a formula for the sum of the first n even integers.

38. Find a formula for the sum of the first n odd integers.

39. Use the result obtained in Problem 36 to find the sum of the first 1000 positive integers.

40. Use the result obtained in Problem 38 to find the sum of the first 50 odd integers.

Miscellaneous Applications

41. Cookie-Jar Savings A couple decides to set aside $5 each month the first year of their marriage, $15 each month the second year, $25 each month the third year, and so on, increasing the monthly amount by $10 each year. Find the total amount that they will have set aside by the end of the fifteenth year.

42. **Cookie-Jar Savings—Continued** In Problem 41, find a formula for the total amount that the couple will have set aside by the end of the nth year.

43. **Distance Traveled** An automobile accelerating at a constant rate travels 2 m the first second, 6 m the second second, 10 m the third second, and so on, traveling an additional 4 m each second. Find the total distance that the automobile has traveled after 6 seconds.

44. **Total Distance** Find a formula for the total distance traveled by the automobile in Problem 43 after n seconds.

45. **Annuity** If the same amount of money P is invested each year for n years at a rate of interest r compounded annually, then the amount accumulated after the nth payment is given by

$$S = P(1 + r)^{n-1} + P(1 + r)^{n-2} + \cdots + P(1 + r) + P.$$

Such a savings plan is called an **annuity**. Show that the value of the annuity after the nth payment is

$$S = P\left[\frac{(1 + r)^n - 1}{r}\right].$$

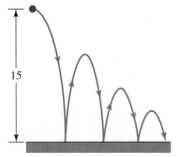

15

46. **Watch the Bouncing Ball** A ball is dropped from an initial height of 15 ft onto a concrete slab. Each time it bounces, it reaches a height of $\frac{2}{3}$ its preceding height. What height does it reach on its third bounce? On its nth bounce? How many times does the ball have to hit the concrete before its height is less than $\frac{1}{2}$ ft? See **FIGURE 10.2.1**.

FIGURE 10.2.1 Bouncing ball in Problem 46

47. **Total Distance** In Problem 46, find the total distance the ball has traveled up to the time when it hits the concrete slab for the seventh time.

48. **Desalinization** A solution of salt water containing 10 kg of salt is passed through a filter that removes 20% of the salt. The resulting solution is then filtered again, removing 20% of the remaining salt. If 20% of the salt is removed during each filtration, find the amount of salt removed from the solution after 10 filtrations.

49. **Drug Accumulation** A patient takes 50 mg of a drug each day and of the amount accumulated, 90% is excreted each day by bodily functions. Determine how much of the drug has accumulated in the body immediately after the eighth dosage.

Display of soup cans

50. **Pyramid Display** A grocer wants to display canned soup in a pyramid with 20 cans on the bottom row, 19 cans on the next row, 18 on the next row, and so on, with a single can at the top. How many cans of soup are required for the display?

For Discussion

Chess board

51. **A Chess Master** According to legend, the king of a Middle Eastern country was so taken with the new game of chess that he queried its peasant inventor on how he might reward him. The inventor's modest request was for the sum of the grains of wheat that would fill the chess board according to the rule: 1 grain on the first square, 2 grains on the second square, 4 on the third square, 8 on the fourth square, and so on, for the entire 64 squares. The king immediately acceded to this request. If an average bushel contains 10^6 grains of wheat, how many bushels did the king owe the inventor? Do you think the peasant lived to see his reward?

10.3 Mathematical Induction

≡ **Introduction** Frequently, a mathematical statement or proposition that depends on the natural numbers or positive integers $N = \{1, 2, 3, \ldots\}$ can be proved using a technique known as **mathematical induction**. Suppose we can show two things:

- *a statement is true for the number* 1; *and*
- *whenever the statement is true for the positive integer k, then it is true for the next positive integer k + 1.*

In other words, suppose we can demonstrate that the

$$\boxed{\text{statement is true for 1}} \qquad (1)$$

and that the

$$\boxed{\text{statement is true for } k} \text{ implies the } \boxed{\text{statement is true for } k + 1.} \qquad (2)$$

What can we conclude from this? From (1) we have that

the statement is true for the number 1,

and by (2)

the statement is true for the number $1 + 1 = 2.$

In addition, it now follows from (2) that

the statement is true for the number $2 + 1 = 3,$
the statement is true for the number $3 + 1 = 4,$
the statement is true for the number $4 + 1 = 5,$

and so on. Symbolically, we can represent this sequence of implications by

$$\boxed{\text{statement is true for 1}} \Rightarrow \boxed{\text{statement is true for 2}} \Rightarrow \boxed{\text{statement is true for 3}} \Rightarrow \cdots$$

It seems clear that the statement must be true for *all* positive integers n. This is precisely the assertion of the following principle.

THEOREM 10.3.1 Principle of Mathematical Induction

Let $S(n)$ be a statement involving a positive integer n such that

  (*i*) $S(1)$ is true, and
 (*ii*) whenever $S(k)$ is true for a positive integer k, then $S(k + 1)$ is also true.

Then $S(n)$ is true for every positive integer.

Although we have stated the Principle of Mathematical Induction as a theorem, it is actually considered to be an *axiom* of the natural numbers.

By way of a physical analogy to the foregoing principle, imagine that we have an endless row of correctly spaced dominoes each standing on its end. Suppose we can demonstrate that whenever a domino (give it a name, say, the kth domino) falls over that its neighboring domino (the $(k + 1)$st domino) also falls over. Then we conclude that all the dominoes must fall over provided we can show one more thing, namely, that the first domino falls over.

We now illustrate the use of induction with several examples. We begin with an example from arithmetic.

Falling dominoes

EXAMPLE 1   Using Mathematical Induction

Prove that the sum of the first n positive integers is given by

$$1 + 2 + 3 + \cdots + n = \frac{n(n + 1)}{2}. \tag{3}$$

Solution Here the statement $S(n)$ is the formula in (3). The first step is to show that $S(1)$ is true, where $S(1)$ is the statement

$$1 = \frac{1 \cdot 2}{2}.$$

Since this is clearly true, condition (*i*) of the Principle of Mathematical Induction is satisfied.

The next step is to verify condition (*ii*). This requires that from the hypothesis "$S(k)$ is true," we prove that "$S(k + 1)$ is true." Thus we assume that the statement $S(k)$,

$$1 + 2 + 3 + \cdots + k = \frac{k(k + 1)}{2}, \tag{4}$$

is true. From this assumption we wish to demonstrate that $S(k + 1)$,

$$1 + 2 + 3 + \cdots + (k + 1) = \frac{(k + 1)[(k + 1) + 1]}{2}, \tag{5}$$

is also true. Now we can obtain a formula for the sum of the first $k + 1$ positive integers by using the equality (4) and some algebra:

$$
\underbrace{1 + 2 + 3 + \cdots + k}_{\text{by (4) this equals } \frac{k(k+1)}{2}} + (k + 1) = \frac{k(k + 1)}{2} + (k + 1)
$$

$$= \frac{k(k + 1) + 2(k + 1)}{2}$$

$$= \frac{(k + 1)(k + 2)}{2}$$

$$= \frac{(k + 1)[(k + 1) + 1]}{2}. \quad \leftarrow \text{this is (5)}$$

Thus we have shown that the statement $S(k + 1)$ is true. It follows from the Principle of Mathematical Induction that $S(n)$ is true for every positive integer n.   ≡

In basic algebra we learned how to factor. In particular, from the factorizations

$$
\begin{aligned}
x - y &= x - y, \\
x^2 - y^2 &= (x - y)(x + y), &&\leftarrow \text{see (1) of Section 1.5} \\
x^3 - y^3 &= (x - y)(x^2 + xy + y^2), &&\leftarrow \text{see (4) of Section 1.5} \\
x^4 - y^4 &= (x^2 - y^2)(x^2 + y^2) = (x - y)(x + y)(x^2 + y^2),
\end{aligned}
$$

a reasonable conjecture is that $x - y$ is always a factor of $x^n - y^n$ for any positive integer n. We now prove that this is so.

EXAMPLE 2   Using Mathematical Induction

Prove that $x - y$ is a factor of $x^n - y^n$ for any positive integer n.

Solution For the statement $S(n)$,

$$x - y \text{ is a factor of } x^n - y^n,$$

we must show that the two conditions (i) and (ii) are satisfied. For $n = 1$ we have the true statement $S(1)$,

$$x - y \text{ is a factor of } x^1 - y^1.$$

Now assume that $S(k)$,

$$x - y \text{ is a factor of } x^k - y^k,$$

is true. Using this assumption, we must show that $S(k + 1)$ is true; that is, $x - y$ is a factor of $x^{k+1} - y^{k+1}$. To this end we perform a bit of cleverness, namely, let's *subtract* and *add* xy^k to $x^{k+1} - y^{k+1}$:

$$x^{k+1} - y^{k+1} = x^{k+1} \overbrace{- xy^k + xy^k}^{0} - y^{k+1} = x\overbrace{(x^k - y^k)}^{\substack{x - y \text{ is assumed} \\ \text{to be a factor of} \\ \text{this term}}} + y^k\overbrace{(x - y)}^{\substack{\text{here is a} \\ \text{factor of} \\ x - y}}. \quad (6)$$

But by hypothesis, $x - y$ is a factor of $x^k - y^k$. Therefore, $x - y$ is a factor of *both* terms on the right-hand side of (6). It follows that $x - y$ is a factor of the right-hand side, and thus we have shown that the statement $S(k + 1)$,

$$x - y \text{ is a factor of } x^{k+1} - y^{k+1},$$

is true. It follows by the Principle of Mathematical Induction that $x - y$ is a factor of $x^n - y^n$ for any positive integer n. ≡

EXAMPLE 3　　Using Mathematical Induction

Prove that $8^n - 1$ is divisible by 7 for all positive integers n.

Solution We let $S(n)$ be the statement "$8^n - 1$ is divisible by 7 for all positive integers n." With $n = 1$ we see that $8^1 - 1 = 7$ is obviously divisible by 7.

Therefore $S(1)$ is true. Now let us assume that $S(k)$ is true; that is, $8^k - 1$ is divisible by 7 for some positive integer k. Using that assumption we must show that $8^{k+1} - 1$ is divisible by 7. Consider

$$
\begin{aligned}
8^{k+1} - 1 &= 8^k 8 - 1 \\
&= 8^k(1 + 7) - 1 \qquad \leftarrow \text{rearrange terms} \\
&= \underbrace{(8^k - 1)}_{\substack{\text{assumed to be} \\ \text{divisible by 7}}} + \underbrace{7 \cdot 8^k}_{\text{divisible by 7}}.
\end{aligned}
$$

The last equality proves $S(k + 1)$ is true because both $8^k - 1$ and $7 \cdot 8^k$ are divisible by 7. It follows from the Principle of Mathematical Induction that $S(n)$ is true for every positive integer n. ≡

10.3 Exercises Answers to selected odd-numbered problems begin on page ANS-29.

In Problems 1–20, use the Principle of Mathematical Induction to prove that the given statement is true for every positive integer n.

1. $2 + 4 + 6 + \cdots + 2n = n^2 + n$
2. $1 + 3 + 5 + \cdots + (2n - 1) = n^2$
3. $1^2 + 2^2 + 3^2 + \cdots + n^2 = \frac{1}{6}n(n + 1)(2n + 1)$
4. $1^3 + 2^3 + 3^3 + \cdots + n^3 = \frac{1}{4}n^2(n + 1)^2$

5. $\displaystyle\sum_{k=1}^{n} \frac{1}{2^k} + \frac{1}{2^n} = 1$

6. $\displaystyle\sum_{k=1}^{n} (4k - 5) = n(2n - 3)$

7. $\displaystyle\frac{1}{1 \cdot 2} + \frac{1}{2 \cdot 3} + \frac{1}{3 \cdot 4} + \cdots + \frac{1}{n(n + 1)} = \frac{n}{n + 1}$

8. $\displaystyle\frac{1}{2 \cdot 3} + \frac{1}{3 \cdot 4} + \frac{1}{4 \cdot 5} + \cdots + \frac{1}{(n + 1)(n + 2)} = \frac{n}{2n + 4}$

9. $1 + 4 + 4^2 + \cdots + 4^{n-1} = \frac{1}{3}(4^n - 1)$

10. $10 + 10^2 + 10^3 + \cdots + 10^n = \frac{1}{9}(10^{n+1} - 10)$

11. $n^3 + 2n$ is divisible by 3   **12.** $n^2 + n$ is divisible by 2

13. 4 is a factor of $5^n - 1$ **14.** 6 is a factor of $n^3 - n$

15. 7 is a factor of $3^{2n} - 2^n$ **16.** $x + y$ is a factor of $x^{2n-1} + y^{2n-1}$

17. If $a \geq -1$, then $(1 + a)^n \geq 1 + na$.   **18.** $2n \leq 2^n$

19. If $r > 1$, then $r^n > 1$.   **20.** If $0 < r < 1$, then $0 < r^n < 1$.

For Discussion

21. If we assume that

$$2 + 4 + 6 + \cdots + 2n = n^2 + n + 1$$

is true for $n = k$, show that the formula is true for $n = k + 1$. Show, however, that the formula itself is false. Explain why this does not violate the Principle of Mathematical Induction.

10.4 The Binomial Theorem

≡ **Introduction** When $(a + b)^n$ is expanded for an arbitrary positive integer n, the exponents of a and b follow a definite pattern. For example, from the expansions

$$(a + b)^2 = a^2 + 2ab + b^2$$
$$(a + b)^3 = a^3 + 3a^2b + 3ab^2 + b^3$$
$$(a + b)^4 = a^4 + 4a^3b + 6a^2b^2 + 4ab^3 + b^4,$$

we see that the exponents of a *decrease* by 1, starting with the first term, whereas the exponents of b *increase* by 1, starting with the second term. In the case of $(a + b)^4$, we have

powers decreasing by 1
↓
$$(a + b)^4 = a^4 + 4a^3b^1 + 6a^2b^2 + 4a^1b^3 + b^4.$$
↑
powers increasing by 1

To extend this pattern, we consider the first and last terms to be multiplied by b^0 and a^0, respectively; that is,

$$(a + b)^4 = a^4b^0 + 4a^3b^1 + 6a^2b^2 + 4a^1b^3 + a^0b^4. \qquad (1)$$

We also note that the sum of the exponents in each of the five terms of the expansion

$$4 = 3 + 1$$

$(a + b)^4$ is 4. For example, in the second term we have $4\overbrace{a^3b^1}$.

EXAMPLE 1   Using (1)

Expand $(y^2 - 1)^4$.

Solution With the identifications $a = y^2$ and $b = -1$, it follows from (1) and the laws of exponents that

$$
\begin{aligned}
(y^2 - 1)^4 &= (y^2 + (-1))^4 \\
&= (y^2)^4 + 4(y^2)^3(-1) + 6(y^2)^2(-1)^2 + 4(y^2)(-1)^3 + (-1)^4 \\
&= y^8 - 4y^6 + 6y^4 - 4y^2 + 1.
\end{aligned}
$$

≡

□ **The Coefficients** The coefficients in the expansion of $(a + b)^n$ also follow a pattern. To illustrate, we display the coefficients in the expansions of $(a + b)^0$, $(a + b)^1$, $(a + b)^2$, $(a + b)^3$, and $(a + b)^4$ in a triangular array

$$
\begin{array}{ccccccccc}
& & & & 1 & & & & \\
& & & 1 & & 1 & & & \\
& & 1 & & 2 & & 1 & & \\
& 1 & & 3 & & 3 & & 1 & \\
1 & & 4 & & 6 & & 4 & & 1
\end{array}
\tag{2}
$$

Observe that each number in the interior of this array is the *sum* of the two numbers directly above it. Thus the next line in the array can be obtained as follows:

$$
\begin{array}{ccccccccccc}
& 1 & & 4 & & 6 & & 4 & & 1 & \\
1 & & 5 & & 10 & & 10 & & 5 & & 1.
\end{array}
$$

As you might expect, these numbers are the coefficients of the powers of a and b in the expansion of $(a + b)^5$; that is,

$$
(a + b)^5 = 1a^5 + 5a^4b + 10a^3b^2 + 10a^2b^3 + 5ab^4 + 1b^5. \tag{3}
$$

The array obtained by continuing in this manner is called **Pascal's triangle** after the French philosopher and mathematician **Blaise Pascal** (1623–1662).

EXAMPLE 2   Using (3)

Expand $(3 - x)^5$.

Solution From (3), with $a = 3$ and $b = -x$, we can write

$$
\begin{aligned}
(3 - x)^5 &= (3 + (-x))^5 \\
&= 1(3)^5 + 5(3)^4(-x) + 10(3)^3(-x)^2 + 10(3)^2(-x)^3 + 5(3)(-x)^4 + 1(-x)^5 \\
&= 243 - 405x + 270x^2 - 90x^3 + 15x^4 - x^5.
\end{aligned}
$$

≡

□ **Factorial Notation** Before we give a general formula for the expansion of $(a + b)^n$, it will be helpful to introduce **factorial notation**. The symbol $r!$ is defined for any positive integer r as the product

$$
r! = r \cdot (r - 1) \cdot (r - 2) \cdots 3 \cdot 2 \cdot 1, \tag{4}
$$

◀ See Problem 63 in Exercises 2.1.

and is read "r factorial." For example, $1! = 1$ and $4! = 4 \cdot 3 \cdot 2 \cdot 1 = 24$. Also, it is convenient to define

$$
0! = 1.
$$

EXAMPLE 3   A Simplification

Simplify $\dfrac{r!(r + 1)}{(r - 1)!}$, where r is a positive integer.

Solution Using the definition of $r!$ in (4) we can write the numerator as

$$r!(r + 1) = (r + 1)r! = (r + 1)r(r - 1) \cdots 2 \cdot 1 = (r + 1)r(r - 1)!.$$

Thus, $\dfrac{r!(r + 1)}{(r - 1)!} = \dfrac{(r + 1)r(r - 1)!}{(r - 1)!} = (r + 1)r.$ ≡

☐ **The Binomial Theorem** The general formula for the expansion of $(a + b)^n$ is given in the following result, known as the **Binomial Theorem**.

THEOREM 10.4.1   Binomial Theorem

For any positive integer n,

$$(a + b)^n = a^n + \frac{n}{1!}a^{n-1}b + \frac{n(n - 1)}{2!}a^{n-2}b^2$$

$$+ \cdots + \frac{n(n - 1) \cdots (n - r + 1)}{r!}a^{n-r}b^r + \cdots + b^n. \tag{5}$$

By paying attention to the increasing powers on b in (5) we see that the expression

$$\frac{n(n - 1) \cdots (n - r + 1)}{r!}a^{n-r}b^r \tag{6}$$

is the $(r + 1)$st term in the expansion of $(a + b)^n$. For $r = 0, 1, \ldots, n$, the numbers

$$\frac{n(n - 1) \cdots (n - r + 1)}{r!} \tag{7}$$

are called **binomial coefficients** and are, of course, the same as those obtained from Pascal's triangle. Before proving the Binomial Theorem by mathematical induction, we consider some examples.

EXAMPLE 4   Using (5)

Expand $(a + b)^4$.

Solution We use the Binomial Theorem (5) with coefficients given by (7). With $n = 4$ we obtain:

$$(a + b)^4 = a^4 + \frac{4}{1!}a^{4-1}b + \frac{4 \cdot 3}{2!}a^{4-2}b^2 + \frac{4 \cdot 3 \cdot 2}{3!}a^{4-3}b^3 + \frac{4 \cdot 3 \cdot 2 \cdot 1}{4!}b^4$$

$$= a^4 + 4a^3b + \frac{12}{2}a^2b^2 + \frac{24}{6}ab^3 + \frac{24}{24}b^4$$

$$= a^4 + 4a^3b + 6a^2b^2 + 4ab^3 + b^4.$$ ≡

EXAMPLE 5   Finding the Sixth Term

Find the sixth term in the expansion of $(x^2 - 2y)^7$.

Solution Since (6) gives the $(r + 1)$st term in the expansion of $(a + b)^n$, the sixth term in the expansion of $(x^2 - 2y)^7$ corresponds to $r = 5$ (that is, $r + 1 = 5 + 1 = 6$).

With the identifications $n = 7, r = 5, a = x^2$, and $b = -2y$, it follows from (6) that the sixth term is

$$\frac{7 \cdot 6 \cdot 5 \cdot 4 \cdot 3}{5!}(x^2)^{7-5}(-2y)^5 = 21x^4(-32y^5)$$
$$= -672x^4y^5. \qquad \equiv$$

☐ **An Alternative Form** The binomial coefficients can be written in a more compact manner using factorial notation. If r is any integer such that $0 \leq r \leq n$, then

$$n(n-1)\cdots(n-r+1) = \frac{n(n-1)\cdots(n-r+1)}{1} \cdot \overbrace{\frac{(n-r)(n-r-1)\cdots 3 \cdot 2 \cdot 1}{(n-r)(n-r-1)\cdots 3 \cdot 2 \cdot 1}}^{\text{this fraction is 1}}$$
$$= \frac{n(n-1)\cdots(n-r+1)(n-r)(n-r-1)\cdots 3 \cdot 2 \cdot 1}{(n-r)(n-r-1)\cdots 3 \cdot 2 \cdot 1}$$
$$= \frac{n!}{(n-r)!}.$$

Thus the binomial coefficients of $a^{n-r}b^r$ for $r = 0, 1, \ldots, n$ given in (7) are the same as $n!/r!(n-r)!$. This latter quotient is usually denoted by the symbol $\binom{n}{r}$. That is, the **binomial coefficients** are

$$\binom{n}{r} = \frac{n!}{r!(n-r)!}. \qquad (8)$$

Hence the Binomial Theorem (5) can be written in the alternative form

$$(a+b)^n = \binom{n}{0}a^n + \binom{n}{1}a^{n-1}b + \cdots + \binom{n}{r}a^{n-r}b^r + \cdots + \binom{n}{n}b^n. \qquad (9)$$

It is this form that we will use to prove (5).

☐ **Summation Notation** The Binomial Theorem can be expressed in a compact manner by using summation notation. Using (6) and (8), the sums in (5) and (9) can be written as

$$(a+b)^n = \sum_{k=0}^{n} \frac{n(n-1)\cdots(n-k+1)}{k!} a^{n-k}b^k$$

or

$$(a+b)^n = \sum_{k=0}^{n} \binom{n}{k} a^{n-k}b^k,$$

respectively. From these forms it is apparent that since the index of summation starts at 0 and ends at n, a binomial expansion contains $n + 1$ terms.

The following property of the binomial coefficient $\binom{n}{r}$ will play a pivotal role in the proof of the Binomial Theorem. For any integer $r, 0 < r \leq n$, we have

$$\binom{n}{r-1} + \binom{n}{r} = \binom{n+1}{r}. \qquad (10)$$

We leave the verification of (10) as an exercise (see Problem 63 in Exercises 10.4).

☐ **Proof of Theorem 10.4.1** We now prove the Binomial Theorem by mathematical induction. Substituting $n = 1$ into (9) gives a true statement,

$$(a+b)^1 = \binom{1}{0}a^1 + \binom{1}{1}b^1 = a + b,$$

since
$$\binom{1}{0} = \frac{1!}{0!1!} = 1 \quad \text{and} \quad \binom{1}{1} = \frac{1!}{1!0!} = 1.$$

This completes the verification of the first condition of the Principle of Mathematical Induction.

For the second condition, we assume that (9) is true for some positive integer $n = k$:

$$(a + b)^k = \binom{k}{0}a^k + \binom{k}{1}a^{k-1}b + \cdots + \binom{k}{r}a^{k-r}b^r + \cdots + \binom{k}{k}b^k. \quad (11)$$

Using this assumption we then must show that (9) is also true for $n = k + 1$. To do this we multiply both sides of (11) by $(a + b)$:

$$(a + b)(a + b)^k = (a + b)\left[\binom{k}{0}a^k + \binom{k}{1}a^{k-1}b + \cdots + \binom{k}{r}a^{k-r}b^r + \cdots \binom{k}{k}b^k\right]$$

$$= \binom{k}{0}(a^{k+1} + a^k b) + \binom{k}{1}(a^k b + a^{k-1}b^2) + \cdots + \binom{k}{r}(a^{k-r+1}b^r + a^{k-r}b^{r+1}) + \cdots + \binom{k}{k}(ab^k + b^{k+1}) \quad (12)$$

$$= \binom{k}{0}a^{k+1} + \left[\binom{k}{0} + \binom{k}{1}\right]a^k b + \left[\binom{k}{1} + \binom{k}{2}\right]a^{k-1}b^2 + \cdots + \left[\binom{k}{r-1} + \binom{k}{r}\right]a^{k-r+1}b^r + \cdots + \binom{k}{k}b^{k+1}.$$

Using (10) to rewrite the coefficient of the $(r + 1)$st term in (12) as

$$\binom{k}{r-1} + \binom{k}{r} = \binom{k+1}{r}$$

and the facts that $(a + b)(a + b)^k = (a + b)^{k+1}$,

$$\binom{k}{0} = 1 = \binom{k+1}{0} \quad \text{and} \quad \binom{k}{k} = 1 = \binom{k+1}{k+1},$$

the last line in (12) becomes

$$(a + b)^{k+1} = \binom{k+1}{0}a^{k+1} + \binom{k+1}{1}a^k b + \cdots + \binom{k+1}{r}a^{k+1-r}b^r + \cdots + \binom{k+1}{k+1}b^{k+1}.$$

Because this is (9) with n replaced by $k + 1$, the proof is complete by the Principle of Mathematical Induction.

≡

10.4 Exercises   Answers to selected odd-numbered problems begin on page ANS-29.

In Problems 1–12, evaluate the given expression.

1. $3!$

2. $5!$

3. $\dfrac{2!}{5!}$

4. $\dfrac{6!}{3!}$

5. $3!4!$

6. $0!5!$

7. $\dbinom{5}{3}$

8. $\dbinom{6}{3}$

9. $\dbinom{7}{6}$

10. $\dbinom{9}{9}$

11. $\dbinom{4}{1}$

12. $\dbinom{4}{0}$

In Problems 13–16, simplify the given expression.

13. $\dfrac{n!}{(n-1)!}$

14. $\dfrac{(n-1)!}{(n-3)!}$

15. $\dfrac{n!(n+1)!}{(n+2)!(n+3)!}$

16. $\dfrac{(2n+1)!}{(2n)!}$

In Problems 17–26, use factorial notation to rewrite the given product.

17. $5 \cdot 4 \cdot 3 \cdot 2 \cdot 1$

18. $7 \cdot 6 \cdot 5 \cdot 4 \cdot 3 \cdot 2 \cdot 1$

19. $100 \cdot 99 \cdot 98 \cdots 3 \cdot 2 \cdot 1$

20. $t(t-1)(t-2) \cdots 3 \cdot 2 \cdot 1$

21. $(4 \cdot 3 \cdot 2 \cdot 1)(5 \cdot 4 \cdot 3 \cdot 2 \cdot 1)$

22. $(6 \cdot 5 \cdot 4 \cdot 3 \cdot 2 \cdot 1)/(3 \cdot 2 \cdot 1)$

23. $4 \cdot 3$

24. $10 \cdot 9 \cdot 8$

25. $n(n-1), n \geq 2$

26. $n(n-1)(n-2) \cdots (n-r+1), n \geq r$

In Problems 27–32, answer true or false.

27. $5! = 5 \cdot 4!$ _____

28. $3! + 3! = 6!$ _____

29. $\dfrac{8!}{4!} = 2!$ _____

30. $\dfrac{8!}{4} = 2$ _____

31. $n!(n+1) = (n+1)!$ _____

32. $\dfrac{n!}{n} = (n-1)!$ _____

In Problems 33–42, use the Binomial Theorem to expand the given expression.

33. $(x^2 - 5y^4)^2$

34. $(x^{-1} + y^{-1})^3$

35. $(x^2 - y^2)^3$

36. $(x^{-2} + 1)^4$

37. $(x^{1/2} + y^{1/2})^4$

38. $(3 - y^2)^4$

39. $(x^2 + y^2)^5$

40. $\left(2x + \dfrac{1}{x}\right)^5$

41. $(a - b - c)^3$

42. $(x + y + z)^4$

43. By referring to Pascal's triangle, determine the coefficients in the expansion of $(a + b)^n$ for $n = 6$ and $n = 7$.

44. If $f(x) = x^n$, where n is a positive integer, use the Binomial Theorem to simplify the difference quotient:
$$\frac{f(x + h) - f(x)}{h}.$$

In Problems 45–54, find the indicated term in the expansion of the given expression.

45. Sixth term of $(a + b)^6$

46. Second term of $(x - y)^5$

47. Fourth term of $(x^2 - y^2)^6$

48. Third term of $(x - 5)^5$

49. Fifth term of $(4 + x)^7$

50. Seventh term of $(a - b)^7$

51. Tenth term of $(x + y)^{14}$

52. Fifth term of $(t + 1)^4$

53. Eighth term of $(2 - y)^9$

54. Ninth term of $(3 - z)^{10}$

55. Find the coefficient of the constant term in $(x + 1/x)^{10}$.

56. Find the first five terms in the expansion of $(x^2 - y)^{11}$.

57. Use the first four terms in the expansion of $(1 - 0.01)^5$ to find an approximation to $(0.99)^5$. Compare with the answer obtained from a calculator.

58. Use the first four terms in the expansion of $(1 + 0.01)^{10}$ to find an approximation to $(1.01)^{10}$. Compare with the answer obtained from a calculator.

59. Without adding the terms, determine the value of $\displaystyle\sum_{k=0}^{4}\binom{4}{k}4^k$.

60. If $\displaystyle\sum_{k=0}^{5}\binom{5}{k}x^{5-k}=0$, what is x?

61. Use the Binomial Theorem to show that

$$\sum_{k=0}^{n}(-1)^k\binom{n}{k}=0.$$

62. Use the Binomial Theorem to show that

$$\binom{n}{0}+\binom{n}{1}+\cdots+\binom{n}{n}=2^n.$$

63. Prove that

$$\binom{n}{r-1}+\binom{n}{r}=\binom{n+1}{r},\qquad 0<r\le n.$$

64. Prove that

$$\binom{n}{r+1}=\frac{n-r}{r+1}\binom{n}{r},\qquad 0\le r<n.$$

10.5 Principles of Counting

≡ **Introduction** A wide variety of practical problems involve counting the number of ways in which something can occur. For example, the telephone prefix at a certain university is 642. If the prefix is followed by four digits, how many telephone numbers are possible before a second prefix is needed? We will be able to solve this problem (see Example 2) and others using the counting techniques discussed in this section.

☐ **Tree Diagram** We begin by considering a more abstract problem. How many different arrangements can be made of the three letters a, b, and c using two letters at a time? One way to solve this problem is to list all the possible arrangements. As shown in FIGURE 10.6.1, a **tree diagram** can be used to illustrate all the possibilities. From the point labeled "Start," line segments lead to each of the three possible choices for a first letter. From each of these, a line segment leads to each of the possible choices for a second letter. Each possible arrangement corresponds to a path, or **branch** of the tree, beginning at the "Start" and traveling to the right through the tree. We see that there are 6 different arrangements of the three letters:

$$ab,\quad ac,\quad ba,\quad bc,\quad ca,\quad cb.$$

Another way to solve the foregoing problem is to recognize that each arrangement consists of a selection of letters to fill the two blank positions indicated by the red lines:

$$\underset{\text{first letter}}{\underline{\quad\quad}}\ \underset{\text{second letter}}{\underline{\quad\quad}}.$$

Any one of the *three* letters a, b, or c can be chosen for the first position. Once this choice is made, any one of the *two* remaining letters can be chosen for the second position.

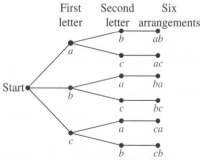

FIGURE 10.5.1 Tree diagram for number of arrangements of a, b, c taken two at a time

Since each of the three letters for the first position can be associated with either of the remaining two letters, the total number of arrangements is given by the *product*

$$\underset{\substack{\text{first}\\\text{letter}}}{3} \cdot \underset{\substack{\text{second}\\\text{letter}}}{2} = 6.$$

This simple example illustrates the **Fundamental Counting Principle**.

THEOREM 10.5.1 Fundamental Counting Principle

If one event can occur in m different ways and, after it has happened, a second event can occur in n different ways, then the total number of ways in which both events can take place is the product mn.

The Fundamental Counting Principle can be extended to three or more events in an obvious way:

Simply multiply the number of ways each event can occur.

EXAMPLE 1 Number of Outfits

A college student has 5 shirts, 3 pairs of slacks, and 2 pairs of shoes. How many different outfits can he wear consisting of a shirt, a pair of slacks, and a pair of shoes?

Solution Three selections or events are to occur, with 5 choices for the first event (choosing a shirt), 3 choices for the second event (choosing a pair of slacks), and 2 choices for the third event (choosing a pair of shoes). By the Fundamental Counting Principle, the number of different outfits is the product $5 \cdot 3 \cdot 2 = 30$. ≡

We now return to the problem given in the introduction.

EXAMPLE 2 Telephone Numbers

The telephone prefix at a certain university is 642. If the prefix is followed by four digits, how many different telephone numbers are possible before a second prefix is needed?

Solution Four events are to occur: selecting the first digit after the prefix, selecting the second digit after the prefix, and so on. Since repeated digits are allowed in telephone numbers, any one of the 10 digits 0, 1, 2, 3, 4, 5, 6, 7, 8, 9 can be selected for each position. Hence there are $10 \cdot 10 \cdot 10 \cdot 10 = 10{,}000$ possible different phone numbers with the single prefix 642. ≡

EXAMPLE 3 Arrangements of Letters

How many different ways are there to arrange the letters in the word RANDOM?

Solution Since RANDOM has 6 distinct letters, there are 6 events: choosing the first letter, choosing the second letter, and so on. Any one of the 6 letters can be chosen for the first position, then any of the *remaining* 5 letters can be chosen for the second position, then any of the *remaining* 4 letters can be chosen for the third position, and so on. The total number of arrangements is $6 \cdot 5 \cdot 4 \cdot 3 \cdot 2 \cdot 1 = 720$. ≡

☐ **Permutations** A **permutation** is an arrangement that is made by using some or all of the elements of a set *without repetition*. This means that no element of the set appears more than once in the arrangement. For example, 312 is a permutation of the digits in the set $\{1, 2, 3\}$, but 112 is not. In Example 3, each of the rearrangements of the six letters in the word RANDOM (for instance, MODRAN) is a permutation. More generally, we have the following definition.

DEFINITION 10.5.1 Permutation

An ordered arrangement of r elements selected from a set of n distinct elements is called a **permutation** of n elements taken r at a time ($n \geq r$).

☐ **Notation** We will use the symbol $P(n, r)$ to denote the number of permutations of n distinct objects taken r at a time. Using the notation $P(n, r)$, we write the number of permutations of 5 objects taken 3 at a time as $P(5, 3)$.

It is possible to find an explicit formula for $P(n, r)$, that is, the number of permutations of n distinct objects taken r at a time for $0 \leq r \leq n$. For $r \geq 1$, we can think of the process of forming a permutation of n objects taken r at a time as r events: choose the first object, choose the second object, and so on. When we make the first choice, there are n objects available; when we make the second choice, there are $n - 1$ objects; for the third choice, there are $n - 2$ objects; and so on. When we choose the rth object, there are $n - (r - 1)$ objects to choose from. Thus from Theorem 10.5.1,

$$P(n, r) = \overbrace{n(n - 1)(n - 2) \cdots (n - (r - 1))}^{r \text{ factors}}$$

or
$$P(n, r) = n(n - 1)(n - 2) \cdots (n - r + 1). \tag{1}$$

An alternative expression for $P(n, r)$ involving factorial notation can be found by multiplying the right-hand side of (1) by

$$\frac{(n - r)!}{(n - r)!} = 1.$$

The result is

$$P(n, r) = \frac{n(n - 1)(n - 2) \cdots (n - r + 1)\overbrace{(n - r)(n - r - 1) \cdots 2 \cdot 1}^{(n-r)!}}{(n - r)!},$$

or
$$P(n, r) = \frac{n!}{(n - r)!}. \tag{2}$$

When $r = n$, formula (2) reduces to

$$P(n, n) = \frac{n!}{0!} = n!,$$

because 0! is defined to be 1. This result is the same as that obtained by using the counting principle in Theorem 10.5.1:

$$P(n, n) = n(n - 1)(n - 2) \cdots 2 \cdot 1 = n!, \tag{3}$$

since any one of the n objects can be chosen first, any one of the remaining objects can be chosen second, and so on. In Example 3, the number of 6-letter arrangements of the words RANDOM is the number of permutations of the 6 letters taken 6 at a time, that is, $P(6, 6) = 6! = 720$.

If $r = 0$, we define $P(n, 0) = 1$, which is consistent with (2).

EXAMPLE 4　　　　　**Using (2) and (3)**

Evaluate **(a)** $P(5, 3)$　**(b)** $P(5, 1)$　**(c)** $P(5, 5)$.

Solution In (a) and (b) we use formula (2):

(a) $P(5, 3) = \dfrac{5!}{(5 - 3)!} = \dfrac{5!}{2!} = \dfrac{5 \cdot 4 \cdot 3 \cdot \overbrace{2 \cdot 1}^{2!}}{2!} = 60,$

(b) $P(5, 1) = \dfrac{5!}{(5 - 1)!} = \dfrac{5!}{4!} = \dfrac{5 \cdot \overbrace{4 \cdot 3 \cdot 2 \cdot 1}^{4!}}{4!} = 5.$

(c) From formula (3) we find that

$$P(5, 5) = 5! = 5 \cdot 4 \cdot 3 \cdot 2 \cdot 1 = 120. \qquad\equiv$$

EXAMPLE 5　　　　**Awarding Medals**

At a track meet 6 athletes are entered in the 100 m dash. In how many ways can gold, silver, and bronze metals be awarded?

Solution We wish to count the number of ways of arranging 3 of the 6 athletes in the winning positions. The solution is given by the number of permutations of 6 things (athletes) taken 3 at a time:

$$P(6, 3) = \frac{6!}{(6 - 3)!} = \frac{6!}{3!} = 120.$$

This problem can also be solved using the Fundamental Counting Principle. Since there are 3 choices to be made, with 6 athletes available for the gold medal, 5 for the silver, and 4 for the bronze, we find $6 \cdot 5 \cdot 4 = 120$. $\qquad\equiv$

EXAMPLE 6　　　　**Arrangements of Books**

How many arrangements are possible for 10 different books on a bookshelf?

Solution We wish to find the number of permutations of 10 objects taken 10 at a time, or $P(10, 10) = 10! = 3,628,800$. $\qquad\equiv$

Ten books on a shelf

☐ **Combinations** In the preceding discussion we were concerned with the number of ways of arranging or choosing r elements from a set of n elements, where the order in which they were arranged or chosen was considered. However, in certain applications the order of the elements is not important. For example, if a committee of two is to be chosen from the four students Angie, Brandon, Cecilia, and David, the committee formed by choosing Angie and Brandon is the same as the committee formed by choosing Brandon and Angie. A selection of objects in which the order does not make any difference is called a **combination**.

DEFINITION 10.5.2　Combination

A subset of r elements of a set of n distinct elements is called a **combination** of n elements taken r at a time ($n \geq r$).

☐ **Notation** We use the symbol $C(n, r)$ to denote the number of combinations of n distinct objects taken r at a time. By using (2) it is possible to derive a formula for

$C(n, r)$. At the beginning of this section we saw that there are 6 arrangements (permutations) of the 3 letters a, b, and c taken 2 at a time:

$$\underbrace{\overbrace{ab}^{\text{same combination}} \quad \overbrace{ac \quad ba}^{} \quad \overbrace{bc \quad ca}^{\text{same combination}} \quad cb.}_{\text{same combination}} \tag{4}$$

In (4) we see that if we disregard the order in which the letters are listed, then there are only 3 combinations of the letters: ab, ac, and bc. Thus, $C(3, 2) = 3$. We see that each of these combinations can be arranged in 2! ways to yield the list of permutations in (4). By the Fundamental Counting Principle,

$$P(3, \ 2) = 6 = 2! \, C(3, 2).$$

In general, for $0 < r \leq n$, each of the $C(n, r)$ combinations can be rearranged in $r!$ different ways, so that

$$P(n, r) = r! \, C(n, r),$$

or

$$C(n, r) = \frac{P(n, r)}{r!} = \frac{\dfrac{n!}{(n - r)!}}{r!}.$$

Thus,

$$C(n, r) = \frac{n!}{(n - r)! \, r!}. \tag{5}$$

For $r = 0$, we define $C(n, 0) = 1$, which is consistent with formula (5).

Note that $C(n, r)$ is identical to the binomial coefficient $\dbinom{n}{r}$ in the expansion of $(a + b)^n$, where n is a nonnegative integer. See (7) and (8) of Section 10.4.

EXAMPLE 7   Using Formula (5)

Evaluate **(a)** $C(5, 3)$ **(b)** $C(5, 1)$ **(c)** $C(5, 5)$.

Solution Using formula (5), we have the following:

(a) $C(5, 3) = \dfrac{5!}{(5 - 3)!3!} = \dfrac{5!}{2! \, 3!} = \dfrac{5 \cdot 4 \cdot \overbrace{3!}^{3 \cdot 2 \cdot 1}}{2! \, 3!} = 10,$

(b) $C(5, 1) = \dfrac{5!}{(5 - 1)!1!} = \dfrac{5!}{4! \, 1!} = \dfrac{5 \cdot \overbrace{4!}^{4 \cdot 3 \cdot 2 \cdot 1}}{4! \, 1!} = 5,$

(c) $C(5, 5) = \dfrac{5!}{(5 - 5)!5!} = \dfrac{5!}{0! \, 5!} = \dfrac{1}{\underset{1}{\underbrace{0!}}} = 1.$ $\equiv$

EXAMPLE 8   Number of Card Hands

How many different 7-card hands can be dealt from a deck of 52 cards?

Solution Since a hand is the same regardless of the order of the cards, we are talking about the number of combinations of 52 cards taken 7 at a time. Using (5), the solution is

$$\begin{aligned} C(52, 7) &= \frac{52!}{45! \, 7!} = \frac{52 \cdot 51 \cdot 50 \cdot 49 \cdot 48 \cdot 47 \cdot 46 \cdot 45!}{45! \, 7!} \\ &= \frac{52 \cdot 51 \cdot 50 \cdot 49 \cdot 48 \cdot 47 \cdot 46}{7!} = 133{,}784{,}560. \end{aligned}$$

Note that we cancelled the larger of the two factorials 45! and 7! to simplify the calculations of $C(52, 7)$. $\equiv$

EXAMPLE 9 **Organizing a Club**

A card club has 8 members.
(a) In how many ways can 3 members be chosen to be president, secretary, and treasurer?
(b) In how many ways can a committee of 3 members be chosen?

Solution In choosing officers, order *does* matter, whereas in choosing a committee, the order of the selection does not affect the resulting committee. Thus in (a) we are counting permutations and in (b) we are counting combinations. We find

(a) $P(8, 3) = \dfrac{8!}{5!} = 336,$

(b) $C(8, 3) = \dfrac{8!}{5!3!} = 56.$ ☰

In deciding whether to use the formula for $P(n, r)$ or $C(n, r)$, consider the following two informal rules. ◄ Note of Caution

- Permutations are involved if you are considering arrangements in which *different orderings* of the same objects *are to be counted*.
- Combinations are involved if you are considering ways of choosing objects in which the *order* of the chosen objects *makes no difference*.

EXAMPLE 10 **Choosing Reporters**

A college newspaper staff has 6 junior reporters and 8 senior reporters. In how many ways can 2 junior and 3 senior reporters be chosen for a special assignment?

Solution Two events are to occur: the selection of 2 junior reporters and the selection of 3 senior reporters. Because the order in which the 2 junior reporters are chosen makes no difference, we count combinations. Therefore, the number of ways of choosing 2 junior reporters is

$$C(6, 2) = \frac{6!}{4!2!} = 15.$$

Likewise in selecting the 3 senior reporters order does not matter, so we again count combinations:

$$C(8, 3) = \frac{8!}{5!3!} = 56.$$

Thus we choose the junior reporters in 15 ways and, for each of these selections, there are 56 ways of selecting the senior reporters. Applying the Fundamental Counting Principle gives

$$C(6, 2) \cdot C(8, 3) = 15 \cdot 56 = 840$$

ways to make the choices for the special assignment. ☰

EXAMPLE 11 **Selecting a Display**

A cheese store has 10 varieties of domestic cheese and 8 varieties of imported cheese. In how many ways can a selection of 6 cheeses, consisting of 2 domestic and 4 imported varieties, be placed on a display shelf?

Solution The domestic varieties can be chosen in $C(10, 2)$ ways and the imported varieties in $C(8, 4)$ ways. Thus by the Fundamental Counting Principle the 6 cheeses can be selected in $C(10, 2) \cdot C(8, 4)$ ways. Up to this point in the solution, order has not been important in making the *selection* of the cheeses. Now we observe that *each* selection of 6 cheeses can be placed or *arranged* on the shelf in $P(6, 6)$ ways. Thus the total number of ways the cheese can be displayed is

$$C(10, 2) \cdot C(8, 4) \cdot P(6, 6) = \frac{10!}{8! \, 2!} \cdot \frac{8!}{4! \, 4!} \cdot \frac{6!}{(6 - 6)!}$$
$$= 2{,}268{,}000.$$

≡

10.5 **Exercises** Answers to selected odd-numbered problems begin on page ANS-30.

In Problems 1–4, use a tree diagram.

1. List all possible arrangements of the letters a, b, and c.
2. If a coin is tossed 4 times, list all possible sequences of heads (H) and tails (T).
3. If a red die and a black die are rolled, list all possible results.
4. If a coin is tossed and then a die is rolled, list all possible results.

In Problems 5–8, use the Fundamental Counting Principle.

5. **Number of Meals** A cafeteria offers 8 salads, 6 entrees, 4 vegetables, and 3 desserts. How many different meals are possible if one item is selected from each category?
6. **Number of Systems** How many different stereo systems consisting of speakers, receiver, and CD player can be purchased if a store carries 6 models of speakers, 4 of receivers, and 2 of CD players?
7. **Number of Prefixes** How many different 3-digit telephone prefixes are possible if neither 0 nor 1 can occupy the first position?
8. **Number of License Plates** If a license plate consists of 3 letters followed by 3 digits, how many license plates are possible if the first letter cannot be O or I?

In Problems 9–16, evaluate $P(n, r)$.

9. $P(6, 3)$ 10. $P(6, 4)$
11. $P(6, 1)$ 12. $P(4, 0)$
13. $P(100, 2)$ 14. $P(4, 4)$
15. $P(8, 6)$ 16. $P(7, 6)$

In Problems 17–24, evaluate $C(n, r)$.

17. $C(4, 2)$ 18. $C(4, 1)$
19. $C(50, 2)$ 20. $C(2, 2)$
21. $C(13, 11)$ 22. $C(8, 2)$
23. $C(2, 0)$ 24. $C(7, 4)$

Miscellaneous Applications

In Problems 25–28, use permutations to solve the given problem.

25. **Family Portrait** In how many ways can a family of four line up in a row to have their family portrait taken?
26. **Volunteer Work** As part of a fund-raising drive, a volunteer is given 5 names to contact. In how many different orders can the volunteer complete the task?

A family of four

CHAPTER 10 SEQUENCES AND SERIES

27. **Scrabble** A *Scrabble* game player has the following 7 letters: A, T, E, L, M, Q, F.
    (a) How many different 7-letter "words" can be considered?
    (b) How many different 5-letter "words"?
28. **Politics** From a class of 24, elections are held for president, vice president, secretary, and treasurer. In how many ways can the offices be filled?

The game *Scrabble*®

In Problems 29–32, use combinations to solve the given problem.

29. **Good Luck!** A student must answer any 10 questions on a 12-question exam. In how many different ways can the student select the questions?
30. **Chem Lab** For a chemistry lab class, a student must correctly identify 3 "unknown" samples. In how many ways can the 3 samples be chosen from 10 chemicals?
31. **Volunteers** In how many ways can 5 subjects be chosen from a group of 10 volunteers for a psychology experiment?
32. **Potpourri** In how many ways can 4 herbs be chosen from 8 available herbs to make a potpourri?

In Problems 33–44, use one or more of the techniques discussed in this section to solve the given counting problem.

33. **Spelling Bee** If 10 students enter a spelling bee, in how many different ways can first- and second-place awards be made?
34. **Show Business** A theater company has a repertoire consisting of 8 dramatic skits, 6 comedies, and 4 musical numbers. In how many ways can a program be selected consisting of a dramatic skit followed by either a comedy or a musical number?
35. **Take Your Pick** A pediatrician allows a well-behaved child to select any 2 of 5 small plastic toys to take home. How many different selections of toys are possible?
36. **Tournament Rankings** If 8 teams enter a soccer tournament, in how many different ways can first, second, and third place be decided, assuming ties are not allowed?
37. **Another Jackson Pollock** If 8 colors are available to make an abstract spatter-paint picture, how many different color combinations are possible if only 3 colors are chosen?
38. **Seating Arrangements** Three couples have reserved seats in a row at the theater. In how many different ways can they be seated
    (a) if there are no restrictions?
    (b) if each couple wishes to sit together?
    (c) if the 3 women and 3 men wish to sit together in 2 groups?
39. **Mastermind** In a popular board game that originated in England called *Mastermind*, one player creates a secret "code" by filling 4 slots with any one of 6 colors. How many codes are possible
    (a) if repetitions are not allowed?
    (b) if repetitions are allowed?
    (c) if repetitions and blank slots are allowed?
40. **Super Mastermind** Some advertisements for the game *Super Mastermind* (a more difficult version of the *Mastermind* game described in Problem 39) claim that up to 59,000 codes are possible. If *Super Mastermind* involves filling 5 slots with any one of 8 colors and if blanks and repetitions are allowed, is the claim correct?
41. **Playing with Letters** From 5 different consonants and 3 different vowels, how many 5-letter "words" can be made consisting of 3 different consonants and 2 different vowels?

The game *Mastermind*®

42. Defective Lights A box contains 24 Christmas tree bulbs, 4 of which are defective. In how many ways can 4 bulbs be chosen so that
   (a) all 4 are defective?
   (b) all 4 are good?
   (c) 2 are good and 2 are defective?
   (d) 3 are good and 1 is defective?

43. More Playing with Letters How many 3-letter "words" can be made from 4 different consonants and 2 different vowels
   (a) if the middle letter must be a vowel?
   (b) if the first letter cannot be a vowel? Assume that repeated letters are not allowed.

44. Store Display A wine store has 12 different California wines and 8 different French wines. In how many ways can 6 bottles of wine consisting of 4 California and 2 French wines
   (a) be selected for display?
   (b) be placed in a row on a display shelf?

10.6 Introduction to Probability

≡ **Introduction** As we mentioned in the chapter introduction, the development of the mathematical theory of **probability** was initially motivated by questions arising in the seventeenth century about games of chance. Today, applications of probability are found in medicine, sports, law, business, and many other areas. In this section we present a brief introduction to this fascinating subject.

☐ **Terminology** Consider an experiment that has a finite number of possible results or **outcomes**. The set S of all possible outcomes of a particular experiment is called the **sample space** of the experiment. For our purposes we will assume that each outcome is *equally likely* to occur. Thus, if the experiment consists of tossing, or flipping, a fair coin, there are two possible equally likely outcomes: obtaining a head or obtaining a tail. If the outcome of obtaining a head is denoted by H and the outcome of obtaining a tail is denoted by T, then the sample space of the experiment can be written in set notation as

$$S = \{H, T\}. \tag{1}$$

Any subset E of a sample space S is called an **event**. Generally, an event E is one or more outcomes of an experiment. For example,

$$E = \{H\} \tag{2}$$

is the event of obtaining a head when a coin is tossed.

There are two possible outcomes in tossing a coin

A die showing 4 is one of six possible outcomes

EXAMPLE 1   Sample Space and Two Events

On a single roll of a fair die, there is an equal chance of obtaining a 1, 2, 3, 4, 5, or 6. Thus the sample space of the experiment of rolling a fair die is the set

$$S = \{1, 2, 3, 4, 5, 6\}. \tag{3}$$

(a) The event E_1 of obtaining a 4 on a roll of the die is the subset $E_1 = \{4\}$ of S.
(b) The event E_2, consisting of obtaining an odd number on a roll of the die, is the subset $E_2 = \{1, 3, 5\}$ of S. ≡

We shall use the notation $n(S)$ to denote the number of outcomes in a sample space S and $n(E)$ to denote the number of outcomes associated with an event E. Thus, in Example 1 we have $n(S) = 6$; in parts (a) and (b) of the example we have $n(E_1) = 1$ and $n(E_2) = 3$, respectively.

The definition of the probability $P(E)$ of an event E is expressed in terms of $n(S)$ and $n(E)$.

DEFINITION 10.6.1 Probability of an Event

Let S be the sample space of an experiment and let E be an event. If each outcome of the experiment is equally likely, then the **probability** of the event E is given by

$$P(E) = \frac{n(E)}{n(S)}, \qquad (4)$$

where $n(E)$ and $n(S)$ denote the number of outcomes in the sets E and S, respectively.

EXAMPLE 2   **The Probability of Tossing a Head**

Find the probability of obtaining a head if a coin is tossed.

Solution From (1) and (2), $E = \{H\}$, $S = \{H, T\}$, and so $n(E) = 1$ and $n(S) = 2$. From (4) of Definition 10.6.1 the probability of obtaining a head is

$$P(E) = \frac{n(E)}{n(S)} = \frac{1}{2}. \qquad \equiv$$

EXAMPLE 3   **Three Probabilities**

On a single roll of a fair die, find the probability
(a) of obtaining a 4, (b) of obtaining an odd number, (c) of obtaining a number that is not a 4.

Solution Let the symbols E_1, E_2, and E_3 denote, respectively, the events in parts (a), (b), and (c) of this example. Also, in each part we have $S = \{1, 2, 3, 4, 5, 6\}$.
(a) From part (a) of Example 1, $E_1 = \{4\}$, and so $n(E_1) = 1$ and $n(S) = 6$. From (4), the probability of obtaining a 4 when a die is rolled is then

$$P(E_1) = \frac{n(E_1)}{n(S)} = \frac{1}{6}.$$

(b) From part (b) of Example 1, $E_2 = \{1, 3, 5\}$, so $n(E_2) = 3$ and $n(S) = 6$. Again from (4), the probability of rolling an odd number is

$$P(E_2) = \frac{n(E_2)}{n(S)} = \frac{3}{6} = \frac{1}{2}.$$

(c) The event of obtaining a number that is not a 4 on a roll of a die is the subset $E_3 = \{1, 2, 3, 5, 6\}$ of S. Using $n(E_3) = 5$ and $n(S) = 6$ the probability of obtaining a number that is not a 4 is

$$P(E_3) = \frac{n(E_3)}{n(S)} = \frac{5}{6}. \qquad \equiv$$

EXAMPLE 4   Probability of 7

Find the probability of obtaining a total of 7 when two dice are rolled.

Solution Since there are 6 numbers on each die, we conclude from the Fundamental Counting Principle of Section 10.5 that there are $6 \cdot 6 = 36$ possible outcomes in the sample space S; that is, $n(S) = 36$. In the accompanying table, we have listed the possible ways of obtaining a total of 7.

E = two dice showing a total of 7						
First die	1	2	3	4	5	6
Second die	6	5	4	3	2	1

From the table we see that $n(E) = 6$. Hence from (4) the probability of throwing a 7 with two dice is

$$P(E) = \frac{n(E)}{n(S)} = \frac{6}{36} = \frac{1}{6}.$$ ≡

EXAMPLE 5   Using Combinations

A bag contains 5 white marbles and 3 red marbles. A person reaches into the bag and randomly withdraws 3 marbles. What is the probability that all the marbles will be white?

Solution The sample space S of the experiment is the set of all possible combinations of 3 marbles drawn from the 8 marbles in the bag. The number of ways of choosing 3 marbles from a bag of 8 marbles is the number of *combinations* of 8 objects taken 3 at a time; that is, $n(S) = C(8, 3)$. Similarly, the number of ways of choosing 3 white marbles from 5 white marbles is the number of combinations $n(E) = C(5, 3)$. Since the event E is "all marbles are white," we have

$$P(E) = \frac{n(E)}{n(S)} = \frac{C(5, 3)}{C(8, 3)} = \frac{\dfrac{5!}{3!2!}}{\dfrac{8!}{3!5!}} = \frac{5}{28}.$$ ≡

☐ **Bounds on the Probability of an Event** Since any event E is a subset of a sample space S, it follows that $0 \le n(E) \le n(S)$. By dividing the last inequality by $n(S)$ we see that

$$0 \le \frac{n(E)}{n(S)} \le \frac{n(S)}{n(S)}$$

or $$0 \le P(E) \le 1.$$

If $E = S$, then $n(E) = n(S)$ and $P(E) = n(S)/n(S) = 1$; whereas if E has no elements, we take $E = \varnothing$, $n(\varnothing) = 0$, and $P(E) = n(\varnothing)/n(S) = 0/n(S) = 0$. If $P(E) = 1$, then E always happens and E is called a **certain event**. On the other hand, if $P(E) = 0$, then E is an **impossible event**, that is, E never happens.

EXAMPLE 6   Rolling a Die

Suppose a fair die is rolled once.
(a) What is the probability of obtaining a 7?
(b) What is the probability of obtaining a number less than 7?

One of six possible ways of throwing a 7 on a pair of dice

Solution (a) Because the number 7 is not in the set S of all possible outcomes (3) the event E of "obtaining a 7" is an impossible event; that is, $E = \varnothing, n(\varnothing) = 0$. Therefore,

$$P(E) = \frac{n(\varnothing)}{n(S)} = \frac{0}{6} = 0.$$

(b) Because the outcomes of rolling a fair die are all positive integers less than 7 we have $E = \{1, 2, 3, 4, 5, 6\} = S$. Thus E is a certain event and

$$P(E) = \frac{n(E)}{n(S)} = \frac{6}{6} = 1. \qquad\qquad \equiv$$

☐ **Complement of an Event** The set of all outcomes in the sample space S that do not belong to an event E is called the **complement of E** and is denoted by the symbol E'. For example, in rolling a die, if E is the event of "obtaining a 4," then E' is the event of "obtaining any number *except* 4." Because events are sets, we can describe the relationship between an event E and its complement E' using the operations of union and intersection:

$$E \cup E' = S \qquad \text{and} \qquad E \cap E' = \varnothing.$$

In view of the foregoing properties we can write $n(E) + n(E') = n(S)$. Dividing both sides of the last equality by $n(S)$ we see that the probabilities of E and E' are related by

$$\frac{n(E)}{n(S)} + \frac{n(E')}{n(S)} = \frac{n(S)}{n(S)}$$

or

$$P(E) + P(E') = 1. \qquad\qquad (5)$$

For instance, the complement of the event $E_1 = \{4\}$ in part (a) of Example 3 is the set $E_1' = E_3 = \{1, 2, 3, 5, 6\}$ in part (c). Observe in accordance with (5), we have $P(E_1) + P(E_3) = P(E_1) + P(E_1') = \frac{1}{6} + \frac{5}{6} = 1$.
    The relationship (5) is useful in either of the two forms:

$$P(E) = 1 - P(E') \qquad \text{or} \qquad P(E') = 1 - P(E). \qquad (6)$$

The second of the two formulas in (6) allows us to find the probability of an event if we know the probability of its complement. Sometimes it is easier to calculate $P(E')$ than it is to calculate $P(E)$. Also, it is interesting to note that the equation $P(E) + P(E') = 1$ can be interpreted as saying that *something* must happen.

EXAMPLE 7   **Probability of an Ace**

If 5 cards are drawn from a well-shuffled 52-card deck without replacement, find the probability of obtaining at least one ace.

Solution We let E be the event of obtaining at least one ace. Since E consists of all 5-card hands that contain 1, 2, 3, or 4 aces, it is actually easier to consider E'; that is, all 5-card hands that contain no aces. The sample space S consists of all possible 5-card hands. From Section 10.5 we have that $n(S) = C(52, 5)$. Since 48 of the 52 cards are *not* aces we find $n(E') = C(48, 5)$. By (4) the probability of drawing 5 cards where none of the cards are aces is given by

$$P(E') = \frac{C(48, 5)}{C(52, 5)} = \frac{1{,}712{,}304}{2{,}598{,}960}.$$

From the first formula in (5) the probability of drawing 5 cards where at least one of them is an ace is

$$P(E) = 1 - P(E') = 1 - \frac{1,712,304}{2,598,960} \approx 0.3412.$$ ≡

Up to this point we have considered the probability of a single event. In the discussion that follows, we examine the probability of two or more events.

□ **Union of Two Events** Two events E_1 and E_2 are said to be **mutually exclusive** if they have no outcomes, or elements, in common. In other words the events E_1 and E_2 cannot occur at the same time. In terms of sets, E_1 and E_2 are **disjoint** sets; that is, $E_1 \cap E_2 = \varnothing$. Recall, the set $E_1 \cup E_2$ consists of the elements that are in E_1 or in E_2. In this case of mutually exclusive events the number of outcomes in the set $E_1 \cup E_2$ is given by

Review the notions of the union and intersection of two sets in Section 1.1

$$n(E_1 \cup E_2) = n(E_1) + n(E_2). \tag{7}$$

By dividing (7) by $n(S)$ we obtain

$$\frac{n(E_1 \cup E_2)}{n(S)} = \frac{n(E_1)}{n(S)} + \frac{n(E_2)}{n(S)}.$$

In view of (4), the foregoing expression is the same as

$$P(E_1 \cup E_2) = P(E_1) + P(E_2). \tag{8}$$

In the next example we return to the results in Example 3.

EXAMPLE 8   **Mutually Exclusive Events**

On a single roll of a fair die, find the probability of obtaining a 4 or an odd number.

Solution From Example 3 the two events are $E_1 = \{4\}$, $E_2 = \{1, 3, 5\}$, and the sample space is again $S = \{1, 2, 3, 4, 5, 6\}$. The events of rolling a 4 and rolling an odd number are mutually exclusive: $E_1 \cap E_2 = \{4\} \cap \{1, 3, 5\} = \varnothing$. Thus by (8) the probability $P(E_1$ or $E_2)$ of rolling a 4 or an odd number is given by

$$P(E_1 \cup E_2) = P(E_1) + P(E_2) = \frac{1}{6} + \frac{3}{6} = \frac{4}{6} = \frac{2}{3}.$$

Alternative Solution From $E_1 \cup E_2 = \{1, 3, 4, 5\}$, $n(E_1 \cup E_2) = 4$, and so (4) of Definition 10.6.1 yields

$$P(E_1 \cup E_2) = \frac{n(E_1 \cup E_2)}{n(S)} = \frac{4}{6} = \frac{2}{3}.$$ ≡

The additive property in (8) extends to the probability of three or more mutually exclusive events. See Problems 31 and 32 in Exercises 10.6.

□ **Addition Rule** Formula (8) is just a special case of a more general rule. In (8) there were no outcomes in common in the events E_1 and E_2. Of course, this need not be the case. For example, in the experiment of rolling a single fair die, the events $E_1 = \{1\}$ and $E_2 = \{1, 3, 5\}$ are not mutually exclusive because the number 1 is an element in both sets. When two sets E_1 and E_2 have a nonempty intersection, the number of outcomes in $n(E_1 \cup E_2)$ is not given by (7) but rather by the formula

$$n(E_1 \cup E_2) = n(E_1) + n(E_2) - n(E_1 \cap E_2). \tag{9}$$

Dividing (9) by $n(S)$ yields

$$\frac{n(E_1 \cup E_2)}{n(S)} = \frac{n(E_1)}{n(S)} + \frac{n(E_2)}{n(S)} - \frac{n(E_1 \cap E_2)}{n(S)}$$

or

$$P(E_1 \cup E_2) = P(E_1) + P(E_2) - P(E_1 \cap E_2). \qquad (10)$$

The result in (10) is called the **addition rule** of probability.

■ EXAMPLE 9   Probability of a Union of Two Events

On a single roll of a fair die, find the probability of obtaining a 1 or an odd number.

Solution The sets are $E_1 = \{1\}$, $E_2 = \{1, 3, 5\}$, and $S = \{1, 2, 3, 4, 5, 6\}$. Now $\{1\} \cap \{1, 3, 5\} = \{1\}$ so that $n(E_1 \cap E_2) = 1$. Thus by (10) the probability of rolling a 1 or an odd number is given by

$$P(E_1 \cup E_2) = P(E_1) + P(E_2) - P(E_1 \cap E_2) = \frac{1}{6} + \frac{3}{6} - \frac{1}{6} = \frac{3}{6} = \frac{1}{2}.$$

Alternative Solution Since E_1 is a subset of E_2, $E_1 \cup E_2 = E_2 = \{1, 3, 5\}$, and $n(E_1 \cup E_2) = 3$. From (4) of Definition 10.6.1,

$$P(E_1 \cup E_2) = \frac{n(E_1 \cup E_2)}{n(S)} = \frac{3}{6} = \frac{1}{2}. \qquad ≡$$

It might help if you think of the symbols $P(E_1 \cup E_2)$ and $P(E_1 \cap E_2)$ in (10) as $P(E_1$ or $E_2)$ and $P(E_1$ and $E_2)$, respectively.

◀ Note

■ EXAMPLE 10   Probability of a Union of Two Events

A single card is drawn from a well-shuffled standard deck. Find the probability of obtaining either an ace or a heart.

Solution As shown in the photo to the left, a standard deck contains 52 cards divided into 4 suits with 13 cards in each suit. Thus the sample space S of this experiment consists of the 52 cards. The event E_1 of drawing an ace consists of the 4 aces and so the probability of drawing an ace is $P(E_1) = \frac{4}{52}$. The event E_2 of drawing a card that is a heart consists of the 13 hearts in that suit and so the probability of drawing a heart is $P(E_2) = \frac{13}{52}$. Since one of the hearts is an ace, $n(E_1 \cap E_2) = 1$, and so $P(E_1 \cap E_2) = \frac{1}{52}$. Therefore, from (10)

$$P(\overbrace{E_1 \cup E_2}^{\text{ace or a heart}}) = P(\overbrace{E_1}^{\text{ace}}) + P(\overbrace{E_2}^{\text{heart}}) - P(\overbrace{E_1 \cap E_2}^{\text{ace and a heart}})$$

$$= \frac{4}{52} + \frac{13}{52} - \frac{1}{52} = \frac{16}{52} = \frac{4}{13}. \qquad ≡$$

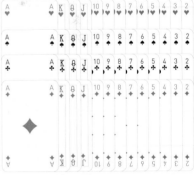

52-card deck in Example 10

10.6   Exercises   Answers to selected odd-numbered problems begin on page ANS-30.

In Problems 1–4, use set notation to write the sample space S of the given experiment.

1. Two coins are tossed.
2. Three coins are tossed.
3. A die is rolled and then a coin is tossed.
4. Two dice are rolled.

In Problems 5–12, find the probability of the given event.

5. Drawing a face card (jack, queen, or king) from a deck of 52 cards
6. Drawing a heart from a deck of 52 cards
7. Rolling a 2 with a single die
8. Rolling a number less than 3 with a single die
9. Rolling snake eyes (a total of 2) with two dice
10. Rolling a total of 7 or 11 with two dice
11. Obtaining all heads when 3 coins are tossed
12. Obtaining exactly 1 head when 3 coins are tossed

In Problems 13–16, find the probability of obtaining the indicated hand by drawing 5 cards without replacement from a well-shuffled 52-card deck.

13. Four of a kind (such as 4 aces)
14. A straight (5 cards in sequence, such as 4, 5, 6, 7, 8, where an ace can count as a 1 or an ace)
15. A flush (5 cards, all of the same suit)
16. A royal flush (10, jack, queen, king, and ace, all of the same suit)

In Problems 17–20, use the first formula in (5) to find the probability of the given event.

17. Obtaining at least 1 heart if 5 cards are drawn without replacement from a 52-card deck
18. Obtaining at least 1 face card if 5 cards are drawn without replacement from a 52-card deck
19. Obtaining at least 1 head in 10 tosses of a coin
20. Obtaining at least one 6 when 3 dice are rolled

Miscellaneous Applications

21. Family Planning Assume that the probability of having a girl equals the probability of having a boy. Find the probability that a family with 4 children has at least 1 girl.
22. Thank You! OOPS! After Joshua writes personalized thank-you notes to each of his 3 aunts for their birthday gifts, his sister randomly inserts them into preaddressed envelopes. Find the probability that **(a)** each aunt receives the correct thank-you note, **(b)** at least one aunt receives the correct thank-you note.
23. Now Hiring Five male and eight female applicants are found to be qualified for 3 identical positions as bank tellers. If 3 of the applicants are selected at random, find the probability that
(a) only women are hired, **(b)** at least one woman is hired.
24. Forming a Committee A committee of 6 people is to be chosen at random from a group of 4 administrators, 7 faculty members, and 8 staff members. Find the probability that all 4 administrators and no faculty members are on the committee.
25. Just Guessing On a 10-question true–false examination, find the probability of scoring 100% if a student guesses the answer for each question.
26. Got Caramel? In a box of 20 chocolates of the same shape and appearance, 10 are known to have caramel centers. Four chocolates are selected at random from the box. Find the probability that all four will have caramel centers.

In Problems 27–30, proceed as in Example 8 to find the indicated probability.

27. **A Natural** In the dice game *craps*, a player rolls two dice and wins on the first roll if a total of 7 or 11 is obtained. Find the probability of winning on the first roll.

28. **Black or Red** A drawer contains 8 black socks, 4 white socks, and 2 red socks. If 1 sock is drawn at random, find the probability that it is either black or red.

29. **You Want to Bet?** At the beginning of the baseball season, an oddsmaker estimates that the probability of the Dodgers winning the World Series is $\frac{1}{10}$ and the probability of the Mets winning is $\frac{1}{20}$. On the basis of these probabilities determine the probability that either the Dodgers or the Mets will win the World Series.

30. **Trying for a Good Grade** A student estimates that his probability of earning an A in a certain math course is $\frac{3}{10}$, a B is $\frac{2}{5}$, a C is $\frac{1}{5}$, and a D is $\frac{1}{10}$. What is the probability that he earns either an A or a B?

31. **Rolling Dice** Two dice are rolled. Find the probability that the total showing on the dice is at most 4.

32. **Tossing a Coin** A coin is tossed 5 times. Let E_1 be the event of obtaining 3 tails, E_2 be the event of obtaining 4 tails, and E_3 be the event of obtaining 5 tails. Intuitively, which of the following probabilities
    (a) $P(E_1 \text{ or } E_2)$ (b) $P(E_2 \text{ or } E_3)$ (c) $P(E_1 \text{ or } E_2 \text{ or } E_3)$
    is the least number? Now compute each probability in parts (a)–(c).

33. **Raindrops Keep Falling** According to the newspaper there is a 40% probability of rain tomorrow. What is the probability that it will not rain tomorrow?

34. **Will She Lose?** A tennis player believes that she has a 75% chance of winning a tournament. Assuming ties are played off, what does she think the probability of losing is?

In Problems 35 and 36, proceed as in Example 10 to find the indicated probability.

35. **More Rolling Dice** Two dice are rolled. Find the probability that the total showing on the dice is either an even number or a multiple of 3.

36. **Choosing at Random** At ABC Plumbing and Heating Company, 30% of the workers are female, 70% are plumbers, and 40% of the workers are female plumbers. If a worker is chosen at random, find the probability that the worker is either female or a plumber.

For Discussion

37. A 12-sided die can be constructed in the form of a regular dodecahedron; each face of the die is a regular pentagon. See FIGURE 10.6.1. When rolled, one of the pentagonal faces will be horizontal to a table top. If each of the numbers from 1 to 6 appears twice on the die, show that the probability of each outcome is the same as that for an ordinary 6-sided die.

FIGURE 10.6.1 12-sided die in Problem 37

38. Suppose a die is a 12-sided regular dodecahedron as in Problem 37, where each face of the die is a regular pentagon. But in this case, suppose that each face bears one of the numbers 1, 2, . . . , 12 as shown in FIGURE 10.6.2.
    (a) If two such dice are rolled, what is the probability of obtaining a total of 13?
    (b) A total of 8?
    (c) A total of 23?
    (d) What number total is least likely to appear?

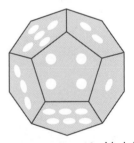

FIGURE 10.6.2 12-sided die in Problem 38

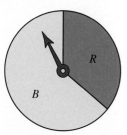

FIGURE 10.6.3 Spinner in
Problem 39

39. For the spinner shown in FIGURE 10.6.3, let S be the sample space for a single spin of the spinner. Let B and R be the events that the pointer lands on blue and red, respectively, so that $S = \{B, R\}$. What, if anything, is wrong with the computation $P(B) = n(B)/n(S) = \frac{1}{2}$ for the probability of the pointer landing on blue?

Project

40. The Birthday Problem Find the probability that in a group of n people at least 2 people have the same birthday. Assume that a year has 365 days. Consider the three cases:
    (a) $n = 10$ **(b)** $n = 25$ **(c)** $n = 90$.

Same birthday

10.7 Convergence of Sequences and Series

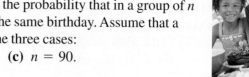 ≡ **Introduction** Sequences and series are important and are studied in depth in a typical course in calculus. In that study, we distinguish between sequences that are convergent or are divergent. In the discussion that follows we examine these concepts from an intuitive point of view. Because the discussion involves the notion of a limit, you are urged to reread Sections 1.5 and 4.11.

☐ **Convergence** The sequence $\left\{\dfrac{n}{n + 1}\right\}$ is an example of a **convergent** sequence. Although it is apparent that the terms of the sequence,

$$\frac{1}{2}, \frac{2}{3}, \frac{3}{4}, \frac{4}{5}, \ldots, \frac{n}{n + 1}, \ldots$$

are increasing as n increases, the values $a_n = \dfrac{n}{n + 1}$ do not increase without bound. This is because $n < n + 1$ and so

$$\frac{n}{n + 1} < 1$$

for all values of n. For example, for $n = 100$, $a_{100} = \frac{100}{101} < 1$. Moreover, it appears that the terms of the sequence can be made closer and closer to 1 by letting the values of n become progressively larger. Using the $\rightarrow$ symbol for the word *approach* as we did in earlier chapters, this is written

$$a_n = \frac{n}{n + 1} \rightarrow 1 \qquad \text{as} \qquad n \rightarrow \infty.$$

We can see the foregoing a little better by dividing the numerator and the denominator of the general term $n/(n + 1)$ by n:

$$\frac{n}{n + 1} = \frac{1}{1 + \dfrac{1}{n}}.$$

As $n \to \infty$ the term $1/n$ in the denominator gets closer and closer to 0 and so

$$\frac{1}{1 + \dfrac{1}{n}} \to \frac{1}{1 + 0} = 1.$$

We write $\lim\limits_{n \to \infty} \dfrac{n}{n + 1} = 1$ and we say that sequence $\left\{ \dfrac{n}{n + 1} \right\}$ **converges** to 1.

☐ **Notation** In general, if the nth term a_n of a sequence $\{a_n\}$ can be made arbitrarily close to a number L for n sufficiently large we say that the sequence $\{a_n\}$ **converges** to L. We indicate that a sequence is convergent to a number L by writing either

$$a_n \to L \qquad \text{as} \qquad n \to \infty \qquad \text{or} \qquad \lim_{n \to \infty} a_n = L.$$

The notions of "arbitrarily close" and "for n sufficiently large" are made precise in a course in calculus. For our purposes, in determining whether a sequence $\{a_n\}$ converges, we will work directly with $\lim\limits_{n \to \infty} a_n$ and proceed as we did in the examination of $\lim\limits_{x \to a} f(x)$ in Section 1.5.

We summarize the discussion.

DEFINITION 10.7.1   Convergent Sequence

(*i*) A sequence $\{a_n\}$ is said to be **convergent** if

$$\lim_{n \to \infty} a_n = L. \qquad (1)$$

The number L is said to be the **limit of the sequence**.

(*ii*) If $\lim\limits_{n \to \infty} a_n$ does not exist, then the sequence is said to be **divergent**.

If a sequence $\{a_n\}$ converges, then its limit L is a unique number.

If a_n either increases or decreases without bound as $n \to \infty$, then $\{a_n\}$ is necessarily divergent and we write, respectively,

$$\lim_{n \to \infty} a_n = \infty \qquad \text{or} \qquad \lim_{n \to \infty} a_n = -\infty. \qquad \blacktriangleleft \text{ In each case, the limits do not exist.}$$

In the first case we say that $\{a_n\}$ **diverges to infinity** and in the second, $\{a_n\}$ **diverges to negative infinity**. For example, the sequence $1, 2, 3, \ldots, n, \ldots$ diverges to infinity.

To determine whether a sequence converges or diverges we often have to rely on analytic procedures (such as algebra) or on previously proven theorems. So in this brief discussion we will accept without proof the following three results:

$$\lim_{n \to \infty} c = c, \text{ where } c \text{ is any real constant}, \qquad (2)$$

$$\lim_{n \to \infty} \frac{1}{n^r} = 0, \text{ where } r \text{ is a positive rational number}, \qquad (3)$$

$$\lim_{n \to \infty} r^n = 0, \text{ for } |r| < 1, \, r \text{ a nonzero real number}. \qquad (4)$$

EXAMPLE 1   **Three Convergent Sequences**

(a) The constant sequence $\{\pi\}$,

$$\pi, \pi, \pi, \pi, \ldots$$

converges to π because of (2), $\lim\limits_{n \to \infty} \pi = \pi$.

(b) The sequence $\left\{\dfrac{1}{\sqrt{n}}\right\}$,

When $n = 1,000,000$, the laws of exponents ▶ shows that $\frac{1}{\sqrt{1,000,000}} = 0.001$.

$$\frac{1}{\sqrt{1}}, \frac{1}{\sqrt{2}}, \frac{1}{\sqrt{3}}, \frac{1}{\sqrt{4}}, \dots \qquad \text{or} \qquad 1, \frac{1}{\sqrt{2}}, \frac{1}{\sqrt{3}}, \frac{1}{2}, \dots$$

converges to 0. With the identification $r = \frac{1}{2}$ in (3), we have

$$\lim_{n\to\infty} \frac{1}{\sqrt{n}} = \lim_{n\to\infty} \frac{1}{n^{1/2}} = 0.$$

(c) The sequence $\left\{\left(\frac{1}{2}\right)^n\right\}$,

The 20th term of the sequence is ▶ approximately $a_{20} \approx 0.00000095$.

$$\frac{1}{2}, \frac{1}{2^2}, \frac{1}{2^3}, \frac{1}{2^4}, \dots \qquad \text{or} \qquad \frac{1}{2}, \frac{1}{4}, \frac{1}{8}, \frac{1}{18}, \dots$$

converges to 0. With the identifications $r = \frac{1}{2}$ and $|r| = \frac{1}{2} < 1$ in (4), we see that $\lim_{n\to\infty} \left(\frac{1}{2}\right)^n = 0$. ≡

▮ EXAMPLE 2   Divergent Sequences

(a) The sequence $\{(-1)^{n-1}\}$,

$$1, -1, 1, -1, \dots$$

is divergent. As $n \to \infty$, the terms of the sequence oscillate between 1 and -1. Thus $\lim_{n\to\infty} (-1)^n$ does not exist because $a_n = (-1)^n$ does not approach a *single* constant L for large values of n.

(b) The first four terms of the sequence $\left\{\left(\frac{5}{2}\right)^n\right\}$ are

$$\frac{5}{2}, \frac{25}{4}, \frac{125}{8}, \frac{625}{16}, \dots \qquad \text{or} \qquad 2.5, 6.25, 15.625, 39.0625, \dots.$$

The 20th term of the sequence is ▶ approximately $a_{20} \approx 90,949,470.2$.

Because the general term $a_n = \left(\frac{5}{2}\right)^n$ increases without bound as $n \to \infty$, we conclude that $\lim_{n\to\infty} a_n = \infty$; in other words, the sequence diverges to infinity. ≡

Expanding on (4) and part (b) of Example 2, it can be proved that

- *The sequence $\{r^n\}$ converges to 0 for $|r| < 1$, and diverges for $|r| > 1$.*   (5)

It follows from (5) that every geometric sequence $\{ar^{n-1}\}$ for which $|r| < 1$ converges to 0.

It is often necessary to manipulate the general term of a sequence to demonstrate convergence of the sequence.

▮ EXAMPLE 3   Convergent Sequence

Determine whether the sequence $\left\{\sqrt{\dfrac{n}{9n+1}}\right\}$ converges.

Solution By dividing the numerator and denominator by n it follows that

$$\frac{1}{9 + \dfrac{1}{n}} \to \frac{1}{9}$$

as $n \to \infty$. Thus, we can write

$$\lim_{n\to\infty} \sqrt{\frac{n}{9n+1}} = \lim_{n\to\infty} \sqrt{\frac{1}{9 + \dfrac{1}{n}}} = \sqrt{\frac{1}{9}} = \frac{1}{3}.$$

The sequence converges to $\frac{1}{3}$. ≡

CHAPTER 10 SEQUENCES AND SERIES

EXAMPLE 4   **Convergent Sequence**

Determine whether the sequence $\left\{ \dfrac{12e^n - 5}{3e^n + 2} \right\}$ converges.

Solution Since $e > 1$, a fast inspection of the general term may lead you to the false conclusion that the sequence is divergent because $12e^n - 5 \to \infty$ and $3e^n + 2 \to \infty$ as $n \to \infty$. But if we divide the numerator and denominator by e^n and then use $12 - 5e^{-n} \to 12$ and $3 + 2e^{-n} \to 3$ as $n \to \infty$, we can write

$$\lim_{n\to\infty} \frac{12e^n - 5}{3e^n + 2} = \lim_{n\to\infty} \frac{12 - 5e^{-n}}{3 + 2e^{-n}} = \frac{12 - 0}{3 + 0} = 4.$$

◀ Note that $e^{-n} = \left(\frac{1}{e}\right)^n$. Since $1/e < 1$, it follows from (4) that $\left(\frac{1}{e}\right)^n \to 0$ as $n \to \infty$.

The sequence converges to 4.   ≡

☐ **Infinite Series** Under certain conditions it is possible to assign a numerical value to an **infinite series**. In Section 10.2 we saw that we could add terms of a sequence using summation notation. Associated with every sequence $\{a_n\}$ is another sequence called the **sequence of partial sums** $\{S_n\}$, where S_1 is the first term, S_2 is the sum of the first two terms, S_3 is the sum of the first three terms, and so on. In symbols:

sequence: $a_1, a_2, a_3, \ldots, a_n, \ldots$
sequence of partial sums: $a_1, a_1 + a_2, a_1 + a_2 + a_3, \ldots, a_1 + a_2 + a_3 + \cdots + a_n, \ldots$

In other words, the sequence of partial sums for $\{a_n\}$ is the sequence $\{S_n\}$, where the general term can be written $S_n = \sum_{k=1}^{n} a_k$. Just as we can ask whether a sequence $\{a_n\}$ converges, we now ask whether a sequence of partial sums can converge.

This question is answered in the next definition.

DEFINITION 10.7.2   Convergent Infinite Series

(*i*) If $a_1, a_2, a_3, \ldots, a_n, \ldots$ is an **infinite sequence**, we say that

$$\sum_{k=1}^{\infty} a_k = a_1 + a_2 + a_3 + \cdots + a_n + \cdots$$

is an **infinite series**.

(*ii*) An infinite series $\sum_{k=1}^{\infty} a_k$ is said to be **convergent** if the sequence of partial sums $\{S_n\}$ converges, that is,

$$\lim_{n\to\infty} S_n = \lim_{n\to\infty} \sum_{k=1}^{n} a_k = S.$$

The number S is called the **sum** of the infinite series.

(*iii*) If $\lim_{n\to\infty} S_n$ does not exist, the infinite series is said to be **divergent**.

Although the proper place for digging deeper into the above concepts is a course in calculus, we can readily illustrate the notion of convergence of an infinite series using geometric series.

Suspend, for the sake of illustration, ▶
that you know this rational number.

Every student of mathematics knows that

$$0.333\ldots \tag{6}$$

is the decimal representation of a well-known rational number. The decimal in (6) is the same as the infinite series

$$0.3 + 0.03 + 0.003 + \cdots = \frac{3}{10} + \frac{3}{100} + \frac{3}{1000} + \cdots$$

$$= \frac{3}{10} + \frac{3}{10^2} + \frac{3}{10^3} + \cdots = \sum_{k=1}^{\infty} \frac{3}{10^k}. \tag{7}$$

If we consider the geometric sequence

$$\frac{3}{10}, \frac{3}{10^2}, \frac{3}{10^3}, \ldots,$$

it is possible to find a formula for the general term of the associated sequence of partial sums:

$$
\begin{aligned}
S_1 &= \frac{3}{10} = 0.3 \\
S_2 &= \frac{3}{10} + \frac{3}{10^2} = 0.33 \\
S_3 &= \frac{3}{10} + \frac{3}{10^2} + \frac{3}{10^3} = 0.333 \\
&\vdots \\
S_n &= \frac{3}{10} + \frac{3}{10^2} + \frac{3}{10^3} + \cdots + \frac{3}{10^n} = \overset{n\ 3's}{\overbrace{0.333\ldots 3.}}
\end{aligned}
\tag{8}
$$

In view (8) of Section 10.2 with the identifications $a = \frac{3}{10}$ and $r = \frac{1}{10}$ we can write the general term S_n of the sequence (8):

$$S_n = \frac{3}{10} \frac{1 - \left(\frac{1}{10}\right)^n}{1 - \frac{1}{10}}. \tag{9}$$

We now let n increase without bound, that is, $n \to \infty$. From (4) and (5) we know that $\left(\frac{1}{10}\right)^n \to 0$ as $n \to \infty$ and so the limit of (9) is

$$\lim_{n\to\infty} S_n = \lim_{n\to\infty} \frac{3}{10} \frac{1 - \left(\frac{1}{10}\right)^n}{1 - \frac{1}{10}} = \frac{\frac{3}{10}}{\frac{9}{10}} = \frac{3}{9} = \frac{1}{3}.$$

Thus, $\frac{1}{3}$ is the sum of the infinite series in (7):

$$\frac{1}{3} = \sum_{k=1}^{\infty} \frac{3}{10^k} \quad \text{or} \quad \frac{1}{3} = 0.333\ldots.$$

☐ **Geometric Series** In general, the sum of an **infinite geometric series**

$$\sum_{k=1}^{\infty} ar^{k-1} = a + ar + ar^2 + \cdots + ar^{n-1} + \cdots \tag{10}$$

is defined whenever $|r| < 1$. To see why this is so, recall from Section 10.2 that

$$S_n = \sum_{k=1}^{n} ar^{k-1} = a + ar + ar^2 + \cdots + ar^{n-1} = \frac{a(1 - r^n)}{1 - r}. \tag{11}$$

By letting $n \to \infty$ and using $r^n \to 0$ whenever $|r| < 1$, we see that

$$\lim_{n\to\infty} S_n = \lim_{n\to\infty} \frac{a(1 - r^n)}{1 - r} = \frac{a}{1 - r}.$$

Therefore for $|r| < 1$ we define the sum of the infinite geometric series in (10) to be $a/(1 - r)$.

THEOREM 10.7.1   Sum of a Geometric Series

(*i*)   An infinite geometric series $\sum_{k=1}^{\infty} ar^{k-1}$ **converges** for $|r| < 1$. The **sum** of the series is then

$$\sum_{k=1}^{\infty} ar^{k-1} = a + ar + ar^2 + \cdots + ar^{n-1} + \cdots = \frac{a}{1 - r}. \qquad (12)$$

(*ii*)   An infinite geometric series $\sum_{k=1}^{\infty} ar^{k-1}$ **diverges** for $|r| \geq 1$.

A divergent geometric series $\sum_{k=1}^{\infty} ar^{k-1}$ has no sum.

Formula (12) gives a method for converting a repeating decimal to a quotient of integers. We use the fact that:

Every repeating decimal is the sum of an infinite geometric series.

Before giving another example of this, let's be clear that a **repeating decimal** is a decimal number that after a finite number of decimal places has a sequence of one or more digits that repeats endlessly.

◀ Recall from Section 1.1, a **rational number** is one that is either a terminating decimal or a repeating decimal. An **irrational number** is one that is neither a terminating nor a repeating decimal.

EXAMPLE 5   **Repeating Decimal**

Write $0.232323 \ldots$ as a quotient of integers.

Solution Written as an infinite geometric series, the repeating decimal is the same as

$$\frac{23}{100} + \frac{23}{100^2} + \frac{23}{100^3} + \cdots = \sum_{k=1}^{\infty} \frac{23}{100^k}.$$

With the identifications $a = \frac{23}{100}$ and $|r| = \left|\frac{1}{100}\right| < 1$ it follows from (12) that

$$\sum_{k=1}^{\infty} \frac{23}{100^k} = \frac{\frac{23}{100}}{1 - \frac{1}{100}} = \frac{\frac{23}{100}}{\frac{99}{100}} = \frac{23}{99}. \qquad\qquad \equiv$$

EXAMPLE 6   **Repeating Decimal**

Write $0.72555 \ldots$ as a quotient of integers.

Solution The repeating digit 5 does not appear until the third decimal place so we write the number as the sum of a terminating decimal and a repeating decimal:

$$
\begin{aligned}
0.72555 \ldots &= 0.72 + \overbrace{0.00555 \ldots}^{\text{geometric series}} \\
&= \tfrac{72}{100} + \left(\tfrac{5}{1000} + \tfrac{5}{10{,}000} + \tfrac{5}{100{,}000} + \cdots\right) \\
&= \tfrac{72}{100} + \left(\tfrac{5}{10^3} + \tfrac{5}{10^4} + \tfrac{5}{10^5} + \cdots\right) \qquad \leftarrow a = \tfrac{5}{10^3},\, r = \tfrac{1}{10} \\
&= \tfrac{72}{100} + \frac{\tfrac{5}{10^3}}{1 - \tfrac{1}{10}} \qquad\qquad\qquad\quad \leftarrow \text{from (12)} \\
&= \tfrac{72}{100} + \tfrac{5}{900}.
\end{aligned}
$$

Combining the last two rational numbers by a common denominator we find

$$0.72555\ldots = \frac{653}{900}.$$

≡

Every repeating decimal number (rational number) is a geometric series, but do not get the impression that the sum of every convergent geometric series need be a quotient of integers.

EXAMPLE 7   Sum of a Geometric Series

The infinite series $1 - \dfrac{1}{e} + \dfrac{1}{e^2} - \dfrac{1}{e^3} + \cdots$ is a convergent geometric series because $|r| = |-1/e| = 1/e < 1$. By (12) the sum of the series is the number

$$\frac{1}{1 - (-1/e)} = \frac{e}{e + 1}.$$

≡

EXAMPLE 8   A Divergent Geometric Series

The infinite series

$$2 - 3 + \frac{3^2}{2} - \frac{3^3}{2^2} + \cdots$$

is a divergent geometric series because $|r| = \left|-\frac{3}{2}\right| = \frac{3}{2} > 1$.

≡

NOTES FROM THE CLASSROOM

(i) When written in terms of summation notation, a geometric series may not be immediately recognizable, or if it is, the values of a and r may not be apparent. For example, to see whether $\sum_{n=3}^{\infty} 4\left(\frac{1}{2}\right)^{n+2}$ is a geometric series it is a good idea to write out two or three terms:

$$\sum_{n=3}^{\infty} 4\left(\frac{1}{2}\right)^{n+2} = \overbrace{4\left(\frac{1}{2}\right)^{5}}^{a} + \overbrace{4\left(\frac{1}{2}\right)^{6}}^{ar} + \overbrace{4\left(\frac{1}{2}\right)^{7}}^{ar^2} + \cdots.$$

From the right side of the last equality, we can make the identifications $a = 4\left(\frac{1}{2}\right)^5$ and $|r| = \frac{1}{2} < 1$. Consequently, the sum of the series is $\dfrac{4\left(\frac{1}{2}\right)^5}{1 - \frac{1}{2}} = \frac{1}{4}$. If desired, although there is no real need to do this, we can express $\sum_{n=3}^{\infty} 4\left(\frac{1}{2}\right)^{n+2}$ in the more familiar form $\sum_{k=1}^{\infty} ar^{k-1}$ by letting $k = n - 2$. The result is

$$\sum_{n=3}^{\infty} 4\left(\frac{1}{2}\right)^{n+2} = \sum_{k=1}^{\infty} 4\left(\frac{1}{2}\right)^{k+4} = \sum_{k=1}^{\infty} \overbrace{4\left(\frac{1}{2}\right)^{5}}^{a} \overbrace{\left(\frac{1}{2}\right)^{k-1}}^{r^{k-1}}.$$

(ii) In general, it is very difficult to find the sum of a convergent infinite series using the sequence of partial sums. In most cases it is impossible to find a formula for the general term $S_n = \sum_{k=1}^{n} a_k$ of this sequence. The geometric series is, of course, an important exception. But there is another type of infinite series whose sum can be found by finding the limit of the sequence $\{S_n\}$. If interested, see Problems 37 and 38 in Exercises 10.7.

In Problems 1–20, determine whether the given sequence converges.

1. $\left\{\dfrac{10}{n}\right\}$

2. $\left\{1 + \dfrac{1}{n^2}\right\}$

3. $\left\{\dfrac{1}{5n + 6}\right\}$

4. $\left\{\dfrac{4}{2n + 7}\right\}$

5. $\left\{\dfrac{3n - 2}{6n + 1}\right\}$

6. $\left\{\dfrac{n}{1 - 2n}\right\}$

7. $\left\{\dfrac{3n(-1)^{n-1}}{n + 1}\right\}$

8. $\left\{\left(-\tfrac{1}{3}\right)^n\right\}$

9. $\left\{\dfrac{n^2 - 1}{2n}\right\}$

10. $\left\{\dfrac{7n}{n^2 + 1}\right\}$

11. $\left\{\sqrt{\dfrac{2n + 1}{n}}\right\}$

12. $\left\{\dfrac{n}{\sqrt{n + 1}}\right\}$

13. $\{\cos n\pi\}$

14. $\{\sin n\pi\}$

15. $\left\{\dfrac{5 - 2^{-n}}{6 + 4^{-n}}\right\}$

16. $\left\{\dfrac{2^n}{3^n + 1}\right\}$

17. $\left\{\dfrac{10e^n - 3e^{-n}}{2e^n + e^{-n}}\right\}$

18. $\left\{4 + \dfrac{3^n}{2^n}\right\}$

19. $2, \tfrac{2}{3}, \tfrac{2}{9}, \tfrac{2}{27}, \ldots$

20. $1 + \tfrac{1}{2}, \tfrac{1}{2} + \tfrac{1}{3}, \tfrac{1}{3} + \tfrac{1}{4}, \tfrac{1}{4} + \tfrac{1}{5}, \ldots$

In Problems 21–26, write the given repeating decimal as a quotient of integers.

21. $0.222\ldots$

22. $0.555\ldots$

23. $0.616161\ldots$

24. $0.393939\ldots$

25. $1.314314\ldots$

26. $0.5262626\ldots$

In Problems 27–36, determine whether the given infinite geometric series converges. If convergent, find its sum.

27. $2 + 1 + \dfrac{1}{2} + \cdots$

28. $1 + \dfrac{1}{3} + \dfrac{1}{9} + \cdots$

29. $\dfrac{2}{3} - \dfrac{4}{9} + \dfrac{8}{27} - \cdots$

30. $1 + 0.1 + 0.01 + \cdots$

31. $9 + 2 + \dfrac{4}{9} + \cdots$

32. $\dfrac{1}{81} - \dfrac{1}{54} + \dfrac{1}{36} - \cdots$

33. $\displaystyle\sum_{k=1}^{\infty} \dfrac{1}{(\sqrt{3} - \sqrt{2})^{k-1}}$

34. $\displaystyle\sum_{k=1}^{\infty} \left(\dfrac{\sqrt{5}}{1 + \sqrt{5}}\right)^{k-1}$

35. $\displaystyle\sum_{k=1}^{\infty} (-3)^k 7^{-k}$

36. $\displaystyle\sum_{k=1}^{\infty} \pi^k \left(\tfrac{1}{3}\right)^{k-1}$

37. The infinite series $\displaystyle\sum_{k=1}^{\infty} \dfrac{1}{k(k + 1)}$ is an example of a **telescoping series**. For such series it is possible to find a formula for the general term S_n of the sequence of partial sums.

(a) Use the partial fraction decomposition

$$\frac{1}{k(k+1)} = \frac{1}{k} - \frac{1}{k+1}$$

as an aid in finding a formula for S_n. This will also explain the meaning of the word *telescoping*.

(b) Use part (a) to find the sum of the infinite series.

38. Use the procedure in part (a) of Problem 37 to find the sum of the infinite series $\sum_{k=1}^{\infty} \frac{1}{(k+1)(k+2)}$.

Miscellaneous Applications

39. Distance Traveled A ball is dropped from an initial height of 15 ft onto a concrete slab. Each time the ball bounces, it reaches a height of $\frac{2}{3}$ its preceding height. Use an infinite geometric series to determine the distance the ball travels before it comes to rest.

40. Drug Accumulation A patient takes 15 mg of a drug at the same time each day. If 80% of the drug accumulated is excreted each day by bodily functions, how much of the drug will accumulate in the patient's body after a long period of time, that is, as $n \to \infty$? (Assume that the measurement of the accumulation is made immediately after each dose.)

Calculator/Computer Problems

41. It can be proved that the terms of the sequence $\{a_n\}$ defined recursively by the formula

$$a_{n+1} = \frac{1}{2}\left(a_n + \frac{r}{a_n}\right), \quad r > 0,$$

converges when $a_1 = 1$ and $r = 3$. Use a calculator to find the first 10 terms of the sequence. Conjecture the limit of the sequence.

42. The sequence

$$\left\{1 + \tfrac{1}{2} + \tfrac{1}{3} + \cdots + \tfrac{1}{n} - \ln n\right\}$$

is known to converge to a number γ called **Euler's constant**. Calculate at least the first 10 terms of the sequence. Conjecture the limit of the sequence.

For Discussion

43. Use algebra to show that the sequence $\{\sqrt{n}(\sqrt{n+1} - \sqrt{n})\}$ converges.

44. Use the graph of the inverse tangent function to show the sequence

$$\left\{\frac{\pi}{4} - \arctan(n)\right\} \text{ converges.}$$

45. The infinite series $\sum_{k=1}^{\infty} \frac{2^k - 1}{4^k}$ is known to be convergent. Discuss how the sum of the series can be found. State any assumptions that you make.

46. Find the values of x for which the infinite series $\sum_{k=1}^{\infty} \left(\frac{x}{2}\right)^{k-1}$ converges.

47. The infinite series

$$1 + 1 + 1 + \cdots$$

is a divergent geometric series with $r = 1$. Note that formula (5) does not yield the general term for the sequence of partial sums. Find a formula for S_n and use that formula to argue that the infinite series is divergent.

48. Consider the rational function $f(x) = 1/(1 - x)$. Show that

$$\frac{1}{1 - x} = 1 + x + x^2 + \cdots.$$

For what values of x is the equality true?

49. Discuss whether the equality $1 = 0.999\ldots$ is true or false.

50. The Trains and the Fly At a specified time two trains T_1 and T_2, 20 miles apart on the same track, start on a collision course at a rate of 10 mi/h. Suppose that at the precise instant the trains start a fly leaves the front of train T_1, flies at a rate of 20 mi/h in a straight line to the front of the engine of train T_2, then flies back to T_1 at 20 mi/h, then back to T_2, and so on. Use geometric series to find the total distance traversed by the fly when the trains collide (and the fly is squashed). Then use common sense to find the total distance the fly flies. See FIGURE 10.7.1.

FIGURE 10.7.1 Trains and fly in Problem 50

51. Embedded Squares In FIGURE 10.7.2 the square shown in red is 1 unit on a side. A second blue square is constructed inside the first square by connecting the midpoints of the first one. A third green square is constructed by connecting the midpoints of the sides of the second square, and so on.

(a) Find a formula for the area A_n of the nth inscribed square.

(b) Make a conjecture about the convergence of the sequence $\{A_n\}$.

(c) Consider the sequence $\{S_n\}$, where $S_n = A_1 + A_2 + \cdots + A_n$. Calculate the numerical values of the first 10 terms of this sequence.

(d) Make a conjecture about the convergence of the sequence $\{S_n\}$.

52. Length of a Polygonal Path In FIGURE 10.7.3, there are twelve blue rays emanating from the origin and the angle between each pair of consecutive rays is $30°$. The line segment AP_1 is perpendicular to ray L_1, the line segment P_1P_2 is perpendicular to ray L_2, and so on.

(a) Show that the length of the red polygonal path $AP_1P_2P_3\ldots$ is the infinite series

$$AP_1 + P_1P_2 + P_2P_3 + P_3P_4 + \cdots$$
$$= \sin 30° + (\cos 30°)\sin 30° + (\cos 30°)^2\sin 30° + (\cos 30°)^3\sin 30° + \cdots.$$

(b) Find the sum of the infinite series in part (a).

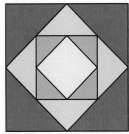

FIGURE 10.7.2 Embedded squares in Problem 51

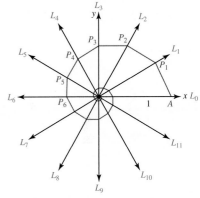

FIGURE 10.7.3 Polygonal path in Problem 52

CHAPTER 10 | **Review Exercises** Answers to selected odd-numbered problems begin on page ANS-30.

A. Fill in the Blanks

In Problems 1–22, fill in the blanks.

1. The next three terms in the arithmetic sequence $2x + 1, 2x + 4, \ldots$ are _____ .

2. $\dfrac{x}{2} + \dfrac{x^2}{4} + \dfrac{x^3}{6} + \cdots + \dfrac{x^{10}}{20} = \sum_{-}^{-}$ _____

3. The fifth term of the sequence $\left\{ \sum_{k=1}^{n} \dfrac{1}{k} \right\}$ is _____ .

4. The twentieth term of the arithmetic sequence $-2, 3, 8, \ldots$ is _____.

5. The common ratio r of the geometric sequence $\left\{ \dfrac{2^{n+1}}{5^{n-1}} \right\}$ is _____.

6. The common difference d of the arithmetic sequence $\left\{ 8 - \dfrac{n}{2} \right\}$ is _____.

7. $\displaystyle\sum_{k=1}^{50} (3 + 2k) =$ _____.

8. $\displaystyle\sum_{k=1}^{100} (-1)^k =$ _____.

9. $3 - 1 + \frac{1}{3} - \frac{1}{9} + \cdots =$ _____.

10. For $|x| > 1$, $\displaystyle\sum_{k=0}^{\infty} \dfrac{1}{x^k} =$ _____.

11. $\displaystyle\sum_{k=1}^{10} 3\left(\tfrac{1}{2}\right)^{k-1} =$ _____.

12. $\displaystyle\sum_{k=1}^{\infty} 3\left(\tfrac{1}{2}\right)^{k-1} =$ _____.

13. $\dbinom{100}{100} =$ _____.

14. $\dbinom{100}{0} =$ _____.

15. For the sequence $1, 2, 3, \ldots$,
$$1 + 2 + 3 + \cdots + 299 + 300 = \underline{\hspace{2cm}}.$$

16. If a sequence is defined recursively by $a_{n+1} = (-1)^n a_n + 1$, $a_1 = 1$, then $a_8 =$ _____.

17. If $C(n + 1, n) = 5$, then $n =$ _____.

18. $C(5, 3)/C(8, 3) =$ _____.

19. The sequence $\left\{ \dfrac{1 - 2n}{4n + 5} \right\}$ converges to _____.

20. The fifth term of an arithmetic sequence is -1 and its twelfth term is 13. The general term a_n of the sequence is _____.

21. If E_1 and E_2 are mutually exclusive events such that $P(E_1) = \frac{1}{5}$ and $P(E_2) = \frac{1}{3}$, then $P(E_1 \cup E_2) =$ _____.

22. If $P(E_1) = 0.3$, $P(E_2) = 0.8$, and $P(E_1 \cap E_2) = 0.7$, then $P(E_1 \cup E_2) =$ _____.

B. True/False

In Problems 1–20, answer true or false.

1. $2(8!) = 16!$ _____

2. $\dfrac{10!}{9!} = 10$ _____

3. $(n - 1)!n = n!$ _____

4. $2^{10} < 10!$ _____

5. There is no constant term in the expansion of $\left(x + \dfrac{1}{x^2} \right)^{20}$. _____

6. There are exactly 100 terms in the expansion of $(a + b)^{100}$. _____

7. A sequence that is defined recursively by $a_{n+1} = (-1)a_n$ is a geometric sequence. _____

8. $\{\ln 5^n\}$ is an arithmetic sequence. _____

9. $\displaystyle\sum_{k=1}^{5} \ln k = \ln 120$ _____

10. $3 = 2.999\ldots$ _____

11. $0! = 1$ _____

12. $P(n, n) = n!$ _____

13. The sequence $\{n \sin n\pi\}$ is convergent. _____

14. The series $\displaystyle\sum_{k=1}^{\infty} \left(\tfrac{1000}{1001} \right)^k$ is divergent. _____

15. The sequence defined recursively by $a_{n+1} = 2a_n + 1$, $a_1 = -1$, is a constant sequence. _____

16. The quotient of two nonterminating repeating decimals is always a rational number. _____

17. If $\frac{8}{3}$ and $\frac{16}{9}$ are the ninth and tenth terms of a geometric sequence, then the seventh term of the sequence is 6. _____

18. The geometric sequence in Problem 17 is divergent. _____

19. The sequence defined recursively by $a_{n+1} = na_n$, $a_1 = 1$, is the same as the sequence $\{n!\}$. _____

20. A math professor's salary in her first year of teaching was \$15,000. If she received a raise of 4.5% each year, then in her 10th year of teaching her salary was $15,000(1.045)^9$. _____

C. Review Exercises

In Problems 1–4, list the first five terms of the given sequence.

1. $\{6 - 3(n - 1)\}$

2. $\{-5 + 4n\}$

3. $\{(-1)^n n\}$

4. $\left\{ \dfrac{(-1)^n 2^n}{n + 3} \right\}$

5. List the first five terms of the sequence defined by $a_1 = 1$, $a_2 = 3$, and $a_n = (n + 1)a_{n-1} + 2$.

6. Find the seventeenth term of the arithmetic sequence with first term 3 and third term 11.

7. Find the first term of the geometric sequence with third term $-\frac{1}{2}$ and fourth term 1.

8. Find the sum of the first 30 terms of the sequence defined by $a_1 = 4$ and $a_{n+1} = a_n + 3$.

9. Find the sum of the first 10 terms of the geometric series with first term 2 and common ratio $-\frac{1}{2}$.

10. Write $2.515151\ldots$ as an infinite geometric series and express the sum as a quotient of integers.

11. Best Gift Determine the best gift from the following choices:
A: \$10 each month for 10 years.
B: 10¢ the first month, 20¢ the second month, 30¢ the third month, and so on, receiving an increase of 10¢ each month for 10 years.
C: 1¢ the first month, 2¢ the second month, 4¢ the third month, and so on, doubling the amount received each month for 2 years.

12. Distance Traveled Galileo discovered that the distance a mass moves down an inclined plane in consecutive time intervals is proportional to an odd integer. Therefore, the total distance D that a mass will move down the inclined plane in n seconds is proportional to $1 + 3 + 5 + \cdots + (2n - 1)$. Show that D is proportional to n^2.

13. Annuity If an annual rate of interest r is compounded continuously, then the amount S accrued in an annuity immediately after the nth deposit of P dollars is given by

$$S = P + Pe^r + Pe^{2r} + \cdots + Pe^{(n-1)r}.$$

Show that

$$S = P\frac{1 - e^{rn}}{1 - e^r}.$$

14. Number of Sales In 2009 a new high-tech firm projects that its sales will double each year for the next 5 years. If its sales in 2009 are \$1,000,000, what does it expect its sales to be in 2014?

In Problems 15–20, use the Principle of Mathematical Induction to prove that the given statement is true for every positive integer n.

15. $n^2(n + 1)^2$ is divisible by 4

16. $\displaystyle\sum_{k=1}^{n} (2k + 6) = n(n + 7)$

17. $1(1!) + 2(2!) + \cdots + n(n!) = (n + 1)! - 1$

18. 9 is a factor of $10^{n+1} - 9n - 10$

19. $\left(1 + \dfrac{1}{1}\right)\left(1 + \dfrac{1}{2}\right)\left(1 + \dfrac{1}{3}\right)\cdots\left(1 + \dfrac{1}{n}\right) = n + 1$

20. $(\cos\theta + i\sin\theta)^n = \cos n\theta + i\sin n\theta$, where $i^2 = -1$

In Problems 21–26, evaluate the given expression.

21. $\dfrac{6!}{4! - 3!}$

22. $\dfrac{6!4!}{10!}$

23. $C(7, 2)$

24. $P(9, 6)$

25. $\dfrac{(n + 3)!}{n!}$

26. $\dfrac{(2n + 1)!}{(2n - 1)!}$

In Problems 27–30, use the Binomial Theorem to expand the given expression.

27. $(a + 4b)^4$

28. $(2y - 1)^6$

29. $(x^2 - y)^5$

30. $(4 - (a + b))^3$

In Problems 31–34, find the indicated term in the expansion of the given expression.

31. Fourth term of $(5a - b^3)^8$

32. Tenth term of $(8y^2 - 2x)^{11}$

33. Fifth term of $(xy^2 + z^3)^{10}$

34. Third term of $\left(\dfrac{10}{a} - 3bc\right)^7$

35. A multiple of x^2 occurs as which term in the expansion of $(x^{1/2} + 1)^{40}$?

36. Solve for x:

$$\sum_{k=0}^{n} \binom{n}{k} x^{2n-2k}(-4)^k = 0.$$

37. If the first term of an infinite geometric series is 10 and the sum of the series is $\frac{25}{2}$, then what is the value of the common ratio r?

38. Consider the sequence $\{a_n\}$ whose first four terms are

$$1, \quad 1 + \tfrac{1}{2}, \quad 1 + \dfrac{1}{2 + \frac{1}{2}}, \quad 1 + \dfrac{1}{2 + \dfrac{1}{2 + \frac{1}{2}}}, \ldots$$

    (a) With $a_1 = 1$, find a recursion formula that defines the sequence.
    (b) What are the fifth and sixth terms of the sequence?

In Problems 39 and 40, conjecture whether the given sequence converges.

39. $\left\{\dfrac{2^n}{n!}\right\}$

40. $\sqrt{3}, \sqrt{3\sqrt{3}}, \sqrt{3\sqrt{3\sqrt{3}}}, \ldots$

41. If a coin is tossed 3 times, use a tree diagram to find all possible sequences of heads (H) and tails (T).

42. List all possible 3-digit numbers using only the digits 2, 4, 6, and 8.

43. **Ice Cream** If 32 different flavors of ice cream are available, in how many ways can a double scoop cone be ordered:
   (a) if both scoops must be different flavors?
   (b) if both scoops can be the same flavor?
      [*Hint*: Assume that the order in which the scoops are placed on the cone does not matter.]

44. **More Ice Cream** At a dessert bar there are 3 flavors of ice cream, 6 different toppings, 2 kinds of nuts, and whipped cream. How many different sundaes can be made consisting of 1 flavor of ice cream with 1 topping:
   (a) if nuts and whipped cream are required?
   (b) if nuts are optional, but whipped cream is required?
   (c) if both nuts and whipped cream are optional?

45. **Build Your Own** Domingo's Pizza offers 10 extra toppings. How many different pizzas can be made using just 3 of the toppings?

46. **Poker Hand** In a certain poker game a hand consists of 5 cards drawn from a standard 52-card deck with 4 suits.
   (a) How many 5-card hands are possible?
   (b) A *full house* is a 5-card hand consisting three of a kind and a pair. How many full houses are possible?
   (c) How many full houses are there consisting of 2 kings and 3 aces?

Full house

47. **Rearrangements** In making up a scrambled word puzzle, how many rearrangements of the letters in the word *shower* are possible?

48. **Time to Plant** Burtee's seed catalog offers 9 varieties of tomatoes. In how many ways can a gardener choose 3 to order?

49. **Modeling** There are 10 casual and 12 formal outfits to be modeled one at a time in a fashion show. In how many different orders can they be shown:
   (a) if all the casual outfits are grouped together and all the formal outfits are grouped together?
   (b) if there are no restrictions on the order?

50. **At the Races** In how many ways can win, place, and show (that is, first-, second-, and third-place finish) be decided if 10 horses are entered in a race? Assume that there are no ties.

In Problems 51 and 52, use set notation to write the sample space of the given experiment.

51. The spinner in FIGURE 10.R.1 is spun twice.
52. The spinner in Figure 10.R.1 is spun once and then a coin is tossed.

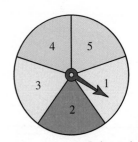

FIGURE 10.R.1 Spinner in Problems 51 and 52

53. **Drawing Cards** If two cards are drawn from a well-shuffled 52-card deck, what is the probability that both are black?

54. **Choosing a Pen** Five pens are selected at random from a batch of 100 Pic pens. If 90% of this batch of Pic pens will write the first time, what is the probability that
   (a) all 5 of the pens selected will write the first time?
   (b) none of them will write the first time?
   (c) at least 1 of them will write the first time?

55. **Family Planning** Assume that the probability of giving birth to a female baby equals the probability of giving birth to a male baby. In a family of 4 children, which is more likely: (*i*) all the same sex, (*ii*) 2 of each sex, (*iii*) 3 of one sex and 1 of the other?

56. **Average Young Woman** Statistics indicate that the probability of death in the next year for a 20-year-old female is 0.0006. What is the probability that an "average" 20-year-old female will live through the next year?

B	I	N	G	O
1	16	33	52	72
12	20	41	47	65
2	22	FREE	55	68
7	30	36	60	74
8	28	45	49	61

FIGURE 10.R.2 Bingo card in Problem 58

57. Feeling Lucky? A drawer contains 8 black socks and 4 white socks. If 2 socks are drawn at random, what is the probability that:
 (a) a black pair is obtained?
 (b) a white pair is obtained?
 (c) a matching pair is obtained?

58. Bingo A Bingo card has 5 rows and 5 columns. See FIGURE 10.R.2. Any five of the numbers 1 through 15 appear in the first column (designated B); any five of the numbers 16 through 30 appear in the second column (I); any four of 31 through 45 appear in the third column (N), where the center square marked "FREE" is found; any five of 46 through 60 appear in the fourth column (G); and any five of 61 through 75 appear in the last column (O). How many different Bingo cards are possible? (Consider 2 cards to be different if any 2 corresponding entries are different.)

59. More Bingo One version of Bingo requires a player to cover all the numbers on the card as numbers are called out at random. See Problem 58.
 (a) What is the minimum number of calls before there can be a winner in this version?
 (b) Assume that there is a winner at the minimum number of calls obtained in part (a). What is the probability that the card being played is a winning card at that point?

60. Areas Let $\{A_n\}$ be the sequence of areas of the isosceles triangles shown in FIGURE 10.R.3. Find the sum of the infinite series $A_1 + A_2 + A_3 + \cdots$.

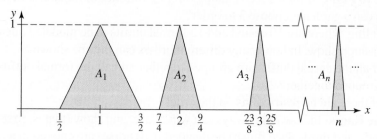

FIGURE 10.R.3 Isosceles triangles in Problem 60

Final Examination
Answers to all questions are given in the *Student Resource Manual*.

A. Fill in the Blanks

In Problems 1–20, fill in the blanks.

1. Completing the square in x for $2x^2 + 6x + 5$ gives _____.
2. In the binomial expansion of $(1 - 2x)^3$ the coefficient of x^2 is _____.
3. In interval notation, the solution set of the inequality $\dfrac{x(x^2 - 9)}{x^2 - 25} \geq 0$ is _____.
4. If $a - 3$ is a negative number, then $|a - 3|$ = _____.
5. If $|5x| = 80$, then x = _____.
6. If (a, b) is a point in the third quadrant, then $(-a, b)$ is a point in the _____ quadrant.
7. The point $(1, 7)$ is on a graph in the Cartesian plane. Give the coordinates of another point on the graph if the graph is:
   (a) symmetric with respect to the x-axis. _____
   (b) symmetric with respect to the y-axis. _____
   (c) symmetric with respect to the origin. _____
8. The lines $6x + 2y = 1$ and $kx - 9y = 5$ are parallel if k = _____. The lines are perpendicular if k = _____.
9. The complete factorization of the function $f(x) = x^3 - 2x^2 - 6x$ is _____.
10. The only potential rational zeros of $f(x) = x^3 + 4x + 2$ are _____.
11. The phase shift of the graph of $y = 5\sin(4x + \pi)$ is _____.
12. If $f(x) = x^4 \arctan(x/2)$, then the exact value of $f(-2)$ is _____ .
13. If $\sin x = \frac{3}{5}$, $\pi/2 < x < \pi$, then $\sin 2x$ = _____ .
14. If $\left(\frac{1}{3}\right)^x = 81$, then x = _____.
15. $\arccos\left(-\frac{\sqrt{3}}{2}\right) =$ _____.
16. $5\ln 2 - \ln\frac{2}{3} = \ln$ _____
17. The graph of $y = \ln(2x + 5)$ has the vertical asymptote $x =$ _____.
18. The domain of the function $y = \ln(x^2 - 2x)$ is _____.
19. The number of five element subsets that can be formed from the set of letters in the English alphabet is _____.
20. If the first three terms of an arithmetic sequence are $a_1 = 10$, $a_2 = 6.5$, and $a_3 = 3$, then a_{11} = _____.

B. True/False

In Problems 1–20, answer true of false.

1. The absolute value of any real number x is positive. _____
2. The inequality $|x| > -1$ has no solutions. _____
2. For any function f, if $f(a) = f(b)$, then $a = b$. _____
4. The graph of $y = f(x + c)$, $c > 0$, is the graph of $y = f(x)$ shifted c units to the right. _____
5. The points $(1, 3)$, $(3, 11)$, and $(5, 19)$ are collinear. _____
6. The function $f(x) = x^5 - 4x^3 + 2$ is an odd function. _____
7. $x + \frac{1}{4}$ is a factor of the function $f(x) = 64x^4 + 16x^3 + 48x^2 - 36x - 12$. _____

8. If $b^2 - 4ac < 0$, the graph of $f(x) = ax^2 + bx + c$, $a \neq 0$, does not cross the x-axis. _____

9. $f(x) = \dfrac{\sqrt{x}}{2x + 1}$ is a rational function. _____

10. If $f(x) = x^5 + 3x - 1$, then there exists a number c in $[-1, 1]$ such that $f(c) = 0$. _____

11. The graph of the function $f(x) = \dfrac{1}{x - 1} + \dfrac{1}{x - 2}$ has no x-intercepts. _____

12. $x = 0$ is a vertical asymptote for the graph of the rational function
$$f(x) = \frac{x^2 - 2x}{x}. \text{_____}$$

13. The graph of $y = \cos(x/6)$ is the graph of $y = \cos x$ stretched horizontally. _____

14. $f(x) = \csc x$ is not defined at $x = \pi/2$. _____

15. The function $f(x) = e^{-4x^2}$ is not one-to-one. _____

16. The exponential function $f(x) = \left(\frac{3}{2}\right)^x$ increases on the interval $(-\infty, \infty)$. _____

17. The domain of the function $f(x) = \ln x + \ln(x - 4)$ is $(4, \infty)$. _____

18. The solutions of the equation $\ln x^2 = \ln 3x$ are $x = 0$ and $x = 3$. _____

19. $\cos^2 x + \cos^2(x - \pi/2) = 1$ _____

20. $\ln|\csc x| + \ln|\sin x| = 0$ _____

C. Exercises _____

1. Match the given interval with the appropriate inequality.

    (*i*) $[2, 4]$ (*ii*) $[2, 4)$

    (*iii*) $(2, 4)$ (*iv*) $(2, 4]$

    (a) $|x - 3| \leq 1$ **(b)** $1 < x - 1 \leq 3$

    (c) $-2 < 2 - x \leq 0$ **(d)** $|x - 3| < 1$

2. Write the solution of the absolute-value inequality $|3x - 1| > 7$ using interval notation.

3. The answer to a problem given in the back of a mathematics text is $1 + \sqrt{3}$, but your answer is $2/(\sqrt{3} - 1)$. Are you correct?

4. In which quadrants in the Cartesian plane is the quotient x/y negative?

5. Which one of the following equations best describes a circle that passes through the origin? The symbols a, b, c, d, and e stand for different nonzero real constants.

    (a) $ax^2 + by^2 + cx + dy + e = 0$ **(b)** $ax^2 + ay^2 + cx + dy + e = 0$

    (c) $ax^2 + ay^2 + cx + dy = 0$ **(d)** $ax^2 + by^2 + cx + dy = 0$

    (e) $ax^2 + ay^2 + e = 0$ **(f)** $ax^2 + ay^2 + cx + e = 0$

6. Match the given rational function f with the most appropriate phrase.

    (*i*) $f(x) = \dfrac{x^4}{x^2 - 2}$ (*ii*) $f(x) = \dfrac{x^2}{x^2 + 2}$

    (*iii*) $f(x) = \dfrac{x^5}{x^2 + 2}$ (*iv*) $f(x) = \dfrac{x^3}{x^2 + 2}$

    (a) slant asymptote   **(b)** no asymptotes

    (c) horizontal asymptote   **(d)** vertical asymptote

7. What is the range of the rational function $f(x) = \dfrac{10}{x^2 + 1}$?

8. What is the domain of the function $f(x) = \dfrac{\sqrt{x+2}}{x^2}$?

9. Find an equation of the line that passes through the origin and through the point of intersection of the graphs of $x + y = 1$ and $2x - y = 7$.

10. Find a quadratic function f whose graph has the y-intercept $(0, -6)$ and the vertex of the graph is $(1, 4)$.

In calculus you are often required to rewrite a function either in a simpler form or in a form that is more helpful in solving the problem. In Problems 11–16, rewrite each function by following the given instruction. In calculus you would be expected to recognize what to do from the context of the actual problem.

11. $f(x) = \sqrt{x^6 + 4} - x^3$. Express f as a quotient using rationalization and simplification.

12. $f(x) = \dfrac{5x^3 - 4x^2\sqrt{x} + 8}{\sqrt[3]{x}}$. Carry out the indicated division and express each term as a power of x.

13. $f(x) = \dfrac{7x^2 - 7x - 6}{x^3 - x^2}$. Decompose f into partial fractions.

14. $f(x) = \dfrac{1}{1 + \sin x}$. Express f in terms of $\sec x$ and $\tan x$.

15. $f(x) = e^{3\ln x}$. Express f as a power of x.

16. $f(x) = |x^2 - 3x|$. Express f without absolute value signs.

In calculus you are often required to find zeros of a function. In Problems 17 and 18, solve the equation $f(x) = 0$ by following the given instruction.

17. $f(x) = x^2\frac{1}{2}(4 - x^2)^{-1/2}(-2x) + 2x\sqrt{4 - x^2}$. Rewrite f as a single expression without negative exponents.

18. $f(x) = 2\sin x\cos x - \sin x$. Find the zeros of f on the interval $[-\pi, \pi]$.

In Problems 19 and 20, compute and simplify the difference quotient $\dfrac{f(x + h) - f(x)}{h}$ for the given function.

19. $f(x) = \dfrac{3x}{2x + 5}$

20. $f(x) = -x^3 + 10x^2$

21. Consider the trigonometric function $y = -8\sin(\pi x/3)$. What is the amplitude of the function? Give an interval over which one cycle of the graph is completed.

22. If $\tan\theta = \sqrt{5}$ and $\pi < \theta < 3\pi/2$, then what is the value of $\cos\theta$?

23. Suppose $f(x) = \sin x$ and $f(c) = 0.7$. What is the value of

$$2f(-c) + f(c + 2\pi) + f(c - 6\pi)?$$

24. Suppose $f(x) = \sin x$ and $g(x) = \ln x$. Solve $(f \circ g)(x) = 0$.

25. Find the x- and y-intercepts of the parabola whose equation is

$$(y + 4)^2 = 4(x + 1).$$

26. Find the center, foci, vertices, and endpoints of the minor axis of the ellipse whose equation is

$$x^2 + 2y^2 + 2x - 20y + 49 = 0.$$

27. The slant asymptotes of a hyperbola are $y = -5x + 2$ and $y = 5x - 8$. What is the center of the hyperbola?

28. From a point 220 ft from the base of a cell-phone antenna a person measures a $30°$ angle of inclination from the ground to the top of the antenna. What is the angle of inclination to the top of the antenna if the person moves 100 ft closer to its base?

29. Iodine-131 is radioactive and is used in certain medical procedures. Assume that iodine-131 decays exponentially. If the half-life of I-131 is 8 days, then how much of a 5-gram sample remains at the end of 15 days?

30. The polar coordinate equation $r = 3\cos 4\theta$ is a rose curve with eight petals. Find all radian-measure angles satisfying $0 \le \theta \le 2\pi$ for which $|r| = 3$.

31. Give the three Pythagorean trigonometric identities.

32. Without the aid of a calculator, find the exact value of

$$\cos 80° \cos 50° + \sin 80° \sin 50°.$$

33. Give the point that is common to the graphs of all exponential functions $f(x) = b^x, b > 0, b \ne 1$.

34. Give the y-intercept, the x-intercept, and horizontal asymptote for the graph of $f(x) = 4^x - 3$.

35. Describe how the graph of $y = \ln(-x)$ can be obtained from the graph of $y = \ln x$.

36. Find the asymptotes of the hyperbola

$$-x^2 + 10x + 9y^2 - 54y + 47 = 0.$$

37. Sketch the graph of the given function.

(a) $f(x) = \sqrt{4 - x^2}$ (b) $f(x) = -\frac{1}{2}\sqrt{4 - x^2}$

(c) $f(x) = \sqrt{x^2 - 4}$ (d) $f(x) = \sqrt{x^2 + 4}$

38. Without doing any work, describe in detail the graph of

$$r = \frac{10}{3 + 2\sin(\theta + 3\pi/4)}.$$

39. Solve the linear system

$$\begin{cases} x - 2y + 3z = 1 \\ x + y - z = 5 \\ 4x - 5y + 8z = 8 \end{cases}$$

and interpret the solution geometrically.

40. Solve the equation $\begin{vmatrix} x & 0 & 4 \\ 0 & x & 0 \\ 4 & 0 & x \end{vmatrix} = 0$ for x.

41. Here is a nonlinear system of equations taken from a calculus text:

$$\begin{cases} 2x\lambda = -4 \\ 2y\lambda = 2y \\ x^2 + y^2 = 9. \end{cases}$$

Solve for x, y, and λ.

42. Graph the system of inequalities:

$$\begin{cases} y \le 2^x \\ 6y - 7x \ge 10 \\ x \ge 0. \end{cases}$$

In Problems 43–46, answer the given question about the sequence

$$128, 64, 32, 16, \ldots.$$

43. What is the eighth term of the sequence?
44. What is the sum S_8 of the first eight terms of the sequence?
45. Is the sequence convergent or divergent?
46. Does the infinite series

$$128 + 64 + 32 + 16 + \cdots$$

have a sum S?

In Problems 47–50, find the nth term a_n of the given sequence.

47. $-2, -1, 0, 1, \ldots$　　　　　**48.** $0, 3, 8, 15, \ldots$
49. $1000, -100, 10, -1, \ldots$　　**50.** $1, \frac{1}{7}, \frac{1}{13}, \frac{1}{19}, \ldots$

51. If d is a digit (any numeral 0 through 9), find a rational number whose decimal representation is $0.ddd\ldots.$

52. How many ten-digit telephone numbers are possible within a given three-digit area code if the last seven digits of the telephone number cannot start with 0 or 1?

Appendix A
Complex Numbers

A.1   Arithmetic Operations and Properties

≡ Introduction No one person "invented" complex numbers, but controversies surrounding the use of these numbers existed in the sixteenth century. In their quest to solve polynomial equations by formulas involving radicals early dabblers in mathematics were forced to admit that there were other kinds of numbers besides positive integers. Equations such as $x^2 - 2x + 2 = 0$ and $x^3 + 6x^2 + 11x = 0$ that yielded "solutions" $1 + \sqrt{-1}$ and $-3 - \sqrt{-2}$ caused particular consternation within the community of fledgling mathematical scholars because *everyone* knew that there are no numbers such as $\sqrt{-1}$ and $\sqrt{-2}$, numbers whose square is negative. Such "numbers" exist only in one's imagination, or as one philosopher opined "the imaginary, the bosom child of complex mysticism." Over time these "imaginary numbers" did not go away, mainly because mathematicians as a group are tenacious and some are even practical. A famous mathematician held that even though "they exist in our imagination . . . nothing prevents us from . . . employing them in calculations." Mathematicians also hate to throw anything away. After all, a collective memory still lingered that negative numbers at first were branded "fictitious." The concept of *number* evolved over centuries; gradually the set of numbers grew from just positive integers to include rational numbers, negative numbers, and irrational numbers. But in the eighteenth century the number concept took a gigantic evolutionary step forward when the German mathematician Carl Friedrich Gauss put the so-called *imaginary numbers*, or as they were now beginning to be called *complex numbers*, on a logical and consistent footing by treating them as an extension of the real number system.

Our goal in this first appendix section is to examine some basic definitions and the arithmetic of complex numbers.

☐ The Imaginary Unit Even after gaining wide respectability, through the seminal works of Carl Friedrich Gauss and the French mathematician **Augustin Louis Cauchy** (1789–1857), the unfortunate name "imaginary" has survived down the centuries. The symbol i was originally used as a disguise for the embarrassing symbol $\sqrt{-1}$. We now say that i is the **imaginary unit** and define it by the property $i^2 = -1$. Nevertheless, it is still common practice to write $i = \sqrt{-1}$. Indeed, using the last symbol we are able to define the **principal square of a negative number** as follows.

DEFINITION A.1.1   Principal Square Root

If c is a positive real number, then the **principal square root** of $-c$ is defined by
$$\sqrt{-c} = \sqrt{c(-1)} = \sqrt{c}\sqrt{-1} = \sqrt{c}\,i. \qquad (1)$$

EXAMPLE 1   Principal Square Roots

Find the principal square root of **(a)** $\sqrt{-4}$ and **(b)** $\sqrt{-5}$.

Solution From (1) of Definition A.1.1,

(a) $\sqrt{-4} = \sqrt{(-1)(4)} = \sqrt{-1}\sqrt{4} = i(2) = 2i$
(b) $\sqrt{-5} = \sqrt{(-1)(5)} = \sqrt{-1}\sqrt{5} = i\sqrt{5} = \sqrt{5}i$.   ≡

□ **Terminology** The complex number system contains the imaginary unit i, all real numbers, products such as bi, b real, and sums such as $a + bi$, where a and b are real numbers. In particular, a **complex number** is defined to be any expression of the form

$$z = a + bi, \tag{2}$$

where a and b are real numbers and $i^2 = -1$. The form given in (2) is called the **standard form** of a complex number. The numbers a and b are called the **real** and **imaginary parts** of z, respectively. A complex number of the form $0 + bi$ is said to be a **pure imaginary number**. Note that by choosing $b = 0$ in (2), we obtain a **real number**. Thus the set R of real numbers is a subset of the set C of complex numbers.

◄ Be careful here, the imaginary part of $a + bi$, is *not bi*; it is the real number b.

EXAMPLE 2   Real and Imaginary Parts

(a) The complex number $z = 4 + (-5)i$ is written as $z = 4 - 5i$. The real part of z is 4 and its imaginary part is -5.
(b) $z = 10i$ is a pure imaginary number.
(c) $z = 6 + 0i = 6$ is a real number.   ≡

EXAMPLE 3   Writing in the Standard Form $a + bi$

Express each of the following in the standard form $a + bi$.

(a) $-3 + \sqrt{-7}$ **(b)** $2 - \sqrt{-25}$

Solution Using (1) of Definition A.1.1, we can write

(a) $-3 + \sqrt{-7} = -3 + i\sqrt{7} = -3 + \sqrt{7}i$,
(b) $2 - \sqrt{-25} = 2 - i\sqrt{25} = 2 - 5i$.   ≡

In order to solve certain equations involving complex numbers, it is necessary to specify when two complex numbers are equal.

DEFINITION A.1.2   Equality of Complex Numbers

Two complex numbers are equal if and only if their real parts are equal and imaginary parts are equal. That is, if $z_1 = a + bi$ and $z_2 = c + di$,

$$z_1 = z_2 \text{ if and only if } a = c \text{ and } b = d.$$

EXAMPLE 4   A Simple Equation

Solve for x and y:

$$(2x + 1) + (-2y + 3)i = 2 - 4i.$$

Solution By Definition A.1.2 we must have

$$2x + 1 = 2 \quad \text{and} \quad -2y + 3 = -4.$$

Solving each equation yields $x = \frac{1}{2}$ and $y = \frac{7}{2}$. $\equiv$

Addition and multiplication for complex numbers are defined as follows.

DEFINITION A.1.3 Sum, Difference, and Product

If $z_1 = a + bi$ and $z_2 = c + di$, then

(i) their **sum** is given by $\qquad\qquad z_1 + z_2 = (a + c) + (b + d)i$
(ii) their **difference** is given by $\qquad z_1 - z_2 = (a - c) + (b - d)i$
(iii) and their **product** is given by $\qquad z_1 z_2 = (ac - bd) + (bc + ad)i$

□ **Properties of Complex Numbers** Using the definition of addition and multiplication of complex numbers, it can be shown that many of the basic properties of the real number system also apply to the complex number system. In particular, the associative, commutative, and distributive laws hold for complex numbers. We further observe that in Definition A.1.3(*i*):

- *The **sum** of two complex numbers is obtained by adding their corresponding real and imaginary parts.*

Similarly, Definition A.1.3(*ii*) shows that

- *The **difference** of two complex numbers is obtained by subtracting their corresponding real and imaginary parts.*

Also, rather than memorizing (*iii*) of Definition A.1.3:

- *The **product** of two complex numbers can be obtained by using the associative, commutative, and distributive laws and the fact that $i^2 = -1$.*

Applying this approach, we find that

$$
\begin{aligned}
(a + bi)(c + di) &= (a + bi)c + (a + bi)di && \leftarrow \text{distributive law} \\
&= ac + (bc)i + (ad)i + (bd)i^2 && \leftarrow \text{distributive law} \\
&= ac + (bc)i + (ad)i + (bd)(-1) \\
&= ac + (bd)(-1) + (bc)i + (ad)i && \leftarrow \text{factor out } i \\
&= (ac - bd) + (bc + ad)i.
\end{aligned}
$$

This is the same result as the product given by Definition A.1.3(*iii*). These techniques are illustrated in the following example.

EXAMPLE 5   **Sum, Difference, and Product**

If $z_1 = 5 - 6i$ and $z_2 = 2 + 4i$, find **(a)** $z_1 + z_2$, **(b)** $z_1 - z_2$, and **(c)** $z_1 z_2$.

Solution (a) The colors in the diagram below show how to add z_1 and z_2:

add the real parts

$$z_1 + z_2 = (5 - 6i) + (2 + 4i) = (5 + 2) + (-6 + 4)i = 7 - 2i.$$

add the imaginary parts

(b) Analogous to part (a) we now subtract the real and imaginary parts:

$$z_1 - z_2 = (5 - 6i) - (2 + 4i) = (5 - 2) + (-6 - 4)i = 3 - 10i.$$

(c) Using the distributive law, we write the product $(5 - 6i)(2 + 4i)$ as

$$(5 - 6i)(2 + 4i) = (5 - 6i)2 + (5 - 6i)4i \leftarrow \text{distributive law}$$
$$= 10 - 12i + 20i - 24i^2 \leftarrow \begin{array}{l}\text{factor } i \text{ from the two middle} \\ \text{terms and replace } i^2 \text{ by } -1\end{array}$$
$$= 10 - 24(-1) + (-12 + 20)i$$
$$= 34 + 8i. \qquad \equiv$$

Note of Caution ▶ Not all the properties of the real number system hold for complex numbers. In particular, the property of radicals $\sqrt{a}\sqrt{b} = \sqrt{ab}$ is *not* true when both a and b are negative. To see this, consider that

$$\sqrt{-1}\sqrt{-1} = ii = i^2 = -1 \qquad \text{whereas} \qquad \sqrt{(-1)(-1)} = \sqrt{1} = 1.$$

Thus, $\sqrt{-1}\sqrt{-1} \neq \sqrt{(-1)(-1)}$. However, if *only one of a or b* is negative, then we do have $\sqrt{a}\sqrt{b} = \sqrt{ab}$.

In the set C of complex numbers, the **additive identity** is the number $0 = 0 + 0i$, and the **multiplicative identity** is the number $1 = 1 + 0i$. The number $-z = -a - bi$ is called the **additive inverse of** $z = a + bi$ because

$$z + (-z) = z - z = (a - a) + (b - b)i = 0 + 0i = 0.$$

In order to obtain the **multiplicative inverse** of a nonzero complex number $z = a + bi$, we introduce the concept of the **conjugate** of a complex number.

DEFINITION A.1.4   Conjugate

If $z = a + bi$ is a complex number, then the number $\bar{z} = a - bi$ is called the **conjugate** of z.

In other words, the conjugate of a complex number $z = a + bi$ is the complex number obtained by changing the sign of its imaginary part. For example, the conjugate of $8 + 13i$ is $8 - 13i$, and the conjugate of $-5 - 2i$ is $-5 + 2i$.

The following computations show that both the sum and the product of a complex number z and its conjugate $\bar{z}$ are *real* numbers:

$$z + \bar{z} = (a + bi) + (a - bi) = 2a \qquad (3)$$
$$z\bar{z} = (a + bi)(a - bi) = a^2 - b^2i^2 = a^2 + b^2. \qquad (4)$$

The latter property makes conjugates very useful in finding the multiplicative inverse $1/z, z \neq 0$, and in dividing two complex numbers.

We summarize the procedure.

- To **divide** a complex number z_1 by a complex number z_2, multiply the numerator and denominator of z_1/z_2 by the conjugate of the denominator z_2. That is,

$$\frac{z_1}{z_2} = \frac{z_1}{z_2} \cdot \frac{\bar{z}_2}{\bar{z}_2} = \frac{z_1\bar{z}_2}{z_2\bar{z}_2}$$

and then use the fact that the product $z_2\bar{z}_2$ is the sum of the squares of the real and imaginary parts of z_2.

▮ EXAMPLE 6   Division

For $z_1 = 3 - 2i$ and $z_2 = 4 + 5i$, express the given complex number in the form $a + bi$.

(a) $\dfrac{1}{z_1}$ **(b)** $\dfrac{z_1}{z_2}$

Solution In each case, we multiply both the numerator and the denominator by the conjugate of the denominator and simplify.

(a) $\dfrac{1}{z_1} = \dfrac{1}{3-2i} = \dfrac{1}{3-2i} \cdot \dfrac{3+2i}{3+2i} = \dfrac{3+2i}{3^2+(-2)^2} = \dfrac{3}{13} + \dfrac{2}{13}i$

conjugate of z_1 ↓ ; from (4) ; standard form $a + bi$

(b) $\dfrac{z_1}{z_2} = \dfrac{3-2i}{4+5i} = \dfrac{3-2i}{4+5i} \cdot \dfrac{4-5i}{4-5i} = \dfrac{12-8i-15i+10i^2}{4^2+5^2}$

$\qquad = \dfrac{2-23i}{41} = \dfrac{2}{41} - \dfrac{23}{41}i \leftarrow$ standard form $a + bi$

≡

From the definition of addition and subtraction of two complex numbers, it is readily shown that the conjugate of a sum and difference of two complex numbers is the sum and difference of the conjugates. This property, along with three other properties of the conjugate are summarized as a theorem.

THEOREM A.1.1   Properties of the Conjugate

Let z_1 and z_2 be any two complex numbers. Then

(i) $\overline{z_1 \pm z_2} = \bar{z}_1 \pm \bar{z}_2$ (ii) $\overline{z_1 z_2} = \bar{z}_1 \bar{z}_2$

(iii) $\overline{\left(\dfrac{z_1}{z_2}\right)} = \dfrac{\bar{z}_1}{\bar{z}_2}, z_2 \neq 0$ (iv) $\bar{\bar{z}} = z$

Of course, the conjugate of any finite sum (product) of complex numbers is the sum (product) of the conjugates.

☐ **Quadratic Equations** Complex numbers make it possible to solve quadratic equations $ax^2 + bx + c = 0$ when the discriminant $b^2 - 4ac$ is negative. We now see that the solutions from the quadratic formula

$$x_1 = \dfrac{-b - \sqrt{b^2 - 4ac}}{2a} \quad \text{and} \quad x_2 = \dfrac{-b + \sqrt{b^2 - 4ac}}{2a} \tag{5}$$

represent complex numbers. Note that in fact the solutions are conjugates of each other. As the next example shows these solutions can be written in standard form.

EXAMPLE 7   **Complex Solutions**

Solve $x^2 - 8x + 25 = 0$.

Solution From the quadratic formula, we obtain

$$x = \dfrac{-(-8) \pm \sqrt{(-8)^2 - 4(1)(25)}}{2(1)} = \dfrac{8 \pm \sqrt{-36}}{2}.$$

Using $\sqrt{-36} = 6i$ we obtain

$$x = \dfrac{8 \pm 6i}{2} = 4 \pm 3i.$$

Thus, the solution set of the equation is $\{4 - 3i, 4 + 3i\}$.

≡

□ **Conjugate Solutions** As we already know from Theorem 3.3.4 on page 151, if a polynomial function $f(x)$ has real coefficients, then complex roots of the polynomial equation $f(x) = 0$ appear in conjugate pairs. Observe in Example 7 that if $x_1 = 4 - 3i$ and $x_2 = 4 + 3i$, then $\bar{x}_2 = x_1$. Moreover, it is easily seen that $\bar{x}_1 = x_2$.

A.1 | Exercises | Answers to selected odd-numbered problems begin on page ANS-31.

In Problems 1–10, find the indicated power of i.

1. i^3 **2.** i^4 **3.** i^5 **4.** i^6 **5.** i^7

6. i^8 **7.** i^{-1} **8.** i^{-2} **9.** i^{-3} **10.** i^{-6}

In Problems 11–56, perform the indicated operation. Write the answer in standard form $a + bi$.

11. $\sqrt{-100}$ **12.** $-\sqrt{-8}$

13. $-3 - \sqrt{-3}$ **14.** $\sqrt{-5} - \sqrt{-125} + 5$

15. $(3 + i) - (4 - 3i)$ **16.** $(5 + 6i) + (-7 + 2i)$

17. $2(4 - 5i) + 3(-2 - i)$ **18.** $-2(6 + 4i) + 5(4 - 8i)$

19. $i(-10 + 9i) - 5i$ **20.** $i(4 + 13i) - i(1 - 9i)$

21. $3i(1 + i) - 4(2 - i)$ **22.** $i + i(1 - 2i) + i(4 + 3i)$

23. $(3 - 2i)(1 - i)$ **24.** $(4 + 6i)(-3 + 4i)$

25. $(7 + 14i)(2 + i)$ **26.** $\left(-5 - \sqrt{3}i\right)\left(2 - \sqrt{3}i\right)$

27. $(4 + 5i) - (2 - i)(1 + i)$ **28.** $(-3 + 6i) + (2 + 4i)(-3 + 2i)$

29. $i(1 - 2i)(2 + 5i)$ **30.** $i\left(\sqrt{2} - i\right)\left(1 - \sqrt{2}i\right)$

31. $(1 + i)(1 + 2i)(1 + 3i)$ **32.** $(2 + i)(2 - i)(4 - 2i)$

33. $(1 - i)[2(2 - i) - 5(1 + 3i)]$ **34.** $(4 + i)[i(1 + 3i) - 2(-5 + 3i)]$

35. $(4 + i)^2$ **36.** $(3 - 5i)^2$

37. $(1 - i)^2(1 + i)^2$ **38.** $(2 + i)^2(3 + 2i)^2$

39. $\dfrac{1}{4 - 3i}$ **40.** $\dfrac{5}{3 + i}$ **41.** $\dfrac{4}{5 + 4i}$

42. $\dfrac{1}{-1 + 2i}$ **43.** $\dfrac{i}{1 + i}$ **44.** $\dfrac{i}{4 - i}$

45. $\dfrac{4 + 6i}{i}$ **46.** $\dfrac{3 - 5i}{i}$ **47.** $\dfrac{1 + i}{1 - i}$

48. $\dfrac{2 - 3i}{1 + 2i}$ **49.** $\dfrac{4 + 2i}{2 - 7i}$ **50.** $\dfrac{\frac{1}{2} - \frac{7}{2}i}{4 + 2i}$

51. $i\left(\dfrac{10 - i}{1 + i}\right)$ **52.** $i\left(\dfrac{1 - 2\sqrt{3}i}{1 + \sqrt{3}i}\right)$ **53.** $(1 + i)\dfrac{2i}{1 - 5i}$

54. $(5 - 3i)\dfrac{1 - i}{2 - i}$ **55.** $4 - 9i + \dfrac{25i}{2 + i}$ **56.** $i\left(-6 + \dfrac{11}{5}i\right) + \dfrac{2 + i}{2 - i}$

In Problems 57–64, use Definition A.1.2 to solve for x and y.

57. $2(x + yi) = i(3 - 4i)$ **58.** $(x + yi) + 4(1 - i) = 5 - 7i$

59. $i(x + yi) = (1 - 6i)(2 + 3i)$ **60.** $10 + 6yi = 5x + 24i$

61. $(1 + i)(x - yi) = i(14 + 7i) - (2 + 13i)$

62. $i^2(1 - i)(1 + i) = 3x + yi + i(y + xi)$

63. $x + yi = \dfrac{i^3}{2 - i}$ **64.** $25 - 49i = x^2 - y^2i$

In Problems 65–76, solve the given equation.

65. $x^2 + 9 = 0$

66. $x^2 + 8 = 0$

67. $2x^2 = -5$

68. $3x^2 = -1$

69. $2x^2 - x + 1 = 0$

70. $x^2 - 2x + 10 = 0$

71. $x^2 + 8x + 52 = 0$

72. $3x^2 + 2x + 5 = 0$

73. $4x^2 - x + 2 = 0$

74. $x^2 + x + 2 = 0$

75. $x^4 + 3x^2 + 2 = 0$

76. $2x^4 + 9x^2 + 4 = 0$

77. The two square roots of the complex number i are the two numbers z_1 and z_2 that are solutions of the equation $z^2 = i$. Let $z = x + iy$ and find z^2. Then use Definition A.1.2 to find z_1 and z_2.

78. Proceed as in Problem 77 to find two numbers z_1 and z_2 that satisfy the equation $z^2 = -3 + 4i$.

For Discussion

In Problems 79–82, prove the given properties involving the conjugates of $z_1 = a + bi$ and $z_2 = c + di$.

79. $\bar{z}_1 = z_1$ if and only if z_1 is a real number.

80. $\overline{z_1 + z_2} = \bar{z}_1 + \bar{z}_2$ **81.** $\overline{z_1 \cdot z_2} = \bar{z}_1 \cdot \bar{z}_2$ **82.** $\overline{z_1^2} = (\bar{z}_1)^2$

A.2 Trigonometric Form of Complex Numbers

≡ **Introduction** A complex number $z = a + bi$ is uniquely determined by an *ordered pair* of real numbers (a, b). The first and second entries of the ordered pairs correspond, in turn, with the real and imaginary parts of the complex number. For example, the ordered pair $(2, -3)$ corresponds to the complex number $z = 2 - 3i$. Conversely, the complex number $z = 2 - 3i$ determines the ordered pair $(2, -3)$. The numbers 10, i, and $-5i$ are equivalent to $(10, 0)$, $(0, 1)$, and $(0, -5)$, respectively. In this manner we are able to associate a complex number $z = a + bi$ with a *point* (a, b) in a rectangular coordinate system.

□ **Complex Plane** Because of the correspondence between a complex number $z = a + bi$ and one and only one point $P(a, b)$ in a coordinate plane we shall use the terms *complex number* and *point* interchangeably. The coordinate plane illustrated in FIGURE A.2.1 is called the **complex plane** or simply the **z-plane**. The horizontal or x-axis is called the **real axis** because each point on that axis represents a real number. The vertical or y-axis is called the **imaginary axis** because a point on that axis represents a pure imaginary number.

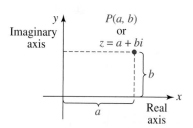

FIGURE A.2.1 Complex plane

| EXAMPLE 1 | Graphing Complex Numbers |

Graph the complex numbers

$$z_1 = 5 + 4i, \quad z_2 = -2i, \quad z_3 = -2 - 3i, \quad \text{and} \quad z_4 = -4 + 2i.$$

Solution We identify the complex numbers z_1, z_2, z_3, z_4 with the points $(5, 4)$, $(0, -2)$, $(-2, -3)$, $(-4, 2)$, respectively. These points are, in turn, the red, blue, green, and orange dots in FIGURE A.2.2.   ≡

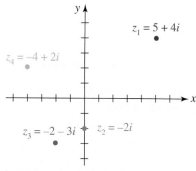

FIGURE A.2.2 The complex numbers in Example 1 interpreted as points

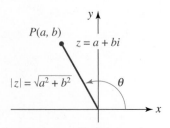

FIGURE A.2.3 Modulus and argument of a complex number z

◻ Trigonometric Form If $z = a + bi$ is a nonzero complex number and $P(a, b)$ is its geometric representation, as shown in FIGURE A.2.3, then the distance from P to the origin is given by $\sqrt{a^2 + b^2}$. This distance is called the **modulus**, **magnitude**, or **absolute value** of z and is denoted by $|z|$,

$$|z| = \sqrt{a^2 + b^2}. \tag{1}$$

For example, if $z = i$, then $|i| = \sqrt{0^2 + 1^2} = \sqrt{1^2} = 1$; if $z = 3 - 4i$, then $|3 - 4i| = \sqrt{3^2 + (-4)^2} = \sqrt{25} = 5$. From (4) of Section A.1, we know that if $\bar{z} = a - bi$ is the conjugate of $z = a + bi$, then $z\bar{z} = a^2 + b^2$. Hence (1) can also be written as

$$|z| = \sqrt{z\bar{z}}.$$

If we let θ be the angle in standard position whose terminal side passes through $P(a, b)$ and $r = |z|$, then $\cos\theta = a/r$ and $\sin\theta = b/r$, from which we obtain $a = r\cos\theta$ and $b = r\sin\theta$. Substituting these expressions for a and b in $z = a + bi$, we obtain $z = a + bi = (r\cos\theta) + (r\sin\theta)i$ or

$$z = r(\cos\theta + i\sin\theta). \tag{2}$$

We say that (2) is the **trigonometric form**, or **polar form**, of the complex number z. The angle θ is called the **argument** of z and satisfies $\tan\theta = b/a$. However, θ is not necessarily $\arctan(b/a)$ since θ is not restricted to the interval $(-\pi/2, \pi/2)$. See Examples 2 and 3 that follow. Also, the argument θ is *not uniquely determined*, since $\cos\theta = \cos(\theta + 2k\pi)$ and $\sin\theta = \sin(\theta + 2k\pi)$ for any integer k. If $z = a + bi = 0$, then $a = b = 0$. In this case, $r = 0$ and we can take any angle θ as the argument.

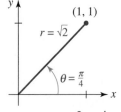

FIGURE A.2.4 Complex number in part (a) of Example 2

▮ EXAMPLE 2   Trigonometric Form

Write the complex numbers in trigonometric form: **(a)** $1 + i$ **(b)** $1 - \sqrt{3}i$.

Solution (a) If we identify $a = 1$ and $b = 1$, then the modulus of $1 + i$ is

$$r = |1 + i| = \sqrt{(1)^2 + (1)^2} = \sqrt{2}.$$

Because $\tan\theta = b/a = 1$ and the point $(1, 1)$ lies in the first quadrant, we can take the argument of the complex number to be $\theta = \pi/4$, as shown in FIGURE A.2.4. Thus,

$$z = \sqrt{2}\left(\cos\frac{\pi}{4} + i\sin\frac{\pi}{4}\right).$$

(b) In this case,

$$r = |1 - \sqrt{3}i| = \sqrt{1^2 + (-\sqrt{3})^2} = \sqrt{4} = 2.$$

From $\tan\theta = -\sqrt{3}/1 = -\sqrt{3}$ and the fact that $(1, -\sqrt{3})$ lies in the fourth quadrant, we take $\theta = \tan^{-1}(-\sqrt{3}) = -\pi/3$, as shown in FIGURE A.2.5. Thus,

$$z = 2\left[\cos\left(-\frac{\pi}{3}\right) + i\sin\left(-\frac{\pi}{3}\right)\right]. \qquad \equiv$$

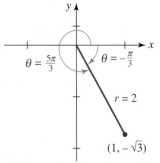

FIGURE A.2.5 Complex number in part (b) of Example 2

Note ▶

Following convention, in the remainder of the discussion as well as in Exercises A.2 we will take the argument θ of a complex number z either as an angle in measured radians in the interval $[0, 2\pi)$ or an angle measured in degrees that satisfies $0 \le \theta < 360°$. For example, the answer in part (b) of Example 2 can be written in the alternative form

$$z = 2\left(\cos\frac{5\pi}{3} + i\sin\frac{5\pi}{3}\right).$$

The argument of $1 - \sqrt{3}i$ that lies in the interval $[0, 2\pi)$ is $\theta = 5\pi/3$, as is shown in Figure A.2.5.

▮ EXAMPLE 3　　　　**Trigonometric Form**

Express the complex number

$$z = 2\sqrt{2}\left(\cos\frac{7\pi}{4} + i\sin\frac{7\pi}{4}\right)$$

in the standard form $z = a + bi$.

Solution　By using the reference angle concept discussed in Section 4.2, we find $\cos(7\pi/4) = \sqrt{2}/2$ and $\sin(7\pi/4) = -\sqrt{2}/2$. Therefore,

$$z = 2\sqrt{2}\left(\cos\frac{7\pi}{4} + i\sin\frac{7\pi}{4}\right) = 2\sqrt{2}\left(\frac{\sqrt{2}}{2} - i\frac{\sqrt{2}}{2}\right)$$

or $z = 2 - 2i$.　　　　　　　　　　　　　　　　　　　　　　　≡

▮ EXAMPLE 4　　　　**Trigonometric Form**

Find the trigonometric form of $z = -4 + 5i$.

Solution　The modulus of $z = -4 + 5i$ is

$$r = |-4 + 5i| = \sqrt{16 + 25} = \sqrt{41}.$$

Because the point $(-4, 5)$ lies in the second quadrant, we must take care to adjust the value of the angle obtained from $\tan\theta = -\frac{5}{4}$ and a calculator so that our final answer is a quadrant II angle. See FIGURE A.2.6. One approach is to use a calculator set in radian mode to obtain the reference angle $\theta' = \tan^{-1}\frac{5}{4} \approx 0.8961$ radian. The desired second-quadrant angle is then $\theta = \pi - \theta' \approx 2.2455$. Thus,

$$z \approx \sqrt{41}(\cos 2.2455 + i\sin 2.2455).$$

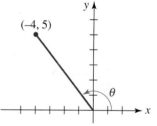

FIGURE A.2.6 Complex number in Example 4

Alternatively the foregoing trigonometric form can be written using a degree-measured angle. With the calculator set in degree mode, we would obtain $\theta' \approx 51.34°$ and $\theta = 180° - \theta' \approx 128.66°$, from which it follows that

$$z \approx \sqrt{41}(\cos 128.66° + i\sin 128.66°).$$　　　≡

▮ EXAMPLE 5　　　　**Modulus and Argument of a Product**

Find the modulus and the argument of $z_1 z_2$, where $z_1 = 2i$ and $z_2 = 1 + i$.

Solution　The product is

$$z_1 z_2 = 2i(1 + i) = -2 + 2i,$$

and hence the modulus is

$$r = |z_1 z_2| = |-2 + 2i| = \sqrt{8} = 2\sqrt{2}.$$

By identifying $a = -2$ and $b = 2$, we have $\tan\theta = -1$. Since θ is a second quadrant angle, we conclude that the argument of $z_1 z_2$ is $\theta = 3\pi/4$. See FIGURE A.2.7.　≡

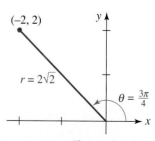

FIGURE A.2.7 The product in Example 5

☐ **Multiplication and Division** In Example 5 notice that the modulus $r = 2\sqrt{2}$ of the product $z_1 z_2$ is the *product* of the modulus $r_1 = 2$ of z_1 and the modulus $r_2 = \sqrt{2}$ of z_2. Also, the argument $\theta = 3\pi/4$ of $z_1 z_2$ is the *sum* of the arguments $\theta_1 = \pi/2$ and $\theta_2 = \pi/4$ of z_1 and z_2, respectively. We have illustrated a particular case of the following theorem, which describes how to multiply and divide complex numbers when they are written in trigonometric form.

THEOREM A.2.1 Product and Quotient

If $z_1 = r_1(\cos\theta_1 + i\sin\theta_1)$ and $z_2 = r_2(\cos\theta_2 + i\sin\theta_2)$, then

$$z_1 z_2 = r_1 r_2[\cos(\theta_1 + \theta_2) + i\sin(\theta_1 + \theta_2)] \qquad (3)$$

$$\frac{z_1}{z_2} = \frac{r_1}{r_2}[\cos(\theta_1 - \theta_2) + i\sin(\theta_1 - \theta_2)], \quad r_2 \neq 0 \qquad (4)$$

PROOF: We will prove only (4) of Theorem A.2.1; the proof of (3) is very similar. If we multiply the numerator and the denominator of

$$\frac{z_1}{z_2} = \frac{r_1(\cos\theta_1 + i\sin\theta_1)}{r_2(\cos\theta_2 + i\sin\theta_2)}$$

by $\cos\theta_2 - i\sin\theta_2$, we obtain

$$\frac{z_1}{z_2} = \frac{r_1}{r_2} \frac{(\cos\theta_1 + i\sin\theta_1)(\cos\theta_2 - i\sin\theta_2)}{\cos^2\theta_2 + \sin^2\theta_2} \qquad \leftarrow \text{ denominator equals 1}$$

$$= \frac{r_1}{r_2}(\cos\theta_1 + i\sin\theta_1)(\cos\theta_2 - i\sin\theta_2).$$

Performing the multiplication and then using the difference formulas from Section 4.6, we have

$$\frac{z_1}{z_2} = \frac{r_1}{r_2}[\overbrace{(\cos\theta_1\cos\theta_2 + \sin\theta_1\sin\theta_2)}^{\text{see (2) of Theorem 4.6.1}} + i\overbrace{(\sin\theta_1\cos\theta_2 - \cos\theta_1\sin\theta_2)}^{\text{see (5) of Theorem 4.6.2}}]$$

$$= \frac{r_1}{r_2}[\cos(\theta_1 - \theta_2) + i\sin(\theta_1 - \theta_2)]. \qquad \equiv$$

▮ **EXAMPLE 6**   **Product and Quotient**

If $z_1 = 4(\cos 75° + i\sin 75°)$ and $z_2 = \frac{1}{2}(\cos 45° + i\sin 45°)$, find **(a)** $z_1 z_2$ **(b)** z_1/z_2. Express each answer in the standard form $a + bi$.

Solution (a) From (3) of Theorem A.2.1 we can write the product as

$$z_1 z_2 = 4 \cdot \overbrace{\frac{1}{2}}^{\substack{\text{multiply} \\ \text{moduli}}}[\cos\overbrace{(75° + 45°)}^{\substack{\text{add} \\ \text{arguments}}} + i\sin(75° + 45°)$$

$$= 2[\cos 120° + i\sin 120°] = 2\left[-\frac{1}{2} + \frac{\sqrt{3}}{2}i\right]$$

and so $z_1 z_2 = -1 + \sqrt{3}i$.

(b) Now from (4) of Theorem A.2.1 the quotient is

$$\frac{z_1}{z_2} = \overset{\substack{\text{divide}\\\text{moduli}}}{\frac{4}{\frac{1}{2}}} \overset{\substack{\text{subtract}\\\text{arguments}}}{[\cos(75° - 45°) + i\sin(75° - 45°)]}$$

$$= 8[\cos 30° + i\sin 30°] = 8\left[\frac{\sqrt{3}}{2} + \frac{1}{2}i\right]$$

or $z_1/z_2 = 4\sqrt{3} + 4i$.   ≡

A.2 | **Exercises** Answers to selected odd-numbered problems begin on page ANS-31.

In Problems 1–10, graph the given complex number(s) and evaluate and graph the indicated complex number.

1. $z_1 = 2 + 5i$; $\bar{z}_1$
2. $z_1 = -8 - 4i$; $\frac{1}{4}\bar{z}_1$
3. $z_1 = 1 + i, z_2 = 2 - 2i$; $z_1 + z_2$
4. $z_1 = 4i, z_2 = -4 + i$; $z_1 - z_2$
5. $z_1 = 6 - 3i, z_2 = -i$; $\bar{z}_1 + z_2$
6. $z_1 = 5 + 2i, z_2 = -1 + 2i$; $z_1 + \bar{z}_2$
7. $z_1 = -2i, z_2 = 1 - i$; $z_1 z_2$
8. $z_1 = 1 + i, z_2 = 2 - i$; $z_1 z_2$
9. $z_1 = 2\sqrt{3} + 2i, z_2 = 1 - \sqrt{3}i$; $\dfrac{z_1}{z_2}$
10. $z_1 = i, z_2 = 1 - i$; $\dfrac{z_1}{z_2}$

In Problems 11–22, find the modulus and an argument of the given complex number.

11. $z = \dfrac{1}{2} - \dfrac{\sqrt{3}}{2}i$
12. $z = 4 + 3i$
13. $z = \sqrt{2} - 4i$
14. $z = -5 + 2i$
15. $z = \dfrac{3}{4} - \dfrac{1}{4}i$
16. $z = -8 - 2i$
17. $z = 3 + 3i$
18. $z = -1 - i$
19. $z = \sqrt{3} + i$
20. $z = 2 - 2\sqrt{3}i$
21. $z = 2 - i$
22. $z = 4 + 8i$

In Problems 23–32, write the given complex number in trigonometric form.

23. $z = -4i$
24. $z = 15i$
25. $z = 5\sqrt{3} + 5i$
26. $z = 3 + i$
27. $z = -2 + 5i$
28. $z = 2 + 2\sqrt{3}i$
29. $z = 3 - 5i$
30. $z = -10 + 6i$
31. $z = -2 - 2i$
32. $z = 1 - i$

In Problems 33–42, write the given complex number in the standard form $z = a + bi$. Do not use a calculator.

33. $z = \sqrt{2}\left(\cos\dfrac{\pi}{4} + i\sin\dfrac{\pi}{4}\right)$
34. $z = 6\left(\cos\dfrac{7\pi}{4} + i\sin\dfrac{7\pi}{4}\right)$
35. $z = 10(\cos 210° + i\sin 210°)$
36. $z = \sqrt{5}(\cos 420° + i\sin 420°)$
37. $z = 2(\cos 30° + i\sin 30°)$
38. $z = 7\left(\cos\dfrac{7\pi}{12} + i\sin\dfrac{7\pi}{12}\right)$

39. $z = \cos\dfrac{5\pi}{3} + i\sin\dfrac{5\pi}{3}$ **40.** $z = \dfrac{3}{2}\left(\cos\dfrac{5\pi}{4} + i\sin\dfrac{5\pi}{4}\right)$

41. $z = 4[\cos(\tan^{-1}2) + i\sin(\tan^{-1}2)]$ **42.** $z = 20\left[\cos\left(\tan^{-1}\dfrac{3}{5}\right) + i\sin\left(\tan^{-1}\dfrac{3}{5}\right)\right]$

In Problems 43–48, find $z_1 z_2$ and z_1/z_2 in trigonometric form by first writing z_1 and z_2 in trigonometric form.

43. $z_1 = 3i, z_2 = 6 + 6i$ **44.** $z_1 = 1 + i, z_2 = -1 + i$

45. $z_1 = 1 + \sqrt{3}i, z_2 = 2\sqrt{3} + 2i$ **46.** $z_1 = 5i, z_2 = -10i$

47. $z_1 = \sqrt{3} + i, z_2 = 5 - 5i$

48. $z_1 = -\sqrt{2} + \sqrt{2}i, z_2 = \dfrac{5\sqrt{2}}{2} + \dfrac{5\sqrt{2}}{2}i$

In Problems 49–52, find $z_1 z_2$ and z_1/z_2. Write the answer in the standard form $z = a + bi$.

49. $z_1 = \sqrt{6}\left(\cos\dfrac{\pi}{3} + i\sin\dfrac{\pi}{3}\right), z_2 = \sqrt{2}\left(\cos\dfrac{7\pi}{4} + i\sin\dfrac{7\pi}{4}\right)$

50. $z_1 = 10\left(\cos\dfrac{7\pi}{6} + i\sin\dfrac{7\pi}{6}\right), z_2 = \dfrac{1}{2}\left(\cos\dfrac{\pi}{6} + i\sin\dfrac{\pi}{6}\right)$

51. $z_1 = 3\left(\cos\dfrac{\pi}{4} + i\sin\dfrac{\pi}{4}\right), z_2 = 4\left(\cos\dfrac{15\pi}{8} + i\sin\dfrac{15\pi}{8}\right)$

52. $z_1 = \cos 57° + i\sin 57°, z_2 = 7(\cos 73° + i\sin 73°)$

A.3   Powers and Roots of Complex Numbers

≡ **Introduction**   The trigonometric form of a product $z_1 z_2$ given in (3) of Theorem A.2.1 of the last section also gives a means of computing *powers* of a complex number, that is, z^n, where n is a positive integer. In this section we also show how to find the n distinct nth *roots* of a complex number z.

    We begin the discussion with an example.

☐ **Powers of a Complex Number**   Suppose $z = 1 + i$ and we wish to compute z^3. Of course, there are several ways of proceeding. We can carry out the multiplications

See (*iii*) of Definition A.1.3 on page APP-3. ▶

zz and $(zz)z$ using the standard forms of the numbers or we can treat the number $z = 1 + i$ as a binomial and use the binomial expansion

$$(a + b)^3 = a^3 + 3a^2 b + 3ab^2 + b^3$$

with $a = 1$ and $b = i$. Alternatively, we can use the trigonometric form

$$z = 1 + i = \sqrt{2}\left(\cos\frac{\pi}{4} + i\sin\frac{\pi}{4}\right).$$

With $z = z_1 = z_2$ in (3) of Theorem A.2.1, we obtain the square of z:

$$z^2 = z \cdot z = (\sqrt{2})(\sqrt{2})\left[\cos\left(\frac{\pi}{4} + \frac{\pi}{4}\right) + i\sin\left(\frac{\pi}{4} + \frac{\pi}{4}\right)\right]$$

$$= (\sqrt{2})^2\left[\cos 2\left(\frac{\pi}{4}\right) + i\sin 2\left(\frac{\pi}{4}\right)\right].$$

Then (3) of Theorem A.2.1 also gives

$$z^3 = z^2 \cdot z = (\sqrt{2})^2(\sqrt{2})\left[\cos\left(\frac{2\pi}{4} + \frac{\pi}{4}\right) + i\sin\left(\frac{2\pi}{4} + \frac{\pi}{4}\right)\right]$$

$$= (\sqrt{2})^3\left[\cos 3\left(\frac{\pi}{4}\right) + i\sin 3\left(\frac{\pi}{4}\right)\right] \qquad (1)$$

$$= 2\sqrt{2}\left(\cos\frac{3\pi}{4} + i\sin\frac{3\pi}{4}\right).$$

After simplifying the last expression we get $z^3 = -2 + 2i$.

The result in (1),

$$z^3 = (\sqrt{2})^3\left[\cos 3\left(\frac{\pi}{4}\right) + i\sin 3\left(\frac{\pi}{4}\right)\right], \qquad (2)$$

illustrates a particular case of the following theorem named after the French mathematician **Abraham DeMoivre** (1667–1754). The formal proof of this theorem requires mathematical induction, which is discussed in Section 10.3.

THEOREM A.3.1 DeMoivre's Theorem

If $z = r(\cos\theta + i\sin\theta)$ and n is a positive integer, then

$$z^n = r^n(\cos n\theta + i\sin n\theta) \qquad (3)$$

Inspection of (2) shows that the result is DeMoivre's theorem with $z = 1 + i$, $r = \sqrt{2}, \theta = \pi/4$ in blue, and $n = 3$ in red.

EXAMPLE 1 **Power of a Complex Number**

Evaluate $(\sqrt{3} + i)^8$.

Solution First, the modulus of $\sqrt{3} + i$ is $r = \sqrt{(\sqrt{3})^2 + 1^2} = 2$. Then from $\tan\theta = 1/\sqrt{3}$, an argument of the number is $\theta = \pi/6$ since $(\sqrt{3}, 1)$ lies in quadrant I. Hence from DeMoivre's theorem,

$$(\sqrt{3} + i)^8 = 2^8\left[\cos 8\left(\frac{\pi}{6}\right) + i\sin 8\left(\frac{\pi}{6}\right)\right] = 256\left(\cos\frac{4\pi}{3} + i\sin\frac{4\pi}{3}\right)$$

$$= 256\left(-\frac{1}{2} - \frac{\sqrt{3}}{2}i\right) = -128 - 128\sqrt{3}i. \qquad \equiv$$

☐ **Roots of a Complex Number** Recall from algebra that -2 and 2 are said to be square roots of the number 4 because $(-2)^2 = 4$ and $2^2 = 4$. In other words, the two square roots of 4 are distinct solutions of the equation $w^2 = 4$. In like manner we say $w = 3$ is a cube root of 27 since $w^3 = 3^3 = 27$. In general, we say that a number $w = a + bi$ is a complex **nth root** of a nonzero complex number z if $w^n = (a + bi)^n = z$, where n is a positive integer. For example, you are urged to verify that $w_1 = \frac{1}{2}\sqrt{2} + \frac{1}{2}\sqrt{2}i$ and $w_2 = -\frac{1}{2}\sqrt{2} - \frac{1}{2}\sqrt{2}i$ are the two square roots of the complex number $z = i$ because $w_1^2 = i$ and $w_2^2 = i$. See also Problem 77 in Exercises A.1.

We will now demonstrate that there are exactly n solutions of the equation $w^n = z$.

Let the modulus and the argument of w be ρ and ϕ, respectively, so that $w = \rho(\cos\phi + i\sin\phi)$. If w is an nth root of the complex number $z = r(\cos\theta + i\sin\theta)$, then $w^n = z$. DeMoivre's theorem enables us to write the last equation as

$$\rho^n(\cos n\phi + i\sin n\phi) = r(\cos\theta + i\sin\theta).$$

When two complex numbers are equal, their moduli are necessarily equal. Thus we have

$$\rho^n = r \qquad \text{or} \qquad \rho = r^{1/n}$$

and $$\cos n\phi + i\sin n\phi = \cos\theta + i\sin\theta.$$

Equating the real and imaginary parts in this equation gives

$$\cos n\phi = \cos\theta, \quad \sin n\phi = \sin\theta,$$

from which it follows that $n\phi = \theta + 2k\pi$, or

$$\phi = \frac{\theta + 2k\pi}{n},$$

where k is any integer. As k takes on the successive integer values $0, 1, 2, \ldots, n-1$, we obtain n distinct roots of z. For $k \geq n$, the values of $\sin\phi$ and $\cos\phi$ repeat the values obtained by letting $k = 0, 1, 2, \ldots, n-1$. To see this, suppose that $k = n + m$, where $m = 0, 1, 2, \ldots$. Then

$$\phi = \frac{\theta + 2(n + m)\pi}{n} = \frac{\theta + 2m\pi}{n} + 2\pi.$$

Since the sine and cosine each have period 2π, we have

$$\sin\phi = \sin\left(\frac{\theta + 2m\pi}{n}\right) \quad \text{and} \quad \cos\phi = \cos\left(\frac{\theta + 2m\pi}{n}\right),$$

and so no new roots are obtained for $k \geq n$. Summarizing these results gives the following theorem.

THEOREM A.3.2 Complex Roots

If $z = r(\cos\theta + i\sin\theta)$ and n is a positive integer, then n distinct complex nth roots of z are given by

$$w_k = r^{1/n}\left[\cos\left(\frac{\theta + 2k\pi}{n}\right) + i\sin\left(\frac{\theta + 2k\pi}{n}\right)\right] \qquad (4)$$

where $k = 0, 1, 2, \ldots, n - 1$.

We will denote the n roots by $w_0, w_1, \ldots, w_{n-1}$ corresponding to $k = 0, 1, \ldots, n-1$, respectively, in (4).

EXAMPLE 2   Three Cube Roots

Find the three cube roots of i.

Solution In the trigonometric form for i, $r = 1$ and $\theta = \pi/2$, so that

$$i = \cos\frac{\pi}{2} + i\sin\frac{\pi}{2}.$$

With $n = 3$ we find from (4) of Theorem A.3.2 that

$$w_k = 1^{1/3}\left[\cos\left(\frac{\pi/2 + 2k\pi}{3}\right) + i\sin\left(\frac{\pi/2 + 2k\pi}{3}\right)\right], \quad k = 0, 1, 2.$$

Now for

$$k = 0, \quad w_0 = \cos\frac{\pi}{6} + i\sin\frac{\pi}{6}$$

$$k = 1, \quad w_1 = \cos\left(\frac{\pi}{6} + \frac{2\pi}{3}\right) + i\sin\left(\frac{\pi}{6} + \frac{2\pi}{3}\right)$$

$$= \cos\frac{5\pi}{6} + i\sin\frac{5\pi}{6}$$

$$k = 2, \quad w_2 = \cos\left(\frac{\pi}{6} + \frac{4\pi}{3}\right) + i\sin\left(\frac{\pi}{6} + \frac{4\pi}{3}\right)$$

$$= \cos\frac{3\pi}{2} + i\sin\frac{3\pi}{2}.$$

Therefore, in standard form the three cube roots of i are $w_0 = \frac{1}{2}\sqrt{3} + \frac{1}{2}i$, $w_1 = -\frac{1}{2}\sqrt{3} + \frac{1}{2}i$. and $w_2 = -i$. $\equiv$

The three cube roots of i found in Example 2 are plotted in FIGURE A.3.1. We note that they are equally spaced around a circle of radius 1 centered at the origin. In general, the n distinct nth roots of a nonzero complex number z are equally spaced on the circumference of the circle of radius $|z|^{1/n}$ with center at the origin.

As the next example shows, the roots of a complex number do not have to be "nice" numbers as in Example 2.

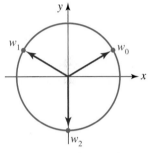

FIGURE A.3.1 Three cube roots of i in Example 2

EXAMPLE 3   Solving an Equation

Solve the equation $z^4 = 1 + i$.

Solution Solving this equation is equivalent to finding the four complex fourth roots of the number $1 + i$. In this case, the modulus and an argument of $1 + i$ are $r = \sqrt{2}$ and $\theta = \pi/4$, respectively. From (4) with $n = 4$ and the symbol z_k playing the part of w_k we obtain

$$z_k = (\sqrt{2})^{1/4}\left[\cos\left(\frac{\pi/4 + 2k\pi}{4}\right) + i\sin\left(\frac{\pi/4 + 2k\pi}{4}\right)\right]$$

$$= \sqrt[8]{2}\left[\cos\left(\frac{\pi/4 + 2k\pi}{4}\right) + i\sin\left(\frac{\pi/4 + 2k\pi}{4}\right)\right], \quad k = 0, 1, 2, 3,$$

$$k = 0, \quad z_0 = \sqrt[8]{2}\left(\cos\frac{\pi}{16} + i\sin\frac{\pi}{16}\right)$$

$$k = 1, \quad z_1 = \sqrt[8]{2}\left(\cos\frac{9\pi}{16} + i\sin\frac{9\pi}{16}\right)$$

$$k = 2, \quad z_2 = \sqrt[8]{2}\left(\cos\frac{17\pi}{16} + i\sin\frac{17\pi}{16}\right)$$

$$k = 3, \quad z_3 = \sqrt[8]{2}\left(\cos\frac{25\pi}{16} + i\sin\frac{25\pi}{16}\right).$$

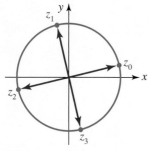

FIGURE A.3.2 Four fourth roots of $1 + i$ in Example 3

With the aid of a calculator we find the approximate standard forms,

$$z_0 \approx 1.0696 + 0.2127i$$
$$z_1 \approx -0.2127 + 1.0696i$$
$$z_2 \approx -1.0696 - 0.2127i$$
$$z_3 \approx 0.2127 - 1.0696i.$$

As shown in FIGURE A.3.2 the four roots lie on a circle centered at the origin of radius $r = \sqrt[8]{2} \approx 1.09$ and are spaced at equal angular intervals of $2\pi/4 = \pi/2$ radians beginning with the root whose argument is $\pi/16$. ≡

A.3 **Exercises** Answers to selected odd-numbered problems begin on page ANS-31.

In Problems 1–10, use DeMoivre's theorem to calculate the given power. Write your answer in the standard form $z = a + bi$. If necessary, use a calculator.

1. $\left(\cos\dfrac{5\pi}{8} + i\sin\dfrac{5\pi}{8} \right)^{24}$

2. $\left(\cos\dfrac{\pi}{10} + i\sin\dfrac{\pi}{10} \right)^5$

3. $\left[\sqrt{2}\left(\cos\dfrac{\pi}{16} + i\sin\dfrac{\pi}{16} \right) \right]^4$

4. $\left[\sqrt{3}\left(\cos\dfrac{7\pi}{16} + i\sin\dfrac{7\pi}{16} \right) \right]^4$

5. $[\sqrt{3}(\cos 21° + i\sin 21°)]^{10}$

6. $\left[\sqrt{3}\left(\cos\dfrac{\pi}{24} + i\sin\dfrac{\pi}{24} \right) \right]^8$

7. $[\sqrt{5}(\cos 13.5° + i\sin 13.5°)]^6$

8. $[2(\cos 67° + i\sin 67°)]^3$

9. $[3.2(\cos 12° + i\sin 12°)]^3$

10. $[\frac{1}{2}(\cos 24° + i\sin 24°)]^5$

In Problems 11 and 12, use (3) of this section and (4) of Section A.2 to simplify the given complex number. Write your answer in the standard form $z = a + bi$.

11. $\dfrac{\left[2\left(\cos\dfrac{\pi}{16} + i\sin\dfrac{\pi}{16} \right) \right]^{10}}{\left[4\left(\cos\dfrac{3\pi}{8} + i\sin\dfrac{3\pi}{8} \right) \right]^3}$

12. $\dfrac{\left(\cos\dfrac{\pi}{9} + i\sin\dfrac{\pi}{9} \right)^{12}}{\left[\dfrac{1}{2}\left(\cos\dfrac{\pi}{6} + i\sin\dfrac{\pi}{6} \right) \right]^5}$

In Problems 13–24, use the trigonometric form of a complex number along with DeMoivre's theorem to calculate the given power. Write your answer in the standard form $z = a + bi$.

13. i^{30}

14. $-i^{15}$

15. $(1 + i)^6$

16. $(1 - i)^9$

17. $(-2 + 2i)^4$

18. $(-4 - 4i)^3$

19. $(\sqrt{3} + i)^5$

20. $(-\sqrt{3} + i)^{10}$

21. $\left(\dfrac{\sqrt{2}}{2} - \dfrac{\sqrt{6}}{2}i \right)^9$

22. $\left(\dfrac{\sqrt{3}}{6} + \dfrac{1}{2}i \right)^8$

23. $(1 + 2i)^4$

24. $(\frac{1}{2} + \frac{1}{2}i)^{20}$

In Problems 25–34, find the indicated roots. Write your answer in the standard form $z = a + bi$.

25. The three cube roots of -8

26. The three cube roots of 1

27. The four fourth roots of i

28. The two square roots of i

29. The four fourth roots of $-1 - \sqrt{3}i$

30. The two square roots of $-1 + \sqrt{3}i$

31. The two square roots of $1 + i$

32. The three cube roots of $-2\sqrt{3} + 2i$

33. The six sixth roots of $64(\cos 54° + i\sin 54°)$

34. The two square roots of $81\left(\cos\dfrac{5\pi}{3} + i\sin\dfrac{5\pi}{3}\right)$

In Problems 35 and 36, find the indicated roots. Proceed as in Example 3 and plot these roots on an appropriate circle.

35. The six sixth roots of 1

36. The eight eighth roots of 1

37. For what positive integers n will $\left(\sqrt{2}/2 + \sqrt{2}i/2\right)^n$ be equal to 1? Equal to i? Equal to $-\sqrt{2}/2 - \sqrt{2}i/2$? Equal to $\sqrt{2}/2 + \sqrt{2}i/2$?

38. (a) Verify that $(4 + 3i)^2 = 7 + 24i$.
   (b) Use part (a) to find the two values of $(7 + 24i)^{1/2}$.

In Problems 39–42, solve the given equation. Write your answer in the standard form $z = a + bi$.

39. $z^4 + 1 = 0$

40. $z^3 - 125i = 0$

41. $z^2 + 8 + 8\sqrt{3}i = 0$

42. $z^2 - 8z + 18 = 8i$

For Discussion

43. DeMoivre's theorem implies

$$(\cos\theta + i\sin\theta)^2 = \cos 2\theta + i\sin 2\theta.$$

Use this information to derive trigonometric identities for $\cos 2\theta$ and $\sin 2\theta$ by multiplying out the left-hand side of the equation and then equating real and imaginary parts.

44. Use a procedure analogous to that outlined in Problem 43 to find trigonometric identities of $\cos 3\theta$ and $\sin 3\theta$.

Appendix B
Descartes' Rule of Signs

≡ **Introduction** Suppose that $y = f(x)$ is a polynomial function with real coefficients and is arranged in the usual manner of descending powers of x. From this form it is possible to determine the maximum number of positive zeros and the maximum number of negative zeros by examining the variations of sign in $f(x)$. We say that a **variation of sign** occurs when two consecutive terms have opposite signs. For example, in the polynomial

$$f(x) = 9x^6 - 7x^4 - 8x^3 + 2x - 14, \qquad (1)$$

sign change   sign change   sign change

there are three variations of sign: between the first and second terms, between the third and fourth terms, and between the fourth and fifth terms.

We state the following rule without proof.

THEOREM B.1.1   Descartes' Rule of Signs

Let $y = f(x)$ be a polynomial function with real coefficients that is arranged in descending powers of x.

   (*i*) The number of *positive zeros* of $f(x)$ is either equal to the number of variations of signs of $f(x)$ or less than this number by an even integer.
   (*ii*) The number of *negative zeros* of $f(x)$ is either equal to the number of variations of signs of $f(-x)$ or less than this number by an even integer.

▰ EXAMPLE 1   Maximum Number of Positive Zeros

Suppose that a polynomial $y = f(x)$ has five variations of sign. Descartes' rule stipulates that possibilities for the number of positive zeros of $f(x)$ is five, three, or one. Thus the maximum number of positive zeros is five.   ≡

▰ EXAMPLE 2   Number of Zeros

The polynomial function $f(x) = x^3 - 3x - 1$ has one variation of sign. From Descartes' rule we can conclude that $f(x)$ has precisely one positive zero. Notice that the number 1 reduced by an even integer is negative, and we cannot have a negative number of zeros. Now inspection of

$$f(-x) = -x^3 + 3x - 1$$

reveals two variations of sign. Therefore, $f(x)$ has either two or no negative zeros.   ≡

Because the polynomial $f(x) = x^2 - 10x + 25$ has two variations of sign, we know by Descartes' rule that $f(x)$ has either two or no positive zeros. But from

$$f(x) = x^2 - 10x + 25 = (x - 5)^2$$

we see that 5 is a positive zero of multiplicity two. This leads us to an important point: In the application of Descartes' rule, we must count a zero of multiplicity k as k zeros.

EXAMPLE 3   Equation (1) Revisited

In (1) we saw that polynomial function $f(x) = 9x^6 - 7x^4 - 8x^3 + 2x - 14$ has three variations of sign. Descartes' rule stipulates that the number of positive zeros of $f(x)$ is either three or one. Because

$$f(-x) = 9x^6 - 7x^4 + 8x^3 - 2x - 14$$

also has three variations of sign $f(x)$ has either three or one negative zeros.   ≡

EXAMPLE 4   Section 3.4 Revisited

In Example 1 of Section 3.4 we discovered that polynomial function

$$f(x) = 3x^4 - 10x^3 - 3x^2 + 8x - 2$$

has four real zeros

$$2 + \sqrt{2}, 2 - \sqrt{2}, \tfrac{1}{3}, -1 \tag{2}$$

by using the Rational Zeros Theorem. Had we used Descartes' rule before applying Theorem 3.4.2, we could have determined that $f(x)$ has three or one positive zeros and one negative zero. Observe in (2) that there are three positive numbers and one negative number.   ≡

> **B.1**   Exercises   Answers to selected odd-numbered problems begin on page ANS-32.

In Problems 1–10, use Descartes' rule of signs to determine the possibilities for the number of positive and negative zeros of the given polynomial function.

1. $f(x) = 8x^2 + 2x - 3$
2. $f(x) = x^2 + 4x + 4$
3. $f(x) = 7x^3 - 6x^2 + x - 5$
4. $f(x) = 10x^3 - 8x - 2$
5. $f(x) = x^3 + 4x^2 + 6x + 1$
6. $f(x) = x^3 - 2$
7. $f(x) = -x^4 + 8x^3 - 5x - 9$
8. $f(x) = x^5 - 12x^4 + 2x^2 + 7x - 16$
9. $f(x) = x^5 + x^4 + x^3 - x^2 - x + 1$
10. $f(x) = 3x^6 + 5x^3 + x + 8$

For Discussion

11. Consider the polynomial function

$$f(x) = x^6 + x^5 + 3x^4 + 5x^3 - x^2 + 10x + 5.$$

Based on the information obtained from Descartes' rule, construct a table that lists all the possible combinations in which the positive, negative, and complex zeros of $f(x)$ can occur. Do not attempt to find the zeros.

Appendix C
Formulas from Geometry

Area A, Circumference C, Volume V, Surface Area S

RECTANGLE	PARALLELOGRAM	TRAPEZOID

$A = lw, \ C = 2l + 2w$	$A = bh$	$A = \frac{1}{2}(a + b)h$

RIGHT TRIANGLE	TRIANGLE	EQUILATERAL TRIANGLE

Pythagorean Theorem: $c^2 = a^2 + b^2$	$A = \frac{1}{2}bh, \ C = a + b + c$	$h = \frac{\sqrt{3}}{2}s, \ A = \frac{\sqrt{3}}{4}s^2$

CIRCLE	CIRCULAR RING	CIRCULAR SECTOR
$A = \pi r^2, \ C = 2\pi r$	$A = \pi(R^2 - r^2)$	$A = \frac{1}{2}r^2\theta, \ s = r\theta$

ELLIPSE	ELLIPSOID	SPHERE
$A = \pi ab$	$V = \frac{4}{3}\pi abc$	$V = \frac{4}{3}\pi r^3, \ S = 4\pi r^2$

RIGHT CYLINDER	RIGHT CIRCULAR CYLINDER	RECTANGULAR PARALLELEPIPED
$V = Bh, \ B$ area of base	$V = \pi r^2 h, \ S = 2\pi rh$ (lateral side)	$V = lwh, \ S = 2(hl + lw + hw)$

CONE	RIGHT CIRCULAR CONE	FRUSTUM OF A CONE
		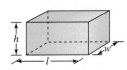
$V = \frac{1}{3}Bh, \ B$ area of base	$V = \frac{1}{3}\pi r^2 h, \ S = \pi r\sqrt{r^2 + h^2}$	$V = \frac{1}{3}\pi h(r_1^2 + r_1 r_2 + r_2^2)$

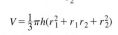

Answers to Selected Odd-Numbered Problems

Exercises 1.1   Page 8

1. $a + 2 > 0$ **3.** $a + b \geq 0$ **5.** $2b + 4 \geq 100$

7. $(-\infty, 0)$

9. $[5, \infty)$

11. $(8, 10]$

13. $[-2, 4]$

15. $-7 \leq x \leq 9$ **17.** $x < 2$

19. $(-5, \infty)$

21. $(-\infty, 4]$

23. $\left(-\infty, \frac{3}{4}\right)$

25. $\left(-\infty, \frac{5}{2}\right]$

27. $(-10, 6)$

29. $(-5, 3)$

31. $[-10, -8)$

33. $[0, 6)$

35. $(-3, 3)$

37. $(-\infty, 0] \cup [5, \infty)$

39. $(2, 6)$

41. $\left[-\frac{3}{2}, 6\right]$

43. $(-\infty, -1) \cup (2, 4)$

45. $[-2, -1] \cup [1, 2]$

47. $(-\infty, -8)$

49. $(-\infty, -5] \cup (0, \infty)$

51. $\left(-\infty, \frac{1}{3}\right) \cup (1, \infty)$

53. $(-5, 0] \cup [1, \infty)$

55. $(-\infty, -1) \cup [1, 2]$

57. $(-\infty, -3) \cup (-1, 1)$

59. If x is the number, then $x < 8$.   **61.** $n > 10$
63. If x denotes the width, then $x > 7$.   **65.** $R > \frac{10}{3}$
67. $(12, 20)$

Exercises 1.2   Page 15

1. $4 - \pi$ **3.** $8 - \sqrt{63}$ **5.** 4 **7.** $-h$
9. $-x + 6$ **11.** 0 **13.** $-2x + 7$ **15.** 3
17. $2x - 2$ **19.** 4 **21.** $4; 5$ **23.** $3; 0$
25. $a = 2, b = 8$ **27.** $m = 4 + \pi, b = 4 + 2\pi$
29. $-\frac{1}{4}, \frac{3}{4}$ **31.** $-\frac{1}{2}, \frac{5}{6}$ **33.** $\frac{2}{3}, 2$
35. $\left(-\frac{4}{5}, \frac{4}{5}\right)$

37. $(-\infty, -10) \cup (4, \infty)$

39. $[3, 4]$

41. $\left(-\infty, -1 - \sqrt{2}\right] \cup \left[1 - \sqrt{2}, \infty\right)$

43. $\left(-\frac{7}{3}, 3\right)$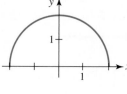

45. (4.99, 5.01)

47. $|x - 4| < 7$ **49.** $|x - 5| > 4$
51. $|x + 3| \geq 2, (-\infty, -5] \cup [-1, \infty)$
53. $|A_B - A_M| \leq 3$ **55.** (11.95, 12.05)

Exercises 1.3   Page 21

1. *(graph with points (4, 5), (2, 3), (0, 2), (−1, −3))*

3. *(graph with points (3, 3), $\left(-1, \frac{4}{3}\right)$, (0, 0), $\left(-\frac{1}{2}, -2\right)$)*

5. II **7.** III **9.** II **11.** I
13. III **15.** IV
17. *(graph with points $(-b, -a)$, (a, b), $(b, -a)$, $(-a, -a)$, $(-b, b)$, $(-a, b)$, $(a, -b)$, $(b, -b)$, $(-a, -b)$, (a, a), $(-b, a)$, (b, a), $(-a, a)$)* **19.** (3, 6)

21. *(graph showing $x = 0$, $y = 0$)*

23. *(graph showing region $y = 2$, $y = -2$, $x = -1$, $x = 1$)*

25. *(graph showing region between $x = -4$ and $x = 4$)*

27. $2\sqrt{5}$ **29.** 10

31. 5 **33.** not a right triangle
35. a right triangle **37.** an isosceles triangle
39. (a) $2x + y - 5 = 0$
     (b) The points (x, y) lie on the perpendicular bisector of the
         line segment joining A and B.
41. (6, 8) and (6, −4) **43.** $\left(1, \frac{5}{2}\right)$
45. $\left(-\frac{9}{2}, \frac{5}{2}\right)$ **47.** $\left(3a, -\frac{3}{2}b\right)$ **49.** (5, −1) **51.** (−7, −10)
53. 6   **55.** (2, −5) **57.** $\left(\frac{7}{2}, \frac{13}{2}\right), (4, 7), \left(\frac{9}{2}, \frac{15}{2}\right)$

Exercises 1.4   Page 30

1. center (0, 0), radius $\sqrt{5}$ **3.** center (0, 3), radius 7
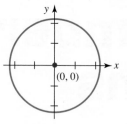 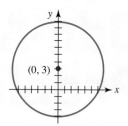

5. center $\left(\frac{1}{2}, \frac{3}{2}\right)$, radius 1
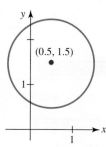

7. center (0, −4), radius 4 **9.** center (−1, 2), radius 3
11. center (10, −8), radius 6 **13.** center (−1, −4), radius $\sqrt{\frac{33}{2}}$
15. $x^2 + y^2 = 1$ **17.** $x^2 + (y - 2)^2 = 2$
19. $(x - 1)^2 + (y - 6)^2 = 8$ **21.** $x^2 + y^2 = 5$
23. $(x - 5)^2 + (y - 6)^2 = 36$

25. 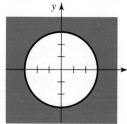 **27.** *(graph)*

29. $y = 3 + \sqrt{4 - x^2}; x = \sqrt{4 - (y - 3)^2}$
31. 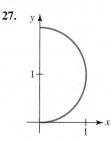 **33.**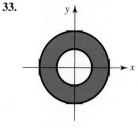

35. $\left(3 - \sqrt{13}, 0\right), \left(3 + \sqrt{13}, 0\right), \left(0, -6 - 2\sqrt{10}\right),$
     $\left(0, -6 + 2\sqrt{10}\right)$
37. (0, 0), origin **39.** $(-1, 0), \left(0, \frac{1}{2}\right)$, no symmetry
41. (0, 0), x-axis **43.** (−2, 0), (2, 0), (0, −4), y-axis
45. $\left(1 - \sqrt{3}, 0\right), \left(1 + \sqrt{3}, 0\right), (0, -2)$, no symmetry
47. (0, 0), $\left(-\sqrt{3}, 0\right), \left(\sqrt{3}, 0\right)$, origin
49. (0, −4), (0, 4), x-axis
51. (0, −3), (0, 3), x-axis, y-axis, and origin
53. $\left(-\sqrt{7}, 0\right), \left(\sqrt{7}, 0\right)$, origin
55. (−4, 0), (5, 0), $\left(0, -\frac{10}{3}\right)$, no symmetry
57. (9, 0), (0, −3), no symmetry **59.** (9, 0), (0, 9), no symmetry
61. (−4, 0), (4, 0), (0, −4), (0, 4), x-axis, y-axis, and origin
63. x-axis, y-axis, and origin **65.** y-axis

67.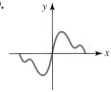

69.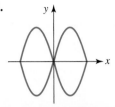

71.

Exercises 1.5   Page 42

1. (a) $x + 5$ **(b)** 10   **3. (a)** $x - 6$ **(b)** -5

5. (a) $\dfrac{x + 3}{x - 3}$ **(b)** -5 **7. (a)** $x^2 + x + 1$ **(b)** 3

9. (a) $\dfrac{x^2 + x + 1}{x + 4}$ **(b)** $\frac{3}{5}$ **11. (a)** $\dfrac{x + 1}{x^2 - x + 1}$ **(b)** 0

13. (a) $4 + h$ **(b)** 4   **15. (a)** $4(x + 2)$ **(b)** 12

17. (a) $3 + 3x + x^2$ **(b)** 3   **19. (a)** $2h^2 + h - 4$ **(b)** -4

21. (a) $\dfrac{1}{x + 4}$ **(b)** $\frac{1}{6}$ **23. (a)** $\dfrac{1}{x + 10}$ **(b)** $\frac{1}{20}$

25. (a) $-\dfrac{4 + h}{4(2 + h)^2}$ **(b)** $-\frac{1}{4}$ **27. (a)** $\dfrac{1}{\sqrt{x} + 3}$ **(b)** $\frac{1}{6}$

29. (a) $\sqrt{7 + x} + \sqrt{7}$ **(b)** $2\sqrt{7}$ **31. (a)** $5 + \sqrt{t}$ **(b)** 10

33. (a) $4(\sqrt{y^2 + y + 1} + \sqrt{y + 1})$ **(b)** 8

35. $\dfrac{ax - 1}{ax}$ **37.** $2(2x - 3)^3(2x - 1)(9x + 10)$

39. $\dfrac{6x(2 - x)}{(-4x + 6)^{3/2}}$ **41.** $y' = \dfrac{x + y}{3y^2 - x}$

43. $y' = \dfrac{2x}{1 - 2y}$ **45.** $y' = \dfrac{x^2 - 2xy + 2y + y^2}{2x}$

Chapter 1 Review Exercises   Page 45

A. 1. $x \leq 9$ **3.** II

   5. $(2, -3)$ **7.** $(x + 2)^2 + (y + 5)^2 = 36$

   9. $\sqrt{10}$ **11.** $\left(-\frac{5}{2}, 0\right), \left(\frac{5}{2}, 0\right), (0, -5)$

  13. center $(8, 0)$, radius 8   **15.** $(-3, 4), (-3, -4)$

  17. $x^2 + y^2 > 36$ **19.** $|x - \sqrt{2}| > 3$

B. 1. false   **3.** true   **5.** false   **7.** true

   9. true   **11.** true   **13.** false   **15.** true

  17. true   **19.** true

C. 1. $a^2 < ab$ **3.** $a < a + b$

   5. 10   **7.** $\leq$

   9. $-4 \leq x \leq 3$ **11.** $a = 4, b = 6$

  13. (a) $-6 < x \leq 2$ **(b)** $(-6, 2]$

  15. (a) $x \geq -4$ or $-4 \leq x < \infty$ **(b)** $[-4, \infty)$

  17. $(-\infty, -3]$ **19.** $(4, 12)$

  21. $(-\infty, -10) \cup (10, \infty)$ **23.** $\left(-\frac{1}{3}, 3\right)$

  25. $\left[-1, \frac{5}{2}\right]$ **27.** $(-1, 0) \cup (1, \infty)$

  29. $(0, 1) \cup (1, \infty)$

  31. $(x - 1)^2 + (y - 1)^2 = 1$, $(x - 2)^2 + (y - 2)^2 = 4$

  33. $x^2 + (y + 1)^2 = 1$, $x^2 + (y + 2)^2 = 4$ **35.** $\frac{3}{10} < d_o < \frac{3}{4}$

  37. (a) $\dfrac{1}{2x + 1}$ **(b)** $\frac{1}{2}$ **39. (a)** $(\sqrt{x} + 2)(x + 4)$ **(b)** 32

Exercises 2.1   Page 55

1. $24, 2, 8, 35$ **3.** $0, 1, 2, \sqrt{6}$

5. $-\frac{3}{2}, 0, \frac{3}{2}, \sqrt{2}$

7. $-2x^2 + 3x, -8a^2 + 6a, -2a^4 + 3a^2, -50x^2 - 15x,$
   $-8a^2 - 2a + 1, -2x^2 - 4xh - 2h^2 + 3x + 3h$

9. $-2, 2$ **11.** $\left[\frac{1}{2}, \infty\right)$

13. $(-\infty, 1)$ **15.** $\{x \mid x \neq 0, x \neq 3\}$

17. $\{x \mid x \neq 5\}$ **19.** $(-\infty, \infty)$

21. $[-5, 5]$ **23.** $(-\infty, 0] \cup [5, \infty)$

25. $(-2, 3]$ **27.** not a function

29. function   **31.** $[-4, 4], [0, 5]$

33. $[1, 9], [1, 6]$ **35.** $-\frac{6}{5}$

37. $2, 3$ **39.** $0, \frac{1}{3}, -9$

41. $-1, 1$ **43.** $(8, 0), (0, -4)$

45. $\left(\frac{3}{2}, 0\right), \left(\frac{5}{2}, 0\right), (0, 15)$ **47.** $\left(0, -\frac{1}{4}\right)$

49. $(-2, 0), (2, 0), (0, 3)$

51. $f_1(x) = \sqrt{x + 5}, f_2(x) = -\sqrt{x + 5}; [-5, \infty)$

53. $f(x) = 4x - 11$ **55.** $f(x) = -\frac{1}{4}x^3 - x + 1$

57. $f(x) = \dfrac{2x - 10}{x}$ **59.** $0, -3.4, 0.3, 2, 3.8, 2.9; (0, 2)$

61. $3.6, 2, 3.3, 4.1, 2, -4.1; (-3.2, 0), (2.3, 0), (3.8, 0)$

63. (a) $2; 6; 120; 5040$ **(c)** $(n + 1)(n + 2)$

Exercises 2.2   Page 65

1. even   **3.** neither even nor odd

5. odd   **7.** even   **9.** even   **11.** odd

13. neither even nor odd

15. (a)   **(b)**

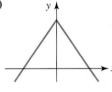

17. (a)   **(b)**

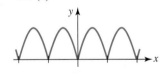

19. $f(2) = 4, f(-3) = 7$ **21.** $g(1) = 5, g(-4) = -8$

23. $(-2, 3), (3, -2)$ **25.** $(-8, 1), (-3, -4)$

27. $(-6, 2), (-1, -3)$ **29.** $(2, 1), (-3, -4)$

31. $(-2, 15), (3, -60)$

33. (a)   **(b)**

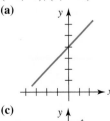

(c)   **(d)**

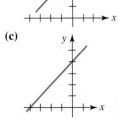

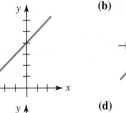

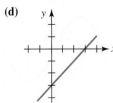

(e)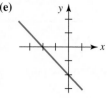

(f)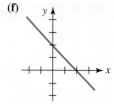

(f)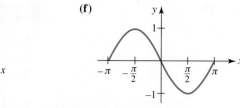

35. (a)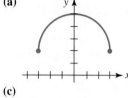

(b)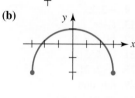

(g)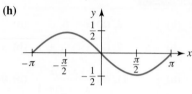

(c)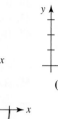

(d)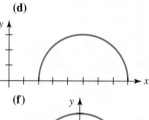

(h)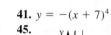

(e)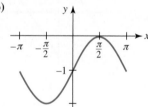

(f)

39. $y = (x - 1)^3 + 5$ **41.** $y = -(x + 7)^4$

43. 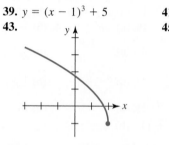   **45.**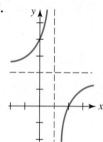

37. (a)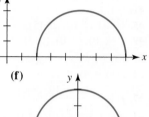

(b)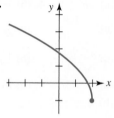

1. $-\frac{7}{2}$; 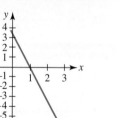   **3.** 5;

(c)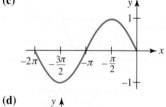

5. -1; 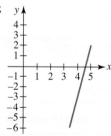   **7.** $-\frac{5}{12}$

(d)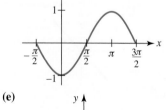

9. $\frac{3}{4}$; $(-4, 0)$, $(0, 3)$; 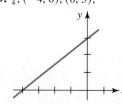   **11.** $\frac{2}{3}$; $\left(\frac{9}{2}, 0\right)$, $(0, -3)$;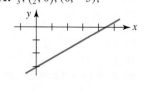

(e)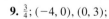

13. $-\frac{2}{5}$; $(4, 0)$, $\left(0, \frac{8}{5}\right)$;

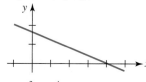

15. $-\frac{2}{3}$; $\left(\frac{3}{2}, 0\right)$, $(0, 1)$

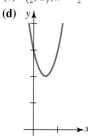

13. (a) $(0, 3)$
(b) $y = 4\left(x - \frac{1}{2}\right)^2 + 2$
(c) $\left(\frac{1}{2}, 2\right)$, $x = \frac{1}{2}$
(d)

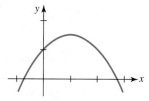

17. $y = \frac{2}{3}x + \frac{4}{3}$
19. $y = 2$
21. $y = -x + 3$
23. $y = -2x + 7$
25. $y = 1$
27. $x = -2$
29. $y = -3x - 2$
31. $x = 5$
33. $y = -4x + 11$ **35.** $y = -\frac{1}{5}x - 5$ **37.** $\left(\frac{63}{16}, \frac{31}{8}\right)$
39. (a) and **(c)** are parallel, **(b)** and **(e)** are parallel; **(a)** and **(c)** are perpendicular to **(b)** and **(e)**; **(d)** is perpendicular to **(f)**
41. (a) and **(d)** are perpendicular, **(b)** and **(c)** are perpendicular, **(e)** and **(f)** are perpendicular
43. $f(x) = \frac{1}{2}x + \frac{11}{2}$
45. $\left(-\frac{5}{6}, \frac{8}{3}\right)$;

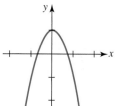

47. $(-1, 3)$;

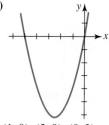

15. (a) $\left(1 - \sqrt{3}, 0\right)$, $\left(1 + \sqrt{3}, 0\right)$, $(0, 1)$
(b) $y = -\frac{1}{2}(x - 1)^2 + \frac{3}{2}$
(c) $\left(1, \frac{3}{2}\right)$, $x = 1$
(d)

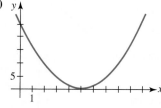

49. -9 **51.** $y = x + 3$ **53. (a)** $T_F = \frac{9}{5}T_C + 32$
55. 1,680; approximately 35.3 years

Exercises 2.4   Page 82

1.

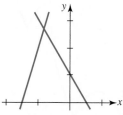

3.

17. (a) $(5, 0)$, $(0, 25)$
(b) $y = (x - 5)^2$
(c) $(5, 0)$, $x = 5$
(d)

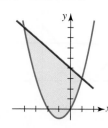

5.

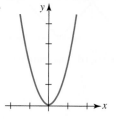

7. (a) $(0, 0)$, $(-5, 0)$
(b) $y = \left(x + \frac{5}{2}\right)^2 - \frac{25}{4}$
(c) $\left(-\frac{5}{2}, -\frac{25}{4}\right)$, $x = -\frac{5}{2}$
(d)

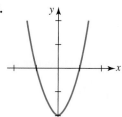

19. Minimum functional value is $f\left(\frac{4}{3}\right) = -\frac{13}{3}$; $\left[-\frac{13}{3}, \infty\right)$
21. Increasing on $[0, \infty)$, decreasing on $(-\infty, 0]$
23. Increasing on $(-\infty, -3]$, decreasing on $[-3, \infty)$
25. The graph of $y = x^2$ is shifted horizontally ten units to the right.
27. The graph of $y = x^2$ is compressed vertically, followed by a reflection in the x-axis, followed by a horizontal shift of four units to the left, followed by vertical shift of nine units upward.
29. As the equation is given it can be interpreted as the graph of $y = x^2$ shifted horizontally six units to the right, followed by a reflection in the y-axis, followed by a vertical shift of four units downward.
31. $y = (x + 2)^2$ **33.** $y = -x^2 - 1$
35. $y = -(x - 1)^2 + 5$ **37.** $f(x) = 2x^2 + 3x + 5$
39. $f(x) = 4(x - 1)^2 + 2$
41. $(-4, 8)$, $(1, 3)$,   **43.** $(-2, 2)$, $(0, 2)$,

9. (a) $(-1, 0)$, $(3, 0)$, $(0, 3)$
(b) $y = -(x - 1)^2 + 4$
(c) $(1, 4)$, $x = 1$
(d)

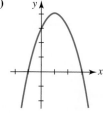

11. (a) $(1, 0)$, $(2, 0)$, $(0, 2)$
(b) $y = \left(x - \frac{3}{2}\right)^2 - \frac{1}{4}$
(c) $\left(\frac{3}{2}, -\frac{1}{4}\right)$, $x = \frac{3}{2}$
(d)

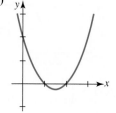

45. 19
47. (a) $d^2 = 5x^2 - 10x + 25$ **(b)** $(1, 2)$
49. (a) $s(t) = -16t^2 + 64t + 6$, $v(t) = -32t + 64$
(b) 70 ft, 0 ft/s   **(c)** 4 s, -64 ft/s

51. (a) 117.6 m, −9.8 m/s
(b) in 5 seconds
(c) −49 m/s
53. (a) The graph of $R(D) = -kD^2 + kPD$ is a parabola with vertex at $-b/2a = (-kP)/(-2k) = P/2$. Since k is positive, the graph opens downward, and so $R(D)$ is a maximum at this value. Since $R(D)$ measures the rate at which the disease spreads, we conclude that the disease spreads most rapidly when exactly one-half the population is infected.
(b) 3×10^{-5}
(c) approximately 48
(d) approximately 62, 79, 102, and 130

| Exercises 2.5 | Page 89 |

1. $2, 4, -5$
3. $3, 0, 8, 2 + 2\sqrt{2}$
5. (a) 1 **(b)** 1 **(c)** 0 **(d)** 1
(e) 1 **(f)** 0
7. (a) 3 **(b)** $-1, \sqrt{2}$ **(c)** $\sqrt[3]{-2}, 1$ **(d)** $\sqrt[3]{-3}, 0$
(e) $\sqrt{3}$ **(f)** -2
9. $(0, 0)$, continuous,

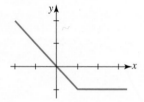

11. $(0, 0)$, continuous,

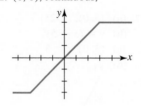

13. x-intercepts are the points $(x, 0)$, where $-2 \leq x < -1$, y-intercept is $(0, 2)$, the function is discontinuous at every integer value of x,

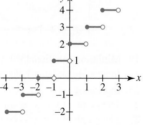

15. x-intercepts are the points $(x, 0)$, where $0 \leq x < 1$, y-intercept is $(0, 0)$, the function is discontinuous at every integer value of x,

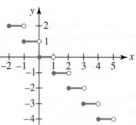

17. $(-3, 0)$, $(0, 3)$, continuous,

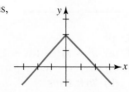

19. $(-2, 0)$, $(2, 0)$, $(0, 2)$, continuous,

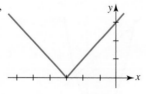

21. $(-3, 0)$, $(1, 0)$, $(0, -1)$, continuous,

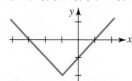

23. $\left(\frac{5}{3}, 0\right)$, $(0, -5)$, continuous,

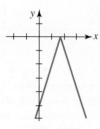

25. $(-1, 0)$, $(1, 0)$, $(0, 1)$, continuous,

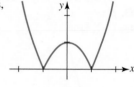

27. $(0, 0)$, $(2, 0)$, continuous,

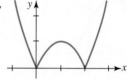

29. $(-2, 0)$, $(2, 0)$, $(0, 2)$, continuous,

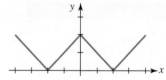

31. $(1, 0)$, $(0, 1)$, continuous, **33.** $(1, 0)$, $(0, 1)$, continuous,

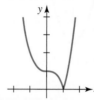 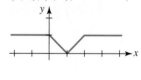

35. $\{-1, 1\}$
37. $f(x) = \begin{cases} x + 2, & x < 0 \\ -2x + 2, & 0 \leq x < 2 \\ -2, & x \geq 2 \end{cases}$
39. $f(x) = \begin{cases} -x, & x < -3 \\ \sqrt{9 - x^2}, & -3 \leq x < 3 \\ x, & x \geq 3 \end{cases}$

41.

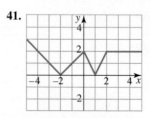

43. $f(x) = \begin{cases} 1, & x > 0 \\ -1, & x < 0 \end{cases}$ **45.** $k = 1$

47. $g(x) = \lceil x \rceil = \begin{cases} -2, & -3 < x \le -2 \\ -1, & -2 < x \le -1 \\ 0, & -1 < x \le 0 \\ 1, & 0 < x \le 1 \\ 2, & 1 < x \le 2 \\ 3, & 2 < x \le 3 \end{cases}$

1. $(f + g)(x) = 3x^2 - x + 1$, domain: $(-\infty, \infty)$
$(f - g)(x) = -x^2 + x + 1$, domain: $(-\infty, \infty)$
$(fg)(x) = 2x^4 - x^3 + 2x^2 - x$, domain: $(-\infty, \infty)$
$(f/g)(x) = (x^2 + 1)/(2x^2 - x)$,
domain: real numbers except $x = 0$ and $x = \frac{1}{2}$
3. $(f + g)(x) = x + \sqrt{x - 1}$, domain: $[1, \infty)$,
$(f - g)(x) = x - \sqrt{x - 1}$, domain: $[1, \infty)$,
$(fg)(x) = x\sqrt{x - 1}$, domain: $[1, \infty)$,
$(f/g)(x) = x/\sqrt{x - 1}$, domain: $(1, \infty)$
5. $(f + g)(x) = 3x^3 - 3x^2 + 3x + 1$, domain: $(-\infty, \infty)$,
$(f - g)(x) = 3x^3 - 5x^2 + 7x - 1$, domain: $(-\infty, \infty)$,
$(fg)(x) = 3x^5 - 10x^4 + 16x^3 - 14x^2 + 5x$,
domain: $(-\infty, \infty)$,
$(f/g)(x) = (3x^3 - 4x^2 + 5x)/(1 - x)^2$,
domain: real numbers except $x = 1$
7. $(f + g)(x) = \sqrt{x + 2} + \sqrt{5 - 5x}$, domain: $[-2, 1]$,
$(f - g)(x) = \sqrt{x + 2} - \sqrt{5 - 5x}$, domain: $[-2, 1]$,
$(fg)(x) = \sqrt{5(x + 2)(1 - x)}$, domain: $[-2, 1]$,
$\left(\dfrac{f}{g}\right)(x) = \sqrt{\dfrac{x + 2}{5 - 5x}}$, domain: $[-2, 1)$
9. $10, 8, -1, 2, 0$
11. $(f \circ g)(x) = x$, domain: $[1, \infty)$,
$(g \circ f)(x) = \sqrt{x^2} = |x|$, domain: $(-\infty, \infty)$
13. $(f \circ g)(x) = \dfrac{1}{2x^2 + 1}$, domain: $(-\infty, \infty)$,
$(g \circ f)(x) = \dfrac{4x^2 - 4x + 2}{4x^2 - 4x + 1}$,
domain: real numbers except $x = \frac{1}{2}$
15. $(f \circ g)(x) = x, (g \circ f)(x) = x$
17. $(f \circ g)(x) = \dfrac{x^3 + 1}{x}, (g \circ f)(x) = \dfrac{x^2}{x^3 + 1}$
19. $(f \circ g)(x) = x + 1 + \sqrt{x - 1}, (g \circ f)(x) = x + 1 + \sqrt{x}$
21. $(f \circ f)(x) = 4x + 18, \left(f \circ \dfrac{1}{f}\right)(x) = \dfrac{6x + 19}{x + 3}$
23. $(f \circ f)(x) = x^4, \left(f \circ \dfrac{1}{f}\right)(x) = \dfrac{1}{x^4}$
25. $(f \circ g \circ h)(x) = |x - 1|$ **27.** $(f \circ g \circ g)(x) = 54x^4 + 7$
29. $(f \circ f \circ f)(x) = 8x - 35$ **31.** $f(x) = x^5, g(x) = x^2 - 4x$
33. $f(x) = x^2 + 4\sqrt{x}, g(x) = x - 3$
35.

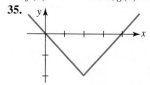
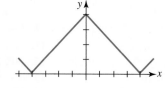

37. $y = \begin{cases} x + 3, & -3 \le x < -2 \\ x + 2, & -2 \le x < -1 \\ x + 1, & -1 \le x < 0 \\ x, & 0 \le x < 1 \\ x - 1, & 1 \le x < 2 \\ x - 2, & 2 \le x < 3 \end{cases}$

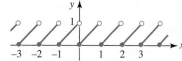

39.

41.

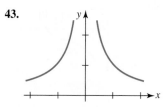

43.

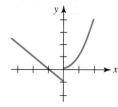

45. **(a)** $(-2, 3), (1, 0)$ **(b)** $d = -x^2 - x + 2$ **(c)** $\frac{9}{4}$
47. $d = \sqrt{10,000 + 250,000t^2}$; approximately 2,502 ft

1. not one-to-one   **3.** not one-to-one
5. one-to-one
7. one-to-one,   **9.** not one-to-one,

25. domain is $[4, \infty)$; range is $[0, \infty)$

27. $f^{-1}(x) = \dfrac{4}{x^2}, (0, \infty), (0, \infty)$

29. $f^{-1}(x) = -\frac{1}{2}x + 3$,   **31.** $f^{-1}(x) = \sqrt[3]{x - 2}$,

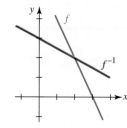
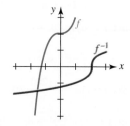

33. $f^{-1}(x) = (2 - x)^2, (-\infty, 2]$

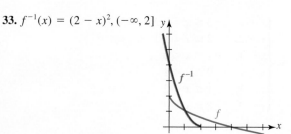

35. $f^{-1}(x) = \dfrac{x + 1}{2x}$, domain of f^{-1} is the set of real numbers except $x = 0$, range of f^{-1} is the set of real numbers except $y = \frac{1}{2}$, range of f is the set of real numbers except $y = 0$

37. $f^{-1}(x) = \dfrac{3x}{2x - 7}$, domain of f^{-1} is the set of real numbers except $x = \frac{7}{2}$, range of f^{-1} is the set of real numbers except $y = \frac{3}{2}$, range of f is the set of real numbers except $y = \frac{7}{2}$

39. $(20, 2)$ **41.** $(12, 9)$

43.

45.

47. $f^{-1}(x) = \frac{1}{2}\sqrt{x - 2}, [2, \infty)$

49. $f^{-1}(x) = 2\sqrt{1 - x^2}, [0, 1]$

Exercises 2.8 Page 112

1. $S(x) = x + \dfrac{50}{x}; (0, \infty)$ **3.** $S(x) = 3x^2 - 4x + 2; [0, 1]$

5. $A(x) = 100x - x^2; [0, 100]$ **7.** $A(x) = 2x - \frac{1}{2}x^2; [0, 4]$

9. $d(x) = \sqrt{2x^2 + 8}; (-\infty, \infty)$ **11.** $P(A) = 4\sqrt{A}; (0, \infty)$

13. $d(C) = C/\pi; (0, \infty)$ **15.** $A(h) = \dfrac{1}{\sqrt{3}}h^2; (0, \infty)$

17. $A(x) = \dfrac{1}{4\pi}x^2; (0, \infty)$ **19.** $s(h) = \dfrac{30h}{25 - h}; [0, 25)$

21. $S(w) = 3w^2 + \dfrac{1200}{w}; (0, \infty)$

23. $d(t) = 20\sqrt{13t^2 + 8t + 4}; (0, \infty)$

25. $V(h) = \begin{cases} 120h^2, & 0 \le h < 5 \\ 1200h - 3000, & 5 \le h \le 8 \end{cases}; [0, 8]$

27. $f(x) = x - x^2; (-\infty, \infty)$ **29.** $F(x) = 2x + \dfrac{16,000}{x}; (0, \infty)$

31. $C(x) = 4x + \dfrac{640,000}{x}; (0, \infty)$

33. $A(x) = \frac{1}{2}xp - x^2; [0, \frac{1}{2}p]$

35. (a) $A(x) = x^2 + \dfrac{128,000}{x}; (0, \infty)$

  (b) $A(x) = 2x^2 + \dfrac{128,000}{x}; (0, \infty)$

37. $V(x) = 20x - 40x^2; [0, \frac{1}{2}]$

39. $A(x) = 40 + 4x + \dfrac{64}{x}; (0, \infty)$

41. $L(x) = x + \dfrac{8x}{\sqrt{x^2 - 64}}; (8, \infty)$

43. $L(x) = \dfrac{1}{4\pi}(L^2x - x^3); [0, L]$

45. $V(x) = 5x\sqrt{64 - x^2}; [0, 8]$

47. $T(x) = \frac{1}{3}\sqrt{x^2 + 1} + \frac{1}{2}\sqrt{x^2 - 8x + 17}; [0, 4]$

Exercises 2.9 Page 122

1. (a) $6 + h$ **(b)** 6 **(c)** $y = 6x - 15$

3. (a) $-1 + h$ **(b)** -1 **(c)** $y = -x - 1$

5. (a) $-23 - 12h - 2h^2$ **(b)** -23 **(c)** $y = -23x + 32$

7. (a) $\dfrac{1}{2(-1 + h)}$ **(b)** $-\frac{1}{2}$ **(c)** $y = -\frac{1}{2}x - 1$

9. (a) $\dfrac{1}{\sqrt{4 + h} + 2}$ **(b)** $\frac{1}{4}$ **(c)** $y = \frac{1}{4}x + 1$

11. (a) 0 **(b)** $f'(x) = 0$

13. (a) $-8x - 4h$ **(b)** $f'(x) = -8x$

15. (a) $6x + 3h - 1$ **(b)** $f'(x) = 6x - 1$

17. (a) $3x^2 + 3xh + h^2 + 5$ **(b)** $f'(x) = 3x^2 + 5$

19. (a) $\dfrac{1}{(4 - x)(4 - x - h)}$ **(b)** $f'(x) = \dfrac{1}{(4 - x)^2}$

21. (a) $\dfrac{-1}{(x - 1)(x + h - 1)}$ **(b)** $f'(x) = \dfrac{-1}{(x - 1)^2}$

23. (a) $1 - \dfrac{1}{x(x + h)}$ **(b)** $f'(x) = 1 - \dfrac{1}{x^2}$

25. (a) $\dfrac{2}{\sqrt{x + h} + \sqrt{x}}$ **(b)** $f'(x) = \dfrac{1}{\sqrt{x}}$

27. $(2, 17); 11; y = 11x - 5$

29. $(1, 2); 8; y = 8x - 6$ **31.** $\left(\frac{1}{2}, \frac{5}{2}\right); -3; y = -3x + 4$

33. (a) $3x + 3a$ **(b)** $f'(a) = 6a$

35. (a) $10(x^2 + ax + a^2)$ **(b)** $f'(a) = 30a^2$

37. (a) $\dfrac{-1}{ax}$ **(b)** $f'(a) = \dfrac{-1}{a^2}$

39. (a) $\dfrac{\sqrt{7}}{\sqrt{x} + \sqrt{a}}$ **(b)** $f'(a) = \dfrac{1}{2}\sqrt{\dfrac{7}{a}}$

Chapter 2 Review Exercises Page 124

A. 1. $-\frac{1}{3}$ **3.** $(-\infty, 5)$

5. $x = 0, x = 2$ **7.** $k = -\frac{6}{5}$

9. $m = -\frac{3}{2}$ **11.** $\left(1 - \sqrt{2}, 0\right), \left(1 + \sqrt{2}, 0\right), (0, -1)$

13. $f(x) = \frac{7}{4}(x + 2)^2$ **15.** $(10, 2)$

17. $(0, 5)$ **19.** $a = -\frac{1}{4}$ and $a = 8$

21. second **23.** $[0.5, 7]$

25. approximately $(1.4, 0), (2.7, 0),$ and $(6, 0)$ **27.** $[2, 4], [6, 7]$

28. approximately the interval $(1.4, 2.7)$ **31.** 1.2

B. 1. true   **3.** false   **5.** true   **7.** true
   9. true   **11.** true   **13.** true   **15.** false
   17. true   **19.** true   **21.** false

C. 1. $f(x) = x^2, g(x) = \dfrac{3x - 5}{x}$

   3. (a) $y = (x + 3)^3 - 2$ **(b)** $y = x^3 - 7$
      (c) $y = (x - 1)^3$ **(d)** $y = -x^3 + 2$
      (e) $y = -x^3 - 2$ **(f)** $y = 3x^3 - 6$

   5. domain: $(\pi/2, 3\pi/2)$, range: $(-\infty, \infty)$

   7. $f(x) = \begin{cases} x + 1, & x < 0 \\ -x + 1, & 0 \le x < 1, \\ x - 1, & x \ge 1 \end{cases}$ **9.** $(-\infty, \infty)$

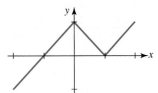

   11. $\{x \mid x \ne -3, x \ne 2\}$ **13.** $f^{-1}(x) = -1 + \sqrt[3]{x}$
   15. $A(h) = h^2(1 - \pi/4)$ **17.** $d(s) = \sqrt{3}s$
   19. (a) $d(t) = 6t$ **(b)** $d(t) = \sqrt{90^2 + (90 - 6t)^2}$

   21. $S(x) = 20x + \dfrac{5}{x}, x > 0$

   23. $A(x) = 2x(1 - \pi x)$, where x is the radius of the semicircle
   25. $f'(x) = -6x + 16, y = 4x + 24$

   27. $f'(x) = \dfrac{1}{x^3}, y = 8x - 6$ **29.** $(0, 0), \left(\tfrac{2}{3}, -\tfrac{4}{27}\right)$

Exercises 3.1   Page 138

1.
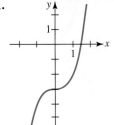

3.
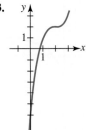

5.
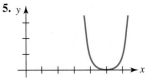

7.
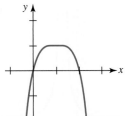

9. odd   **11.** neither even nor odd
13. (f)   **15.** (e)   **17.** (b)
19. $f(x) = 2x^4 + 3x^2 - 6$ **21.** $f(x) = -7x(x - 1)(x + 3)^2$
23.
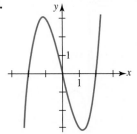

25.
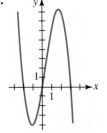

27.
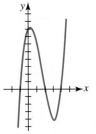

29.
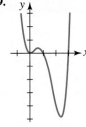

31.
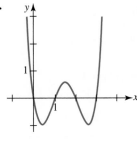

33.
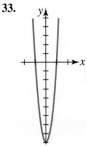

35.
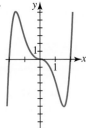

37.
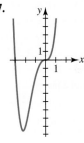

39.
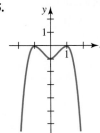

41.
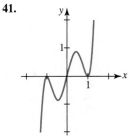

43.
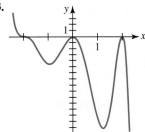

45. (b) $f(x) = (x - 1)^2(x + 2)$
47. $k = -\tfrac{7}{16}$ **49.** $k = -\tfrac{10}{3}$
51. the odd positive integers
53. $[0, 15]$;
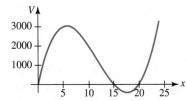

55. $V(h) = \tfrac{1}{3}\pi(R^2h - h^3)$

1. $f(x) = x^2 \cdot 8 + 4x - 7$
3. $f(x) = (x^2 + x - 1) \cdot (5x - 12) + 21x - 11$
5. $f(x) = (x + 2)^2 \cdot (2x - 4) + 5x + 21$
7. $f(x) = (3x^2 - x) \cdot (9x + 3) + 4x - 2$
9. $f(x) = (6x^2 + 4x + 1) \cdot (x^3 - 2) + 12x^2 + 8x + 2$
11. $r = 6$ **13.** $r = \frac{29}{8}$ **15.** $r = 76$ **17.** $f(2) = 2$
19. $f(-5) = -74$ **21.** $f\left(\frac{1}{2}\right) = \frac{303}{16}$
23. $q(x) = 2x + 3, r = 11$
25. $q(x) = x^2 - 4x + 12, r = -34$
27. $q(x) = x^3 + 2x^2 + 4x + 8, r = 32$
29. $q(x) = x^4 - 4x^3 + 16x^2 - 8x + 32, r = -132$
31. $q(x) = x^2 - 2x + \sqrt{3}, r = 0$
33. $f(-3) = 51$ **35.** $f(1) = 1$
37. $f(4) = 5369$ **39.** $k = -1$ **41.** $k = -\frac{1}{5}$ **43.** $k = -4$

1. $f(x) = 4\left(x - \frac{1}{4}\right)(x - 1)^2$ **3.** 5 is not a zero
5. $f(x) = 3\left(x + \frac{2}{3}\right)(x - 2 + \sqrt{2})(x - 2 - \sqrt{2})$
7. $f(x) = 4(x + 3)(x - 5)\left(x - \frac{1}{2}\right)\left(x + \frac{1}{2}\right)$
9. $f(x) = 9(x - 1)\left(x + \frac{1}{3}\right)^2(x + 8)$
11. $f(x) = (x - 1)^3(x + 5)(x - 5)$
13. $x - 5$ is not a factor
15. $f(x) = (x - 1)\left(x + \frac{1}{2} + \frac{1}{2}\sqrt{7}i\right)\left(x + \frac{1}{2} - \frac{1}{2}\sqrt{7}i\right)$
17. $x - \frac{1}{3}$ is not a factor
19. $f(x) = (x - 1)(x - 2)(x - 2i)(x + 2i)$
21. $f(x) = 2(x - 1)^2(x + 1)\left(x + \frac{3}{2}\right)$
23. $f(x) = 3\left(x - \frac{5}{3}\right)(x + 2i)(x - 2i)$
25. $f(x) = 5\left(x - \frac{2}{5}\right)(x + 1 - i)(x + 1 + i)$
27. $f(x) = (x - 3)(x + 3)(x - 1 + 2i)(x - 1 - 2i)$
29. $f(x) = (x - 2)(x - 1)(x + 3)^2$
$= x^4 + 3x^3 - 7x^2 - 15x + 18$
31. $f(x) = x^5 - 6x^4 + 10x^3$ **33.** $f(x) = x^2 - 2x + 37$
35. 0 is a simple zero, $\frac{5}{4}$ is a zero of multiplicity two, $\frac{1}{2}$ is a zero of multiplicity three
37. $-\frac{2}{3}$ is a zero of multiplicity two, $\frac{2}{3}$ is a zero of multiplicity two
39. $k = -36, f(x) = 2(x - 3)\left(x + 1 - \sqrt{5}i\right)\left(x + 1 + \sqrt{5}i\right)$
41. $f(x) = -\frac{1}{16}(x - 4)(x + 2)^2$

1. $\frac{2}{5}$ **3.** 3
5. $\frac{1}{2}$ (multiplicity 2)   **7.** no rational zeros
9. $\frac{1}{3}, \frac{3}{2}$ **11.** 0, 1   **13.** $-3, 0, 2$ **15.** $\frac{3}{2}$
17. $-\frac{1}{5}$ **19.** $-\frac{1}{2}$ (multiplicity two), $\frac{1}{3}$ (multiplicity two)
21. $\frac{3}{8}, -\frac{1}{2} - \frac{1}{2}\sqrt{5}, -\frac{1}{2} + \frac{1}{2}\sqrt{5};$
$f(x) = (8x - 3)\left(x + \frac{1}{2} + \frac{1}{2}\sqrt{5}\right)\left(x + \frac{1}{2} - \frac{1}{2}\sqrt{5}\right)$
23. $\frac{4}{5}, \frac{5}{2}, -\sqrt{2}, \sqrt{2};$
$f(x) = (5x - 4)(2x - 5)\left(x + \sqrt{2}\right)\left(x - \sqrt{2}\right)$
25. $-4, -1, 1, -\sqrt{5}, \sqrt{5};$
$f(x) = (x + 4)(x + 1)(x - 1)\left(x + \sqrt{5}\right)\left(x - \sqrt{5}\right)$
27. $0, 1, 3, -1 - \sqrt{2}, -1 + \sqrt{2};$
$f(x) = 4x(x - 1)(x - 3)\left(x + 1 + \sqrt{2}\right)\left(x + 1 - \sqrt{2}\right)$
29. $-1, \frac{1}{4}$ (multiplicity 2); $f(x) = (x + 1)(4x - 1)^2(x^2 - 2x + 3)$
31. $-\frac{1}{2}$ **33.** $-\frac{3}{2}, 2, -2 - \sqrt{3}, -2 + \sqrt{3}$
35. 1 (mutliplicity 3)

37. $f(x) = 3x^4 - x^3 - 39x^2 + 49x - 12$
39. $\frac{3}{4}$ **41.** $f(x) = -\frac{1}{6}(x - 1)(x - 2)(x - 3)$
43. 3 inches or $\frac{1}{2}(7 - \sqrt{33}) \approx 0.63$ inches

1. -1.531 **3.** -1.314
5. $1.611; 3.820$ **7.** $-1.141; 1.141$
9. 1.730 in.

1.

x	3.1	3.01	3.001	3.0001	3.00001
$f(x)$	62	602	6,002	60,002	600,002
x	2.9	2.99	2.999	2.9999	2.99999
$f(x)$	-58	-598	$-5,998$	$-59,998$	$-599,998$

3. Asymptotes: $x = 2, y = 0$ **5.** Asymptotes: $x = -1, y = 1$
Intercepts: $\left(0, -\frac{1}{2}\right)$ Intercepts: $(0, 0)$

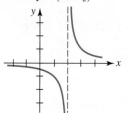
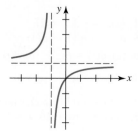

7. Asymptotes: $x = -\frac{3}{2}, y = 2$ **9.** Asymptotes: $x = -1, y = -1$
Intercepts: $\left(\frac{9}{4}, 0\right), (0, -3)$ Intercepts: $(1, 0), (0, 1)$

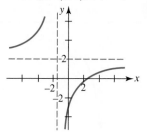
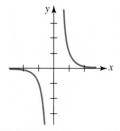

11. Asymptotes: $x = 1, y = 0$ **13.** Asymptotes: $x = 0, y = 0$
Intercepts: $(0, 1)$ Intercepts: none

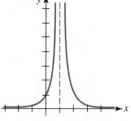

15. Asymptotes: $x = 1, x = -1,$ **17.** Asymptotes: $x = 0, x = 2,$
$y = 0$ $y = 0$
Intercepts: $(0, 0)$ Intercepts: none

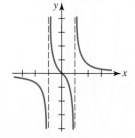
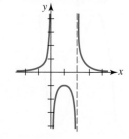

19. Asymptotes: $x = 0$, $y = -1$
Intercepts: $(-1, 0)$, $(1, 0)$

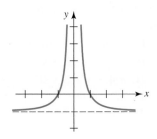

21. Asymptotes: $x = 1$, $y = -2$
Intercepts: $(-2, 0)$, $(2, 0)$, $(0, 8)$

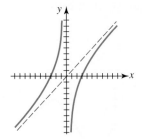

23. Asymptotes: $x = 0$, $y = x$
Intercepts: $(-3, 0)$, $(3, 0)$

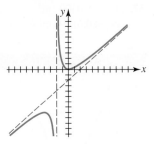

25. Asymptotes: $x = -2$,
$y = x - 2$
Intercepts: $(0, 0)$

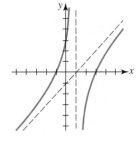

27. Asymptotes: $x = 1$, $y = x - 1$
Intercepts: $(3, 0)$, $(-1, 0)$, $(0, 3)$

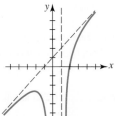

29. Asymptotes: $x = 1$, $x = 0$, $y = x + 1$
Intercepts: $(2, 0)$

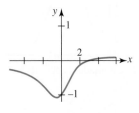

31. $(3, 0)$;

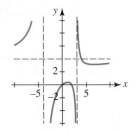

33. $(4, 4)$;

35. $(-3, -6)$

37. $y = \dfrac{x - 5}{x - 2}$

39. $y = \dfrac{3x(x - 3)}{(x + 1)(x - 2)}$

41. Hole in the graph at $x = 1$
Intercepts: $(-1, 0)$, $(0, 1)$

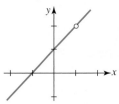

43. Hole in the graph at $x = -1$
Intercepts: none

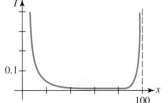

45. $R \to 5$ as $r \to \infty$;

47. $I(x) \to \infty$ as $x \to 0^+$; $I(x) \to \infty$ as $x \to 100^-$

Exercises 3.7 — Page 183

1. $\frac{7}{16}$

3. $\frac{1}{2}$

5. (a) $\frac{27}{4}$ **(b)** $\frac{33}{4}$

7. (a) 20 **(b)** 20

9. $\frac{85}{4}$

11. 6.85; 7.15

13. 9.32 acres; 8.48 acres

Chapter 3 Review Exercises — Page 185

A. 1. $(1, 0)$; $(0, 0)$, $(5, 0)$

3. $f(x) = x^4$

5. $k = \frac{2}{3}$

7. $x = 1$, $x = 4$

9. $y = -\frac{1}{2}$

11. $n = 0$, $n = 1$, $n = 2$

13. $-i$, $-1 + i$, $-1 - i$

15. 4

17. $y = 2x + 4$, $x = 0$, $x = 1$

19. the origin

B. 1. true **3.** true **5.** true **7.** true

9. true **11.** false **13.** false **15.** false

17. true **19.** true

C. 1. $3x^3 - \frac{1}{2}x + 1 + \dfrac{-\frac{1}{2}x + 5}{2x^2 - 1}$

3. $7x^3 + 14x^2 + 22x + 53 + \dfrac{109}{x - 2}$

5. $r = f(-3) = -198$ **7.** n an odd positive integer

9. $\pm 1, \pm 3, \pm 5, \pm 15, \pm \frac{1}{2}, \pm \frac{3}{2}, \pm \frac{5}{2}, \pm \frac{15}{2}, \pm \frac{1}{4}, \pm \frac{3}{4}, \pm \frac{5}{4}, \pm \frac{15}{4}, \pm \frac{1}{8}, \pm \frac{3}{8}, \pm \frac{5}{8}, \pm \frac{15}{8}$

11. $f(x) = (x - 2)\left(x - \frac{7}{2} + \frac{1}{2}\sqrt{3}i\right)\left(x - \frac{7}{2} - \frac{1}{2}\sqrt{3}i\right)$

13. $k = -\frac{21}{2}$ **15.** $k = \frac{3}{2}$

17. $f(x) = 3x^2(x + 2)^2(x - 1)$ **19.** $f(x) = \dfrac{4(x + 2)}{(x - 2)(x + 4)}$

21. (f) **23.** (d)

25. (h) **27.** (c)

29. (b)

31. $y = 0, x = -4, x = 2, (-2, 0), \left(0, -\frac{1}{4}\right),$

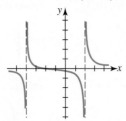

1.

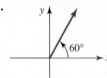

60°

3.

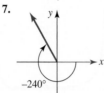

135°

5.

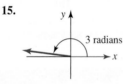

1140°

7.

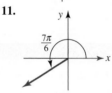

−240°

9.

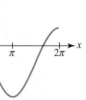

$\frac{\pi}{3}$

11.

$\frac{7\pi}{6}$

13.

$-\frac{\pi}{6}$

15.

3 radians

17. 10.6547°　　　　**19.** 5.17°
21. 210° 46′48″　　　**23.** 30°48′36″
25. $\pi/18$　　　　　**27.** $\pi/4$
29. $3\pi/2$　　　　　**31.** $-23\pi/18$
33. 40°　　　　　　**35.** 120°
37. 225°　　　　　**39.** 177.62°
41. 155°　　　　　**43.** 110°
45. −205°　　　　**47.** $7\pi/4$
49. 1.3π　　　　**51.** $2\pi - 4 \approx 2.28$
53. $-\pi/4$
55. (a) 41.75°　　　　　　**(b)** 131.75°
57. (a) The given angle is greater than 90°.　**(b)** 81.6°
59. (a) $\pi/4$　　　　　　**(b)** $3\pi/4$
61. (a) The given angle is greater than $\pi/2$.　**(b)** $\pi/3$
63. (a) 216°, 1.2π　　　　**(b)** −1845°, -10.25π
65. because the hour hand moves clockwise: −60°, $-\pi/3$
67. (a) 16 h　　　　　　**(b)** 2 h
69. (a) 9　　　　　　　**(b)** 15
71. (a) 1.5　　　　　　**(b)** 85.94°
75. (a) 0.000072921 rad/s　**(b)** 3.074641 km/s
77. 1.15 statute miles
79. (a) 3π rad/s　　　　**(b)** 300π cm/s
81. (a) 711.1 rev/min　　**(b)** 4468 rad/min

1. $\sqrt{21}/5$　　　　　　**3.** $-\sqrt{5}/3$
5. $\pm 3\sqrt{5}/7$　　　　**7.** $\pm 2\sqrt{6}/5 \approx \pm 0.98$
9. $\sin t = \pm 1/\sqrt{5},\ \cos t = \pm 2/\sqrt{5}$
11. (a) −1　　　**(b)** 0　　**13. (a)** 0　　　　**(b)** 1
15. $\pi/3, \sqrt{3}/2, -\frac{1}{2}$　　　　**17.** $\pi/4, -\sqrt{2}/2, -\sqrt{2}/2$
19. $\pi/6, -\frac{1}{2}, \sqrt{3}/2$　　　　**21.** $\pi/4, -\sqrt{2}/2, \sqrt{2}/2$
23. $\pi/6, -\frac{1}{2}, -\sqrt{3}/2$　　**25.** $\pi/3, \sqrt{3}/2, \frac{1}{2}$
27. $\sqrt{3}/2$　　　**29.** $\sqrt{2}/2$　　**31.** −1
33. $\sin(t + 2\pi) = \sin t$ for $t = \pi$
35. $\sin(-t) = -\sin t$ for $t = 3 + \pi$
37. $\cos(-t) = \cos t$ for $t = 0.43$　　　**39.** $\sqrt{2}/2$
41. $-\sqrt{3}/2$　**43.** $\sqrt{3}/2$　**45.** $-\sqrt{3}/2$　**47.** $0, \pi$
49. $\pi/4, 7\pi/4$　**51.** 30°, 330°　**53.** 225°, 315°　**55.** 4.81 m
57. (a) 978.0309 cm/s²　　　**(b)** 983.21642 cm/s²
　　(c) 980.61796 cm/s²

1.

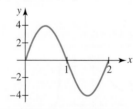

3.

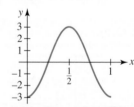

5.

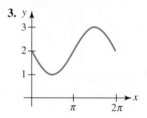

7. $y = -3\sin x$

9. $y = 1 - 3\cos x$
11. $(n, 0)$, where n is an integer
13. $((2n + 1)\pi, 0)$, where n is an integer
15. $(\pi/4 + n\pi, 0)$, where n is an integer
17. $(\pi/2, 0)$; $(\pi/2 + 2n\pi, 0)$, where n is an integer
19. $y = 3\sin 2x$　　　　　　**21.** $y = \frac{1}{2}\cos \pi x$
23. $y = -\sin \pi x$
25. amplitude: 4; period: 2　　**27.** amplitude: 3; period: 1

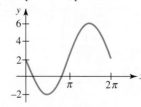

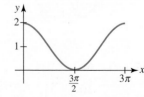

29. amplitude: 4; period: 2π　　**31.** amplitude: 1; period: 3π

33. amplitude: 1; period: 2π

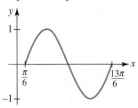

35. amplitude: 1; period: 2π

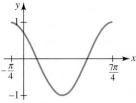

37. amplitude: 4; period: π

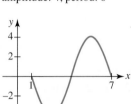

39. amplitude: 3; period: 4π

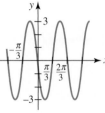

41. amplitude: 4; period: 6

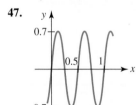

43. $y = -5 + 3\cos\left(6x + \dfrac{3\pi}{2}\right)$

45.

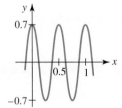

$y = 3\sin(3x - \pi)$ $y = 3\cos(3x - \pi)$

47.

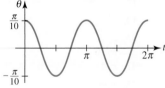

$y = 0.7\sin[4\pi(x - 4)]$ $y = 0.7\cos[4\pi(x - 4)]$

51.

53.

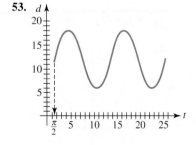

1.

x	$\frac{2\pi}{3}$	$\frac{3\pi}{4}$	$\frac{5\pi}{6}$	π	$\frac{7\pi}{6}$	$\frac{5\pi}{4}$	$\frac{4\pi}{3}$	$\frac{3\pi}{2}$	$\frac{5\pi}{3}$	$\frac{7\pi}{4}$	$\frac{11\pi}{6}$	2π
$\tan x$	$-\sqrt{3}$	-1	$-\dfrac{1}{\sqrt{3}}$	0	$\dfrac{1}{\sqrt{3}}$	1	$\sqrt{3}$	$-$	$-\sqrt{3}$	-1	$-\dfrac{1}{\sqrt{3}}$	0
$\cot x$	$-\dfrac{1}{\sqrt{3}}$	-1	$-\sqrt{3}$	$-$	$\sqrt{3}$	1	$\dfrac{1}{\sqrt{3}}$	0	$-\dfrac{1}{\sqrt{3}}$	-1	$-\sqrt{3}$	$-$

3. $\sqrt{3}$ **5.** undefined   **7.** $-2/\sqrt{3}$ **9.** -1

11. -2 **13.** undefined   **15.** -2 **17.** $\sqrt{2}$

19. $\cot x = -\dfrac{1}{2}$, $\sec x = -\sqrt{5}$, $\cos x = -\dfrac{1}{\sqrt{5}}$,

     $\sin x = \dfrac{2}{\sqrt{5}}$, $\csc x = \dfrac{\sqrt{5}}{2}$

21. $\sin x = \frac{3}{4}$, $\cos x = \dfrac{\sqrt{7}}{4}$, $\tan x = \dfrac{3}{\sqrt{7}}$,

     $\cot x = \dfrac{\sqrt{7}}{3}$, $\sec x = \dfrac{4}{\sqrt{7}}$

23. $\csc x = 3$, $\cos x = -\dfrac{2\sqrt{2}}{3}$, $\sec x = -\dfrac{3}{2\sqrt{2}}$,

     $\tan x = -\dfrac{1}{2\sqrt{2}}$, $\cot x = -2\sqrt{2}$

25. $\sec x = \frac{13}{12}$, $\sin x = -\frac{5}{13}$, $\csc x = -\frac{13}{5}$,

     $\tan x = -\frac{5}{12}$, $\cot x = -\frac{12}{5}$

27. $\tan x = 3$, $\cot x = \frac{1}{3}$, $\sec x = \pm\sqrt{10}$, $\csc x = \pm\dfrac{\sqrt{10}}{3}$

29. period: 1; x-intercepts: $(n, 0)$, where n is an integer;

     asymptotes: $x = \dfrac{2n + 1}{2}$, n an integer;

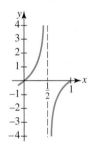

31. period: $\pi/2$;

     x-intercepts: $\left(\dfrac{2n + 1}{4}\pi, 0\right)$,

     where n is an integer;

     asymptotes: $x = n\pi/2$, n an integer;

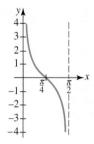

33. period: 2π;

     x-intercepts: $\left(\dfrac{\pi}{2} + 2n\pi, 0\right)$,

     where n is an integer;

     asymptotes: $x = \dfrac{3\pi}{2} + 2n\pi$,

     n an integer;

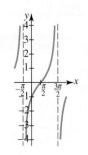

35. period: 1;

x-intercepts: $\left(\frac{1}{4} + n, 0\right)$,

where n is an integer;

asymptotes: $x = n$,

n an integer;

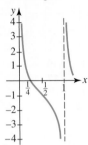

39. period: 2;

asymptotes: $x = n$,

n an integer;

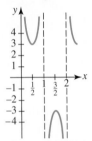

43. period: π;

asymptotes: $x = \dfrac{2n - 1}{4}\pi$,

n an integer;

37. period: 2π;

asymptotes: $x = \dfrac{2n + 1}{2}\pi$,

n an integer;

41. period: $2\pi/3$;

asymptotes: $x = \dfrac{n\pi}{3}$,

n an integer;

45. $\cot x = -\tan\left(x - \dfrac{\pi}{2}\right)$

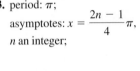

Exercises 4.5   Page 231

1. 1

3. 1

5. −1

7. 0

9. 1

11. $\sin t$

13. $\tan\alpha$

15. $\csc x$

17. $\sec\alpha$

19. $\sin t$

21. $2\sec^2 t$

65. $a\sin\theta$

67. $a\tan\theta$

69. $\tan\theta$

71. $\frac{\sqrt{3}}{3}\sin\theta\cos\theta$

73. $\frac{\sqrt{7}}{7}\cos\theta$

Exercises 4.6   Page 240

1. $\frac{\sqrt{2}}{4}\left(1 + \sqrt{3}\right)$

3. $\frac{\sqrt{2}}{4}\left(1 + \sqrt{3}\right)$

5. $\frac{\sqrt{2}}{4}\left(1 + \sqrt{3}\right)$

7. $2 + \sqrt{3}$

9. $\frac{\sqrt{2}}{4}\left(1 - \sqrt{3}\right)$

11. $\frac{\sqrt{2}}{4}\left(\sqrt{3} - 1\right)$

13. $-\frac{\sqrt{2}}{4}\left(1 + \sqrt{3}\right)$

15. $-2 + \sqrt{3}$

17. $\frac{\sqrt{2}}{4}\left(1 - \sqrt{3}\right)$

19. $\frac{\sqrt{2}}{4}\left(\sqrt{3} + 1\right)$

---

21. $-\frac{\sqrt{2}}{4}\left(1 + \sqrt{3}\right)$

23. $\sin 2\beta$

25. $\cos(2\pi/5)$

27. $\frac{1}{2}\tan 6t$

29. (a) $\frac{5}{9}$ **(b)** $-\frac{2\sqrt{14}}{9}$ **(c)** $-\frac{2\sqrt{14}}{5}$

31. (a) $\frac{3}{5}$ **(b)** $\frac{4}{5}$ **(c)** $\frac{4}{3}$

33. (a) $-\frac{119}{169}$ **(b)** $-\frac{120}{169}$ **(c)** $\frac{120}{119}$

35. $\frac{1}{2}\sqrt{2 + \sqrt{3}}$

37. $\frac{1}{2}\sqrt{2 + \sqrt{2}}$

39. $\frac{1}{2}\sqrt{2 - \sqrt{2}}$

41. $-2 - \sqrt{3}$

43. $-2\sqrt{2 + \sqrt{3}}$

45. (a) $\frac{2\sqrt{13}}{13}$ **(b)** $\frac{3\sqrt{13}}{13}$ **(c)** $\frac{3}{2}$

47. (a) $-\sqrt{(5 - \sqrt{5})/10}$ **(b)** $\sqrt{(5 + \sqrt{5})/10}$

(c) $-\frac{1}{2}\left(1 + \sqrt{5}\right)$

49. (a) $\frac{\sqrt{30}}{6}$ **(b)** $\frac{\sqrt{6}}{6}$ **(c)** $\frac{\sqrt{5}}{5}$

61. $y = 2\cos 2x$, amplitude 2, period π

63. $y = \sin 4x$, amplitude 1, period $\pi/2$

65. (a) $-\frac{2}{9}\left(\sqrt{10} + 1\right)$ **(b)** $\frac{1}{9}\left(\sqrt{5} - 4\sqrt{2}\right)$

(c) $\frac{2}{9}\left(1 - \sqrt{10}\right)$ **(d)** $\frac{1}{9}\left(\sqrt{5} + 4\sqrt{2}\right)$

67. $2\sqrt{2 + \sqrt{3}} \approx 3.86$

Exercises 4.7   Page 245

1. $\frac{1}{2}(\cos\theta + \cos 7\theta)$

3. $\frac{1}{2}(\cos 3x - \cos 7x)$

5. $\frac{1}{2}\left(\cos x + \cos\dfrac{5x}{3}\right)$

7. $\frac{1}{2}(\sin 20x - \sin 2x)$

9. $\sin 4\beta - \sin 2\beta$

11. $-\cos 2x$

13. $\frac{1}{2} - \frac{\sqrt{3}}{4}$

15. $\frac{1}{4}$

17. $\frac{\sqrt{2}}{4} + \frac{\sqrt{3}}{4}$

19. $-2\cos 3y\sin 2y$

21. $-2\sin\dfrac{5x}{2}\sin 2x$

23. $2\cos 4x\cos 2x$

25. $2\sin\dfrac{\omega_1 + \omega_2}{2}t\cos\dfrac{\omega_1 - \omega_2}{2}t$ **27.** $-\sin 3t\sin 2t$

29. 0

31. 1

33. $-\frac{\sqrt{6}}{2}$

35. $\frac{\sqrt{2}}{2}$

37. $f(t) = 0.06\sin 750\pi t\cos 250\pi t$

Exercises 4.8   Page 253

1. 0

3. π

5. $\pi/3$

7. $-\pi/3$

9. $\pi/4$

11. $-\pi/6$

13. $-\pi/4$

15. $\frac{4}{5}$

17. $-\frac{\sqrt{5}}{2}$

19. $\frac{\sqrt{5}}{5}$

21. $\frac{5}{3}$

23. $\frac{1}{5}$

25. 1.2

27. $\pi/16$

29. 0

31. $\pi/4$

33. $\dfrac{x}{\sqrt{1 + x^2}}$

35. $\dfrac{x}{\sqrt{1 - x^2}}$

37. $\dfrac{\sqrt{1 - x^2}}{x}$

39. $\dfrac{\sqrt{1 + x^2}}{x}$

41.

43.

45.

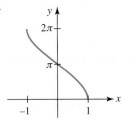

47.

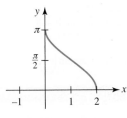

49. domain: $(-\infty, \infty)$,
range: $(0, \pi)$

51. domain: $(-\infty, -1] \cup [1, \infty)$,
range: $[0, \pi/2) \cup (\pi/2, \pi]$

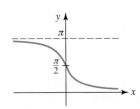

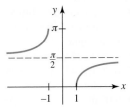

53. (a) x in $(-\infty, \infty)$ (b) x in $(0, \pi)$
55. (a) x in $(-\infty, -1] \cup [1, \infty)$ (b) x in $[0, \pi/2) \cup (\pi/2, \pi]$
59. 0.9273   **61.** -0.7297
63. 2.5559   **65.** $19.9°, 70.1°$
67. $5.76°$ **69.** $\phi = 0.5404$ radian $\approx 31°$

43. $x = \dfrac{\pi}{6} + 2n\pi$, $x = \dfrac{5\pi}{6} + 2n\pi$, where n is an integer
45. $\theta = 90° + 180°n$, $\theta = 360°n$, where n is an integer
47. $\{0, \pi/10, 3\pi/10, \pi/2, 7\pi/10, 9\pi/10, \pi, 11\pi/10, 13\pi/10, 3\pi/2, 17\pi/10, 19\pi/10\}$
49. $\{0, \pm 2\pi/5, \pm 2\pi/3, \pm 4\pi/5\}$
51. $\{0, \pm \pi/3, \pm 2\pi/3, \pm \pi/2\}$
53. $(\pi/3, 0), (2\pi/3, 0), (\pi, 0)$ **55.** $\left(\frac{2}{3}, 0\right), \left(\frac{10}{3}, 0\right), \left(\frac{14}{3}, 0\right)$
57. $(\pi, 0), (2\pi, 0), (3\pi, 0)$ **59.** $(\pi/3, 0), (\pi, 0), (5\pi/3, 0)$
61. The equation has infinitely many solutions.   **63.** The equation has infinitely many solutions.

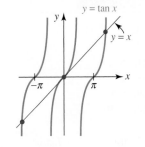

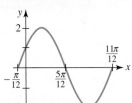

65. 1.37, 1.82   **67.** $-1.02, 0.55$
69. 0.58, 1.57, 1.81   **71.** $30°, 150°$
73. $60°$
75. $t = \frac{1}{120}\left(\frac{1}{6} + n\right)$, where n is an integer.
77. (a) 36.93 million square kilometers   (b) $w = 31$ weeks
(c) August

Exercises 4.9   Page 262

1. $x = \dfrac{\pi}{3} + 2n\pi$ or $x = \dfrac{2\pi}{3} + 2n\pi$, where n is an integer
3. $x = \dfrac{\pi}{4} + 2n\pi$ or $x = \dfrac{7\pi}{4} + 2n\pi$, where n is an integer
5. $x = \dfrac{5\pi}{6} + n\pi$, where n is an integer
7. $x = \pi + 2n\pi = (2n + 1)\pi$, where n is an integer
9. $x = n\pi$, where n is an integer
11. $x = \dfrac{3\pi}{2} + 2n\pi$, where n is an integer
13. $\theta = 60° + 360°n$, $\theta = 120° + 360°n$, where n is an integer
15. $\theta = 135° + 180°n$, where n is an integer
17. $\theta = 120° + 360°n$, $\theta = 240° + 360°n$, where n is an integer
19. $x = n\pi$, where n is an integer
21. no solutions
23. $\theta = 120° + 360°n$, $\theta = 240° + 360°n$, where n is an integer
25. $\theta = 90° + 180°n$, $\theta = 135° + 180°n$, where n is an integer
27. $x = \dfrac{\pi}{2} + n\pi$, where n is an integer
29. $\theta = 10° + 120°n$, $\theta = 50° + 120°n$, where n is an integer
31. $x = \dfrac{\pi}{2} + 2n\pi$, where n is an integer
33. $x = n\pi$, $x = \dfrac{2\pi}{3} + 2n\pi$, $x = \dfrac{4\pi}{3} + 2n\pi$, where n is an integer
35. $\theta = 30° + 360°n$, $\theta = 150° + 360°n$, $\theta = 270° + 360°n$, where n is an integer
37. $x = \dfrac{\pi}{2} + n\pi$, where n is an integer
39. $x = n\pi$, where n is an integer
41. $\theta = 2n\pi$, $\theta = \dfrac{\pi}{2} + 2n\pi$, where n is an integer

---

Exercises 4.10   Page 267

1. $y = \sqrt{2}\sin(\pi x + 3\pi/4)$;
amplitude: $\sqrt{2}$; period: 2;
phase shift: $\frac{3}{4}$; one cycle of the graph is

3. $y = 2\sin(2x + \pi/6)$;
amplitude: 2; period: π;
phase shift: $\pi/12$; one cycle of the graph is

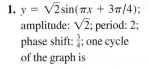

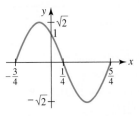

5. $y = \sin(x + 5\pi/4)$;
amplitude: 1; period: 2π;
phase shift: $5\pi/4$; one cycle of the graph is

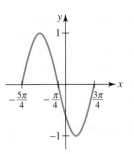

7. $y = \frac{\sqrt{13}}{4}\sin(2t + 0.5880)$; amplitude: $\frac{\sqrt{13}}{4}$ feet;
period: π seconds; frequency: $1/\pi$ cycles per second
9. $y = \frac{\sqrt{5}}{2}\sin(4t + 4.2487)$; amplitude: $\frac{\sqrt{5}}{2}$ feet;
period: $\pi/2$ seconds; frequency: $2/\pi$ cycles per second
11. $y_0 = -\frac{5\sqrt{3}}{4}$, $v_0 = \frac{5}{2}$
13. $(\pi/12, 0), (7\pi/12, 0), (13\pi/12, 0)$

Exercises 4.11   Page 272

1. $\frac{1}{2}$ **3.** -1
5. $-\frac{\sqrt{3}}{2}$ **7.** 5
9. 0   **11.** -2
13. 0   **15.** 1
17. 1   **19.** $y = x$
21. $y = \frac{1}{2} + \frac{\sqrt{3}}{2}\left(x - \frac{\pi}{6}\right)$ **23.** $f'(x) = -\sin x$
25. $f'(x) = 5\cos 5x$

Chapter 4 Review Exercises   Page 274

A. 1. $36°$ **3.** $\left(-\frac{\sqrt{3}}{2}, \frac{1}{2}\right)$
5. $\sqrt{3}$ **7.** $\frac{3}{2\sqrt{2}}$
9. $(0, -2)$ **11.** $\frac{2}{\sqrt{5}}$
13. $\sin 2\pi x$ **15.** $\sqrt{2}$
17. 10   **19.** $\pi/5$
21. 3   **23.** $(5, 0)$
25. 10

B. 1. true   **3.** true   **5.** true   **7.** true
9. true   **11.** true   **13.** true   **15.** false
17. true   **19.** false   **21.** true   **23.** false
25. true

C. 1. $\pm 6\sin(\pi x/2)$ **3.** $3\sin(4x - \pi), -3\sin(4x + \pi)$
5. $\{\pi/2, \pi\}$ **7.** $\{\pi/6, 5\pi/6, 7\pi/6, 11\pi/6\}$
9. $\{\pi/4, 3\pi/4, 5\pi/4, 7\pi/4\}$
11. $\{0, \pi/6, \pi/2, 5\pi/6, \pi, 7\pi/6, 3\pi/2, 11\pi/6, 2\pi\}$
13. $\{\pi/4, 5\pi/4\}$ **15.** $2\pi/3$
17. $\frac{3}{\sqrt{7}}$ **19.** 0
21. $\frac{12}{13}$ **23.** $\sqrt{1 - x^2}$
25. $y = -\sin x; y = \cos(x + \pi/2)$
27. $y = 1 + \frac{1}{2}\sin(x + \pi/2); y = 1 + \frac{1}{2}\cos x$

Exercises 5.1   Page 282

1. $\sin\theta = \frac{4}{5}$, $\cos\theta = \frac{3}{5}$, $\tan\theta = \frac{4}{3}$, $\csc\theta = \frac{5}{4}$, $\sec\theta = \frac{5}{3}$, $\cot\theta = \frac{3}{4}$
3. $\sin\theta = 3\sqrt{10}/10$, $\cos\theta = \sqrt{10}/10$, $\tan\theta = 3$, $\csc\theta = \sqrt{10}/3$, $\sec\theta = \sqrt{10}$, $\cot\theta = \frac{1}{3}$
5. $\sin\theta = \frac{2}{5}$, $\cos\theta = \sqrt{21}/5$, $\tan\theta = 2\sqrt{21}/21$, $\csc\theta = \frac{5}{2}$, $\sec\theta = 5\sqrt{21}/21$, $\cot\theta = \sqrt{21}/2$
7. $\sin\theta = \frac{1}{3}$, $\cos\theta = 2\sqrt{2}/3$, $\tan\theta = \sqrt{2}/4$, $\csc\theta = 3$, $\sec\theta = 3\sqrt{2}/4$, $\cot\theta = 2\sqrt{2}$
9. $\sin\theta = y/\sqrt{x^2 + y^2}$, $\cos\theta = x/\sqrt{x^2 + y^2}$, $\tan\theta = y/x$, $\csc\theta = \sqrt{x^2 + y^2}/y$, $\sec\theta = \sqrt{x^2 + y^2}/x$, $\cot\theta = x/y$
11. $b = 2.04, c = 4.49$ **13.** $a = 11.71, c = 14.19$
15. $\alpha = 60°, \beta = 30°, a = 2.6$
17. $\alpha = 21.8°, \beta = 68.2°, c = 10.8$
19. $\alpha = 48.6°, \beta = 41.4°, b = 7.9$ **21.** $a = 8.5, c = 21.7$
23. 36.53

Exercises 5.2   Page 287

1. 52.1 m   **3.** 66.4 ft
5. 409.7 ft   **7.** height: 15.5 ft; distance: 12.6 ft
9. 8.7°   **11.** 34,157 ft ≈ 6.5 mi
13. 20.2 ft   **15.** 6617 ft
17. Yes, since the altitude of the storm is 6.3 km.

21. 227,100 mi   **23.** $h = c/(\cot\alpha + \cot\beta)$
25. approximately 4.8   **27.** 80.87°
29. $h(\theta) = 1.25\tan\theta$
31. $\theta(x) = \arctan\left(\frac{1}{x}\right) - \arctan\left(\frac{1}{2x}\right)$, where x is measured in meters

Exercises 5.3   Page 294

1. $\gamma = 80°, a = 20.16, c = 20.16$
3. $\alpha = 92°, b = 3.01, c = 3.89$
5. $\alpha = 79.61°, \gamma = 28.39°, a = 12.41$
7. no solution
9. $\alpha = 24.46°, \beta = 140.54°, b = 12.28$; $\alpha = 155.54°, \beta = 9.46°, b = 3.18$
11. no solution
13. $\alpha = 45.58°, \gamma = 104.42°, c = 13.56$; $\alpha = 134.42°, \gamma = 15.58°, c = 3.76$
15. no solution   **17.** 15.80 ft
19. 9.07 m   **21.** 10.35 ft
23. 8.81°   **25.** the vessel sailing at 5 knots

Exercises 5.4   Page 298

1. $\alpha = 37.59°, \beta = 77.41°, c = 7.43$
3. $\alpha = 52.62°, \beta = 83.33°, \gamma = 44.05°$
5. $\alpha = 25°, \beta = 57.67°, c = 7.04$
7. $\alpha = 76.45°, \beta = 57.10°, \gamma = 46.45°$
9. $\alpha = 27.66°, \beta = 40.54°, \gamma = 111.80°$
11. $\alpha = 36.87°, \beta = 53.13°, \gamma = 90°$
13. $\alpha = 26.38°, \beta = 36.34°, \gamma = 117.28°$
15. $\beta = 10.24°, \gamma = 147.76°, a = 6.32$
17. 35.94 nautical miles
19. (a) S33.66° W   **(b)** S2.82° E
21. $\alpha = 119.45°, \beta = 67.98°$
23. 91.77, 176.18

Exercises 5.5   Page 312

1. $|\mathbf{v}| = 2, \theta = 11\pi/6$ **3.** $|\mathbf{v}| = 5, \theta = 2\pi$

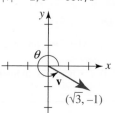

5. $|\mathbf{v}| = 8, \theta = 2\pi/3$ **7.** $|\mathbf{v}| = 10\sqrt{2}, \theta = 3\pi/4$

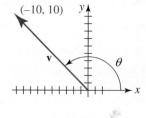

9. $\langle 3, 2\rangle, \langle 1, 4\rangle, \langle -6, -9\rangle, \langle 2, 13\rangle$
11. $\langle 0, 3\rangle, \langle -8, 1\rangle, \langle 12, -6\rangle, \langle -28, 2\rangle$
13. $\langle -\frac{9}{2}, -\frac{29}{4}\rangle, \langle -\frac{11}{2}, -\frac{27}{4}\rangle, \langle 15, 21\rangle, \langle -17, -20\rangle$
15. $-31\mathbf{i} - 14\mathbf{j}, 42\mathbf{i} + 11\mathbf{j}$ **17.** $-\frac{15}{2}\mathbf{i} - \frac{3}{2}\mathbf{j}, 11\mathbf{i} - 3\mathbf{j}$
19. $5.8\mathbf{i} + 8.5\mathbf{j}, -6.6\mathbf{i} - 10.3\mathbf{j}$

21.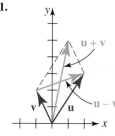

23.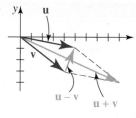

25.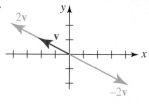

27.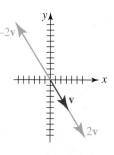

29. horizontal component: 4, vertical component: −6

31. horizontal component: −10, vertical component: 8

33. (a) $2\left(\cos\dfrac{3\pi}{4}\mathbf{i} + \sin\dfrac{3\pi}{4}\mathbf{j}\right)$ (b) $-\sqrt{2}\mathbf{i} + \sqrt{2}\mathbf{j}$

35. (a) $6\left(\cos\dfrac{5\pi}{6}\mathbf{i} + \sin\dfrac{5\pi}{6}\mathbf{j}\right)$ (b) $-3\sqrt{3}\mathbf{i} + 3\mathbf{j}$

37. (a) $\langle 1/\sqrt{2}, 1/\sqrt{2}\rangle$ (b) $\langle -1/\sqrt{2}, -1/\sqrt{2}\rangle$

39. (a) $\langle 0, -1\rangle$ (b) $\langle 0, 1\rangle$ **41.** $\langle\frac{5}{13}, \frac{12}{13}\rangle$

43. 10.16, 30.26°   **45.** 10

47. 1   **49.** −17

51. 12   **53.** 13

55. −68   **57.** $\frac{13}{2}$

59. 8   **61.** $\langle\frac{17}{26}, -\frac{85}{26}\rangle$

63. $25\sqrt{2}$ **65.** 102.53°

67. 63.43°   **69.** not orthogonal

71. orthogonal   **73.** $c = 3$

77. $-\frac{\sqrt{10}}{5}$ **79.** $-3\sqrt{2}$

81. (a) $-\frac{21}{5}\mathbf{i} + \frac{28}{5}\mathbf{j}$ (b) $-\frac{7}{2}\mathbf{i} + \frac{7}{2}\mathbf{j}$

83. $\frac{72}{25}\mathbf{i} + \frac{96}{25}\mathbf{j}$ **85.** 1000 ft-lb

87. $\frac{78}{5}$ ft-lb   **89.** $2\mathbf{i} - 10\mathbf{j}$

91. Actual course makes an angle of 71.6° from the original heading.

93. 52.9 mi/h

Chapter 5 Review Exercises   Page 315

A. 1. Sines   **3.** Cosines

5. approximately 77.06°   **7.** approximately 3.87 mi

9. $\langle\frac{12}{13}, -\frac{5}{13}\rangle$ **11.** 90°

B. 1. false   **3.** false

5. true   **7.** true

9. false   **11.** true

C. 1. $\gamma = 80°, a = 5.32, c = 10.48$

3. $a = 15.76, \beta = 99.44°, \gamma = 29.56°$

5. 42.61 m   **7.** 118 ft

9. (a) 42.35°   **11.** 16,927.6 km

13. front: 18.88°, back: 35.12°   **15.** 233 ft, 182 ft

17. $V(\theta) = 160\sin 2\theta$ **19.** $L(\theta) = 3\csc\theta + 4\sec\theta$

21. $V(\theta) = 360 + 75\cot\theta$

23. $A(\phi) = 100\cos\phi + 50\sin 2\phi$

25. (a) $d = \sqrt{149 - 140\cos\theta}$ (b) $\theta \approx 109.62°$

27. $13\mathbf{i} - 12\mathbf{j}$ **29.** $3\mathbf{i} - 7\mathbf{j}$

31. $7\mathbf{i} - 13\mathbf{j}$ **33.** $\frac{-3}{\sqrt{17}}$

35. $\mathbf{i} + \mathbf{j}$ **37.** $\sqrt{13} + 2\sqrt{2}$

39. $2\sqrt{2}\left(\cos\dfrac{\pi}{4}\mathbf{i} + \sin\dfrac{\pi}{4}\mathbf{j}\right)$ **41.** $-\frac{1}{\sqrt{17}}\mathbf{i} + \frac{4}{\sqrt{17}}\mathbf{j}$

43. $5\sqrt{2}$, 135°

Exercises 6.1   Page 326

1. $(0, 1); y = 0$; decreasing

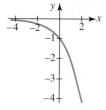

3. $(0, -1); y = 0$; decreasing

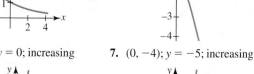

5. $(0, 2); y = 0$; increasing

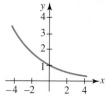

7. $(0, -4); y = -5$; increasing

9. $(0, 2); y = 3$; increasing

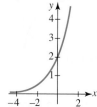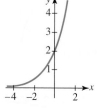

11. $(0, -1 + e^{-3}); y = -1$; increasing

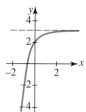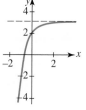

13. $f(x) = 6^x$ **15.** $f(x) = (e^{-2})^x = e^{-2x}$

17. $f(x) = 3^{-x}$ **19.** $(5, \infty)$

21. $(-2, \infty)$ **23.** $(2, 0), (0, -3)$

25. $(-10, 0), (0, 10)$ **27.** $(-2, 0), (-3, 0), (0, 0)$

29. $x > 4$ **31.** $x < 2$

33. 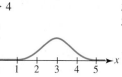   **35.**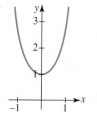

37.   **39.**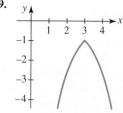

ANSWERS TO SELECTED ODD-NUMBERED PROBLEMS   ANS-17

Answers to Selected Odd-Numbered Problems, CHAPTER 6

41.

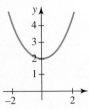

43.

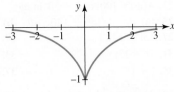

45. $y = \frac{26}{9}x + \frac{29}{9}$

47. approximately $-0.77, 2, 4$

1. $-\frac{1}{2} = \log_4 \frac{1}{2}$

3. $4 = \log_{10} 10,000$

5. $-s = \log_t v$

7. $2^7 = 128$

9. $(\sqrt{3})^8 = 81$

11. $b^v = u$

13. -7

15. 3

17. e

19. 36

21. $\frac{1}{7}$

23. $f(x) = \log_7 x$

25. $(0, \infty); (1, 0), x = 0$

27. $(-\infty, 0); (-1, 0), x = 0$

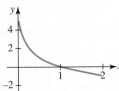

29. $(-3, \infty); (5, 0), x = -3$

31. $(0, \infty); (e, 0), x = 0$

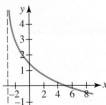

33. $-1 < x < 0$

35. $f(x) = \begin{cases} \ln x, & x > 0 \\ \ln(-x), x < 0 \end{cases}, (-1, 0), (1, 0), x = 0$

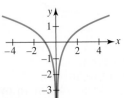

37.

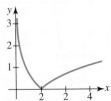

39. $\left(\frac{3}{2}, \infty\right)$

41. the interval $(-3, 3)$

43 $(1, \infty)$

45. $\log_{10} 50$

47. $\ln(x^2 - 2)$

49. $\ln 1 = 0$

51. 0.3011

53. 1.8063

55. 0.3495

57. 0.2007

59. -0.0969

61. 1.6609

61. $\ln y = 10\ln x + \frac{1}{2}\ln(x^2 + 5) - \frac{1}{3}\ln(8x^3 + 2)$

63. $\ln y = 5\ln(x^3 - 3) + 8\ln(x^4 + 3x^2 + 1)$
   $\qquad -\frac{1}{2}\ln x - 9\ln(7x + 5)$

1. 2

3. 2

5. 1.0802

7. 2

9. 3

11. 0.3495

13. 2.7712

15. ± 4

17. ± 3

19. -0.8782

21. 32

23. 45

25. $\pm\frac{1}{10}$

27. 81

29. $\frac{3}{2}$

31. 100

33. $2, 8$

35. 1

37. 4

39. $\frac{7}{2}$

41. $0, 2$

43. $1, 16$

45. $\log_5(1 + \sqrt{2}) = \dfrac{\ln(1 + \sqrt{2})}{\ln 5}$

47. e^{-2}, e

49. 0

51. $(-3, 0)$

53. $\left(1 + \frac{\ln 3}{\ln 4}, 0\right) \approx (1.7925, 0)$

55. $(-1.0397, 0)$

57. $(2.3219, 0)$

59. approximately -0.6606

61. approximately
   $-0.4481, 1.5468$

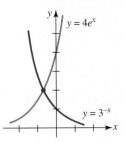

63. $\sqrt{10} \approx 3.1623$

65. e^{-3}, e^3

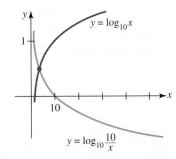

67. 9

69. $x \approx 2.1944$

1. (a) $P(t) = P_0 e^{0.3466t}$ (b) $5.66P_0$ (c) 8.64 h

3. $2,344$ **5.** 201

7. (a) 82 (b) 8.53 days (c) 2000

(d)

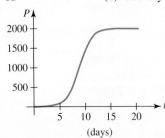

9. $A(t) = 200e^{-0.005077t}$; 177 mg

11. approximately 100 g, 50 g, 25 g
13. approximately $k = -0.08664$; 34.58 days
15. 0.6730 A_0; 0.2264 A_0
17. approximately 16,253 years old
19. approximately 92%
21. (a) 220.2° F   **(b)** 9.2 minutes   **(c)** 80° F
23. 8.74 minutes
25. approximately 1.6 hours
27. \$4,851,651.95; \$2.35 $\times 10^{15}$
29. \$3080.37 in interest
31. (a) 6 days   **(b)** 19.84%
33. $t = -\dfrac{L}{R}\ln\left(1 - \dfrac{IR}{E}\right)$
35. approximately 25 times stronger
37. 5.5
39. 6
41. 7.6
43. 5×10^{-4}
45. 2.5×10^{-7}
47. 10 times as acidic
49. 158.5 times as acidic
51. (a) 5.62×10^{25} ergs
53. 10^{-2} watts/cm^2
55. 65 dB
57. (a) 2.46 mm   **(b)** 0.79 mm, 0.19 mm   **(c)** 7.7×10^{-6} mL

Exercises 6.5   Page 358

3. $\dfrac{1}{x}\log_{10} e = \dfrac{1}{x\ln 10}$
5. $\dfrac{1}{x}$
7. $e^{5x}\left(\dfrac{e^{5h} - 1}{h}\right)$
15. (a) $\dfrac{\sqrt{13}}{2}$ **(b)** $\dfrac{-3}{\sqrt{13}}, -\dfrac{\sqrt{13}}{3}, \dfrac{2}{\sqrt{13}}, -\dfrac{2}{3}$
21. domain: $(-1, 1)$, range: $(-\infty, \infty)$
23. $5e^{5x}$

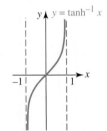

Chapter 6 Review Exercises   Page 360

A. 1. $(0, 5), y = 6$
3. $(-3, 0), x = -4$
5. -1
7. 1000
9. 6
11. $\frac{1}{9}$
13. 3
15. $3 + e$
17. -7.8577
19. 64
21. 8

B. 1. true
3. true
5. true
7. false
9. true
11. false
13. true
15. true
17. true
19. true
21. false

C. 1. $\log_5 0.2 = -1$
3. $9^{1.5} = 27$
5. -2
7. $-\frac{1}{2}$
9. $1 - \log_2 7 = 1 - \dfrac{\ln 7}{\ln 2}$
11. $-2 + \ln 6$

13. $m = -\dfrac{1}{r}\ln(P/S)$

15.

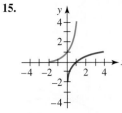

17. C, D, A, B
19. $\dfrac{3^{1-h} - 3}{h}$
21. (ii)
23. (iv)
25. (iii)
27. upward shift of 1 unit
29. $f(x) = 5e^{(-\frac{1}{6}\ln 5)x} = 5e^{-0.2682x}$
31. $f(x) = 5 + \left(\frac{1}{2}\right)^x$
33. After doubling, it will take another 9 hours to double again. In other words, it will take a total of 18 hours for the population to grow to 4 times the initial population.
35. $0.0625A_0$ or $6\frac{1}{4}\%$ of the initial quantity A_0
37. 4.3%
39. $t = \dfrac{1}{c}[\ln b - \ln(\ln a - \ln y)]$
41. \$51,955.78
43. $y = x - 3$
45. $y = 8x$
47. $y = 4x^2 - 4x + 6$
49. $y = 3(x^2 + 1)$
51. $y = 5e^{-3x}$

Exercises 7.1   Page 372

1. Vertex: $(0, 0)$
Focus: $(1, 0)$
Directrix: $x = -1$
Axis: $y = 0$

$y^2 = 4x$

3. Vertex: $(0, 0)$
Focus: $\left(-\frac{1}{3}, 0\right)$
Directrix: $x = \frac{1}{3}$
Axis: $y = 0$

$y^2 = -\frac{4}{3}x$

5. Vertex: $(0, 0)$
Focus: $(0, -4)$
Directrix: $y = 4$
Axis: $x = 0$

$x^2 = -16y$

7. Vertex: $(0, 0)$
Focus: $(0, 7)$
Directrix: $y = -7$
Axis: $x = 0$

$x^2 = 28y$

9. Vertex: $(0, 1)$
Focus: $(4, 1)$
Directrix: $x = -4$
Axis: $y = 1$

$(y - 1)^2 = 16x$

11. Vertex: $(-5, -1)$
Focus: $(-5, -2)$
Directrix: $y = 0$
Axis: $x = -5$

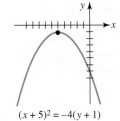

$(x + 5)^2 = -4(y + 1)$

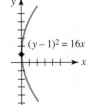

Answers to Selected Odd-Numbered Problems, CHAPTER 7

13. Vertex: $(-5, -6)$
Focus: $(-4, -6)$
Directrix: $x = -6$
Axis: $y = -6$

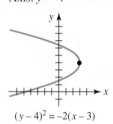

$(y + 6)^2 = 4(x + 5)$

15. Vertex: $\left(-\frac{5}{2}, -1\right)$
Focus: $\left(-\frac{5}{2}, -\frac{15}{16}\right)$
Directrix: $y = -\frac{17}{16}$
Axis: $x = -\frac{5}{2}$

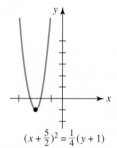

$\left(x + \frac{5}{2}\right)^2 = \frac{1}{4}(y + 1)$

17. Vertex: $(3, 4)$
Focus: $\left(\frac{5}{2}, 4\right)$
Directrix: $x = \frac{7}{2}$
Axis: $y = 4$

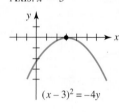

$(y - 4)^2 = -2(x - 3)$

19. Vertex: $(0, 0)$
Focus: $\left(0, \frac{1}{8}\right)$
Directrix: $y = -\frac{1}{8}$
Axis: $x = 0$

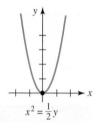

$x^2 = \frac{1}{2}y$

21. Vertex: $(3, 0)$
Focus: $(3, -1)$
Directrix: $y = 1$
Axis: $x = 3$

$(x - 3)^2 = -4y$

23. Vertex: $(-2, 1)$
Focus: $(-1, 1)$
Directrix: $x = -3$
Axis: $y = 1$

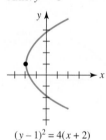

$(y - 1)^2 = 4(x + 2)$

25. $x^2 = 28y$ **27.** $y^2 = -16x$ **29.** $y^2 = 10x$
31. $(x - 2)^2 = 12y$ **33.** $(y - 4)^2 = -12(x - 2)$
35. $(x - 1)^2 = 32(y + 3)$ **37.** $(y + 3)^2 = 32x$
39. $x^2 = 7y$ **41.** $(x - 5)^2 = -24(y - 1)$
43. $x^2 = \frac{1}{2}y$ **45.** $(3, 0), (0, -2), (0, -6)$
47. $\left(-3\sqrt{2}, 0\right), \left(3\sqrt{2}, 0\right), (0, 9)$
49. At the focus 6 in. from the vertex
51. $y = -2$ **53.** 27 ft
55. 4.5 ft **57.** (a) 8

Exercises 7.2   Page 380

1. Center: $(0, 0)$
Foci: $(\pm 4, 0)$
Vertices: $(\pm 5, 0)$
Minor axis endpoints: $(0, \pm 3)$
Eccentricity: $\frac{4}{5}$

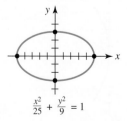

$\frac{x^2}{25} + \frac{y^2}{9} = 1$

3. Center: $(0, 0)$
Foci: $\left(0, \pm \sqrt{15}\right)$
Vertices: $(0, \pm 4)$
Minor axis endpoints: $(\pm 1, 0)$
Eccentricity: $\frac{\sqrt{15}}{4}$

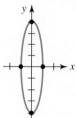

$\frac{x^2}{1} + \frac{y^2}{16} = 1$

5. Center: $(0, 0)$
Foci: $\left(\pm \sqrt{7}, 0\right)$
Vertices: $(\pm 4, 0)$
Minor axis endpoints: $(0, \pm 3)$
Eccentricity: $\frac{\sqrt{7}}{4}$

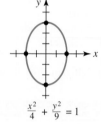

$\frac{x^2}{16} + \frac{y^2}{9} = 1$

7. Center: $(0, 0)$
Foci: $\left(0, \pm \sqrt{5}\right)$
Vertices: $(0, \pm 3)$
Minor axis endpoints: $(\pm 2, 0)$
Eccentricity: $\frac{\sqrt{5}}{3}$

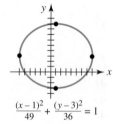

$\frac{x^2}{4} + \frac{y^2}{9} = 1$

9. Center: $(1, 3)$
Foci: $\left(1 \pm \sqrt{13}, 3\right)$
Vertices: $(-6, 3), (8, 3)$
Minor axis endpoints: $(1, -3), (1, 9)$
Eccentricity: $\frac{\sqrt{13}}{7}$

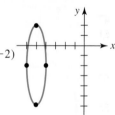

$\frac{(x - 1)^2}{49} + \frac{(y - 3)^2}{36} = 1$

11. Center: $(-5, -2)$
Foci: $\left(-5, -2 \pm \sqrt{15}\right)$
Vertices: $(-5, -6), (-5, 2)$
Minor axis endpoints: $(-6, -2), (-4, -2)$
Eccentricity: $\frac{\sqrt{15}}{4}$

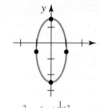

$\frac{(x + 5)^2}{1} + \frac{(y + 2)^2}{16} = 1$

13. Center: $\left(0, -\frac{1}{2}\right)$
Foci: $\left(0, -\frac{1}{2} \pm \sqrt{3}\right)$
Vertices: $\left(0, -\frac{5}{2}\right), \left(0, \frac{3}{2}\right)$
Minor axis endpoints: $\left(-1, -\frac{1}{2}\right), \left(1, -\frac{1}{2}\right)$
Eccentricity: $\frac{\sqrt{3}}{2}$

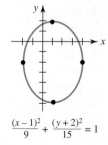

$\frac{x^2}{1} + \frac{\left(y + \frac{1}{2}\right)^2}{4} = 1$

15. Center: $(1, -2)$
Foci: $\left(1, -2 \pm \sqrt{6}\right)$
Vertices: $\left(1, -2 \pm \sqrt{15}\right)$
Minor axis endpoints: $(-2, -2), (4, -2)$
Eccentricity: $\sqrt{\frac{2}{5}}$

$\frac{(x - 1)^2}{9} + \frac{(y + 2)^2}{15} = 1$

ANS-20   ANSWERS TO SELECTED ODD-NUMBERED PROBLEMS

17. Center: $(2, -1)$
Foci: $(2, -5), (2, 3)$
Vertices: $(2, -6), (2, 4)$
Minor axis endpoints: $(-1, -1), (5, -1)$
Eccentricity: $\frac{4}{5}$

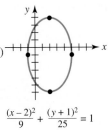

$$\frac{(x-2)^2}{9} + \frac{(y+1)^2}{25} = 1$$

19. Center: $(0, -3)$
Foci: $(\pm\sqrt{6}, -3)$
Vertices: $(-3, -3), (3, -3)$
Minor axis endpoints: $(0, -3 \pm \sqrt{3})$
Eccentricity: $\frac{\sqrt{6}}{3}$

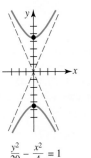

$$\frac{x^2}{9} + \frac{(y+3)^2}{3} = 1$$

21. $\dfrac{x^2}{25} + \dfrac{y^2}{16} = 1$

23. $\dfrac{x^2}{8} + \dfrac{y^2}{9} = 1$

25. $\dfrac{x^2}{1} + \dfrac{y^2}{9} = 1$

27. $\dfrac{(x-1)^2}{16} + \dfrac{(y+3)^2}{4} = 1$

29. $\dfrac{x^2}{9} + \dfrac{y^2}{13} = 1$

31. $\dfrac{x^2}{11} + \dfrac{y^2}{9} = 1$

33. $\dfrac{x^2}{3} + \dfrac{y^2}{12} = 1$

35. $\dfrac{x^2}{16} + \dfrac{(y-1)^2}{4} = 1$

37. $\dfrac{(x-1)^2}{7} + \dfrac{(y-3)^2}{16} = 1$

39. $\dfrac{x^2}{45} + \dfrac{(y-2)^2}{9} = 1$

41. $f(x) = -\frac{3}{5}\sqrt{25 - x^2}$, domain is $[-5, 5]$

43. $f(x) = 3 + \frac{6}{7}\sqrt{49 - (x-1)^2}$, domain is $[-6, 8]$

45. greatest distance is 43.5 millions miles; least distance is 28.5 million miles

47. approximately 0.97   **49.** 12 ft

51. The piece of string should be 4 ft long. The tacks should be placed $\sqrt{7}/2$ ft from the center of the rectangle on the major axis of the ellipse.

53. on the major axis, 12 ft to either side from the center of the room

55. $5x^2 - 4xy + 8y^2 - 24x - 48y = 0$

Exercises 7.3   Page 389

1. Center: $(0, 0)$
Foci: $(\pm\sqrt{41}, 0)$
Vertices: $(\pm 4, 0)$
Asymptotes: $y = \pm\frac{5}{4}x$
Eccentricity: $\frac{\sqrt{41}}{4}$

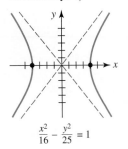

$$\frac{x^2}{16} - \frac{y^2}{25} = 1$$

3. Center: $(0, 0)$
Foci: $(0, \pm\sqrt{73})$
Vertices: $(0, \pm 8)$
Asymptotes: $y = \pm\frac{8}{3}x$
Eccentricity: $\frac{\sqrt{73}}{8}$

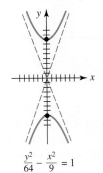

$$\frac{y^2}{64} - \frac{x^2}{9} = 1$$

5. Center: $(0, 0)$
Foci: $(\pm 2\sqrt{5}, 0)$
Vertices: $(\pm 4, 0)$
Asymptotes: $y = \pm\frac{1}{2}x$
Eccentricity: $\frac{\sqrt{5}}{2}$

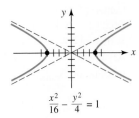

$$\frac{x^2}{16} - \frac{y^2}{4} = 1$$

7. Center: $(0, 0)$
Foci: $(0, \pm 2\sqrt{6})$
Vertices: $(0, \pm 2\sqrt{5})$
Asymptotes: $y = \pm\sqrt{5}x$
Eccentricity: $\sqrt{\frac{6}{5}}$

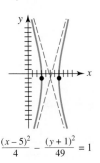

$$\frac{y^2}{20} - \frac{x^2}{4} = 1$$

9. Center: $(5, -1)$
Foci: $(5 \pm \sqrt{53}, -1)$
Vertices: $(3, -1), (7, -1)$
Asymptotes: $y = -1 \pm \frac{7}{2}(x - 5)$
Eccentricity: $\frac{\sqrt{53}}{2}$

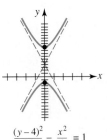

$$\frac{(x-5)^2}{4} - \frac{(y+1)^2}{49} = 1$$

11. Center: $(0, 4)$
Foci: $(0, 4 \pm \sqrt{37})$
Vertices: $(0, -2), (0, 10)$
Asymptotes: $y = 4 \pm 6x$
Eccentricity: $\frac{\sqrt{37}}{6}$

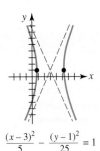

$$\frac{(y-4)^2}{36} - \frac{x^2}{1} = 1$$

13. Center: $(3, 1)$
Foci: $(3 \pm \sqrt{30}, 1)$
Vertices: $(3 \pm \sqrt{5}, 1)$
Asymptotes: $y = 1 \pm \sqrt{5}(x - 3)$
Eccentricity: $\sqrt{6}$

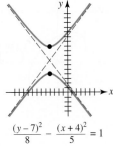

$$\frac{(x-3)^2}{5} - \frac{(y-1)^2}{25} = 1$$

15. Center: $(-4, 7)$
Foci: $(-4, 7 \pm \sqrt{13})$
Vertices: $(-4, 7 \pm 2\sqrt{2})$
Asymptotes: $y = 7 \pm \sqrt{\frac{8}{5}}(x + 4)$
Eccentricity: $\sqrt{\frac{13}{8}}$

$$\frac{(y-7)^2}{8} - \frac{(x+4)^2}{5} = 1$$

17. Center: $(2, 1)$
Foci: $(2 \pm \sqrt{11}, 1)$
Vertices: $(2 \pm \sqrt{6}, 1)$
Asymptotes: $y = 1 \pm \sqrt{\frac{5}{6}}(x - 2)$
Eccentricity: $\sqrt{\frac{11}{6}}$

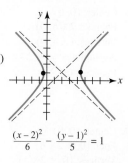

$$\frac{(x-2)^2}{6} - \frac{(y-1)^2}{5} = 1$$

19. Center: $(1, 3)$
Foci: $(1, 3 \pm \frac{1}{2}\sqrt{5})$
Vertices: $(1, 2), (1, 4)$
Asymptotes: $y = 3 \pm 2(x - 1)$
Eccentricity: $\frac{\sqrt{5}}{2}$

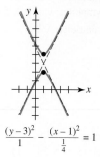

$$\frac{(y-3)^2}{1} - \frac{(x-1)^2}{\frac{1}{4}} = 1$$

21. $\dfrac{x^2}{9} - \dfrac{y^2}{16} = 1$

23. $\dfrac{y^2}{4} - \dfrac{x^2}{12} = 1$

25. $\dfrac{x^2}{9} - \dfrac{y^2}{7} = 1$

27. $\dfrac{y^2}{\frac{25}{4}} - \dfrac{x^2}{\frac{11}{4}} = 1$

29. $\dfrac{x^2}{4} - \dfrac{y^2}{5} = 1$

31. $\dfrac{y^2}{64} - \dfrac{x^2}{16} = 1$

33. $\dfrac{x^2}{4} - \dfrac{y^2}{\frac{64}{9}} = 1$

35. $\dfrac{(y+3)^2}{4} - \dfrac{(x-1)^2}{5} = 1$

37. $\dfrac{(x+1)^2}{4} - \dfrac{(y-2)^2}{5} = 1$

39. $\dfrac{x^2}{4} - \dfrac{y^2}{8} = 1$

41. $\dfrac{(y-3)^2}{1} - \dfrac{(x+1)^2}{4} = 1$

43. $\dfrac{(y-4)^2}{1} - \dfrac{(x-2)^2}{4} = 1$

45. $f(x) = \frac{5}{4}\sqrt{x^2 - 16}$, domain is $(-\infty, -4] \cup [4, \infty)$

47. $f(x) = 4 + 6\sqrt{x^2 + 1}$, domain is $(-\infty, \infty)$

49. $(-7, 12)$

51. $7y^2 - 24xy + 24x + 82y + 55 = 0$

Exercises 7.4   Page 396

1. $(4\sqrt{2}, -2\sqrt{2})$

3. $\left(-\frac{1}{2} - \frac{\sqrt{3}}{2}, \frac{\sqrt{3}}{2} - \frac{1}{2}\right)$

5. $(4 + \sqrt{3}, 1 - 4\sqrt{3})$

7. $(-4, 0)$

9. approximately $(2.31, 6.83)$

11. ellipse rotated $45°$, $3x'^2 + y'^2 = 8$

13. parabola rotated $45°$, $y'^2 = 4\sqrt{2}x'$

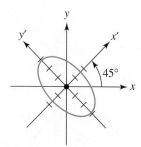

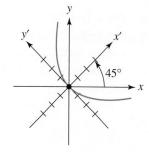

---

15. hyperbola rotated approximately $27°$, $2x'^2 - 3y'^2 = 6$

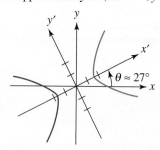

17. ellipse, $3(x' + \sqrt{5})^2 + 8y'^2 = 24$

19. parabola, $(y' - 1)^2 = -(x' - \frac{3}{2})$

21. (a) $y' = x'^2$

(b) $x'y'$-coordinates: $(0, \frac{1}{4})$; xy-coordinates: $(-\frac{1}{8}, \frac{\sqrt{3}}{8})$

(c) $y' = -\frac{1}{4}$, $2x - 2\sqrt{3}y = 1$

23. hyperbola   **25.** parabola   **27.** ellipse

Exercises 7.5   Page 404

1, 3, 5.

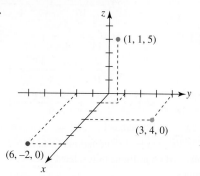

7. The set $\{(x, y, 5) \mid x, y \text{ real numbers}\}$ is a plane perpendicular to the z-axis, 5 units above the xy-plane.

9. The set $\{(2, 3, z) \mid z \text{ a real number}\}$ is a line perpendicular to the xy-plane at $(2, 3, 0)$.

11. $(2, 0, 0), (2, 5, 0) (2, 0, 8), (2, 5, 8), (0, 5, 0), (0, 5, 8), (0, 0, 8),$
$(0, 0, 0)$

13. (a) $(-2, 5, 0), (-2, 0, 4), (0, 5, 4)$ **(b)** $(-2, 5, -2)$ **(c)** $(3, 5, 4)$

15. The union of the three coordinate planes

17. The point $(-1, 2, -3)$

19. The union of the planes $z = 5$ and $z = -5$

21. $\sqrt{70}$ **23. (a)** 7 **(b)** 5

25. right triangle   **27.** isosceles triangle

29. collinear   **31.** not collinear

33. 6 or -2 **35.** $(4, \frac{1}{2}, \frac{3}{2})$ **37.** $(-4, -11, 10)$

39.   **41.**

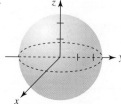
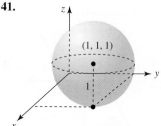

43. center $(-4, 3, 2)$, radius 6

45. center $(0, 0, 8)$, radius 8

47. $(x + 1)^2 + (y - 4)^2 + (z - 6)^2 = 3$

49. $(x - 1)^2 + (y - 1)^2 + (z - 4)^2 = 16$

51. $x^2 + (y - 4)^2 + z^2 = 4$ or $x^2 + (y - 8)^2 + z^2 = 4$

53.

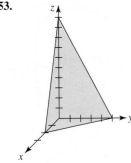

55.

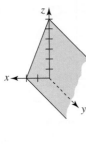

57.

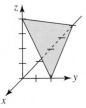

59. $5x - 3y + z = 2$

61. $3x - 2y + 2z = 3$

63. $-x + 3y = 5$

65. $\langle 2, 4, 12 \rangle$

67. $\langle -11, -41, -49 \rangle$

69. $\sqrt{139}$

71. 6

73. -1

75. 15

75. $\langle -\frac{2}{3}, \frac{1}{3}, -\frac{2}{3} \rangle$

79. $4\mathbf{i} - 4\mathbf{j} + 4\mathbf{k}$

81. $-3\mathbf{i} + 19\mathbf{j} + 10\mathbf{k}$

83. $\mathbf{0}$

Chapter 7 Review Exercises Page 407

A. 1. $y^2 = 20x$

3. $(x - 1)^2 = 8(y + 5)$

5. $(0, 0)$

7. $8x^2 = y - 2$

9. $(-3, 0), (-3, -1), (-3, 1)$

11. $(0, -\sqrt{2}), (0, \sqrt{2})$

13. 1

15. $\frac{10}{3}$

17. $x = \frac{1}{2}$

19. $(-7, 4), (9, 4)$

B. 1. true   **3.** false   **5.** true   **7.** true

9. true   **11.** true   **13.** true   **15.** false

17. true   **19.** true

C. 1. Vertex: $(0, 3)$
   Focus: $(-2, 3)$
   Directrix: $x = 2$
   Axis: $y = 3$

3. Vertex: $(1, 0)$
   Focus: $(1, -1)$
   Directrix: $y = 1$
   Axis: $x = 1$

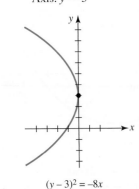

$(y - 3)^2 = -8x$

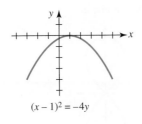

$(x - 1)^2 = -4y$

5. $(x - 1)^2 = 8(y + 5)$

7. $(x - 1)^2 = 3(y - 2)$

9. Center: $(0, -5)$
   Vertices: $(0, -10), (0, 0)$
   Foci: $(0, -5 - \sqrt{22})$,
   $(0, -5 + \sqrt{22})$

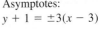

$\frac{x^2}{3} + \frac{(y + 5)^2}{25} = 1$

11. Center: $(-1, 3)$
   Vertices: $(-1, 1), (-1, 5)$
   Foci: $(-1, 3 - \sqrt{3})$,
   $(-1, 3 + \sqrt{3})$

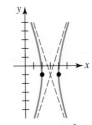

$(x + 1)^2 + \frac{(y - 3)^2}{4} = 1$

13. $\frac{x^2}{41} + \frac{y^2}{16} = 1$

15. $\frac{x^2}{4} + (y + 2)^2 = 1$

17. Center: $(0, 0)$
   Vertices: $(-1, 0), (1, 0)$
   Foci: $(-\sqrt{2}, 0), (\sqrt{2}, 0)$
   Asymptotes: $y = \pm x$

19. Center: $(3, -1)$
   Vertices: $(2, -1), (4, -1)$
   Foci: $(3 - \sqrt{10}, -1)$
   $(3 + \sqrt{10}, -1)$
   Asymptotes:
   $y + 1 = \pm 3(x - 3)$

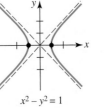

$x^2 - y^2 = 1$

$(x - 3)^2 - \frac{(y + 1)^2}{9} = 1$

21. $\frac{x^2}{36} - \frac{y^2}{28} = 1$

23. $\frac{x^2}{4} - \frac{y^2}{16} = 1$

25.

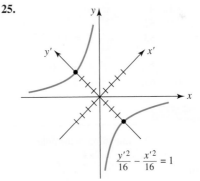

$\frac{y'^2}{16} - \frac{x'^2}{16} = 1$

27. $(x + 5)^2 + \frac{(y - 2)^2}{4} = 1$ **29.** $1.95 \times 10^9 \text{m}$

31. $\frac{25}{7} \text{cm}$

33. $(x - 1)^2 + (y - 4)^2 + (z - 2)^2 = 90$

35. parabola: $y^2 = -8(x - 6)$, ellipse: $x^2/6^2 + y^2/(2\sqrt{5})^2 = 1$

Exercises 8.1 Page 416

1.

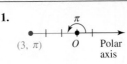

3.

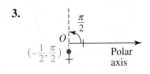

5. $\left(-4, -\frac{\pi}{6}\right)$

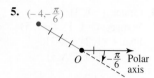

7. (a) $(2, -5\pi/4)$ (b) $(2, 11\pi/4)$
(c) $(-2, 7\pi/4)$ (d) $(-2, -\pi/4)$
9. (a) $(4, -5\pi/3)$ (b) $(4, 7\pi/3)$
(c) $(-4, 4\pi/3)$ (d) $(-4, -2\pi/3)$
11. (a) $(1, -11\pi/6)$ (b) $(1, 13\pi/6)$
(c) $(-1, 7\pi/6)$ (d) $(-1, -5\pi/6)$
13. (a) $(9, -\pi/2)$ (b) $(9, 7\pi/2)$
(c) $(-9, \pi/2)$ (d) $(-9, -3\pi/2)$
15. $\left(-\frac{1}{4}, \frac{\sqrt{3}}{4}\right)$ **17.** $\left(-3, 3\sqrt{3}\right)$
19. $\left(-2\sqrt{2}, -2\sqrt{2}\right)$ **21.** $\left(\frac{\sqrt{3}}{2}, \frac{1}{2}\right)$
23. $(3.696, 1.531)$
25. (a) $\left(2\sqrt{2}, -3\pi/4\right)$ (b) $\left(-2\sqrt{2}, \pi/4\right)$
27. (a) $(2, -\pi/3)$ (b) $(-2, \pi/3)$
29. (a) $(7, 0)$ (b) $(-7, \pi)$
31. (a) $(5, 2.214)$ (b) $(-5, -0.927)$
33.

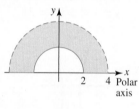

35.

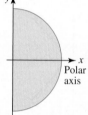

37.

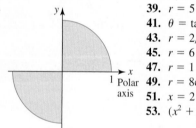

39. $r = 5 \csc\theta$
41. $\theta = \tan^{-1} 7$
43. $r = 2/(1 + \cos\theta)$
45. $r = 6$
47. $r = 1 - \cos\theta$
49. $r = 8\cos\theta$
51. $x = 2$
53. $(x^2 + y^2)^3 = 144x^2y^2$
55. $(x^2 + y^2)^2 = 8xy$ **57.** $x^2 + y^2 + 5y = 0$
59. $8x^2 - 12x - y^2 + 4 = 0$ **61.** $3x + 8y = 5$

Exercises 8.2   Page 425

1. circle

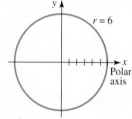

3. line through the pole

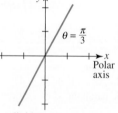

5. spiral

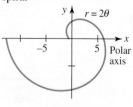

7. cardioid

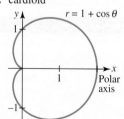

9. cardioid

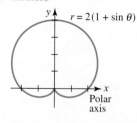

11. limaçon with interior loop

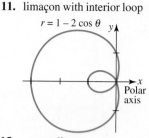

13. dimpled limaçon

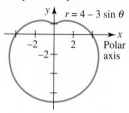

15. convex limaçon

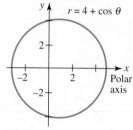

17. rose curve

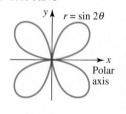

19. rose curve

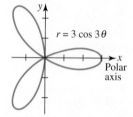

21. rose curve

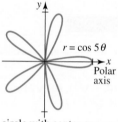

23. circle with center on x-axis

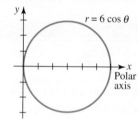

25. circle with center on y-axis

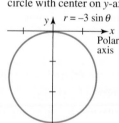

27. lemniscate

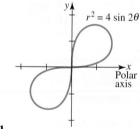

29. lemniscate

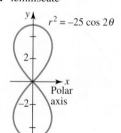

31.

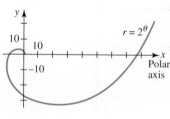

33. $r = \frac{5}{2}$ **35.** $r = 4 - 3\cos\theta$
37. $r = 2\cos 4\theta$ **39.** $(2, \pi/6), (2, 5\pi/6)$
41. $(1, \pi/2), (1, 3\pi/2)$, origin
43. $\left(\sqrt{3}/2, \pi/3\right), \left(\sqrt{3}/2, 2\pi/3\right)$, origin

1. $e = 1$, parabola

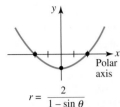

$$r = \frac{2}{1 - \sin\theta}$$

3. $e = \frac{1}{4}$, ellipse

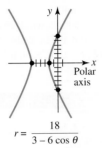

$$r = \frac{16}{4 + \cos\theta}$$

5. $e = 2$, hyperbola

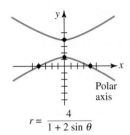

$$r = \frac{4}{1 + 2\sin\theta}$$

7. $e = 2$, hyperbola

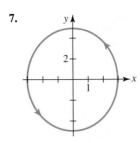

$$r = \frac{18}{3 - 6\cos\theta}$$

9. $e = 1$, parabola

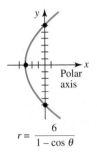

$$r = \frac{6}{1 - \cos\theta}$$

11. $e = 2$, $\dfrac{(y-4)^2}{4} - \dfrac{x^2}{12} = 1$

13. $e = \frac{2}{3}$, $\dfrac{(x - \frac{24}{5})^2}{\frac{1296}{25}} + \dfrac{y^2}{\frac{144}{5}} = 1$

15. $r = \dfrac{3}{1 + \cos\theta}$

17. $r = \dfrac{4}{3 - 2\sin\theta}$

19. $r = \dfrac{12}{1 + 2\cos\theta}$

21. $r = \dfrac{3}{1 + \cos(\theta + 2\pi/3)}$

23. $r = \dfrac{3}{1 - \sin\theta}$

25. $r = \dfrac{1}{1 - \cos\theta}$

27. $r = \dfrac{1}{2 - 2\sin\theta}$

29. parabola, vertex: $(2, \pi/4)$

31. ellipse, vertices: $(10, \pi/3)$ and $\left(\frac{10}{3}, 4\pi/3\right)$

33. $r_p = 8000$ km

35. $r = \dfrac{1.495 \times 10^8}{1 - 0.0167\cos\theta}$

1.

t	-3	-2	-1	0	1	2	3
x	-1	0	1	2	3	4	5
y	$\frac{3}{2}$	2	$\frac{5}{2}$	3	$\frac{7}{2}$	4	$\frac{9}{2}$

intercepts: $(-4, 0), (0, 2)$

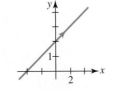

3.

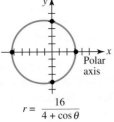

5.

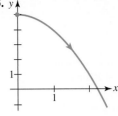

7.

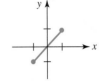

9.

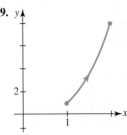

11. $y = x^2 + 3x - 1$, $x \geq 0$

13. $x = 1 - 2y^2$, $-1 \leq x \leq 1$

15. $y = \ln x$, $x > 0$

17. $\dfrac{x^2}{16} + \dfrac{y^2}{4} = 1$

19.

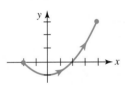

21.

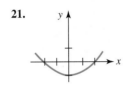

23.

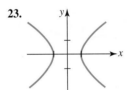
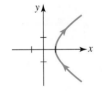

25.

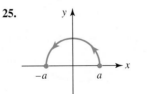
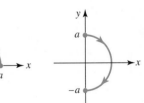

27.

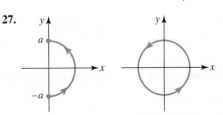

29. $(3, 0); (0, 1), (0, 3)$

31. the line segment between (x_1, y_1) and (x_2, y_2)

33. $x = 95\sqrt{2}\,t$, $y = -16t^2 + 95\sqrt{2}\,t$, $t \geq 0$;
$\left(190\sqrt{2}, 190\sqrt{2} - 64\right) \approx (268.70, 204.70)$

35. $x = \pm\sqrt{r^2 - L^2\sin^2\phi}$, $y = L\sin\phi$

Chapter 8 Review Exercises — Page 442

A. 1. $(1, 1)$
3. $(10, 3\pi/2)$
5. $\frac{1}{2}$
7. hyperbola
9. convex limaçon
11. x-axis
13. $r = 20\sin\theta$
15. line

B. 1. true
3. true
5. false
7. true
9. false
11. false
13. true
15. true
17. false

C. 1. $\left(x - \frac{1}{2}\right)^2 + \left(y - \frac{1}{2}\right)^2 = \frac{1}{2}$
3. $r = 4\sin\theta$
5. $(0, 2), \left(0, -\frac{2}{3}\right)$
7. (a) $\left(\sqrt{6}, -\pi/4\right)$
(b) $\left(-\sqrt{6}, 3\pi/4\right)$
9. circle of radius 5 centered at the origin
11. circle with center on y-axis

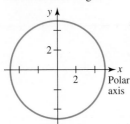

13. cardioid
15. convex limaçon

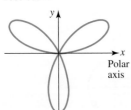
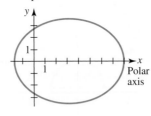

17. rose curve
19. ellipse

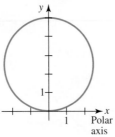
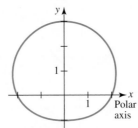

21. $r = 3\sin 10\theta$
23. (a) $r = 2\cos(\theta - \pi/4)$
(b) $x^2 + y^2 = \sqrt{2}x + \sqrt{2}y$
25. (b) center $(b/2, a/2)$, radius $\frac{1}{2}\sqrt{a^2 + b^2}$

Exercises 9.1 — Page 444

1. $(0, 2)$, consistent; independent
3. $\left(-\frac{12}{13}, -\frac{35}{13}\right)$, consistent, independent
5. no solutions, inconsistent
7. $\left(\frac{3}{2}, -\frac{1}{2}\right)$, consistent, independent
9. $(-2\alpha + 4, \alpha)$, α a real number, consistent, dependent
11. $(1, 2, 3)$, consistent, independent
13. $\left(-1, \frac{1}{2}, -3\right)$, consistent, independent
15. no solutions, inconsistent

17. $(0, 0, 0)$, consistent, independent
19. $\left(7, -5, \frac{1}{3}\right)$, consistent, independent
21. $(2\alpha + 3\beta - 2, \beta, \alpha)$, α and β real numbers, consistent, dependent
23. no solutions, inconsistent
25. $(1, -2, 4, 8)$, consistent, independent
27. $\left(-\frac{3}{7}\alpha + \frac{3}{7}, \frac{25}{7}\alpha - \frac{4}{7}, \alpha\right)$, α a real number
29. no solution
31. $2Na + 2H_2O \rightarrow 2NaOH + H_2$
33. $Fe_3O_4 + 4C \rightarrow 3Fe + 4CO$
35. $3Cu + 8NHO_3 \rightarrow 3Cu(NO_3)_2 + 4H_2O + 2NO$
37. $x = 2, y = 3$
39. $x = 10^3, y = 10^{-7}$
41. $T_1 = \dfrac{200\cos 15°}{\sin 40°} \approx 300.54$, $T_2 = \dfrac{200\cos 25°}{\sin 40°} \approx 281.99$
43. plane: 575 mi/h, wind: 25 mi/h
45. 50 gal from the first tank, 40 gal from the second tank
47. P_1: 4 h, P_2: 12 h, P_3: 6 h
49. 25
51. 20 A's, 5 B's, 15 C's

Exercises 9.2 — Page 465

1. $M_{11} = -2, M_{12} = 3, M_{21} = 0, M_{22} = 4; A_{11} = -2, A_{12} = -3, A_{21} = 0, A_{22} = 4$
3. $M_{11} = 5, M_{12} = 10, M_{13} = 3, M_{21} = -35, M_{22} = 29, M_{23} = -21, M_{31} = -8, M_{32} = -16, M_{33} = 15; A_{11} = 5, A_{12} = -10, A_{13} = 3, A_{21} = 35, A_{22} = 29, A_{23} = 21, A_{31} = -8, A_{32} = 16, A_{33} = 15$
5. 27
7. 12
9. $a^2 + b^2$
11. 60
13. -61
15. 0
17. adf
19. $x = 4, y = -3$
21. $x = 2, y = 7$
23. $x = -\frac{5}{3}, y = -\frac{5}{3}$
25. $x = -\frac{1}{6}, y = \frac{2}{3}$
27. $x = 4, y = -4, z = -5$
29. $x = 4, y = 1, z = 2$
31. $x = \frac{1}{4}, y = \frac{3}{4}, z = 1$
33. Cramer's rule not applicable
35. -4
43. $-5, 2$
45. $-2, -1, 3$
47. (b) $v = \sqrt{\dfrac{d_2^2 - d_1^2}{t_2^2 - t_1^2}}; D = \dfrac{1}{2}\sqrt{\dfrac{t_1^2 d_2^2 - t_2^2 d_1^2}{t_2^2 - t_1^2}}$
(c) 947.9 m, 1531 m/s

Exercises 9.3 — Page 471

1. two solutions
3. two solutions

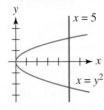
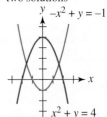

5. one solution
7. $(-1, -1), (2, 2)$

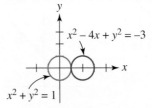

9. $(1, 1)$

11. $(0, 1)$

13. $(\sqrt{3}, 3)$

15. no solutions

17. $(-\sqrt{5}, \sqrt{5}), (\sqrt{5}, \sqrt{5})$

19. $(0, 0), (-2\sqrt{5}, 4), (2\sqrt{5}, 4), (-2\sqrt{3}, -4), (2\sqrt{3}, -4)$

21. no solutions

23. $\left(-\frac{2}{5}\sqrt{5}, \frac{4}{5}\sqrt{5}\right), \left(\frac{2}{5}\sqrt{5}, -\frac{4}{5}\sqrt{5}\right)$

25. $(-\sqrt{5}, 0), (\sqrt{5}, 0), (-2, -1), (2, -1)$

27. $(-1 - \sqrt{3}, 1 - \sqrt{3}), (-1 + \sqrt{3}, 1 + \sqrt{3}),$
$(1 - \sqrt{3}, -1 - \sqrt{3}), (1 + \sqrt{3}, -1 + \sqrt{3})$

29. $\left\{ \left(\frac{\pi}{4} + 2n\pi, \frac{\sqrt{2}}{2} \right) \Big| n = 0, \pm 1, \dots \right\}$
$\cup \left\{ \left(\frac{\pi}{4} + (2n + 1)\pi, -\frac{\sqrt{2}}{2} \right) \Big| n = 0, \pm 1, \dots \right\}$

31. $\left\{ \left(\frac{\pi}{6} + 2n\pi, 1 \right) \Big| n = 0, \pm 1, \dots \right\}$
$\cup \left\{ \left(\frac{5\pi}{6} + 2n\pi, 1 \right) \Big| n = 0, \pm 1, \dots \right\}$
$\cup \left\{ \left(\frac{7\pi}{6} + 2n\pi, -1 \right) \Big| n = 0, \pm 1, \dots \right\}$
$\cup \left\{ \left(\frac{11\pi}{6} + 2n\pi, -1 \right) \Big| n = 0, \pm 1, \dots \right\}$

33. $(0.1, -1), (100,000, 5)$

35. $(-101, -201), (99, 199)$

37. $(5, \log_3 5)$

39. $(\sqrt{3}, \sqrt{3}, -2\sqrt{3}), (-\sqrt{3}, -\sqrt{3}, 2\sqrt{3})$

41. $(1/\sqrt{3}, \sqrt{2/3}, 1/\sqrt{3}), (-1/\sqrt{3}, \sqrt{2/3}, -1/\sqrt{3}),$
$(1/\sqrt{3}, -\sqrt{2/3}, 1/\sqrt{3}), (-1/\sqrt{3}, -\sqrt{2/3}, -1/\sqrt{3}),$
$(1, 0, 0), (-1, 0, 0)$

43. 50 ft $\times$ 80 ft

45. each radius is 4 cm

47. approximately 7.9 in. $\times$ 12.7 in.

49. 2 ft $\times$ 2 ft $\times$ 8 ft, or approximately 7.06 ft $\times$ 7.06 ft $\times$ 0.64 ft

Exercises 9.4   Page 477

1.
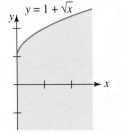

3.
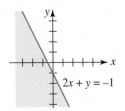

5.
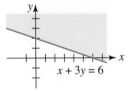

7.
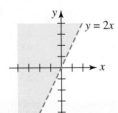

9.
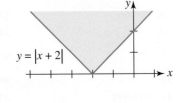

11.
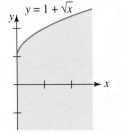

13.
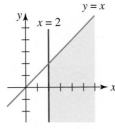

15.
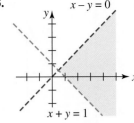

17. no solutions

19.
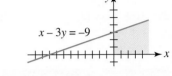

21.
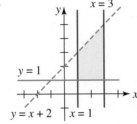

23.
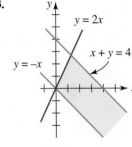

25.
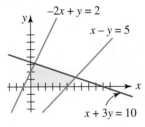

27.
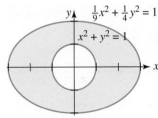

29.
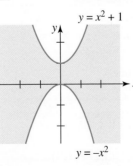

31.
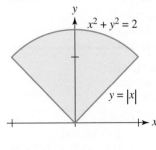

33.
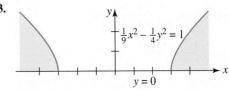

35.

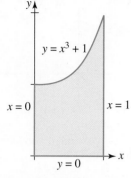

37. $3x + 2y \geq 12$
$x + 2y \geq 8$
$x \geq 0, y \geq 0$

39. $x + y \leq 10$
$1 \leq x \leq 5$
$2 \leq y \leq 6$

Exercises 9.5   Page 484

1. $\dfrac{\frac{1}{2}}{x} - \dfrac{\frac{1}{2}}{x + 2}$

3. $-\dfrac{6}{x + 1} - \dfrac{3}{x - 5}$

5. $\dfrac{\frac{3}{2}}{x + 1} - \dfrac{10}{x + 2} + \dfrac{\frac{21}{2}}{x + 3}$

7. $\dfrac{\frac{3}{2}}{x + 4} + \dfrac{\frac{3}{2}}{x - 4}$

9. $\dfrac{5}{x - 3} + \dfrac{9}{(x - 3)^2}$

11. $-\dfrac{\frac{1}{4}}{x} + \dfrac{\frac{1}{4}}{x^2} + \dfrac{\frac{1}{4}}{x + 2} + \dfrac{\frac{1}{4}}{(x + 2)^2}$

13. $-\dfrac{\frac{11}{27}}{x} - \dfrac{\frac{7}{9}}{x^2} + \dfrac{\frac{1}{3}}{x^3} + \dfrac{\frac{1}{2}}{x - 1} - \dfrac{\frac{5}{54}}{x + 3}$

15. $\dfrac{1}{x - 1} + \dfrac{5x - 2}{x^2 + 9}$

17. $\dfrac{\frac{36}{7}}{2x - 3} + \dfrac{-\frac{4}{7}x + \frac{26}{7}}{x^2 - x + 1}$

19. $-\dfrac{\frac{7}{4}}{t + 1} + \dfrac{\frac{9}{4}}{t - 1} + \dfrac{-\frac{1}{2}t - 4}{t^2 + 1}$

21. $\dfrac{2x}{x^2 + 2} - \dfrac{x}{x^2 + 1}$

23. $\dfrac{1}{x^2 + 1} + \dfrac{2x}{(x^2 + 1)^2}$

25. $x^3 + x + \dfrac{\frac{1}{2}}{x - 1} + \dfrac{\frac{1}{2}}{x + 1}$

27. $\dfrac{1}{2} - \dfrac{\frac{13}{3}}{x + 2} + \dfrac{\frac{13}{6}}{2x + 1}$

29. $x^3 + 2x^2 + 3x + 6 + \dfrac{\frac{64}{5}}{x - 2} + \dfrac{\frac{1}{5}x + \frac{2}{5}}{x^2 + 1}$

Chapter 9 Review Exercises   Page 485

A. 1. $-\frac{3}{2}$
3. -2 or 1
5. half-plane
7. $x = -\frac{1}{3}, y = 5, z = -2$
9. $a = -\frac{1}{2}, b = \frac{3}{2}$

B. 1. true
3. true
5. false
7. true
9. true

C. 1. $(0, 0, 0)$
3. $(-1, 4, -5)$
5. $(-1, 0), (6, -7)$
7. $\left(-\frac{4\sqrt{5}}{5}, -\frac{4\sqrt{5}}{5}\right), \left(-\frac{4\sqrt{5}}{5}, \frac{4\sqrt{5}}{5}\right), \left(\frac{4\sqrt{5}}{5}, -\frac{4\sqrt{5}}{5}\right), \left(\frac{4\sqrt{5}}{5}, \frac{4\sqrt{5}}{5}\right)$
9. $(100, 2)$
11. (e^2, e^{-1})
13. $(1, 0)$
15. 27
17. length of side of square: $\dfrac{4 - \pi}{4(4 + \pi)} \approx 0.03$;

radius of circle: $\dfrac{1}{4 + \pi} \approx 0.14$

19. $\dfrac{\frac{1}{3}}{x} + \dfrac{\frac{1}{4}}{x - 1} - \dfrac{\frac{7}{12}}{x + 3}$

21. $\dfrac{1}{x^2 + 4} - \dfrac{4}{(x^2 + 4)^2}$

23.

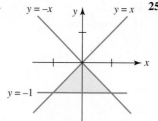

25.

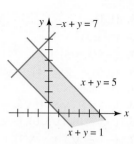

27.

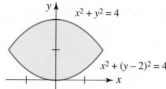

31. $\begin{cases} y \geq x^2 \\ y \geq 2 - x \end{cases}$

33. $\begin{cases} y \leq x^2 \\ y \leq 2 - x \end{cases}$

35. (e)

37. (f)

39. (b)

41. $\begin{cases} x^2 + y^2 \leq 16 \\ x^2 + y^2 \geq 4 \end{cases}$

Exercises 10.1   Page 496

1. $-1, 1, -1, 1, -1, \ldots$
3. $1, 3, 6, 10, 15, \ldots$
5. $\frac{1}{2}, \frac{1}{5}, \frac{1}{10}, \frac{1}{17}, \frac{1}{26}, \ldots$
7. $-1, 2, -3, 4, -5, \ldots$
9. $0, \frac{1}{3}, \frac{2}{13}, \frac{5}{17}, \frac{4}{21}, \ldots$
11. $-2, 4, -8, 16, -32, 36, \ldots$
13. $\frac{1}{49}, 8, \frac{1}{81}$
15. $3, \frac{1}{3}, -3, -\frac{1}{3}, 3, \ldots$
17. $0, 2, 8, 26, 80, \ldots$
19. $1, \frac{1}{2}, \frac{1}{6}, \frac{1}{24}, \frac{1}{120}, \ldots$
21. $7, 9, 11, 13, 15, \ldots$
23. $d = -5; a_n = 4 - 5(n - 1); a_{n+1} = a_n - 5, a_1 = 4$
25. $r = -\frac{3}{4}; a_n = 4\left(-\frac{3}{4}\right)^{n-1}; a_{n+1} = -\frac{3}{4}a_n, a_1 = 4$
27. $d = -11; a_n = 2 - 11(n - 1); a_{n+1} = a_n - 11, a_1 = 2$
29. $r = 0.1y; a_n = 0.1(0.1y)^{n-1}; a_{n+1} = (0.1y)a_n, a_1 = 0.1$
31. $r = -\frac{2}{3}; a_n = \frac{3}{8}\left(-\frac{2}{3}\right)^{n-1}; a_{n+1} = -\frac{2}{3}a_n, a_1 = \frac{3}{8}$
33. 113
35. $\frac{1}{2}$
37. $\frac{1}{8}$
39. 255
41. $4, 7, 10, 13, \ldots$
43. \$3870
45. 6.6%
47. \$145
49. 57,665
51. 32
53. $1, 1, 2, 3, 5, 8, 13, \ldots$

Exercises 10.2   Page 502

1. 14
3. -40
5. $\frac{23}{2}$
7. $1 + \sqrt{2} + \sqrt{3} + 2 + \sqrt{5}$
9. $1 - 1 + 1 - 1$
11. $\displaystyle\sum_{k=1}^{5} (2k + 1)$
13. $\displaystyle\sum_{k=0}^{5} \dfrac{(-1)^k}{3 \cdot 2^k}$
15. $\displaystyle\sum_{k=1}^{5} \dfrac{2^k + 1}{2^k}$
17. 35
19. 564
21. -748
23. $\frac{40}{81}$
25. 66.666
27. $\frac{85}{128}$
29. $a_1 = \frac{9}{2}, a_{10} = \frac{45}{2}$
31. 80
33. 12
35. $\dfrac{x^{15} + y^{15}}{x^{13}y(x + y)}$
37. $n^2 + n$
39. 500,500
41. \$13,500
43. 72 m
47. approximately 69.73 ft
49. approximately 55.6 mg

1. (*i*) $2 = (1)^2 + 1$, is true. (*ii*) Assume that $S(k)$,
$2 + 4 + \cdots + 2k = k^2 + k$, is true. Then

$$
\begin{aligned}
2 + 4 + \cdots + 2k + 2(k + 1) &= k^2 + k + 2(k + 1) \\
&= k^2 + k + 2k + 2 \\
&= (k^2 + 2k + 1) + (k + 1) \\
&= (k + 1)^2 + (k + 1).
\end{aligned}
$$

Thus $S(k + 1)$ is true. By (*i*) and (*ii*) the proof is complete.

3. (*i*) $1^2 = \frac{1}{6}1(1 + 1)[2(1) + 1] = \frac{1}{6}2 \cdot 3$, is true. (*ii*) Assume that
$S(k)$, $1^2 + 2^2 + \cdots + k^2 = \frac{1}{6}k(k + 1)(2k + 1)$, is true. Then

$$
\begin{aligned}
1^2 + 2^2 + \cdots + k^2 + (k + 1)^2 &= \frac{1}{6}k(k + 1)(2k + 1) + (k + 1)^2 \\
&= \frac{k(k + 1)(2k + 1) + 6(k + 1)^2}{6} \\
&= \frac{(k + 1)[k(2k + 1) + 6(k + 1)]}{6} \\
&= \frac{(k + 1)(2k^2 + 7k + 6)}{6} \\
&= \frac{(k + 1)[(k + 2)(2k + 3)]}{6} \\
&= \frac{1}{6}(k + 1)[(k + 1) + 1][2(k + 1) + 1].
\end{aligned}
$$

Thus $S(k + 1)$ is true. By (*i*) and (*ii*) the proof is complete.

5. (*i*) $\frac{1}{2} + \frac{1}{2} = 1$, is true. (*ii*) Assume that $S(k)$,

$$
\left(\frac{1}{2} + \frac{1}{2^2} + \cdots + \frac{1}{2^k} \right) + \frac{1}{2^k} = 1,
$$

is true. Then

$$
\begin{aligned}
\left(\frac{1}{2} + \frac{1}{2^2} + \cdots + \frac{1}{2^k} + \frac{1}{2^{k+1}} \right) + \frac{1}{2^{k+1}} &= \left(\frac{1}{2} + \frac{1}{2^2} + \cdots + \frac{1}{2^k} \right) + \frac{2}{2^{k+1}} \\
&= \left(1 - \frac{1}{2^k} \right) + \frac{2}{2^{k+1}} \\
&= 1 - \frac{1}{2^k} + \frac{1}{2^k} = 1.
\end{aligned}
$$

Thus $S(k + 1)$ is true. By (*i*) and (*ii*) the proof is complete.

7. (*i*) $\dfrac{1}{1 \cdot (1 + 1)} = \dfrac{1}{1 + 1}$, is true. (*ii*) Assume that $S(k)$,

$$
\frac{1}{1 \cdot 2} + \frac{1}{2 \cdot 3} + \cdots + \frac{1}{k(k + 1)} = \frac{k}{k + 1}, \text{ is true. Then}
$$

$$
\begin{aligned}
&\frac{1}{1 \cdot 2} + \frac{1}{2 \cdot 3} + \cdots + \frac{1}{k(k + 1)} + \frac{1}{(k + 1)[(k + 1) + 1]} \\
&= \frac{k}{k + 1} + \frac{1}{(k + 1)[(k + 1) + 1]} \\
&= \frac{k(k + 2) + 1}{(k + 1)(k + 2)} \\
&= \frac{k^2 + 2k + 1}{(k + 1)(k + 2)} \\
&= \frac{(k + 1)^2}{(k + 1)(k + 2)} \\
&= \frac{k + 1}{(k + 1) + 1}.
\end{aligned}
$$

Thus $S(k + 1)$ is true. By (*i*) and (*ii*) the proof is complete.

9. (*i*) $1 = \frac{1}{3}(4 - 1)$ is true. (*ii*) Assume that
$S(k)$, $1 + 4 + 4^2 + \cdots + 4^{k-1} = \frac{1}{3}(4^k - 1)$, is true. Then

$$
\begin{aligned}
1 + 4 + 4^2 + \cdots + 4^{k-1} + 4^k &= \frac{1}{3}(4^k - 1) + 4^k \\
&= \frac{1}{3}4^k + 4^k - \frac{1}{3} \\
&= \frac{4}{3}4^k - \frac{1}{3} \\
&= \frac{1}{3}(4^{k+1} - 1).
\end{aligned}
$$

Thus $S(k + 1)$ is true. By (*i*) and (*ii*) the proof is complete.

11. (*i*) The statement $(1)^3 + 2(1)$ is divisible by 3, is true.
(*ii*) Assume that $S(k)$, $k^3 + 2k$ is divisible by 3, is true; in other
words, $k^3 + 2k = 3x$ for some integer x. Then

$$
\begin{aligned}
(k + 1)^3 + 2(k + 1) &= k^3 + 3k^2 + 3k + 1 + 2k + 2 \\
&= k^3 + 2k + 3k^2 + 3k + 3 \\
&= (k^3 + 2k) + 3(k^2 + k + 1) \\
&= 3x + 3(k^2 + k + 1) \\
&= 3(x + k^2 + k + 1),
\end{aligned}
$$

is divisible by 3. Thus $S(k + 1)$ is true. By (*i*) and (*ii*) the proof
is complete.

13. (*i*) The statement, 4 is a factor of $5 - 1$, is true. (*ii*) Assume that
$S(k)$, 4 is a factor of $5^k - 1$, is true. Then

$$
\begin{aligned}
5^{k+1} - 1 &= 5^k \cdot 5 - 1 \\
&= 5^k \cdot 5 - 5 + 4 \\
&= 5(5^k - 1) + 4.
\end{aligned}
$$

Since 4 is a factor of $5^k - 1$ and of 4, it follows that 4 is a factor
of $5^{k+1} - 1$. Thus $S(k + 1)$ is true. By (*i*) and (*ii*) the proof is
complete.

15. (*i*) The statement, 7 is a factor of $3^2 - 2^1 = 9 - 2$, is true.
(*ii*) Assume that $S(k)$, 7 is a factor of $3^{2k} - 2^k$, is true. Then

$$
\begin{aligned}
3^{2(k+1)} - 2^{k+1} &= 3^{2k} \cdot 3^2 - 2^k \cdot 2 \\
&= 3^{2k} \cdot 9 - 2^k \cdot 2 \\
&= 3^{2k} \cdot (2 + 7) - 2^k \cdot 2 \\
&= 2(3^{2k} - 2^k) + 7 \cdot 3^{2k}.
\end{aligned}
$$

Since 7 is a factor of $3^{2k} - 2^k$ and of $7 \cdot 3^{2k}$, it follows that 7 is a
factor of $3^{2k+2} - 2^{k+1}$. Thus $S(k + 1)$ is true. By (*i*) and (*ii*) the
proof is complete.

17. (*i*) The statement, $(1 + a)^1 \geq 1 + (1)a$ for $a \geq -1$, is true.
(*ii*) Assume that $S(k)$, $(1 + a)^k \geq 1 + ka$ for $a \geq -1$, is true.
Then, for $a \geq -1$,

$$
\begin{aligned}
(1 + a)^{k+1} &= (1 + a)^k(1 + a) \\
&\geq (1 + ka)(1 + a) \\
&= 1 + ka^2 + ka + a \\
&\geq 1 + ka + a \\
&= 1 + (k + 1)a.
\end{aligned}
$$

Thus $S(k + 1)$ is true. By (*i*) and (*ii*) the proof is complete.

19. (*i*) Since $r > 1$, the statement $r^1 > 1$ is true. (*ii*) Assume that $S(k)$.
If $r > 1$, then $r^k > 1$, is true. Then, for $r > 1$,

$$
r^{k+1} = r^k \cdot r > r^k \cdot 1 > 1 \cdot 1 = 1.
$$

Thus $S(k + 1)$ is true. By (*i*) and (*ii*) the proof is complete.

1. 6	**3.** $\frac{1}{60}$	**5.** 144	**7.** 10
9. 7	**11.** 4	**13.** n	
15. $\dfrac{1}{(n + 1)(n + 2)^2(n + 3)}$		**17.** 5!	**19.** 100!

Answers to Selected Odd-Numbered Problems, CHAPTER 10

21. $4!5!$ **23.** $\dfrac{4!}{2!}$ **25.** $\dfrac{n!}{(n-2)!}$ **27.** true

29. false **31.** true **33.** $x^4 - 10x^2y^4 + 25y^8$

35. $x^6 - 3x^4y^2 + 3x^2y^4 - y^6$

37. $x^2 + 4x^{3/2}y^{1/2} + 6xy + 4x^{1/2}y^{3/2} + y^2$

39. $x^{10} + 5x^8y^2 + 10x^6y^4 + 10x^4y^6 + 5x^2y^8 + y^{10}$

41. $a^3 - 3a^2b + 3ab^2 - b^3 - 3a^2c + 6abc - 3b^2c + 3ac^2 - 3bc^2 - c^3$

43. $n = 6$: 1 6 15 20 15 6 1; $n = 7$: 1 7 21 35 35 21 7 1

45. $6ab^5$ **47.** $-20x^6y^6$

49. $2240x^4$ **51.** $2002x^5y^9$

53. $-144y^7$ **55.** 252

57. 0.95099

Exercises 10.5   Page 520

1. *abc, acb, bac, bca, cab, cba*

3. Here (x, y) represents the number x on the red die and the number y on the black die:
$(1, 1), (1, 2), (1, 3), (1, 4), (1, 5), (1, 6), (2, 1), (2, 2), (2, 3),$
$(2, 4), (2, 5), (2, 6), (3, 1), (3, 2), (3, 3), (3, 4), (3, 5), (3, 6),$
$(4, 1), (4, 2), (4, 3), (4, 4), (4, 5), (4, 6), (5, 1), (5, 2), (5, 3),$
$(5, 4), (5, 5), (5, 6), (6, 1), (6, 2), (6, 3), (6, 4), (6, 5), (6, 6)$

5. 576 **7.** 800

9. 120 **11.** 6

13. 9900 **15.** 20, 160

17. 6 **19.** 1225

21. 78 **23.** 1

25. 24 **27.** (a) 5040 (b) 2520

29. 66 **31.** 252

33. 90 **35.** 10

37. 56 **39.** (a) 360 (b) 1296 (c) 2401

41. $C(5, 3) \cdot C(3, 2) \cdot 5! = 3600$ **43.** (a) 40 (b) 80

Exercises 10.6   Page 527

1. {HH, HT, TH, TT}

3. {1H, 2H, 3H, 4H, 6H, 1T, 2T, 3T, 4T, 6T}

5. $\frac{3}{13}$ **7.** $\frac{1}{6}$

9. $\frac{1}{36}$ **11.** $\frac{1}{8}$

13. $\dfrac{13 \cdot 48}{C(52, 5)} \approx 0.00024$ **15.** $\dfrac{4 \cdot C(13, 5)}{C(52, 5)} \approx 0.002$

17. $1 - \dfrac{C(39, 5)}{C(52, 5)} \approx 0.78$ **19.** $1 - \dfrac{1}{2^{10}} \approx 0.999$

21. $\frac{15}{16}$ **23.** (a) $\frac{28}{143}$ (b) $\frac{138}{143}$

25. $\frac{1}{1024}$ **27.** $\frac{6}{36} + \frac{2}{36} = \frac{2}{9}$

29. $\frac{10}{100} + \frac{5}{100} = \frac{3}{20}$ **31.** $\frac{1}{6}$

33. $\frac{3}{5}$ or 60% **35.** $\frac{2}{3}$

Exercises 10.7   Page 537

1. converges to 0 **3.** converges to 0

5. converges to $\frac{1}{2}$ **7.** diverges

9. diverges **11.** converges to $\sqrt{2}$

13. diverges **15.** converges to $\frac{5}{6}$

17. converges to 5 **19.** converges to 0

21. $\frac{2}{9}$ **23.** $\frac{61}{99}$

25. $\frac{1313}{999}$ **27.** 4

29. $\frac{2}{5}$

33. diverges

31. $\frac{81}{7}$

35. $-\frac{3}{10}$

37. (a) $S_n = 1 - \dfrac{1}{n + 1}$ (b) 1

39. 75 ft **41.** $\sqrt{3}$

Chapter 10 Review Exercises   Page 539

A. 1. $2x + 7, 2x + 10, 2x + 13, \ldots$

3. $1 + \frac{1}{2} + \frac{1}{3} + \frac{1}{4} + \frac{1}{5}$ **5.** $\frac{2}{5}$

7. 2700 **9.** $\frac{9}{4}$

11. $\frac{3069}{512}$ **13.** 1

15. 41,150 **17.** 4

19. $-\frac{1}{2}$ **21.** $\frac{8}{15}$

B. 1. false **3.** true

5. true **7.** true

9. true **11.** true

13. true **15.** true

17. true **19.** true

C. 1. $6, 3, 0, -3, -6, \ldots$ **3.** $-1, 2, -3, 4, -5, \ldots$

5. $1, 3, 14, 72, 434, \ldots$ **7.** $-\frac{1}{8}$

9. $\frac{341}{256}$ **11.** C

15. (*i*) $1^2(1 + 1)^2 = 4$ is divisible by 4. (*ii*) Assume $S(k)$, $k^2(k + 1)^2$ is divisible by 4, is true. Then

$$(k + 1)^2(k + 2)^2 = (k + 1)^2(k^2 + 4k + 4)$$
$$= k^2(k + 1)^2 + 4(k + 1)^3$$

is divisible by 4 since each term is divisible by 4. Thus $S(k + 1)$ is true. By (*i*) and (*ii*) the proof is complete.

17. (*i*) $1(1!) = 2! - 1 = 1$, is true. (*ii*) Assume $S(k)$,

$$1(1!) + 2(2!) + \cdots + k(k!) = (k + 1)! - 1,$$

is true. Then

$$1(1!) + 2(2!) + \cdots + k(k!) + (k + 1)(k + 1)!$$
$$= (k + 1)! - 1 + (k + 1)(k + 1)!$$
$$= (k + 1)!(1 + k + 1) - 1$$
$$= (k + 1)!(k + 2) - 1$$
$$= (k + 2)! - 1.$$

Thus $S(k + 1)$ is true. By (*i*) and (*ii*) the proof is complete.

19. (*i*) $\left(1 + \frac{1}{1}\right) = 1 + 1$, is true. (*ii*) Assume $S(k)$,

$$\left(1 + \tfrac{1}{1}\right)\left(1 + \tfrac{1}{2}\right)\left(1 + \tfrac{1}{3}\right)\cdots\left(1 + \tfrac{1}{k}\right) = k + 1,$$

is true. Then

$$\left(1 + \tfrac{1}{1}\right)\left(1 + \tfrac{1}{2}\right)\left(1 + \tfrac{1}{3}\right)\cdots\left(1 + \tfrac{1}{k}\right)\left(1 + \tfrac{1}{k + 1}\right)$$
$$= (k + 1)\left(1 + \tfrac{1}{k + 1}\right)$$
$$= (k + 1 + 1)$$
$$= k + 2.$$

Thus $S(k + 1)$ is true. By (*i*) and (*ii*) the proof is complete.

21. 40 **23.** 21

25. $(n + 1)(n + 2)(n + 3)$

27. $a^4 + 16a^3b + 96a^2b^2 + 256ab^3 + 256b^4$

29. $x^{10} - 5x^8y + 10x^6y^2 - 10x^4y^3 + 5x^2y^4 - y^5$

31. $-175,000a^5b^9$ **33.** $210x^6y^{12}z^{12}$

35. 37th **37.** $\frac{1}{5}$

39. converges to 0

41. HHH, HHT, HTH, HTT, THH, THT, TTH, TTT

43. (a) 496 (b) 328

45. 120 **47.** 720

49. (a) $2 \cdot 10!12! \approx 3.48 \times 10^{15}$
  (b) $22! \approx 1.124 \times 10^{21}$

51. $\{(1, 1), (1, 2), (1, 3), (1, 4), (1, 5), (2, 1), (2, 2), (2, 3), (2, 4),$
  $(2, 5), (3, 1), (3, 2), (3, 3), (3, 4), (3, 5), (4, 1), (4, 2), (4, 3),$
  $(4, 4), (4, 5), (5, 1), (5, 2), (5, 3), (5, 4), (5, 5)\}$

53. $\frac{25}{102}$ **55.** *(iii)*

57. (a) $\frac{14}{33}$ **(b)** $\frac{1}{11}$ **(c)** $\frac{17}{33}$

59. (a) 24 **(b)** $\dfrac{1}{[C(15, 5)]^4 C(15, 4)} \approx \dfrac{1}{11.1 \times 10^{17}} \approx 9 \times 10^{-18}$

Exercises A.1   Page APP-6

1. $-i$ **3.** i
5. $-i$ **7.** $-i$
9. i **11.** $10i$
13. $-3 - \sqrt{3}i$ **15.** $-1 + 4i$
17. $2 - 13i$ **19.** $-9 - 15i$
21. $-11 + 7i$ **23.** $1 - 5i$
25. $35i$ **27.** $1 + 4i$
29. $-1 + 12i$ **31.** -10
33. $-18 - 16i$ **35.** $15 + 8i$
37. 4 **39.** $\frac{4}{25} + \frac{3}{25}i$
41. $\frac{20}{41} - \frac{16}{41}i$ **43.** $\frac{1}{2} + \frac{1}{2}i$
45. $6 - 4i$ **47.** i
49. $-\frac{6}{53} + \frac{32}{53}i$ **51.** $\frac{11}{2} + \frac{9}{2}i$
53. $-\frac{6}{13} - \frac{4}{13}i$ **55.** $9 + i$
57. $x = 2, y = \frac{3}{2}$ **59.** $x = -9, y = -20$
61. $x = -4, y = -5$ **63.** $x = \frac{1}{5}, y = -\frac{2}{5}$
65. $\pm 3i$ **67.** $\pm \frac{\sqrt{10}}{2}i$
69. $\frac{1}{4} - \frac{\sqrt{7}}{4}i, \frac{1}{4} + \frac{\sqrt{7}}{4}i$ **71.** $-4 - 6i, -4 + 6i$
73. $\frac{1}{8} - \frac{\sqrt{31}}{8}i, \frac{1}{8} + \frac{\sqrt{31}}{8}i$ **75.** $\pm i, \pm \sqrt{2}i$
77. $\frac{\sqrt{2}}{2} + \frac{\sqrt{2}}{2}i, -\frac{\sqrt{2}}{2} - \frac{\sqrt{2}}{2}i$

Exercises A.2   Page APP-11

1. $\bar{z}_1 = 2 - 5i$

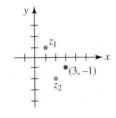

3. $z_1 + z_2 = 3 - i$

5. $\bar{z}_1 + z_2 = 6 + 2i$

7. $z_1 z_2 = -2 - 2i$

---

9. $\dfrac{z_1}{z_2} = 2i$ **11.** $r = 1, \theta = 5\pi/3$

13. $r = 3\sqrt{2}, \theta = 289.47°$ **15.** $r = \frac{\sqrt{10}}{4}, \theta = 341.57°$
17. $r = 3\sqrt{2}, \theta = \pi/4$ **19.** $r = 2, \theta = \pi/6$
21. $r = \sqrt{5}, \theta = 333.43°$

23. $z = 4\left(\cos\frac{3\pi}{2} + i\sin\frac{3\pi}{2}\right)$

25. $z = 10\left(\cos\frac{\pi}{6} + i\sin\frac{\pi}{6}\right)$

27. $z = \sqrt{29}(\cos 111.8° + i\sin 111.8°)$
29. $z = \sqrt{34}(\cos 300.96° + i\sin 300.96°)$

31. $z = 2\sqrt{2}\left(\cos\frac{5\pi}{4} + i\sin\frac{5\pi}{4}\right)$

33. $1 + i$
35. $-5\sqrt{3} - 5i$
37. $\sqrt{3} + i$
39. $\frac{1}{2} - \frac{1}{2}\sqrt{3}i$
41. $\frac{4}{5}\sqrt{5} + \frac{8}{5}\sqrt{5}i$

43. $z_1 z_2 = 18\sqrt{2}\left(\cos\frac{3\pi}{4} + i\sin\frac{3\pi}{4}\right), \dfrac{z_1}{z_2} = \dfrac{\sqrt{2}}{4}\left(\cos\frac{\pi}{4} + i\sin\frac{\pi}{4}\right)$

45. $z_1 z_2 = 8\left(\cos\frac{\pi}{2} + i\sin\frac{\pi}{2}\right), \dfrac{z_1}{z_2} = \dfrac{1}{2}\left(\cos\frac{\pi}{6} + i\sin\frac{\pi}{6}\right)$

47. $z_1 z_2 = 10\sqrt{2}\left(\cos\frac{23\pi}{12} + i\sin\frac{23\pi}{12}\right),$
  $\dfrac{z_1}{z_2} = \dfrac{\sqrt{2}}{5}\left(\cos\frac{5\pi}{12} + i\sin\frac{5\pi}{12}\right)$

49. $z_1 z_2 = 2\sqrt{3}\left(\cos\frac{\pi}{12} + i\sin\frac{\pi}{12}\right) \approx 3.3461 + 0.8966i,$
  $\dfrac{z_1}{z_2} = \sqrt{3}\left(\cos\frac{7\pi}{12} + i\sin\frac{7\pi}{12}\right) \approx -0.4483 + 1.6730i$

51. $z_1 z_2 = 12\left(\cos\frac{\pi}{8} + i\sin\frac{\pi}{8}\right) \approx 11.0866 + 4.5922i,$
  $\dfrac{z_1}{z_2} = \dfrac{3}{4}\left(\cos\frac{3\pi}{8} + i\sin\frac{3\pi}{8}\right) \approx 0.2870 + 0.6929i$

Exercises A.3   Page APP-16

1. -1
3. $2\sqrt{2} + 2\sqrt{2}i$
5. $-\frac{243}{2}\sqrt{3} - \frac{243}{2}i$
7. approximately $19.5543 + 123.4610i$
9. approximately $26.5099 + 19.2605i$
11. $-16i$ **13.** -1
15. $-8i$ **17.** -64
19. $-16\sqrt{3} + 16i$ **21.** $-16\sqrt{2}$
23. $-7 - 24i$ **25.** $1 + \sqrt{3}i, -2, 1 - \sqrt{3}i$
27. $0.9239 + 0.3827i, -0.3827 + 0.9239i,$
  $-0.9239 - 0.3827i, 0.3827 - 0.9239i$
29. $\sqrt[4]{2}\left(\frac{1}{2} + \frac{1}{2}\sqrt{3}i\right), \sqrt[4]{2}\left(-\frac{1}{2}\sqrt{3} + \frac{1}{2}i\right),$
  $\sqrt[4]{2}\left(-\frac{1}{2} - \frac{1}{2}\sqrt{3}i\right), \sqrt[4]{2}\left(\frac{1}{2}\sqrt{3} - \frac{1}{2}i\right)$
31. $\sqrt[4]{2}(0.9239 + 0.3827i), \sqrt[4]{2}(-0.9239 - 0.3827i)$

33. $1.9754 + 0.3129i, 0.7167 + 1.8672i, -1.2586 + 1.5543i,$
$-1.9754 - 0.3129i, -0.7167 - 1.8672i, 1.2586 - 1.5543i$

35. $1, \frac{1}{2} + \frac{1}{2}\sqrt{3}i, -\frac{1}{2} + \frac{1}{2}\sqrt{3}i, -1, -\frac{1}{2} - \frac{1}{2}\sqrt{3}i, \frac{1}{2} - \frac{1}{2}\sqrt{3}i$

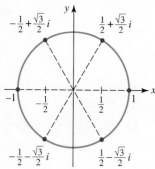

37. $n = 8k, \ k = 1, 2, 3, \ldots; \ n = 2 + 8k, \ k = 0, 1, 2, \ldots;$
$n = 5 + 8k, \ k = 0, 1, 2, \ldots; \ n = 1 + 8k, \ k = 0, 1, 2, \ldots$

39. $\frac{1}{2}\sqrt{2} + \frac{1}{2}\sqrt{2}i, \frac{1}{2}\sqrt{2} - \frac{1}{2}\sqrt{2}i,$
$-\frac{1}{2}\sqrt{2} + \frac{1}{2}\sqrt{2}i, -\frac{1}{2}\sqrt{2} - \frac{1}{2}\sqrt{2}i$

41. $-2 + 2\sqrt{3}i, \ 2 - 2\sqrt{3}i$

1. one positive zero, one negative zero

3. three or one positive zeros, no negative zeros

5. no positive zeros, three or one negative zeros

7. two or no positive zeros, two or no negative zeros

9. two or no positive zeros, three or one negative zeros

Index

Index

Index

K

Kanamori, Hiroo, 347
Kelvin scale, 75
Kepler, Johannes, 367, 379
Kepler's first law of planetary motion, 379
Knots, 297

L

Lascaux cave drawings, 350
Law of Cosines, 295–296
Law of Sines, 291
Law of Universal Gravitation, 379
Laws of exponents, 321–322
Laws of logarithms, 331
Leading coefficient of a polynomial, 131
Left-handed coordinate system, 397
Legs of a right triangle, 279
Leibniz, Wilhelm Gottfried, 182
Lemniscate, 423
Length of major axis, 375
Length of minor axis, 375
Length of transverse axis, 383
Lens equation, 47
Less than ($<$), 1
Less than or equal to ($\leq$), 1–2
Libby, Willard, 345
Liber Abacci, 497
Lighthouse of Alexandria, 290
Limaçon:
    convex, 421
    dimpled, 421
    with an interior loop, 421
Limit(s):
    of a difference quotient, 117
    evaluating, 39–41
    existence of, 38
    indeterminate form of, 39
    notation for, 38
    of a sequence, 531, 533
    of a trigonometric function, 268–271
Line(s):
    family of, 71
    graphs of, 69
    horizontal, 71
    intercepts of, 72
    parallel, 71–72
    perpendicular, 71–72
    point-slope form of, 70
    points of intersection of, 73
    slope of, 69
    slope-intercept form of, 71
    vertical, 71
    with slope, 70
    with no slope, 69
Linear combination of vectors, 305
Linear depreciation, 9
Linear equation:
    in n variables, 447
    in three variables, 400, 449
    in two variables, 68
Linear function:
    definition of, 69, 403
    domain of, 69
Linear inequality in one variable, 5
Linear inequality in two variables:
    definition of, 474
    graphing of, 474
    half-plane, 474
    solution of, 474
    system of, 474, 475
Linear speed, 198
Linear system:
    consistent, 448, 449
    definition of, 447
    equivalent, 447
    inconsistent, 448, 449
    solution of, 447
    in three variables, 449
    in two variables, 447
Local behavior of a function, 134
Local extremum:
    definition of, 134–135
    maximum, 134–135
    minimum, 134–135
Logarithmic equations, 336
Logarithmic function:
    asymptote of, 329
    common, 330
    definition of, 328
    domain of, 328
    graph of, 329
    natural, 330
    one-to-one property of, 336
    properties of, 329, 331
Logarithmic mathematical models, 347
Logarithms, laws of, 331
Logistic function, 343
LORAN, 387
Losing solutions, 258

M

Mach number, 241, 263
Magnitude of a complex number, APP-8
Magnitude of a vector, 301, 402

Index

Index

y-coordinate, 18, 398
y-intercept, 28, 53
yz-plane, 398

z-axis, 398
z-coordinate, 398
Zero(s):
    approximating, 54, 162–163
    bisection method for approximating, 162–163
    complex, 147, 149
    conjugate pairs of, 151
    of cosine function, 210

    finding, 155
    of a function, 53, 135
    of multiplicity *m*, 135, 148
    number of, 148–149
    rational, 155–156
    real, 154
    the real number, 1
    simple, 135, 148
    of sine function, 210
Zero polynomial, 131
Zero solution, 453
Zero vector, 301, 402
z-plane, APP-7

Credits

Chapter 7: **Opener** © Linda Brotkorb/ShutterStock, Inc.; **page 367** Courtesy of NASA/JPL; **page 371** (top) © Popperfoto/Alamy Images; **page 371** (bottom) © Soundsnaps/ShutterStock, Inc.; **page 372** (top) © where-@tiscali.it/ShutterStock, Inc.; **page 372** (bottom) © Corbis; **page 379** © Brand X Pictures/Alamy Images; **page 381** © jiawangkun/ShutterStock, Inc.; **page 389** Courtesy of NASA; **page 403** © Tomas Skopal/ShutterStock, Inc.; **page 410** © David Page/Alamy Images.

Chapter 8: **Opener** © Mircea BEZERGHEANU/ShutterStock, Inc.; **page 419** © Yaroslav/ShutterStock, Inc.; **page 424** © Cristian M/ShutterStock, Inc.; **page 430** Courtesy of Mariner 10, Astrogeology Team, and USGS; **page 432** © Datacraft/age fotostock; **page 439** © Corbis.

Chapter 9: **Opener** © Andrea Danti/ShutterStock, Inc.; **page 484** © Corbis

Chapter 10: **Opener** © Peter Stone/Alamy Images; **page 497** Courtesy of Liam Quin; **page 504** (top) © Melvyn Longhurst/Alamy; **page 504** (bottom) © Fesus Robert/ShutterStock, Inc.; **page 505** © Monkey Business Images/ShutterStock, Inc.; **page 517** © Sergielev/ShutterStock, Inc.; **page 520** © juan carlos tinjaca/ShutterStock, Inc.; **page 521** (top) © Tony Rolls/Alamy; **page 521** (bottom) Courtesy of Pressman Toy Corporation; **page 522** (top) © James Steidl/ShutterStock, Inc.; **page 522** (bottom) © Gjermund Alsos/ShutterStock, Inc.; **page 524** © Sebastian Kaulitzki/ShutterStock, Inc.; **page 527** © alaaddin/ShutterStock, Inc.; **page 530** © Monkey Business Images/ShutterStock, Inc.; **page 536** © Sofos Design/ShutterStock, Inc.; **page 543** © Michael D Brown/ShutterStock, Inc.

Unless otherwise indicated, all photographs and illustrations are under copyright of Jones & Bartlett Learning.

Some images in this book feature models. These models do not necessarily endorse, represent, or participate in the activities represented in the images.

Review of Trigonometry

Unit Circle Definition of Sine and Cosine

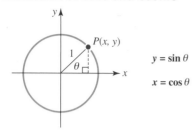

$$y = \sin\theta$$
$$x = \cos\theta$$

Other Trigonometric Functions

$$\tan\theta = \frac{y}{x} = \frac{\sin\theta}{\cos\theta}, \qquad \cot\theta = \frac{x}{y} = \frac{\cos\theta}{\sin\theta}$$

$$\sec\theta = \frac{1}{x} = \frac{1}{\cos\theta}, \qquad \csc\theta = \frac{1}{y} = \frac{1}{\sin\theta}$$

Right Triangle Definition of Sine and Cosine

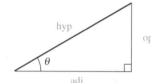

$$\sin\theta = \frac{\text{opp}}{\text{hyp}}$$

$$\cos\theta = \frac{\text{adj}}{\text{hyp}}$$

Other Trigonometric Functions

$$\tan\theta = \frac{\text{opp}}{\text{adj}}, \qquad \cot\theta = \frac{\text{adj}}{\text{opp}}$$

$$\sec\theta = \frac{\text{hyp}}{\text{adj}}, \qquad \csc\theta = \frac{\text{hyp}}{\text{opp}}$$

Signs of Sine and Cosine

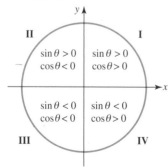

Values of Sine and Cosine for Special Angles

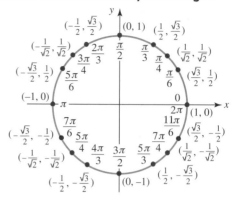

Bounds for Sine and Cosine

$$-1 \le \sin\theta \le 1 \quad \text{and} \quad -1 \le \cos\theta \le 1$$

Periodicity of Trigonometric Functions

$$\sin(\theta + 2\pi) = \sin\theta, \qquad \cos(\theta + 2\pi) = \cos\theta$$

$$\sec(\theta + 2\pi) = \sec\theta, \qquad \csc(\theta + 2\pi) = \csc\theta$$

$$\tan(\theta + \pi) = \tan\theta, \qquad \cot(\theta + \pi) = \cot\theta$$

Cofunction Identities

$$\sin\left(\frac{\pi}{2} - \theta\right) = \cos\theta$$

$$\cos\left(\frac{\pi}{2} - \theta\right) = \sin\theta$$

$$\tan\left(\frac{\pi}{2} - \theta\right) = \cot\theta$$

Pythagorean Identities

$$\sin^2\theta + \cos^2\theta = 1$$

$$1 + \tan^2\theta = \sec^2\theta$$

$$1 + \cot^2\theta = \csc^2\theta$$

Even/Odd Identities

Even	Odd
$\cos(-\theta) = \cos\theta$	$\sin(-\theta) = -\sin\theta$
$\sec(-\theta) = \sec\theta$	$\csc(-\theta) = -\csc\theta$
	$\tan(-\theta) = -\tan\theta$
	$\cot(-\theta) = -\cot\theta$